MATHEMATICAL FOUNDATIONS OF ELASTICITY

JERROLD E. MARSDEN
Department of Mathematics
University of California, Berkeley

THOMAS J. R. HUGHES
Division of Applied Mechanics
Stanford University

DOVER PUBLICATIONS, INC.
New York

Bibliographical Note

This Dover edition, first published in 1994, is an unabridged, corrected republication of the work first published by Prentice-Hall, Inc., Englewood Cliffs, N.J., 1983.

Library of Congress Cataloging in Publication Data

Marsden, Jerrold E.
Mathematical foundations of elasticity / Jerrold E. Marsden, Thomas J. R. Hughes.
p. cm.
Originally published: Englewood Cliffs, N.J. : Prentice-Hall, c1983.
Includes bibliographical references and index.
ISBN-13: 978-0-486-67865-8
ISBN-10: 0-486-67865-2
1. Elasticity. I. Hughes, Thomas J. R. II. Title.
QA931.M42 1993
531'.382'0151—dc20 93–42631
CIP

Manufactured in the United States by LSC Communications
67865210 2022
www.doverpublications.com

To Nancy, Susan, Chris, Emily, Alison, Ian, and Elizabeth

CONTENTS

PREFACE

This book treats parts of the mathematical foundations of three-dimensional elasticity using modern differential geometry and functional analysis. It is intended for mathematicians, engineers, and physicists who wish to see this classical subject in a modern setting and to see some examples of what newer mathematical tools have to contribute.

Disclaimer There are three things that every beginner in elasticity theory should know. The first is that "Kirchhoff" has two h's in it. The second is that Hooke's law will not be found as a basic axiom (it "really" means you are working with the linearized theory). The third is that researchers in elasticity theory are very opinionated, even when they are wrong. During our own work in this field we have refused to fight, and in keeping with this pacifist approach, we now issue these general disclaimers: This book is neither complete nor unbiased. Furthermore, we have not mentioned many deep and highly erudite works, nor have we elucidated alternative approaches to the subject. Any historical comments we make on subjects prior to 1960 are probably wrong, and credits to some theorems may be incorrectly assigned. Excellent historical sketches are available in the works of Truesdell [1968], Sokolnikoff [1956], and Sneddon [1980] cited in the bibliography.

The Two-Track and Box Approach To a mathematician, a tensor $\mathbf{t}$ is a section of a certain bundle over a manifold. To an engineer or physicist, a tensor t^{ijk} is an object dressed in indices. This is one of many unfortunate paper barriers that have retarded the growth of, and interest in, mathematical elasticity. The beginner should learn to speak both languages and to ignore notational disputes.

For example, beginners who are already trained in some geometry and who realize that ∇f is a vector, while df is a one-form, will recognize at once that the deformation gradient $\boldsymbol{F}$ is not a gradient at all, but is simply the derivative of the deformation. They may also recognize that the rate of deformation tensor is just the Lie derivative of the Riemannian metric on space, and that the Cauchy–Green tensor is the pull-back of the Riemannian metric on space by the deformation.

To aid the reader in this linguistic endeavor we have tried to present as many formulas as possible in both languages. This is done through numerous boxes that summarize the important formulas both ways. These boxes are also used to isolate more advanced or optional material.

Subjects Covered The first two chapters cover the background geometry—which is developed as it is needed—and use this discussion to obtain the basic results on kinematics and dynamics of continuous media. Chapter 3 narrows the discussion to elastic materials. Chapter 4 on linearization gives a systematic way to linearize a nonlinear field theory along with a basic mathematical tool—the inverse function theorem. Chapter 5 deals with variational principles. Chapter 6 presents a relatively self-contained account of the use of functional analysis (such as elliptic theory and semigroups) in elasticity. Chapter 7 introduces bifurcation theory. We originally planned to include a chapter on numerical methods as well, but space and timeliness did not allow us to do so.

Level and Background The book is written at the beginning graduate level, with occasional excursions to the research frontier. Some parts, such as the first five chapters and parts of the remainder, are accessible to good undergraduates. To read this book one should have a solid background in advanced calculus (for example, J. Marsden [1974a] is adequate). One should also be prepared to invest considerable time in learning geometry and functional analysis as the book is read. Most of what is needed is in this book, but it may be useful to consult some of the references that follow.

The Use of Geometry and Functional Analysis We have found differential geometry helpful in sorting out the foundations of the subject. Deeper analytical facts about elasticity require a serious knowledge of functional analysis, including partial differential equations. The reader should realize that many researchers understand one or the other of these subjects, but very few understand both because of the large investment of time and effort involved. Therefore, one should adjust one's aspirations and depth of reading accordingly. For example, if one's goal is to get to modern research in the buckling of shells as fast as possible, it may be a mistake to start on page 1. It is obvious that a large part of any book is irrelevant to such a specific endeavor. Rather, one should jump directly into the current literature (for example, see Section 7.2) and use this book to complete the necessary background. On the other hand, if one has the time to go through the requisite geometry, the insights gained into nonlinear elasticity will be worthwhile. Examples of how geometry is used in elasticity are discussed in Section 6 of the introductory chapter. Likewise, abstract functional analysis is often accused of not shedding any light on "practical" problems of elasticity. Recent progress in constitutive inequalities and numerical methods demonstrates that this view is incorrect.

Point of Departure and Interdependence of Chapters Because of the large amount of geometry involved in the first three chapters, we have written an introductory Chapter to enable readers to bypass parts of Chapters 1–3. After studying Sections 1–5 of the introductory chapter, such readers should be ready to undertake Chapters 4–7. These four chapters do contain some dependence on Chapters 1–3, but this dependence is minimal and may be bypassed if one has a background in elasticity obtained from other sources. We also recommend the introductory chapter for readers who intend to seriously study Chapters 1–3 to keep their work in perspective. Chapters 4–7 are in logical order, but it is not necessary to have full mastery of one before proceeding. To this end, ample cross references are given.

Notation We have adopted a reasonably simple system used by some current workers. This is summarized in a brief glossary. Readers should understand that if they hear a lecture on elasticity, the conventions will probably differ from those here or their own. Here boldface type is used to distinguish abstract tensors from their components. For example, $\boldsymbol{\sigma}$ means the abstract Cauchy stress tensor, while σ^{ab} represents its components. The only other nonstandard notation is the use of block boldface for the fourth-order elasticity tensors, such as C, whose components are denoted C^{ABCD}, and A, whose components are denoted A^{aAbB}. Occasionally the same symbol has two meanings in the book, when the intended meaning is clear from the context. We find this preferable to a multitudinous proliferation of alphabets and fonts that are impossible to reproduce in the classroom.

Things We Fuss Over; Things We Don't Most mathematicians, physicists and engineers now agree that the distinction between a linear transformation and a matrix is worth fussing over. We believe that one should also distinguish tensors from tensor components. However, we do not fuss over whether Euclidean space should be written as $\mathbb{R}^3$ or not. To abstract $\mathbb{R}^3$ properly, we believe that manifolds should be used. They are unquestionably the appropriate setting for tensor analysis.

Resistance to the use of abstract manifolds is frequently encountered, simply because most work in elasticity occurs in $\mathbb{R}^3$. In the literature, $\mathbb{R}^3$ is often replaced by abstract vector spaces. This arena is *not* suitable for general tensor analysis.

Indeed, as Einstein has so profoundly taught us, deep insights can be gained by removing one's blinders to see the theory in the grander time-proven context of covariant formulations. This is why we encourage the use of manifolds.

We do not fuss over measure-theoretic questions that are often used to introduce mass and force densities, for example. If one understands the Radon–Nikodym derivative, it takes only a few minutes to understand this even though technical intricacies may be nontrivial. We chose not to go into measure-theoretic formalism because it requires a lengthy exposition that would divert us from our main goal.

Numbering Conventions Within Chapter 1, the eleventh item of the third section is referred to as 3.11 and the third section is referred to as Section 3. In subsequent chapters this item is referred to as 3.11, Chapter 1, and the third section is referred to as Section 1.3. Similar conventions apply to discussions enclosed in boxes. Figures are given their full labels for editorial reasons. Formulas are numbered within a section when it helps the exposition.

References A relatively large bibliography is included at the back of this book. Specific references cited in the text are listed by author and year like this: Ball [1977b]. It is wise for beginning students to consult a few key books regularly. We recommend the following:

(a) an introductory modern text on continuum mechanics, such as Malvern [1969] or Gurtin [1981b];
(b) one of the classical texts on elasticity such as Love [1927], Sokolnikoff [1956], Landau and Lifshitz [1970], Green and Adkins [1970], or Green and Zerna [1968];
(c) the encyclopedic treatise of Truesdell and Noll [1965] (which has a massive bibliography);
(d) a modern book on manifolds and tensor analysis, such as Abraham Marsden and Ratiu [1982], Bishop and Goldberg [1968], Schutz [1980], Spivak [1975], or Warner [1971], and a classical one such as Eisenhart [1926], Schouten [1954], or Synge and Schild [1956];
(e) a book on functional analysis such as Balakrishnan [1976], Oden [1979], or Yosida [1971].

More advanced readers should consult other contemporary works for comparisons and other points of view. For example, we find the following additional references useful:

(a) Kondo [1955] for an early attempt at the use of geometry in elasticity,
(b) Truesdell and Toupin [1960], Rivlin [1966a], and Eringen [1975] on basic principles;
(c) Gurtin [1972a] on linear elasticity;
(d) Knops and Wilkes [1973] on elastic stability;

(e) Fichera [1972a and b] and Knops and Payne [1971] on existence and uniqueness theorems;
(f) Bloom [1979] on the use of geometry in dislocation theory;
(g) Naghdi [1972] on general shell theory and Ciarlet [1983] on the derivation of plate theory from three dimensional elasticity;
(h) Antman [1972a], [1983] on rod theory and bifurcations in elasticity.

Acknowledgments The main part of this book grew out of a course given by us at Berkeley in 1975–76. A preliminary set of notes by us was published in Volume II of *Nonlinear Analysis and Mechanics*, edited by R. Knops (Pitman, 1978). We are indebted to Professor Knops for encouraging this publication and to the readers who sent us comments.

The support of the National Science Foundation, the Army Research Office, the University of California, the Carnegie and Killam Foundations and the Miller Institute is gratefully acknowledged.

We thank many colleagues directly and indirectly for their comments and criticisms, especially Stuart Antman, David Bao, John Ball, Iris Bloomer, Jack Carr, Paul Chernoff, David Chillingworth, Constantine Dafermos, Georges Duvaut, Marcelo Epstein, Jerry Ericksen, Marty Golubitsky, Morton Gurtin, Philip Holmes, Tosio Kato, Robin Knops, Jill Mesirov, George Oster, Miguel Ortiz, John Pierce, Peter Pinsky, Karl Pister, Miles Rubin, Gloria Sanchez, David Schaeffer, Reuven Segev, Marshall Slemrod, Juan Simo, Steve Wan, and Nigel Wilkes.

We are indirectly indebted to the founding masters of the modern theory, especially Ronald Rivlin and Clifford Truesdell. Their works and those of Stuart Antman, John Ball, Morton Gurtin, Robin Knops, Paul Naghdi, and Walter Noll have had a large influence on our development of the subject.

A BRIEF GLOSSARY OF CONVENTIONS AND NOTATIONS

CONVENTIONS: As far as possible, we shall use the following:

(a) sets and manifolds—lightface script; examples: $\mathcal{B}$, $\mathcal{S}$, $\mathfrak{M}$

(b) points, point mappings, scalars—lightface italic; examples: $X \in \mathcal{B}$, $\phi: \mathcal{B} \to \mathcal{S}$

(c) vectors and vector fields, 2-tensors—boldface italic; examples: $\boldsymbol{v}$, $\boldsymbol{V}$, $\boldsymbol{F}$, $\boldsymbol{C}$, $\boldsymbol{\sigma}$ $\boldsymbol{P}$, $\boldsymbol{S}$

(d) higher-order tensors—boldface block letters; examples: **A, C**

(e) material quantities—upper case

(f) spatial quantities—lower case

NOTATIONS

$\mathbb{R}^n$	Euclidean n-space
$\mathcal{B}$	reference configuration of a body
X	point in $\mathcal{B}$
$\mathcal{S}$	the space in which the body moves (usually $\mathcal{S} = \mathbb{R}^3$)
x	point in $\mathcal{S}$
$\{X^A\}$	coordinates on $\mathcal{B}$
$\{x^a\}$	coordinates on $\mathcal{S}$
Z^I, z^i	Euclidean coordinates on $\mathcal{B} \subset \mathbb{R}^3$, $\mathcal{S} = \mathbb{R}^3$
G_{AB} or $\boldsymbol{G}$	Riemannian metric on $\mathcal{B}$
g_{ab} or $\boldsymbol{g}$	Riemannian metric on $\mathcal{S}$
Γ^A_{BC}	Christoffel symbols for $\boldsymbol{G}$
γ^a_{bc}	Christoffel symbols for $\boldsymbol{g}$
$T\mathfrak{M}$	tangent bundle of a manifold $\mathfrak{M}$
ϕ^a or $\phi: \mathcal{B} \to \mathcal{S}$	a configuration (or deformation)

dV	material volume element as a measure
$\mathbf{dV}$	material volume element as a tensor
U^a or $\boldsymbol{U}$	displacement
u^a or $\boldsymbol{u}: \mathcal{B} \to T\mathcal{S}$	linearization of a deformation
$F^a{}_A$ or $\boldsymbol{F} = T\phi$	deformation gradient = tangent of ϕ
v^a or $\boldsymbol{v}$	spatial velocity
V^a or $\boldsymbol{V}$	material velocity
a^a or $\boldsymbol{a}$	spatial acceleration
A^a or $\boldsymbol{A}$	material acceleration
C_{AB} or $\boldsymbol{C}$	(left) Cauchy–Green or deformation tensor
t^a or $\boldsymbol{t}$	Cauchy stress vector
σ^{ab} or $\boldsymbol{\sigma}$	Cauchy stress tensor
P^{aA} or $\boldsymbol{P}$	first Piola–Kirchhoff stress tensor
$\hat{P}^{aA}$ or $\hat{\boldsymbol{P}}$	corresponding constitutive function
S^{AB} or $\boldsymbol{S}$	second Piola–Kirchhoff stress tensor
A^{aAbB} or $\mathbf{A}$	first elasticity tensor
C^{ABCD} or $\mathbf{C}$	second elasticity tensor
c^{abcd} or $\mathbf{c}$	classical elasticity tensor (for the linearized theory)
d_{ab} or $\boldsymbol{d} = L_v \boldsymbol{g}$	rate of deformation tensor
$\rho(x)$	mass density in the current configuration
$\rho_{\mathrm{Ref}}(X)$	mass density in the reference configuration
W or E or e	stored energy function
$\hat{W}$ or $\hat{E}$ or $\hat{e}$	corresponding constitutive function
Ψ or ψ	free energy
$\hat{\Psi}$ or $\hat{\psi}$	corresponding constitutive function
Θ, N	temperature, entropy
$\phi^*\mathbf{t}$	pull-back of a tensor $\mathbf{t}$ by the map ϕ
$\phi_*\mathbf{t} = (\phi^{-1})^*\mathbf{t}$	push-forward by the map ϕ
$\dot{\mathbf{t}}$	material derivative of a tensor $\mathbf{t}$
$L_v\mathbf{t} = \dfrac{\partial \mathbf{t}}{\partial t} + \mathcal{L}_v\mathbf{t}$	Lie derivative of a tensor $\mathbf{t}$ along a vector field v
$SO(3)$	proper orthogonal transformations of $\mathbb{R}^3$
$I_1(C), I_2(C), I_3(C)$	invariants of a symmetric 3×3 matrix C
$f: \mathfrak{X} \to \mathfrak{Y}$	map between Banach spaces
$Df(x_0)$	derivative of f at a point $x_0 \in \mathfrak{X}$
e^{tA}	semigroup generated by an operator A
$\boldsymbol{AB}$	multiplication of linear maps or matrices
Ax or $\boldsymbol{A}\cdot x$	linear operator applied to x
$Df(x)\cdot v$	derivative of f at x in direction v
$\boldsymbol{\sigma}\cdot\boldsymbol{n} = \sigma^{ab}n_b$	contraction of tensors
$\boldsymbol{S}: \boldsymbol{D} = S^{AB}D_{AB}$	double contraction of tensors
$f \circ g$	composition of maps

A POINT OF DEPARTURE

This preliminary chapter provides a quick survey of a few standard topics in elasticity theory from a classical point of view. The treatment is exclusively in Euclidean space $\mathbb{R}^3$, using standard Euclidean coordinates. One of the first tasks we face in the book is to repeat this material in a more general "intrinsic" context. The preview aims to be as elementary as possible, while still getting to a few issues of current interest. It can be read prior to, or in conjunction with, the main body of the text. The only background needed is calculus of several variables and linear algebra; some first-year physics is helpful for motivation.[1]

Warning. This introductory material is not where the book actually starts. This is intended to give certain readers a quick overview of where we are going. It proceeds at a very different tempo from the text and is sometimes chatty and imprecise. The material presented may be good for some readers, especially mathematicians who wish to learn elasticity for the first time. Experienced readers may wish to omit this and turn directly to Chapter 1.

1. KINEMATICS

In continuum mechanics, *kinematics* refers to the mathematical description of the deformation and motion of a piece of material. For example, if the beam shown in (a) of Figure 1.1 is loaded, it will bend. This is an example of a *deforma-*

[1]We are indebted to John Ball, whose lectures "Elementary Elasticity from Scratch" at Berkeley inspired this preview. Parts of the exposition are taken directly from his lectures, but have been rewritten to conform to the notations of this book. Of course any inaccuracies are the responsibility of the authors.

tion of a body (see Figure 1.1(b)). We describe the new static equilibrium state it assumes by mapping points in the unloaded (reference) state to their corresponding positions in the loaded (current) state. An example of a *motion* of a body is the oscillatory behavior of the same beam as it responds to the blow of a sledge hammer (Figure 1.1(c)).

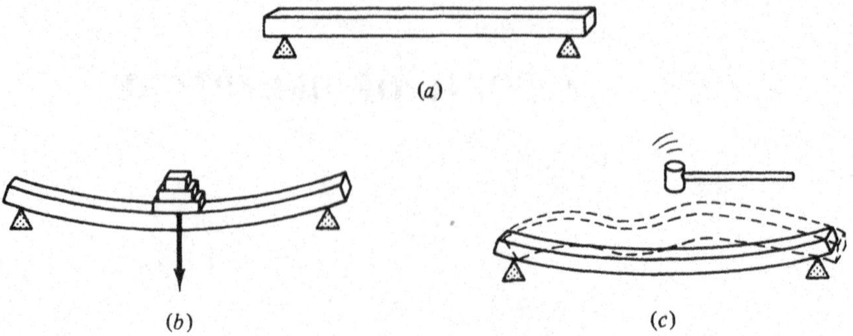

Figure 1.1 (a) Beam (reference configuration). (b) Loaded beam (deformed configuration). (c) Oscillating beam (motion).

Motions and Configurations Abstractly, let a *reference configuration* $\mathcal{B}$ be chosen for our body. By this, we mean the closure of an open set in $\mathbb{R}^3$ with piecewise smooth boundary. A *configuration* of $\mathcal{B}$ is a mapping $\phi: \mathcal{B} \longrightarrow \mathbb{R}^3$ that is sufficiently smooth, orientation preserving, and invertible. Points in $\mathcal{B}$ are denoted $X = (X^1, X^2, X^3) \in \mathcal{B}$ and are called *material points*, while points in $\mathbb{R}^3$ are denoted $x = (x^1, x^2, x^3) \in \mathbb{R}^3$ and are called *spatial points*. We write $x = \phi(X)$.[2] The words "configuration" and "deformation" are used synonymously.

A *motion* of $\mathcal{B}$ is a time-dependent family of configurations, written $x = \phi(X, t)$. The *velocity* of the material point X is defined by $V(X, t) = (\partial/\partial t)\phi(X, t)$ regarded as a vector emanating from the point x. The velocity viewed as a function of (x, t), denoted $v(x, t)$, is called the *spatial velocity*. Thus $V(X, t) = v(x, t)$, where $x = \phi(X, t)$. See Figure 1.2. If ϕ has components ϕ^1, ϕ^2, and ϕ^3—that is, if $\phi = (\phi^1, \phi^2, \phi^3)$—then $V = (V^1, V^2, V^3)$, where $V^i = \partial\phi^i/\partial t$ $(i = 1, 2, 3)$.

Acceleration and Material Derivatives The *material acceleration* of a motion $\phi(X, t)$ is defined by

$$A(X, t) = \frac{\partial^2 \phi}{\partial t^2}(X, t) = \frac{\partial V}{\partial t}(X, t).$$

[2]In the main text, components relative to general coordinates are denoted with letters $a, b, \ldots$ or $A, B, \ldots$ and Euclidean coordinates by $i, j, \ldots$ or $I, J, \ldots$. In this chapter, we employ only Euclidean coordinates and use $i, j, k, \ldots$ for simplicity.

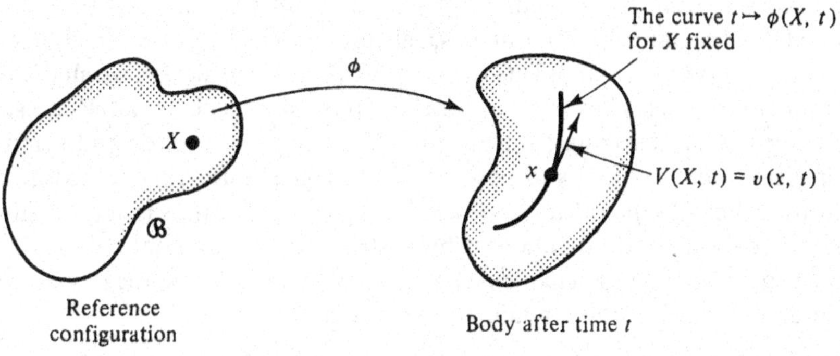

Figure 1.2

By the chain rule,

$$\frac{\partial V}{\partial t} = \frac{\partial v}{\partial t} + (v \cdot \nabla)v; \qquad \text{that is,} \quad \frac{\partial V^i}{\partial t} = \frac{\partial v^i}{\partial t} + \frac{\partial v^i}{\partial x^j} v^j.$$

More generally, if $Q(x, t)$ is a material quantity—a given function of (X, t)—and $q(x, t) = Q(X, t)$ is the same quantity expressed as a function of (x, t), then the chain rule gives

$$\frac{\partial Q}{\partial t} = \frac{\partial q}{\partial t} + (v \cdot \nabla)q; \qquad \text{that is,} \quad \frac{\partial Q}{\partial t} = \frac{\partial q}{\partial t} + \frac{\partial q}{\partial x^j} v^j.$$

The right-hand side is called the *material derivative* of q and is denoted $Dq/Dt = \dot{q}$. Thus Dq/Dt is the derivative of q with respect to t, holding X fixed, while $\partial q/\partial t$ is the derivative of q with respect to t holding x fixed. In particular, the acceleration is given by $\dot{v} = Dv/Dt = \partial V/\partial t$.

Deformation Gradient and Polar Decomposition. Above, V was used for the material velocity, a notation we shall use throughout the book. Below, V will also, according to standard usage, denote a matrix. The latter usage will occur in isolated places at various points in the book. The meaning of this symbol will always be clear from the context.

The 3×3 matrix of partial derivatives of ϕ is denoted $F = D\phi$ and is called the *deformation gradient*. Thus, $F^i{}_J = \partial \phi^i / \partial X^J$. Since ϕ is assumed to preserve orientation, $\det F > 0$. By the polar decomposition from linear algebra, we can uniquely decompose F as

$$F = RU = VR$$

where R is a proper orthogonal matrix called the *rotation*, and U and V—called the right and left *stretch tensors*—are positive-definite and symmetric. Indeed, $U = \sqrt{F^T F}$ and $V = \sqrt{F F^T}$, where F^T is the transpose of F; $(F^T)_i{}^J = F^i{}_J$. We call $C = F^T F = U^2$ the *right Cauchy–Green tensor* and $b = F F^T = V^2$ is the *left Cauchy–Green tensor*. Thus,[3] $C_{ij} = F^k{}_i F^k{}_j$ and $b^{ij} = F^i{}_k F^j{}_k$. In this setting of Euclidean rectangular coordinates, raising or lowering indices does not affect the values of tensor components (e.g., $F^i{}_j = F_{ij}$). Nevertheless, we shall attempt to maintain natural placement as much as possible. Since U and V are similar, their eigenvalues are equal; since U and V are positive-definite and symmetric, their eigenvalues are real and positive. These eigenvalues are denoted $\lambda_1, \lambda_2, \lambda_3$ and are called the *principal stretches*. Thus, a deformation is locally given to first order by a rotation followed by a stretching by amounts λ_i along three orthogonal axes (the *principal directions*) or vice versa. Thus, the deviation of λ_i from unity measures the amount of *strain* in a deformation.

2 BALANCE LAWS

By balancing the rate of change of momentum of a portion of a body with the total force acting on it, the fundamental equations of a continuum are derived. Other laws that fall into the same framework are conservation of mass and the balance of torque with moment of momentum.

Conservation of Mass Given a motion $\phi(X, t)$ of a body $\mathcal{B}$, let $\rho(x, t)$ be the mass density of the deformed body at time t and let $\rho_{\text{Ref}}(X)$ be the mass density in the reference configuration. Let $\mathcal{U} \subset \mathcal{B}$ be a subbody (an open set with piecewise smooth boundary) and let $\mathcal{U}_t = \phi(\mathcal{U}, t)$ be the subbody after time t. Conservation of mass states that

$$\text{mass}(\mathcal{U}) = \text{mass}(\mathcal{U}_t); \qquad \text{that is,} \quad \int_{\mathcal{U}} \rho_{\text{Ref}}(X)\, dX = \int_{\mathcal{U}_t} \rho(x, t)\, dx.$$

By the change of variables formula and the arbitrariness of $\mathcal{U}$, conservation of mass is equivalent to $\rho_{\text{Ref}} = J\rho$, where $J = \det(F)$ is the Jacobian of ϕ with t held constant. The formula $DJ/Dt = J \operatorname{div} v$, where $\operatorname{div} v = \partial v^i/\partial x^i$ is the divergence of v with t held constant, can be verified by direct calculation. Thus, $\rho_{\text{Ref}} = J\rho$ yields the equation of continuity: $D\rho/Dt + \rho \operatorname{div} v = 0$. Therefore, we may interpret $J = \rho v$ as the *mass flux*, and, by virtue of the divergence theorem, rewrite conservation of mass as

$$\int_{\mathcal{U}_t} \frac{\partial \rho}{\partial t}(x, t)\, dx = -\int_{\partial \mathcal{U}_t} J \cdot n\, da$$

where n is the unit outward normal of $\partial \mathcal{U}_t$ and da is its corresponding area element.

Balance of Momentum A body $\mathcal{B}$ undergoing a motion $\phi(X, t)$ will be acted on by applied *body forces* $b(x, t)$ (per unit mass) and applied *surface forces* $\tau(x, t)$ (per unit area), as in Figure 2.1. The pair of forces (b, τ) is called the *load*,

[3] Repeated indices are assumed summed unless otherwise indicated; for example, $F^k{}_i F^k{}_j = \sum_{k=1}^{3} F^k{}_i F^k{}_j$.

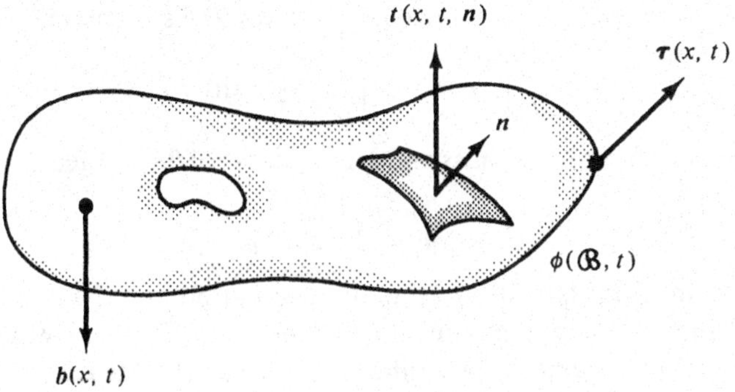

Figure 2.1

and is often given or prescribed in advance. For example, $\boldsymbol{b}$ may be gravity and $\boldsymbol{\tau}$ may be a load as in Figure 1.1(b). In addition, the body generally experiences internal forces of stress across any given surface. Let $\boldsymbol{t}(x, t, \boldsymbol{n})$ be the force per unit area at position x at time t across a surface element with unit normal $\boldsymbol{n}$; in other words, $\boldsymbol{t}$ is the force per unit deformed area engendered by the material outside (the $+\boldsymbol{n}$ direction) acting on the material inside (the $-\boldsymbol{n}$ direction).

The continuum analog of Newton's second law states that for any subbody $\mathcal{U} \subset \mathcal{B}$,

$$\frac{d}{dt}\int_{\mathcal{U}_t} \rho \boldsymbol{v}\, dx = \int_{\partial\mathcal{U}_t} \boldsymbol{t}(x, t, \boldsymbol{n})\, da + \int_{\mathcal{U}_t} \rho \boldsymbol{b}\, dx.$$

This assertion is called *balance of (linear) momentum*. Likewise, balance of *angular momentum* (measured with reference to the origin of coordinates) states that

$$\frac{d}{dt}\int_{\mathcal{U}_t} \boldsymbol{x} \times \rho \boldsymbol{v}\, dx = \int_{\partial\mathcal{U}_t} \boldsymbol{x} \times \boldsymbol{t}(x, t, \boldsymbol{n})\, da + \int_{\mathcal{U}_t} \boldsymbol{x} \times \rho \boldsymbol{b}\, dx$$

where $\boldsymbol{x}$ is the vector from the origin to the point x. (In the present context, $\boldsymbol{x}$ and x may be regarded as identical.)

Cauchy's theorem (proved in Section 2.1) states that if balance of momentum holds, then $\boldsymbol{t}$ depends linearly on $\boldsymbol{n}$; that is, there is a two-tensor $\boldsymbol{\sigma}(x, t)$ such that $\boldsymbol{t}(x, t, \boldsymbol{n}) = \boldsymbol{\sigma}(x, t)\boldsymbol{n}$. Componentwise, $\boldsymbol{\sigma}(x, t)$ is a 3×3 matrix σ^{ij} and $\boldsymbol{t}$ is a vector with components t^i such that $t^i = \sigma^{ij} n_j$ (sum on j). We call $\boldsymbol{\sigma}$ the *Cauchy stress tensor*. Thus, σ^{ij} is the ith component of the force per unit area measured across a surface element with unit normal $\boldsymbol{e}_j$, the jth standard basis vector.

Substituting $\boldsymbol{t} = \boldsymbol{\sigma} \cdot \boldsymbol{n}$ into the balance of (linear) momentum and using the divergence theorem gives

$$\frac{d}{dt}\int_{\mathcal{U}_t} \rho \boldsymbol{v}\, dx = \int_{\mathcal{U}_t} \operatorname{div} \boldsymbol{\sigma}\, dx + \int_{\mathcal{U}_t} \rho \boldsymbol{b}\, dx$$

where $(\operatorname{div} \boldsymbol{\sigma})^i = \partial\sigma^{ij}/\partial x^j$. By the change of variables theorem,

$$\frac{d}{dt}\int_{\mathcal{U}_t} q\,dx = \int_{\mathcal{U}_t} (\dot{q} + q \operatorname{div} v)\,dx.$$

Thus from conservation of mass and arbitrariness of $\mathcal{U}_t$, we get *Cauchy's equation of motion*:

$$\rho\dot{v} = \operatorname{div} \boldsymbol{\sigma} + \rho \boldsymbol{b}.$$

Note that in this derivation, smoothness hypotheses have been implicitly assumed. Such hypotheses are invalid if phenomena like shock waves are studied, and this derivation needs to be studied more carefully (see Chapter 2).

Similar arguments show that the balance of angular momentum leads to symmetry of the stress: $\sigma^{ij} = \sigma^{ji}$.

The Material Picture In continuum mechanics it is often convenient to express things with respect to a reference configuration rather than to the current configuration—that is, to use material quantities rather than spatial ones. When this idea is applied to the Cauchy stress, we get the Piola–Kirchhoff stress. The *first Piola–Kirchhoff stress vector* is the vector $\boldsymbol{T}(X, t, \boldsymbol{N})$, which is parallel to the Cauchy stress $\boldsymbol{t}(x, t, \boldsymbol{n})$, but measures the force per unit *undeformed area*. Here, $x = \phi(X, t)$ and $\boldsymbol{n}$ and $\boldsymbol{N}$ are unit vectors, normal to the corresponding deformed and undeformed area elements. Thus, we desire $\boldsymbol{T}\,dA = \boldsymbol{t}\,da$ (see Figure 2.2).

The *Piola Identity*:

$$\operatorname{DIV}(JF^{-1}) = 0; \qquad \text{that is,} \quad \frac{\partial}{\partial X^J}\left(J\frac{\partial X^J}{\partial x^i}\right) = 0$$

is useful. It follows from the fact that JF^{-1} is the transpose of the matrix of cofactors of F; $(JF^{-1})_k{}^r = \epsilon_{ijk}\epsilon^{pqr}F^i{}_pF^j{}_q$ where ϵ_{ijk} is $+1$ or -1 according as

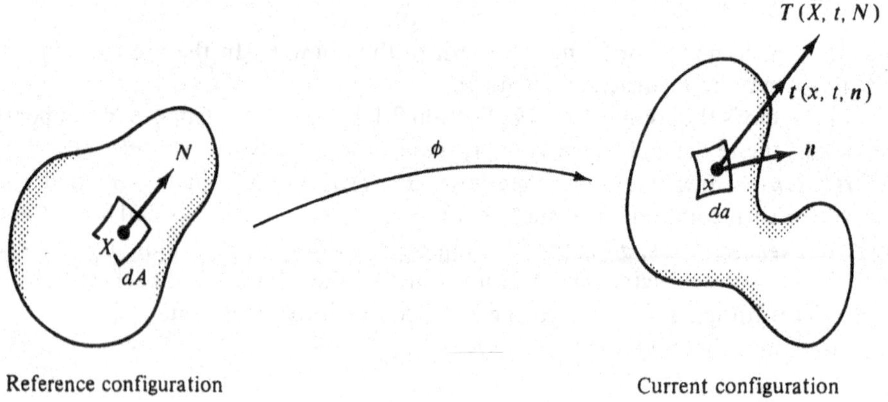

Figure 2.2

i, j, k is an even or odd permutation of 1, 2, 3 and is zero otherwise, and similarly for ϵ^{pqr}. From the Piola identity and the chain rule, we find that if w is a vector function of (x, t) and we define its *Piola transform* W by $W(X, t) = JF^{-1}w(x, t)$—that is, $W^i(X, t) = (\partial X^i/\partial x^j)Jw^j$—then

$$\text{DIV } W = J \text{ div } w; \qquad \text{that is,} \quad \frac{\partial W^i}{\partial X^i} = J\frac{\partial w^i}{\partial x^i}.$$

From the divergence and change of variables theorems, we get

$$\int_{\partial \mathfrak{U}_t} w \cdot n \, da = \int_{\partial \mathfrak{U}} W \cdot N \, dA.$$

Therefore,

$$\int_{\partial \mathfrak{U}_t} w \cdot n \, da = \int_{\partial \mathfrak{U}} w \cdot JF^{-T} N \, dA$$

where $F^{-T} = (F^{-1})^T = (F^T)^{-1}$, and so $n \, da = JF^{-T}N \, dA$. Multiplying by σ gives $\sigma n \, da = t \, da = J\sigma F^{-T} N \, dA$. Thus, we can write $T = PN$, where $P = J\sigma F^{-T}$, that is, $P^{iJ} = J\sigma^{ik}(\partial X^J/\partial x^k)$. This two-tensor P^{iJ} is called the *first Piola–Kirchhoff stress*. In terms of it, balance of momentum becomes

$$\frac{d}{dt}\int_{\mathfrak{U}} \rho_{\text{Ref}} V \, dx = \int_{\partial \mathfrak{U}} PN \, dA + \int_{\mathfrak{U}} \rho_{\text{Ref}} B \, dX$$

where $B(X, t) = b(x, t)$ is the body force expressed as a function of X. The pointwise version is

$$\rho_{\text{Ref}} \frac{\partial V}{\partial t} = \text{DIV } P + \rho_{\text{Ref}} B$$

where $(\text{DIV } P)^i = (\partial/\partial X^J)P^{iJ}$. This form of Cauchy's equations of motion is the most suitable for studying Hamiltonian and variational principles. See Section 4 which follows.

The *second Piola–Kirchhoff stress vector* is obtained by transforming $PN = T$, which is a vector based at x, to a vector based at X by setting $\mathfrak{T} = F^{-1}T$. Thus $\mathfrak{T} = SN$, where $S = F^{-1}P$, that is, $S^{IJ} = (\partial X^I/\partial x^k)P^{kJ}$; S is a 3×3 matrix called the *second Piola–Kirchhoff stress tensor*.

Already the reader will notice the need for a clear systematic index notation. Such notation is given by tensor analysis and is developed in Chapter 1. This tensor notation enables us to work in general coordinates, to remember the formulas, and it occasionally prevents us from asking intrinsically absurd questions. For example, we have remarked that σ is symmetric: $\sigma^{ij} = \sigma^{ji}$. We might ask: is P symmetric? This question seems reasonable to ask at this stage, but when we correctly understand P as a tensor, we will realize that the question does not make sense. It is analogous to asking: is a bilinear form $\beta: \mathcal{V} \times \mathcal{W} \to \mathbb{R}$, where $\mathcal{V}$ and $\mathcal{W}$ are *different* vector spaces, symmetric? This question as posed is meaningless. Rather than P, one should ask: is S symmetric? Indeed, it is.

3 ELASTIC MATERIALS

Newton's second law in particle mechanics cannot be solved until we know how the force depends on the position and velocity of the particle. Likewise, Cauchy's equation of motion in continuum mechanics cannot be solved until we know how the stress depends on the motion.

Elastic and Hyperelastic Materials A material is *elastic* when we can write the first Piola–Kirchhoff stress $\boldsymbol{P}$ as a function $\hat{\boldsymbol{P}}$ of points $X \in \mathcal{B}$ and 3×3 matrices $\boldsymbol{F}$ with $\det \boldsymbol{F} > 0$ such that the stress associated with a configuration ϕ is

$$\boldsymbol{P}(X, t) = \hat{\boldsymbol{P}}(X, \boldsymbol{F}(X, t))$$

where $\boldsymbol{F}$ is the deformation gradient of ϕ, $F^i{}_J = \partial\phi^i/\partial X^J$, as usual. Why we do not include dependence on higher derivatives is an important point that is discussed at length in Chapter 3. By contrast, the viscous stresses in a *fluid* depend on the *velocity* gradients. Specification of a $\hat{\boldsymbol{P}}$ for an elastic material is called a *stress-strain law*, and $\hat{\boldsymbol{P}}$ is called a *constitutive function.*

A useful class of elastic materials are called *hyperelastic*. These are materials for which there is a *stored energy function* $\hat{W}$ depending on points $X \in \mathcal{B}$ and 3×3 matrices $\boldsymbol{F}$ with $\det \boldsymbol{F} > 0$ such that $\hat{\boldsymbol{P}} = \rho_{\text{Ref}}(\partial \hat{W}/\partial \boldsymbol{F})$, that is, $\hat{P}_i{}^J = \rho_{\text{Ref}}(\partial \hat{W}/F^i{}_J)$.

Material Frame Indifference Before giving examples, it is useful to discuss more carefully how $\hat{W}$ depends on $\boldsymbol{F}$. Consider a rigid motion of space given by the transformation $\xi: \mathbb{R}^3 \longrightarrow \mathbb{R}^3$; $\xi(x) = \boldsymbol{Q}x + \boldsymbol{c}$, where $\boldsymbol{Q}$ is a proper orthogonal matrix and $\boldsymbol{c}$ is a constant vector. Such a transformation should just rotate the stress vector $\boldsymbol{T}(X, \boldsymbol{F}, \boldsymbol{N}) = \hat{\boldsymbol{P}}(X, \boldsymbol{F})\boldsymbol{N}$. Thus, we demand the identity $\hat{\boldsymbol{T}}(X, \boldsymbol{QF}, \boldsymbol{N}) = \boldsymbol{Q}\hat{\boldsymbol{T}}(X, \boldsymbol{F}, \boldsymbol{N})$, or, equivalently, $\hat{\boldsymbol{P}}(X, \boldsymbol{QF}) = \boldsymbol{Q}\hat{\boldsymbol{P}}(X, \boldsymbol{F})$. This states simply that if we view the same configuration from a rotated point of view, then the stress transforms by the same rotation. This requirement is called *frame indifference or objectivity*. For a hyperelastic material, frame indifference is equivalent to the identity $\hat{W}(X, \boldsymbol{QF}) = \hat{W}(X, \boldsymbol{F})$ for all 3×3 matrices $\boldsymbol{F}$ with $\det \boldsymbol{F} > 0$ and all proper orthogonal matrices $\boldsymbol{Q}$. Moreover, frame indifference implies balance of angular momentum (symmetry of the stress). (See Section 3.2 for the proofs of these assertions.) If we write $\boldsymbol{F} = \boldsymbol{RU}$ and choose $\boldsymbol{Q} = \boldsymbol{R}^{-1}$, we see that $\hat{W}(X, \boldsymbol{F}) = \hat{W}(X, \boldsymbol{U}) = \hat{W}(X, \sqrt{\boldsymbol{C}})$, where $\boldsymbol{C}$ is the (right) Cauchy–Green tensor. Thus, frame indifference is equivalent to the assertion that $\hat{W}$ depends on $\boldsymbol{F}$ through $\boldsymbol{C}$.

The relationship $\hat{\boldsymbol{P}} = \rho_{\text{Ref}}(\partial \hat{W}/\partial \boldsymbol{F})$ is equivalent to $\hat{\boldsymbol{S}} = 2\rho_{\text{Ref}}(\partial \hat{W}/\partial \boldsymbol{C})$, using the chain rule. In Section 3.3 we shall see how to write $\boldsymbol{\sigma}$ as a derivative of $\hat{W}$ with respect to the metric tensor.

Isotropic Materials Most explicit models of elastic materials are not only frame indifferent, but exhibit symmetry of the actual material as well. An elastic material is *homogeneous* if $\hat{\boldsymbol{P}}$ does not depend explicitly upon X. Thus, different portions of the material exhibit the same mechanical behavior. An elastic material is *isotropic* if $\hat{\boldsymbol{P}}(X, \boldsymbol{F}\boldsymbol{Q}) = \hat{\boldsymbol{P}}(X, \boldsymbol{F})$ for all proper orthogonal 3×3 matrices $\boldsymbol{Q}$. This means that the stress for a given deformation gradient $\boldsymbol{F}$ is unaltered if we *first* rotate the material by $\boldsymbol{Q}$. For hyperelastic materials, isotropy is equivalent to $\hat{W}(X, \boldsymbol{F}\boldsymbol{Q}) = \hat{W}(X, \boldsymbol{F})$ for all proper orthogonal $\boldsymbol{Q}$; that is, $\hat{W}$ depends on $\boldsymbol{F}$ only through $\boldsymbol{V}$.

A hyperelastic material is frame indifferent, homogeneous, and isotropic if and only if $\hat{W}(X, \boldsymbol{F}) = \Phi(\lambda_1, \lambda_2, \lambda_3)$, where Φ is a symmetric function of the principal stretches $\lambda_1, \lambda_2, \lambda_3$. This assertion follows by writing $\boldsymbol{F} = \boldsymbol{R}\boldsymbol{U} = \boldsymbol{R}\boldsymbol{L}\,\mathrm{diag}\,(\lambda_1, \lambda_2, \lambda_3)\boldsymbol{L}^{-1}$ for proper orthogonal matrices $\boldsymbol{R}$ and $\boldsymbol{L}$ and noting that

$$\begin{aligned}\hat{W}(X, \boldsymbol{F}) &= \hat{W}(X, \boldsymbol{R}\boldsymbol{L}\,\mathrm{diag}\,(\lambda_1, \lambda_2, \lambda_3)\boldsymbol{L}^{-1})\\ &= \hat{W}(X, \mathrm{diag}\,(\lambda_1, \lambda_2, \lambda_3)\boldsymbol{L}^{-1}) \quad \text{(by frame indifference)}\\ &= \hat{W}(X, \mathrm{diag}\,(\lambda_1, \lambda_2, \lambda_3)) \quad \text{(by isotropy).}\end{aligned}$$

The argument is now readily completed. In Section 3.5 we show that this is equivalent to $\hat{W}$ being a function only of the principal invariants of $\boldsymbol{C}$. We also show that the general form of the stress for an isotropic elastic material is $\boldsymbol{S} = \alpha_0 \boldsymbol{I} + \alpha_1 \boldsymbol{C} + \alpha_2 \boldsymbol{C}^2$ or $\boldsymbol{\sigma} = \beta_0 \boldsymbol{i} + \beta_1 \boldsymbol{b} + \beta_2 \boldsymbol{b}^2$ ($\boldsymbol{I}, \boldsymbol{i}$ = identity matrices), where $\boldsymbol{C} = \boldsymbol{F}^T\boldsymbol{F}$, $\boldsymbol{b} = \boldsymbol{F}\boldsymbol{F}^T$, and α_i and β_i are scaler functions of the invariants of $\boldsymbol{C}$ and $\boldsymbol{b}$, respectively.

Linear Elasticity The most widely used models have been those in linear elasticity. This theory is appropriate when small displacements from a given deformation are involved. In Chapter 4, we discuss in detail how to obtain this theory by linearizing the nonlinear theory. Briefly, and in a special case, this proceeds as follows. Let $\hat{\boldsymbol{P}}$ be a given constitutive stress function for a (nonlinear) homogeneous elastic material. Suppose $\hat{\boldsymbol{P}}(\boldsymbol{I}) = \boldsymbol{O}$. Thus, the trivial deformation $\phi(X) = X$ has zero stress. Consider a family of motions $\phi_\epsilon(X, t)$ that depend on a small parameter ϵ. Suppose $\phi_0(X, t) = X$, so the motion ϕ_0 satisfies (trivially) the equations of motion $\rho_{\mathrm{Ref}}(\partial \boldsymbol{V}/\partial t) = \mathrm{DIV}\,\hat{\boldsymbol{P}}(\boldsymbol{F})$ with zero body force. Now suppose that ϕ_ϵ satisfies these same equations for each ϵ. Writing $\phi_\epsilon(X, t) = X + \epsilon \boldsymbol{u}(X, t) + O(\epsilon^2)$ we seek to find the equations governing the linear part $\boldsymbol{u}$. Note that

$$\boldsymbol{u} = \frac{\partial}{\partial \epsilon}\phi_\epsilon \bigg|_{\epsilon=0} \quad \text{and} \quad \frac{\partial \boldsymbol{u}}{\partial t} = \frac{\partial}{\partial \epsilon}\boldsymbol{V}_\epsilon \bigg|_{\epsilon=0}.$$

Differentiating

$$\rho_{\mathrm{Ref}}\frac{\partial \boldsymbol{V}_\epsilon}{\partial t} = \mathrm{DIV}\,\hat{\boldsymbol{P}}(\boldsymbol{F}_\epsilon); \qquad \text{that is} \quad \rho_{\mathrm{Ref}}\frac{\partial V^i_\epsilon}{\partial t} = \frac{\partial}{\partial X^j}\left[\hat{P}^{ij}\left(\frac{\partial \phi^k_\epsilon}{\partial X^l}\right)\right]$$

in ϵ at $\epsilon = 0$ gives

$$\rho_{\mathrm{Ref}}\frac{\partial^2 u^i}{\partial t^2} = \frac{\partial}{\partial X^j}\left(\frac{\partial \hat{P}^{ij}}{\partial F^k{}_l}(\boldsymbol{I})\,\frac{\partial u^k}{\partial X^l}\right).$$

The quantity $(\partial \hat{P}^{ij}/\partial F^k{}_l)(\boldsymbol{I}) = \mathrm{c}^{ij}{}_k{}^l(X)$, which depends on the four indices $i, j, k,$

l, is called the *classical elasticity tensor.* For homogeneous materials these are constants, and the equations of linear elasticity take the form

$$\rho_{\text{Ref}} \frac{\partial^2 u^i}{\partial t^2} = \mathrm{c}^{ij}{}_k{}^l \frac{\partial^2 u^k}{\partial X^j \, \partial X^l}.$$

This is a type of vector wave equation examined in Chapters 4 and 6.

In the special case of a homogeneous isotropic material, there are constants λ and μ called *Lamé moduli* such that

$$\mathrm{c}_{ijkl} = \lambda \delta_{ij} \delta_{kl} + \mu(\delta_{ik}\delta_{jl} + \delta_{il}\delta_{jk})$$

where $\delta_{ij} = 1$ if $i = j$ but 0 if $i \neq j$. This is proved in a manner similar to the corresponding assertions for nonlinear elastic materials. Much of the classical literature on elasticity is, in fact, confined to this case (see the references in the preface).

An Elastic Fluid Perhaps the simplest example of a *nonlinear* elastic material is one for which $\hat{W}(F) = h(J)$; that is, $\hat{W}$ is a function only of the determinant of F. This is homogeneous, frame indifferent, and isotropic. Here $\Phi(\lambda_1, \lambda_2, \lambda_3) = h(\lambda_1 \lambda_2 \lambda_3)$. For this $\hat{W}$, the first Piola–Kirchhoff stress tensor is given by

$$\hat{\boldsymbol{P}} = \rho_{\text{Ref}} \frac{\partial \hat{W}}{\partial \boldsymbol{F}} = \rho_{\text{Ref}} h'(J) \cdot \frac{\partial J}{\partial \boldsymbol{F}}.$$

A calculation gives $\partial J / \partial \boldsymbol{F} = J\boldsymbol{F}^{-T}$ (the matrix of cofactors). Thus, the Cauchy stress is $\hat{\boldsymbol{\sigma}} = (1/J)\hat{\boldsymbol{P}}\boldsymbol{F}^T = \rho_{\text{Ref}}\, h'(J)\boldsymbol{i}$. Because of the relative simplicity of $\hat{\boldsymbol{\sigma}}$ over $\hat{\boldsymbol{P}}$, this example is easiest to express spatially. Let $p(\rho) = -\rho_{\text{Ref}} h'(J)$, so $\boldsymbol{\sigma} = -p(\rho)\boldsymbol{i}$. This defines the *pressure.* The Cauchy equations of motion are therefore

$$\rho \frac{D\boldsymbol{v}}{Dt} = -\nabla p + \rho \boldsymbol{b}.$$

These are the Euler equations for a compressible perfect fluid. Thus, we see that elastic materials include perfect fluids as well. (The Navier–Stokes equations require the dependence of $\boldsymbol{\sigma}$ on $\nabla \boldsymbol{v}$, not just on $\boldsymbol{F}$, to obtain the viscous terms.)

In general, solids and liquids can be distinguished by their symmetry groups. At a point $X \in \mathcal{B}$, the *symmetry group* of a material is defined by

$$\mathcal{S}_X = \{A \in \mathfrak{M}^+_{3\times 3} \mid \hat{\boldsymbol{P}}(X, \boldsymbol{F}\boldsymbol{A}) = \hat{\boldsymbol{P}}(X, \boldsymbol{F}) \text{ for all } \boldsymbol{F} \in \mathfrak{M}^+_{3\times 3}\}$$

where $\mathfrak{M}^+_{3\times 3} = \{3 \times 3$ matrices $\boldsymbol{A}$ with $\det \boldsymbol{A} > 0\}$. The elastic material is *solid* when $\mathcal{S}_X \subset \mathrm{SO}(3)$ (the 3×3 proper orthogonal matrices), and *fluid* when $\mathcal{S}_X = \{\boldsymbol{A} \in \mathfrak{M}^+_{3\times 3} \mid \det \boldsymbol{A} = 1\} = \mathrm{SL}(3)$. Isotropy means $\mathrm{SO}(3) \subset \mathcal{S}_X$. Thus, $\hat{W}(\boldsymbol{F}) = h(J)$ is an example of an elastic fluid. *Crystals* generally have a symmetry group lying between SO(3) and SL(3). The classification of materials by means of symmetry groups is not discussed in this book; see Wang and Truesdell [1973].

Incompressibility Incompressible elasticity is obtained by imposing the constraint $J = 1$, or, equivalently, div $\boldsymbol{v} = 0$. Corresponding to this constraint, we have a "Lagrange multiplier" in the stress. The stress for a motion $\phi(X, t)$ is $\boldsymbol{P}(X, t) = -pF^{-T} + \hat{\boldsymbol{P}}(X, \boldsymbol{F}(X, t))$, where $\hat{\boldsymbol{P}}$ is a given constitutive function and p is to be determined by the constraint $J = 1$. For example, $\hat{\boldsymbol{P}} = 0$ corresponds to the Euler equations of a perfect incompressible fluid:

$$\left.\begin{aligned} \rho \frac{D\boldsymbol{v}}{Dt} &= -\nabla p + \rho \boldsymbol{b}, \\ \operatorname{div} \boldsymbol{v} &= 0. \end{aligned}\right\}$$

Mooney–Rivlin Materials Perhaps the nonlinear elastic solid material that is best understood from the point of view of constitutive theory is rubber, which usually behaves like an incompressible, homogeneous, hyperelastic, isotropic solid. A stored energy function for rubberlike materials that is often used is due to Mooney [1940] and Rivlin [1948a]. This is given in terms of the principal stretches λ_1, λ_2 and λ_3 by

$$\Phi(\lambda_1, \lambda_2, \lambda_3) = \alpha(\lambda_1^2 + \lambda_2^2 + \lambda_3^2 - 3) + \beta((\lambda_2\lambda_3)^2 + (\lambda_3\lambda_1)^2 + (\lambda_1\lambda_2)^2 - 3)$$

where α and β are positive constants. The 3s are included so that $\Phi(1, 1, 1) = 0$. The limiting case $\beta = 0$ is called a *neo-Hookean material.* Generalizations where the exponents "2" are replaced by other numbers have been successfully used by Ogden [1972]. The constants in the model are determined by experiment. Such constitutive functions are generally valid for deformations that are not too large; exactly "how large" is hard to specify without detailed experimental analysis.

For the compressible case, a model that is sometimes used is the *Hadamard material,* with

$$\Phi(\lambda_1, \lambda_2, \lambda_3) = \alpha(\lambda_1^2 + \lambda_2^2 + \lambda_3^2 - 3) + h(\lambda_1\lambda_2\lambda_3)$$

where $h(\delta)$ has a minimum at $\delta = 1$ and $h(\delta) \longrightarrow \infty$ as $\delta \longrightarrow 0$ and as $\delta \longrightarrow \infty$.

At this point, these models are largely ad hoc. Much work remains to be done on developing faithful models of real materials. Examples of other constitutive models are found in Truesdell and Noll [1965] and Malvern [1969].

4 BOUNDARY VALUE PROBLEMS

We shall append some initial and boundary conditions to Cauchy's equation of motion.

Boundary Conditions Some boundary conditions often encountered are:

(a) *Displacement.* $\phi(X, t) = \phi_d(X)$ is prescribed for $X \in \partial\mathcal{B}$.

(b) *Traction.* $\boldsymbol{P}(X, t)N(X) = \boldsymbol{\tau}(X)$ is prescribed for $X \in \partial\mathcal{B}$, where $N(X)$ is the outward unit normal to $\partial\mathcal{B}$ at X.

(c) *Pressure.* $\boldsymbol{\sigma}(x, t)\boldsymbol{n}(x) = -p(t)\boldsymbol{n}(x)$ is prescribed for $x \in \partial(\phi(\mathcal{B}, t))$, where p is a given scalar function of t.

Other boundary conditions may be a combination of these. For example, if a steel bar is pulled to a prescribed length, then displacement boundary conditions are used on its ends, and zero traction conditions are used on its sides. (See Figure 4.1.) In boundary condition (b), note that $\boldsymbol{\tau}(X)$ remains parallel throughout the motion. We say that such a load is *dead.* Blowing up a balloon provides an example of (c).

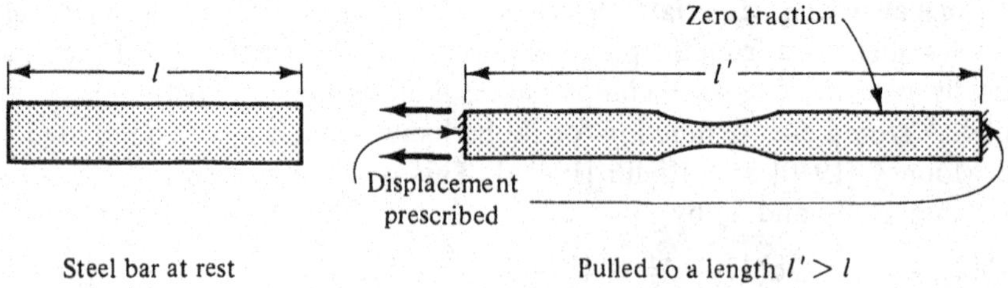

Figure 4.1

Elastostatics The basic mathematical problem in elastostatics is to find a configuration $\phi: \mathcal{B} \longrightarrow \mathbb{R}^3$ such that

$$\mathrm{DIV}(\boldsymbol{P}) + \rho_{\mathrm{Ref}}\boldsymbol{B} = \boldsymbol{0}$$

and satisfying a given set of boundary conditions. If $\boldsymbol{P}(X) = \hat{\boldsymbol{P}}(X, \boldsymbol{F}(X))$ is a given constitutive function for stress and if $\boldsymbol{B}$ is given, then this is a nonlinear boundary value problem for the unknown ϕ. Suppose that the material is hyperelastic and that the displacement is prescribed on $\partial_d \subset \partial\mathcal{B}$ and the traction is prescribed on $\partial_\tau \subset \partial\mathcal{B}$, where $\partial_d \cap \partial_\tau = \varnothing$ and $\overline{\partial_d \cup \partial_\tau} = \partial\mathcal{B}$. Define a functional I on the set $\mathcal{C}$ of all configurations $\phi: \mathcal{B} \longrightarrow \mathbb{R}^3$ satisfying $\phi = \phi_d$ on ∂_d, by means of the potential energy:

$$I(\phi) = \int_{\mathcal{B}} \{\rho_{\mathrm{Ref}}(X)\hat{W}(X, \boldsymbol{F}(X)) - \rho_{\mathrm{Ref}}(X)\boldsymbol{\phi}(X)\cdot\boldsymbol{B}(X)\}\, dX - \int_{\partial_\tau} \boldsymbol{\tau}(X)\cdot\boldsymbol{\phi}(X)\, dA(X).$$

Then, assuming sufficient smoothness, $\phi \in \mathcal{C}$ will be a solution to the boundary value problem if and only if the following *variational principle* holds: $\delta I(\phi) = 0$; that is, for all $\boldsymbol{u}: \mathcal{B} \longrightarrow \mathbb{R}^3$ vanishing on ∂_d,

$$DI(\phi)\cdot\boldsymbol{u} \equiv \frac{d}{d\epsilon} I(\phi + \epsilon\boldsymbol{u})\bigg|_{\epsilon=0} = 0.$$

Indeed, the chain rule gives the *weak form* of the equations:

$$DI(\phi)\cdot \boldsymbol{u} = \int_{\mathcal{B}} \left\{\rho_{\text{Ref}} \frac{\partial \hat{W}}{\partial F}\cdot \nabla \boldsymbol{u} - \rho_{\text{Ref}} \boldsymbol{u}\cdot \boldsymbol{B}\right\} dX - \int_{\partial_\tau} \boldsymbol{\tau}\cdot \boldsymbol{u}\, dA.$$

Integrating by parts and using $\hat{\boldsymbol{P}} = \rho_{\text{Ref}}(\partial \hat{W}/\partial F)$ and $\boldsymbol{u} = \boldsymbol{0}$ on ∂_d, we get

$$DI(\phi)\cdot \boldsymbol{u} = -\int_{\mathcal{B}} \boldsymbol{u}\cdot\{\text{DIV}\, \boldsymbol{P} + \rho_{\text{Ref}} \boldsymbol{B}\}\, dX + \int_{\partial_\tau} \boldsymbol{u}\cdot\{\boldsymbol{P}\boldsymbol{N} - \boldsymbol{\tau}\}\, dA.$$

Since $\boldsymbol{u}$ is arbitrary, this is equivalent to DIV $\boldsymbol{P} + \rho_{\text{Ref}} \boldsymbol{B} = \boldsymbol{0}$ on $\mathcal{B}$ and $\boldsymbol{P}\boldsymbol{N} = \boldsymbol{\tau}$ on ∂_τ. Variational structures are presented in Chapter 5, and existence and uniqueness theorems are studied in Chapter 6.

Elastodynamics In elastodynamics, the problem is to find a motion $\phi(X, t)$ that satisfies Cauchy's equation of motion with prescribed boundary conditions and prescribed initial deformation $\phi(X, 0)$ and velocity $V(X, 0)$. There is a Hamiltonian variational principle for this problem given by

$$\delta L(\phi) = 0 \qquad \text{where} \quad L(\phi) = \int_0^T \left\{\int_{\mathcal{B}} \tfrac{1}{2}\rho_{\text{Ref}} \|V\|^2\, dX - I(\phi)\right\} dt$$

and the variations are now over curves in $\mathcal{C}$ with prescribed end-points.

Less is known about existence and uniqueness for elastodynamics than for elastostatics. Even assuming existence, many key properties are not well understood. In particular, the basic problem of *elastic stability*—"Are minima of the potential I dynamically stable?"—is largely open. Some aspects of what is currently known are discussed in Section 6.6.

A Few Nonrigorous Examples A few basic examples are important to bear in mind when reading the theory. Examples of this type are discussed further in Chapter 7.

(a) *Nonuniqueness of equilibrium solutions* (buckling). See Figure 4.2.
(b) *Nonuniqueness for pure traction* (*dead load, no buckling*). See Figure 4.3.
(c) *Bifurcation to periodic oscillations* (Hopf bifurcation; flutter of venetian blinds). See Figure 4.4.

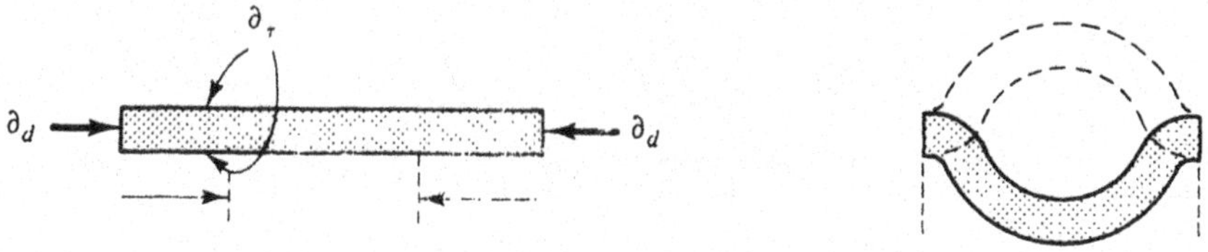

Prescribed displacement

Two solutions in two dimensions; a circle of solutions in three dimensions

Figure 4.2

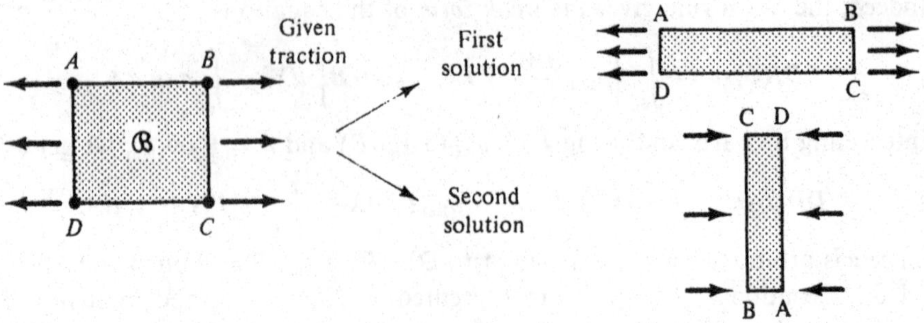

Figure 4.3

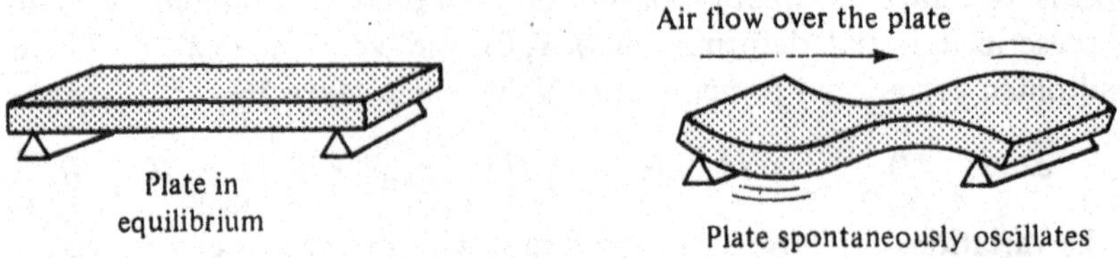

Figure 4.4

Exact Solutions Consider the case of a homogeneous material with $\boldsymbol{B} = \mathbf{0}$. A trivial, but instructive class of equilibrium solutions are the *homogeneous deformations*: $\boldsymbol{x} = \boldsymbol{F}\boldsymbol{X} + \boldsymbol{c}$, where $\boldsymbol{F}$ is a constant 3×3 matrix, and $\boldsymbol{c}$ is a constant vector. For example, consider simple shear (see Figure 4.5).

$$x^1 = X^1 + \kappa X^2, \qquad x^2 = X^2, \qquad x^3 = X^3$$

Here $$\boldsymbol{F} = \begin{bmatrix} 1 & \kappa & 0 \\ 0 & 1 & 0 \\ 0 & 0 & 1 \end{bmatrix}, \qquad \boldsymbol{C} = \boldsymbol{F}^T\boldsymbol{F} = \begin{bmatrix} 1 & \kappa & 0 \\ \kappa & \kappa^2 + 1 & 0 \\ 0 & 0 & 1 \end{bmatrix},$$

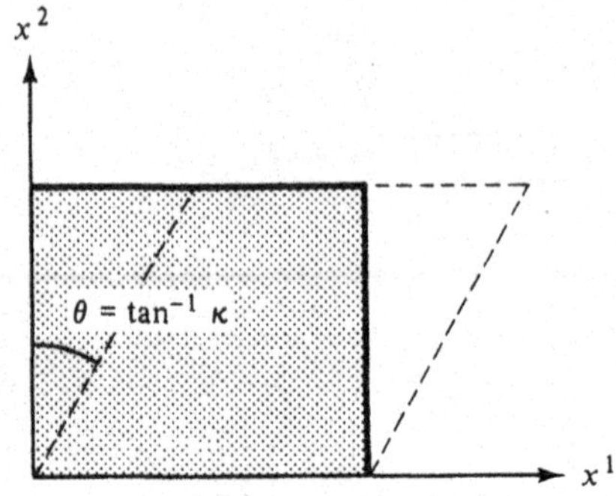

Figure 4.5

and
$$b = \begin{bmatrix} \kappa^2+1 & \kappa & 0 \\ \kappa & 1 & 0 \\ 0 & 0 & 1 \end{bmatrix}.$$

The general form of the Cauchy stress is $\boldsymbol{\sigma} = \beta_0 \boldsymbol{i} + \beta_1 \boldsymbol{b} + \beta_2 \boldsymbol{b}^2$, where β_0, β_1, β_2 are scalar functions of the invariants of $\boldsymbol{b}$, which are $\operatorname{tr}(\boldsymbol{b}) = 3 + \kappa^2$ (the trace of $\boldsymbol{b}$), $\frac{1}{2}[(\operatorname{tr} \boldsymbol{b})^2 - (\operatorname{tr} \boldsymbol{b}^2)] = 3 + \kappa^2$, and $\det \boldsymbol{b} = 1$. We obtain

$$\boldsymbol{\sigma} = \begin{bmatrix} \beta_0 + \beta_1(\kappa^2+1) + \beta_2[(\kappa^2+1)^2 + \kappa^2] & \beta_1\kappa + \beta_2\kappa(\kappa^2+2) & 0 \\ \beta_1\kappa + \beta_2\kappa(\kappa^2+2) & \beta_0 + \beta_1 + \beta_2(\kappa^2+1) & 0 \\ 0 & 0 & \beta_0 + \beta_1 + \beta_2 \end{bmatrix}.$$

The columns of $\boldsymbol{\sigma}$ give the forces acting on planes with normals in the three coordinate directions. Notice that the *normal stresses*, σ_{11}, σ_{22}, and σ_{33} need not vanish, so that simple shear cannot be maintained by a shear stress alone. However, note that if the reference state is unstressed, then $\boldsymbol{\sigma} = 0$ when $\kappa = 0$; that is, $\beta_0 + \beta_1 + \beta_2 = 0$ when $\kappa = 0$. In this case, σ_{33}, σ_{11}, and σ_{22} are $O(\kappa^2)$, that is, are second order in κ. Note too the *universal* relation $\sigma_{11} - \sigma_{22} = \kappa\sigma_{12}$, which holds for all isotropic elastic materials in simple shear.

Relevant to this example is the universal deformation theorem of Ericksen [1954], which states that if a motion $\phi(X, t)$ can be maintained by surface tractions alone in *any* homogeneous compressible isotropic hyperelastic material, then $\phi(X, t) = \boldsymbol{a} + \boldsymbol{G}X + \boldsymbol{\sigma}(\boldsymbol{b} + \boldsymbol{H}X)$, where $\boldsymbol{a}$, $\boldsymbol{b}$ are constant vectors and $\boldsymbol{G}$, $\boldsymbol{H}$ are constant 3×3 matrices. (See Shield [1971] for a simple proof.) Ericksen has also analyzed the universal motions for incompressible isotropic homogeneous hyperelasticity: examples are simple shear, radial motions of spherical and cylindrical shells, and the twisting of cylinders.

Another very interesting example is Rivlin's [1948b] analysis of the equilibrium deformations of an incompressible block of neo-Hookean material under a uniform tension T (Figure 4.6). Surprisingly, one finds that if $T < 3\sqrt[3]{2}\,\alpha$

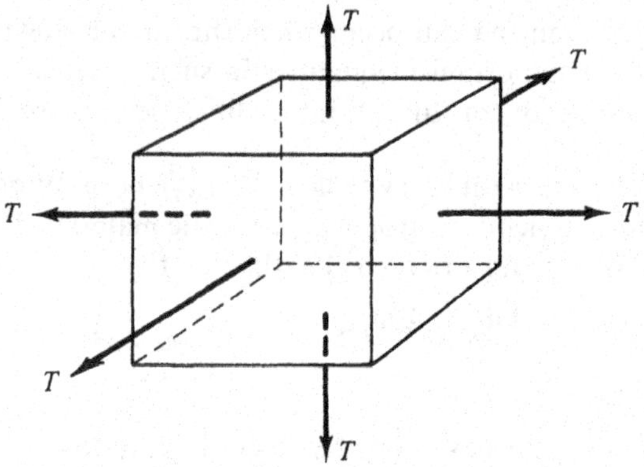

Figure 4.6

where α is the constant in the neo-Hookean constitutive function, then there is just one (stable) solution, whereas if $T > 3\sqrt[3]{2}\,\alpha$, there are seven (homogeneous) solutions. See Section 7.2 for more information.

Linearization Instability A situation of considerable mathematical interest is the pure traction problem in elastostatics for a nonlinear elastic material near a natural state (in which the stress is zero) and for small loads. Somewhat surprisingly, the solutions need not be unique and do not correspond in a simple way to the linearized theory. This led the Italian school, beginning with Signorini in the 1930s, to devote considerable attention to this apparent difficulty. Today we view this as a particularly interesting topic in bifurcation theory and as an example of *linearization instability*—a situation in which the solutions of the linearized problem are not tangent to solution curves in the nonlinear problem (the curve parameters being perturbation parameters). Similar phenomena occur in other basic field theories of mathematical physics, such as general relativity (this is due to Fischer, Marsden, and Moncrief; see Marsden [1981] for an exposition and references).

While the traction problem in elastostatics is linearization unstable, it is formally obvious that the dynamic problem is linearization stable. This remark, due to Capriz and Podio-Guidugli [1974], may be comforting to those who are uneasy about linearization instabilities, but the traction problem is tantalizing nonetheless. Linearization stability is discussed in general terms in Section 4.4, and the traction problem is examined in Section 7.3.

5 CONSTITUTIVE INEQUALITIES

Stored energy functions are subject to several possible restrictions. These restrictions are typified by statements like "stress increases with strain," which eventually lead to inequalities. Such inequalities are best viewed with scrutiny and care. Any proposed inequalities normally have a limited range of validity, and interesting phenomena can occur when they break down. We shall briefly discuss some of the important inequalities in such a critical framework. For a comprehensive survey of constitutive inequalities, see Truesdell and Noll [1965].

Baker-Ericksen Inequalities Let $\Phi(\lambda_1, \lambda_2, \lambda_3)$ be a smooth stored energy function for a homogeneous isotropic hyperelastic material. The *Baker–Ericksen inequalities* (see Baker and Ericksen [1954]) state that

$$\frac{\lambda_i\Phi_i - \lambda_j\Phi_j}{\lambda_i - \lambda_j} > 0 \quad \text{if } \lambda_i \neq \lambda_j \qquad \text{(BE)}_1$$

where $\Phi_i = \partial\Phi/\partial\lambda_i$.

These inequalities are based on the physical situation described as follows. Consider a rectangular block of material with stored energy function Φ, being pulled uniformly on each side. Consider a homogeneous deformation of the form

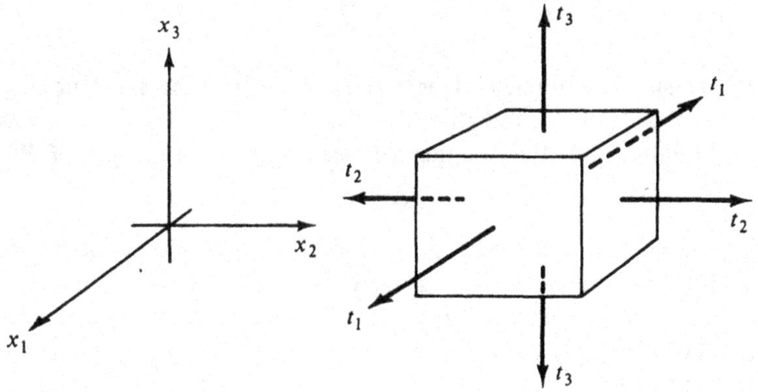

Figure 5.1

$x_i = \lambda_i X_i$, so $\boldsymbol{F} = \text{diag}(\lambda_1, \lambda_2, \lambda_3)$. (See Figure 5.1.) To support such a deformation, we apply a force normal to each face. For convenience, we set $\rho_{\text{Ref}} = 1$. Since $\partial\Phi/\partial\boldsymbol{F} = \text{diag}(\Phi_1, \Phi_2, \Phi_3)$, the first Piola–Kirchhoff stress tensor is $\boldsymbol{P} = \text{diag}(\Phi_1, \Phi_2, \Phi_3)$, so these normal forces are given by $(T_1, T_2, T_3) = (\Phi_1, \Phi_2, \Phi_3)$ on the three pairs of opposite faces.

Recall that the T_i measure the force per unit of reference area. The forces per unit current area are given (through the Piola transform for the Cauchy stress, that is, $\boldsymbol{P} = J\boldsymbol{\sigma}\boldsymbol{F}^{-T}$, or directly) by

$$(t_1, t_2, t_3) = \left(\frac{\lambda_1\Phi_1}{J}, \frac{\lambda_2\Phi_2}{J}, \frac{\lambda_3\Phi_3}{J}\right) \qquad \text{where} \quad J = \det \boldsymbol{F} = \lambda_1\lambda_2\lambda_3.$$

Intuitively, one expects that each T_i is an increasing function of λ_i: the more deformation in each direction, the more force required. This leads to the following convexity requirements[4]:

$$\Phi \text{ is strictly convex in each argument, that is, } \Phi_{ii} > 0 \quad (i = 1, 2, 3) \qquad \text{(BE)}_2$$

One also expects that if λ_1 is larger than λ_2, then t_1 is larger than t_2. Thus, the inequalities

$$(t_i - t_j)(\lambda_i - \lambda_j) > 0 \quad \text{if} \quad \lambda_i \neq \lambda_j \qquad \text{(BE)}_1'$$

are also reasonable. From the formula $t_i = \lambda_i\Phi_i/J$, we see that the inequalities $\text{(BE)}_1'$ are equivalent to the Baker–Ericksen inequalities (BE)_1.

An indication of how these inequalities are related to others and their ranges of validity is given below and is summarized in Table 5.1.

---

[4]A function f is called *convex* when $f(\lambda x + (1 - \lambda)y) \leq \lambda f(x) + (1 - \lambda)f(y)$ for $x \neq y$ and $0 < \lambda < 1$; (f is *strictly convex* if $\leq$ can be replaced by $<$). Geometrically, this means that the chord joining $(x, f(x))$ and $(y, f(y))$ lies above the graph of f. If f is C^2, it is convex if and only if $f'' \geq 0$. This is also equivalent to monotonicity of f': $[f'(x) - f'(y)][x - y] \geq 0$ if $x \neq y$.

Convexity and Monotonicity Inequalities Consider the same homogeneous deformation again. We found that $(t_i - t_j)(\lambda_i - \lambda_j) > 0$ if $\lambda_i \neq \lambda_j$ was a reasonable inequality. What about the corresponding monotonicity of $\boldsymbol{P}$: that is,

$$(T_i - T_j)(\lambda_i - \lambda_j) > 0 \quad \text{if} \quad \lambda_i \neq \lambda_j? \tag{M}$$

Keeping in mind that T_i represents the normal forces per unit *undeformed area*, we see that (M) is less obvious intuitively. In fact, using nearly incompressible materials like rubber, Ball (1977a) has described a reasonable situation for which $(\mathrm{BE})_1$ *is valid, while* (M) *is invalid.*

Notice that strict convexity of Φ in all arguments implies (M). Indeed we take strict convexity of Φ to mean that

$$[\nabla\Phi(\bar{\lambda}_1, \bar{\lambda}_2, \bar{\lambda}_3) - \nabla\Phi(\lambda_1, \lambda_2, \lambda_3)]\cdot(\bar{\lambda}_1 - \lambda_1, \bar{\lambda}_2 - \lambda_2, \bar{\lambda}_3 - \lambda_3) > 0 \tag{SCΦ}$$

Taking $(\bar{\lambda}_1, \bar{\lambda}_2, \bar{\lambda}_3) = (\lambda_2, \lambda_1, \lambda_3)$, (SC$\Phi$) reduces to (M). Thus for the same experiment, (SCΦ) must be violated.

In place of convexity of Φ, one may contemplate the convexity of $\hat{W}$ as a function of $\boldsymbol{F}$. This is unacceptable for three basic reasons:

(1) it implies uniqueness of solutions, so it precludes buckling,
(2) it is incompatible with material frame indifference, and
(3) the domain $\{\boldsymbol{F} | J > 0\}$ of $\hat{W}$ is not convex, and so convexity of $\hat{W}$ precludes the reasonable requirement that $\hat{W}(\boldsymbol{E}) \longrightarrow \infty$ as $J \longrightarrow 0$.

Observation (1) is due to Hill [1957]. Because of (2) and considerations of thermodynamic stability, Coleman and Noll [1959] sought a weaker inequality. Using arguments involving work done on deforming bodies, they proposed the following inequality for the Piola–Kirchhoff stress $\boldsymbol{P}(\boldsymbol{F})$.

$$\left.\begin{array}{l}\text{If } \boldsymbol{\Lambda} \text{ is a positive-definite symmetric matrix}\\ \text{(i.e., a pure stretch) and } \boldsymbol{\Lambda} \neq \boldsymbol{I}, \text{ then}\\ [\boldsymbol{P}(\boldsymbol{\Lambda F}) - \boldsymbol{P}(\boldsymbol{F})][\boldsymbol{\Lambda F} - \boldsymbol{F}] > 0.\end{array}\right\} \tag{CN}$$

For hyperelastic materials, (CN) may be translated to a convexity-type condition on the stored energy function. For homogeneous isotropic hyperelastic materials, (CN) implies that Φ is strictly convex and is subject therefore to the same criticisms as above. That Φ being strictly convex is unreasonable has been pointed out by Hill [1968], [1970], Rivlin [1973], and Sidoroff [1974]. That (CN) implies (SCΦ) was noted by Ogden [1970], [1972]; compare Ball [1977a, b].

Strong Ellipticity If $\hat{\boldsymbol{P}}$ is a given constitutive function for the first Piola–Kirchhoff stress, the four-index tensor A defined by $\mathsf{A}^{iJ}{}_k{}^L = \partial\hat{P}^{iJ}/\partial F^k{}_L$ is called the *elasticity tensor*.

We say that $\hat{\boldsymbol{P}}$ is *strongly elliptic* (at a particular deformation ϕ) if there is an $\epsilon > 0$ such that

$$\mathsf{A}_{iJkL} v^i v^k w^J w^L \geq \epsilon \,\| \boldsymbol{v} \|^2 \,\| \boldsymbol{w} \|^2 \tag{SE}$$

for all vectors $v, w \in \mathbb{R}^3$. If we put $\epsilon = 0$ in (SE), we call this condition *semi-ellipticity.*

If Id is a natural state—that is, $\hat{P} = 0$ at the deformation $\phi = \text{Id}$—then A_{ijkl} coincides with c_{ijkl}, which was considered in our discussion of linear elasticity in Section 3. Omitting body forces, the equations of linear elasticity for small deviations u from a given deformation ϕ are

$$\rho \frac{\partial^2 u^i}{\partial t^2} = \frac{\partial}{\partial x^j} \mathsf{A}^{ij}{}_k{}^l \frac{\partial u^k}{\partial x^l}.$$

If one seeks solutions in the form of traveling waves, i.e., $u(x, t) = v\psi(x \cdot w - ct)$ where v, w are constant vectors with $\|w\| = 1$, one finds the condition $c^2 v^i = \mathsf{A}^i{}_{jkl} w^j v^k w^l$. This states that the matrix $E_{ik} = \mathsf{A}_{ijkl} w^j w^l$ for w fixed, has v^i as an eigenvector, with eigenvalue c^2. Since $c^2 > 0$, multiplying by v^i, summing and rescaling (to remove the restriction $\|w\| = 1$) gives (SE). Thus, *the existence of traveling waves with real wave speeds implies* (SE). This condition was first isolated by Legendre and Hadamard about 1900.

Strong ellipticity is also a convexity condition for hyperelastic materials. It is equivalent to saying that if $K = v \otimes w$—that is, $K^{ij} = v^i w^j$ is a 3×3 rank 1 matrix—then $\hat{W}$ is strictly convex along the line joining F and $F + K$. Indeed, observe that

$$\frac{d^2}{d\epsilon^2} \hat{W}(F + \epsilon K) = \mathsf{A}_{ijkl} v^i v^k w^j w^l.$$

A basic fact is that with enough regularity, A_{ijkl} evaluated at a smooth minimum of the stored energy function, is semi-elliptic. We shall present theorems of this type in Chapter 6.

What are the implications of strong ellipticity? By choosing $F = \text{diag}(\lambda_1, \lambda_2, \lambda_3)$, $v = (1, 0, 0)$, and $w = (1, 0, 0)$, we get $\Phi_{11} > 0$. Similarly, Φ_{22} and Φ_{33} are positive. Thus (SE) implies $(\text{BE})_2$. Taking

$$K = \begin{bmatrix} 0 & 1 & 0 \\ 0 & 0 & 0 \\ 0 & 0 & 0 \end{bmatrix} \quad \text{and} \quad F = \text{diag}(\lambda_1, \lambda_2, \lambda_3),$$

one sees that $(\text{BE})_1$ also holds. Thus, *strong ellipticity implies the Baker–Ericksen inequalities.* In Chapter 6 we shall see that strong ellipticity is just what is needed for a good existence and uniqueness theory for linear elastostatics and elastodynamics.

The violation of strong ellipticity has little or nothing to do with the loss of stability and buckling, as we shall see in Chapter 7 (compare Truesdell and Noll [1965], p. 275). Antman [1973a] has shown that strong ellipticity is compatible with the necking of bars in tension. Ericksen [1975] has suggested that strong ellipticity may fail when "phase transitions" occur. Potentially, this is a deep remark, but its implications are not well understood at present. In their studies of plane strain and crack propagation, Knowles and Sternberg [1977], [1978] have encountered situations in which strong ellipticity fails. Thus, while strong ellipticity holds over wide ranges—including buckling—and is physically rather compelling, it is not necessarily universal.

Polyconvexity Perhaps it is not an accident that what has turned out to be the most basic inequality for the existence theory of linear elasticity (the strong ellipticity condition) also has the most convincing physical basis (reality of wave speeds and the Baker–Ericksen inequalities). Motivated by such a situation, it seems wise to pay attention to conditions that arise in nonlinear existence theory.

Local existence and uniqueness theorems in nonlinear elastostatics—looking for solutions near a given one for perturbed boundary conditions, forces, or constitutive functions (Stoppelli [1954])—and elastodynamics—solutions of the initial value problem for short time (Hughes, Kato, and Marsden [1977])—are based on strong ellipticity. However, no global theorems that are based on strong ellipticity as the main constitutive assumption are known.

The earliest global existence theory for elastostatics (for example, see Beju [1971]) was based on convexity of the stored energy function. As we have seen, such a convexity hypothesis is unreasonable. Using the notion of quasi-convexity due to Morrey [1952], Ball [1977b] proved global existence theorems for elasto-

Table 5.1 Some Constitutive Inequalities

Inequality	Implies	Implied by	Could be violated when
Baker–Ericksen $(BE)_1$ and $(BE)_2$	"stress increases with strain"	strong ellipticity (SE)	phase transitions occur(?)
strict convexity of Φ (SCΦ)	(M) and (SE)	Coleman–Noll inequality (CN)	buckling or large deformations occur
strict convexity of W	(SCΦ) and uniqueness of solutions in elastostatics	?	nonuniqueness of solutions holds or material frame indifference is demanded
Coleman–Noll inequality (CN)	(SCΦ), (M), and (SE)	Coleman's hypotheses on work	large deformations occur
strong ellipticity	(BE)	reality of wave speeds	phase transitions occur(?)
polyconvexity (P)	existence of solutions in elastostatics and (SE)	?	?

statics. His main requirement was that the stored energy function $\hat{W}$ be *polyconvex*—that is, that there be a convex function g with arguments $\boldsymbol{F}$, adj $\boldsymbol{F}$ (the matrix of cofactors of $\boldsymbol{F}$) and det $\boldsymbol{F}$ such that

$$\hat{W}(\boldsymbol{F}) = g(\boldsymbol{F}, \operatorname{adj} \boldsymbol{F}, \det \boldsymbol{F}). \tag{P}$$

The existence theorem also requires certain growth conditions on g (discussed in Section 6.4). The case where $\hat{W}$ is given by $\hat{W}(\boldsymbol{F}) = h(\det \boldsymbol{F})$ for h a convex function (see Section 3), shows that polyconvexity of $\hat{W}$ does not imply convexity of $\hat{W}$. Polyconvexity holds for materials with a stored energy function of the Mooney–Rivlin type (Ball [1977a], Section 6). While strict polyconvexity does imply strong ellipticity, its physical meaning is not well understood. Some of this theory is presented in Section 6.4. The situation is summarized in Table 5.1 and Figure 5.3.

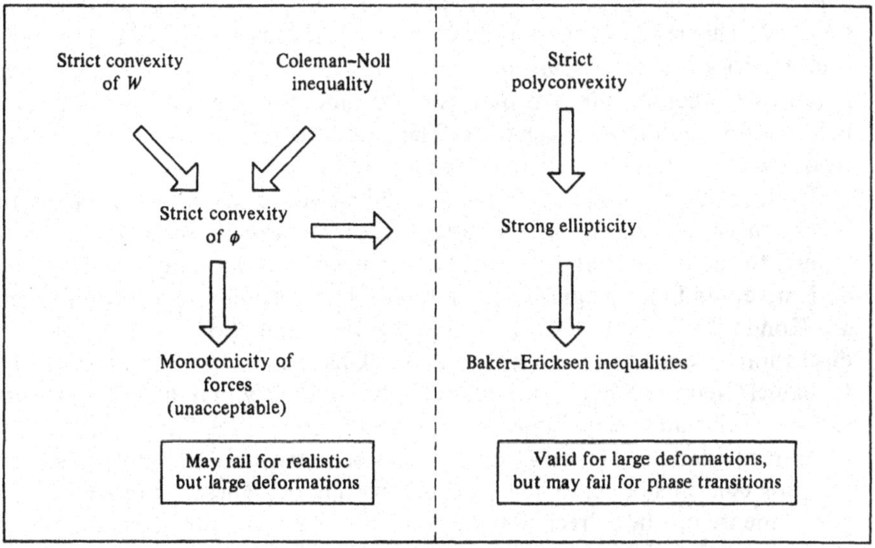

Figure 5.3

6 THE ROLE OF GEOMETRY AND FUNCTIONAL ANALYSIS

These two large branches of mathematics will be applied to the theory of elasticity in this book. This section is devoted to explaining how and where they are applied to give the reader perspective and motivation, and also to avoid disappointments. They are very useful subjects, but the reader must be warned that both take major efforts to master. It is wise to examine the possible benefits to see if this effort is relevant to one's needs.

Several examples of how geometric concepts are used in elasticity are discussed first. Perhaps the most obvious one is in rod and shell theory. Since a rod or shell is usually represented by a curve or a surface in $\mathbb{R}^3$—that is, a one- or two-dimensional manifold in $\mathbb{R}^3$—geometric concepts such as the curvature of a curve or surface enter in a fundamental way. For example, the work done in bending a flexible rod is, according to the classical theory of Bernoulli and Euler, proportional to the integral of the square of the curvature along the rod. As another example, we can ask what is the analog for shells of the equation of continuity derived in Section 2, namely, $D\rho/Dt + \rho \operatorname{div} \boldsymbol{v} = 0$. Let the velocity of the shell be decomposed into tangential $\boldsymbol{v}_\parallel$ and normal components $v_n \boldsymbol{n}$ (where $\boldsymbol{n}$ is the unit normal to the shell). Then the correct equation for conservation of mass is

$$\frac{D\rho}{Dt} + \rho \, \overline{\operatorname{div}} \, \boldsymbol{v}_\parallel + \rho v_n \kappa = 0$$

where κ is the mean curvature of the shell and $\overline{\operatorname{div}}$ is the divergence operator on the shell. This relationship is proved in Box 5.1, Chapter 1. Although rod and shell theories are not discussed in any detail in this book, this example shows that a clear understanding of differential geometry is crucial for understanding this area of continuum mechanics. The standard references for rod and shell theory are Antman [1972a] and Naghdi [1972].

The important concept of the Lie derivative occurs throughout elasticity theory in computations such as stress rates. Nowadays such things are well known to many workers in elasticity, but it was not so long ago that the Lie derivative was first recognized to be relevant to elasticity (two early references are Kondo [1955] and Guo Zhong-Heng [1963]). [See Box 6.1, Chapter 1 for a discussion of objective rates and Box 3.1, Chapter 3 for an application to the Duhamel–Neumann hypothesis of constitutive theory.] Numerous other applications of this and related concepts are given throughout Chapters 1–3.

A third example of how geometry has been used in the book is in understanding concepts of "covariance" and "invariance." These are related to Nolls' work theorem, the Green–Rivlin–Naghdi energy theorem, material frame indifference, framings, inertial frames, spacetime slicings, principles of general covariance for relativistic elasticity, and so on. We found much of the literature on these topics aesthetically unsatisfactory. The use of geometry in attempting to isolate the basic principles that are covariant—that is, that make intrinsic tensorial sense independent of a preferred coordinate system—automatically clears up several basic issues. For example, balance of linear momentum does not make tensorial sense as it stands. However, one can make covariant sense out of balance of energy principles with no reference to rigid body motions. That is, the Noll and Green–Rivlin–Naghdi programs can be done covariantly, although this is not obvious given the existing literature. It is doubtful that we could have seen this without the aid of abstract geometry. These topics are discussed in Boxes 4.1 (Chapter 1), 3.1, 3.2, and 3.3 (Chapter 2), and in Section 3.3.

Finally, we mention a fourth use of geometry discussed in Chapter 5. That chapter is devoted to the variational or Hamiltonian structure of elasticity. We show in a precise sense that nonlinear elastodynamics is a Hamiltonian system. Here abstract geometric thinking makes this rather transparent. Furthermore, conservation laws can all be obtained by the straightforward application of well established conservation principles (symmetry and Noether theorems). In the literature, these were discovered essentially by ad hoc methods. For complex continua (such as director theories), it is important to have a systematic method. The general theory of Hamiltonian systems is one way to help in this endeavor.

Functional analysis needed in the book is developed in Section 6.1 and 6.2. The uses of analysis start with the obvious one: existence and uniqueness theorems. Existence theorems play an important role in obtaining satisfactory numerical algorithms (see for example Oden and Reddy [1976a] and Chorin et al. [1978]). Another example is Ball's work which has led to a deeper appreciation of constitutive inequalities (see Section 5 above and Section 6.4).

Functional analysis also contributes to linear elasticity. For example, the use of semigroup theory in linear elastodynamics clarifies notions of "well-posedness" and "dynamic stability." Indeed, it allows one to improve the known theorems in the subject (see Section 6.3).

In bifurcation theory (Chapter 7), functional analysis and geometry are used together. Functional analysis is used to analyze solvability, nonuniqueness, and stability questions, while geometry and topology are used to count the number of solutions (using, for example, degree theory) and give us devices that reveal when enough imperfection parameters have been included (using, for example, singularity theory). Anyone who wishes to follow the mainstream of current research in bifurcation theory and its applications to elasticity must be literate, or better fluent, in both geometry and functional analysis.

Geometry and functional analysis go together very well. Their blending is often called "nonlinear analysis" or "global analysis." It is our conviction that in future years these subjects will become basic tools for many researchers in elasticity. This is why we have used them in this book.

CHAPTER 1

GEOMETRY AND KINEMATICS OF BODIES

Normally a solid continuum body occupies an open subset of three-dimensional Euclidean space (or the closure of an open subset if the boundary is counted). However,shells and rods may be modeled using surfaces and curves. "Exotic" materials, such as liquid crystals, may require higher dimensional spaces for their description. To have a unified approach, as well as for conceptual clarity, it is useful to think geometrically and to represent bodies in terms of manifolds.

This chapter concerns the description of bodies, their motions, and their configurations. The laws of dynamics which govern the motion of bodies, are discussed in the next chapter. We begin here with bodies in $\mathbb{R}^3$ and gradually work up to a description in terms of manifolds. We assume the reader is familiar with advanced calculus on $\mathbb{R}^n$. For the moment we do not distinguish between "Euclidean space" and "$\mathbb{R}^3$."

1.1 MOTIONS OF SIMPLE BODIES[1]

1.1 Definition A *simple body* is an open set $\mathcal{B} \subset \mathbb{R}^3$. A *configuration* of $\mathcal{B}$ is a mapping $\phi : \mathcal{B} \to \mathbb{R}^3$. The set of all configurations of $\mathcal{B}$ is denoted $\mathcal{C}$, or by $\mathcal{C}(\mathcal{B})$ if there is danger of confusion. Points in $\mathcal{B}$ are denoted by capital letters $X, Y, \ldots$.

A configuration represents a deformed state of the body, as in Figure 1.1.1. As the body moves, we obtain a family of configurations, depending on time t.

[1]The use of the word "simple" here is not to be confused with Noll's use of the term "simple material." (See Truesdell and Noll [1965], p. 60 and Malvern [1969], p. 391.)

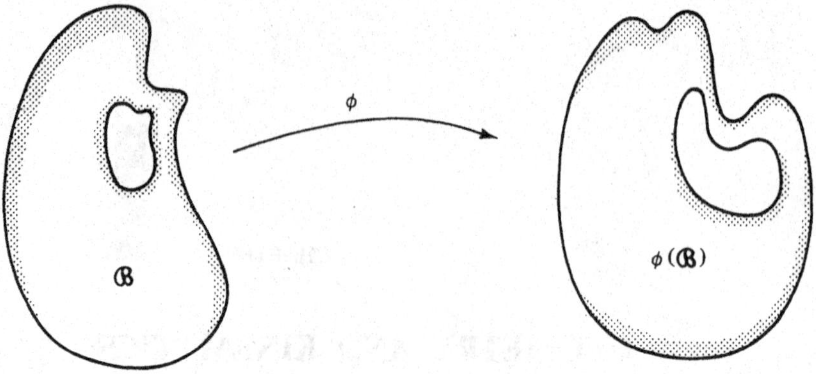

Figure 1.1.1

1.2 Definition A *motion* of a body $\mathcal{B}$ is a curve in $\mathcal{C}$; that is, a mapping $t \mapsto \phi_t \in \mathcal{C}$ of $\mathbb{R}$ to $\mathcal{C}$ (or some open interval of $\mathbb{R}$ to $\mathcal{C}$). For $t \in \mathbb{R}$ fixed, we write $\phi_t(X) = \phi(X, t)$. Likewise, if we wish to hold $X \in \mathcal{B}$ fixed, we write $\phi_X(t) = \phi(X, t)$. The map $V_t\colon \mathcal{B} \to \mathbb{R}^3$ defined by

$$V_t(X) = V(X, t) = \frac{\partial \phi(X, t)}{\partial t} = \frac{d}{dt}\phi_X(t)$$

(assuming the derivative exists) is called the *material velocity* of the motion.

If $c(t)$ is a curve in $\mathbb{R}^3$, the tangent to $c(t)$ is defined by $c'(t) = \text{limit}_{h \to 0}\,(c(t+h) - c(t))/h$. If the standard Euclidean coordinates of $c(t)$ are $(c^1(t), c^2(t), c^3(t))$, then

$$c'(t) = \left(\frac{dc^1}{dt}, \frac{dc^2}{dt}, \frac{dc^3}{dt}\right).$$

To avoid confusion with other coordinate systems we shall write ϕ_z^i, and so on, for the Euclidean components of ϕ. Since $\mathbb{R}^3$ is the set of all real triples, denoted $z = (z^1, z^2, z^3)$, and for fixed $X \in \mathcal{B}$, $\phi(X, t)$ is a curve in $\mathbb{R}^3$, we get

$$\begin{aligned} V(X, t) &= (V_z^1(X, t), V_z^2(X, t), V_z^3(X, t)) \\ &= \left(\frac{\partial \phi_z^1}{\partial t}(X, t), \frac{\partial \phi_z^2}{\partial t}(X, t), \frac{\partial \phi_z^3}{\partial t}(X, t)\right). \end{aligned}$$

We regard $V(X, t)$ as a vector based at the point $\phi(X, t)$; if $\phi(X, 0) = X$, then $V(X, t)$ is the velocity at time t of the particle that started out at X.[2] See Figure 1.1.2.

1.3 Definition The *material acceleration* of a motion is defined by

$$A_t(X) = A(X, t) = \frac{\partial}{\partial t}V(X, t) = \frac{d}{dt}V_X(t) \quad \text{(if the derivative exists).}$$

---

[2]If we regard the map $t \mapsto \phi_t$ as a curve in $\mathcal{C}$, then the notation $V_t = d\phi_t/dt$ is appropriate. However, we shall use $\partial\phi_t/\partial t$ for both roles for simplity.

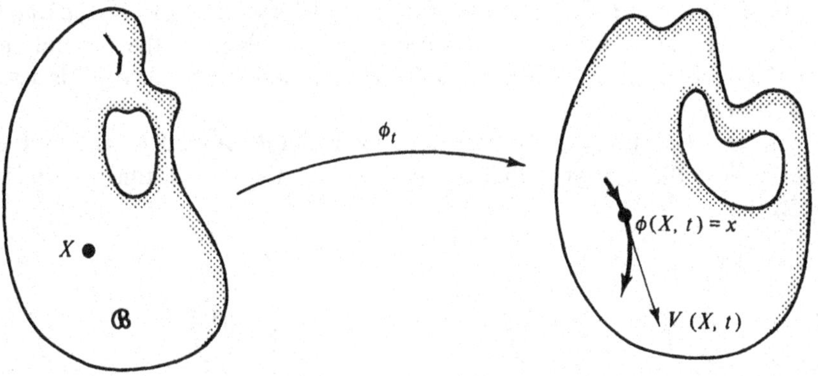

Figure 1.1.2

In Euclidean coordinates,

$$A^i_t(X, t) = \frac{\partial V^i_t}{\partial t}(X, t) = \frac{\partial^2 \phi^i_t}{\partial t^2}(X, t).$$

1.4 Definition A motion ϕ_t of $\mathcal{B}$ is called *regular* or *invertible* if each $\phi_t(\mathcal{B})$ is open and ϕ_t has an inverse $\phi_t^{-1}: \phi_t(\mathcal{B}) \to \mathcal{B}$. A C^r *regular motion* is a C^r motion [i.e., $\phi(X, t)$ is a C^r function of (X, t)] such that ϕ_t^{-1} is also C^r. (The inverse function theorem, recalled later, is relevant here.)

Intuitively, a regular motion is one for which nothing "catastrophic" like ripping, pinching, or interpenetration of matter has occurred.[3] Some of the commonly encountered quantities of continuum mechanics are not well defined if ϕ is not regular, whereas others remain well defined. Since there are physically important cases that are not regular—such as contact problems in which $\mathcal{B}$ may consist of two disconnected pieces that ϕ brings together—it is important to differentiate between quantities that may or may not be applicable to the formulation of this class of problems.

1.5 Definition Let ϕ_t be a C^1 regular motion of $\mathcal{B}$. The *spatial velocity* of the motion is defined by[4]

$$\boldsymbol{v}_t: \phi_t(\mathcal{B}) \to \mathbb{R}^3, \qquad \boldsymbol{v}_t = V_t \circ \phi_t^{-1}.$$

If ϕ_t is a C^2 regular motion, we define the *spatial acceleration* by

$$\boldsymbol{a}_t: \phi_t(\mathcal{B}) \to \mathbb{R}^3, \qquad \boldsymbol{a}_t = A_t \circ \phi_t^{-1}.$$

---

[3]Notice that if a body folds and undergoes self-contact but no interpenetration, then the motion can still be C^r regular as we have defined it. However, ϕ_t cannot be extended to include the boundary of $\mathcal{B}$ and still be regular.

[4]By the local existence theory for vector fields, ϕ_t will be regular exactly when $\boldsymbol{v}_t$ remains defined and is at least C^1. See Section 1.6.

We do not like to use phrases like "material coordinates" or "spatial coordinates" because there are no "coordinates" involved in the general setting. Another technique for considering velocity, the "convective picture," is described in Section 1.2.

Figure 1.1.3 shows the different domains for $\boldsymbol{V}$, $\boldsymbol{A}$ and $\boldsymbol{v}$, $\boldsymbol{a}$. Observe that $\boldsymbol{v}$, $\boldsymbol{a}$ "follow the motion," that is, they are defined on the time-dependent set $\phi_t(\mathcal{B})$.

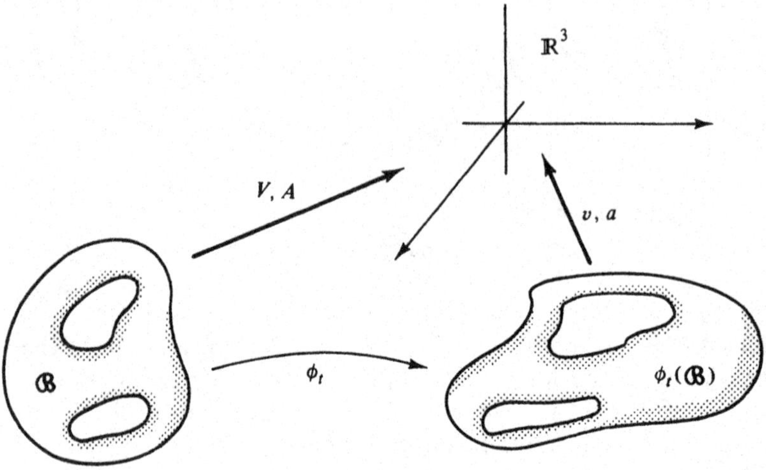

Figure 1.1.3

Now we want to work out expressions for $\boldsymbol{V}$, $\boldsymbol{v}$, $\boldsymbol{A}$, and $\boldsymbol{a}$ in general coordinate systems.

1.6 Definitions, Notation, and Conventions[5] A *coordinate system* $\{x^a\}$ $(a = 1, 2, 3)$ on $\mathbb{R}^3$ is a C^∞ mapping (x^1, x^2, x^3) of an open set $\mathcal{U}_z \subset \mathbb{R}^3$ to $\mathbb{R}^3$ such that

(i) the range is an open set $\mathcal{U}_x \subset \mathbb{R}^3$, and
(ii) the mapping

$$(z^1, z^2, z^3) \mapsto (x^1(z^1, z^2, z^3), x^2(z^1, z^2, z^3), x^3(z^1, z^2, z^3))$$

of $\mathcal{U}_z$ to $\mathcal{U}_x$ has a C^∞ inverse, whose components are denoted $z^i(x^1, x^2, x^3)$.

---

[5]We use, as far as possible, upper case letters to refer to the body $\mathcal{B}$ and lower case for space $\mathcal{S}$. Thus $\{x^a\}$ is a coordinate system in space and $\{X^A\}$ is one on the body. Another common conventions for indices are those of Truesdell and Noll: $\{x^i\}$ for space coordinates and $\{X^\alpha\}$ for body coordinates. Our notation on the indices follows Rivlin [1974a] and Eringen [1975], but otherwise is largely that of Truesdell and Noll.

Coordinate lines are the curves $c_1(t)$, $c_2(t)$, $c_3(t)$ whose components in Euclidean coordinates are $z^i(c_1(t)) = z^i(t, x^2, x^3)$, where x^2 and x^3 are fixed. Similar definitions hold for c_2 and c_3. The tangents to these curves are the coordinate basis vectors; thus

$$\boldsymbol{e}_a = \frac{\partial z^i}{\partial x^a}\hat{\boldsymbol{\imath}}_i$$

where $\hat{\boldsymbol{\imath}}_i$ $(i = 1, 2, 3)$ are the standard basis vectors in $\mathbb{R}^3$. Note that $\boldsymbol{e}_a \in \mathbb{R}^3$ and is a function of x^1, x^2, x^3; that is, $\boldsymbol{e}_a \colon \mathcal{U}_x \to \mathbb{R}^3$. We always use the *summation convention*: summation on repeated indices is understood.

For example, spherical coordinates (where $\mathcal{U}_z$ is $\mathbb{R}^3$ minus a plane) define a coordinate system in $\mathbb{R}^3$.

Because of condition (ii) above, the Jacobian of the transformation $z^i \mapsto x^a(z^i)$ is nonsingular, so $\{\boldsymbol{e}_a\}$ is a basis of $\mathbb{R}^3$ for each (x^1, x^2, x^3). Later we shall require an orientation on the vectors $\boldsymbol{e}_1$, $\boldsymbol{e}_2$, and $\boldsymbol{e}_3$, but this is not needed now. Figure 1.1.4 depicts the coordinate lines and basis vectors for a general coordinate system in $\mathbb{R}^3$.

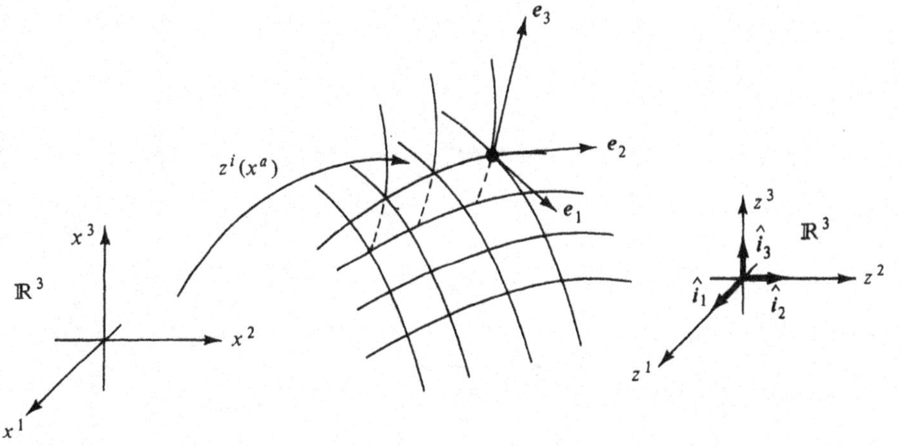

Figure 1.1.4

Coordinate systems on the ambient space $\mathbb{R}^3$ are here denoted $\{x^a\}$ while those on $\mathcal{B}$ are denoted $\{X^A\}$. The corresponding Euclidean coordinates are denoted $\{z^i\}$ and $\{Z^I\}$. The basis vectors for the systems $\{X^A\}$, $\{z^i\}$, and $\{Z^I\}$ are denoted, respectively, $\boldsymbol{E}_A$, $\hat{\boldsymbol{\imath}}_i$, and $\hat{\boldsymbol{I}}_I$.

1.7 Proposition *Let $\phi(X, t)$ be a C^1 motion of $\mathcal{B}$ and let V be the material velocity. Let $\{x^a\}$ be a coordinate system on $\mathbb{R}^3$ and let $\phi^a(X, t) = x^a(\phi(X, t))$. Then*

$$V^a(X, t) = \frac{\partial \phi^a}{\partial t}(X, t)$$

where $V^a(X, t)$ are the components[6] of V relative to e_a at the point $\phi(X, t)$ [*with coordinates $x^b(\phi(X, t))$*].

Proof Consider $((\partial/\partial t)\phi^a)e_a$ [where ϕ^a stands for $\phi^a(X, t) = x^a(\phi(X, t))$ and where e_a stands for $e_a(x^b(\phi(X, t)))$]. We have

$$\left(\frac{\partial}{\partial t}\phi^a\right)e^a = \frac{\partial x^a}{\partial z^i}\frac{\partial \phi^i_z}{\partial t}e_a = \frac{\partial x^a}{\partial z^i}V^i_z\frac{\partial z^j}{\partial x^a}\hat{\imath}_j$$

$$= \delta^j_i V^i_z \hat{\imath}_j = V^i_z \hat{\imath}_i = V = V^a e_a$$

since $\partial x^a/\partial z^i$ and $\partial z^i/\partial x^a$ are inverse matrices. ∎

Suppose $\bar{x}^a$ is another coordinate system on $\mathbb{R}^3$ with, say, the same domain $\mathfrak{U}_z$. By composition we can form the change of coordinate functions $\bar{x}^a\,(x^b)$ and $x^a\,(\bar{x}^b)$; see Figure 1.1.5.

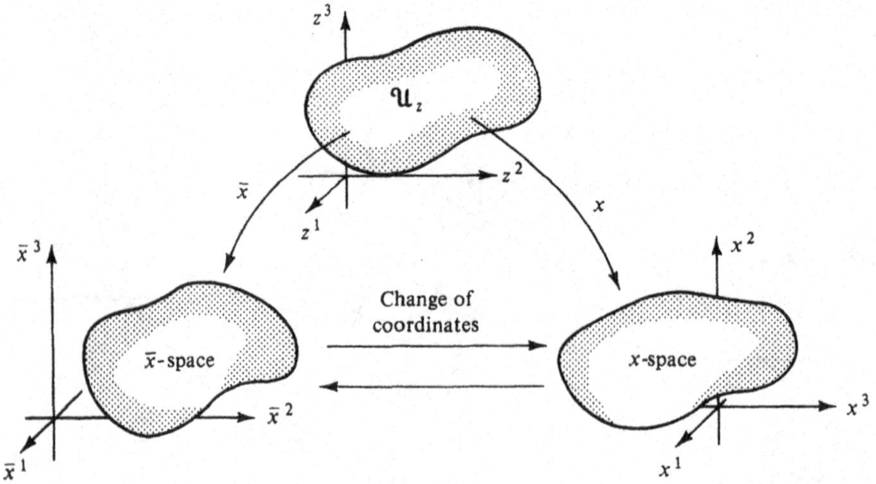

Figure 1.1.5

The transformation property of V is worked out next; one says V transforms "like a vector." We let $\bar{V}^a$ denote the components[7] of V in the basis $\bar{e}_a$ associated with $\{\bar{x}^a\}$.

---

[6]If there is danger of confusion, we write V^a_x for the components of V in the x-coordinate system.

[7]More consistently, we should write $V^a_{\bar{x}}$ for $\bar{V}^a$, but it is conventional in many tensor analysis texts to write $\bar{V}^a$.

1.8 Proposition *$\bar{V}^a = (\partial\bar{x}^a/\partial x^b)V^b$, where $\bar{V}^a$ stands for $\bar{V}^a(X, t)$, and so forth.*

Proof By the chain rule and 1.7,

$$\bar{V}^a = \frac{\partial}{\partial t}\bar{\phi}^a = \frac{\partial \bar{x}^a}{\partial x^b}\frac{\partial}{\partial t}\phi^b = \frac{\partial \bar{x}^a}{\partial x^b}V^b.$$

An alternative proof is obtained by nothing that $\bar{e}_a = (\partial x^b/\partial\bar{x}^a)e_b$ and then equating $V^a e_a$ with $\bar{V}^a\bar{e}_a$. ∎

To work out the components of the acceleration in a general coordinate system we recall a few notations from calculus on $\mathbb{R}^n$. If $\mathcal{U} \subset \mathbb{R}^n$ is an open set in $\mathbb{R}^n$ and f maps $\mathcal{U}$ to $\mathbb{R}^m$, the *derivative* of f at $x_0 \in \mathcal{U}$ is a linear transformation $Df(x_0)\colon \mathbb{R}^n \longrightarrow \mathbb{R}^m$. Formally, for every $\epsilon > 0$ there must be a $\delta > 0$ such that $h \in \mathbb{R}^n$ and $(\|h\| < \delta)$ imply

$$\|f(x_0 + h) - f(x_0) - Df(x_0)\cdot h\| \leq \epsilon\|h\|.$$

Notice that only the norm and the linear structure of Euclidean space are involved. Coordinates are not mentioned. If we represent the linear transformation $Df(x_0)$ in the standard basis, we get the matrix $\partial f^i_z/\partial z^j$, the Jacobian matrix.

The usual calculus rules hold. For instance, the chain rule states that $D(g\circ f)(x_0) = Dg(f(x_0))\circ Df(x_0)$; the second "$\circ$" stands for composition of linear maps—that is, matrix multiplication. The *inverse function theorem*[8] says that if $f\colon \mathcal{U} \subset \mathbb{R}^n \longrightarrow \mathbb{R}^n$ is C^r $(r \geq 1)$ and $Df(x_0)$ is invertible, then f has a local C^r inverse f^{-1}, mapping some neighborhood of $f(x_0)$ to some neighborhood of x_0 and $Df^{-1}(f(x_0)) = [Df(x_0)]^{-1}$.

Returning to the context of 1.6, let us work out the derivative of $e_a = (\partial z^i/\partial x^a)\hat{\imath}_i$. Clearly,

$$\frac{\partial e_a}{\partial x^b} = \frac{\partial^2 z^i}{\partial x^a\,\partial x^b}\hat{\imath}_i = \frac{\partial^2 z^i}{\partial x^a\,\partial x^b}\frac{\partial x^c}{\partial z^i}e_c.$$

This object arises frequently so is given a name:

1.9 Definition The *Christoffel symbols* of the coordinate system $\{x^a\}$ on $\mathbb{R}^3$ are defined by

$$\gamma^c_{ab} = \frac{\partial^2 z^i}{\partial x^a\,\partial x^b}\frac{\partial x^c}{\partial z^i}$$

which are regarded as functions of x^d. The Christoffel symbols of a coordinate system $\{X^A\}$ are denoted Γ^A_{BC}.

Note that the γ's are symmetric in the sense that $\gamma^c_{ab} = \gamma^c_{ba}$.

---

[8]This theorem is proved in a more general infinite-dimensional context in Section 4.1.

1.10 Proposition *Let $\phi(X, t)$ be a C^2 motion of $\mathcal{B}$, and V and A the material velocity and acceleration. Then the components A^a of A in the basis $\mathbf{e}_a$ of a coordinate system $\{x^a\}$ are given by*

$$\boxed{A^a = \frac{\partial V^a}{\partial t} + \gamma^a_{bc} V^b V^c.}$$

Proof

$$A = \frac{\partial V}{\partial t} = \frac{\partial V^i_z}{\partial t}\hat{\imath}_i = \frac{\partial}{\partial t}\left(\frac{\partial z^i}{\partial x^c} V^c\right)\hat{\imath}_i$$

$$= \frac{\partial^2 z^i}{\partial x^b \partial x^c}\frac{\partial x^b}{\partial t} V^c \hat{\imath}_i + \frac{\partial z^i}{\partial x^a}\frac{\partial V^a}{\partial t}\hat{\imath}_i = \frac{\partial^2 z^i}{\partial x^b \partial x^c} V^b V^c \frac{\partial x^a}{\partial z^i}\mathbf{e}_a + \frac{\partial V^a}{\partial t}\mathbf{e}_a$$

Comparison with $A = A^a \mathbf{e}_a$ yields the proposition. ∎

Noting that $\bar{\mathbf{e}}_a = (\partial x^b/\partial \bar{x}^a)\mathbf{e}_b$ and comparing $A = A^a\mathbf{e}_a = \bar{A}^b \bar{\mathbf{e}}_b$ yields:

1.11 Proposition *A^a transforms as a vector; that is, $\bar{A}^b = (\partial \bar{x}^b/\partial x^a)A^a$.*

1.12 Definition Let $\mathbf{v}$ and $\mathbf{w}$ be two vector fields on $\mathbb{R}^3$—that is, maps of open sets in $\mathbb{R}^3$ to $\mathbb{R}^3$. Assume $\mathbf{v}$ is C^1. Thus $D\mathbf{v}(x)$ is a linear map of $\mathbb{R}^3$ to $\mathbb{R}^3$, so $D\mathbf{v}(x)\cdot\mathbf{w}(x)$ is a vector field on $\mathbb{R}^3$. It is called the *covariant derivative* of $\mathbf{v}$ along $\mathbf{w}$ and is denoted $\nabla_{\mathbf{w}}\mathbf{v}(x)$ or $\mathbf{w}\cdot\nabla\mathbf{v}$.

1.13 Proposition *In a coordinate system $\{x^a\}$,*

$$\boxed{(\nabla_{\mathbf{w}}\mathbf{v})^a = \frac{\partial v^a}{\partial x^b} w^b + \gamma^a_{bc} w^b v^c}$$

and $(\nabla_{\mathbf{w}}\mathbf{v})^a$ transforms as a vector.

Proof Using Euclidean coordinates, and the matrix $\partial v^i_z/\partial z^j$ of $D\mathbf{v}$ in the standard basis,

$$\nabla_{\mathbf{w}}\mathbf{v} = \frac{\partial v^i_z}{\partial z^j} w^j_z \hat{\imath}_i = \left[\frac{\partial}{\partial z^j}\left(\frac{\partial z^i}{\partial x^c} v^c\right)\right]\left(w^d \frac{\partial z^j}{\partial x^d}\right)\left(\frac{\partial x^a}{\partial z^i}\mathbf{e}_a\right)$$

$$= \frac{\partial^2 z^i}{\partial x^b \partial x^c}\frac{\partial x^b}{\partial z^j} v^c w^d \frac{\partial z^j}{\partial x^d}\frac{\partial x^a}{\partial z^i}\mathbf{e}_a + \frac{\partial z^i}{\partial x^c}\frac{\partial v^c}{\partial x^b}\frac{\partial x^b}{\partial z^j} w^d \frac{\partial z^j}{\partial x^d}\frac{\partial x^a}{\partial z^i}\mathbf{e}_a$$

$$= \gamma^a_{bc}\delta^b_d v^c w^d \mathbf{e}_a + \delta^b_d \frac{\partial v^c}{\partial x^b}\delta^a_c w^d \mathbf{e}_a = \gamma^a_{bc} v^c w^b \mathbf{e}_a + \frac{\partial v^a}{\partial x^b} w^b \mathbf{e}_a. \quad ∎$$

One writes $v^a{}_{|b} = \partial v^a/\partial x^b + \gamma^a_{bc} v^c$ so that $(\nabla_{\mathbf{w}}\mathbf{v})^a = v^a{}_{|b} w^b$.

Recall that for a regular motion, $V_t = \mathbf{v}_t \circ \phi_t$, that is, $V(X, t) = \mathbf{v}(\phi(X, t), t)$. In a coordinate system $\{x^a\}$, $V^a(X, t) = v^a(\phi(X, t), t)$, and so $\partial V^a/\partial t = \partial v^a/\partial t + (\partial v^a/\partial x^b)V^b$. Since $A^a = \partial V^a/\partial t + \gamma^a_{bc} V^b V^c$, we get

$$A^a(X, t) = \frac{\partial v^a}{\partial t}(\phi(X, t), t) + \frac{\partial v^a}{\partial x^b}(\phi(X, t), t)V^b(X, t) + \gamma^a_{bc}(\phi(X, t), t)V^b(X, t)V^c(X, t).$$

Substituting $A^a(X, t) = a^a(\phi(X, t), t)$ and $x = \phi(X, t)$, we get

$$a^a(x, t) = \frac{\partial v^a}{\partial t}(x, t) + \frac{\partial v^a}{\partial x^b}(x, t)v^b(x, t) + \gamma^a_{bc}(x, t)v^b(x, t)v^c(x, t).$$

The following proposition summarizes the situation:

1.14 Proposition *For a C^2 motion we have*

$$\boxed{\boldsymbol{a} = \frac{\partial \boldsymbol{v}}{\partial t} + \nabla_{\boldsymbol{v}} \boldsymbol{v}.}$$

In a coordinate system $\{x^a\}$, this reads

$$\boxed{a^a = \frac{\partial v^a}{\partial t} + \frac{\partial v^a}{\partial x^b}v^b + \gamma^a_{bc}v^b v^c.}$$

We call $\dot{\boldsymbol{v}} = (\partial \boldsymbol{v}/\partial t) + \nabla_{\boldsymbol{v}} \boldsymbol{v}$ the *material time derivative* of $\boldsymbol{v}$. Thus $\dot{\boldsymbol{v}} = \boldsymbol{a}$. The general definition follows.

1.15 Definition Let $g_t\colon \phi_t(\mathcal{B}) \longrightarrow \mathbb{R}^3$ be a given C^1 mapping. We call

$$\dot{g}(x, t) = \frac{\partial}{\partial t} g(x, t) + Dg_t(x) \cdot \boldsymbol{v}_t(x)$$

the *material time derivative* of g. Sometimes $\dot{g}$ is denoted Dg/Dt.

If we define $G_t = g_t \circ \phi_t$, then an application of the chain rule gives

$$\frac{\partial}{\partial t} G(X, t) = \dot{g}(\phi_t(X, t)).$$

This formula justifies the terminology "material time derivative." Again, $Dg_t(x) \cdot \boldsymbol{v}_t(x)$ is called the *covariant derivative* of g along $\boldsymbol{v}_t$.

Box 1.1 ***Summary of Important Formulas for Section 1.1***

Motion

$\phi_t\colon \mathcal{B} \longrightarrow \mathbb{R}^3$	$\phi^a_t = x^a \circ \phi_t$
$\phi_t(X) = \phi(X, t)$	

Velocity

$V = \dfrac{\partial \phi}{\partial t}$	$V^a = \dfrac{\partial \phi^a}{\partial t}$
$\boldsymbol{v}_t = V_t \circ \phi_t^{-1}$	$v^a_t = V^a_t \circ \phi_t^{-1}$

Covariant Derivative

$D\boldsymbol{v} \cdot \boldsymbol{w} = \nabla_{\boldsymbol{w}} \boldsymbol{v}$	$(\nabla_{\boldsymbol{w}} \boldsymbol{v})^a = \dfrac{\partial v^a}{\partial x^b} w^b + \gamma^a_{bc} w^b v^c$

Christoffel Symbols

$$\gamma^a_{bc} = \frac{\partial^2 z^l}{\partial x^b \partial x^c} \frac{\partial x^a}{\partial z^l}$$

Acceleration

$$A = \frac{\partial V}{\partial t} \qquad A^a = \frac{\partial V^a}{\partial t} + (\gamma^a_{bc} \circ \phi_t) V^b V^c$$

$$a = A_t \circ \phi_t^{-1} = \dot{v} = \frac{\partial v}{\partial t} + \nabla_v v \qquad a^a = \frac{\partial v^a}{\partial t} + \frac{\partial v^a}{\partial x^b} v^b + \gamma^a_{bc} v^b v^c$$

Coordinate Change

$$\bar{\phi}^a_t = \bar{x}^a \circ \phi_t$$

$$\bar{V}^a = \left(\frac{\partial \bar{x}^a}{\partial x^b} \circ \phi_t\right) V^b, \quad \bar{v}^a = \frac{\partial \bar{x}^a}{\partial x^b} v^b$$

$$\bar{A}^a = \left(\frac{\partial \bar{x}^a}{\partial x^b} \circ \phi_t\right) A^b, \quad \bar{a}^a = \frac{\partial \bar{x}^a}{\partial x^b} a^b$$

Figure 1.1.6 goes with these formulas.

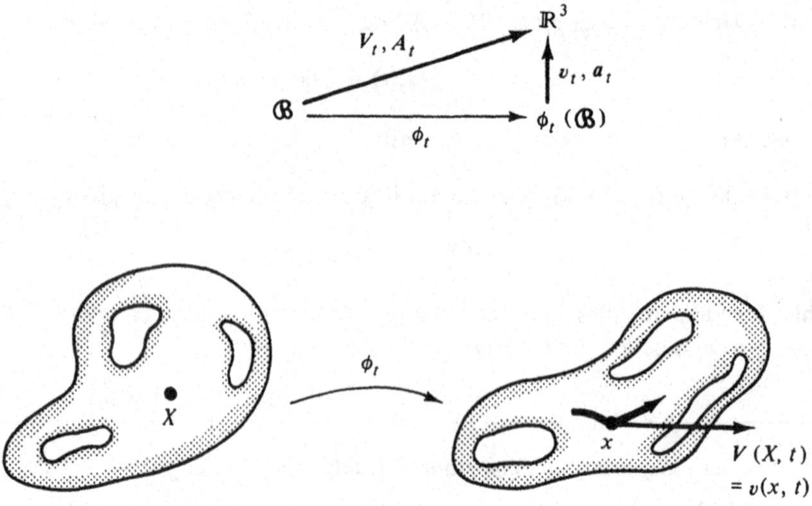

Figure 1.1.6

Problem 1.1 Let $\{x^a\}$ and $\{X^A\}$ denote cylindrical coordinate systems on $\mathbb{R}^3$, that is,

$$x^1 = r, \qquad x^2 = \theta, \qquad x^3 = z;$$

$$X^1 = R, \qquad X^2 = \Theta, \qquad X^3 = Z.$$

Let $\mathcal{B} = \mathbb{R}^3$ and consider a rigid counterclockwise motion about the z-axis given by

$$\phi_t^1(R, \Theta, Z) = R, \qquad \phi_t^2(R, \Theta, Z) = \Theta + 2\pi t, \quad \text{and} \quad \phi_t^3(R, \Theta, Z) = Z$$

where $\phi_t^a = x^a \circ \phi_t$. Compute V_t^a, v_t^a, A_t^a, and a_t^a.

Let $\{\bar{x}^a\}$ denote a Cartesian coordinate system for $\mathbb{R}^3$ given by

$$\bar{x}^1 = x^1 \cos x^2, \qquad \bar{x}^2 = x^1 \sin x^2, \quad \text{and} \quad \bar{x}^3 = x^3.$$

Determine $\bar{V}_t^a$, $\bar{v}_t^a$, $\bar{A}_t^a$, and $\bar{a}_t^a$ from the change-of-coordinate formulas given in the summary, and by directly differentiating $\bar{\phi}_t^a$; compare.

Problem 1.2 For a map $g_t\colon \phi_t(\mathcal{B}) \to \mathbb{R}^3$, work out a formula for the covariant derivative $Dg_t(x)\cdot v_t(x)$ relative to a general spatial coordinate system on $\mathbb{R}^3$.

1.2 VECTOR FIELDS, ONE-FORMS, AND PULL-BACKS

We shall now start using the terminology of manifolds. We begin by giving the general definition of a manifold and then revert to our special case of open sets in $\mathbb{R}^3$ to allow the reader time to become acquainted with manifolds. The basic guidelines and manifold terminology we will use later in the book are given in this section.

2.1 Definition A *smooth n-manifold* (or a manifold modeled on $\mathbb{R}^n$) is a set[9] $\mathfrak{M}$ such that: (1) For each $P \in \mathfrak{M}$ there is a subset $\mathcal{U}$ of $\mathfrak{M}$ containing P, and a one-to-one mapping, called a *chart* or *coordinate system*, $\{x^\alpha\}$ from $\mathcal{U}$ onto an open set $\mathcal{V}$ in $\mathbb{R}^n$; x^α will denote the components of this mapping ($\alpha = 1, 2, \ldots, n$). (2) If x^α and $\bar{x}^\alpha$ are two such mappings, the change of coordinate functions $\bar{x}^\alpha(x^1, \ldots, x^n)$ are C^∞.

If $\{\bar{\bar{x}}^\alpha\}$ maps a set $\mathcal{U} \subset \mathfrak{M}$ one-to-one onto an open set in $\mathbb{R}^n$, and if the change of coordinate functions with the given coordinate functions are C^∞, then $\{\bar{\bar{x}}^\alpha\}$ will also be called a *chart* or *coordinate system*.

For instance, an open set $\mathfrak{M} \subset \mathbb{R}^n$ is a manifold. We take a single chart, $\{Z^I\}$, to be the identity map to define the manifold structure. By allowing all possible coordinate systems that are C^∞ functions of the Z^I, we enlarge our set of coordinate systems. We could also start with all of these coordinate systems at the outset, according to taste.

Thus, a manifold embodies the idea of allowing general coordinate systems (see Figure 1.1.5); it allows us to consider curved objects like surfaces (two-manifolds) in addition to open sets in Euclidean space. The manifold $\mathfrak{M}$ becomes a topological space by declaring the $\mathcal{U}$'s to be open sets. In our work we will

[9]One can start with a topological space $\mathfrak{M}$ or, using the differential structure, make $\mathfrak{M}$ into a topological space later. We chose the latter since it minimizes the number of necessary concepts.

consider both the body $\mathcal{B}$ and the containing space $\mathcal{S}$ to be special cases of manifolds. The descriptions of some realistic bodies *require* this generality, such as shells and liquid crystals (see Sections 1.5 and 2.2). However, the description of any body *benefits from* manifold terminology. Examples occur in the study of covariance and relativistic elasticity.

An important mathematical discovery made around the turn of the century was that one could define the tangent space to a manifold without using a containing space, as one might naively expect. Unfortunately, the abstraction necessary to do this causes confusion to those trying to learn the subject. To help guide the reader, we list three possible approaches:

(a) *Derivations.* The idea here is that in order to specify a vector tangent to $\mathfrak{M}$, we can give a rule defining the derivatives of all real-valued functions in that direction; such directional derivatives are derivations (see Bishop and Goldberg [1968], pp. 47–48).

(b) *Curves.* We intuitively think of tangent vectors as velocities of curves; therefore, vectors can be specified as equivalence classes of curves, two curves being equivalent if they have the same tangent vector in some, and hence in every, chart (see Abraham and Marsden [1978], p. 43).

(c) *Local transformation properties.* We can use the transformation rules for vectors found in Section 1.1 to define a local vector in a coordinate chart as a vector in $\mathbb{R}^n$ and then use the coordinate transformation rules to define an equivalence relation on pairs of charts and local vectors. Each equivalence class obtained is a tangent vector to $\mathfrak{M}$ (see Lang [1972], pp. 26, 47).

For $\mathfrak{M}$ open in $\mathbb{R}^n$, the tangent space is easy to define directly:

2.2 Definition Let $\mathfrak{M} \subset \mathbb{R}^n$ be an open set and let $P \in \mathfrak{M}$. The *tangent space* to $\mathfrak{M}$ at P is simply the vector space $\mathbb{R}^n$ regarded as vectors emanating from P; this tangent space is denoted $T_P\mathfrak{M}$. The *tangent bundle* of $\mathfrak{M}$ is the product $T\mathfrak{M} = \mathfrak{M} \times \mathbb{R}^n$ consisting of pairs $(P, \mathbf{w})$ of base points P and tangent vectors at P. The map π (or $\pi_{\mathfrak{M}}$ if there is danger of confusion) from $T\mathfrak{M}$ to $\mathfrak{M}$ mapping a tangent vector $(P, \mathbf{w})$ to its base point P is called the *projection.* We may write $T_P\mathfrak{M} = \{P\} \times \mathbb{R}^n$ as a set in order to keep the different tangent spaces distinguished, or denote tangent vectors by $\mathbf{w}_P = (P, \mathbf{w})$ to indicate the base point P which is meant.

The idea of the tangent bundle, then, is to think of a tangent vector as being a vector equipped with a base point to which it is attached (see Figure 1.2.1).

Once the tangent bundle of a manifold is defined, one makes it into a manifold by introducing the coordinates of vectors as in Section 1.1. For the special case in which $\mathfrak{M} = \mathcal{B}$ is an open set in $\mathbb{R}^n$ this is easy. For the rest of this section we confine ourselves to this case and we use notation adapted to it; that is, points in $\mathcal{B}$ are denoted X and coordinate systems are written $\{X^A\}$. However, the reader should be prepared to apply the ideas to general manifolds after finishing the section.

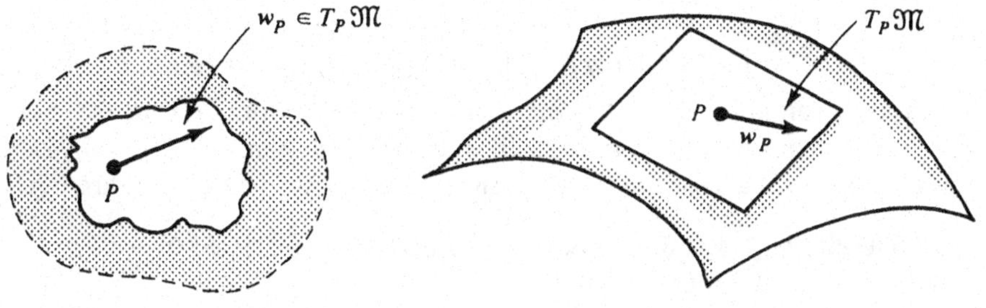

$\mathfrak{M}$ an open set in $\mathbb{R}^3$ $\qquad$ $\mathfrak{M}$ a surface

Figure 1.2.1

2.3 Definition Let $\mathcal{B} \subset \mathbb{R}^n$ be open and $T\mathcal{B} = \mathcal{B} \times \mathbb{R}^n$ be its tangent bundle. Let $\{X^A\}$ be a coordinate system on $\mathcal{B}$. The corresponding *coordinate system induced on* $T\mathcal{B}$ is defined by mapping $W_X = (X, W)$ to $(X^A(X), W^A)$, where $X \in \mathcal{B}$ and $W^A = (\partial X^A/\partial Z^I)W^I{}_Z$ are the components of W in the coordinate system $\{X^A\}$, as explained in Section 1.1.

For $\mathcal{B} \subset \mathbb{R}^3$ open, $T\mathcal{B}$ is a six-dimensional manifold. In general if $\mathcal{B}$ is an n-manifold, $T\mathcal{B}$ is a $2n$-manifold.

In Euclidean space we know what is meant by a C^r map. A mapping of manifolds is C^r if it is C^r when expressed in local coordinates.

2.4 Proposition *(a) Let $\mathcal{B} \subset \mathbb{R}^n$ be open and let $f\colon \mathcal{B} \to \mathbb{R}$ be a C^1 function. Let $W_X = (X, W) \in T_X\mathcal{B}$. Let $W_X[f]$ denote the derivative of f at X in the direction W_X—that is, $W_X[f] = Df(X)\cdot W$. If $\{X^A\}$ is any coordinate system on $\mathcal{B}$, then $W_X[f] = (\partial f/\partial X^A)W^A$, where it is understood that $\partial f/\partial X^A$ is evaluated at X.*

(b) If $c(t)$ is a C^1 curve in $\mathcal{B}$, $c(0) = X$, and $W_X = (X, W) = (X, c'(0))$ is the tangent to $c(t)$ at $t = 0$, then, in any coordinate system $\{X^A\}$,

$$W^A = \frac{dc^A}{dt}, \qquad \text{where} \quad c^A(t) = X^A(c(t)).$$

Proof

(a) $$Df(X)\cdot W = \frac{\partial f}{\partial Z^I}W^I{}_Z = \frac{\partial f}{\partial X^A}\frac{\partial X^A}{\partial Z^I}W^I{}_Z = \frac{\partial f}{\partial X^A}W^A.$$

(b) $$\frac{d}{dt}c^A(t) = \frac{\partial X^A}{\partial Z^I}W^I{}_Z = W^A \text{ (evaluated at } c(t)). \quad \blacksquare$$

Following standard practice, we let $c'(0)$ stand both for $(X, c'(0))$ and $c'(0) \in \mathbb{R}^n$; there is normally no danger of confusion.

The above proposition gives a correspondence between the "transformation of coordinate" definition and the other methods of defining the tangent space. Observe that the mapping $f \mapsto W_X[f]$ is a *derivation*; that is, it satisfies

$$W_X[f + g] = W_X[f] + W_X[g] \quad \text{(sum rule)}$$

and

$$W_X[fg] = fW_X[g] + gW_X[f] \quad \text{(product rule).}$$

In a coordinate system $\{X^A\}$ the basis vectors $\boldsymbol{E}_A = (\partial Z^I/\partial X^A)\boldsymbol{I}_I$ (see Section 1.1) are sometimes written $\partial/\partial X^A$, since for any function f, $\boldsymbol{E}_A[f] = \partial f/\partial X^A$, by (a) in 2.4 (i.e., the coordinates of $\boldsymbol{E}_A$ in the coordinate system $\{X^B\}$ are δ^B_A).

2.5 Definition Let $\mathcal{B}$ be open in $\mathbb{R}^n$ and let $\mathcal{S} = \mathbb{R}^n$. If $\phi\colon \mathcal{B} \to \mathcal{S}$ is C^1, the *tangent map* of ϕ is defined as follows:

$$T\phi\colon T\mathcal{B} \to T\mathcal{S}, \qquad \text{where} \quad T\phi(X, W) = (\phi(X), D\phi(X)\cdot W).$$

For $X \in \mathcal{B}$, we let $T_X\phi$ denote the restriction of $T\phi$ to $T_X\mathcal{B}$, so $T_X\phi$ becomes the linear map $D\phi(X)$ when base points are dropped.

We notice that the following diagram commutes (see Figure 1.2.2):

$$\begin{array}{ccc} T\mathcal{B} & \xrightarrow{\;T\phi\;} & T\mathcal{S} \\ \downarrow{\scriptstyle \pi_{\mathcal{B}}} & & \downarrow{\scriptstyle \pi_{\mathcal{S}}} \\ \mathcal{B} & \xrightarrow{\;\phi\;} & \mathcal{S} \end{array}$$

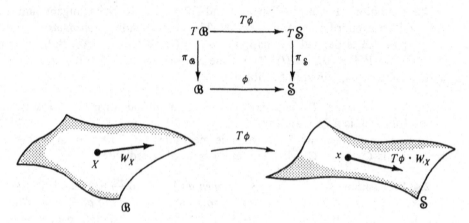

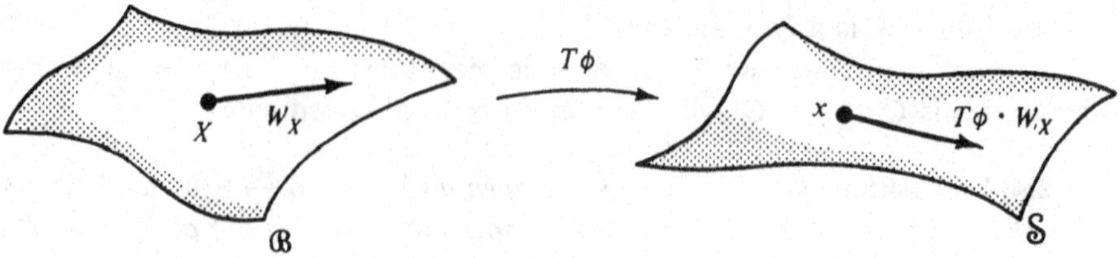

Figure 1.2.2

The vector, $T\phi\cdot W_X$ is called the *push-forward* of W_X by ϕ and is sometimes denoted $\phi_* W_X$. (This use of the word "push-forward" is not to be confused with its use for vector fields defined in 2.9 below.) The next proposition is the essence of the fact that $T\phi$ makes intrinsic sense on manifolds.

2.6 Proposition *(a) If $c(t)$ is a curve in $\mathcal{B}$ and $W_X = c'(0)$, then*

$$T\phi\cdot W_X = \frac{d}{dt}\phi(c(t))\Big|_{t=0}$$

(the base points X of W_X and $\phi(X)$ of $T\phi\cdot W_X$ being understood).

(b) If $\{X^A\}$ is a coordinate chart on $\mathcal{B}$ and $\{x^a\}$ is one on $\mathcal{S}$, then, for $W \in T_X\mathcal{B}$,

$$(T\phi\cdot W_X)^a = \frac{\partial\phi^a}{\partial X^A} W^A$$

that is, in coordinate charts the matrix of $T_X\phi$ is the Jacobian matrix of ϕ evaluated at X.[10]

[10]Sometimes $\partial\phi^a/\partial X^A$ is written $\partial x^a/\partial X^A$, but this can cause confusion.

Proof

(*a*) By the chain rule, $\left.\frac{d}{dt}\phi(c(t))\right|_{t=0} = D\phi(X)\cdot c'(0)$.

(*b*)
$$\begin{aligned} D\phi(X)\cdot W &= \frac{\partial \phi^i_z}{\partial Z^I} W^I_Z \hat{\imath}_i \quad \text{(representation in the standard basis)} \\ &= \frac{\partial \phi^a}{\partial Z^I}\frac{\partial z^i}{\partial x^a} W^I_Z \hat{\imath}_i \quad \text{(chain rule)} \\ &= \frac{\partial \phi^a}{\partial X^A}\left(\frac{\partial X^A}{\partial Z^I} W^I_Z\right)\left(\frac{\partial z^i}{\partial x^a}\hat{\imath}_i\right) \quad \text{(chain rule)} \\ &= \frac{\partial \phi^a}{\partial X^A} W^A e_a. \quad \blacksquare \end{aligned}$$

Later we shall examine the sense in which $T\phi$ is a tensor. For now we note the following transformation rule:

$$\frac{\partial \bar{\phi}^a}{\partial \bar{X}^A} = \frac{\partial X^B}{\partial \bar{X}^A}\frac{\partial \phi^b}{\partial X^B}\frac{\partial \bar{x}^a}{\partial x^b}.$$

The chain rule can be expressed in terms of tangents as follows:

2.7 Proposition *Let $\phi\colon \mathcal{B} \to \mathcal{S}$ and $\psi\colon \mathcal{S} \to \mathcal{U}$ be C^r maps of manifolds $(r \geq 1)$. Then $\psi\circ\phi$ is a C^r map and $T(\psi\circ\phi) = T\psi\circ T\phi$.*

Proof Each side evaluated on (X, W) gives $(\psi(\phi(X)), D\psi(\phi(X))\cdot(D\phi(X)\cdot W))$ by the chain rule. $\blacksquare$

This "T" formulation keeps track of the base points automatically. The reader should draw the commutative diagram that goes with this as an exercise.

Next we formulate vector fields and the spatial and material velocities in manifold language.

2.8 Definitions If $\mathcal{Q}$ is a manifold (e.g., either $\mathcal{B}$ or $\mathcal{S}$), a *vector field* on $\mathcal{Q}$ is a mapping $\boldsymbol{v}\colon \mathcal{Q} \to T\mathcal{Q}$ such that $\boldsymbol{v}(q) \in T_q\mathcal{Q}$ for all $q \in \mathcal{Q}$.

If $\mathcal{B}$ and $\mathcal{S}$ are manifolds and $\phi\colon \mathcal{B} \to \mathcal{S}$ is a mapping, a *vector field covering* ϕ is a mapping $V\colon \mathcal{B} \to T\mathcal{S}$ such that $V(X) \in T_{\phi(X)}\mathcal{S}$ for all $X \in \mathcal{B}$.

These diagrams commute (where $i\colon \mathcal{Q} \to \mathcal{Q}$ is the identity map):

and we have the corresponding pictures shown in Figure 1.2.3.

A vector field covering the identity mapping is just a vector field. Also, if V is a vector field covering an invertible map ϕ, then $\boldsymbol{v} = V\circ\phi^{-1}$ is a vector field on $\mathcal{Q} = \phi(\mathcal{B})$.

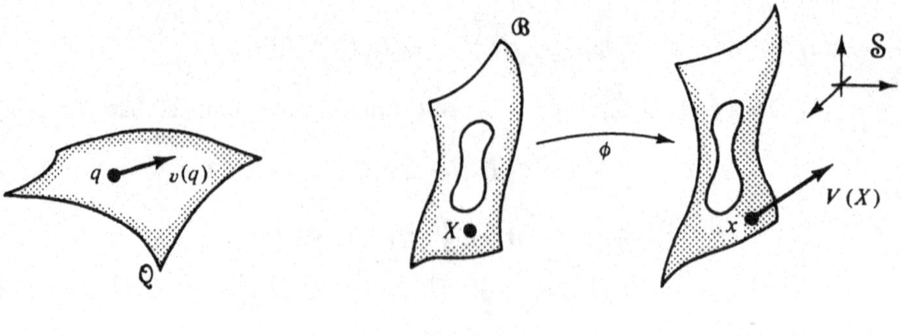

Figure 1.2.3

2.9 Definitions If Y is a vector field on $\mathcal{B}$ and $\phi: \mathcal{B} \longrightarrow \mathcal{S}$ is a C^1 mapping, then $V = T\phi \circ Y$, a vector field covering ϕ, is called the *tilt* of Y by ϕ.

If ϕ is regular, then $\phi_* Y = T\phi \circ Y \circ \phi^{-1}$, a vector field on $\phi(\mathcal{B})$ is called the *push-forward* of Y by ϕ.

If v is a vector field on $\phi(\mathcal{B})$ and ϕ is regular (i.e., $\phi(\mathcal{B})$ is open in $\mathcal{S}$ and $\phi: \mathcal{B} \longrightarrow \phi(\mathcal{B})$ has a C^1 inverse), $\phi^* v = T(\phi^{-1}) \circ v \circ \phi$, a vector field on $\mathcal{B}$ is called the *pull-back* of v by ϕ.

If V is a vector field covering ϕ, we say that $V \circ \phi^{-1}$ (a vector field on $\phi(\mathcal{B})$) is *V expressed in the spatial picture* and $T\phi^{-1} \circ V$ (a vector field on $\mathcal{B}$) is *V expressed in the convected picture.*

The diagram relevant to these concepts is as follows:

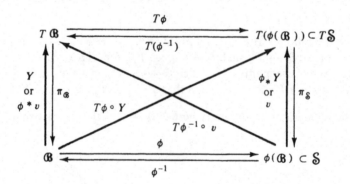

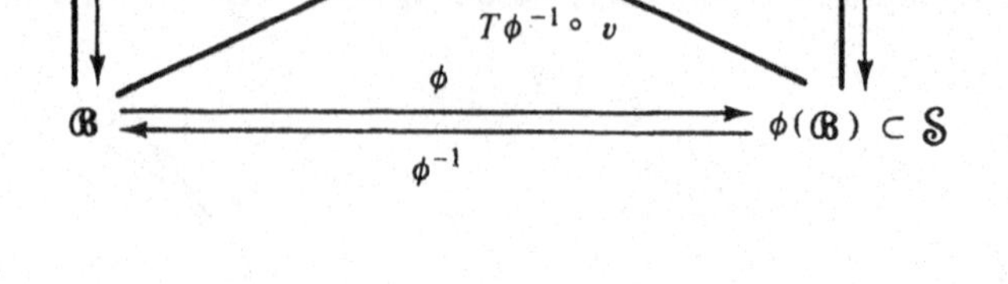

From 2.6(b) we have the coordinate description of these objects. For example, for the tilt,

$$(T\phi \circ Y(X))^a = \frac{\partial \phi^a}{\partial X^A}(X) \cdot Y^A(X),$$

and for the push-forward,

$$(\phi_* Y(X))^a = \frac{\partial \phi^a}{\partial X^A}(X) \cdot Y^A(X), \qquad \text{where} \quad x = \phi(X).$$

2.10 Definition[11] Let ϕ_t be a regular motion of $\mathcal{B}$ in $\mathcal{S}$ and let V_t and v_t be the material and spatial velocities defined in Section 1.1. We call $\phi_t^* v_t = \mathbf{v}_t$ the *convective velocity*; it is a vector field on $\mathcal{B}$ for each t.

The physical meaning of the convective velocity is explained by the use of "convected coordinates" as follows.

2.11 Proposition *Let ϕ_t be a regular motion of $\mathcal{B}$ in $\mathcal{S}$, and V_t, v_t, and $\mathbf{v}_t$ the material, spatial, and convective velocities, respectively. Suppose $\{X^A\}$ is a coordinate system on $\mathcal{B}$ and $\{x^a\}$ is one on $\mathcal{S}$. Let χ^A be the coordinate system on $\phi_t(\mathcal{B})$ defined by $\chi_t^A = X^A \circ \phi_t^{-1}$. Then the components of $\mathbf{v}_t$ with respect to X^A equal those of V_t (or v_t) with respect to χ_t^A (evaluated at the appropriate points).*

Proof Let V_χ^A be the components of V with respect to χ_t^A and ν^A those of $\mathbf{v}$ with respect to X^A. Thus

$$\mathbf{v} = \nu^A E_A, \qquad V = V_\chi^A \Xi_A$$

where $E_A = (\partial Z^I/\partial X^A)\hat{I}_I$ and $\Xi_A = (\partial z^i/\partial \chi^A)\hat{\imath}_i$. By definition, $V(X, t) = T\phi \cdot \mathbf{v}(X, t)$, so from 2.6(b), $V = (\partial \phi_\Xi^B/\partial X^A)\nu^A \Xi_B$. However, $\phi_\Xi^B(X, t) = (X^B \circ \phi_t^{-1})(\phi_t(X)) = X^B(X)$, so $\partial \phi_\Xi^B/\partial X^A = \delta^B{}_A$, and $V = \nu^A \Xi_A$ as required. ∎

Problem 2.1 Prove that $\Xi_A = T\phi \cdot E_A$. From this, obtain an alternative proof of 2.11.

Problem 2.2 Define the convective acceleration $\boldsymbol{\alpha}$ by $\boldsymbol{\alpha}_t = \phi_t^* a$ where a is the spatial acceleration.

(a) Show that the components of $\boldsymbol{\alpha}_t$ with respect to $\{X^A\}$ equal those of A_t with respect to $\{\chi^A\}$.
(b) Find the relationship between the convected acceleration and the convected velocity in components.

The coordinates χ^A may be thought of as being convected by the motion, or being scribed on $\mathcal{B}$ and carried with it as it moves. See Figure 1.2.4. A dual procedure may also be constructed; that is, one could convect the coordinates $\{x^a\}$ on $\phi_t(\mathcal{B})$ back to $\mathcal{B}$ by composition with ϕ_t.

The operations of pull-back and push-forward may be performed on tensor fields on a general manifold $\mathcal{Q}$. We pause here briefly to consider the situation for one-forms—that is, for "covariant" vectors as opposed to "contravariant" vectors.

2.12 Definitions Let $\mathcal{Q}$ be an n-manifold and $q \in \mathcal{Q}$. A *one-form at q* is a linear mapping $\boldsymbol{\alpha}_q \colon T_q\mathcal{Q} \longrightarrow \mathbb{R}$; the vector space of one-forms at q is denoted $T_q^*\mathcal{Q}$. The *cotangent bundle* of $\mathcal{Q}$ is the disjoint union of the sets $T_q^*\mathcal{Q}$ (made into

[11]In the general case in which $\mathcal{B}$ and $\mathcal{S}$ have different dimensions, v_t need not be tangent to $\phi_t(\mathcal{B})$, but only to $\mathcal{S}$. Thus $\phi_t^* v_t$ will not make sense, even if ϕ_t is regular. See Section 1.5 for details. Here there is no trouble since we are in the special case $\mathcal{B}$ open in $\mathbb{R}^n$ and $\mathcal{S} = \mathbb{R}^n$.

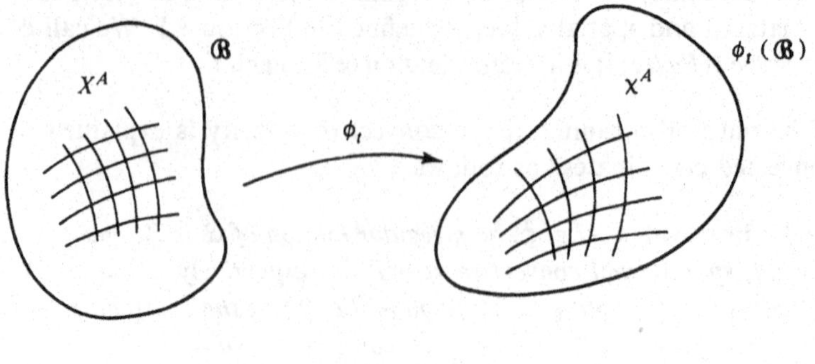

Figure 1.2.4

a manifold, as was the tangent bundle). A *one-form on* $\mathbb{Q}$ is a map $\boldsymbol{\alpha}: \mathbb{Q} \longrightarrow T^*\mathbb{Q}$ such that $\boldsymbol{\alpha}_q = \boldsymbol{\alpha}(q) \in T_q^*\mathbb{Q}$ for all $q \in \mathbb{Q}$.

If $\{x^i\}$ are coordinates on $\mathbb{Q}$, we saw that they induce coordinates on $T\mathbb{Q}$; if $\mathbb{Q} \subset \mathbb{R}^n$ is open, $T\mathbb{Q} = \mathbb{Q} \times \mathbb{R}^n$ and $(q, \boldsymbol{v})$ is mapped to $(x^i(q), v^j)$, where $\boldsymbol{v} = v^j\boldsymbol{e}_j$. For $T^*\mathbb{Q} = (\mathbb{Q} \times \mathbb{R}^{n*})$, we map $(q, \boldsymbol{\alpha})$ to $(x^i(q), \alpha_j)$, where $\boldsymbol{\alpha} = \alpha_j\boldsymbol{e}^j$, $\boldsymbol{e}^j$ being the basis dual to $\boldsymbol{e}_j$; that is, $\boldsymbol{e}^j(\boldsymbol{e}_i) = \delta^j{}_i$.

2.13 Definition Let $f: \mathbb{Q} \longrightarrow \mathbb{R}$ be C^1 so that $Tf: T\mathbb{Q} \longrightarrow T\mathbb{R} = \mathbb{R} \times \mathbb{R}$. The second factor (the "vector part") is called $\boldsymbol{df}$, the *differential* of f. Thus $\boldsymbol{df}$ is a one-form[12] on $\mathbb{Q}$.

The reader may verify the following.

2.14 Proposition *If $\{x^i\}$ is a coordinate system on $\mathbb{Q}$, then $\boldsymbol{e}^i = \boldsymbol{dx}^i$; that is, $\boldsymbol{dx}^i$ is the dual basis of $\partial/\partial x^i$; furthermore,*

$$\boxed{\boldsymbol{df} = \frac{\partial f}{\partial x^i}\boldsymbol{dx}^i.}$$

The transformation rule for one-forms is easy to work out. If $\boldsymbol{\alpha}$ has components α_i relative to $\{x^i\}$ and $\bar{\alpha}_i$ relative to $\{\bar{x}^i\}$, then $\bar{\alpha}_i = (\partial x^j/\partial \bar{x}^i)\alpha_j$.

2.15 Definitions If $\phi: \mathfrak{B} \longrightarrow \mathbb{S}$ is a mapping, a *one-form over* ϕ is a mapping $\boldsymbol{\Lambda}: \mathfrak{B} \longrightarrow T^*\mathbb{S}$ such that for $X \in \mathfrak{B}$, $\boldsymbol{\Lambda}_X \equiv \boldsymbol{\Lambda}(X) \in T^*_{\phi(X)}\mathbb{S}$.

If ϕ is C^1 and $\boldsymbol{\beta}$ is a one-form on $\mathbb{S}$, then $\phi^*\boldsymbol{\beta}$—the one-form on $\mathfrak{B}$ defined by $(\phi^*\boldsymbol{\beta})_X \cdot \boldsymbol{W}_X = \boldsymbol{\beta}_{\phi(X)} \cdot (T\phi \cdot \boldsymbol{W}_X)$ for $X \in \mathfrak{B}$ and $\boldsymbol{W}_X \in T_X\mathfrak{B}$—is called the *pull-back* of $\boldsymbol{\beta}$ by ϕ. (In contrast to vector fields, the pull-back of a one-form does not

---

[12]Do not confuse $\boldsymbol{df}$ with the gradient of f, introduced later, which is a *vector field* on $\mathbb{Q}$. In this book all tensor fields and tensor operations are denoted by boldface characters. Points, coordinates, components, mappings, and scalar fields are denoted by lightface characters.

use the inverse of ϕ, so does not require ϕ to be regular.) If ϕ is regular, we can define the *push-forward* of a one-form γ on $\mathcal{B}$ by $\phi_*\gamma = (\phi^{-1})^*\gamma$.

Problem 2.3 Let $T_x^*\phi: T_x^*\mathcal{S} \to T_X^*\mathcal{B}$ be the dual of $T_X\phi$. Show that

$$(\phi^*\boldsymbol{\beta})_X = T_x^*\phi \circ \boldsymbol{\beta} \circ \phi^{-1}$$

The coordinate expression for pull-back follows.

2.16 Proposition *If $\{X^A\}$ are coordinates on $\mathcal{B}$ and $\{x^a\}$ are coordinates on $\mathcal{S}$, then*

$$(\phi^*\boldsymbol{\beta}(X))_A = \frac{\partial \phi^a}{\partial X^A}(X) \cdot \beta_a(\phi(X)).$$

Proof

$$(\phi^*\boldsymbol{\beta})_X \cdot \boldsymbol{E}_A = \boldsymbol{\beta}_{\phi(X)} \cdot T\phi \cdot \boldsymbol{E}_A = \boldsymbol{\beta}_{\phi(X)} \cdot \frac{\partial \phi^a}{\partial X^A} \boldsymbol{e}_a = \frac{\partial \phi^a}{\partial X^A} \boldsymbol{\beta}_{\phi(X)} \cdot \boldsymbol{e}_a$$

so the result follows since $\boldsymbol{\beta} \cdot \boldsymbol{e}_a = \beta_a$ ∎

Using the chain rule one can easily prove the following properties of pull-back:

2.17 Proposition *(a) Let $\phi: \mathcal{B} \to \mathcal{S}$ be C^1 and $f: \mathcal{S} \to \mathbb{R}$ be C^1. Set $\phi^*f = f \circ \phi$. Then $\phi^*(df) = d(\phi^*f)$.*
(b) If $\psi: \mathcal{S} \to \mathcal{Q}$ is C^1 and γ is a one-form on $\mathcal{Q}$, then $(\psi \circ \phi)^*\gamma = \phi^*(\psi^*\gamma)$

In coordinates, pull-back by ϕ acts the same way as if ϕ were a coordinate transformation. The following shows that the rule in 2.17(a) embodies the coordinate expression:

$$\phi^*\boldsymbol{\beta} = \phi^*(\beta_a dx^a) = (\beta_a \circ \phi)\phi^* dx^a = (\beta_a \circ \phi) d\phi^a = (\beta_a \circ \phi) \frac{\partial \phi^a}{\partial X^B} dX^B$$

so $\quad (\phi^*\boldsymbol{\beta})_B = \beta_a \circ \phi \dfrac{\partial \phi^a}{\partial X^B}$.

One can think of one-forms as row vectors and vectors as column vectors in the sense of matrix algebra. The natural contraction $\boldsymbol{\beta} \cdot \boldsymbol{Y}$ defined by $(\boldsymbol{\beta} \cdot \boldsymbol{Y})(X) = \boldsymbol{\beta}_X(\boldsymbol{Y}(X))$, where $\boldsymbol{\beta}$ is a one-form on $\mathcal{B}$ and $\boldsymbol{Y}$ a vector field, is just matrix multiplication (i.e. $\beta_A Y^A$). If $\boldsymbol{\beta} = df$, then $\boldsymbol{\beta} \cdot \boldsymbol{Y}$ is just $\boldsymbol{Y}[f]$ discussed earlier.

Box 2.1 below makes some further connections between "classical" and "modern" tensor analysis, and summarizes the situation.

Problem 2.4 Consider the motion ϕ_t and coordinate systems of Problem 1.1.

(i) Let $\boldsymbol{w}$ be a vector field on $\mathbb{R}^3$. Calculate $\phi_t^*\boldsymbol{w}$.
(ii) What if $\boldsymbol{w}$ is the spatial velocity?
(iii) Let $\boldsymbol{\alpha}$ be a one form on $\mathbb{R}^3$. Calculate $\phi_t^*\boldsymbol{\alpha}$.
(iv) Let $\boldsymbol{Y}(Z^1, Z^2, Z^3) = (0, -Z^3, Z^2)$. Calculate the ϕ_t-tilt of $\boldsymbol{Y}$.

Box 2.1 ***Notational Relationship between Classical Tensor Analysis and Analysis on Manifolds***

Classical Tensor Analysis Let $\{x^a\}$ denote a (curvilinear) coordinate system defined on an open subset of $\mathbb{R}^n$. Let z^i and $\hat{\boldsymbol{\imath}}_i$ denote the Cartesian coordinate functions of $\mathbb{R}^n$ and the corresponding unit basis vectors (the standard basis of $\mathbb{R}^n$), respectively. We may view the z^i as functions of x^a and vice-versa. The coordinate basis vectors $\boldsymbol{e}_a$ (thought of as column vectors) corresponding to x^a, are defined by $\boldsymbol{e}_a = (\partial z^i/\partial x^a)\hat{\boldsymbol{\imath}}_i$ and are tangent to their respective coordinate curves x^a. The dual basis $\boldsymbol{e}^a$ (throught of as row vectors) is defined by the inner product $\boldsymbol{e}^a \cdot \boldsymbol{e}_b = \delta^a{}_b$. If we define the *metric tensor* g_{ab} by

$$g_{ab} = \frac{\partial z^i}{\partial x^a}\frac{\partial z^j}{\partial x^b}\delta_{ij}$$

and let g^{ab} denote the inverse matrix of g_{ab} (so $g^{ac}g_{cb} = \delta^a{}_b$), then $\boldsymbol{e}^a = g^{ab}\boldsymbol{e}_b$. (This is an easy verification; see Section 1.3 for details.) Both $\boldsymbol{e}_a$ and $\boldsymbol{e}^a$ are usually viewed as vectors although they satisfy different transformation laws; that is, if $\bar{x}^a$ denotes another coordinate system, then

$$\bar{\boldsymbol{e}}_a = \frac{\partial x^b}{\partial \bar{x}^a}\boldsymbol{e}_b \quad \text{and} \quad \bar{\boldsymbol{e}}^a = \frac{\partial \bar{x}^a}{\partial x^b}\boldsymbol{e}^b.$$

To distinguish between the two types of vectors, the terminologies covariant and contravariant are applied to $\boldsymbol{e}_a$ and $\boldsymbol{e}^a$, respectively. (Actually, $\boldsymbol{e}^a$ is not a vector, but a one-form; the use of the metric is the source of confusion.)

Tensor Analysis on Manifolds Let $\{x^a\}$ denote a coordinate system defined on an open set in a manifold $\mathcal{S}$. The coordinate basis corresponding to $\{x^a\}$ is denoted $\partial/\partial x^a$, and the dual basis is denoted $\boldsymbol{d}x^a$. The terminology "vector" is reserved for $\partial/\partial x^a$ and the terminology "one-form" (or covector) is used for $\boldsymbol{d}x^a$. The transformation rules for $\partial/\partial x^a$ and $\boldsymbol{d}x^a$ are suggested by the differential operator notations; that is, if $\bar{x}^a$ denotes another coordinate system on $\mathcal{S}$, then

$$\frac{\partial}{\partial \bar{x}^a} = \frac{\partial x^b}{\partial \bar{x}^a}\frac{\partial}{\partial x^b} \quad \text{and} \quad \boldsymbol{d}\bar{x}^a = \frac{\partial \bar{x}^a}{\partial x^b}\boldsymbol{d}x^b.$$

One sees from these expressions that $\partial/\partial x^a$ and $\boldsymbol{d}x^a$ are the analogs of $\boldsymbol{e}_a$ and $\boldsymbol{e}^a$, respectively. If we define $\boldsymbol{e}^a$ as the dual basis in the proper sense (i.e., $\boldsymbol{e}^a$ is a linear functional) without use of the metric, then $\boldsymbol{e}_a = \partial/\partial x^a$ and $\boldsymbol{e}^a = \boldsymbol{d}x^a$.

Tensor analysis on manifolds is developed without assuming the structure of a "background" $\mathbb{R}^n$. Thus Cartesian coordinate systems and standard bases are unavailable, so relations of the form $\boldsymbol{e}_a = (\partial z^i/\partial x^a)\hat{\boldsymbol{\imath}}_i$ are not allowed. The notion of a vector can be defined operationally via the directional derivative as follows. Let $\boldsymbol{v} = v^a(\partial/\partial x^a) = v^a\boldsymbol{e}_a$ denote a vector field on $\mathcal{S}$. Then the directional derivative of the function $f: \mathcal{S} \to \mathbb{R}$ in the direction v, denoted $v[f]$, is given by $v[f] = v^a(\partial f/\partial x^a)$, which is independent of the coordinate system. The analogous quantity for vectors in $\mathbb{R}^n$ is given by

$$v[f] \stackrel{\text{def}}{=} \frac{d}{d\epsilon} f(z^i + \epsilon v^i)\bigg|_{\epsilon=0} = v^i \frac{\partial f}{\partial z^i} = v^a \frac{\partial f}{\partial x^a}.$$

Thus the end result is the same.

The differential of f, denoted df, can be defined by $df(v) = v[f]$. From this, the duality relation follows immediately: $dx^a(\partial/\partial x^b) = \partial x^a/\partial x^b = \delta^a{}_b$. This is the analog of the relationship $e^a \cdot e_b = \delta^a{}_b$.

Manifolds are important in the formulation of physical theories. For example, in relativity space–time is modeled as a four-dimensional, pseudo-Riemannian manifold. There is simply no Cartesian structure to fall back on. In continuum mechanics, with a manifold as a basic notion of a body, a geometric theory of structured materials becomes a possibility. Even for a simple body, tensor analysis on manifolds clarifies the basic theory. For instance, using manifold ideas we can see clearly how to formulate the pull-back and push-forward and hence to clarify the meaning of convective velocity (and many more things that are considered in the following sections). The metric is not required to talk about e_a and e^a. The confusion arises through unnecessary identifications brought about by the presence of the Cartesian structure. If introduced in the manifold context, we can see exactly how the metric is needed (see Section 1.3).

The following table summarizes the relations and notations of classical tensor analysis on $\mathbb{R}^n$ and tensor analysis on manifolds.

Classical Tensor Analysis		*Tensor Analysis on Manifolds*
$\{x^a\}$	Coordinates	$\{x^a\}$
$e_a = \dfrac{\partial z^i}{\partial x^a}\hat{i}_i$	coodinate basis vectors	$\dfrac{\partial}{\partial x^a} = e_a$
$e^a = g^{ab} e_b$	dual basis	$dx^a = e^a$
$e^a \cdot e_b = \delta^a{}_b$	duality relation	$dx^a(\partial/\partial x^b) = \delta^a{}_b$
$\left.\begin{array}{l}\bar{e}_a = \dfrac{\partial x^b}{\partial \bar{x}^a} e_b \\ \bar{e}^a = \dfrac{\partial \bar{x}^a}{\partial x^b} e^b\end{array}\right\}$	change of coordinates	$\left\{\begin{array}{l}\dfrac{\partial}{\partial \bar{x}^a} = \dfrac{\partial x^b}{\partial \bar{x}^a}\dfrac{\partial}{\partial x^b} \\ d\bar{x}^a = \dfrac{\partial \bar{x}^a}{\partial x^b} dx^b\end{array}\right.$
$\left.\begin{array}{l}v = v^a e_a \\ v^a = e^a \cdot v\end{array}\right\}$	coordinate representation of a vector	$\left\{\begin{array}{l}v = v^a(\partial/\partial x^a) \\ v^a = dx^a(v)\end{array}\right.$
$\left.\begin{array}{l}\alpha = \alpha_a e^a \\ \alpha_a = e_a \cdot \alpha\end{array}\right\}$	coordinate representation of a one-form	$\left\{\begin{array}{l}\alpha = \alpha_a dx^a \\ \alpha_a = \alpha(\partial/\partial x^a)\end{array}\right.$
$v[f] = v^a \dfrac{\partial f}{\partial x^a}$	directional derivative	$v[f] = v^a \dfrac{\partial f}{\partial x^a}$

In classical tensor analysis, a vector and one-form are indistinguishable in the sense that we have the representations

$$\mathbf{v} = v^a \mathbf{e}_a = v_a \mathbf{e}^a, \quad \text{and} \quad \boldsymbol{\alpha} = \alpha_a \mathbf{e}^a = \alpha^a \mathbf{e}_a.$$

In tensor analysis on manifolds one avoids confusing vectors and one-forms.

Box 2.2 ***Summary of Important Formulas for Section 1.2***

Tangent Map of $\phi: \mathcal{B} \to \mathcal{S}$

$$T\phi: T\mathcal{B} \to T\mathcal{S},$$

$$T\phi \cdot W_X = T\phi(X, W) = (\phi(X), D\phi(X) \cdot W), \qquad (T\phi \cdot W_X)^a = \frac{\partial \phi^a}{\partial X^A} W^A$$

Projection

$$\pi_{\mathcal{S}}(T\phi \cdot W_X) = \phi(X) \qquad (\pi_{\mathcal{S}}(T\phi \cdot W_X))^a = \phi^a(X)$$

Tilt of a Vector Field $\mathbf{Y}$

$$V = T\phi \circ Y \qquad V^a = \frac{\partial \phi^a}{\partial X^A} Y^A$$

Push-Forward of Vector Field $\mathbf{Y}$

$$\phi_* Y = T\phi \circ Y \circ \phi^{-1} \qquad (\phi_* Y)^a = \left(\frac{\partial \phi^a}{\partial X^A} \circ \phi^{-1}\right)(Y^A \circ \phi^{-1})$$

Pull-Back of Vector Field $\mathbf{v}$

$$\phi^* \mathbf{v} = T\phi^{-1} \circ \mathbf{v} \circ \phi \qquad (\phi^* \mathbf{v})^A = \left(\frac{\partial (\phi^{-1})^A}{\partial x^a} \circ \phi\right)(v^a \circ \phi)$$

Pull-Back of One-Form $\boldsymbol{\beta}$

$$\phi^* \boldsymbol{\beta} = (\boldsymbol{\beta} \circ \phi) \cdot T\phi \qquad (\phi^* \boldsymbol{\beta})_A = (\beta_a \circ \phi) \frac{\partial \phi^a}{\partial X^A}$$

Push-Forward of One-Form $\boldsymbol{\gamma}$

$$\phi_* \boldsymbol{\gamma} = (\phi^{-1})^* \boldsymbol{\gamma} \qquad (\phi_* \boldsymbol{\gamma})_a = (\gamma_A \circ \phi^{-1}) \frac{\partial (\phi^{-1})^A}{\partial x^a}$$

Change of Coordinates

$$\frac{\partial \bar{\phi}^a}{\partial \bar{X}^A} = \frac{\partial \bar{x}^a}{\partial x^b} \frac{\partial \phi^b}{\partial X^B} \frac{\partial X^B}{\partial \bar{X}^A}$$

$$\bar{\beta}_a = \frac{\partial x^b}{\partial \bar{x}^a} \beta_b$$

Differential of a Function

df = second component of Tf $\qquad df = \dfrac{\partial f}{\partial x^a} dx^a$

$dx^a = \mathbf{e}^a$, dual basis of $\mathbf{e}_a$

Directional Derivatives

$$Y[f] = df \cdot Y \qquad Y[f] = \frac{\partial f}{\partial X^A} Y^A$$

Pull-Back of a Differential

$$\phi^* df = d(\phi^* f) \qquad \left(\frac{\partial f}{\partial x^a} \circ \phi\right) \frac{\partial \phi^a}{\partial X^A} = \frac{\partial (f \circ \phi)}{\partial X^A}$$

Chain Rules

$$T(\psi \circ \phi) = T\psi \circ T\phi \qquad \frac{\partial (\psi \circ \phi)^\alpha}{\partial X^A} = \frac{\partial \psi^\alpha}{\partial x^a} \frac{\partial \phi^a}{\partial X^A}$$

$$(\psi \circ \phi)^* = \phi^* \circ \psi^*$$

Convective Velocity

$$\mathbf{V}_t = \phi_t^* v_t \qquad \mathbf{V}^A = \left(\frac{\partial (\phi^{-1})^A}{\partial x^a} \circ \phi\right)(v^a \circ \phi)$$

1.3 THE DEFORMATION GRADIENT

The derivative of the configuration of a body is called the deformation gradient. This object plays a fundamental role in the subsequent theory, so we shall devote this section to its detailed study. Several of the notions here serve as motivation for the geometric considerations of Section 1.4. Reciprocally, the general framework of that section will help give perspective and a deeper understanding of the results here.

3.1 Definition Let $\phi: \mathcal{B} \longrightarrow \mathcal{S}$ be a C^1 configuration of $\mathcal{B}$ in $\mathcal{S}$ ($\mathcal{B}$ and $\mathcal{S}$ can be general manifolds here). The tangent of ϕ is denoted $\boldsymbol{F}$ and is called the *deformation gradient* of ϕ; thus $\boldsymbol{F} = T\phi$. For $X \in \mathcal{B}$, we let $\boldsymbol{F}_X$ or $\boldsymbol{F}(X)$ denote the restriction of $\boldsymbol{F}$ to $T_X\mathcal{B}$. Thus $\boldsymbol{F}(X): T_X\mathcal{B} \longrightarrow T_{\phi(X)}\mathcal{S}$ is a linear transformation for each $X \in \mathcal{B}$.

In Proposition 2.6 we worked out the coordinate description of $\boldsymbol{F}$. We recall this for reference.

3.2 Proposition *Let $\{X^A\}$ and $\{x^a\}$ denote coordinate systems on $\mathcal{B}$ and $\mathcal{S}$, respectively. Then the matrix of $\boldsymbol{F}(X)$ with respect to the coordinate bases $\boldsymbol{E}_A(X)$ and $\boldsymbol{e}_a(x)$ [where $x = \phi(X)$] is given by*

$$F^a{}_A(X) = \frac{\partial \phi^a}{\partial X^A}(X).$$

If we have a motion $\phi(X, t)$, we shall write the components of the deformation gradient for each t as $F^a{}_A(X, t)$, or merely as $F^a{}_A$ if we are suppressing the arguments. The deformation gradient $\boldsymbol{F}$ is an important example of a *two-point tensor*. These objects will be discussed in general in Section 1.4. Notice that the coordinate expression for $F^a{}_A$ does not involve any covariant derivatives. This is because ϕ *is not a vector, but rather is a point mapping of* $\mathcal{B}$ *to* $\mathcal{S}$. (Sometimes $x = \phi(X)$ is represented by an "arrow," but this can be a source of confusion.)

For the remainder of this section we will assume $\mathcal{S} = \mathbb{R}^n$ and $\mathcal{B} \subset \mathbb{R}^n$ is a simple body. (After the reader has digested Section 1.4, $\mathcal{B}$ and $\mathcal{S}$ may be replaced by Riemannian manifolds.)

3.3 Notation We let $\langle , \rangle_x$ denote the standard inner product in $\mathbb{R}^n$ for vectors based at $x \in \mathcal{S}$ and similarly let $\langle , \rangle_X$ be the standard inner product in $\mathbb{R}^n$ at $X \in \mathcal{B}$. For a vector $\boldsymbol{v} \in T_x\mathcal{S}$ we let $\|\boldsymbol{v}\|_x = \langle \boldsymbol{v}, \boldsymbol{v}\rangle_x^{1/2}$ be the length of $\boldsymbol{v}$. Similarly the length of $\boldsymbol{W} \in T_X\mathcal{B}$ is denoted $\|\boldsymbol{W}\|_X$. (If there is no danger of confusion, the subscripts may be dropped.)

Let $\boldsymbol{A}\colon T_X\mathcal{B} \longrightarrow T_x\mathcal{S}$ be a linear transformation. Then the *transpose*, or *adjoint* of $\boldsymbol{A}$, written $\boldsymbol{A}^T$, is the linear transformation

$$\boldsymbol{A}^T\colon T_x\mathcal{S} \longrightarrow T_X\mathcal{B} \qquad \text{such that} \quad \langle \boldsymbol{A}\,\boldsymbol{W}, \boldsymbol{v}\rangle_x = \langle \boldsymbol{W}, \boldsymbol{A}^T\boldsymbol{v}\rangle_X$$

for all $\boldsymbol{W} \in T_X\mathcal{B}$ and $\boldsymbol{v} \in T_x\mathcal{S}$. If $\boldsymbol{B}\colon T_x\mathcal{S} \longrightarrow T_x\mathcal{S}$ is a linear transformation, it is called *symmetric* if $\boldsymbol{B} = \boldsymbol{B}^T$.

In a coordinate system $\{x^a\}$ on $\mathcal{S}$, let the metric tensor g_{ab} be defined by $g_{ab}(x) = \langle \boldsymbol{e}_a, \boldsymbol{e}_b\rangle_x$ and similarly define $G_{AB}(X)$ on $\mathcal{B}$ by $G_{AB}(X) = \langle \boldsymbol{E}_A, \boldsymbol{E}_B\rangle_X$. We let g^{ab} and G^{AB} denote the inverse matrices of g_{ab} and G_{AB}; these exist since g_{ab} and G_{AB} are nonsingular.

In Euclidean space, $\boldsymbol{e}_a = (\partial z^i/\partial x^a)\hat{\boldsymbol{\imath}}_i$, so we have the expression

$$g_{ab} = \frac{\partial z^i}{\partial x^a}\frac{\partial z^j}{\partial x^b}\delta_{ij}.$$

Similarly,

$$G_{AB} = \frac{\partial Z^I}{\partial X^A}\frac{\partial Z^J}{\partial X^B}\delta_{IJ}.$$

3.4 Proposition

(i) *For* $\boldsymbol{v}, \boldsymbol{w} \in T_x\mathcal{S}$ *and a coordinate system* $\{x^a\}$, *we have*

$$\boxed{\langle \boldsymbol{v}, \boldsymbol{w}\rangle_x = g_{ab}v^a w^b.}$$

(ii) *If* $\{x^a\}$ *and* $\{X^A\}$ *are coordinate systems on* $\mathcal{S}$ *and* $\mathcal{B}$, *respectively, and* $\phi\colon \mathcal{B} \longrightarrow \mathcal{S}$ *is a* C^1 *configuration of* $\mathcal{B}$, *then the matrix of* $\boldsymbol{F}^T$ *is given by*

$$\boxed{(\boldsymbol{F}^T(x))^A{}_a = g_{ab}(x)F^b{}_B(X)G^{AB}(X),}$$

where $x = \phi(X)$.

Proof

(i) This follows from the definition of g_{ab} and the expressions $v = v^a e_a$ and $w = w^b e_b$:

$$\langle v, w\rangle_x = \langle v^a e_a, w^b e_b\rangle = v^a w^b \langle e_a, e_b\rangle = v^a w^b g_{ab}.$$

(ii) By definition,

$$\langle F^T w, W\rangle_X = \langle FW, w\rangle_x; \qquad \text{that is,} \quad (F^T)^B{}_b w^b W^A G_{AB} = F^a{}_A W^A w^b g_{ab}$$

for all $W \in T_X\mathcal{B}$ and $w \in T_x\mathcal{S}$, where F^T and F have their arguments suppressed. Since W and w are arbitrary, $(F^T)^B{}_b G_{AB} = F^a{}_A g_{ab}$. Multiplying by G^{AC} and using $G_{AB}G^{AC} = \delta^C{}_B$ gives the result. (In order for the map $F^T: T(\phi(\mathcal{B})) \rightarrow T\mathcal{B}$ to be well defined, ϕ must be regular.) ∎

Problem 3.1 Define $F^*(x): T^*_x\mathcal{S} \rightarrow T^*_X\mathcal{B}$ by $(F^*(x)\cdot\boldsymbol{\beta})(W) = \boldsymbol{\beta}(F(x)\cdot W)$ for $\boldsymbol{\beta} \in T^*_x\mathcal{S}$ and $W \in T_X\mathcal{B}$. Show that the matrix representative of F^* with respect to the dual bases e^a and E^A is the transpose of $F^a{}_A$, i.e. dropping x, $F^*(e^a) = F^a{}_A E^A$.

3.5 Definition The (Green) *deformation tensor* (also called the right Cauchy–Green tensor) C is defined by:

$$\boxed{C(X): T_X\mathcal{B} \rightarrow T_X\mathcal{B}, \qquad C(X) = F(X)^T F(X)}$$

Or, for short, $C = F^T F$.

If C is invertible, we let $B = C^{-1}$, called the *Piola deformation tensor.*

The following proposition gives some of the basic properties of C.

3.6 Proposition *Let ϕ be a C^1 configuration of $\mathcal{B}$ in $\mathcal{S}$ and C the deformation tensor.*

(i) *If $\{x^a\}$ and $\{X^A\}$ are coordinate systems on $\mathcal{S}$ and $\mathcal{B}$, respectively, then*

$$\boxed{C^A{}_B = (F^T)^A{}_a F^a{}_B = g_{ab} G^{AC} \frac{\partial \phi^b}{\partial X^C} \frac{\partial \phi^a}{\partial X^B}.}$$

(ii) *C is symmetric and positive-semidefinite; that is, $\langle CW, W\rangle_X \geq 0$, and if each F_X is one-to-one, then C is invertible and positive-definite; that is, $\langle CW, W\rangle > 0$ if $W \neq 0$. (Note that F is one-to-one if ϕ is regular.)*

Proof

(i) follows from the definition of C and 3.4(ii).

(ii) Let $W_1, W_2 \in T_X\mathcal{B}$. Then $\langle CW_1, W_2\rangle_X = \langle F^T F W_1, W_2\rangle_X = \langle FW_1, FW_2\rangle_X = \langle W_1, F^T F W_2\rangle_X = \langle W_1, CW_2\rangle_X$, so C is symmetric. Clearly, $\langle CW, W\rangle_X = \langle FW, FW\rangle_X \geq 0$, so C is positive-semidefinite.

If $\boldsymbol{F}$ is one-to-one and if $\langle \boldsymbol{C}\boldsymbol{W}, \boldsymbol{W}\rangle = \langle \boldsymbol{F}\boldsymbol{W}, \boldsymbol{F}\boldsymbol{W}\rangle$ is zero, then $\boldsymbol{F}\boldsymbol{W} = \boldsymbol{0}$ and hence $\boldsymbol{W} = \boldsymbol{0}$. In particular, $\boldsymbol{C}$ is one-to-one and hence is invertible. ▮

Symmetry of $\boldsymbol{C}$ means that $C_{AB} = C_{BA}$, where $C_{AB} = G_{AC}C^C{}_B$. We call C_{AB} the *associated components* of $\boldsymbol{C}$. We pause briefly to consider this notion.

3.7 Definition Let $\boldsymbol{\alpha}$ be a one form on $\mathcal{S}$, with components α_a in a coordinate system $\{x^a\}$; that is, $\boldsymbol{\alpha} = \alpha_a \boldsymbol{e}^a$. The *associated vector field* $\boldsymbol{\alpha}^\sharp$ is defined to have components $\alpha^a = g^{ab}\alpha_b$; that is, $\boldsymbol{\alpha}^\sharp = \alpha^a \boldsymbol{e}_a = g^{ab}\alpha_b \boldsymbol{e}_a$.

If $\boldsymbol{v}$ is a vector field, the *associated one-form* $\boldsymbol{v}^\flat$ is defined by $\boldsymbol{v}^\flat = v_a \boldsymbol{e}^a$, where $v_a = g_{ab}v^b$. If $\boldsymbol{\sigma}$ is a tensor with components σ_{ab}, then the tensors with components $\sigma_a{}^b = \sigma_{ac}g^{cb}$, $\sigma^a{}_b = \sigma_{cb}g^{ca}$, and $\sigma^{ab} = \sigma_{cd}g^{ca}g^{db}$, are called *associated tensors.* In general, a rank N tensor $\mathbf{t}$ has 2^N associated tensors defined in a similar way. If we write $\mathbf{t}^\flat$, we mean the tensor associated to $\mathbf{t}$ that has all its indices lowered. Similarly, $\mathbf{t}^\sharp$ means $\mathbf{t}$ with all its indices raised.

We wish to emphasize that associated tensors are different objects. Specifically, $v^a\boldsymbol{e}_a$ and $v_a\boldsymbol{e}^a$ are *not* equal. These tensors are related via the metric tensor, but they are *not the same* tensor. This point can become confused because of an over-reliance on the Cartesian structure. The situation will be clarified in the next section. Another thing Section 1.4 will clarify is the very definition of $\boldsymbol{C}$ itself. We shall, in fact, see that $\boldsymbol{C}^\flat = \phi^*\boldsymbol{g}$, the pull-back of the metric $\boldsymbol{g}$ on $\mathcal{S}$ to $\mathcal{B}$. In a similar way, we can push the metric $\boldsymbol{G}$ on $\mathcal{B}$ forward to $\mathcal{S}$. This leads to a new tensor $\boldsymbol{b} = \boldsymbol{c}^{-1}$ with $\boldsymbol{c}^\flat = \phi_*\boldsymbol{G}$. The explicit definition follows.

3.8 Definition Let ϕ be a regular C^1 configuration of $\mathcal{B}$ in $\mathcal{S}$. Then the *Finger deformation tensor* (also called the left Cauchy–Green tensor) is defined on $\phi(\mathcal{B})$ by

$$\boldsymbol{b}(x) = \boldsymbol{F}(X)(\boldsymbol{F}(X))^T : T_x\phi(\mathcal{B}) \longrightarrow T_x\phi(\mathcal{B})$$

[where $X = \phi^{-1}(x)$], or $\boldsymbol{b} = \boldsymbol{F}\boldsymbol{F}^T$ for short. Also, define $\boldsymbol{c} = \boldsymbol{b}^{-1}$.

Note that $\boldsymbol{C}$ is defined whether or not ϕ is regular, but $\boldsymbol{c}$ and $\boldsymbol{b}$ require ϕ^{-1} to exist to be defined. As in 3.6, one can prove the following.

3.9 Proposition *If ϕ is C^1 and regular, then*

(i) $b^a{}_b = g_{bc}G^{AB}\dfrac{\partial\phi^a}{\partial X^A}\dfrac{\partial\phi^c}{\partial X^B}$ *and*

(ii) *$\boldsymbol{b}$ is symmetric and positive-definite.*

Now we shall use a result from linear algebra.

3.10 Lemma *Let $\mathcal{V}$ be a finite-dimensional inner product space and let $\boldsymbol{A}$: $\mathcal{V} \longrightarrow \mathcal{V}$ be a symmetric positive-definite linear transformation; that is, $\boldsymbol{A}^T = \boldsymbol{A}$ and $\langle \boldsymbol{A}\boldsymbol{v}, \boldsymbol{v}\rangle > 0$ if $\boldsymbol{v} \in \mathcal{V}$ ($\boldsymbol{v} \neq 0$). Then there exists a unique symmetric positive-definite linear transformation $\boldsymbol{B}: \mathcal{V} \longrightarrow \mathcal{V}$ such that $\boldsymbol{B}^2 = \boldsymbol{A}$.*

Let us recall the proof of existence (The summation convention is temporarily suspended): There is an orthonormal basis $\boldsymbol{\psi}_1, \ldots, \boldsymbol{\psi}_n$ of eigenvectors for $\boldsymbol{A}$. Let $\lambda_1, \ldots, \lambda_n$ be the corresponding eigenvalues. Then $\boldsymbol{A}\boldsymbol{\psi}_i = \lambda_i \boldsymbol{\psi}_i$, so $\langle \boldsymbol{A}\boldsymbol{\psi}_i, \boldsymbol{\psi}_i \rangle = \lambda_i \|\boldsymbol{\psi}_i\|^2 > 0$ and hence $\lambda_i > 0$. Define $\boldsymbol{B}$ by $\boldsymbol{B}\boldsymbol{\psi}_i = \sqrt{\lambda_i}\, \boldsymbol{\psi}_i$. Thus $\boldsymbol{B}^2\boldsymbol{\psi}_i$ agrees with $\boldsymbol{A}\boldsymbol{\psi}_i$, so $\boldsymbol{B}^2 = \boldsymbol{A}$. This shows $\boldsymbol{B}$ exists. Note that the eigenvalues of $\boldsymbol{B}$ are the square roots of those of $\boldsymbol{A}$.

3.11 Definition Let ϕ be regular and let $\boldsymbol{C}$ and $\boldsymbol{b}$ be defined as above. Let $\boldsymbol{U}$ and $\boldsymbol{V}$ denote the unique symmetric, positive-definite square roots of $\boldsymbol{C}$ and $\boldsymbol{b}$, respectively. We call $\boldsymbol{U}$ and $\boldsymbol{V}$ the *right and left stretch tensors*, respectively. [For each $x \in \mathcal{S}$, $\boldsymbol{V}(x)\colon T_x\mathcal{S} \to T_x\mathcal{S}$ and for each $X \in \mathcal{B}$, $\boldsymbol{U}(X)\colon T_X\mathcal{B} \to T_X\mathcal{B}$.] The eigenvalues of $\boldsymbol{U}$ are called the *principal stretches.*

Warning. Note that $\boldsymbol{V}$ also denotes the material velocity. However the meaning will be clear from the context.

3.12 Proposition *Let ϕ be regular. For each $X \in \mathcal{B}$ there exists an orthogonal transformation $\boldsymbol{R}(X)\colon T_X\mathcal{B} \to T_x\mathcal{S}$ [i.e., $\boldsymbol{R}(X)^T\boldsymbol{R}(X) = \boldsymbol{I}$ (the identity on $T_X\mathcal{B}$) and $\boldsymbol{R}(X)\boldsymbol{R}(X)^T = \boldsymbol{i}$ (the identity on $T_x\mathcal{S}$)] such that*

$$\boldsymbol{F} = \boldsymbol{R}\boldsymbol{U} \quad [\textit{that is, } \boldsymbol{F}(X) = \boldsymbol{R}(X)\circ \boldsymbol{U}(X)]$$

and

$$\boldsymbol{F} = \boldsymbol{V}\boldsymbol{R} \quad [\textit{that is, } \boldsymbol{F}(X) = \boldsymbol{V}(x)\circ \boldsymbol{R}(X)].$$

Moreover, each of these decompositions is unique: if $\boldsymbol{F} = \bar{\boldsymbol{R}}\bar{\boldsymbol{U}}$, where $\bar{\boldsymbol{R}}$ is orthogonal and $\bar{\boldsymbol{U}}$ is symmetric positive-definite, then $\bar{\boldsymbol{R}} = \boldsymbol{R}$ and $\bar{\boldsymbol{U}} = \boldsymbol{U}$. We call $\boldsymbol{R}$ the rotation matrix *and refer to $\boldsymbol{F} = \boldsymbol{R}\boldsymbol{U}$ and $\boldsymbol{F} = \boldsymbol{V}\boldsymbol{R}$ as the* right and left polar decompositions, *respectively.*

Proof Define $\boldsymbol{R}(X) = \boldsymbol{F}(X)\boldsymbol{U}(X)^{-1}$. Then $\boldsymbol{R}^T\boldsymbol{R} = \boldsymbol{U}^{-1}\boldsymbol{F}^T\boldsymbol{F}\boldsymbol{U}^{-1} = \boldsymbol{U}^{-1}\boldsymbol{C}\boldsymbol{U}^{-1} = \boldsymbol{U}^{-1}\boldsymbol{U}^2\boldsymbol{U}^{-1} = \boldsymbol{I}$ since $\boldsymbol{C} = \boldsymbol{U}^2$ by definition of $\boldsymbol{U}$, and since $(\boldsymbol{U}^{-1})^T = \boldsymbol{U}^{-1}$ as $\boldsymbol{U}$ is symmetric positive-definite. Also, $\boldsymbol{R}\boldsymbol{R}^T = \boldsymbol{R}(\boldsymbol{R}^T\boldsymbol{R})\boldsymbol{R}^T = (\boldsymbol{R}\boldsymbol{R}^T)^2$, so $\boldsymbol{R}\boldsymbol{R}^T = \boldsymbol{I}$, since $\boldsymbol{R}\boldsymbol{R}^T$ is nonsingular. Thus the right polar decomposition is established. To establish its uniqueness, let $\boldsymbol{F} = \bar{\boldsymbol{R}}\bar{\boldsymbol{U}} = \boldsymbol{R}\boldsymbol{U}$. Then $\boldsymbol{F}^T\boldsymbol{F} = \bar{\boldsymbol{U}}^T\bar{\boldsymbol{R}}^T\bar{\boldsymbol{R}}\bar{\boldsymbol{U}} = \bar{\boldsymbol{U}}^2$ and $\boldsymbol{F}^T\boldsymbol{F} = \boldsymbol{U}^2$. Since symmetric positive square roots are unique, $\bar{\boldsymbol{U}} = \boldsymbol{U}$. Hence $\bar{\boldsymbol{R}} = \boldsymbol{R}$.

Let $\boldsymbol{F} = \boldsymbol{V}\hat{\boldsymbol{R}}$ be the left polar decomposition, established in the same manner as the right polar decomposition. We prove $\boldsymbol{R} = \hat{\boldsymbol{R}}$. Indeed, $\boldsymbol{F} = \boldsymbol{V}\hat{\boldsymbol{R}} = (\hat{\boldsymbol{R}}\hat{\boldsymbol{R}}^T)\boldsymbol{V}\hat{\boldsymbol{R}} = \hat{\boldsymbol{R}}(\hat{\boldsymbol{R}}^T\boldsymbol{V}\hat{\boldsymbol{R}})$. This has the form of the right decomposition, so by uniqueness, $\hat{\boldsymbol{R}} = \boldsymbol{R}$ and $\boldsymbol{U} = \boldsymbol{R}^T\boldsymbol{V}\boldsymbol{R}$. ∎

From this proof we see that $\boldsymbol{U}$ and $\boldsymbol{V}$ are similar, and thus have the same eigenvalues. Therefore, the principal stretches can be defined in terms of either $\sqrt{\boldsymbol{F}^T\boldsymbol{F}}$ or $\sqrt{\boldsymbol{F}\boldsymbol{F}^T}$. The following commutative diagram summarizes the situation. Notice that, in general it does not make sense to ask that $\boldsymbol{U} = \boldsymbol{V}$ since they map on different spaces. In particular, $\boldsymbol{U}$ and $\boldsymbol{R}$ do not commute. (However, if $\boldsymbol{A}$ is a *symmetric* transformation of an inner product space to itself, the components of its polar decomposition do commute and the left and right decompositions coincide.)

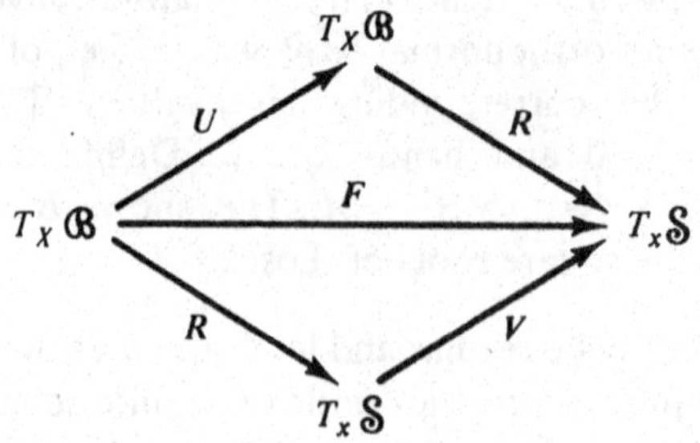

Notice that U and V operate within each fixed tangent space; that is, $U(X)$: $T_X\mathcal{B} \longrightarrow T_X\mathcal{B}$ and $V(x)$: $T_x\mathcal{S} \longrightarrow T_x\mathcal{S}$. On the other hand, R maps $T_X\mathcal{B}$ to $T_x\mathcal{S}$; that is, it shifts the base point as well as rotating.

3.13 Algorithm for Computing the Polar Decomposition Let X, x be fixed, and let F be given. To compute R and U, let $C = F^TF$, and let $\psi^1, \ldots, \psi^n$ be orthonormal eigenvectors of C with eigenvalues $\lambda_1, \ldots, \lambda_n$. Let

$$\Lambda = \begin{bmatrix} \lambda_1 & & 0 \\ & \ddots & \\ 0 & & \lambda_n \end{bmatrix} \quad \text{and} \quad \Psi = (\psi^1, \ldots, \psi^n)$$

so that $\Lambda = \Psi^T C \Psi$. Let $U = \Psi \Lambda^{1/2} \Psi^T$, where

$$\Lambda^{1/2} = \begin{bmatrix} \sqrt{\lambda_1} & & 0 \\ & \ddots & \\ 0 & & \sqrt{\lambda_n} \end{bmatrix} \qquad (\sqrt{\lambda_1}, \ldots, \sqrt{\lambda_n} \text{ are the principal stretches}),$$

and set $R = FU^{-1}$. Use a similar procedure for the left decomposition or let $V = RUR^T$. An explicit formula for U, R in the two-dimensional case is worked out in 3.15.

Observe that $b = V^2 = (RUR^T)(RUR^T) = RU^2R^T = RCR^T$. Thus the Finger deformation tensor b and the deformation tensor C are conjugate under the rotation matrix.

3.14 An Example of the Polar Decomposition[13] Let $\mathcal{B}$ be the unit circular cylinder contained in $\mathbb{R}^3$ and let $\{z^i\}$ and $\{Z^I\}$ denote coincident Cartesian coordinate systems for $\mathbb{R}^3$. Then $\mathcal{B}$ can be written as

$$\mathcal{B} = \{X \mid [Z^1(X)]^2 + [Z^2(X)]^2 \leq 1\}.$$

Consider the configuration $\phi: \mathcal{B} \longrightarrow \mathbb{R}^3$ defined explicitly by

$$z^1(X) = \sqrt{3}\, Z^1(X) + Z^2(X), \qquad z^2(X) = 2Z^2(X), \quad \text{and} \quad z^3(X) = Z^3(X).$$

[13]The data for this example come from Jaunzemis [1967], pp. 78–79.

This configuration may be described as biaxial stretching the in z^1, z^2-plane. The boundary of $\mathcal{B}$, given by $\partial\mathcal{B} = \{X \mid [Z^1(X)]^2 + [Z^2(X)]^2 = 1\}$, is deformed under $\boldsymbol{\phi}$ into an ellipse: $\partial\boldsymbol{\phi}(\mathcal{B}) = \{x \mid [z^1(x)]^2 + z^1(x)z^2(x) + [z^2(x)]^2 = 3\}$. In the coordinate system $\{\bar{z}^i\}$ defined by $\bar{z}^1 = \sqrt{2}\,(z^1 + z^2)$, $\bar{z}^2 = \sqrt{2}\,(z^2 - z^1)$, and $\bar{z}^3 = z^3$, the boundary of $\boldsymbol{\phi}(\mathcal{B})$ can be represented by the equation $[\bar{z}^1(x)]^2/6 + [\bar{z}^2(x)]^2/2 = 1$. Thus the coordinate axes $\bar{z}^1$ and $\bar{z}^2$, which are rotated 45° counterclockwise with respect to z^1 and z^2, coincide with the major and minor axes of the ellipse.

The matrices pertient to the polar decomposition are listed below.

$$\boldsymbol{F} = \begin{bmatrix} \sqrt{3} & 1 & 0 \\ 0 & 2 & 0 \\ 0 & 0 & 1 \end{bmatrix}$$

$$\boldsymbol{C} = \begin{bmatrix} 3 & \sqrt{3} & 0 \\ \sqrt{3} & 5 & 0 \\ 0 & 0 & 1 \end{bmatrix}$$

$$\boldsymbol{\Lambda} = \begin{bmatrix} 6 & 0 & 0 \\ 0 & 2 & 0 \\ 0 & 0 & 1 \end{bmatrix}$$

$$\boldsymbol{\Psi} = \begin{bmatrix} 1/2 & \sqrt{3}/2 & 0 \\ \sqrt{3}/2 & -1/2 & 0 \\ 0 & 0 & 1 \end{bmatrix}$$

$$\boldsymbol{U} = \frac{1}{2\sqrt{2}} \begin{bmatrix} 3 + \sqrt{3} & 3 - \sqrt{3} & 0 \\ 3 - \sqrt{3} & 1 + 3\sqrt{3} & 0 \\ 0 & 0 & 2\sqrt{2} \end{bmatrix}$$

$$\boldsymbol{U}^{-1} = \frac{1}{4\sqrt{6}} \begin{bmatrix} 1 + 3\sqrt{3} & \sqrt{3} - 3 & 0 \\ \sqrt{3} - 3 & 3 + \sqrt{3} & 0 \\ 0 & 0 & 4\sqrt{6} \end{bmatrix}$$

$$\boldsymbol{R} = \frac{1}{2\sqrt{2}} \begin{bmatrix} \sqrt{3} + 1 & \sqrt{3} - 1 & 0 \\ 1 - \sqrt{3} & \sqrt{3} + 1 & 0 \\ 0 & 0 & 2\sqrt{2} \end{bmatrix}$$

$$\boldsymbol{V} = \frac{1}{\sqrt{2}} \begin{bmatrix} \sqrt{3} + 1 & \sqrt{3} - 1 & 0 \\ \sqrt{3} - 1 & \sqrt{3} + 1 & 0 \\ 0 & 0 & \sqrt{2} \end{bmatrix}$$

The physical interpretation of these results follows (see Fig. 1.3.1).

Right decomposition $\boldsymbol{U}$ maps the unit disk in the Z^1, Z^2-plane into an ellipse with major and minor axes rotated 60° counterclockwise with respect to the Z^1

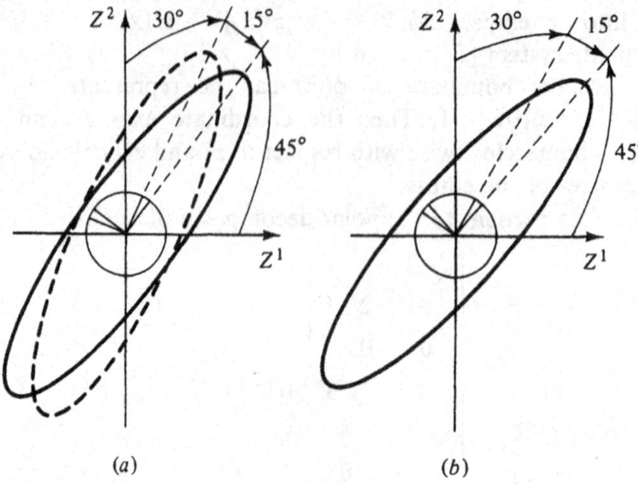

Figure 1.3.1 (a) Right decomposition of $\boldsymbol{F}$: pure deformation (dashed curve) followed by a clockwise rotation through 15°. (b) Left decomposition of $\boldsymbol{F}$: clockwise rotation of the circle through 15°, followed by the pure deformation $\boldsymbol{V}$.

and Z^2 axes, respectively. $\boldsymbol{R}$ then rigidly rotates the elliptic cylinder 15° clockwise about the Z^3-axis into its final position.

Left decomposition $\boldsymbol{R}$ maps the unit disk rigidly into itself. Then $\boldsymbol{V}$ maps the unit disk into an ellipse with major and minor axes coinciding with the $\bar{z}^1$ and $\bar{z}^2$ axes, respectively.

Note that $\boldsymbol{U}$ leaves three orthogonal directions unrotated (the directions are defined by its eigenvectors). Similarly, $\boldsymbol{V}$ leaves three coordinate directions unrotated (these are given by its eigenvectors, which coincide with the axes of the coordinate system $\{\bar{z}^i\}$). The principal stretches—that is, 1, $\sqrt{6}$ and $\sqrt{2}$—which are the eigenvalues of $\boldsymbol{U}$ (or $\boldsymbol{V}$) determine the "size" of the deformation. For example, the major axis of the deformed ellipse is $\sqrt{6}$ and minor axis is $\sqrt{2}$.

Problem 3.2 Find the polar decompositions of

$$\boldsymbol{F} = \begin{bmatrix} 1 & \kappa \\ 0 & 1 \end{bmatrix}.$$

Employ the substitution $\kappa = 2\tan\alpha$ and check that

$$\boldsymbol{U} = \begin{bmatrix} \cos\alpha & \sin\alpha \\ \sin\alpha & (1+\sin^2\alpha)/\cos\alpha \end{bmatrix}.$$

3.15 Proposition *In the two-dimensional case, let*

$$I = \begin{bmatrix} 1 & 0 \\ 0 & 1 \end{bmatrix}$$

$\mathrm{I}_C = \operatorname{trace} C$, *and* $\mathrm{II}_C = \det C$.

Then $$U = \frac{1}{\sqrt{(\mathrm{I}_C + 2\sqrt{\mathrm{II}_C})}}(C - \sqrt{\mathrm{II}_C}I) \quad \text{and} \quad R = FU^{-1}.$$

Proof By the Cayley–Hamilton theorem from linear algebra,

$$U^2 - \mathrm{I}_U U + \mathrm{II}_U I = 0.$$

But $C = U^2$, and $\det C = (\det U)^2$, so $C - \mathrm{I}_U U + \sqrt{\mathrm{II}_C} I = 0$. Taking the trace gives $\mathrm{I}_C - (\mathrm{I}_U)^2 + 2\sqrt{\mathrm{II}_C} = 0$, so $\mathrm{I}_U = \sqrt{\mathrm{I}_C + 2\sqrt{\mathrm{II}_C}}$. Solving $C - \mathrm{I}_U U + \sqrt{\mathrm{II}_C} I = 0$ for U and substituting for I_U gives the result. ∎

The reader might wish to re-work Problem 3.2 using 3.15.

Problem 3.3 The Cayley–Hamilton theorem for 3×3 matrices U states that

$$-U^3 + \mathrm{I}_U U^2 - \mathrm{II}_U U + \mathrm{III}_U I = 0$$

where $\quad \mathrm{I}_U = \operatorname{tr} U, \quad \mathrm{II}_U = \det U(\operatorname{tr} U^{-1}), \quad$ and $\quad \mathrm{III}_U = \det U$

are the principal invariants of U. Use this to work out an explicit formula for U in terms of C and its principal invariants.

Next we study how the deformation tensor and the stretch tensors measure changes in lengths and angles. Recall that if $\sigma : [a, b] \to \mathcal{B}$ is a C^1 (or piecewise C^1) curve, its *length* is given by

$$l(\sigma) = \int_a^b \|\sigma'(\lambda)\| \, d\lambda.$$

3.16 Proposition *Let σ be a C^1 curve in $\mathcal{B}$ and let ϕ be a C^1 configuration of $\mathcal{B}$ in $\mathcal{S}$. Let $\tilde{\sigma} = \phi \circ \sigma$ be the image of σ under ϕ. Then the length of $\tilde{\sigma}$ depends only on σ and on the stretch tensor U.*

Proof From the chain rule $\tilde{\sigma}'(\lambda) = T\phi(\sigma(\lambda))\sigma'(\lambda) = F_{\sigma(\lambda)}\sigma'(\lambda)$. Hence

$$\begin{aligned} \|\tilde{\sigma}'(\lambda)\| &= \langle F_{\sigma(\lambda)}\sigma'(\lambda), F_{\sigma(\lambda)}\sigma'(\lambda)\rangle^{1/2} \\ &= \langle \sigma'(\lambda), F^T_{\sigma(\lambda)}F_{\sigma(\lambda)}\sigma'(\lambda)\rangle^{1/2} \\ &= \langle \sigma'(\lambda), C_{\sigma(\lambda)}\sigma'(\lambda)\rangle^{1/2} \\ &= \langle \sigma'(\lambda), U^2_{\sigma(\lambda)}\sigma'(\lambda)\rangle^{1/2} = \|U_{\sigma(\lambda)}\sigma'(\lambda)\|. \end{aligned}$$

Thus $$l(\tilde{\sigma}) = \int_a^b \|U_{\sigma(\lambda)}\sigma'(\lambda)\| \, d\lambda. \quad ∎$$

We call $\tilde{\sigma}$ the *deformation* of σ under ϕ (see Figure 1.3.2).

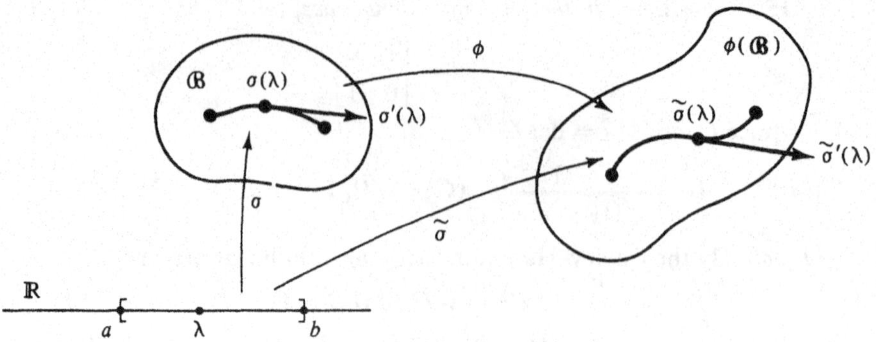

Figure 1.3.2

Recall that the angle θ between two vectors $v_1, v_2 \in T_x\mathcal{S}$ is given by $\cos\theta = \langle v_1, v_2 \rangle / \|v_1\| \cdot \|v_2\|$, and that the angle between two C^1 curves is the angle between their tangent vectors at a point of intersection.

3.17 Proposition *Let σ_1 and σ_2 be two C^1 curves in $\mathcal{B}$ that intersect at $X \in \mathcal{B}$ [i.e., $\sigma_1(\lambda_1) = \sigma_2(\lambda_2) = X$]. Then the angle between the deformed curves $\tilde{\sigma}_1$ and $\tilde{\sigma}_2$ depends only upon σ_1, σ_2, and U_X.*

Proof As in 3.16, $\tilde{\sigma}_i'(\lambda) = F_{\sigma_i(\lambda)}\sigma_i'(\lambda)$ and $\|\tilde{\sigma}_i'(\lambda)\| = \|U_{\sigma_i(\lambda)}\sigma_i'(\lambda)\|$. Now

$$\begin{aligned}\langle \tilde{\sigma}_1'(\lambda_1), \tilde{\sigma}_2'(\lambda_2)\rangle_x &= \langle F_X\sigma_1'(\lambda_1), F_X\sigma_2'(\lambda_2)\rangle_x \\ &= \langle R_X U_X\sigma_1'(\lambda_1), R_X U_X\sigma_2'(\lambda_2)\rangle_x = \langle U_X\sigma_1'(\lambda_1), U_X\sigma_2'(\lambda_2)\rangle_X.\end{aligned}$$

The angle θ between $\tilde{\sigma}_1'(\lambda_1)$ and $\tilde{\sigma}_2'(\lambda_2)$ is thus

$$\cos\theta = \frac{\langle U_X\sigma_1'(\lambda_1), U_X\sigma_2'(\lambda_2)\rangle_X}{\|U_X\sigma_1'(\lambda_1)\|_X \cdot \|U_X\sigma_2'(\lambda_2)\|_X}. \quad \blacksquare$$

Thus U (or C) measures the distortion, as manifested by changes in lengths and angles, due to the configuration ϕ. Note that R, being orthogonal, has no effect on lengths or angles.

Previously we defined the mappings $C: T\mathcal{B} \to T\mathcal{B}$ and $b: T\mathcal{S} \to T\mathcal{S}$. In the ensuing developments we deal with the associated mappings $C^\flat: T\mathcal{B} \to T^*\mathcal{B}$ and $b^\flat: T\mathcal{S} \to T^*\mathcal{S}$ which are obtained by lowering the first index of C and b, respectively. The components of C and b are

$$C^A{}_B = G^{AC} g_{ab} F^b{}_C F^a{}_B \quad \text{and} \quad b^a{}_b = G^{AB} g_{bc} F^a{}_A F^c{}_B$$

and the components of $C^\flat$ and $b^\flat$ are given by

$$C_{AB} = G_{AD} C^D{}_B \quad \text{and} \quad b_{ab} = g_{ac} b^c{}_b.$$

(In the next section we provide an abstract treatment of raising and lowering indices.)

3.18 Definition The *material* (or Lagrangian) *strain tensor* E: $T\mathcal{B} \to T\mathcal{B}$ is defined by $2E = C - I$. The associated tensor $E^\flat$ is given by $2E^\flat = C^\flat - I^\flat$.

In components, $2E^A{}_B = C^A{}_B - \delta^A{}_B$ and $2E_{AB} = C_{AB} - G_{AB}$. Thus $\boldsymbol{E} = \mathbf{0}$ is equivalent to $\boldsymbol{C} = \boldsymbol{I}$, which implies that points in $\mathcal{B}$ experience no relative motion under ϕ. The reason for the factor "2" in the definition of $\boldsymbol{E}$ will be evident when we consider the linearized theory.

3.19 Definition Let ϕ be regular. The *spatial* (or Eulerian) *strain tensor* $\boldsymbol{e}$: $T\mathbb{S} \to T\mathbb{S}$ is defined by $2\boldsymbol{e} = \boldsymbol{i} - \boldsymbol{c}$, where $\boldsymbol{c} = \boldsymbol{b}^{-1}: T\mathbb{S} \to T\mathbb{S}$.

The associated mapping and component forms of $\boldsymbol{e}$ are given by

$$2\boldsymbol{e}^\flat = \boldsymbol{i}^\flat - \boldsymbol{c}^\flat, \qquad 2e_{ab} = g_{ab} - c_{ab}, \quad \text{and} \quad 2e^a{}_b = \delta^a{}_b - c^a{}_b.$$

As we shall see in Section 1.4, the various tensors defined here can be redefined in terms of pull-backs and push-forwards:

$$\begin{aligned} \boldsymbol{C}^\flat &= \phi^*(\boldsymbol{g}) \qquad & \boldsymbol{c}^\flat &= \phi_*(\boldsymbol{G}) \\ \boldsymbol{B}^\sharp &= \phi^*(\boldsymbol{g}^\sharp) \qquad & \boldsymbol{b}^\sharp &= \phi_*(\boldsymbol{G}^\sharp) \\ \boldsymbol{E}^\flat &= \phi^*(\boldsymbol{e}^\flat) \qquad & \boldsymbol{e}^\flat &= \phi_*(\boldsymbol{E}^\flat). \end{aligned}$$

The component forms of these definitions (with $x = \phi(X)$, as usual) are:

$$\begin{aligned} C_{AB}(X) &= g_{ab}(x)F^a{}_A(X)F^b{}_B(X) \qquad & c_{ab}(x) &= G_{AB}(X)(F(X)^{-1})^A{}_a(F(X)^{-1})^B{}_b \\ B^{AB}(X) &= g^{ab}(x)(F(X)^{-1})^A{}_a(F(X)^{-1})^B{}_b \qquad & b^{ab}(x) &= G^{AB}(X)F^a{}_A(X)F^b{}_B(X) \\ E^{AB}(X) &= e_{ab}(x)F^a{}_A(X)F^b{}_B(X) \qquad & e_{ab}(x) &= E_{AB}(X)(F(X)^{-1})^A{}_a(F(X)^{-1})^B{}_b. \end{aligned}$$

The following conditions are equivalent:

$$\begin{aligned} \boldsymbol{C} &= \boldsymbol{I} \qquad & \boldsymbol{c} &= \boldsymbol{i} \\ \boldsymbol{B} &= \boldsymbol{I} \qquad & \boldsymbol{b} &= \boldsymbol{i} \\ \boldsymbol{E} &= \mathbf{0} \qquad & \boldsymbol{e} &= \mathbf{0}. \end{aligned}$$

Box 3.1 ***Shifters and Displacements***

In the special case of Euclidean space, there are some additional concepts that are sometimes useful.

3.20 Definition Let $\mathcal{B} \subset \mathbb{S} = \mathbb{R}^n$ be a simple body and let ϕ be a C^1 configuration. The map

$$\boldsymbol{S}: T\mathcal{B} \to T\mathbb{S}, \qquad \boldsymbol{S}(X, \boldsymbol{W}) = (\phi(X), \boldsymbol{W})$$

is called the *shifter*. If ϕ_t is a motion, we write $\boldsymbol{S}_t(X, \boldsymbol{W}) = (\phi_t(X), \boldsymbol{W})$, and let $\boldsymbol{S}_X$ be the restriction of $\boldsymbol{S}$ to $T_X\mathcal{B}$ so that $\boldsymbol{S}_X: T_X\mathcal{B} \to T_x\mathbb{S}$.

Notice that $\boldsymbol{S}$ parallel transports vectors emanating from X to vectors emanating from $\phi(X)$. See Figure 1.3.3.

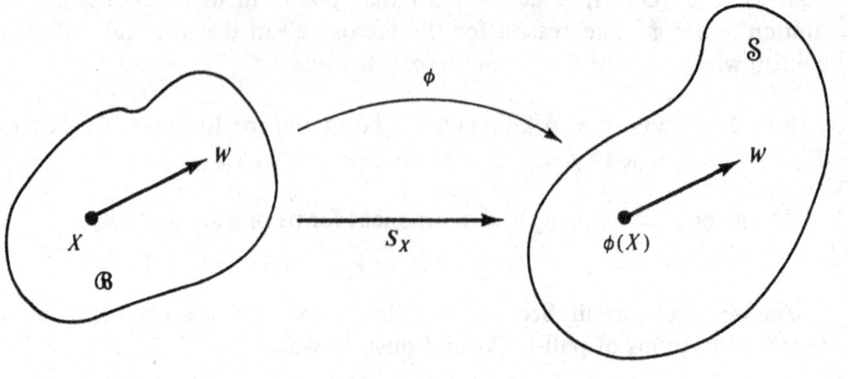

Figure 1.3.3

On general Riemannian manifolds, the notion of shifter still makes sense if a motion is under consideration: $S_t(W_X)$ parallel translates W_X along the curve $\phi_t(X)$. However, shifters have been used primarily in the $\mathbb{R}^n$ context.[14]

3.21 Proposition

(i) *Let $\{x^a\}$ and $\{X^A\}$ be coordinate charts on $\mathcal{S}$ and $\mathcal{B}$, respectively. Then the components of S are given by*

$$S^a{}_A(X) = \frac{\partial x^a}{\partial z^i}(x)\frac{\partial Z^I}{\partial X^A}(X)\,\delta^i{}_I$$

(ii) *S is orthogonal: $S^T = S^{-1}$.*

Proof

(i) Since $S(X, W) = (\phi(X), W)$, S_X as a linear transformation is "the identity": $S^a{}_A W^A = W^a$. But $W^A = (\partial X^A/\partial Z^I)W^I$ and $W^a = (\partial x^a/\partial z^i)W^i$. Substituting these gives the result.

(ii) follows from $\langle S_X W_1, S_X W_2\rangle_x = \langle W_1, W_2\rangle_X$ ▮

Recall that the rotation tensor R rotates as well as moves base points. Using the shifter we can break up R into two parts by defining $R_1 = RS^{-1}$ and $R_2 = S^{-1}R$. Thus we have $R = R_1 S$ (shift and then rotate), and $R = SR_2$ (rotate and then shift). In general, $R_1 \neq R_2$. The diagram that goes with this follows.

[14]For instance, in shell theory $\mathcal{S} = \mathbb{R}^3$, $\mathcal{B}$ is a surface, and shifters are defined as above by ordinary parallel translation in $\mathbb{R}^3$.

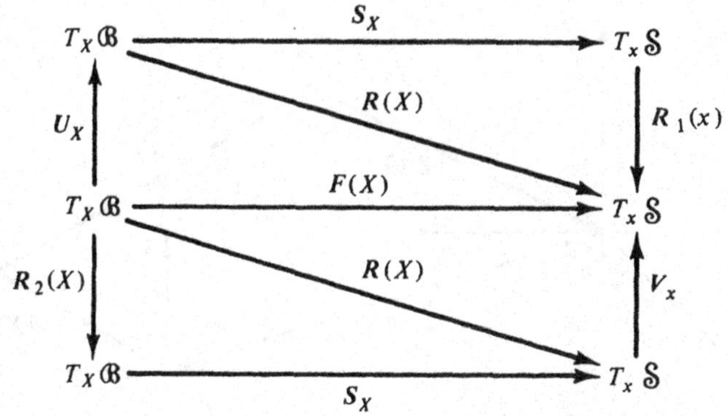

3.22 Definition Let $\phi: \mathcal{B} \to \mathcal{S} = \mathbb{R}^n$ be a C^1 configuration of $\mathcal{B}$. We consider $\mathbb{R}^n$ a vector space; thus it makes sense to write $\phi(X) - X$. The vector field on $\mathcal{B}$, defined by $U(X) = (X, \phi(X) - X)$, is called the *displacement*. See Figure 1.3.4.

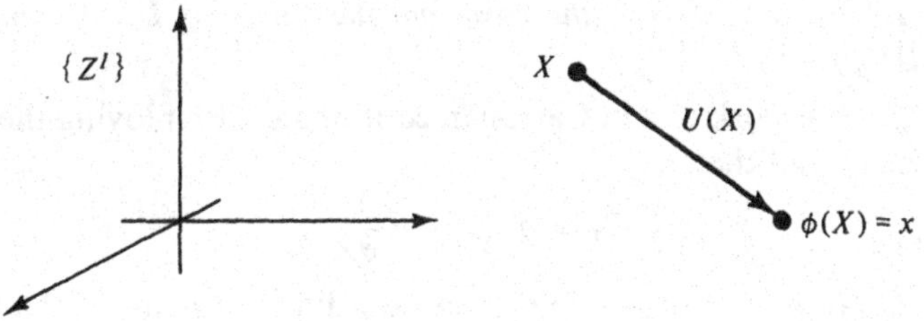

Figure 1.3.4

In the general case in which $\mathcal{S}$ is not a linear space, the displacement is *not defined.* It may be used only when the Cartesian structure of the ambient space is available.

Warning: Transforming U as a vector to new coordinates is *not compatible* with transforming $\phi(X)$ and X as points to the new components (i.e., *the equation* $U = x - X$ *is not a tensorial equation*).

3.23 Proposition *Let* $\{Z^I\}$ *and* $\{z^i\}$ *denote collinear Cartesian coordinate systems for* $\mathbb{R}^n$, *and assume that the origins of these frames are connected by a vector* $Y = Y^I\hat{I}_I$. *Then* $U^I(X) = Y^I - Z^I(X) + \delta^I{}_i\phi^i(X)$. *See Figure 1.3.5.*

Proof If $x \in \mathbb{R}^n$, then $Z^I(x) = Y^I + \delta^I{}_i z^i(x)$, by definition of the coordinates. Let the vector components of U be denoted U^I; then $U^I(X) = \phi^I(X) - Z^I(X)$. Combining these results yields the assertion. ∎

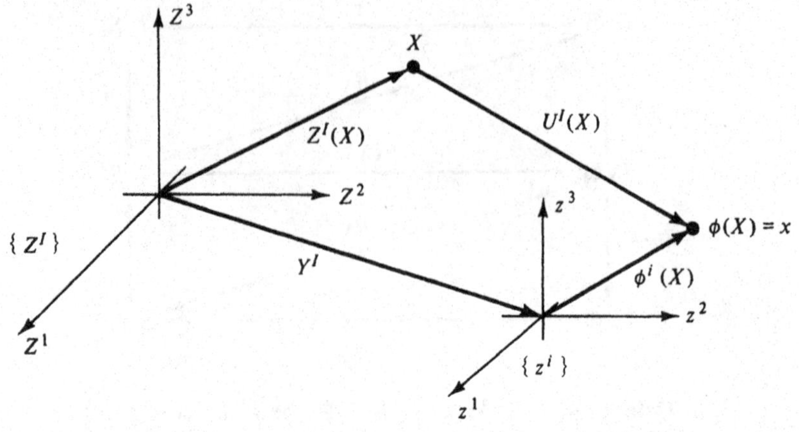

Figure 1.3.5

3.24 Proposition *The deformation gradient is given by* $\boldsymbol{F} = \boldsymbol{S}(\boldsymbol{I} + D\boldsymbol{U})$. *(Here* $D\boldsymbol{U}$ *is the covariant derivative of* $\boldsymbol{U}$*; in component form,* $U^A{}_{|B}$.*)*

Proof We work first in Cartesian coordinates. Employing the result of 3.23, we find that

$$F^i{}_J = \delta^i{}_I\left(\delta^I{}_J + \frac{\partial U^I}{\partial Z^J}\right).$$

Using 3.21(i) and transforming to the coordinate systems $\{X^A\}$ and $\{x^a\}$ yields $F^a{}_B = S^a{}_A(\delta^A{}_B + U^A{}_{|B})$. ∎

3.25 Corollary $\boldsymbol{C} = \boldsymbol{I} + D\boldsymbol{U} + (D\boldsymbol{U})^T + (D\boldsymbol{U})^T D\boldsymbol{U}$
and $2\boldsymbol{E} = D\boldsymbol{U} + (D\boldsymbol{U})^T + (D\boldsymbol{U})^T D\boldsymbol{U}$.

If ϕ is regular, then a "dual displacement" to $\boldsymbol{U}$ can be defined by $\boldsymbol{u} = \boldsymbol{S}\boldsymbol{U}\circ\phi^{-1}$. See Figure 1.3.6. Analogous to the previous development, the following formulas are easily established:

$$\boldsymbol{F}^{-1} = \boldsymbol{S}^{-1}(\boldsymbol{i} - D\boldsymbol{u}), \qquad (D\boldsymbol{u})^a{}_b = u^a{}_{|b}$$
$$\boldsymbol{c} = \boldsymbol{i} - D\boldsymbol{u} - (D\boldsymbol{u})^T + (D\boldsymbol{u})^T(D\boldsymbol{u})$$
$$2\boldsymbol{e} = D\boldsymbol{u} + (D\boldsymbol{u})^T - (D\boldsymbol{u})^T(D\boldsymbol{u}).$$

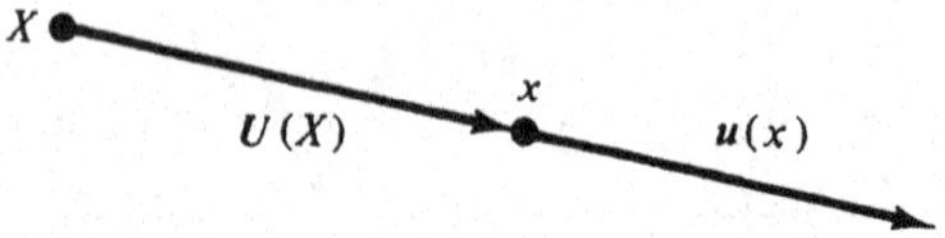

Figure 1.3.6

Problem 3.4 Show that $e^a = S^a{}_A E^A$ and $e_a = S_a{}^A E_A$, where $S_a{}^A = g_{ab} S^b{}_B G^{AB}$.

Problem 3.5 Show that the (unique) solution of the equations $\partial \phi_t / \partial t = v_t \circ \phi_t$ and $Dv_t/Dt = 0$ with $\phi_0 = I$, $v_0 = V$ is $\phi_t = I + tV$, $v_t = S_t(V)$. Give an explicit example on $\mathbb{R}$ to show that ϕ_t can fail to be regular in a finite time.

Problem 3.6 Given a motion ϕ_t and $W \in T_X \mathcal{B}$, show that $(\partial/\partial t)\, S_t(W) = -\nabla_{v_t} S_t(W)$.

Suppose that a motion ϕ_t is under consideration. We want to calculate the rate of change of the various deformation tensors.

3.26 Definition Let ϕ_t be a C^1 motion of $\mathcal{B}$ in $\mathcal{S}$, and let C be the deformation tensor. The material (or Lagrangian) *rate of deformation tensor* $\boldsymbol{D}$ is defined by

$$2D(X, t) = \frac{\partial}{\partial t} C(X, t).$$

Note that $C(X, t)\colon T_X\mathcal{B} \to T_X\mathcal{B}$ for each t, so it makes sense to differentiate $C(X, t)$ in t to obtain another linear transformation. The reason for the factor "2" will become clear when we study the linearized theory; see Chapter 4.

To compute D in coordinates, we introduce the covariant derivative of V. (It will be considered abstractly in terms of covariant differentiation of two-point tensors in Section 1.4.)

3.27 Definition The *covariant derivative* of the material velocity is defined in coordinates $\{x^a\}$ and $\{X^A\}$ on $\mathcal{S}$ and $\mathcal{B}$, respectively, by

$$V^a{}_{|B} = \frac{\partial V^a}{\partial X^B} + \gamma^a_{bd} V^b F^d{}_B$$

or, putting in the arguments,

$$V^a{}_{|B}(X, t) = \frac{\partial V^a}{\partial X^B}(X, t) + \gamma^a_{bd}(\phi(X, t)) V^b(X, t) F^d{}_B(X, t).$$

Problem 3.7 If $\mathcal{B} \subset \mathcal{S} = \mathbb{R}^n$ is a simple body, show that $V^a{}_{|B}$ are simply the matrix entries of the linear map $DV(X)$ relative to the coordinate systems $\{x^a\}$ and $\{X^A\}$. (Imitate the proof of 1.13 and use the chain rule at the appropriate point.) Prove that if ϕ_t is regular, $V^a{}_{|B}(X, t) = v^a{}_{|b}(x, t) F^b{}_B(X, t)$.

3.28 Proposition *In coordinates, the following are the components of* $\mathbf{D}$:

$$2D^A{}_B(X) = G^{CA}(X)g_{ac}(x)\{V^c{}_{|C}(X,t)F^a{}_B(X,t) + V^a{}_{|B}(X,t)F^c{}_C(X,t)\}.$$

The proof requires an observation about the Christoffel symbols γ^a_{bc}, which, in fact, was at the historical origins of Riemannian geometry: γ^a_{bc} depends only on g_{ab}.

3.29 Lemma *We have the identities*

$$\frac{\partial g_{ac}}{\partial x^d} = g_{ba}\gamma^b_{dc} + g_{bc}\gamma^b_{da}$$

and

$$2g_{ab}\gamma^a_{dc} = \left(\frac{\partial g_{cb}}{\partial x^d} + \frac{\partial g_{db}}{\partial x^c} - \frac{\partial g_{dc}}{\partial x^b}\right).$$

Proof The first follows at once from the explicit definitions of g_{ab} and γ^a_{bc} in 3.4 and 1.8, recalling that $(\partial z^i/\partial x^a)(\partial x^a/\partial z^j) = \delta^i_j$. The second identity follows by substituting the formula for g_{ab} into its right-hand side, remembering g_{bc} and γ^a_{bc} are symmetric in (b, c). ▮

Proof of 3.28 We have $G_{AC}(X)C^A{}_B(X,t) = g_{ac}(\phi(X,t))F^c{}_C(X,t)F^a{}_B(X,t)$, and so, differentiating in t,

$$2G_{AC}D^A{}_B = \frac{\partial g_{ac}}{\partial x^d}V^dF^c{}_CF^a{}_B + g_{ac}\frac{\partial V^c}{\partial X^C}F^a{}_B + g_{ac}F^c{}_C\frac{\partial V^a}{\partial X^B}.$$

Using the first identity from 3.29, this equals (after changing two dummy summation variables to b)

$$\left(g_{ab}\gamma^b_{dc}V^dF^c{}_C + g_{ab}\frac{\partial V^b}{\partial X^C}\right)F^a{}_B + \left(g_{bc}\gamma^b_{da}V^dF^a{}_B + g_{bc}\frac{\partial V^b}{\partial X^B}\right)F^c{}_C$$
$$= g_{ab}V^b{}_{|C}F^a{}_B + g_{bc}V^b{}_{|B}F^c{}_C. \quad ▮$$

In Section 1.6, we will recognize $\mathbf{D}^\flat$ as $\phi^*(L_v g)$, the pull-back of the Lie derivative of the metric.

3.30 Proposition *Let ϕ_t be a C^2 motion of $\mathcal{B}$. Let $W_1, W_2 \in T_X\mathcal{B}$ and let $w_i(t) = F(X,t)W_i$ $(i = 1, 2)$. Then*

$$\frac{d}{dt}\langle w_1(t), w_2(t)\rangle_{\phi(X,t)} = 2\langle W_1, D(X,t)W_2\rangle_X.$$

Proof

$$\frac{d}{dt}\langle w_1(t), w_2(t)\rangle_{\phi(X,t)} = \frac{d}{dt}\langle W_1, F(X,t)^T F(X,t) W_2\rangle_X$$

$$= \frac{d}{dt}\langle W_1, C(X,t)W_2\rangle_X = \langle W_1, 2D(X,t)W_2\rangle_X. \quad \blacksquare$$

This proposition shows that $D(X, t)$ measures the rate of change of the inner product of vectors as they are pushed forward by the motion.

Combined with 3.28, 3.30 yields:

3.31 Corollary *In coordinates,*

$$\frac{d}{dt}\{g_{ab}(\phi(X,t))F^a{}_A(X,t)W_1^A F^b{}_B(X,t)W_2^B\}$$
$$= W_1^C W_2^B g_{ac}(\phi(X,t))\{V^c{}_{|C}F^a{}_B + V^a{}_{|B}F^c{}_C\}(X,t).$$

Problem 3.8 Work out the metric and Christoffel symbols for the spherical coordinate system (ρ, θ, ϕ) on $\mathbb{R}^3$.

Problem 3.9 Consider the motion of $\mathbb{R}^2$ in $\mathbb{R}^2$ given by

$$\phi(Z^1, Z^2, t) = \begin{bmatrix} 1 & t\kappa \\ 0 & 1 \end{bmatrix}\begin{bmatrix} Z^1 \\ Z^2 \end{bmatrix}$$

(a steady shearing). Calculate the deformation tensor and the rate of deformation tensor.

Box 3.2 ***Summary of Formulas for Section 3.1***

Covariant Derivatives for Simple Bodies $\mathcal{B}$ in $\mathcal{S} = \mathbb{R}^3$

$U: \mathcal{B} \to T\mathcal{B}$;	$DU: T\mathcal{B} \to T\mathcal{B}$	$U^A{}_{\|B} = \dfrac{\partial U^A}{\partial X^B} + \Gamma^A_{BC}U^C$
$v: \mathcal{S} \to T\mathcal{S}$;	$Dv: T\mathcal{S} \to T\mathcal{S}$	$v^a{}_{\|b} = \dfrac{\partial v^a}{\partial x^b} + \gamma^a_{bc}v^c$
$V: \mathcal{B} \to T\mathcal{S}$;	$DV: T\mathcal{B} \to T\mathcal{S}$	$V^a{}_{\|B} = \dfrac{\partial V^a}{\partial X^B} + (\gamma^a_{bc}\circ\phi)V^b F^c{}_B$
$V = v\circ\phi$;	$DV = Dv\circ F$	$V^a{}_{\|A} = (v^a{}_{\|b}\circ\phi)F^b{}_A$

Metric Tensors

$G: T\mathcal{B} \to T^*\mathcal{B}$	$G_{AB} = \dfrac{\partial Z^I}{\partial X^A}\dfrac{\partial Z^J}{\partial X^B}\delta_{IJ}$
$g: T\mathcal{S} \to T^*\mathcal{S}$	$g_{ab} = \dfrac{\partial z^i}{\partial x^a}\dfrac{\partial z^j}{\partial x^b}\delta_{ij}$
$G^\sharp = G^{-1}$	$(G^{AB}) = (G_{AB})^{-1}$
$g^\sharp = g^{-1}$	$(g^{ab}) = (g_{ab})^{-1}$

Deformation Gradient

$F: T\mathcal{B} \to T\mathcal{S};$ $\quad F(X, W) = (\phi(X), D\phi(X)\cdot W)$ $\qquad F^a{}_A = \dfrac{\partial \phi^a}{\partial X^A}$

$F^T: T\mathcal{S} \to T\mathcal{B};$ $\quad F^T(x, w) = (\phi^{-1}(x), D\phi(\phi^{-1}(x))^T \cdot w)$ $\qquad (F^T)^A{}_a = g_{ab}(F^b{}_B \circ \phi^{-1})(G^{BA} \circ \phi^{-1})$

Polar Decomposition (***U*** and ***V*** are the stretch tensors under this heading)

$F = RU = VR$ $\qquad F^a{}_A = R^a{}_B U^B{}_A = (V^a{}_b \circ \phi) R^b{}_A$

$R: T\mathcal{B} \to T\mathcal{S} \qquad R^{-1} = R^T$

$U: T\mathcal{B} \to T\mathcal{B}$ (symmetric positive-definite) $\qquad G_{AC} U^C{}_B = G_{BC} U^C{}_A$

$V: T\mathcal{S} \to T\mathcal{S}$ (symmetric positive-definite) $\qquad g_{ac} V^c{}_b = g_{bc} V^c{}_a$

Deformation Tensors

$C(X, W) = (X, D\phi(X)^T D\phi(X) W)$

$C = F^T F \qquad C^A{}_B = G^{AC}(g_{ab} \circ \phi) F^b{}_C F^a{}_B$

$B = C^{-1} \qquad (B^A{}_B) = (C^A{}_B)^{-1}$

$2E = C - I \qquad 2E^A{}_B = C^A{}_B - \delta^A{}_B$

$b = FF^T \qquad b^a{}_b = (G^{AB} \circ \phi^{-1}) g_{bc} (F^a{}_A \circ \phi^{-1})(F^c{}_B \circ \phi^{-1})$

$c = b^{-1} \qquad c^a{}_b = (b^a{}_b)^{-1}$

$2e = c - i \qquad 2e^a{}_b = c^a{}_b - \delta^a{}_b$

$C^\flat = \phi^*(g) \qquad C_{AB} = (g_{ab} \circ \phi) F^a{}_A F^b{}_B$

$B^\sharp = \phi^*(g^\sharp) \qquad B^{AB} = (g^{ab} \circ \phi)((F^{-1})^A{}_a \circ \phi)((F^{-1})^B{}_b \circ \phi)$

$E^\flat = \phi^*(e) \qquad E_{AB} = (e_{ab} \circ \phi) F^a{}_A F^b{}_B$

$c^\flat = \phi_*(G) \qquad c_{ab} = (G_{AB} \circ \phi^{-1})(F^{-1})^A{}_a (F^{-1})^B{}_b$

$b^\sharp = \phi_*(G^\sharp) \qquad b^{ab} = (G^{AB} \circ \phi^{-1})(F^a{}_A \circ \phi^{-1})(F^b{}_B \circ \phi^{-1})$

$e^\flat = \phi_*(E^\flat) \qquad e_{ab} = (E_{AB} \circ \phi^{-1})(F^{-1})^A{}_a (F^{-1})^B{}_b$

Shifter in $\mathbb{R}^n$

$S: T\mathcal{B} \to T\mathcal{S};$ $\quad S(X, W) = (\phi(X), W)$ $\qquad S^a{}_A = \left(\dfrac{\partial x^a}{\partial z^i} \circ \phi\right) \delta^i{}_I \dfrac{\partial Z^I}{\partial X^A}$

$S^T: T\mathcal{S} \to T\mathcal{B};$ $\quad S^{-1} = S^T$ $\qquad (S^T)^A{}_a = \left(\dfrac{\partial X^A}{\partial Z^I} \circ \phi^{-1}\right) \delta^I{}_i \dfrac{\partial z^i}{\partial x^a}$

$(S^{-1})^A{}_a = (S^T)^A{}_a$

Displacements in $\mathbb{R}^n$

$U: \mathcal{B} \to T\mathcal{B};$

$U(X) = (X, \phi(X) - X) \qquad U^I = \delta^I{}_i \phi^i - Z^I$

$F = S(I + DU) \qquad F^a{}_A = S^a{}_B(\delta^B{}_A + U^B{}_{|A})$

$C = I + DU + DU^T + DU^T DU \qquad C^A{}_B = \delta^A{}_B + U^A{}_{|B} + G^{AD} G_{BC} U^C{}_{|D} + G^{AD} G_{CE} U^E{}_{|D} U^C{}_{|B}$

$2E = DU + DU^T + DU^T DU \qquad 2E^A{}_B = U^A{}_{|B} + G^{AD} G_{BC} U^C{}_{|D} + G^{AD} G_{CE} U^E{}_{|D} U^C{}_{|B}$

$u = SU \circ \phi^{-1} \qquad u^a = (S^a{}_A \circ \phi^{-1})(U^A \circ \phi^{-1})$

$F^{-1} = S^T(i - Du) \qquad (F^{-1})^A{}_a = (S^T)^A{}_b(\delta^b{}_a - u^b{}_{|a})$

$c = i - Du - Du^T + Du^T Du$	$c^a{}_b = \delta^a{}_b - u^a{}_{\vert b} - g^{ad}g_{bc}u^c{}_{\vert d} + g^{ad}g_{ce}u^e{}_{\vert d}u^c{}_{\vert b}$
$2e = Du + Du^T - Du^T Du$	$2e^a{}_b = u^a{}_{\vert b} + g^{ad}g_{bc}u^c{}_{\vert d} - g^{ad}g_{ce}u^e{}_{\vert d}u^c{}_{\vert b}$
Rate-of-Deformation Tensors	
$D_t\colon T\mathcal{B} \to T\mathcal{B}$;	
$2D_t(X, W) = \left(\frac{d}{dt}C_X(t)\right)W$	$2D^A{}_B = G^{CA}g_{ac}(V^c{}_{\vert C}F^a{}_B + V^a{}_{\vert B}F^c{}_C) = 2\frac{\partial}{\partial t}E^A{}_B$
$d^\flat = \phi_*(D^\flat) = \frac{1}{2}L_v g$ (See Section 1.6.)	$d_{ab} = (D_{AB}\circ\phi^{-1})(F^{-1})^A{}_a(F^{-1})^B{}_b = \frac{1}{2}(v_{a\vert b} + v_{b\vert a}) = \frac{\partial}{\partial t}e_{ab}$

1.4 TENSORS, TWO-POINT TENSORS, AND THE COVARIANT DERIVATIVE

The previous section introduced a number of important geometric concepts. This section will examine these objects in further detail from both the abstract and the computational point of view. By this time the reader should be comfortable enough with manifolds so that we can treat the general case with no more effort than the case of open sets in $\mathbb{R}^3$.

Our notation throughout is as above: $\mathcal{B}$ and $\mathcal{S}$ are manifolds, points in $\mathcal{B}$ are denoted by X and those in $\mathcal{S}$ by x. The tangent spaces are written $T_X\mathcal{B}$ and $T_x\mathcal{S}$. Coordinate systems are denoted $\{X^A\}$ and $\{x^a\}$ for $\mathcal{B}$ and $\mathcal{S}$, respectively, with corresponding bases E_A and e_a and dual bases E^A and e^a. The summation convention is enforced.

4.1 Definition A *tensor* of type $\binom{p}{q}$ at $X \in \mathcal{B}$ is a multilinear mapping

$$\mathbf{T}\colon \underbrace{T_X^*\mathcal{B} \times \cdots \times T_X^*\mathcal{B}}_{p \text{ copies}} \times \underbrace{T_X\mathcal{B} \times \cdots \times T_X\mathcal{B}}_{q \text{ copies}} \to \mathbb{R}$$

(multilinear means linear in each variable separately). The *components* of $\mathbf{T}$ are defined by

$$\mathbf{T}^{A_1A_2\cdots A_p}{}_{B_1B_2\cdots B_q} = \mathbf{T}(E^{A_1}, E^{A_2}, \ldots, E^{A_p}, E_{B_1}, E_{B_2}, \ldots, E_{B_q})$$

so that $\quad \mathbf{T}(\alpha^1, \ldots, \alpha^p, W_1, \ldots, W_q) = \mathbf{T}^{A_1\cdots A_p}{}_{B_1\cdots B_q}\alpha^1_{A_1}\cdots\alpha^p_{A_p}W_1^{B_1}\cdots W_q^{B_q}$

where $\alpha^i \in T_X^*\mathcal{B}$, $\alpha^i = \alpha^i_A E^A$, $W_j \in T_X\mathcal{B}$, and $W_j = W_j^A E_A$. We say that $\mathbf{T}$ is *contravariant* of rank p (p indices up) and *covariant* of rank q (q indices down).

A *tensor field* of type $\binom{p}{q}$ is an assignment $\mathbf{T}(X)$ of a tensor of type $\binom{p}{q}$

at X for each $X \in \mathcal{B}$. The components are functions of X and are denoted

$$\mathsf{T}^{A_1 \cdots A_p}{}_{B_1 \cdots B_q}(X).$$

We can make the space of all tensors on $\mathcal{B}$ into a vector bundle $T^p_q(\mathcal{B})$ over $\mathcal{B}$ (as with the tangent bundle) and a tensor field $\mathbf{T}$ may be regarded as a *section* of this bundle; that is, $\mathbf{T}$ assigns to each base point $X \in \mathcal{B}$ an element of the fiber of $T^p_q(\mathcal{B})$ over X.

A tensor field is said to be of *class* C^r if its components are C^r functions of X^A in any coordinate system.

We can regard vectors as $\binom{1}{0}$ tensors,[15] one-forms as $\binom{0}{1}$ tensors, and functions as $\binom{0}{0}$ tensors.

We can, by a similar process, position the indices in other slots. For instance, a fourth-order tensor that is a multilinear mapping

$$\mathbf{T}: T^*_X\mathcal{B} \times T_X\mathcal{B} \times T^*_X\mathcal{B} \times T_X\mathcal{B} \to \mathbb{R}$$

has indices positioned like this:

$$\mathsf{T}^A{}_B{}^C{}_D.$$

A tensor or tensor field $\mathbf{T}$ has a separate existence from its components $\mathsf{T}^{A_1 A_2 \cdots A_p}{}_{B_1 B_2 \cdots B_q}$, and it makes sense to write down $\mathbf{T}$ without any indices, just as for vectors.

4.2 Proposition *Let $\{X^A\}$ and $\{\bar{X}^A\}$ be two coordinate systems on $\mathcal{B}$ and $\mathbf{T}$ a tensor field of type $\binom{p}{q}$. Then the components of $\mathbf{T}$ in these two systems are related by*

$$\bar{\mathsf{T}}^{A_1 \cdots A_p}{}_{B_1 \cdots B_q} = \frac{\partial \bar{X}^{A_1}}{\partial X^{C_1}} \cdots \frac{\partial \bar{X}^{A_p}}{\partial X^{C_p}} \mathsf{T}^{C_1 \cdots C_p}{}_{D_1 \cdots D_q} \frac{\partial X^{D_1}}{\partial \bar{X}^{B_1}} \cdots \frac{\partial X^{D_q}}{\partial \bar{X}^{B_q}}.$$

Proof This follows from the definition of components, multilinearity, and the formulas

$$\bar{E}^A = \frac{\partial \bar{X}^A}{\partial X^B} E^B, \qquad \bar{E}_A = \frac{\partial X^B}{\partial \bar{X}^A} E_B. \quad \blacksquare$$

4.3 Definition Let $\mathbf{T}$ be a tensor of type $\binom{p}{q}$ and $\mathbf{S}$ a tensor of type $\binom{r}{s}$. Then the *tensor product* $\mathbf{T} \otimes \mathbf{S}$ is the tensor of type $\binom{p+r}{q+s}$ defined by

$$(\mathbf{T} \otimes \mathbf{S})(X): \overbrace{(T^*_X\mathcal{B} \times \cdots \times T^*_X\mathcal{B})}^{p \text{ copies}} \times \overbrace{(T_X\mathcal{B} \times \cdots \times T_X\mathcal{B})}^{q \text{ copies}}$$
$$\times \overbrace{(T^*_X\mathcal{B} \times \cdots \times T^*_X\mathcal{B})}^{r \text{ copies}} \times \overbrace{(T_X\mathcal{B} \times \cdots \times T_X\mathcal{B})}^{s \text{ copies}} \to \mathbb{R};$$

---

[15] If W is a vector at X, we set $W(\alpha) = \alpha(W)$ for $\alpha \in T^*_X\mathcal{B}$.

$$(\mathbf{T} \otimes \mathbf{S})(X)(\alpha^1, \ldots, \alpha^p, W_1, \ldots, W_q, \beta^1, \ldots, \beta^r, Y_1, \ldots, Y_s)$$
$$= \mathbf{T}(X)(\alpha^1, \ldots, \alpha^p, W_1, \ldots, W_q) \cdot \mathbf{S}(X)(\beta^1, \ldots, \beta^r, Y_1, \ldots, Y_s).$$

Thus
$$(\mathbf{T} \otimes \mathbf{S})^{A_1 \cdots A_p}{}_{B_1 \cdots B_q}{}^{C_1 \cdots C_r}{}_{D_1 \cdots D_s}(X)$$
$$= \mathsf{T}^{A_1 \cdots A_p}{}_{B_1 \cdots B_q}(X) \cdot \mathsf{S}^{C_1 \cdots C_r}{}_{D_1 \cdots D_s}(X).$$

Addition and scalar multiplication of tensors are defined in the obvious way.

Problem 4.1 In a coordinate system $\{X^A\}$ on $\mathcal{B}$, show that the matrix of components of $E_A \otimes E_B$ is a matrix of 0's with the exception of a 1 in the (A, B) slot.

The following is easily proven from the definitions.

4.4 Proposition *The following holds for a tensor* $\mathbf{T}$ *of type* $\binom{p}{q}$:

$$\mathbf{T}(X) = \mathsf{T}^{A_1 \cdots A_p}{}_{B_1 \cdots B_q}(X) E_{A_1} \otimes \cdots \otimes E_{A_p} \otimes E^{B_1} \otimes \cdots \otimes E^{B_q}.$$

Next, we define contractions of tensors.

4.5 Definition The *contraction* of a one-form α and a vector field W is defined by

$$\alpha \cdot W(X) = \alpha(X)(W(X)) = \alpha_A(X) W^A(X).$$

If $\mathbf{T}$ and $\mathbf{S}$ are tensors, the contraction of the ith covariant index of $\mathbf{T}$ and jth contravarient index of $\mathbf{S}$ is defined by fixing all the other slots (or indices) and contracting $\mathbf{T}$ in the ith covariant index as a one-form with the jth contravarient index of $\mathbf{S}$ as a vector field.

The operation of contracting the ith contravariant index of a tensor $\mathbf{T}$ with the jth covariant index of the same tensor is defined in a similar fashion. The contraction of a $\binom{1}{1}$ tensor is called its *trace*. (Thus to contract two indices, one upper and one lower, we merely set the indices equal and sum.)

When two indices are simultaneously contracted, we shall use two dots. For example, if R and S are tensors of types $\binom{2}{0}$ and $\binom{0}{2}$, respectively, $R : S$ is the scalar defined by $R^{AB} S_{AB}$.

The ideas of pull-back and push-forward, defined in Section 1.2 for one forms and vector fields, extend easily to tensors as follows.

4.6 Definition Let $\phi : \mathcal{B} \to \mathcal{S}$ be a regular mapping. If $\mathbf{T}$ is a tensor of type $\binom{p}{q}$, its *push-forward* $\phi_* \mathbf{T}$ is a tensor of type $\binom{p}{q}$ on $\phi(\mathcal{B})$ defined by

$$\boxed{(\phi_* \mathbf{T})(x)(\alpha^1, \ldots, \alpha^p, v_1, \ldots, v_q) = \mathbf{T}(X)(\phi^*(\alpha^1), \ldots, \phi^*(\alpha^p), \phi^*(v_1), \ldots, \phi^*(v_q)),}$$

where $\alpha^i \in T^*_x\mathcal{S}$, $v_i \in T_x\mathcal{S}$, $X = \varphi^{-1}(x)$, $\phi^*(\alpha^i)\cdot v = \alpha^i\cdot(T\phi\cdot v)$ and $\phi^*(v_j) = T(\varphi^{-1})\cdot v_j$. The *pull-back*[16] of a tensor $\mathbf{t}$ defined on $\phi(\mathcal{B})$ is given by $\phi^*\mathbf{t} = (\phi^{-1})_*\mathbf{t}$.

In coordinates we have the following relations which result from the definitions and the corresponding formulas for one-forms and vector fields (see 2.9 and 2.16).

4.7 Proposition *Letting $F(X) = T_X\phi$ and $x = \phi(X)$, we have*

$$(\phi_*\mathbf{T})^{a_1\cdots a_p}{}_{b_1\cdots b_q}(x)$$
$$= F^{a_1}{}_{A_1}(X)\cdots F^{a_p}{}_{A_p}(X)\cdot \mathbf{T}^{A_1\cdots A_p}{}_{B_1\cdots B_q}(X)\cdot(F^{-1})^{B_1}{}_{b_1}(x)\cdots(F^{-1})^{B_q}{}_{b_q}(x)$$

and

$$(\phi^*\mathbf{t})^{A_1\cdots A_p}{}_{B_1\cdots B_q}(X)$$
$$= (F^{-1})^{A_1}{}_{a_1}(X)\cdots(F^{-1})^{A_p}{}_{a_p}(x)\cdot \mathbf{t}^{a_1\cdots a_p}{}_{b_1\cdots b_q}(x)\cdot F^{b_1}{}_{B_1}(X)\cdots F^{b_q}{}_{B_q}(X).$$

Before going on to two-point tensors, it is useful to consider first some of the extra structure that enters when we have a Riemannian metric.

4.8 Definition A *Riemannian metric* on $\mathcal{B}$ is a C^∞ covariant two-tensor $\boldsymbol{G}$ $\left(\text{i.e., } \boldsymbol{G} \text{ is a tensor of type } \binom{0}{2}\right)$ such that for each $X \in \mathcal{B}$:

(i) $\boldsymbol{G}(X)$ is symmetric; that is, for $\boldsymbol{W}_1, \boldsymbol{W}_2 \in T_X\mathcal{B}$, $\boldsymbol{G}(X)(\boldsymbol{W}_1, \boldsymbol{W}_2) = \boldsymbol{G}(X)(\boldsymbol{W}_2, \boldsymbol{W}_1)$.

(ii) $\boldsymbol{G}(X)$ is positive-definite: $\boldsymbol{G}(X)(\boldsymbol{W}, \boldsymbol{W}) > 0$ for $\mathbf{0} \neq \boldsymbol{W} \in T_X\mathcal{B}$; in other words, $\boldsymbol{G}(X)$ is an inner product on $T_X\mathcal{B}$. If there is no danger of confusion, we often write

$$\boldsymbol{G}(X)(\boldsymbol{W}_1, \boldsymbol{W}_2) = \langle \boldsymbol{W}_1, \boldsymbol{W}_2\rangle_X.$$

(A Riemannian metric on $\mathcal{S}$ will be denoted $\boldsymbol{g}$.)

Notice that symmetry of $\boldsymbol{G}$ means $G_{AB} = G_{BA}$ and that $\langle \boldsymbol{W}_1, \boldsymbol{W}_2\rangle_X = G_{AB}(X)W_1^A W_2^B$. Condition (ii) states that the matrix G_{AB} is positive-definite.

Warning: In classical tensor analysis, G (or g if we use $\mathcal{S}$) is often used to denote $\sqrt{\det[G_{AB}]}$ (or $\sqrt{\det[g_{ab}]}$).

Next we describe associated tensors (i.e., new tensors obtained by raising and lowering indices) in greater detail than was considered in Section 1.3.

4.9 Definition For $X \in \mathcal{B}$, define the linear transformation $\boldsymbol{G}^\flat(X)$: $T_X\mathcal{B} \to T^*_X\mathcal{B}$ by $\boldsymbol{G}^\flat(X)(\boldsymbol{W}_1) = \boldsymbol{\alpha}(X)$, where $\boldsymbol{\alpha}(X)(\boldsymbol{W}_2) = \langle \boldsymbol{W}_1, \boldsymbol{W}_2\rangle_X$; that is, $\boldsymbol{G}^\flat(X)(\boldsymbol{W}_1) = \langle \boldsymbol{W}_1, \cdot\rangle_X$. Since $\boldsymbol{G}(X)$ is positive-definite $\boldsymbol{G}^\flat(X)$ is invertible; its inverse is denoted $\boldsymbol{G}^\sharp(X)$: $T^*_X\mathcal{B} \to T_X\mathcal{B}$.

---

[16]Regularity is not needed to pull back the covariant components. Pull-back of contravariant components and push-forward of both covariant and contravariant components require regularity. See Box 2.2.

4.10 Proposition *$G^\sharp(X)\cdot E^A = G^{AB}(X)\cdot E_B$, where $G^{AB}(X)$ is the inverse matrix of $G_{AB}(X)$.*

Proof $G^\flat(X)(G^{AB}(X)\cdot E_B)(E_C) = G^{AB}(X)\langle E_B, E_C\rangle_X = G^{AB}(X)G_{BC}(X) = \delta^A{}_C = E^A(E_C)$. Thus $G^\flat(X)(G^{AB}(X)\cdot E_B) = E^A$, so the result holds. ∎

We may regard $G^\sharp$ as a $\binom{2}{0}$ tensor on $\mathcal{B}$; 4.10 states that its components are G^{AB}.

4.11 Definition Let $\mathbf{T}$ be a tensor of type $\binom{p}{q}$ on $\mathcal{B}$. The 2^{p+q} *associated tensors* are defined by applying $G^\flat$ or $G^\sharp$ to each slot. For instance, lowering the first index on $\mathbf{T}$ gives the tensor $\mathbf{T}_1$ defined by

$$\mathbf{T}_1 : T_X\mathcal{B} \times \underbrace{T^*_X\mathcal{B} \times \cdots \times T^*_X\mathcal{B}}_{p-1 \text{ copies}} \times \underbrace{T_X\mathcal{B} \times \cdots \times T_X\mathcal{B}}_{q \text{ copies}} \to \mathbb{R}$$

$$\mathbf{T}_1(Y, \boldsymbol{\alpha}_2, \ldots, \boldsymbol{\alpha}_p, W_1, \ldots, W_q) = \mathbf{T}(G^\flat\cdot Y, \boldsymbol{\alpha}_2, \ldots, \boldsymbol{\alpha}_p, W_1, \ldots, W_q).$$

That is,

$$(\mathsf{T}_1)_{A_1}{}^{A_2\cdots A_p}{}_{B_1\cdots B_q} = G_{A_1C}\mathsf{T}^{CA_2\cdots A_p}{}_{B_1\cdots B_q}.$$

These indices are denoted by $\mathsf{T}_{A_1}{}^{A_2\cdots A_p}{}_{B_1\cdots B_q}$ (although they are the components of $\mathbf{T}_1$, which does not equal $\mathbf{T}$). Similarly, the components of the tensor $\mathbf{T}_2$ obtained by raising the first covariant index of $\mathbf{T}$ are denoted

$$\mathsf{T}^{A_1\cdots A_pB_1}{}_{B_2\cdots B_q} = G^{B_1C}\mathsf{T}^{A_1\cdots A_p}{}_{CB_2\cdots B_q}.$$

Invariantly,

$$\mathbf{T}_2(\boldsymbol{\alpha}_1, \ldots, \boldsymbol{\alpha}_p, \boldsymbol{\beta}_1, W_2, \ldots, W_q) = \mathbf{T}(\boldsymbol{\alpha}_1, \ldots, \boldsymbol{\alpha}_p, G^\sharp\cdot\boldsymbol{\beta}_1, W_2, \ldots, W_q).$$

The tensor obtained by raising all the indices of $\mathbf{T}$ is denoted $\mathbf{T}^\sharp$ and that obtained by lowering all the indices is denoted $\mathbf{T}^\flat$.

If F is a real-valued function, the vector field $(dF)^\sharp = \nabla F$ is called the *gradient* of F; that is,

$$\boxed{\nabla F = G^{AB}\frac{\partial F}{\partial X^B}E_A \quad \text{or} \quad (\nabla F)^A = G^{AB}\frac{\partial F}{\partial X^B}.}$$

4.12 Definition If $\mathbf{T}$ and $\mathbf{S}$ are tensors of the same type (or same total order), their *inner product* is defined by

$$\begin{aligned}\langle \mathbf{T}, \mathbf{S}\rangle &= \text{contraction of } \mathbf{T}^\sharp \text{ and } \mathbf{S}^\flat \text{ on all indices}\\ &= \mathsf{T}^{A_1\cdots A_pB_1\cdots B_q}\mathsf{S}_{A_1\cdots A_pB_1\cdots B_q}.\end{aligned}$$

For instance, suppose $\mathbf{T}$ is a $\binom{0}{2}$ tensor and $\mathbf{S}$ is a $\binom{2}{0}$ tensor. Then $\mathbf{T}\cdot\mathbf{S}$ is usually used to denote the contraction $\mathsf{T}_{AB}\mathsf{S}^{BC}$ and $\mathbf{T}:\mathbf{S}$ denotes the double contraction; thus $\mathbf{T}:\mathbf{S} = \langle \mathbf{T}, \mathbf{S}\rangle$ is the trace of $\mathbf{T}\cdot\mathbf{S}$.

Warning: Raising and lowering indices may not, in general, be interchanged with pull-back or push-forward. Thus, $(\phi^*\mathbf{t})^\sharp \neq \phi^*(\mathbf{t}^\sharp)$.

Problem 4.2 Show that $(\phi^*\mathbf{t})^\sharp$ does coincide with $\phi^*(\mathbf{t}^\sharp)$, provided the first $\sharp$ is taken with respect to ϕ^*g rather than G.

4.13 Proposition *The deformation tensor C is given by*

$$\boxed{C^\flat = \phi^* g,}$$

where $\phi\colon \mathcal{B} \longrightarrow \mathcal{S}$ is a configuration and $\mathbf{g}$ is the metric tensor on $\mathcal{S}$.

Proof From Proposition 3.6 we have

$$(C^\flat)_{AB}(X) = C_{AB}(X) = g_{ab}(x)F^a{}_A(X)F^b{}_B(X),$$

but from 4.7 these are exactly the components of ϕ^*g. ▮

Note that ϕ need not be regular for ϕ^*g and hence for C to be defined. As an exercise we ask the reader to show, using Proposition 3.9, that

$$b^\sharp = \phi_* G^\sharp \quad \text{and} \quad B^\sharp = \phi^* g^\sharp,$$

but we again warn that

$$b^\flat \neq \phi_* G = c^\flat \quad \text{and} \quad B^\flat \neq \phi^* g = C^\flat.$$

Next we shall discuss two-point tensors.[17] These objects play an important role in continuum mechanics; a prime example is the deformation tensor $F^a{}_A$.

4.14 Definition A *two-point tensor* T *of type* $\begin{pmatrix} q & l \\ p & m \end{pmatrix}$ *at* $X \in \mathcal{B}$ *over a mapping* $\phi\colon \mathcal{B} \longrightarrow \mathcal{S}$ is a multilinear mapping

$$\mathsf{T}\colon \underbrace{(T_X^*\mathcal{B} \times \cdots \times T_X^*\mathcal{B})}_{p \text{ copies}} \times \underbrace{(T_X\mathcal{B} \times \cdots \times T_X\mathcal{B})}_{q \text{ copies}}$$
$$\times \underbrace{(T_x^*\mathcal{S} \times \cdots \times T_x^*\mathcal{S})}_{l \text{ copies}} \times \underbrace{(T_x\mathcal{S} \times \cdots \times T_x\mathcal{S})}_{m \text{ copies}} \longrightarrow \mathbb{R},$$

where $x = \phi(X)$.

The *components of* T are defined by

$$\mathsf{T}^{A_1\cdots A_p}{}_{B_1\cdots B_q}{}^{a_1\cdots a_l}{}_{b_1\cdots b_m} = \mathsf{T}(E^{A_1}, \ldots, E^{A_p}, E_{B_1}, \ldots, E_{B_q}, e^{a_1}, \ldots, e^{a_l}, e_{b_1}, \ldots, e_{b_m})$$

relative to coordinates $\{X^A\}$ on $\mathcal{B}$ and $\{x^a\}$ on $\mathcal{S}$.

A *two-point tensor field* T assigns a two-point tensor $\mathsf{T}(X)$ to each point $X \in \mathcal{B}$, as with ordinary tensor fields.[18]

---

[17]A main reference for two-point tensors is Ericksen [1960].

[18]In manifold language, a two-point tensor is a section of the bundle $T^p_q(\mathcal{B}) \otimes T^l_m(\phi^*(T\mathcal{S}))$ over $\mathcal{B}$. Here $\phi^*(T\mathcal{S})$ denotes the pull-back bundle whose fiber over $X \in \mathcal{B}$ is $T_{\phi(X)}\mathcal{S}$.

The positioning of the indices can, of course, be altered just as with ordinary tensors. For example, a two-point tensor with indices $\mathsf{T}^a{}_{Ab}$ is a multilinear mapping $\mathsf{T}: T^*_x\mathcal{S} \times T_X\mathcal{B} \times T_x\mathcal{S} \to \mathbb{R}$.

The deformation gradient $F^a{}_A$ may be regarded as a two-point tensor as follows:

$$F(X): T^*_x\mathcal{S} \times T_X\mathcal{B} \to \mathbb{R}, \qquad (\alpha, W) \mapsto \alpha(T_X\phi \cdot W).$$

The components of F are, of course, just $F^a{}_A = \partial\phi^a/\partial X^A$.

Two-point tensors are natural generalizations of vector fields and one forms over maps. Note that a two-point tensor is regarded as a function of $X \in \mathcal{B}$, and not of $x \in \mathcal{S}$. One can think of a two-point tensor as having two tensor "legs," one in $\mathcal{B}$ and one in $\mathcal{S}$. We leave it for the reader to write down the general coordinate transformation rule for two-point tensors.

The operations on tensors generalize naturally to operations on two-point tensors. These include: raising and lowering indices, tensor products, push-forward, pull-back, tilt, and contraction. For instance, suppose we have a two-point tensor $\mathsf{T}^a{}_{Ab}$ and wish to pull back the first contravariant index a from $\mathcal{S}$ to $\mathcal{B}$. Doing this gives a new two-point tensor $\tilde{\mathsf{T}}$ whose components are denoted $\tilde{\mathsf{T}}^B{}_{Ab}$ and is defined in the following ways.

Abstractly:

$$\tilde{\mathsf{T}}(X): T^*_X\mathcal{B} \times T_X\mathcal{B} \times T_x\mathcal{S} \to \mathbb{R}, \qquad \tilde{\mathsf{T}}(X)(\alpha, W, v) = \mathsf{T}(X)(\phi_*(\alpha), W, v),$$

In components:

$$\tilde{\mathsf{T}}^B{}_{Ab}(X) = \mathsf{T}^a{}_{Ab}(X)(F^{-1})^B{}_a(x) = \mathsf{T}^a{}_{Ab}(X)\frac{\partial X^B}{\partial x^a}.$$

Notice that raising and lowering indices on the $\mathcal{S}$ leg is done using g and on the $\mathcal{B}$ leg using G. For instance,

$$\mathsf{T}_{aAb}(X) = g_{ac}(x)\mathsf{T}^c{}_{Ab}(X) \quad \text{and} \quad \mathsf{T}^{aA}{}_b(X) = G^{AB}(X)\mathsf{T}^a{}_{Bb}(X)$$

A number of fundamental two-point tensors will come up in our later work, so it is essential that the reader become familiar with them. (The first Piola–Kirchhoff stress tensor P is an example.)

> ***Problem 4.3*** Consider the motion of Problem 1.1. Calculate the components of the two-point tensors $F \otimes V$ and $F \otimes v$ in the given coordinates.

> ***Problem 4.4*** (a) Let $\mathsf{A}^a{}_A{}^b{}_B$ be a given fourth-order two-point tensor. Write down a formula for its pull-back to $\mathcal{B}$.
> (b) What is the pull-back of the fourth-order tensor $F^T \otimes F$ to $\mathcal{B}$? Of the second-order tensor $F^T \cdot F$ (contraction on one pair of indices)? What are the associated components of $F^T \cdot F$?

Our next task is to consider some further aspects of Riemannian geometry. We want to generalize covariant differentiation from vector fields on $\mathbb{R}^n$ (Section 1.1) to general tensor fields and two-point tensor fields on Riemannian manifolds. We shall develop what is needed briefly and concisely since there are detailed treatments available in standard text books.

4.15 Definition Let W_1 and W_2 be vector fields on a manifold $\mathfrak{M}$. The *Lie bracket* of W_1 and W_2 is defined (as a derivation) by means of the commutator:

$$[W_1, W_2][f] = W_1[W_2[f]] - W_2[W_1[f]].$$

Recall that $W[f] = W^A(\partial f/\partial X^A)$, the derivative of f in direction W. From this and the definition, one sees that

$$[W_1, W_2]^A = W_1^B \frac{\partial W_2^A}{\partial X^B} - W_2^B \frac{\partial W_1^A}{\partial X^B}.$$

That is, in Euclidean notation, $(W_1 \cdot \nabla)W_2 - (W_2 \cdot \nabla)W_1$.

Problem 4.5 Verify Jacobi's identity: $[[X, Y], Z] + [[Z, X], Y] + [[Y, Z], X] = 0$.

Before defining the covariant derivative, it is useful to have the abstract notion of a connection (or covariant derivative) at hand. We shall link it with the metric shortly.

4.16 Definition A *connection* on a manifold $\mathfrak{M}$ is an operation ∇ that associates to each pair of vector fields W, Y on $\mathfrak{M}$ a third vector field denoted $\nabla_W Y$ and called the *covariant derivative* of Y along W, such that:

(i) $\nabla_W Y$ is linear in each of W and Y;
(ii) $\nabla_{fW} Y = f\nabla_W Y$ for scalar functions f; and
(iii) $\nabla_W(fY) = f\nabla_W Y + (W[f])Y$.

These conditions are reasonable requirements for an operation that is supposed to differentiate Y in the direction W. Note that (iii) is analogous to the product rule for differentiation.

4.17 Definition The *Christoffel symbols* of a connection ∇ on $\mathfrak{M}$ are defined on a coordinate system $\{X^A\}$ by $\Gamma^A_{BC}(X)E_A(X) = (\nabla_{E_B} E_C)(X)$. (Consistent with our previous conventions, we denote the Christoffel symbols on $\mathcal{B}$ by Γ^A_{BC} and those on $\mathcal{S}$ by γ^a_{bc}.)

Problem 4.6 If $\{\bar{X}^A\}$ is another coordinate system on $\mathfrak{M}$, show that

$$\bar{\Gamma}^A_{BC} = \frac{\partial \bar{X}^A}{\partial X^D}\Gamma^D_{EF}\frac{\partial X^E}{\partial \bar{X}^B}\frac{\partial X^F}{\partial \bar{X}^C} + \frac{\partial \bar{X}^A}{\partial X^D}\frac{\partial^2 X^D}{\partial \bar{X}^B \partial \bar{X}^C}.$$

Conclude that Γ^A_{BC} are *not* the components of a tensor.

4.18 Proposition *In coordinates, we have*

$$(\nabla_W Y)^A = \frac{\partial Y^A}{\partial X^B} W^B + \Gamma^A_{BC} W^B Y^C.$$

Proof By the properties (i)–(iii) of a connection:

$$\begin{aligned}\nabla_W Y &= \nabla_{W^B E_B}(Y^C E_C) = W^B \nabla_{E_B}(Y^C E_C)\\ &= W^B(Y^C \nabla_{E_B} E_C + E_B[Y^A]E_A)\\ &= \Gamma^A_{BC} W^B Y^C E_A + W^B \frac{\partial Y^A}{\partial X^B} E_A. \quad \blacksquare\end{aligned}$$

From this proposition, observe that $\nabla_W Y(X)$ depends only on the point value of W at X, and not on its derivatives. Thus if σ is a curve, $\nabla_{\dot\sigma} Y$ makes sense. In fact, this is an important object that we now identify.

4.19 Definition We call $\nabla_{\dot\sigma} Y$ the *material derivative* of Y along σ, and write it DY/dt, or if $\sigma(t) = \phi_X(t)$ [where $\phi(X, t)$ is a given motion], by $\dot Y$.

Using coordinates, 4.18, and the chain rule,

$$\boxed{\left(\frac{DY}{dt}\right)^A = \frac{d}{dt}[Y^A(\sigma(t))] + \Gamma^A_{BC}(\sigma(t))Y^B(\sigma(t))\dot\sigma^C.}$$

If Y is time dependent, set $DY/Dt = \partial Y/\partial t + \nabla_{\dot\sigma} Y$. In particular, for a motion ϕ_t of $\mathcal{B}$ in $\mathcal{S}$ we see that the acceleration is

$$a = \frac{\partial v}{\partial t} + \nabla_v v = \dot v.$$

Another consequence of the fact that $\nabla_W Y(X)$ depends only on the point value of W at X is that ∇Y defines a $\binom{1}{1}$ tensor with components $(\nabla Y)^A{}_B = \partial Y^A/\partial X^B + \Gamma^A_{BC} Y^C$.

There are several ways of defining the covariant derivative of a tensor. One may proceed directly (see, Abraham and Marsden [1978]), using the fact that we desire $\nabla(\mathbf{T} \otimes \mathbf{S}) = \nabla\mathbf{T} \otimes \mathbf{S} + \mathbf{T} \otimes \nabla\mathbf{S}$, or one may use parallel translation. Since we shall need this concept, we choose the latter.

4.20 Definition A vector field Y defined along a curve $\sigma(t)$ is called *parallel* if $\nabla_{\dot\sigma} Y = \mathbf{0}$. A vector Y_0 at $\sigma(t_0)$ is called *parallel transported* or *translated along* σ if it is extended to a parallel vector field Y along σ.

A vector can be parallel transported uniquely since, given $\sigma(t)$, $DY/dt = \mathbf{0}$ is a linear differential equation for $Y(\sigma(t))$, which has a unique solution with initial data Y_0 (as long as σ is, say, continuous).

The intuitive idea is that ∇ reflects the local geometry in such a way that an observer carrying an arrow along a curve in what seems to be a parallel manner,

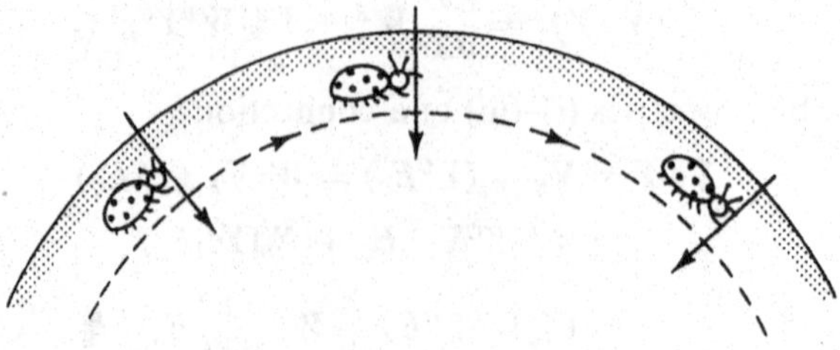

Figure 1.4.1

is, in fact, parallel transporting the arrow. On a sphere, for instance, a parallel transported vector is not parallel transported in space. See Figure 1.4.1.

The following proposition demonstrates a basic link between parallel transport and the covariant derivative.

4.21 Proposition *Let $\sigma(t)$ be a given curve in $\mathfrak{M}$, and Y a vector field defined along σ. Let $S_{t,s}\colon T_{\sigma(s)}\mathfrak{M} \to T_{\sigma(t)}\mathfrak{M}$ denote parallel translation from $\sigma(s)$ to $\sigma(t)$ along σ. (We also call $S_{t,s}$ the* shifter.) *Then*

$$\boxed{\nabla_{\dot\sigma(t)} Y(\sigma(t)) = \frac{d}{ds}(S_{t,s}Y(\sigma(s)))\Big|_{s=t}.}$$

(Note that $S_{t,s}Y(\sigma(t)) \in T_{\sigma(t)}\mathfrak{M}$ for all s.)

Proof By construction, $(S_{s,t_0}Y_0)^A = (S_{s,t_0})^A{}_B Y_0^B = Y_s^A$ satisfies

$$\frac{d}{ds}(S_{s,t_0})^A{}_B Y_0^B + \Gamma^A_{BC}\dot\sigma^C Y_s^B = 0.$$

Thus, since S_{s,t_0} is the identity at $s = t_0$,

$$\frac{d}{ds}(S_{s,t_0})^A{}_B\Big|_{s=t_0} = -\Gamma^A_{BC}\dot\sigma^C.$$

From uniqueness of solution of differential equations, $S_{r,s}\circ S_{s,t} = S_{r,t}$, and so $S_{s,t} = S_{t,s}^{-1}$. Thus

$$\begin{aligned}\left(\frac{d}{ds}S_{t,s}Y(\sigma(s))\Big|_{s=t}\right)^A &= \left(\frac{d}{ds}S_{s,t}^{-1}Y(\sigma(s))\Big|_{s=t}\right)^A\\ &= -\frac{d}{ds}(S_{s,t})^A{}_B\Big|_{s=t} Y^B(\sigma(t)) + \frac{d}{ds}Y^A(\sigma(s))\Big|_{s=t}\\ &= \Gamma^A_{CB}\dot\sigma^C Y^B + \frac{d}{dt}Y^A(\sigma(t)) = (\nabla_{\dot\sigma(t)}Y(\sigma(t)))^A. \quad \blacksquare\end{aligned}$$

4.22 Definition A connection is called *torsion free* if

$$\nabla_W Y - \nabla_Y W = [W, Y]$$

for all vector fields W and Y. The *torsion* of a connection is defined by $\mathrm{Tor}(W, Y) = \nabla_W Y - \nabla_Y W - [W, Y]$ and is a $\begin{pmatrix}1\\2\end{pmatrix}$ tensor; thus a connection is torsion free when its torsion tensor is zero.

4.23 Proposition *A connection is torsion free if and only if* $\Gamma^A_{BC} = \Gamma^A_{CB}$*—that is, the Christoffel symbols are symmetric.*

Proof Write out $\nabla_W Y - \nabla_Y W - [W, Y]$ in coordinates, using the earlier coordinate expressions to see that $(\mathrm{Tor})^A_{BC} = \Gamma^A_{BC} - \Gamma^A_{CB}$. ▮

Now we are ready to find the connection associated with a given metric G on $\mathfrak{M}$. The following is sometimes called the "Fundamental Theorem of Riemannian Geometry," and is based on the early pioneering work of Levi-Civita.

4.24 Theorem *Let $\mathfrak{M}$ be a manifold with a Riemannian metric G. Then there is a unique connection on $\mathfrak{M}$ that is torsion free and for which parallel translation preserves inner products (that is, $S_{t,s}$ is an orthogonal transformation).*

Proof First of all, suppose that such a connection exists. Let $W_1(t)$ and $W_2(t)$ be parallel vector fields along a curve $\sigma(t)$. Therefore, by hypothesis,

$$0 = \frac{d}{dt}\langle W_1(t), W_2(t)\rangle = \frac{d}{dt} G_{AC}(\sigma(t)) W^A_1(t) W^C_2(t)$$

$$= \frac{\partial G_{AC}}{\partial X^D}\dot{\sigma}^D W^A_1 W^C_2 - G_{AC}\Gamma^A_{BD} W^B_1 \dot{\sigma}^D W^C_2 - G_{AC}\Gamma^C_{BD} W^A_1 \dot{\sigma}^D W^B_2$$

since

$$\frac{d}{dt} W^A_i(t) + \Gamma^A_{BC} W^B_i \dot{\sigma}^C = 0,$$

that is, W_i is parallel along $\sigma(t)$. If we take the particular case in which $W_1(0) = E_A$ and $W_2(0) = E_C$ and evaluate at $t = 0$, we get

$$0 = \frac{\partial G_{AC}}{\partial X^D}\dot{\sigma}^D - G_{BC}\Gamma^B_{AD}\dot{\sigma}^D - G_{BA}\Gamma^B_{CD}\dot{\sigma}^D.$$

Since $\dot{\sigma}^D$ is arbitrary,

$$\frac{\partial G_{AC}}{\partial X^D} = G_{BC}\Gamma^B_{AD} + G_{BA}\Gamma^B_{CD}.$$

Note that this is the same as the identity obtained in Euclidean space (see Lemma 3.29). By the same algebra, remembering symmetry of G_{AD} and Γ^A_{BC}, we get

$$\boxed{2G_{AB}\Gamma^A_{DC} = \left(\frac{\partial G_{CB}}{\partial X^D} + \frac{\partial G_{DB}}{\partial X^C} - \frac{\partial G_{DC}}{\partial X^B}\right).}$$

This formula shows that the Christoffel symbols Γ^A_{BC} are uniquely determined. Hence the connection is unique.

For the converse, we can define the Christoffel symbols in any coordinate chart by the preceding formula. Clearly they are symmetric. Then we define $\nabla_W Y$ in terms of the Γ^A_{BC} and check that it is a well-defined vector field (transforms like a vector). Finally, reversing the above computations shows that parallel translation preserves the inner product. ∎

Problem 4.7 Show that the coordinate-free analogs of the identities in the above proof are

$$\frac{d}{dt}\langle W_1(t), W_2(t)\rangle = \left\langle \frac{DW_1}{dt}, W_2\right\rangle + \left\langle W_1, \frac{DW_2}{dt}\right\rangle$$

and
$$\begin{aligned} 2\langle W_1, \nabla_{W_3} W_2\rangle = {} & W_3[\langle W_1, W_2\rangle] + \langle W_3, [W_1, W_2]\rangle \\ & + W_2[\langle W_3, W_1\rangle] - \langle W_2, [W_3, W_1]\rangle \\ & - W_1[\langle W_2, W_3\rangle] - \langle W_1, [W_2, W_3]\rangle. \end{aligned}$$

Given a Riemannian manifold, we shall use the connection above (called the Riemannian or Levi-Civita connection) unless special exception is made.

Problem 4.8 (a) Work out Γ^A_{BC} for $\mathfrak{M}$ the unit sphere S^2 in $\mathbb{R}^3$ with respect to spherical coordinates.

(b) A *geodesic* is a curve $\sigma(t)$ such that $\dot{\sigma}$ is parallel along σ. Show that the geodesics on S^2 are great circles.

(c) Describe parallel translation of a vector around a circle of constant latitude on S^2 relative to (i) the Riemannian manifold $\mathbb{R}^3$ and (ii) the Riemannian manifold S^2.

4.25 Definition Let $S_{t,s}$ be parallel translation along a curve $\sigma(t)$. Extend $S_{t,s}$ to all tensor fields along $\sigma(t)$ in the same manner as push-forward was extended. Explicitly: If $\boldsymbol{\alpha}$ is a one-form at $\sigma(s)$, let

$$(S_{t,s}\boldsymbol{\alpha})(W) = \boldsymbol{\alpha}(S_{s,t}(W)), \quad W \in T_{\sigma(t)}\mathfrak{M}$$

(see Figure 1.4.2). If $\mathbf{T}$ is a tensor of type $\binom{p}{q}$ at $\sigma(s)$, let

$$(S_{t,s}\mathbf{T})(\boldsymbol{\alpha}_1, \ldots, \boldsymbol{\alpha}_p, W_1, \ldots, W_q) = \mathbf{T}(S_{s,t}\boldsymbol{\alpha}_1, \ldots, S_{s,t}\boldsymbol{\alpha}_p, S_{s,t}W_1, \ldots, S_{s,t}W_q).$$

If $W = \dot{\sigma}(t)$, then the *covariant derivative of* $\mathbf{T}$ *along* W at $X = \sigma(t)$ is defined by

$$(\nabla_W \mathbf{T})(X) = \frac{d}{ds}\{S_{t,s}\mathbf{T}(\sigma(s))\}\bigg|_{s=t}.$$

The covariant derivative $\nabla\mathbf{T}$ thereby defines a $\binom{p}{q+1}$ tensor.

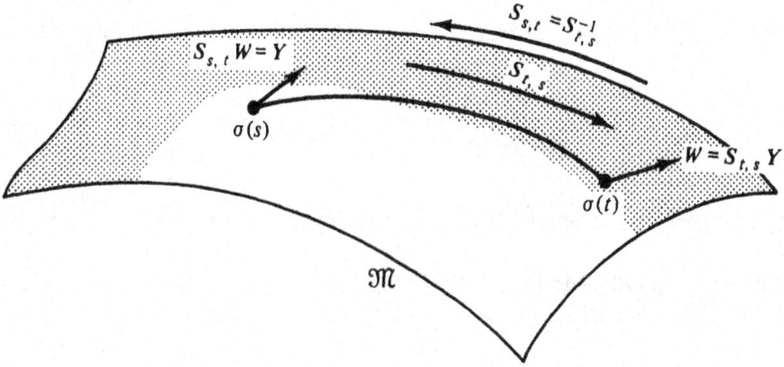

Figure 1.4.2

Notice that $S_{t,s}\mathbf{T}(\sigma(s))$ is a tensor at $\sigma(t)$, so the d/ds derivative occurs within a fixed linear space, just as with vectors in 4.21.

The coordinate expression for covariant derivative may be worked out exactly as for vector fields. That computation leads to the following:

4.26 Proposition *Let* $\mathbf{T}$ *be a tensor of type* $\binom{p}{q}$ *on* $\mathfrak{M}$. *Writing* $(\nabla\mathbf{T})^{A\cdots D}{}_{E\cdots GK} = \mathrm{T}^{A\cdots D}{}_{E\cdots G|K}$, *we have*

$$\mathrm{T}^{A\cdots D}{}_{E\cdots G|K} = \frac{\partial \mathrm{T}^{A\cdots D}{}_{E\cdots G}}{\partial X^K} + \mathrm{T}^{LB\cdots D}{}_{E\cdots G}\Gamma^A_{LK} + (\textit{all upper indices})$$
$$- \mathrm{T}^{A\cdots D}{}_{LF\cdots G}\Gamma^L_{EK} - (\textit{all lower indices})$$

and

$$(\nabla_W\mathbf{T})^{A\cdots D}{}_{E\cdots G} = \mathrm{T}^{A\cdots D}{}_{E\cdots G|K}W^K.$$

Notice that we use a vertical bar to designate the coordinate expression for covariant differentiation.

Problem 4.9 (i) Prove that $\nabla G = 0$ by:

(a) a coordinate calculation;
(b) using the definition and the fact that $S_{t,s}$ preserves inner products.

(ii) Prove the product rule: $\nabla(\mathbf{T}\otimes\mathbf{S}) = \nabla\mathbf{T}\otimes\mathbf{S} + \mathbf{T}\otimes\nabla\mathbf{S}$. What does this assert in coordinates? For $\mathbf{T} = \mathrm{T}^A E_A$?

(iii) If $\phi: \mathfrak{B} \to \mathfrak{S}$ is a regular mapping, show that $\phi^*(\nabla\mathbf{t}) = \nabla^*(\phi^*\mathbf{t})$, where ∇ is the covariant derivative with respect to g on $\mathfrak{S}$, and ∇^* is the covarient derivative with respect to $\phi^*g = C^\flat$ on $\mathfrak{B}$.

(iv) Consult problem 2.2 and write the convected acceleration as a covariant derivative of the convected velocity using the metric $C_t^\flat = \varphi_t^*(g)$.

4.27 Definition The *divergence* of a tensor field $\mathbf{T}$ of type $\binom{p}{q}$ is a tensor of type $\binom{p-1}{q}$ obtained by contracting the last contravariant and covariant indices of $\nabla\mathbf{T}$:

$$(\text{DIV}\,\mathbf{T})^{AB\cdots C}{}_{F\cdots H} = \mathbf{T}^{AB\cdots C}{}_{F\cdots H|C}.$$

(The divergence of a tensor $\mathbf{t}$ on $\mathcal{S}$ is denoted div $\mathbf{t}$.)

For vector fields, the following formula is easy to check:

$$\text{DIV}\,W = W^A{}_{|A} = \frac{1}{\sqrt{\det G}}\frac{\partial}{\partial X^A}(\sqrt{\det G}\,W^A).$$

We shall return to this formula from another point of view in Section 1.7.

Next we consider covariant differentiation of two-point tensors.

4.28 Definition Let $\mathbf{T}$ be a two-point tensor of type $\begin{pmatrix} p & l \\ q & m \end{pmatrix}$ over the map $\phi: \mathcal{B} \to \mathcal{S}$; let $\sigma(t)$ be a curve in $\mathcal{B}$ and $W = \dot{\sigma}(t)$. Let $\tilde{\sigma}(t) = \phi(\sigma(t))$ be the image of σ under ϕ. Let $S_{t,s}$ denote parallel translation along σ in $\mathcal{B}$ and $s_{t,s}$ denote parallel translation along $\tilde{\sigma}$ in $\mathcal{S}$.

Define the two-point tensor $\nabla_W\mathbf{T}$ at $X = \sigma(t)$, a tensor of type $\begin{pmatrix} p & l \\ q & m \end{pmatrix}$, as follows:

$$\nabla_W\mathbf{T} = \frac{d}{ds}(S_{t,s}\mathbf{T}(\sigma(s)))\Big|_{s=t},$$

where

$$(S_{t,s}\mathbf{T})(\boldsymbol{\alpha}^1, \ldots, \boldsymbol{\alpha}^p, W_1, \ldots, W_q; \boldsymbol{\beta}^1, \ldots, \boldsymbol{\beta}^l, v_1, \ldots, v_m)$$
$$= \mathbf{T}(S_{s,t}\boldsymbol{\alpha}^1, \ldots, S_{s,t}\boldsymbol{\alpha}^p, S_{s,t}W_1, \ldots, S_{s,t}W_q;$$
$$s_{s,t}\boldsymbol{\beta}^1, \ldots, s_{s,t}\boldsymbol{\beta}^l, s_{s,t}v_1, \ldots, s_{s,t}v_m),$$

where $\boldsymbol{\alpha}^i \in T^*_X\mathcal{B}$, $W_i \in T_X\mathcal{B}$, $\boldsymbol{\beta}^j \in T^*_x\mathcal{S}$, $v_j \in T_x\mathcal{S}$, $X = \sigma(t)$, and $x = \phi(X)$. This defines, therefore, a two-point tensor $\nabla\mathbf{T}$ of type $\begin{pmatrix} p & l \\ q+1 & m \end{pmatrix}$.

To help understand this definition, we work out the components of $\nabla_W V$ in case V is a vector field over ϕ. By definition,

$$\nabla_W V = \frac{d}{ds}(S_{t,s}V(\sigma(s)))\Big|_{s=t} \qquad \text{where} \quad \dot{\sigma} = W.$$

(See Figure 1.4.3.) Going to a chart, as in 4.21, we get

$$(\nabla_W V)^a = \frac{d}{dt} V^a(\sigma(t)) + \gamma^a_{bc} V^b \frac{d}{dt}\tilde{\sigma}(t)^c.$$

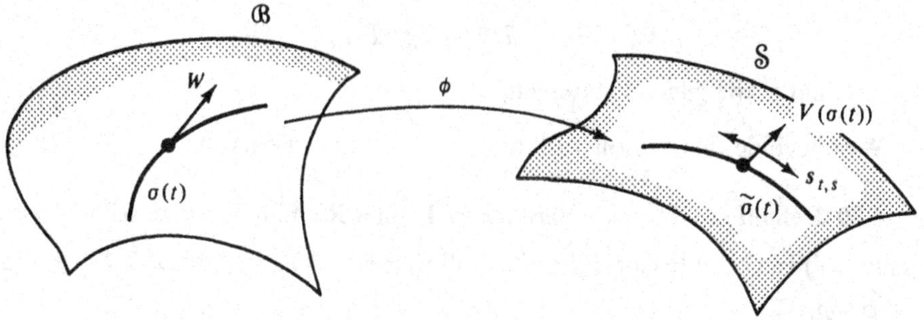

Figure 1.4.3

[Note that $(d/ds)(s_{t,s})^a{}_b = -\gamma^a_{bc}\dot{\tilde{\sigma}}^c$ since $s_{t,s}$ is parallel translation along $\tilde{\sigma}$.] Using the chain rule and the notation $F^a{}_A = \partial\phi^a/\partial X^A$,

$$(\nabla_W V)^a = \frac{\partial V^a}{\partial X^A} W^A + \gamma^a_{bc} V^b F^c{}_A W^A,$$

that is, $$(\nabla V)^a{}_A = V^a{}_{|A} = \frac{\partial V^a}{\partial X^A} + \gamma^a_{bc} V^b F^c{}_A.$$

The reader will recognize this expression from 3.27. A similar computation proves the following:

4.29 Proposition *If* **T** *is a two-point tensor of type* $\begin{pmatrix} p & l \\ q & m \end{pmatrix}$, *then*

$$\begin{aligned} &\mathsf{T}^{AB\cdots D}{}_{E\cdots H}{}^{ab\cdots d}{}_{e\cdots h|K} \\ &= \frac{\partial}{\partial X^K}\mathsf{T}^{AB\cdots D}{}_{E\cdots H}{}^{a\cdots d}{}_{e\cdots h} \\ &\quad + \mathsf{T}^{LB\cdots D}{}_{E\cdots H}{}^{a\cdots d}{}_{e\cdots h}\Gamma^A_{LK} + (\textit{all upper large indices}) \\ &\quad - \mathsf{T}^{A\cdots D}{}_{L\cdots H}{}^{a\cdots d}{}_{e\cdots h}\Gamma^L_{EK} - (\textit{all lower large indices}) \\ &\quad + \mathsf{T}^{A\cdots D}{}_{E\cdots H}{}^{kb\cdots d}{}_{e\cdots h}\gamma^a_{kl}F^l{}_K + (\textit{all upper small indices}) \\ &\quad - \mathsf{T}^{A\cdots D}{}_{E\cdots H}{}^{a\cdots d}{}_{kf\cdots h}\gamma^k_{el}F^l{}_K - (\textit{all lower small indices}). \end{aligned}$$

The divergence of a two-point tensor is defined in the same way as for a tensor:

$$(\text{DIV } \mathbf{T})^{AB\cdots C}{}_{E\cdots H}{}^{a\cdots d}{}_{e\cdots h} = \mathbf{T}^{AB\cdots CD}{}_{E\cdots H}{}^{a\cdots d}{}_{e\cdots h|D}.$$

Problem 4.10 Show that, if $T^{Aa} = g^{ab}T^A{}_b$ (associated tensors), then

$$T^{Aa}{}_{|B} = g^{ab}T^A{}_{b|B}.$$

What is the general statement?

We conclude this section with a brief discussion of curvature.[19]

4.30 Definition The *curvature tensor* $\mathbf{R}$ on a Riemannian manifold $\mathfrak{M}$ (with metric G) is a $\binom{1}{3}$ tensor defined as follows: $\mathbf{R}: T_X^*\mathfrak{M} \times T_X\mathfrak{M} \times T_X\mathfrak{M} \times T_X\mathfrak{M} \longrightarrow \mathbb{R}$, where

$$\begin{aligned}\mathbf{R}(\alpha, W_1, W_2, W_3) &= \alpha(\nabla_{W_1}\nabla_{W_2}W_3 - \nabla_{W_2}\nabla_{W_1}W_3 - \nabla_{[W_1,W_2]}W_3)\\ &= \alpha([\nabla_{W_1}, \nabla_{W_2}]W_3 - \nabla_{[W_1,W_2]}W_3),\end{aligned}$$

$\alpha \in T_X^*\mathfrak{M}$, and $W_i(X) \in T_X\mathfrak{M}$.

One has to verify that $\mathbf{R}$ indeed depends only on the point values of W_i. This can be done by computing the components of $\mathbf{R}$. One gets:

$$\boxed{\mathrm{R}^A{}_{BCD} = \frac{\partial \Gamma^A_{DB}}{\partial X^C} - \frac{\partial \Gamma^A_{CB}}{\partial X^D} + \Gamma^A_{CE}\Gamma^E_{DB} - \Gamma^A_{DE}\Gamma^E_{CB}.}$$

One can substitute the expression for Γ^A_{BC} in terms of G_{AB} to express $\mathrm{R}^A{}_{BCD}$ totally in terms of G_{AB} and $\partial G_{AB}/\partial X^C\, \partial X^D$.

The contraction $R_{BD} = \mathrm{R}^A{}_{BAD}$ is called the *Ricci curvature* and its trace $R^A{}_A = R$ is called the *scalar curvature*.

The geometric significance of the curvature can be illustrated by an example. Let $\Phi: \mathbb{R}^2 \longrightarrow \mathfrak{M}$ be a surface in $\mathfrak{M}$. Then $\mathrm{R}^A{}_{BCD}$ measures the lack of commutativity of second covariant derivatives along the x and y directions:

$$\frac{D}{dx}\frac{D}{dy}W - \frac{D}{dy}\frac{D}{dx}W = \mathbf{R}\cdot\left(\frac{\partial\Phi}{\partial x}, \frac{\partial\Phi}{\partial y}, W\right)$$

where $(x, y) \in \mathbb{R}^2$ are identified with points on the surface via Φ, so D/dx denotes the covariant derivative along the x-coordinate curve, and W is a vector field along the surface.

---

[19]One has to be careful of sign conventions in this topic. The frontispiece of Misner–Thorne–Wheeler [1973] summarizes the situation.

Problem 4.11 Show that on a Riemannian manifold $\mathfrak{M}$,

$$W^A{}_{|B|C} - W^A{}_{|C|B} = R^A{}_{BCD}W^D$$

for a vector field W, by a coordinate calculation.

A manifold with $\mathbf{R} = \mathbf{0}$ is called *flat*. For instance, $\mathbb{R}^3$ is flat: $\mathsf{R}^A{}_{BCD} = 0$ in any coordinate system since it is obviously zero in Cartesian coordinates (note $\Gamma^A_{BC} \neq 0$ in general curvilinear coordinates on $\mathbb{R}^n$).

In books on Riemannian geometry it is proven that a manifold $\mathfrak{M}$ is flat if and only if there are coordinates about each point in which $G_{AB} = \delta_{AB}$: that is, $\mathfrak{M}$ is locally Euclidean, both topologically and metrically.[20]

We conclude with the following, which are sometimes called the *compatibility conditions.* They restrict the motion ϕ of a simple body in terms of its deformation tensor.

4.31 Proposition *Let $\mathcal{B}$ be open in $\mathbb{R}^n$ and $\mathcal{S} = \mathbb{R}^n$. Let $\phi: \mathcal{B} \longrightarrow \mathcal{S}$ be a regular configuration. Let $\mathsf{K}^A{}_{BCD}$ be the "curvature tensor" obtained by using the deformation tensor C_{AB} in place of G_{AB}. Then $\mathsf{K}^A{}_{BCD} = 0$.*

Proof Combining Problem 4.9(iii) with the definition of curvature, we see that $\mathbf{K}$ is the pull-back of $\mathbf{R}$, the curvature tensor of $\mathbb{R}^3$. But $\mathbf{R} = \mathbf{0}$, so $\mathbf{K} = \mathbf{0}$ as well. ∎

There is a related question that is of some interest. This is, given a tensor C_{AB} that is symmetric and positive-definite, when is C_{AB} the deformation tensor of a configuration? The following answers the question locally. (The global question is presumably hard.)

4.32 Proposition *Let $\mathcal{B} \subset \mathbb{R}^n$ be open and $\mathcal{S} = \mathbb{R}^n$. Let C_{AB} be a given positive-definite symmetric two-tensor whose curvature tensor vanishes: $\mathsf{K}^A{}_{BCD} = 0$. Then given any point $x_0 \in \mathcal{B}$, there is a neighborhood $\mathcal{U}$ of x_0 and a regular map $\phi: \mathcal{U} \longrightarrow \mathbb{R}^n$ whose deformation tensor is C_{AB}.*

Proof The hypotheses say that the Riemannian manifold $(\mathcal{B}, C_{AB})$ is flat. Thus ,in a neighborhood $\mathcal{U}$ of x_0, there is a coordinate system $\chi: \mathcal{U} \longrightarrow \mathbb{R}^n$ in which the C_{AB}'s are constants.[21] By following χ with a linear transformation A, we can bring C_{AB} into diagonal form δ_{AB}. Let $\phi = A \circ \chi$. Then $(\phi_* C)_{ab}$ is the Euclidean metric; that is, C_{AB} is the deformation tensor of ϕ. ∎

---

[20] If a metric is flat, one shows that exponential or "canonical" coordinates (which make $\Gamma^A_{BC} = 0$ at one point) do the job. See footnote 21.

[21] This is a theorem of classical Riemannian geometry proved in, for example, the book of Eisenhart [1926]. The map χ is given by the exponential map.

Box 4.1 ***Summary of Important Formulas for Section 1.4***

Tensors of type $\binom{p}{q}$ *on* $\mathfrak{M}$

$$\mathbf{T}: \underbrace{T_X^*\mathfrak{M} \times \cdots \times T_X^*\mathfrak{M}}_{p} \times \underbrace{T_X\mathfrak{M} \times \cdots \times T_X\mathfrak{M}}_{q} \to \mathbb{R}, \qquad \mathsf{T}^{\overbrace{AB\cdots D}^{p \text{ indices}}}{}_{\underbrace{EF\cdots H}_{q \text{ indices}}} = \mathbf{T}(E^A, \ldots, E^D, E_E \cdots E_H)$$

Coordinate Transformations

$$\bar{\mathsf{T}}^{A\cdots D}{}_{E\cdots H} = \frac{\partial \bar{X}^A}{\partial X^I} \cdots \frac{\partial \bar{X}^D}{\partial X^L} \mathsf{T}^{I\cdots L}{}_{M\cdots P} \frac{\partial X^M}{\partial \bar{X}^E} \cdots \frac{\partial X^P}{\partial \bar{X}^H}$$

Tensor Product

$$\mathbf{T} \otimes \mathbf{S}(\boldsymbol{\alpha}^1, \ldots, \boldsymbol{\alpha}^p, W_1, \ldots, W_q, \boldsymbol{\beta}^1, \ldots, \boldsymbol{\beta}^r, Y_1, \ldots, Y_s) = \mathbf{T}(\boldsymbol{\alpha}^1, \ldots, \boldsymbol{\alpha}^p, W_1, \ldots, W_q)\mathbf{S}(\boldsymbol{\beta}^1, \ldots, \boldsymbol{\beta}^r, Y_1, \ldots, Y_s)$$

$$(\mathbf{T} \otimes \mathbf{S})^{A\cdots D}{}_{E\cdots H}{}^{I\cdots L}{}_{M\cdots P} = \mathsf{T}^{A\cdots D}{}_{E\cdots H}\mathsf{S}^{I\cdots L}{}_{M\cdots P}$$

Contraction

Contraction of $\mathbf{T}$ in its last covariant and contravariant slots $\qquad \mathsf{T}^{A\cdots CD}{}_{E\cdots GD}$

(Contraction of $\mathbf{T}$ and $\mathbf{S}$) $= \mathbf{T} \cdot \mathbf{S}$ $\qquad \mathsf{T}^{A\cdots CD}\mathsf{S}_{E\cdots GD}$

(Double contraction of $\mathbf{T}$ and $\mathbf{S}$) $= \mathbf{T} : \mathbf{S}$ $\qquad \mathsf{T}^{AB\cdots CD}\mathsf{S}_{EF\cdots CD}$

(Inner product of $\mathbf{T}$ and $\mathbf{S}$) $= \langle \mathbf{T}, \mathbf{S} \rangle$ $\qquad \mathsf{T}^{AB\cdots CD}\mathsf{S}_{AB\cdots CD}$

Push-Forward

$$(\phi_*\mathbf{T})(\boldsymbol{\alpha}_1, \ldots, \boldsymbol{\alpha}_p, v_1, \ldots, v_q) = \mathbf{T}(\phi^*(\boldsymbol{\alpha}_1), \ldots, \phi^*(\boldsymbol{\alpha}_p), \phi^*(v_1), \ldots, \phi^*(v_q))$$

$$(\phi_*\mathbf{T})^{a\cdots d}{}_{e\cdots h} = F^a{}_A \cdots F^d{}_D \mathsf{T}^{A\cdots D}{}_{E\cdots H}(F^{-1})_e{}^E \cdots (F^{-1})_h{}^H$$

Pull-Back

$$\phi^*\mathbf{t} = \phi^{-1}{}_*\mathbf{t} \qquad (\phi^*\mathbf{t})^{A\cdots D}{}_{E\cdots H} = (F^{-1})^A_a \cdots (F^{-1})^D_d \mathbf{t}^{a\cdots d}{}_{e\cdots h}F^e{}_E \cdots F^h{}_H$$

Riemannian Metric

$$G(X)(W_1, W_2) = \langle W_1, W_2 \rangle_X \qquad \langle W_1, W_2 \rangle_X = G_{AB}(X)W_1{}^A W_2{}^B$$

Associated Tensors

Raise or lower indices by $G^\sharp$ or $G^\flat$ $\qquad$ $\mathrm{T}_A{}^{B\cdots D}{}_{E\cdots H} = G_{AK}\mathrm{T}^{KB\cdots D}{}_{E\cdots H}$, etc.

$\mathbf{T}^\sharp = (\mathbf{T}$ with all indices raised: $\mathrm{T}^{AB\cdots DE\cdots H})$

$\mathbf{T}^\flat = (\mathbf{T}$ with all indices lowered: $\mathrm{T}_{AB\cdots DE\cdots H})$

Two-Point Tensors of Type $\begin{pmatrix} p & l \\ q & m \end{pmatrix}$ *Over* $\phi: \mathcal{B} \to \mathcal{S}$

$$\mathbf{T}: \underbrace{T_X^*\mathcal{B} \times \cdots \times T_X^*\mathcal{B}}_{p} \times \underbrace{T_X\mathcal{B} \times \cdots \times T_X\mathcal{B}}_{q} \times \underbrace{T_x^*\mathcal{S} \times \cdots \times T_x^*\mathcal{S}}_{l} \times \underbrace{T_x\mathcal{S} \times \cdots \times T_x\mathcal{S}}_{m} \to \mathbb{R}$$

$$\mathrm{T}^{A\cdots D}{}_{E\cdots H}{}^{a\cdots d}{}_{e\cdots h} = \mathbf{T}(E^A, \ldots, E^D, E_E, \ldots, E_H, e^a, \ldots, e^d, e_e, \ldots, e_h)$$

Lie Bracket

$[W_1, W_2][f] = W_1[W_2[f]] - W_2[W_1[f]]$ $\qquad$ $[W_1, W_2]^A = W_1^B \dfrac{\partial W_2^A}{\partial X^B} + W_2^B \dfrac{\partial W_1^A}{\partial X^B}$

Christoffel Symbols

Of a connection $\nabla \cdots$ $\qquad$ $\Gamma^A_{BC} E_A = \nabla_{E_B} E_C$

Of a metric $G_{AB} \cdots$ $\qquad$ $2G_{AB}\Gamma^A_{DC} = \left(\dfrac{\partial G_{CB}}{\partial X^D} + \dfrac{\partial G_{DB}}{\partial X^C} - \dfrac{\partial G_{DC}}{\partial X^B}\right)$

Covariant Derivative of a Tensor

$\nabla_W \mathbf{T} = \dfrac{d}{ds} S_{t,s} \mathbf{T}(\sigma(s))\Big|_{s=t}$

$W = \dot\sigma(t)$,

$S_{t,s}$ = parallel translation (shifter)

$$\mathrm{T}^{A\cdots D}{}_{E\cdots G|K} = \frac{\partial \mathrm{T}^{A\cdots D}{}_{E\cdots G}}{\partial X^K} + \mathrm{T}^{LB\cdots D}{}_{E\cdots G}\Gamma^A_{LK} + (\text{all upper indices})$$
$$- \mathrm{T}^{A\cdots D}{}_{LF\cdots G}\Gamma^L_{EK} - (\text{all lower indices})$$

$$(\nabla_W \mathbf{T})^{A\cdots D}{}_{E\cdots G} = \mathrm{T}^{A\cdots D}{}_{E\cdots G|K} W^K$$

Covariant Derivative along a Curve $\sigma(t)$

$\dfrac{D\mathbf{T}}{dt} = \nabla_W \mathbf{T}, \quad W = \dot\sigma(t)$

$$\frac{d}{dt}\mathrm{T}^{A\cdots D}{}_{E\cdots G}(\sigma(t)) + \mathrm{T}^{LB\cdots D}{}_{E\cdots G}\Gamma^A_{LK}W^K + (\text{all upper indices})$$
$$- \mathrm{T}^{A\cdots D}{}_{L\cdots G}\Gamma^L_{EK}W^K - (\text{all lower indices})$$

Covariant Derivative of a Two-Point Tensor **T** *of Type* $\begin{pmatrix} p & l \\ q & m \end{pmatrix}$

$$\nabla_W \mathbf{T} = \frac{d}{ds} S_{t,s} \mathbf{T}(\sigma(s))\Big|_{s=t}$$

$S_{t,s}$ = shifter

$$\mathrm{T}^{A\cdots D}{}_{E\cdots H}{}^{a\cdots d}{}_{e\cdots h|K} = \frac{\partial}{\partial X^K} \mathrm{T}^{A\cdots D}{}_{E\cdots H}{}^{a\cdots d}{}_{e\cdots h}$$
$$+ \mathrm{T}^{L\cdots D}{}_{E\cdots H}{}^{a\cdots d}{}_{e\cdots h} \Gamma^A_{LK} + \text{(all upper large indices)}$$
$$- \mathrm{T}^{A\cdots D}{}_{L\cdots H}{}^{a\cdots d}{}_{e\cdots h} \Gamma^L_{EK} - \text{(all lower large indices)}$$
$$+ \mathrm{T}^{A\cdots D}{}_{E\cdots H}{}^{k\cdots d}{}_{e\cdots h} \gamma^a_{kl} F^l{}_K + \text{(all upper small indices)}$$
$$- \mathrm{T}^{A\cdots D}{}_{E\cdots H}{}^{a\cdots d}{}_{k\cdots h} \gamma^k_{el} F^l{}_K - \text{(all lower small indices)}$$

Divergence

DIV **T** = contraction of $\nabla\mathbf{T}$ on last contravariant index

$$(\mathrm{DIV}\,\mathbf{T})^{A\cdots C}{}_{E\cdots H}{}^{a\cdots d}{}_{e\cdots h} = \mathrm{T}^{A\cdots CD}{}_{E\cdots H}{}^{a\cdots d}{}_{e\cdots h|D}$$

Curvature Tensor

$$\mathbf{R}(\boldsymbol{\alpha}, W_1, W_2, W_3) = \boldsymbol{\alpha}(\nabla_{W_1}\nabla_{W_2}W_3 - \nabla_{W_2}\nabla_{W_1}W_3 - \nabla_{[W_1,W_2]}W_3)$$

$$\mathrm{R}^A{}_{BCD} = \frac{\partial \Gamma^A_{DB}}{\partial X^C} - \frac{\partial \Gamma^A_{CB}}{\partial X^D} + \Gamma^A_{CE}\Gamma^E_{DB} - \Gamma^A_{DE}\Gamma^E_{CB}$$

Compatibility Conditions

Curvature of $C^\flat = \phi^* g$ is zero.

$\mathrm{K}^A{}_{BCD} = 0$, where $\mathrm{K}^A{}_{BCD}$ is the curvature of C_{AB}.

1.5 CONSERVATION OF MASS

We shall use conservation of mass to motivate the geometric ideas in Sections 1.6 and 1.7 and the general balance principles in Section 2.1. We begin with regular motions of simple bodies and then, as a supplement, treat the case of (thin) shells; that is, when $\mathcal{B}$ is a two-manifold, $\mathcal{S} = \mathbb{R}^3$ and the motion consists of embeddings.

We shall begin right away by assuming the existence of a mass density function ρ. We could alternatively assume the existence of a mass measure m and then, assuming m is sufficiently regular, derive ρ by writing $m = \rho\, dv$ (we may use either the Radon–Nikodym theorem or differential forms to do this). We shall bypass the measure theoretic approach, although it is based on the more primitive physical concept of measuring the masses of portions of a body. The reasons are twofold: first, it plays no role in the rest of the book, and second, it is "obvious" how to bridge the gap for those who know measure theory.[22]

5.1 Definition Let $\mathcal{B} \subset \mathbb{R}^n$ be a simple body and let $\phi(X, t)$ be a motion of $\mathcal{B}$. A function $\rho(x, t)$ is said to obey *conservation of mass* provided that for all open sets $\mathcal{U} \subset \mathcal{B}$ with piecewise C^1 boundary (such $\mathcal{U}$ hereafter will be called *nice*),

$$\frac{d}{dt}\int_{\phi_t(\mathcal{U})} \rho(x, t)\, dv = 0$$

where dv denotes the standard Euclidean volume element in $\mathbb{R}^n$.

We call $\rho(x, t)$ the *mass density* (in spatial coordinates) and call $\int_{\mathcal{V}} \rho(x, t)\, dv$ the *mass* of the set $\mathcal{V}$.

Conservation of mass thus states that the mass of any nice material region $\mathcal{U}$ is constant in time. Let us assume that ϕ_t is a regular C^1 motion from now on, and $\phi_0 =$ identity; that is, the body is in the reference configuration at $t = 0$.

5.2 Definition Let $J(X, t)$, called the *Jacobian*, denote the determinant of the linear transformation $F(X, t) = D\phi_t(X)$.

Let us work out $J(X, t)$ in general coordinates.

5.3 Proposition *In (positively oriented)*[23] *coordinates* $\{X^A\}$ *on* $\mathcal{B}$ *and* $\{x^a\}$ *on* $\mathcal{S}$ *we have*

$$\boxed{J(X, t) = \frac{\partial(\phi^1, \ldots, \phi^n)}{\partial(X^1, \ldots, X^n)} \frac{\sqrt{\det g_{ab}(x)}}{\sqrt{\det G_{AB}(X)}},}$$

[22]See, for example, Truesdell [1977].

[23]Coordinates $\{X^A\}$ in $\mathbb{R}^3$ are called positively oriented when $\det(\partial X^A/\partial Z^I) > 0$.

where $\quad x = \phi(X, t) \quad$ *and* $\quad \dfrac{\partial(\phi^1, \ldots, \phi^n)}{\partial(X^1, \ldots, X^n)} = \det\left(\dfrac{\partial \phi^a}{\partial X^A}\right).$

Proof Using Cartesian coordinates, we get

$$J(X, t) = \det\left(\frac{\partial \phi^i}{\partial Z^I}\right) = \det\left[\left(\frac{\partial \phi^a}{\partial X^A}\right)\left(\frac{\partial z^i}{\partial x^a}\right)\left(\frac{\partial X^A}{\partial Z^I}\right)\right] = \det\left(\frac{\partial \phi^a}{\partial X^A}\right)\det\left(\frac{\partial z^i}{\partial x^a}\right)\det\left(\frac{\partial X^A}{\partial Z^I}\right).$$

But

$$\det(g_{ab}) = \det\left(\frac{\partial z^i}{\partial x^a}\frac{\partial z^j}{\partial x^b}\delta_{ij}\right) = \left[\det\left(\frac{\partial z^i}{\partial x^a}\right)\right]^2,$$

with a similar formula for $\det(G_{AB})$. Hence the result follows. ∎

Warning. J is a scalar function, invariant under coordinate transformations. Note that $J \neq \det(F^a{}_A)$; the quantity $\det(F^a{}_A)$ picks up determinant factors under coordinate changes and is sometimes called a "modular" tensor. The factor $(\det g_{ab}/\det G_{AB})^{1/2}$ in the formula for J corrects this, so J is a scalar.

For what follows, we shall need to note that $J(X, t) > 0$. Indeed, $J(X, 0) = 1$, and since ϕ_t is regular, $J(X, t) \neq 0$. Hence $J(X, t)$ is positive by the intermediate value theorem.

5.4 Proposition

$$\boxed{\frac{\partial J}{\partial t} = (\operatorname{div} v)J = (v^a{}_{|a})J.}$$

Proof We shall prove this by a direct calculation. For a proof without coordinates, see Section 1.7. The reader can easily prove the following fact from matrix algebra: if $(a_{ij}(t))$ is a time-dependent matrix, then

$$\frac{d}{dt}\det(a_{ij}(t)) = \frac{da_{ij}}{dt}(\text{Cof})^{ij}$$

(sum on both i and j), where $(\text{Cof})^{ij}$ is the (i, j)th cofactor of (a_{ij}). Therefore,

$$\begin{aligned}\frac{\partial}{\partial t}J(X, t) &= \left(\frac{\partial}{\partial t}\left(\frac{\partial \phi^a}{\partial X^A}\right)(\text{Cof})_a{}^A\right)\sqrt{\frac{\det g_{cd}}{\det G_{CD}}} + \frac{\partial(\phi^1, \ldots, \phi^n)}{\partial(X^1, \ldots, X^n)}\frac{\partial}{\partial t}\sqrt{\frac{\det g_{cd}}{\det G_{CD}}} \\ &= \frac{\partial v^a}{\partial x^b}\frac{\partial \phi^b}{\partial X^A}(\text{Cof})_a{}^A\sqrt{\frac{\det g_{cd}}{\det G_{CD}}} \\ &\quad + \frac{\partial(\phi^1, \ldots, \phi^n)}{\partial(X^1, \ldots, X^n)}\frac{1}{\sqrt{\det G_{CD}}}\frac{\partial}{\partial x^a}(\sqrt{\det g_{cd}})v^a.\end{aligned}$$

The first sum is a sum of determinants with the ath row of $\partial \phi^a/\partial X^A$ replaced by $(\partial v^a/\partial x^b)(\partial \phi^b/\partial X^A)$. Since repeated rows yield a zero determinant, only the term with $b = a$ survives. Thus

$$\begin{aligned}\frac{\partial}{\partial t}J(X,t) &= \frac{\partial v^a}{\partial x^a}J(X,t) + J(X,t)\frac{1}{\sqrt{\det g_{cd}}}\frac{\partial}{\partial x^a}(\sqrt{\det g_{cd}})v^a \\ &= J(X,t)\frac{1}{\sqrt{\det g_{cd}}}\frac{\partial}{\partial x^a}(\sqrt{\det g_{cd}}\, v^a) \\ &= J(X,t)v^a{}_{|a}. \quad \blacksquare\end{aligned}$$

This proof will be thoroughly understood if the reader will repeat it explicitly for $\mathbb{R}^3$, using Cartesian coordinates and the formula

$$\frac{d}{dt}\begin{vmatrix} a_{11} & a_{12} & a_{13} \\ a_{21} & a_{22} & a_{23} \\ a_{31} & a_{32} & a_{33}\end{vmatrix} = \begin{vmatrix} \dot a_{11} & \dot a_{12} & \dot a_{13} \\ a_{21} & a_{22} & a_{23} \\ a_{31} & a_{32} & a_{33}\end{vmatrix} + \begin{vmatrix} a_{11} & a_{12} & a_{13} \\ \dot a_{21} & \dot a_{22} & \dot a_{23} \\ a_{31} & a_{32} & a_{33}\end{vmatrix} + \begin{vmatrix} a_{11} & a_{12} & a_{13} \\ a_{21} & a_{22} & a_{23} \\ \dot a_{31} & \dot a_{32} & \dot a_{33}\end{vmatrix}.$$

5.5 Definition A motion ϕ_t is called *volume preserving* (also called isochoric or incompressible) if volume$[\phi_t(\mathcal{U})]$ = volume$[\mathcal{U}]$ for every nice region $\mathcal{U} \subset \mathcal{B}$.

5.6 Proposition *Assume ϕ_t is a C^1 regular motion. Then the following are equivalent:*

(i) *ϕ_t is volume preserving;*
(ii) $J(X, t) = 1$; *and*
(iii) $\operatorname{div} v = 0$.

Proof (i) is equivalent to (ii) by the change of variables formula, and (ii) is equivalent to (iii) by 5.4 and $J(X, 0) = 1$. $\blacksquare$

Next we use 5.4 to establish the basic *equation of continuity*. We write $\rho_{\text{Ref}}(X) = \rho(X, 0)$, the mass density in the reference configuration (assumed to occur at $t = 0$). Recall that $\dot\rho = \partial\rho/\partial t + d\rho\cdot v = \partial\rho/\partial t + v[\rho]$ is the material derivative of ρ.

5.7 Theorem *Assume ϕ_t is a C^1 regular motion and that $\rho(x, t)$ is a C^1 function. Then the following are equivalent:*

(i) *Conservation of Mass;*
(ii) $\rho(x, t)J(X, t) = \rho_{\text{Ref}}(X)$ [*where* $x = \phi(X, t)$]; *and*
(iii) *the equation of continuity,*

$$\boxed{\dot\rho + \rho \operatorname{div} v = 0; \qquad \text{that is,} \quad \frac{\partial\rho}{\partial t} + \operatorname{div}(\rho v) = 0.}$$

Proof Assume (i); that is, $\int_{\phi_t(\mathcal{U})} \rho(x, t)\, dv = \int_{\mathcal{U}} \rho_{\text{Ref}}(X)\, dV$. By the change of variables formula this is equivalent to $\int_{\mathcal{U}} \rho(x, t)J(X, t)\, dV = \int_{\mathcal{U}} \rho_{\text{Ref}}(X)\, dV$. Since

$\mathcal{U}$ is arbitrary, this is equivalent to (ii). (The proof employs an easy calculus exercise: if $f(x)$ is a continuous function and $\int_{\mathfrak{D}} f(x)\,dv = 0$ for every ball $\mathfrak{D}$, then $f = 0$.) By 5.4, $(\partial/\partial t)(\rho(x, t)J(X, t) = \dot{\rho}J + \rho\dot{J} = \dot{\rho}J + (\rho \operatorname{div} v)J$, so (ii) is equivalent to (iii). ∎

The process of passing from (i) to (iii) is called *localization.* In the volume preserving case, the equation of continuity reduces to $\dot{\rho} = 0$; that is, $\rho(x, t) = \rho(X, 0)$.

Problem 5.1 Omitting the assumption that $\phi_0 =$ identity, show that 5.7 remains true for *some* function ρ_{Ref} on $\mathcal{B}$.

Next we investigate the geometry behind conservation of mass a little further. We want to make dv and dV, the volume elements on $\mathcal{S}$ and $\mathcal{B}$, into respectable tensors, $\boldsymbol{dv}$ and $\boldsymbol{dV}$ respectively.

Warning. These notations should not be confused with the spatial and material velocities.

5.8 Definition The *volume form*, regarded as an antisymmetric $\binom{0}{n}$ tensor, is defined by $\boldsymbol{dv}(w_1, \ldots, w_n) = \sqrt{\det\langle w_i, w_j\rangle}$ for $w_1, \ldots, w_n$ positively oriented; that is, $\det\langle w_i, w_j\rangle > 0$ and is extended to all $w_1, \ldots, w_n$ by skew symmetry and multilinearity.

Recall that the volume of the parallelopiped spanned by n vectors $w_1, \ldots, w_n$ in $\mathbb{R}^n$ is $\sqrt{\det\langle w_i, w_j\rangle}$. The tensor $\boldsymbol{dv}$ is completely antisymmetric to reflect the same property of the determinant: it vanishes if any $w_i = \mathbf{0}$ and therefore if $\boldsymbol{dv}$ is to be linear in each w_i, it must change sign if two w_i's are interchanged.

In terms of a general (oriented) coordinate system in $\mathbb{R}^3$, 5.8 reads

$$\boldsymbol{dv} = \sqrt{\det g_{ab}}\,(\boldsymbol{dx}^1 \otimes \boldsymbol{dx}^2 \otimes \boldsymbol{dx}^3)_{\text{antisymmetrized}} \equiv \sqrt{\det g_{ab}}\,\boldsymbol{dx}^1 \wedge \boldsymbol{dx}^2 \wedge \boldsymbol{dx}^3.$$

(Section 1.7 elaborates on this notation.)

5.9 Definition The *mass form* is defined by $\mathsf{m} = \rho\,\boldsymbol{dv}$ (each of m and ρ depends on (x, t)). Also, let $\mathsf{m}_{\text{Ref}} = \rho_{\text{Ref}}\,\boldsymbol{dV}$, the mass form of the reference configuration.

Using the mass form we can express conservation of mass entirely in terms of pull-backs.

5.10 Proposition *Conservation of mass is equivalent to*

$$\boxed{\phi_t^*\mathsf{m} = \mathsf{m}_{\text{Ref}}.}$$

Proof $\phi^*\mathsf{m} = \phi^*(\rho\,\boldsymbol{dv}) = (\rho\circ\phi)\phi^*\,\boldsymbol{dv}$. (The pull-back of a product is the product of the pull-backs.) We claim that $\phi^*\,\boldsymbol{dv} = J\,\boldsymbol{dV}$, from which it follows that $\phi^* \mathbf{m} = \mathbf{m}_{\text{Ref}}$ is equivalent to $\rho(x, t)J(X, t) = \rho_{\text{Ref}}(X)$—that is, conservation of mass. To prove the claim we use the definitions of pull-back and dv:

$$\begin{aligned}(\phi^* dv)(W_1, W_2, W_3) &= dv(D\phi_t \cdot W_1, D\phi_t \cdot W_2, D\phi_t \cdot W_3)\\ &= [\det\langle D\phi_t \cdot W_i, D\phi_t \cdot W_j\rangle]^{1/2}\\ &= \left[\det\left(g_{ab}\frac{\partial \phi^a}{\partial X^A}W_i^A\frac{\partial \phi^b}{\partial X^B}W_j^B\right)\right]^{1/2}\\ &= \sqrt{\det g_{ab}}\cdot\frac{\partial(\phi^1, \phi^2, \phi^3)}{\partial(X^1, X^2, X^3)}\cdot\sqrt{\frac{\det(G_{AB}W_i^A W_j^B)}{\det(G_{AB})}}\\ &= J(t, X)\, dV(W_1, W_2, W_3). \quad \blacksquare\end{aligned}$$

These ideas are concisely treated using differential forms and the Lie derivative in Section 1.7.

Problem 5.2 Write out the equation of continuity in spherical coordinates in $\mathbb{R}^3$.

Box 5.1 *Conservation of Mass for Shells and the Second Fundamental Form*[24]

The expressions for dv and dV make sense on an arbitrary Riemannian manifold as does the definition of conservation of mass. The equivalence with $\rho J = \rho_{\text{Ref}}$ and $\phi^* \mathbf{m} = \mathbf{m}_{\text{Ref}}$ remains valid. However, the equation of continuity requires revision; indeed, it does not make sense as stated, since div v is not always defined. In fact, v is a vector tangent to $\mathcal{S}$ defined at points of $\phi_t(\mathcal{B})$, so if $\mathcal{B}$ and $\mathcal{S}$ have different dimensions, div v does not make sense.

By definition, a *shell* is a body $\mathcal{B}$ that is a two-manifold in $\mathbb{R}^3 = \mathcal{S}$. Physically, it is a sheet whose thickness is being ignored. The correct form of the equation of continuity depends on the second fundamental form of an embedded hypersurface (in our case, the shell), so we consider first some relevant geometry.

5.11 Definition Let $\mathfrak{M}$ and $\mathfrak{N}$ be Riemannian manifolds, $\mathfrak{M} \subset \mathfrak{N}$ with $\dim \mathfrak{N} = \dim \mathfrak{M} + 1$. The Riemannian structure on $\mathfrak{M}$ is assumed to equal that obtained from $\mathfrak{N}$. The *second fundamental form* of $\mathfrak{M}$ is the $\binom{0}{2}$ tensor $\boldsymbol{k}$ on $\mathfrak{M}$ such that

$$\boxed{\boldsymbol{k}(W_1, W_2) = \langle \nabla_{W_1}\boldsymbol{n}, W_2\rangle_X,}$$

[24] The standard reference for shells is Naghdi [1972].

where $\boldsymbol{\nabla}$ is the covariant derivative on $\mathfrak{N}$, $\boldsymbol{W}_1, \boldsymbol{W}_2 \in T_X\mathfrak{M}$, and $\boldsymbol{n}$ is the unit outward normal of $\mathfrak{M}$. (We assume $\mathfrak{M}$ is oriented; that is, it has a well defined "outward" normal.)

5.12 Proposition *We have*

(i) $\boldsymbol{k}(\boldsymbol{W}_1, \boldsymbol{W}_2) = -\langle \boldsymbol{\nabla}_{\boldsymbol{W}_1}\boldsymbol{W}_2, \boldsymbol{n}\rangle$, (*Weingarten equation*)
(ii) $\boldsymbol{k}$ *is a symmetric tensor, and*
(iii) *in coordinates* $(x^1, \ldots, x^{n+1})$ *for which* x^{n+1} *is the outward normal direction to* $\mathfrak{M}$, *and such that* $\boldsymbol{e}_{n+1} = \boldsymbol{n}$,

$$k_{ab} = -\gamma_{ab}^{n+1} = \frac{1}{2}\frac{\partial g_{ab}}{\partial x^{n+1}} \quad (1 \leq a, b \leq n).$$

Proof (i) Since $\boldsymbol{W}_2$ is parallel to $\mathfrak{M}$, $\langle \boldsymbol{W}_2, \boldsymbol{n}\rangle = 0$. Hence

$$0 = \boldsymbol{\nabla}_{\boldsymbol{W}_1}\langle \boldsymbol{W}_2, \boldsymbol{n}\rangle = \langle \boldsymbol{\nabla}_{\boldsymbol{W}_1}\boldsymbol{W}_2, \boldsymbol{n}\rangle + \langle \boldsymbol{W}_2, \boldsymbol{\nabla}_{\boldsymbol{W}_1}\boldsymbol{n}\rangle.$$

(ii) follows from (iii), so we prove (iii); indeed, by (i),

$$k_{ab} = -\langle \boldsymbol{\nabla}_{\boldsymbol{e}_a}\boldsymbol{e}_b, \boldsymbol{n}\rangle = -\gamma_{ab}^{c}\delta_c^{n+1} = -\gamma_{ab}^{n+1}.$$

Using the explicit formula for γ_{bc}^{a} in terms of g_{ab} (see the proof of 4.24 or 3.29) yields the last equality. ∎

For a surface in space, $\kappa = \operatorname{tr} \boldsymbol{k}$, the trace of $\boldsymbol{k}$, is called the *mean curvature* (sum of the principal curvatures; the eigenvalues of $\boldsymbol{k}$) and $\det \boldsymbol{k}$ (product of the principal curvatures) is called the *Gaussian curvature*. [*Aside*: The second fundamental form provides the link between the connections and curvatures on $\mathfrak{M}$ and $\mathfrak{N}$, through the Gauss–Codazzi equations:

(i) ${}^{(\mathfrak{N})}\mathsf{R}_{abcd} = {}^{(\mathfrak{M})}\mathsf{R}_{abcd} + k_{bd}k_{ac} - k_{ad}k_{bc} \quad (1 \leq a, b, c, d \leq n)$,
(ii) ${}^{(\mathfrak{N})}\boldsymbol{\nabla}_{\boldsymbol{W}_1}\boldsymbol{W}_2 - {}^{(\mathfrak{M})}\boldsymbol{\nabla}_{\boldsymbol{W}_1}\boldsymbol{W}_2 = -\, \boldsymbol{k}(\boldsymbol{W}_1, \boldsymbol{W}_2)\boldsymbol{n}$.

These are not hard to prove: for these and related formulas, see any book in Riemannian geometry, such as Yano [1970].]

Now we return to our main concern—conservation of mass. As in 5.7, the equation of continuity boils down to a computation of $\partial J/\partial t$. The definition of J in general is $J\, \boldsymbol{dV} = \boldsymbol{\phi}^*\, \boldsymbol{dv}$, where $\boldsymbol{dv}$ is the volume element on $\phi_t(\mathfrak{B})$; 5.3 remains valid. Also, $\boldsymbol{\phi}^*\, \boldsymbol{dv}$ = volume element of $(\boldsymbol{\phi}^*\boldsymbol{g} = \boldsymbol{C}^\flat)$ from the definition of pull-back and 5.8. Write $\boldsymbol{\mu}(\boldsymbol{\phi}^*\boldsymbol{g})$ for the volume element of $\boldsymbol{\phi}^*\boldsymbol{g}$. By the computation in 5.4,

$$\frac{d}{dt}\det(g_{ab}(t)) = \operatorname{tr}\left(\frac{\partial g_{ab}}{\partial t}\right)\det(g_{ab}).$$

Thus $$\frac{\partial J}{\partial t}\boldsymbol{dV} = \tfrac{1}{2}\boldsymbol{\mu}(\boldsymbol{\phi}^*\boldsymbol{g})\operatorname{tr}\left(\frac{\partial}{\partial t}\boldsymbol{\phi}_t^*\boldsymbol{g}\right) = \frac{1}{2}\operatorname{tr}\left(\frac{\partial}{\partial t}\boldsymbol{\phi}_t^*\boldsymbol{g}\right)\boldsymbol{\phi}^*\,\boldsymbol{dv}.$$

This calculation proves the following abstract result:

5.13 Proposition *Let $\mathcal{B}$ and $\mathcal{S}$ be general Riemannian manifolds and ϕ_t an embedded C^1 motion of $\mathcal{B}$ in $\mathcal{S}$. Then*

$$\frac{\partial J}{\partial t} = \frac{1}{2}\,\mathrm{tr}_C\left(\frac{\partial}{\partial t}\phi_t^* g\right)J = D_{AB}B^{AB}J,$$

where the trace is taken using the metric $\phi_t^* g$ and $\boldsymbol{B}^\sharp = (\boldsymbol{C}^\flat)^{-1}$.

To apply this to shells, let $\boldsymbol{v}$ be the spatial velocity and $\boldsymbol{n}$ the unit normal. Decompose $\boldsymbol{v}$ into a component parallel to $\phi_t(\mathcal{B})$ and a component normal to it (see Figure 1.5.1):

$$\boldsymbol{v} = \boldsymbol{v}_\parallel + v_n \boldsymbol{n}.$$

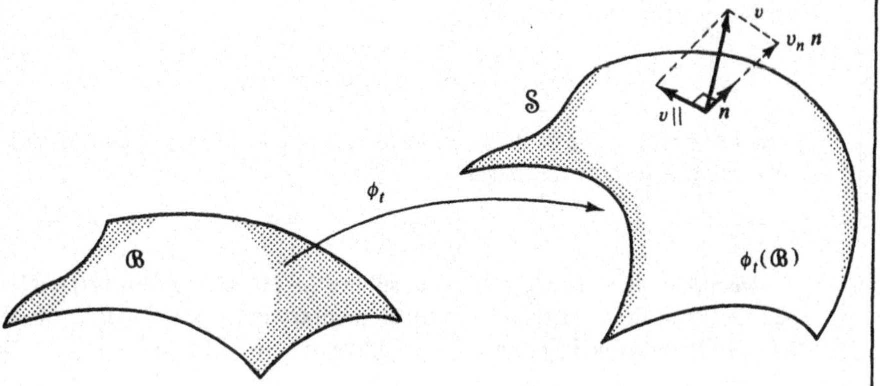

Figure 1.5.1

We must compute $(\partial/\partial t)\phi_t^* g$, that is, $(\partial/\partial t)C_{AB}$. Write $2\boldsymbol{D}^\flat = (\partial/\partial t)\phi_t^* g$ as before.

5.14 Proposition *For shells, the rate of deformation tensor is*

$$2\boldsymbol{D}^\flat = \phi_t^*((\nabla \boldsymbol{v}_\parallel)^\flat + [(\nabla \boldsymbol{v}_\parallel)^\flat]^T + 2v_n \boldsymbol{k}),$$

that is, $\quad 2D_{AB} = (v^\parallel_{a|b} + v^\parallel_{b|a})F^a{}_A F^b{}_B + 2v_n k_{ab} F^a{}_A F^b{}_B,$

where $\boldsymbol{k}$ is the second fundamental form.

Proof $C_{AB} = g_{ab}F^a{}_A F^b{}_B$, so, as in 3.28,

$$2D_{AB} = \left(\frac{\partial g_{ab}}{\partial x^c}v^c + g_{cb}\frac{\partial v^c}{\partial x^a} + g_{ac}\frac{\partial v^c}{\partial x^b}\right)F^a{}_A F^b{}_B$$
$$= (v_{a|b} + v_{b|a})F^a{}_A F^b{}_B.$$

But $\nabla \boldsymbol{v} = \nabla(\boldsymbol{v}_\parallel + v_n \boldsymbol{n}) = \nabla \boldsymbol{v}_\parallel + \nabla v_n \otimes \boldsymbol{n} + v_n \nabla \boldsymbol{n}$. The covariant derivative here is on $\mathcal{S}$. Since $\boldsymbol{n}$ is orthogonal to $\phi_t(\mathcal{B})$, the second term pulls back to zero. Since $(\nabla \boldsymbol{n})^\flat = \boldsymbol{k}$, and $\boldsymbol{k}$ is symmetric, we get the result. ∎

We shall write $\bar{\nabla}$ for the covariant derivative on $\phi_t(\mathcal{B})$ to avoid confusion with the covariant derivative ∇ on $\mathcal{S}$. However, $\nabla \boldsymbol{v}_\parallel$ and $\bar{\nabla} \boldsymbol{v}_\parallel$ differ by $\boldsymbol{k}(\boldsymbol{v}_\parallel, \cdot)\boldsymbol{n}$, which is normal to $\phi_t(\mathcal{B})$ and hence has zero pull-back to $\mathcal{B}$. Thus $\phi_t^*(\nabla \boldsymbol{v}_\parallel) = \phi_t^*(\bar{\nabla} \boldsymbol{v}_\parallel)$. Since the divergence of $\boldsymbol{v}_\parallel$ on $\phi_t(\mathcal{B})$ is given by $\overline{\operatorname{div}}\, \boldsymbol{v}_\parallel = \operatorname{tr}(\bar{\nabla} \boldsymbol{v}_\parallel)$, 5.14 yields

$$\operatorname{tr}_C \boldsymbol{D} = \phi_t^*(\overline{\operatorname{div}}\, \boldsymbol{v}_\parallel) + \phi_t^*(v_n \operatorname{tr} \boldsymbol{k}) = (\overline{\operatorname{div}}\, \boldsymbol{v}_\parallel) \circ \phi_t + (v_n \operatorname{tr} \boldsymbol{k}) \circ \phi_t.$$

Substituting in 5.13 using $\rho J = \rho_{\text{Ref}}$ gives the following:

5.15 Theorem *Conservation of mass for shells is equivalent to the equation of continuity for shells:*

$$\boxed{\dot{\rho} + \rho\, \overline{\operatorname{div}}\, \boldsymbol{v}_\parallel + \rho v_n \operatorname{tr} \boldsymbol{k} = 0}$$

where $\dot{\rho} = \partial \rho / \partial t + \nabla \rho \cdot \boldsymbol{v}$, $\boldsymbol{k}$ *is the second fundamental form of the surface* $\phi_t(\mathcal{B})$, *and* $\operatorname{tr} \boldsymbol{k}$ *is its mean curvature.*

Problem 5.3 Suppose $\phi_t(\mathcal{B})$ is a sphere in $\mathbb{R}^3$ with radius $r(t)$. Given ρ_{Ref}, derive a formula for $\rho(x, t)$ by elementary considerations. Verify that 5.15 holds by explicitly computing $\boldsymbol{k}$ for a sphere of radius $r(t)$.

Ans: $k_{ab} = \begin{bmatrix} r & 0 \\ 0 & r \sin^2 \phi \end{bmatrix}$ in spherical coordinates.

Problem 5.4 Derive the equation of continuity for a wire—that is, when $\mathcal{B} \subset \mathbb{R}^3$ is a 1-manifold. You will need the Frénet formulas from a geometry text.

Problem 5.5 If $\mathcal{B} \subset \mathcal{S}$ is open, but the metric on $\mathcal{S}$ is time dependent, show that the equation of continuity becomes

$$\dot{\rho} + \rho \operatorname{div} \boldsymbol{v} + \frac{1}{2} \operatorname{tr}\left(\frac{\partial g}{\partial t}\right) = 0.$$

Box 5.2 ***Summary of Important Formulas for Section 1.5***

Conservation of Mass for Simple Bodies

$$\frac{d}{dt} \int_{\phi_t(\mathcal{U})} \rho(x, t)\, dv = 0 \qquad (\mathcal{U} \subset \mathcal{B})$$

Jacobian $J(X, t)$	
$\phi^* \, dv = J \, dV$	$J = \dfrac{\partial(\phi^1, \ldots, \phi^n)}{\partial(X^1, \ldots, X^n)} \sqrt{\dfrac{\det g_{ab}}{\det G_{AB}}}$, where $\dfrac{\partial(\phi^1, \ldots, \phi^n)}{\partial(X^1, \ldots, X^n)} = \det\left(\dfrac{\partial \phi^a}{\partial X^A}\right)$
Volume Element	
$dv(w_1, \ldots, w_n) = \sqrt{\det\langle w_i, w_j\rangle}$	$dv = \sqrt{\det g_{ab}} \, dx^1 \wedge \cdots \wedge dx^n$
Mass Form	
$\mathsf{m} = \rho \, dv$	$\mathsf{m} = \rho\sqrt{\det g_{ab}} \, dx^1 \wedge \cdots \wedge dx^n$
Conservation of Mass	
$\phi_t^* \mathsf{m} = \mathsf{m}_{\text{Ref}}$	$(\rho \circ \phi) J = \rho_{\text{Ref}}$
Rate of Change of Jacobian	
Simple Bodies:	
$\dfrac{\partial J}{\partial t} = (\operatorname{div} v) J$	$\dfrac{\partial J}{\partial t} = (v^a{}_{\vert a}) J$
Shells:	
$\dfrac{\partial J}{\partial t} = \dfrac{1}{2} \operatorname{tr}_c\left(\dfrac{\partial}{\partial t} \phi_t^* g\right) J = (\overline{\operatorname{div}}\, v_\parallel + v_n \operatorname{tr} k) J$; $v = v_\parallel + v_n n$	$\dfrac{\partial J}{\partial t} = ((v_\parallel)^a{}_{\vert a} + v_n k^a{}_a) J$, where coordinates are on $\phi_t(\mathcal{B})$ $(a = 1, 2)$; $\begin{cases} v^a = v_\parallel^a & (a = 1, 2) \\ v^3 = v_n \end{cases}$ and the third coordinate is orthogonal to $\phi_t(\mathcal{B})$.
Second Fundamental Form:	
$k(W_1, W_2) = \langle \nabla_{W_1} n, W_2 \rangle = -\langle \nabla_{W_1} W_2, n \rangle$	$k_{ab} = -\gamma_{ab}^{n+1} = \dfrac{1}{2} \dfrac{\partial g_{ab}}{\partial x^{n+1}}$
Equation of Continuity	
Simple Bodies:	
$\dot{\rho} + \rho \operatorname{div} v = 0$	$\dfrac{\partial \rho}{\partial t} + \dfrac{\partial \rho}{\partial x^a} v^a + \rho v^a{}_{\vert a} = 0$
Shells:	
$\dot{\rho} + \rho(\overline{\operatorname{div}}\, v_\parallel + v_n \operatorname{tr} k) = 0$	$\dfrac{\partial \rho}{\partial t} + \dfrac{\partial \rho}{\partial x^a} v^a + \rho(v^a_{\parallel \vert a} + v_n k^a{}_a) = 0$

1.6 FLOWS AND LIE DERIVATIVES

We shall begin by defining the notion of a tangent vector to a curve c within the context of manifold theory. Heretofore the tangent has been denoted dc/dt. Strictly speaking, this is not an appropriate notation for a tangent vector in a manifold since it omits the base point of the vector. If $\mathcal{S}$ is an n-dimensional differentiable manifold and $c\colon \mathcal{I} \longrightarrow \mathcal{S}$ is a smooth curve, where $\mathcal{I}$ is an open interval of the real line, then in a coordinate chart $\{x^a\}$, $dc/dt = (dc^1/dt, \ldots, dc^n/dt)$, the "vector part" of the tangent vector. To develop the appropriate notion of tangent vector we apply the tangent map T to c. (The interval $\mathcal{I}$ may be viewed as a differentiable manifold and thus $Tc\colon T\mathcal{I} \longrightarrow T\mathcal{S}$.) By definition of the tangent of a map, $Tc(r, s) = (c(r), (dc/dt)(r)s)$, where $(r, s) \in \mathcal{I} \times \mathbb{R} = T\mathcal{I}$ is a tangent vector to $\mathcal{I}$. Since we ultimately want the vector part of the tangent vector to be simply dc/dt, we evaluate Tc on the unit vector based at r: $Tc(r, 1) = (c(r), (dc/dt)(r))$. Thus $Tc(\cdot, 1)\colon \mathcal{I} \longrightarrow T\mathcal{S}$ is what we meant when we discussed the tangent vector to the curve c. The use of dc/dt in place of $Tc(\cdot, 1)$ is a frequent abuse of notation in manifold theory, which we shall follow.

6.1 Definition Let $w\colon \mathcal{U} \longrightarrow T\mathcal{S}$ be a vector field, where $\mathcal{U}$ is an open subset of $\mathcal{S}$. A curve $c\colon \mathcal{I} \longrightarrow \mathcal{S}$, where $\mathcal{I}$ is an open interval, is called an *integral curve* of w if for every $r \in \mathcal{I}$

$$\frac{dc}{dt}(r) = w(c(r)).$$

6.2 Proposition *Let w be a C^k vector field on $\mathcal{U}$ (open in $\mathcal{S}$), where $k \geq 1$. Then for each $x \in \mathcal{U}$ there exists an $\epsilon > 0$ and a C^{k+1} integral curve c_x of w defined on $(-\epsilon, \epsilon)$ such that $c_x(0) = x$. Furthermore, c_x is unique in the sense that if $\tilde{c}_x$ satisfies the same conditions, but is defined on an interval $(-\tilde{\epsilon}, \tilde{\epsilon})$ $(\tilde{\epsilon} > 0)$, then $c_x = \tilde{c}_x$ on the intersection of the intervals.*

Proof Let $\{x^a\}$ be a coordinate system about a point $p \in \mathcal{S}$. Then $c_x^a = x^a(c(r))$, so by the chain rule,

$$\frac{dc_x^a}{dt}(r) = dx^a(Tc_x(r, 1)) = dx^a(w(c_x(r))) = w^a(c_x(r)),$$

for each $r \in (-\epsilon, \epsilon)$. Also, $c_x^a(0) = x^a(c_x(0)) = x^a(p)$. This is a system of n ordinary differential equations with given initial conditions. Since w is C^k, so are the w^a's (this is what w being C^k means), and thus by the fundamental existence theorem of ordinary differential equations,[25] there exist C^{k+1} functions c_x^a satisfying the differential equations and initial conditions. These functions are unique in the sense spelled out in the hypotheses. ∎

6.3 Generalization If w depends upon the "time variable" r itself—that is, if $w\colon \mathcal{U} \times (-\epsilon, \epsilon) \longrightarrow T\mathcal{S}$—we define an integral curve by

$$\boxed{\frac{dc}{dt} = w(c(t), t).}$$

[25]See any of the standard references on analysis or ordinary differential equations, for example, Marsden [1974a] or Hartman [1973].

If w depends upon t in a C^k manner, then the conclusions of the previous proposition still hold.

6.4 Example Let v: $\phi_t(\mathcal{B}) \times (-\epsilon, \epsilon) \to T\mathcal{S}$ denote the spatial (Eulerian) velocity vector field of a regular motion ϕ_t with ϕ_0 = identity. An integral curve of v through $X \in \mathcal{B} \subset \mathcal{S}$ is given by the motion; that is, $\phi_X(t)$ is an integral curve of v:

$$\frac{d\phi_X}{dt} = T\phi_X(t, 1) = \left(\phi_X(t), \frac{d}{dt}\phi_X(t)\right) = v(\phi_X(t), t)$$

and $\phi_X(0) = X$.

6.5 Definition Let w: $\mathcal{S} \times \mathcal{I} \to T\mathcal{S}$ be a C^k vector field with $\mathcal{I}$ and $\mathcal{S}$ as before. The collection of maps $\psi_{t,s}$ such that for each s and x, $t \mapsto \psi_{t,s}(x)$ is an integral curve of w and $\psi_{s,s}(x) = x$ is called the *flow* or *evolution operator* of w.

The flow may be thought of as the totality of integral curves of w. The basic existence theory implies that the flow of a C^k vector field is jointly C^k where defined. Normally one takes $\psi_{r,s}(x)$ to mean the maximally extended flow—that is, with the largest domain possible.

6.6 Example Let v_t be the spatial velocity vector field of a motion ϕ_t. Then the collection $\{\phi_{t,s} \mid \phi_{t,s} = \phi_t \circ \phi_s^{-1}\colon \phi_s(\mathcal{B}) \to \phi_t(\mathcal{B})\}$ is the flow of v. It is clear from this definition that $\phi_{t,s} \circ \phi_{s,r} = \phi_{t,r}$ and that $\phi_{t,t}$ = identity, for all $r, s, t \in \mathbb{R}$ for which the flow is defined. These are properties of flows in general that may be verified using uniqueness of integral curves.

6.7 Remark If the vector field w is time independent, then $\psi_{t,s}$ depends only on the difference $t - s$. In this case, one writes $\psi_{t-s} = \psi_{t,s}$, and the flow constitutes a local one-parameter group—that is, $\psi_t \circ \psi_s = \psi_{t+s}$.

The tangent map of $\psi_{t,s}$ is defined in the usual way, as are the push-forward and pull-back induced by $\psi_{t,s}$, by merely holding t and s fixed and regarding $\psi_{t,s}$ as a mapping of $\mathcal{S}$ to $\mathcal{S}$.

6.8 Definition Let w be a C^1 (time-dependent) vector field on $\mathcal{S}$ and let $\psi_{t,s}$ denote its flow. If $\mathbf{t}$ is a C^1 (possibly time-dependent) tensor field on $\mathcal{S}$, then the *Lie derivative*[26] of $\mathbf{t}$ with respect to w is defined by

$$\boxed{L_w \mathbf{t} = \left(\frac{d}{dt}\psi_{t,s}^* \mathbf{t}_t\right)\Big|_{t=s}.}$$

---

[26]The first uses of the Lie derivative in mechanics that we know of are Cartan [1922] and Slebodzinski [1931]. In continuum mechanics the first comprehensive reference we know of is Guo Zhong-Heng [1963].

If we hold t fixed in $\mathbf{t}_t$, we obtain the *autonomous Lie derivative*:

$$\boxed{\mathfrak{L}_w \mathbf{t} = \frac{d}{dt}(\psi^*_{t,s}\mathbf{t}_s)\Big|_{t=s}.}$$

Thus $L_w \mathbf{t} = \partial \mathbf{t}/\partial t + \mathfrak{L}_w \mathbf{t}$.

6.9 Remark Note that the Lie derivative does not depend on a metric or a connection on $\mathcal{S}$.

The differentiation d/dt in the definition of the Lie derivative makes sense since for s fixed, the pull-back of $\mathbf{t}$ is a curve of tensors in a fixed linear space.

Before enumerating general properties of the Lie derivative, we compute it in coordinates for a few simple cases.

6.10 Examples

(1) Consider the Lie derivative of a function f on $\mathcal{S}$. In this case, $(\psi^*_{t,s}f_t)(x) = f(\psi_{t,s}(x), t)$, and, therefore,

$$(L_w f)(x, t) = \left(\frac{\partial f}{\partial t}(\psi_{t,s}(x), t) + \frac{\partial f}{\partial x^a}(\psi_{t,s}(x), t)\frac{\partial \psi^a_{t,s}}{\partial t}(x)\right)\Big|_{t=s}.$$

Since $\psi_{s,s}(x) = \psi_x(s) = x$ and $\dfrac{\partial \psi^a_{t,s}}{\partial t}(x) = \dfrac{d}{dt}\psi^a_x(t) = w^a(\psi_x(t), t)$, we have

$$(L_w f)(x, t) = \frac{\partial f}{\partial t}(x, t) + \frac{\partial f}{\partial x^a}(x, t)w^a(x, t),$$

that is, $L_w f = \partial f/\partial t + w[f]$.

(2) Let $\boldsymbol{\alpha}$ be a one-form. Then

$$(\psi^*_{t,s}\boldsymbol{\alpha}_t)(x) = \boldsymbol{\alpha}_a(\psi_{t,s}(x), t)\frac{\partial \psi^a_{t,s}}{\partial x^b}(x)e^b(x).$$

Thus $$(L_w\boldsymbol{\alpha})(x, t) = \left[\left(\frac{\partial}{\partial t}\alpha_a + \frac{\partial \alpha_a}{\partial x^c}\frac{\partial \psi^c_{t,s}}{\partial t}\right)\frac{\partial \psi^a_{t,s}}{\partial x^b} + \alpha_a\frac{\partial^2 \psi^a_{t,s}}{\partial t\,\partial x^b}\right]\Big|_{t=s} e^b.$$

Note that from $\psi_{s,s}(x) = x$, we have $(\partial \psi^a_{s,s}/\partial x^b)(x) = (\partial x^a/\partial x^b)(x) = \delta^a{}_b$. We also need to note that if w is a C^1 vector field,

$$\frac{\partial^2 \psi^a_{t,s}}{\partial t\,\partial x^b} = \frac{\partial}{\partial x^b}\frac{\partial \psi^a_{t,s}}{\partial t} = \frac{\partial w^a}{\partial x^b}.$$

[This uses a general fact: if $f(x, y)$ is a C^1 function of two variables and $\partial f/\partial x$ is also C^1, then the mixed partial derivatives exist and commute, and $\partial^2 f/\partial x\,\partial y$ is continuous.[27]] Combining these results enables us to write

$$L_w\boldsymbol{\alpha} = \left(\frac{\partial \alpha_a}{\partial t} + \frac{\partial \alpha_a}{\partial x^b}w^b + \alpha_b\frac{\partial w^b}{\partial x^a}\right)e^a.$$

Note that throughout, the basis one-form e^a is fixed at x and thus is un-

[27]For a proof, see, for example, Apostol [1974].

affected by the differentiation. (This is another way of understanding that a metric structure of $\mathcal{S}$ is not involved in the notion of Lie derivative.) Also observe that $\boldsymbol{L}_w\boldsymbol{\alpha}$ depends on the values of w in a neighborhood of the point in question since $\boldsymbol{L}_w\boldsymbol{\alpha}$ involves partial derivatives of the components of w. This is to be contrasted with covariant differentiation, for which $\nabla_w\boldsymbol{\alpha}$ depends solely on the value of w at the point in question.

(3) Finally, we consider the case of a vector field $\boldsymbol{v}$. Here

$$(\psi_{t,s}^*\boldsymbol{v}_t)(x) = v^a(\psi_{t,s}(x), t)\frac{\partial(\psi_{t,s}^{-1})^b}{\partial x^a}(\psi_{t,s}(x))\boldsymbol{e}_b(x).$$

To compute the Lie derivative we use the formula

$$\frac{\partial}{\partial t}\left(\frac{\partial(\psi_{t,s}^{-1})^b}{\partial x^a}(\psi_{t,s}(x))\right)\bigg|_{t=s} = -\frac{\partial w^b}{\partial x^a}$$

which is obtained in the same way as computing the derivative of the inverse of a matrix depending upon a parameter. Thus

$$\boldsymbol{L}_w\boldsymbol{v} = \left(\frac{\partial v^a}{\partial t} + \frac{\partial v^a}{\partial x^b}w^b - v^b\frac{\partial w^a}{\partial x^b}\right)\boldsymbol{e}_a$$

or, in coordinate-free notation,

$$\boldsymbol{L}_w\boldsymbol{v} = \frac{\partial \boldsymbol{v}}{\partial t} + [w, \boldsymbol{v}]$$

where [,] is the Lie bracket of vector fields (see Section 1.2).

A general coordinate expression can be given for the Lie derivative of a tensor of arbitrary type, namely,

$$\begin{aligned}(\boldsymbol{L}_w\mathbf{t})^{ab\cdots c}{}_{de\cdots f} &= \frac{\partial}{\partial t}\mathrm{t}^{ab\cdots c}{}_{de\cdots f} + \frac{\partial}{\partial x^g}\mathrm{t}^{ab\cdots c}{}_{de\cdots f}\,w^g \\ &\quad - \mathrm{t}^{gb\cdots c}{}_{de\cdots f}\frac{\partial w^a}{\partial x^g} - (\text{all upper indices}) \\ &\quad + \mathrm{t}^{ab\cdots c}{}_{ge\cdots f}\frac{\partial w^g}{\partial x^d} + (\text{all lower indices}).\end{aligned}$$

This follows by using the computations in the preceding example applied to each index.

The following proposition enables us to express the Lie derivative in terms of the covariant derivative.

6.11 Proposition *For a torsion-free connection the partial derivatives of the preceding general formula may be replaced by covariance derivatives; that is,*

$$\begin{aligned}(\boldsymbol{L}_w\mathbf{t})^{ab\cdots c}{}_{de\cdots f} &= \frac{\partial}{\partial t}\mathrm{t}^{gb\cdots c}{}_{de\cdots f} + \mathrm{t}^{ab\cdots c}{}_{de\cdots f|g}\,w^g \\ &\quad - \mathrm{t}^{gb\cdots c}{}_{de\cdots f}\,w^a{}_{|g} - (\text{all upper indices}) \\ &\quad + \mathrm{t}^{ab\cdots c}{}_{ge\cdots f}\,w^g{}_{|d} + (\text{all lower indices}).\end{aligned}$$

Recall that the torsion, Tor, of a connection ∇ is the (0, 2) tensor defined by $\text{Tor}(v, w) = \nabla_v w - \nabla_w v - [v, w]$. Torsion-free simply means that Tor = 0, from which it follows that the Christoffel symbols γ^a_{bc} are symmetric; that is, $\gamma^a_{bc} = \gamma^a_{cb}$. With this fact the proof of the proposition is a simple manipulative exercise, which we omit.

6.12 Corollary *(i) Let v be the spatial velocity vector field of a motion ϕ_t and $\mathbf{g}$ denote a Riemannian metric on $\mathcal{S}$. Then $\frac{1}{2}\mathbf{L}_v\mathbf{g} = \mathbf{d}$, the spatial* rate-of-deformation tensor.

(ii) Let f be a function on $\mathcal{S}$. Then the material derivative is $\dot{f} = \mathbf{L}_v f$. In particular, $\mathbf{L}_v J = \dot{J} =$ (div v)J, where J is the Jacobian.

Proof From the general formula of Proposition 6.11,

$$(\mathbf{L}_v\mathbf{g})_{ab} = \frac{\partial g_{ab}}{\partial t} + g_{ab|c}v^c + g_{cb}v^c{}_{|a} + g_{ac}v^c{}_{|b}.$$

(The vertical bar is understood to signify covariant differentiation with respect to the Riemannian connection.) Since $g_{ab|c} = 0$, we have $(\mathbf{L}_v\mathbf{g})_{ab} = v_{b|a} + v_{a|b}$, from which the result (i) is immediate; (ii) follows similarly. ∎

6.13 Proposition *Let $\boldsymbol{\mu} = \boldsymbol{\mu}(g)$ denote the volume form of g on $\mathcal{S}$. [Recall that $\boldsymbol{\mu}(g) = (\det g_{ab})^{1/2}\, dx^1 \wedge dx^2 \wedge \cdots \wedge dx^n$, where the wedge notation refers to the skew symmetrization of $dx^1 \otimes dx^2 \otimes \cdots \otimes dx^n$.] Then $\nabla_v\boldsymbol{\mu} = \mathbf{0}$ and $\mathbf{L}_v\boldsymbol{\mu} =$ (div v)$\boldsymbol{\mu}$.*

This proposition is another way of formulating $\dot{J} =$ (div v)J. We return to this point and give more details in Section 1.7.

We now enumerate some general properties of the Lie derivative.

6.14 Proposition

(i) *If w and $\mathbf{t}$ are C^k ($k \geq 1$), then $\mathbf{L}_w\mathbf{t}$ is a C^{k-1} tensor field of the same type as $\mathbf{t}$.*

(ii) *Symmetries and skew symmetries of $\mathbf{t}$ are also possessed by $\mathbf{L}_w\mathbf{t}$.*

(iii) *Let $\mathbf{t}_1$ and $\mathbf{t}_2$ be tensor fields of the same type. Then $\mathbf{L}_w(\mathbf{t}_1 + \mathbf{t}_2) = \mathbf{L}_w\mathbf{t}_1 + \mathbf{L}_w\mathbf{t}_2$.*

(iv) *$\mathbf{L}_w$ is a derivation: $\mathbf{L}_w(\mathbf{t}_1 \otimes \mathbf{t}_2) = \mathbf{L}_w\mathbf{t}_1 \otimes \mathbf{t}_2 + \mathbf{t}_1 \otimes \mathbf{L}_w\mathbf{t}_2$.*

(v) *Let w_1 and w_2 be vector fields. Then $\mathfrak{L}_{w_1+w_2} = \mathfrak{L}_{w_1} + \mathfrak{L}_{w_2}$.*

(vi) *$\mathbf{L}_w$ commutes with contractions.*

(vii) *$\mathbf{L}_w w = \partial w/\partial t$, that is, $\mathfrak{L}_w w = \mathbf{0}$.*

(viii) *Let f be a C^2 function and let $\mathbf{d}f$ denote the differential of f; that is, $\mathbf{d}f = (\partial f/\partial x^a)\, dx^a$. Then $\mathbf{L}_w\mathbf{d}f = \mathbf{d}(\mathbf{L}_w f)$.*

(ix) *If $\mathbf{t}$ is a general tensor and ψ is a diffeomorphism, then $\psi^*(\mathfrak{L}_w\mathbf{t}) = \mathfrak{L}_{\psi^* w}(\psi^*\mathbf{t})$.*

(x) *The following holds: $(d/dt)(\psi^*_{t,s}\mathbf{t}_t) = \psi^*_{t,s}(\mathbf{L}_w\mathbf{t})$.*

The proofs are relatively simple verifications, using the definitions or coordinate formulas, so we leave them as exercises.

6.15 Definition A map $\psi: \mathcal{S} \to \mathcal{S}$ is called an *isometry* of a metric g, if $\psi^* g = g$. A vector field w is a *Killing vector field*[28] (or *infinitesimal isometry*) if each map $\psi_{t,s}$ of the flow of w is an isometry is $\mathcal{S}$.

6.16 Proposition *If w is a Killing vector field, then $L_w g = 0$, and conversely.*

Proof If w is a Killing field, then

$$L_w g = \left(\frac{d}{dt} \psi_{t,s}^* g\right)\Big|_{t=s} = 0$$

since $\psi_{t,s}$ is an isometry. The converse follows from 6.14(x). ∎

6.17 Remark A Killing vector field w therefore satisfies *Killing's Equation* $w_{a|b} + w_{b|a} = 0$ (its covariant derivative is skew symmetric).

Problem 6.1 Let $\mathcal{S} = \mathbb{R}^3$ with the usual metric. Show that $\mathcal{S}$ possesses six linearly independent Killing vector fields: namely, $\partial/\partial z^i$ ($i = 1, 2, 3$) and $z^i\, \partial/\partial z^j - z^j\, \partial/\partial z^i$ ($i, j = 1, 2, 3$; $i \neq j$). Show that the most general form of a time-dependent isometry corresponding to these Killing vectors is

$$\psi_t^i(x) = c^i(t) + Q^i{}_j(t) z^j(x),$$

where the c^i's are components of a C^∞ curve $c: \mathbb{R} \to \mathcal{S}$ and the $Q^i{}_j$'s are the components of a C^∞ curve of orthogonal matrices.

Problem 6.2 Consider the vector fields $\boldsymbol{t} = x\boldsymbol{i} - xy\boldsymbol{j}$ and $\boldsymbol{w} = x^2\boldsymbol{i} - y\boldsymbol{j}$ in the plane. Calculate $\mathfrak{L}_w \mathbf{t}$ using 6.11. Also calculate the flow of w explicitly and verify that Definition 6.8 holds.

Problem 6.3 Define $\mathfrak{L}_w \nabla$ for a connection ∇ [see Problem 4.8(iii)]. Show that $\mathfrak{L}_w \nabla = \nabla\nabla w + w \cdot \mathbf{R}$; that is, $(\mathfrak{L}_w \nabla)^a_{bc} = w^a{}_{|b|c} + w^d \mathrm{R}^a{}_{dbc}$. (This can be used as a basis for the introduction of curvature.)

Box 6.1 ***Objective Rates: An Application of Lie Derivatives***

The subject of objective rates has been a controversial one in continuum mechanics. All so-called objective rates of second-order tensors are in fact Lie derivatives. Observe in the following formulas that the operations of raising and lowering indices—which commute

---

[28]After W. Killing [1892].

with $\partial/\partial t$ and ∇—do *not* commute with Lie differentiation. This fact is important in understanding this topic. Let $\boldsymbol{\sigma}_1 = \sigma^{ab}e_a \otimes e_b$ be a given tensor field on $\mathcal{S}$ and let $\boldsymbol{\sigma}_2 = \sigma_a{}^b e^a \otimes e_b$, $\boldsymbol{\sigma}_3 = \sigma^a{}_b e_a \otimes e^b$, and $\boldsymbol{\sigma}_4 = \sigma_{ab}e^a \otimes e^b$, denote its associated tensors (e.g., $\sigma^{ab} = g^{ac}\sigma_c{}^b$, etc.). Then the following relations hold: *

$$(L_v\boldsymbol{\sigma}_1)^{ab} = \dot{\sigma}^{ab} - \sigma^{cb}v^a{}_{|c} - \sigma^{ac}v^b{}_{|c}$$

$$g^{ac}(L_v\boldsymbol{\sigma}_2)_c{}^b = \dot{\sigma}^{ab} - \sigma^{ad}v^b{}_{|d} + \sigma^{db}v_d{}^{|a}$$

$$(L_v\boldsymbol{\sigma}_3)^a{}_c g^{cb} = \dot{\sigma}^{ab} - \sigma^{db}v^a{}_{|d} + \sigma^{ad}v_d{}^{|b}$$

$$g^{ac}(L_v\boldsymbol{\sigma}_4)_{cd}g^{db} = \dot{\sigma}^{ab} + \sigma^{cb}v_c{}^{|a} + \sigma^{ac}v_c{}^{|b}$$

and $$(L_v(\boldsymbol{\sigma}_1 \otimes \boldsymbol{\mu}))^{ab} = ((L_v\boldsymbol{\sigma}_1)^{ab} + \sigma^{ab}\operatorname{div} v)\boldsymbol{\mu}$$

where $\dot{\sigma}^{ab} = \partial\sigma^{ab}/\partial t + \sigma^{ab}{}_{|c}v^c$.

The tensor $L_v\boldsymbol{\sigma}_1$ has been associated with the name Oldroyd (see Oldroyd [1950]) and $L_v(\boldsymbol{\sigma}_1 \otimes \boldsymbol{\mu})$ with the name Truesdell (see Truesdell [1955a, b]). We see that all of these tensors are different manifestations of the Lie derivative of $\boldsymbol{\sigma}$.

Any linear combination of the preceding formulas also qualifies as an "objective flux"; for example,

$$\tfrac{1}{2}((L_v\boldsymbol{\sigma}_3)^a{}_c g^{cb} + g^{ac}(L_v\boldsymbol{\sigma}_2)_c{}^b) = \dot{\sigma}^{ab} + \sigma^{ad}\omega_d{}^b - \sigma^{db}\omega^a{}_d$$

where $\omega^a{}_b$ are associated components of the spin $\omega_{ab} = \frac{1}{2}(v_{a|b} - v_{b|a})$. This tensor is associated with the name Jaumann (see Jaumann [1911]).

We note in passing that, like the Lie derivative in general, the right-hand sides may be expressed without using covariant derivatives. For example,

$$(L_v\boldsymbol{\sigma}_1)^{ab} = \frac{\partial\sigma^{ab}}{\partial t} + \frac{\partial\sigma^{ab}}{\partial x^c}v^c - \sigma^{cb}\frac{\partial v^a}{\partial x^c} - \sigma^{ac}\frac{\partial v^b}{\partial x^c}.$$

And if $\sigma^{ab} = \sigma^{ba}$, then

$$(L_v\boldsymbol{\sigma}_1)^{ab} = \frac{\partial\sigma^{ab}}{\partial t} + \frac{\partial\sigma^{ab}}{\partial x^c}v^c - 2 \times \text{symmetric part of } \left(\sigma^{ac}\frac{\partial v^b}{\partial x^c}\right).$$

(One seems never to see this in practice, but it could yield savings in numerical computations in that it is unnecessary to compute the Christoffel symbols.)

We conclude this box with a general discussion of what we mean by objective.

6.18 Definition Let $\mathbf{t}$ be a tensor field (or tensor density) on a manifold $\mathcal{S}$ and ψ a diffeomorphism of $\mathcal{S}$ to another manifold $\mathcal{S}'$. We say that $\mathbf{t}' = \psi_*\mathbf{t}$ is the *objective transformation* of $\mathbf{t}$ (i.e., $\mathbf{t}$ transforms in the usual way under the map ψ).

*Some of these results were known to Sedov around 1960.

6.19 Theorem *Let ϕ_t be a regular motion of $\mathcal{B}$ in $\mathcal{S}$ with spatial velocity field v_t. Let ξ_t be a motion of $\mathcal{S}$ in $\mathcal{S}'$, and let $\phi'_t = \xi_t \circ \phi_t$ be the* superposed motion[29] *of $\mathcal{B}$ in $\mathcal{S}'$.*

Let $\mathbf{t}$ be a given time-dependent tensor field on $\mathcal{S}$ and let $\mathbf{t}' = \xi_* \mathbf{t}$, that is, transform $\mathbf{t}$ objectively.

Let v' be the velocity field of ϕ'_t. Then

$$L_{v'}\mathbf{t}' = \xi_*(L_v \mathbf{t}).$$

Thus, "objective tensors (or tensor densities) have objective Lie derivative." This is remarkable since v itself is *not* objective as we shall see immediately in the proof.

Proof We first note that $v'_t = w_t + \xi_{t*}v_t$ where w_t is the spatial velocity of ξ_t. This follows by differentiating $\phi'_t(X) = \xi_t(\phi_t(X))$ in t. Now we compute

$$\begin{aligned}
L_{v'}\mathbf{t}' &= L_{w+\xi_* v}(\xi_*\mathbf{t}) = \mathfrak{L}_{w+\xi_* v}(\xi_*\mathbf{t}) + \frac{\partial}{\partial t}(\xi_*\mathbf{t}) \\
&= \xi_*(\mathfrak{L}_v\mathbf{t}) + \mathfrak{L}_w(\xi_*\mathbf{t}) + \frac{\partial}{\partial t}(\xi_*\mathbf{t}) = \xi_*(\mathfrak{L}_v\mathbf{t}) + L_w(\xi_*\mathbf{t}) \\
&= \xi_*(\mathfrak{L}_v\mathbf{t}) + \frac{d}{dt}\xi^*_{t,s}(\xi_{t*}\mathbf{t}_t)\Big|_{t=s} \\
&= \xi_*(\mathfrak{L}_v\mathbf{t}) + \frac{d}{dt}(\xi_t \circ \xi_s^{-1})^*(\xi_{t*}\mathbf{t}_t)\Big|_{t=s} \qquad (\text{see } 6.6) \\
&= \xi_*(\mathfrak{L}_v\mathbf{t}) + \frac{d}{dt}\xi_{s*}\mathbf{t}_t\Big|_{t=s} \\
&= \xi_*\left(\mathfrak{L}_v\mathbf{t} + \frac{d}{dt}\mathbf{t}_t\Big|_{t=s}\right) = \xi_*(L_v\mathbf{t}). \quad \blacksquare
\end{aligned}$$

In order to master the proof fully, a coordinate computation may also be done. We do so in case $\mathbf{t}$ is a vector field t. Let $\{x^a\}$ be coordinates on $\mathcal{S}$, $\{\xi^\alpha\}$ be coordinates on $\mathcal{S}'$, and write ξ^α for the coordinates of ξ. From $\phi'^\alpha = \xi^\alpha(\phi^a(X, t))$ we get $v'^\alpha = w^\alpha + (\partial\xi^\alpha/\partial x^a)v^a$. Since t^a is objective, $t'^\alpha = (\partial\xi^\alpha/\partial x^a)t^a$. From our coordinate formulas for the Lie derivative,

$$(L_{v'}t')^\alpha = \frac{\partial t'^\alpha}{\partial t} + \frac{\partial t'^\alpha}{\partial \xi^\beta}v'^\beta - t'^\beta\frac{\partial v'^\alpha}{\partial \xi^\beta}$$

[29]If the reader prefers, think of ξ_t passively as a time-dependent chart of $\mathcal{S}$ so that ϕ'_t is the same motion viewed at in a different chart (an observer transformation). This is the same idea as representing a matrix as either a linear transformation (active) or as a change of basis (passive). See Box 4.1, Section 1.4.

$$= \frac{\partial}{\partial t}\left(\frac{\partial \xi^\alpha}{\partial x^a} t^a\right) + \frac{\partial}{\partial \xi^\beta}\left(\frac{\partial \xi^\alpha}{\partial x^a} t^a\right)\left(w^\beta + \frac{\partial \xi^\beta}{\partial x^b} v^b\right)$$

$$- \left(\frac{\partial \xi^\beta}{\partial x^a} t^a\right) \frac{\partial}{\partial \xi^\beta}\left(w^\alpha + \frac{\partial \xi^\alpha}{\partial x^b} v^b\right)$$

$$= \left(\frac{\partial}{\partial x^a} w^\alpha\right) t^a + \left(-\frac{\partial}{\partial \xi^\beta} t'^\alpha w^\beta + \frac{\partial \xi^\alpha}{\partial x^a}\frac{\partial t^a}{\partial t}\right) + \frac{\partial}{\partial \xi^\beta}\left(\frac{\partial \xi^\alpha}{\partial x^a} t^a\right) w^\beta$$

$$- \frac{\partial \xi^\beta}{\partial x^a} t^a \frac{\partial w^\alpha}{\partial \xi^\beta} + \frac{\partial}{\partial \xi^\beta}\left(\frac{\partial \xi^\alpha}{\partial x^a} t^a\right) \frac{\partial \xi^\beta}{\partial x^b} v^b - \frac{\partial \xi^\beta}{\partial x^a} t^a \frac{\partial}{\partial \xi^\beta}\left(\frac{\partial \xi^\alpha}{\partial x^b} v^b\right).$$

The second term is there because t'^α has an explicit time dependence through ξ_t^{-1}. All terms cancel except those for $(\partial \xi^\beta/\partial x^a)(L_v t)^a$—that is, $\xi_*(L_v t)$—so we get the result. The reader may do the same computation for one-forms or two-tensors.

As a corollary, all the "objective fluxes" discussed earlier are objective tensors with this proviso: if the metric tensor g_{ab}, or g^{ab} appears explicitly on the left hand side and the rates are to transform like tensors with the *same* g_{ab} resulting after the transformation, ξ must be an isometry at the point of interest. In this sense, there are two levels of objectivity, objectivity with respect to diffeomorphisms, and objectivity with respect to isometries. The Oldroyd and Truesdell rates are objective with respect to diffeomorphisms while the remaining rates discussed above are objective with respect to isometries. Rates which are objective with respect to diffeomorphisms are called *covariant*. This subject of covariance is taken up in Sections 2-4 and 3-3.

Box 6.2 ***Summary of Important Formulas for Section 1.6***

Tangent Vector to a Curve

$c: \mathcal{I} \to \mathcal{S}$

$$\frac{dc}{dt} = Tc(t, 1) \in T_{c(t)}\mathcal{S} \qquad \left(\frac{dc}{dt}\right)^a = \frac{dc^a}{dt}$$

Integral Curve of w

$$\frac{dc}{dt} = w(c(t)) \qquad \frac{dc^a}{dt} = w^a(c^1(t), \ldots, c^n(t))$$

$$(a = 1, 2, \ldots, n)$$

Flow (Evolution operator) $\psi_{t,s}$ *of* w

$$\frac{d\psi_{t,s}}{dt}(x) = w(\psi_{t,s}(x))$$

$(\psi_{t,s}(x))^a = c^a(t)$, where c^a is an integral curve with $c^a(s) = x^a$

$$\psi_{s,s}(x) = x$$

Flow Associated wtih a Motion ϕ_t

$\phi_{t,s} = \phi_t \circ \phi_s^{-1}$ = flow of the spatial velocity v_t

Lie Derivative of a Time-Dependent Tensor $\mathbf{t}$

$$\mathbf{L}_w \mathbf{t} = \frac{\partial \mathbf{t}}{\partial t} + \mathfrak{L}_w \mathbf{t}$$

$$\mathbf{L}_w \mathbf{t} = \frac{d}{dt}(\psi_{t,s}^* \mathbf{t}_t)\bigg|_{t=s}$$

$$\mathfrak{L}_w \mathbf{t} = \frac{d}{dt}(\psi_{t,s}^* \mathbf{t}_s)\bigg|_{t=s},$$

where $\psi_{t,s}$ is the flow of w

$$\begin{aligned}(\mathbf{L}_w \mathbf{t})^{ab\cdots c}{}_{de\cdots f} &= \frac{\partial}{\partial t}\mathbf{t}^{ab\cdots c}{}_{de\cdots f} + \frac{\partial}{\partial x^g}\mathbf{t}^{ab\cdots c}{}_{de\cdots f}w^g \\ &\quad - \mathbf{t}^{gb\cdots c}{}_{de\cdots f}\frac{\partial w^a}{\partial x^g} \\ &\quad - \text{(all upper indices)} \\ &\quad + \mathbf{t}^{ab\cdots c}{}_{ge\cdots f}\frac{\partial w^g}{\partial x^d} \\ &\quad + \text{(all lower indices)} \\ &= \frac{\partial}{\partial t}\mathbf{t}^{ab\cdots c}{}_{de\cdots f} + \mathbf{t}^{ab\cdots c}{}_{de\cdots f|g}w^g \\ &\quad - \mathbf{t}^{gb\cdots c}{}_{de\cdots f}w^a{}_{|g} \\ &\quad - \text{(all upper indices)} \\ &\quad + \mathbf{t}^{ab\cdots c}{}_{ge\cdots f}w^g{}_{|d} \\ &\quad + \text{(all lower indices)}\end{aligned}$$

(using any torsion-free connection).

For a vector field, $\mathbf{L}_w v = \dfrac{\partial v}{\partial t} + [w, v] = \dfrac{\partial v}{\partial t} + (\nabla_w v - \nabla_v w)$.

For a differential form $\boldsymbol{\alpha}$, $\mathbf{L}_w \boldsymbol{\alpha} = \mathbf{d}\mathbf{i}_w \boldsymbol{\alpha} + \mathbf{i}_w \mathbf{d}\boldsymbol{\alpha}$ (see Section 1.7).

Rate of Deformation Tensor

$$\mathbf{d} = \tfrac{1}{2}\mathbf{L}_v \mathbf{g} \qquad d_{ab} = \tfrac{1}{2}(v_{a|b} + v_{b|a}) = \tfrac{1}{2}(\mathbf{L}_v \mathbf{g})_{ab}$$

Material Derivative

$$\dot{f} = \mathbf{L}_v f \qquad \dot{f} = \frac{\partial f}{\partial t} + \frac{\partial f}{\partial x^a}v^a$$

Jacobians and Lie Derivatives

$$\mathbf{L}_v J = \dot{J} = (\operatorname{div} v)J \qquad \dot{J} = (v^a{}_{|a})J$$

J = Jacobian determinant of ϕ_t $\quad J = \det(F^a{}_A)$

Properties of Lie Differentiation

$$\mathbf{L}_w(\mathbf{t}_1 + \mathbf{t}_2) = \mathbf{L}_w \mathbf{t}_1 + \mathbf{L}_w \mathbf{t}_2$$

$$\mathbf{L}_w(\mathbf{t}_1 \otimes \mathbf{t}_2) = \mathbf{L}_w \mathbf{t}_1 \otimes \mathbf{t}_2 + \mathbf{t}_1 \otimes \mathbf{L}_w \mathbf{t}_2$$

$$\mathbf{L}_w \mathbf{d}f = \mathbf{d}\mathbf{L}_w f$$

$$\mathfrak{L}_{w_1+w_2}\mathbf{t} = \mathfrak{L}_{w_1}\mathbf{t} + \mathfrak{L}_{w_2}\mathbf{t}$$

$$\psi^*(\mathfrak{L}_w \mathbf{t}) = \mathfrak{L}_{\psi^* w}\psi^* \mathbf{t}$$

$$\frac{d}{dt}(\psi_{t,s}^* \mathbf{t}_t) = \psi_{t,s}^*(\mathbf{L}_w \mathbf{t})$$

Killing's Equations

$$\mathbf{L}_w \mathbf{g} = \mathbf{0} \qquad w_{a|b} + w_{b|a} = 0$$

(The flow of $\boldsymbol{w}$ consists of isometries if and only if Killing's equations hold.)

Objective Tensors

Transformation rule

$$\mathbf{t}' = \xi_* \mathbf{t} \qquad (\mathbf{t}')^{\alpha\beta\cdots\gamma}{}_{\epsilon\mu\cdots\nu} = \frac{\partial \xi^\alpha}{\partial x^a} \cdots \frac{\partial \xi^\gamma}{\partial x^c} \mathbf{t}^{ab\cdots c}{}_{ef\cdots g} \frac{\partial x^e}{\partial \xi^\epsilon} \cdots \frac{\partial x^g}{\partial \xi^\nu}$$

Lie Derivatives

$$L_{v'} \mathbf{t}' = \xi_* L_v \mathbf{t}$$

(The Lie derivative of an objective tensor is objective.)

1.7 DIFFERENTIAL FORMS AND THE PIOLA TRANSFORMATION

Skew-symmetric covariant tensors are called differential forms. They have a rich algebraic and differential structure with many applications to the physical sciences.[30] We shall consider some of these here. One of the principal applications is to the Piola transformation—a fundamental operation relating the material and spatial descriptions of a continuous medium. We shall consider a few other applications as well, to Hamiltonian systems and variational principles; these are treated with more specific reference to continuum mechanics in Chapter 5. We shall also consider volume elements, integration, Stokes' theorem and Gauss' theorem in the language of differential forms.

7.1 Definition A *k-form* on a manifold $\mathfrak{M}$ is a $\binom{0}{k}$ tensor $\boldsymbol{\alpha}$ on $\mathfrak{M}$ that is skew symmetric; that is, for $X \in \mathfrak{M}$,

$$\boldsymbol{\alpha}_X : \underbrace{T_X\mathfrak{M} \times \cdots \times T_X\mathfrak{M}}_{k \text{ copies}} \to \mathbb{R}$$

is a multilinear mapping and

$$\boldsymbol{\alpha}_X(\boldsymbol{W}_{\pi(1)}, \boldsymbol{W}_{\pi(2)}, \ldots, \boldsymbol{W}_{\pi(k)}) = (\operatorname{sgn} \pi)\boldsymbol{\alpha}_X(\boldsymbol{W}_1, \ldots, \boldsymbol{W}_k)$$

for any $\boldsymbol{W}_1, \ldots, \boldsymbol{W}_k \in T_X\mathfrak{M}$ and any permutation π on $\{1, \ldots, k\}$, where $\operatorname{sgn} \pi$ is the sign of π ($+1$ or -1 according to whether π is an even or odd permutation). One also refers to a k-form as a *differential k-form* or simply as a *differential form*. We write $\boldsymbol{\alpha}(X) = \boldsymbol{\alpha}_X$ for the value of $\boldsymbol{\alpha}$ at $X \in \mathfrak{M}$, and leave off the argument X when convenient.

In terms of components, skew symmetry means the following: if any *two* indices of the components $\alpha_{A_1 \cdots A_k}$ are interchanged, then $\alpha_{A_1 \cdots A_k}$ changes sign.

---

[30]See, for example, Cartan [1922], Gallissot [1958], Flanders [1963], and Abraham and Marsden [1978].

If we skew-symmetrize the tensor product, we get the following basic algebraic operation on differential forms.

7.2 Definition Let $\boldsymbol{\alpha}$ be a k-form and $\boldsymbol{\beta}$ an l-form. Define the $(k+l)$-form $\boldsymbol{\alpha} \wedge \boldsymbol{\beta}$, called their *wedge* or *exterior product*, by

$$(\boldsymbol{\alpha} \wedge \boldsymbol{\beta})(W_1, \ldots, W_k, W_{k+1}, \ldots, W_{k+l}) = \frac{1}{k!\,l!} \sum_{\text{all } \pi} (\operatorname{sgn} \pi) \boldsymbol{\alpha}(W_{\pi(1)}, \ldots, W_{\pi(k)}) \boldsymbol{\beta}(W_{\pi(k+1)}, \ldots, W_{\pi(k+l)}).$$

There are several possible choices of normalization factor.[31] The one chosen here gives the equivalent formula

$$(\boldsymbol{\alpha} \wedge \boldsymbol{\beta})(W_1, \ldots, W_k, W_{k+1}, \ldots, W_{k+l}) = \sum{}' (\operatorname{sign} \pi) \boldsymbol{\alpha}(W_{\pi(1)}, \ldots, W_{\pi(k)}) \boldsymbol{\beta}(W_{\pi(k+1)}, \ldots, W_{\pi(k+l)})$$

where $\sum'$ denotes the sum over permutations satisfying $\pi(1) < \ldots < \pi(k)$ and $\pi(k+1) < \ldots < \pi(k+l)$.

The wedge product $\wedge$ is associative, $\boldsymbol{\alpha} \wedge (\boldsymbol{\beta} \wedge \boldsymbol{\gamma}) = (\boldsymbol{\alpha} \wedge \boldsymbol{\beta}) \wedge \boldsymbol{\gamma}$, and satisfies the following commutation relation: $\boldsymbol{\alpha} \wedge \boldsymbol{\beta} = (-1)^{kl} \boldsymbol{\beta} \wedge \boldsymbol{\alpha}$. These are algebraic verifications that we shall omit. (If you get stuck, consult, for example, Abraham and Marsden [1978].) Some algebraic relationships which help in understanding the wedge product are contained in the following problem.

Problem 7.1

(i) If $\boldsymbol{\alpha}$ is a two-form and $\boldsymbol{\beta}$ a one-form, show that

$$(\boldsymbol{\alpha} \wedge \boldsymbol{\beta})(W_1, W_2, W_3) = \boldsymbol{\alpha}(W_1, W_2)\boldsymbol{\beta}(W_3) - \boldsymbol{\alpha}(W_1, W_3)\boldsymbol{\beta}(W_2) + \boldsymbol{\alpha}(W_2, W_3)\boldsymbol{\beta}(W_1).$$

(ii) If $\boldsymbol{\alpha}$, $\boldsymbol{\beta}$, and $\boldsymbol{\gamma}$ are one-forms,

$$\begin{aligned}(\boldsymbol{\alpha} \wedge \boldsymbol{\beta} \wedge \boldsymbol{\gamma})(W_1, W_2, W_3) &= \boldsymbol{\alpha}(W_1)\boldsymbol{\beta}(W_2)\boldsymbol{\gamma}(W_3) + \boldsymbol{\alpha}(W_2)\boldsymbol{\beta}(W_3)\boldsymbol{\gamma}(W_1) \\ &\quad + \boldsymbol{\alpha}(W_3)\boldsymbol{\beta}(W_1)\boldsymbol{\gamma}(W_2) - \boldsymbol{\alpha}(W_2)\boldsymbol{\beta}(W_1)\boldsymbol{\gamma}(W_3) \\ &\quad - \boldsymbol{\alpha}(W_1)\boldsymbol{\beta}(W_3)\boldsymbol{\gamma}(W_2) - \boldsymbol{\alpha}(W_3)\boldsymbol{\beta}(W_2)\boldsymbol{\gamma}(W_1).\end{aligned}$$

(iii) In coordinates, if $\boldsymbol{\alpha}$ is a k-form, show that

$$\begin{aligned}\boldsymbol{\alpha} &= \frac{1}{k!} \alpha_{A_1 \cdots A_k}\, dX^{A_1} \wedge \cdots \wedge dX^{A_k} \\ &= \sum_{A_1 < \cdots < A_k} \alpha_{A_1 \cdots A_k}\, dX^{A_1} \wedge \cdots \wedge dX^{A_k}\end{aligned}$$

---

[31]Some books (e.g. Kobayashi and Nomizu [1963]) use the factor $1/(k+l)!$ here. This convention leads to awkward factors in subsequent formulas however.

where the summation is over all indices $A_1, \ldots, A_k$ satisfying $A_1 < \cdots < A_k$.

(iv) In $\mathbb{R}^3$, if $\boldsymbol{\alpha}$ and $\boldsymbol{\beta}$ are one-forms, show that the coefficients of $\boldsymbol{\alpha} \wedge \boldsymbol{\beta}$ in the standard basis are components of the cross product of $\boldsymbol{\alpha}$ and $\boldsymbol{\beta}$.

(v) For mappings $\phi: \mathfrak{M} \to \mathfrak{N}$ and $\psi: \mathfrak{L} \to \mathfrak{M}$, and differential forms $\boldsymbol{\alpha}$ and $\boldsymbol{\beta}$ on $\mathfrak{N}$, verify that $\phi^*(\boldsymbol{\alpha} \otimes \boldsymbol{\beta}) = \phi^*\boldsymbol{\alpha} \otimes \phi^*\boldsymbol{\beta}$, $\phi^*(\boldsymbol{\alpha} \wedge \boldsymbol{\beta}) = \phi^*\boldsymbol{\alpha} \wedge \phi^*\boldsymbol{\beta}$, and $(\phi \circ \psi)^*\boldsymbol{\alpha} = \psi^*(\phi^*\boldsymbol{\alpha})$.

7.3 Definition If W is a vector field on $\mathfrak{M}$ and $\boldsymbol{\alpha}$ is a k-form, the contraction of W with the first index of $\boldsymbol{\alpha}$ is called the *interior product* and is denoted by

$$i_W\boldsymbol{\alpha} \quad \text{or} \quad W \lrcorner\, \boldsymbol{\alpha}.$$

Thus $i_W\boldsymbol{\alpha}$ is a $(k-1)$-form given by

$$\boxed{(i_W\boldsymbol{\alpha})(W_2, \ldots, W_k) = \boldsymbol{\alpha}(W, W_2, \ldots, W_k), \quad \text{or} \quad (i_W\boldsymbol{\alpha})_{A_2\cdots A_k} = W^{A_1}\alpha_{A_1\cdots A_k}.}$$

Some properties of this contraction are given in the next problem:

Problem 7.2 (i) On $\mathbb{R}^3$ and in Euclidean coordinates, let $d\boldsymbol{v}$ be the volume element ($d\boldsymbol{v} = dx \wedge dy \wedge dz$), and let $\boldsymbol{\alpha} = \alpha_x\, dx + \alpha_y\, dy + \alpha_z\, dz$ and $\boldsymbol{\beta} = \beta_x\, dx + \beta_y\, dy + \beta_z\, dz$. Let $\boldsymbol{v} = (\alpha_y\beta_z - \alpha_z\beta_y)\boldsymbol{i} - (\alpha_x\beta_z - \alpha_z\beta_x)\boldsymbol{j} + (\alpha_x\beta_y - \alpha_y\beta_x)\boldsymbol{k}$, be the cross product. Show that $\boldsymbol{\alpha} \wedge \boldsymbol{\beta} = i_v\, d\boldsymbol{v}$.

(ii) Prove that if $\boldsymbol{\alpha}$ is a k-form, $i_W(\boldsymbol{\alpha} \wedge \boldsymbol{\beta}) = (i_W\boldsymbol{\alpha}) \wedge \boldsymbol{\beta} + (-1)^k\boldsymbol{\alpha} \wedge i_W\boldsymbol{\beta}$.

(iii) Show that

$$i_W\boldsymbol{\alpha} = \frac{1}{(k-1)!} W^b\alpha_{ba_2\cdots a_k}\, dx^{a_2} \wedge \cdots \wedge dx^{a_k}.$$

The three most important types of differentiation in tensor analysis are: covariant differentiation ∇, Lie differentiation $\mathfrak{L}_W$, and exterior differentiation $\boldsymbol{d}$. We have already met the first two and now turn to $\boldsymbol{d}$. The exterior derivative generalizes the notion of gradient, divergence, and curl to differential forms. We want it to capture the identities div curl $= 0$ and curl grad $= \mathbf{0}$ as well as $\boldsymbol{d}$ being a derivative.

7.4 Theorem *Given a manifold $\mathfrak{M}$, there is a unique linear operator $\boldsymbol{d}$ taking (smooth) k-forms $\boldsymbol{\alpha}$ on $\mathfrak{M}$ to (smooth)$(k+1)$-forms $\boldsymbol{d\alpha}$ on $\mathfrak{M}$ such that:*

(i) $\boldsymbol{d}(\boldsymbol{d\alpha}) = 0$;

(ii) $\boldsymbol{d}(\boldsymbol{\alpha} \wedge \boldsymbol{\beta}) = (\boldsymbol{d\alpha}) \wedge \boldsymbol{\beta} + (-1)^k\boldsymbol{\alpha} \wedge \boldsymbol{d\beta}$;

(iii) *for functions f, $\boldsymbol{d}f$ coincides with the differential of f as defined in Section 1.2; and*

(iv) *if $\mathfrak{U} \subset \mathfrak{M}$ is open, then $\boldsymbol{d}(\boldsymbol{\alpha}|\mathfrak{U}) = (\boldsymbol{d\alpha})|\mathfrak{U}$.*

Proof First we show uniqueness by using these properties to derive a formula for $\mathbf{d}\boldsymbol{\alpha}$. Let

$$\boldsymbol{\alpha} = \frac{1}{k!}\alpha_{A_1\cdots A_k}\, \mathbf{d}X^{A_1} \wedge \cdots \wedge \mathbf{d}X^{A_k}.$$

(we can pass to a local chart by virtue of (iv)). Then since $\mathbf{d}\,\mathbf{d}X^A = \mathbf{0}$ we get

$$\begin{aligned}\mathbf{d}\boldsymbol{\alpha} &= \frac{1}{k!}\mathbf{d}\alpha_{A_1\cdots A_k} \wedge \mathbf{d}X^{A_1} \wedge \cdots \wedge \mathbf{d}X^{A_k}\\ &= \frac{1}{k!}\frac{\partial \alpha_{A_1\cdots A_k}}{\partial X^B}\, \mathbf{d}X^B \wedge \mathbf{d}X^{A_1} \wedge \cdots \wedge \mathbf{d}X^{A_k}.\end{aligned}$$

Thus, if there is such a $\mathbf{d}$, it is unique. To show that $\mathbf{d}$ exists, define it by this formula. One has to show that $\mathbf{d}\boldsymbol{\alpha}$ is a tensor (i.e., transforms properly) and satisfies (i), (ii), and (iii). These are all straightforward. For example, to prove (i), we use the above formula for $\mathbf{d}$ twice:

$$\mathbf{d}\,\mathbf{d}\boldsymbol{\alpha} = \frac{1}{k!}\frac{\partial^2 \alpha_{A_1\cdots A_k}}{\partial X^C\, \partial X^B}\, \mathbf{d}X^C \wedge \mathbf{d}X^B \wedge \mathbf{d}X^{A_1} \wedge \cdots \wedge \mathbf{d}X^{A_k}$$

But

$$\frac{\partial^2 \alpha_{A_1\cdots A_k}}{\partial X^C\, \partial X^B}$$

is symmetric in B and C (by the equality of mixed partial derivatives) and $\mathbf{d}X^C \wedge \mathbf{d}X^B$ is skew symmetric. Therefore, the sum vanishes. ∎

There is a useful coordinate-free formula for $\mathbf{d}$ (due to Palais [1954]): If $\boldsymbol{W}_0, \boldsymbol{W}_1, \ldots, \boldsymbol{W}_k$ are vector fields, then

$$\begin{aligned}\mathbf{d}\boldsymbol{\alpha}(\boldsymbol{W}_0, \ldots, \boldsymbol{W}_k) &= \sum_{i=0}^{k}(-1)^i \boldsymbol{W}_i[\boldsymbol{\alpha}(\underbrace{\boldsymbol{W}_0, \ldots, \boldsymbol{W}_{i-1}, \boldsymbol{W}_{i+1}, \ldots, \boldsymbol{W}_k}_{\boldsymbol{W}_i \text{ missing}})]\\ &\quad + \sum_{i<j}(-1)^{i+j}\boldsymbol{\alpha}([\boldsymbol{W}_i, \boldsymbol{W}_j], \underbrace{\boldsymbol{W}_0, \ldots, \boldsymbol{W}_k}_{\boldsymbol{W}_i, \boldsymbol{W}_j \text{ omitted}}).\end{aligned}$$

This can be verified using the coordinate expressions of each side; we omit the details.

Problem 7.3 In Euclidean coordinates (x, y, z) on $\mathbb{R}^3$, establish:

(i) $\operatorname{grad} f = (\mathbf{d}f)^\sharp$;
(ii) $\operatorname{div} \boldsymbol{v}$ is such that $\mathbf{d}(\boldsymbol{i}_{\boldsymbol{v}}\, \mathbf{d}v) = (\operatorname{div} \boldsymbol{v})\, \mathbf{d}v$, where $\mathbf{d}v = \mathbf{d}x \wedge \mathbf{d}y \wedge \mathbf{d}z$; and
(iii) $\mathbf{d}(\boldsymbol{v}^\flat) = \boldsymbol{i}_{\nabla\times\boldsymbol{v}}\, \mathbf{d}v$.

Use $\mathbf{dd} = 0$ to show that curl grad $= 0$ and div curl $= 0$.

Problem 7.4 On $\mathbb{R}^4$ with coordinates (t, x, y, z) let $\boldsymbol{F} = \boldsymbol{E}^\flat \wedge \boldsymbol{dt} + \boldsymbol{i}_B(\boldsymbol{dx} \wedge \boldsymbol{dy} \wedge \boldsymbol{dz})$ (the Faraday two-form), and $*\boldsymbol{F} = -\boldsymbol{B}^\flat \wedge \boldsymbol{dt} + \boldsymbol{i}_E(\boldsymbol{dx} \wedge \boldsymbol{dy} \wedge \boldsymbol{dz})$ (the Maxwell two-form). Show that Maxwell's equations may be written $\boldsymbol{dF} = \boldsymbol{0}$, $\boldsymbol{d}(*\boldsymbol{F}) = 4\pi*\boldsymbol{J}$, where $*\boldsymbol{J} = \boldsymbol{i}_J(\boldsymbol{dt} \wedge \boldsymbol{dx} \wedge \boldsymbol{dy} \wedge \boldsymbol{dz})$ and $\boldsymbol{J}^\flat = \rho\, \boldsymbol{dt} + j_a\, \boldsymbol{dx}^a$.

7.5 Proposition *If $\phi: \mathfrak{M} \to \mathfrak{N}$ is a (smooth) mapping and $\boldsymbol{\alpha}$ is a k-form on $\mathfrak{N}$, then $\phi^* \boldsymbol{d\alpha} = \boldsymbol{d}\phi^*\boldsymbol{\alpha}$; that is, pull-back commutes with exterior differentiation.*

Proof First we verify this for functions f. But $\phi^*f = f\circ\phi$ and so, by the chain rule, $\boldsymbol{d}(\phi^*f) = \boldsymbol{d}(f\circ\phi) = \boldsymbol{d}f\cdot T\phi = \phi^*\boldsymbol{d}f$. In general, let

$$\boldsymbol{\alpha} = \frac{1}{k!}\alpha_{a_1\cdots a_k}\, \boldsymbol{dx}^{a_1} \wedge \cdots \wedge \boldsymbol{dx}^{a_k}.$$

Since $\phi^*(\boldsymbol{\alpha} \wedge \boldsymbol{\beta}) = \phi^*\boldsymbol{\alpha} \wedge \phi^*\boldsymbol{\beta}$ and $\boldsymbol{d}\phi^*f = \phi^*\boldsymbol{d}f$,

$$\phi^*\boldsymbol{\alpha} = \frac{1}{k!}\alpha_{a_1\cdots a_k}\circ\phi\, \boldsymbol{d}\phi^{a_1} \wedge \cdots \wedge \boldsymbol{d}\phi^{a_k}$$

(where ϕ^a stands for $x^a\circ\phi$). Using properties of $\boldsymbol{d}$,

$$\begin{aligned}\boldsymbol{d}\phi^*\boldsymbol{\alpha} &= \frac{1}{k!}\frac{\partial\alpha_{a_1\cdots a_k}}{\partial x^b}\frac{\partial\phi^b}{\partial X^A}\boldsymbol{d}X^A \wedge \boldsymbol{d}\phi^{a_1} \wedge \cdots \wedge \boldsymbol{d}\phi^{a_k}\\ &= \frac{1}{k!}\frac{\partial\alpha_{a_1\cdots a_k}}{\partial x^b}\phi^*(\boldsymbol{dx}^b) \wedge \phi^*\, \boldsymbol{dx}^{a_1} \wedge \cdots \wedge \phi^*\, \boldsymbol{dx}^{a_k}\\ &= \phi^*\left(\frac{1}{k!}\frac{\partial\alpha_{a_1\cdots a_k}}{\partial x^b}\boldsymbol{dx}^b \wedge \boldsymbol{dx}^{a_1} \wedge \cdots \wedge \boldsymbol{dx}^{a_k}\right) = \phi^*\boldsymbol{d\alpha}. \quad \blacksquare\end{aligned}$$

Next we shall establish the basic link between the exterior derivative $\boldsymbol{d}$ and the Lie derivative $\mathfrak{L}$.

7.6 Theorem *Let $\boldsymbol{\alpha}$ be a k-form on $\mathfrak{N}$ and $\boldsymbol{w}$ a vector field on $\mathfrak{N}$. Then*

$$\boxed{\mathfrak{L}_w\boldsymbol{\alpha} = \boldsymbol{d}\boldsymbol{i}_w\boldsymbol{\alpha} + \boldsymbol{i}_w\boldsymbol{d\alpha}, \qquad \text{that is,} \quad \mathfrak{L}_w = \boldsymbol{d}\boldsymbol{i}_w + \boldsymbol{i}_w\boldsymbol{d}.}$$

This formula is one of several "magic formulas" of Cartan [1922].

Proof From Section 1.6 we have

$$(\mathfrak{L}_w\boldsymbol{\alpha})_{a_1\cdots a_k} = \frac{\partial\alpha_{a_1\cdots a_k}}{\partial x^b}w^b + \alpha_{ba_2\cdots a_k}\frac{\partial w^b}{\partial x^{a_1}} + \text{(all lower indices)}.$$

On the other hand, from our formulas for $\boldsymbol{d}$ and $\boldsymbol{i}_w$,

$$\begin{aligned}(\boldsymbol{d}\boldsymbol{i}_w\boldsymbol{\alpha} + \boldsymbol{i}_w\boldsymbol{d\alpha}) = \boldsymbol{d}&\left(\frac{1}{(k-1)!}w^b\alpha_{ba_2\cdots a_k}\,\boldsymbol{dx}^{a_2} \wedge \cdots \wedge \boldsymbol{dx}^{a_k}\right)\\ &+ \frac{1}{k!}\left(w^b\frac{\partial\alpha_{a_1\cdots a_k}}{\partial x^b}\boldsymbol{dx}^{a_1} \wedge \cdots \wedge \boldsymbol{dx}^{a_k}\right.\\ &- w^{a_1}\frac{\partial\alpha_{a_1\cdots a_k}}{\partial x^b}\boldsymbol{dx}^b \wedge \boldsymbol{dx}^{a_2} \wedge \cdots \wedge \boldsymbol{dx}^{a_k}\\ &\left.+ w^{a_2}\frac{\partial\alpha_{a_1a_2\cdots a_k}}{\partial x^b}\boldsymbol{dx}^b \wedge \boldsymbol{dx}^{a_1} \wedge \boldsymbol{dx}^{a_3} \wedge \cdots \wedge \boldsymbol{dx}^{a_k} - \cdots + \cdots\right).\end{aligned}$$

In this expression the derivative from α in the first term cancels the last group of k terms (which are all equal). Therefore, this simplifies to

$$\frac{1}{(k-1)!}\frac{\partial w^b}{\partial x^{a_1}}\alpha_{ba_2\cdots a_k}\,dx^{a_1}\wedge dx^{a_2}\wedge\cdots\wedge dx^{a_k} + \frac{1}{k!}w^b\frac{\partial\alpha_{a_1\cdots a_k}}{\partial x^b}\,dx^{a_1}\wedge dx^{a_2}\wedge\cdots\wedge dx^{a_k}.$$

Since the expression for $(\mathfrak{L}_w\alpha)_{a_1\cdots a_k}$ is skew symmetric in $a_1, \ldots, a_k$,

$$\begin{aligned}\mathfrak{L}_w\alpha &= \frac{1}{k!}w^b\frac{\partial\alpha_{a_1\cdots a_k}}{\partial x^b}\,dx^{a_1}\wedge\cdots\wedge dx^{a_k}\\ &\quad+\frac{1}{k!}\left(\alpha_{ba_2\cdots a_k}\frac{\partial w^b}{\partial x^{a_1}}+\text{(all lower indices)}\right)dx^{a_1}\wedge\cdots\wedge dx^{a_k}\\ &= \frac{1}{k!}w^b\frac{\partial\alpha_{a_1\cdots a_k}}{\partial x^b}\,dx^{a_1}\wedge\cdots\wedge dx^{a_k}\\ &\quad+\frac{1}{(k-1)!}\frac{\partial w^b}{\partial x^{a_1}}\alpha_{ba_2\cdots a_k}\,dx^{a_1}\wedge\cdots\wedge dx^{a_k}\end{aligned}$$

since all the lower index terms are equal. Thus the two expressions agree. ∎

7.7 Corollary $\mathfrak{L}_w d = d\mathfrak{L}_w$.

Proof

$$\mathfrak{L}_w\,d\alpha = di_w(d\alpha) + i_w\,d(d\alpha) = di_w(d\alpha) = d(i_w\,d\alpha + di_w\alpha) = d\mathfrak{L}_w\alpha. \quad ∎$$

Box 7.1 ***Summary of Identities Relating d, i_w, and $\mathfrak{L}_w$ for Differential Forms***

1. (a) $d\circ d = 0$, (b) $i_w i_w = 0$
2. (a) $d(\alpha\wedge\beta) = d\alpha\wedge\beta + (-1)^k\alpha\wedge d\beta$, α a k-form
   (b) $i_w(\alpha\wedge\beta) = i_w\alpha\wedge\beta + (-1)^k\alpha\wedge i_w\alpha$
   (c) $\mathfrak{L}_w(\alpha\wedge\beta) = \mathfrak{L}_w\alpha\wedge\beta + \alpha\wedge\mathfrak{L}_w\beta$
3. (a) $\phi^*(\alpha\wedge\beta) = \phi^*\alpha\wedge\phi^*\beta$
   (b) $\phi^*d\alpha = d\phi^*\alpha$
   (c) $\phi^*i_w\alpha = i_{\phi^*w}\phi^*\alpha$ (if ϕ is regular)
   (d) $\phi^*\mathfrak{L}_w\alpha = \mathfrak{L}_{\phi^*w}\phi^*\alpha$ (if ϕ is regular)
4. (a) $$d\alpha(w_0, w_1, \ldots, w_k) = \sum_{i=0}^{k}(-1)^i w_i[\alpha(\underbrace{w_0, \ldots, w_k}_{w_i \text{ missing}})] + \sum_{i<j}(-1)^{i+j}\alpha([w_i, w_j], \underbrace{w_0, \ldots, w_k}_{w_i,\, w_j \text{ missing}})$$

   and in a chart,

   (b) $$(d\alpha)_x(w_0, \ldots, w_k) = \sum_{i=0}^{k}(-1)^i D\alpha_x\cdot w_i\cdot(\underbrace{w_0, \ldots, w_k}_{w_i \text{ missing}})$$

5. $(\mathcal{L}_w\boldsymbol{\alpha})(w_1, \ldots, w_k) = w[\boldsymbol{\alpha}(w_1, \ldots, w_k)] - \sum_{i=1}^{k} \boldsymbol{\alpha}(w_1, \ldots, [w, w_i], \ldots, w_k)$ ($[w, w_i]$ is in the ith slot of $\boldsymbol{\alpha}$).
6. (a) $\mathcal{L}_w = i_w d + d i_w$
   (b) $\mathcal{L}_w d = d\mathcal{L}_w$
   (c) $\mathcal{L}_{[w,v]}\boldsymbol{\alpha} = \mathcal{L}_w\mathcal{L}_v\boldsymbol{\alpha} - \mathcal{L}_v\mathcal{L}_w\boldsymbol{\alpha}$
   (d) $i_{[w,v]}\boldsymbol{\alpha} = \mathcal{L}_w i_v\boldsymbol{\alpha} - i_v\mathcal{L}_w\boldsymbol{\alpha}$
   (e) $\mathcal{L}_{fw}\boldsymbol{\alpha} = f\mathcal{L}_w\boldsymbol{\alpha} + df \wedge i_w\boldsymbol{\alpha}$

In the boxes that follow we present two applications of differential forms which are of interest in mechanics.

Box 7.2 ***The Poincaré Lemma and Variational Principles***

A differential form $\boldsymbol{\alpha}$ is called *closed* if $d\boldsymbol{\alpha} = \mathbf{0}$ and *exact* if $\boldsymbol{\alpha} = d\boldsymbol{\beta}$ for a form $\boldsymbol{\beta}$. Since $d^2 = \mathbf{0}$, every exact form is closed. The converse is not true. (For example, the form $\boldsymbol{\alpha} = x\,dy - y\,dx$ restricted to the unit circle is closed but is not exact, for its line integral—in the sense of advanced calculus—is $2\pi \neq 0$. Locally $\boldsymbol{\alpha} = d\theta$, where θ is angular measure.)

7.8 Poincaré Lemma *Let $\boldsymbol{\alpha}$ be a closed k-form on a manifold $\mathfrak{M}$. Then in some neighborhood of each point, $\boldsymbol{\alpha}$ is exact.*

Proof (*cf.* Moser [1965]) Using coordinates, it suffices to prove the lemma on a ball $\mathcal{U}$ centered at the origin in $\mathbb{R}^n$. Consider the radial motion given by $\phi_t(X) = tX$. For $t > 0$, ϕ_t is a regular mapping of $\mathbb{R}^n$ to itself. The velocity field of ϕ_t is $v_t(x) = \boldsymbol{x}/t$. ($\boldsymbol{x}$ means the vector from the origin to the point x.) From $\mathcal{L}_v = i_v d + d i_v$ and the flow definition of Lie derivatives,

$$\begin{aligned}\frac{d}{dt}\phi_t^*\boldsymbol{\alpha} &= \phi_t^*\mathcal{L}_v\boldsymbol{\alpha} = \phi_t^*(di_v\boldsymbol{\alpha} + i_v d\boldsymbol{\alpha}) = \phi_t^*(di_v\boldsymbol{\alpha}) \qquad (\text{since } d\boldsymbol{\alpha} = \mathbf{0})\\ &= d(\phi_t^* i_v\boldsymbol{\alpha}) \quad (\text{by 7.5}).\end{aligned}$$

Integrating from $t = 0+$ to $t = 1$, noting $\phi_1 = \text{Identity}$, gives $\boldsymbol{\alpha} = d\int_0^1 (\phi_t^* i_v\boldsymbol{\alpha})\,dt$, so we can choose $\boldsymbol{\beta} = \int_0^1 \phi_t^* i_v\boldsymbol{\alpha}\,dt$. Explicitly,

$$\boldsymbol{\beta}_x(w_1, \ldots, w_{k-1}) = \int_0^1 t^{k-1}\boldsymbol{\alpha}_{tx}(\boldsymbol{x}, w_1 \ldots, w_{k-1})\,dt. \quad \blacksquare$$

It follows that in $\mathbb{R}^3$, if $\nabla \times \boldsymbol{v} = \mathbf{0}$, then $\boldsymbol{v} = \nabla f$ for some f and that if div $\boldsymbol{v} = 0$, then $\boldsymbol{v} = \nabla \times \boldsymbol{w}$ for some $\boldsymbol{w}$; these are well-known results in vector calculus.

Note. We proved the Poincaré lemma for finite dimensional manifolds; essentially the same proof works in infinite dimensional spaces as well.

We now apply the Poincaré lemma to a problem in the calculus of variations. (We return to this topic in Chapter 5.) Let us first recall how the Euler–Lagrange equations for a Lagrangian $\mathcal{L}(q, \dot{q})$ may be regarded as the equations for a critical point. Let $\mathcal{Q}$ denote the space of all paths $q(t) \in \mathbb{R}^n$ with $0 \leq t \leq T$ and $q(0)$ and $q(T)$ fixed at specified values. Then $\mathcal{Q}$ is an infinite-dimensional space, but let us apply calculus to it in any case. (This points out the need to generalize our ideas to infinite dimensions which we do in Chapter 4.) The tangent space to $\mathcal{Q}$ is obtained by differentiating a curve q_λ in $\mathcal{Q}$ and thus consists of all vector functions $\boldsymbol{h}(\cdot)$, which are zero at $t = 0$ and $t = T$. Define

$$L\colon \mathcal{Q} \longrightarrow \mathbb{R} \quad \text{by} \quad L(q) = \int_0^T \mathcal{L}(q(t), \dot{q}(t))\, dt.$$

7.9 Proposition *A curve $q(\cdot) \in \mathcal{Q}$ is a critical point of L; that is, $DL(q) = 0$ if and only if the Euler–Lagrange equations hold:*

$$\frac{d}{dt}\frac{\partial \mathcal{L}}{\partial \dot{q}^i} - \frac{\partial \mathcal{L}}{\partial q^i} = 0 \quad (i = 1, \ldots, n).$$

(All functions appearing are assumed continuous.)

Proof Let us differentiate L in the direction $\boldsymbol{h}$ using the chain rule:

$$\begin{aligned} DL(q)\cdot \boldsymbol{h} &= \frac{d}{d\lambda} L(q + \lambda \boldsymbol{h})_{|\lambda=0} \\ &= \frac{d}{d\lambda}\int_0^T \mathcal{L}(q(t) + \lambda \boldsymbol{h}(t), \dot{q}(t) + \lambda \dot{\boldsymbol{h}}(t))\, dt_{|\lambda=0} \\ &= \int_0^T \left(\frac{\partial \mathcal{L}}{\partial q}\cdot \boldsymbol{h} + \frac{\partial \mathcal{L}}{\partial \dot{q}}\cdot \dot{\boldsymbol{h}}\right) dt. \end{aligned}$$

Since $\boldsymbol{h}$ vanishes at $t = 0$ and $t = T$, we can integrate by parts to get

$$DL(q)\cdot \boldsymbol{h} = \int_0^T \left(\frac{\partial \mathcal{L}}{\partial q} - \frac{d}{dt}\frac{\partial \mathcal{L}}{\partial \dot{q}}\right)\cdot \boldsymbol{h}(t)\, dt.$$

Thus if the Euler–Lagrange equations hold, obviously $DL(q) = 0$. Conversely, if $DL(q)\cdot \boldsymbol{h} = 0$ for all $\boldsymbol{h}$ and if $\partial\mathcal{L}/\partial q - (d/dt)(\partial \mathcal{L}/\partial \dot{q}) \neq 0$, then choose a neighborhood about a point t_0 where $\partial\mathcal{L}/\partial q - (d/dt)(\partial \mathcal{L}/\partial \dot{q})$ is nowhere zero (by continuity) and choose a parallel vector $\boldsymbol{h}(t) \neq \mathbf{0}$ in this neighborhood and zero outside; then $DL(q)\cdot \boldsymbol{h} \neq 0$. Therefore, if $DL(q) = 0$, the Euler–Lagrange equations must hold. ∎

Abstracting this, let $\mathfrak{X}$ be a Banach space, let $\langle\ ,\ \rangle$ be a bilinear form (e.g., an inner product) on $\mathfrak{X}$, and let $A: \mathfrak{X} \to \mathfrak{X}$ be a given (nonlinear) operator.

7.10 Definition We say A is a *potential operator* if there is a function $L: \mathfrak{X} \to \mathbb{R}$ such that $dL(x)\cdot v = \langle A(x), v\rangle$ for all x in $\mathfrak{X}$ and $v \in \mathfrak{X}$.

In view of 7.9, $A(x) = 0$ represents the Euler–Lagrange equations for $x \in \mathfrak{X}$, in abstract form. The next theorem is due to Vainberg [1964] (although the present proof is due to the authors—See Hughes and Marsden [1977]).

7.11 Theorem *A given operator A is a potential operator if and only if for each $x \in \mathfrak{X}$, v_1 and $v_2 \in \mathfrak{X}$,*

$$\langle DA(x)\cdot v_1, v_2\rangle = \langle DA(x)\cdot v_2, v_1\rangle.$$

If $\langle\ ,\ \rangle$ is symmetric, this is equivalent to saying $DA(x)$ is a symmetric linear operator on $\mathfrak{X}$.

Proof Consider the one-form $\boldsymbol{\alpha}(x)\cdot v = \langle A(x), v\rangle$. By definition, A is a potential operator if and only if $\boldsymbol{\alpha}$ is exact. By the Poincaré lemma, this is the case if and only if $d\boldsymbol{\alpha} = \mathbf{0}$. But by Formula 4(b) in Box 7.1,

$$d\boldsymbol{\alpha}(x)\cdot(v_1, v_2) = \langle DA(x)\cdot v_1, v_2\rangle - \langle DA(x)\cdot v_2, v_1\rangle,$$

so the result follows immediately. ∎

It may be instructive for the reader to write out an explicit proof using $L(x) = \int_0^1 \langle A(tx), x\rangle\, dt$.

This result gives necessary and sufficient conditions for a given set of equations to be the Euler–Lagrange equations for some Lagrangian.[32]

Box 7.3 ***A Geometric Formulation of Hamiltonian Mechanics***

In Chapter 5 we study elastodynamics as an infinite-dimensional Hamiltonian system. Here we indicate briefly how differential forms and Lie derivatives can be used in classical mechanics (see Arnold [1978] and Abraham and Marsden [1978] for further details).

[32]To ensure that L comes from a Lagrangian density requires the further assumption that the operator A is a local operator. Then L will be a density and A will be the usual Euler–Lagrange operator if $\langle\ ,\ \rangle$ is the L^2 inner product. For related work, see Lawruk and Tulczyjew [1977] and Takens [1977].

Consider a given Hamiltonian function $H(q, \boldsymbol{p})$ for $q \in \mathbb{R}^n$, $\boldsymbol{p} \in \mathbb{R}^{n*}$ (i.e., $\boldsymbol{p} = p_i\, dx^i$ is to be regarded as a one-form), and the associated Hamiltonian equations

$$\frac{dq^i}{dt} = \frac{\partial H}{\partial p_i}, \qquad \frac{dp_i}{dt} = -\frac{\partial H}{\partial q^i} \quad (i = 1, 2, \ldots, n)$$

(Recall that if $H = (1/2m)\langle \boldsymbol{p}, \boldsymbol{p} \rangle + V(q)$, these reduce to Newton's second law: $dq^i/dt = p_i/m$ and $m(d^2q^i/dt^2) = -\partial V/\partial q^i$.)

Let $\boldsymbol{X}_H$ be the vector field for Hamilton's equations,

$$\boldsymbol{X}_H = \left(\frac{\partial H}{\partial p_i}, -\frac{\partial H}{\partial q^i}\right).$$

Then $(q(t), \boldsymbol{p}(t))$ is an integral curve of $\boldsymbol{X}_H$ if and only if Hamilton's equations hold.

Let $\boldsymbol{\omega} = \boldsymbol{d}q^i \wedge \boldsymbol{d}p_i$ denote the fundamental two-form. Observe that it has the matrix of components given by the skew-symmetric matrix $\begin{bmatrix} 0 & I \\ -I & 0 \end{bmatrix}$. We have the following key identity: $\boldsymbol{i}_{X_H}\boldsymbol{\omega} = \boldsymbol{d}H$. (*Proof*

$$\begin{aligned} \boldsymbol{i}_{X_H}\boldsymbol{\omega} &= \boldsymbol{i}_{X_H}(\boldsymbol{d}q^i \wedge \boldsymbol{d}p_i) \\ &= (\boldsymbol{i}_{X_H}\boldsymbol{d}q^i) \wedge \boldsymbol{d}p_i - \boldsymbol{d}q^i \wedge \boldsymbol{i}_{X_H}\boldsymbol{d}p_i \quad \text{[see identity 2(b), Box 7.1]} \\ &= \frac{\partial H}{\partial p_i}\boldsymbol{d}p_i + \frac{\partial H}{\partial q^i}\boldsymbol{d}q^i = \boldsymbol{d}H. \end{aligned}$$

Note that $\boldsymbol{\omega}$ is *closed:* $\boldsymbol{d\omega} = \boldsymbol{0}$. In fact, $\boldsymbol{\omega}$ is *exact:* $\boldsymbol{\omega} = \boldsymbol{d\theta}$, where $\boldsymbol{\theta} = -p_i\, \boldsymbol{d}q^i$.

Now we can easily prove a number of key results about Hamiltonian systems:

Conservation of energy If F_t is the flow of $\boldsymbol{X}_H$ (see Section 1.6), then $H \circ F_t = H$. (*Proof:*

$$\begin{aligned} \frac{d}{dt} H \circ F_t &= \frac{d}{dt} F_t^* H = F_t^* \mathfrak{L}_{X_H} H = F_t^*(\boldsymbol{i}_{X_H}\boldsymbol{d}H) \\ &= F_t^*(\boldsymbol{i}_{X_H}\boldsymbol{i}_{X_H}\boldsymbol{\omega}) = 0, \end{aligned}$$

thus $\quad H \circ F_t = H \circ F_0 = H.$)

Each F_t *is a canonical transformation;* that is, $F_t^*\boldsymbol{\omega} = \boldsymbol{\omega}$. (*Proof*

$$\begin{aligned} \frac{d}{dt}F_t^*\boldsymbol{\omega} &= F_t^*\mathfrak{L}_{X_H}\boldsymbol{\omega} = F_t^*(\boldsymbol{d}\boldsymbol{i}_{X_H}\boldsymbol{\omega} + \boldsymbol{i}_{X_H}\boldsymbol{d\omega}) \\ &\quad \text{(Cartan's "Magic Formula")} \\ &= F_t^*(\boldsymbol{d}\boldsymbol{i}_{X_H}\boldsymbol{\omega}) \quad (\text{since } \boldsymbol{d\omega} = \boldsymbol{0}) = F_t^*\boldsymbol{ddH} = \boldsymbol{0}.) \end{aligned}$$

Liouville's Theorem F_t preserves the phase volume:

$$\boldsymbol{\mu} = \boldsymbol{d}q^1 \wedge \cdots \wedge \boldsymbol{d}q^n \wedge \boldsymbol{d}p_1 \wedge \cdots \wedge \boldsymbol{d}p_n.$$

(*Proof* By computation, one finds that

$$\underbrace{\boldsymbol{\omega} \wedge \boldsymbol{\omega} \wedge \cdots \wedge \boldsymbol{\omega}}_{n \text{ times}} = n!\,(-1)^{[n/2]}\boldsymbol{\mu}.$$

Here $[n/2]$ denotes the largest integer $\leq n/2$. Since $F_t^*\boldsymbol{\omega} = \boldsymbol{\omega}$, and $F_t^*(\boldsymbol{\omega} \wedge \boldsymbol{\omega}) = F_t^*\boldsymbol{\omega} \wedge F_t^*\boldsymbol{\omega} = \boldsymbol{\omega} \wedge \boldsymbol{\omega}$, we get $F_t^*\boldsymbol{\mu} = \boldsymbol{\mu}$.)

One can establish other basic facts like conservation laws in the presence of a symmetry group (Noether's theorem) in a similar way. We shall return to this in Chapter 5.

Next we summarize, without proofs, a few key facts about integration on manifolds. These are familiar results in $\mathbb{R}^n$ which are converted to the language of manifolds.[33] Let $\mathfrak{M}$ be a manifold, possibly with a (piecewise smooth) boundary. We assume $\mathfrak{M}$ is oriented; that is, $\mathfrak{M}$ has a covering by charts such that the Jacobian of the change of coordinates between any two of these coordinate systems is positive.

Let $\boldsymbol{\mu}$ be an n-form on $\mathfrak{M}$ (n is the dimension of $\mathfrak{M}$). In any coordinate chart $\{X^A\}$, $\boldsymbol{\mu}$ has the form $\boldsymbol{\mu} = f\,\boldsymbol{d}X^1 \wedge \cdots \wedge \boldsymbol{d}X^n$. If we integrate f in this chart and do the same for a covering of $\mathfrak{M}$ by charts not counting overlaps twice, we get a well-defined *number*, $\int_{\mathfrak{M}} \boldsymbol{\mu}$ (the change of variables formula and transformation properties of $\boldsymbol{\mu}$ guarantee that this number is independent of the way in which $\mathfrak{M}$ is sliced up).

7.12 Theorem (*Change of Variables*) *If $\phi\colon \mathfrak{M} \to \mathfrak{N}$ is a regular C^1 mapping that is orientation preserving, and $\boldsymbol{\mu}$ is an n-form on $\phi(\mathfrak{M})$, then*

$$\boxed{\int_{\mathfrak{M}} \phi^*\boldsymbol{\mu} = \int_{\phi(\mathfrak{M})} \boldsymbol{\mu}.}$$

In Section 1.4 we described how to obtain a volume element on an oriented Riemannian manifold $(\mathfrak{M}, G)$: $\boldsymbol{d}V = \sqrt{\det G_{AB}}\,\boldsymbol{d}X^1 \wedge \cdots \wedge \boldsymbol{d}X^n$, and the Jacobian of ϕ by $\phi^*\,\boldsymbol{d}v = J\,\boldsymbol{d}V$. Using this notation, 7.12 can be written

$$\int_{\mathfrak{M}} (f \circ \phi) J\, dV = \int_{\phi(\mathfrak{M})} f\, dv$$

for a scalar function f on $\phi(\mathfrak{M})$ by writing $\boldsymbol{\mu} = f\,\boldsymbol{d}v$. Under an integral sign we shall usually write dV instead of $\boldsymbol{d}V$ in accordance with usage in integration theory.

[33]The proofs of the following theorems are omitted; they may be found in one of the standard references (e.g., Spivak [1975], Lang [1972], or Abraham and Marsden [1978]).

7.13 Theorem (*Stokes' Theorem*) *If $\partial\mathfrak{M}$ is positively oriented*[34] *and* $\boldsymbol{\alpha}$ *is an $(n-1)$-form on $\mathfrak{M}$, then*

$$\boxed{\int_{\mathfrak{M}} d\boldsymbol{\alpha} = \int_{\partial\mathfrak{M}} \boldsymbol{\alpha}.}$$

This version of Stokes' theorem includes as special cases, the usual theorems of Green, Gauss, and Stokes. We shall obtain the divergence theorem on a Riemannian manifold as a special case after a few preparatory results.

7.14 Proposition *If W is a vector field on $\mathfrak{M}$, then $\mathfrak{L}_W\, dV = (\mathrm{DIV}\, W)\, dV$.*

Proof Observe that dV is closed (any n-form on an n-manifold is closed). Thus by the magic formula,

$$\begin{aligned}
\mathfrak{L}_W\, dV &= d(i_W\, dV) = d(i_W \sqrt{\det G_{AB}}\, dX^1 \wedge \cdots \wedge dX^n) \\
&= d[\sqrt{\det G_{AB}}(W^1\, dX^2 \wedge \cdots \wedge dX^n \\
&\qquad - W^2\, dX^1 \wedge dX^3 \wedge \cdots \wedge dX^n + \cdots)] \\
&= \frac{\partial}{\partial X^C}(\sqrt{\det G_{AB}} W^C)\, dX^1 \wedge \cdots \wedge dX^n \\
&= \frac{1}{\sqrt{\det G_{AB}}} \frac{\partial}{\partial X^C}(\sqrt{\det G_{AB}} W^C)\, dV = (\mathrm{div}\, W)\, dV. \quad \blacksquare
\end{aligned}$$

This gives an easy proof of the formula for the rate of change of the Jacobian, which was proved directly in Section 1.5.

7.15 Corollary $\partial/\partial t\, J = J \cdot (\mathrm{div}\, v) \circ \phi$.

Proof
$$\begin{aligned}
\frac{\partial}{\partial t}(\phi_t^*\, dv) &= \phi_t^* \mathfrak{L}_v\, dv = \phi_t^*(\mathrm{div}\, v\, dv) = (\mathrm{div}\, v \circ \phi_t)\phi_t^*\, dv \\
&= J(\mathrm{div}\, v \circ \phi)\, dV. \quad \blacksquare
\end{aligned}$$

To get the divergence theorem we need one more important observation.

7.16 Proposition *Let N be the unit outward normal to $\partial\mathfrak{M}$ and W a vector field on $\mathfrak{M}$. Then on $\partial\mathfrak{M}$, $W \cdot N\, dA = i_W\, dV$, where dA is the "area" element on $\partial\mathfrak{M}$ (i.e., dA is the "dV" for the $(n-1)$-manifold $\partial\mathfrak{M}$).*

Proof Let $\{X^A\}$ be coordinates for $\mathfrak{M}$ in which $\partial\mathfrak{M}$ is the plane $X^1 = 0$ and for which $N = (1, 0, 0, \ldots)$, and $\mathfrak{M}$ is described by $X^1 < 0$. We can arrange for N to be normal to this plane at any particular point by a linear transformation

---

[34]In an oriented coordinate chart $\{X^1, \ldots, X^n\}$ for $\mathfrak{M}$ in which $\partial\mathfrak{M}$ is the plane $X^1 = 0$, $\mathfrak{M}$ must be represented as the half-space $X^1 < 0$. This agrees with the usual choice of normals in Green's, Gauss', and Stokes' theorems in vector calculus.

of coordinates; that is $G_{1A} = 0$ $(A = 2, \ldots, n)$ and $G_{11} = 1$. The metric tensor on $\partial\mathfrak{M}$ is G_{AB} $(A, B = 2, \ldots, n)$. As above,

$$
\begin{aligned}
i_W\, dV &= i_W \sqrt{\det G_{AB}}\, dX^1 \wedge \cdots \wedge dX^n \\
&= \sqrt{\det G_{AB}} \sum_{i=1}^{n} (-1)^{i-1} W^i \underbrace{dX^1 \wedge \cdots \wedge dX^n}_{dX^i \text{ missing}}
\end{aligned}
$$

If we evaluate this expression at a point satisfying $X^1 = 0$, we get only the first term in this sum:

$$i_W\, dV = \sqrt{\det G_{AB}} W^1\, dX^2 \wedge \cdots \wedge dX^n = (W \cdot N)\, dA. \quad \blacksquare$$

7.17 Theorem (*Divergence Theorem*) *If W is a vector field on $\mathfrak{M}$, then*

$$\boxed{\int_{\mathfrak{M}} \operatorname{DIV} W\, dV = \int_{\partial\mathfrak{M}} W \cdot N\, dA.}$$

Proof Let $\boldsymbol{\alpha} = i_W\, dV$. Then $\int_{\partial\mathfrak{M}} \boldsymbol{\alpha} = \int_{\partial\mathfrak{M}} W \cdot N\, dA$ by 7.16. Also, $d\boldsymbol{\alpha} = d i_W\, dV = (\operatorname{DIV} W)\, dV$ by 7.14. Hence the result follows from Stokes' theorem. $\blacksquare$

Problem 7.5 (i) Explicitly recover the classical Stokes' theorem for oriented surfaces from 7.13.

(ii) Show that, on a Riemannian manifold $(\mathfrak{N}, \boldsymbol{g})$,

$$\omega_{ab} = \frac{1}{2}(v_{a|b} - v_{b|a}) = \frac{1}{2}\left(\frac{\partial v_a}{\partial x^b} - \frac{\partial v_b}{\partial x^a}\right)$$

are the components of the spin $\boldsymbol{\omega} = \frac{1}{2} \boldsymbol{d}(\boldsymbol{v}^\flat)$. Formulate a Stokes' theorem for a general oriented surface in a Riemannian manifold.

Problem 7.6 Set S^{AB} be a 2-tensor on $\mathfrak{M}$ and $\operatorname{DIV} S = S^{AB}{}_{|B}$, its divergence. Let α_A be a one-form on $\mathfrak{M}$. Prove the following *integration by parts formula* (in index notation):

$$\int_{\mathfrak{M}} \alpha_A S^{AB}{}_{|B}\, dV = -\int_{\mathfrak{M}} \alpha_{A|B} S^{AB}\, dV + \int_{\partial\mathfrak{M}} \alpha_A S^A{}_B N^B\, dA.$$

Now we turn our attention to the Piola transformation. This concept is of fundamental importance in the subsequent chapters. The Piola transformation is analogous to pull-back, except that there is a Jacobian present as well. This is indicative that volume or area forms are being transformed. We begin by defining the Piola transform of vector fields, and revert to $\mathcal{B}$, $\mathcal{S}$ notation in place of $\mathfrak{M}$ and $\mathfrak{N}$ because we have configurations of bodies in mind.

7.18 Definition Let $\boldsymbol{y}$ be a vector field on $\mathcal{S}$ and $\phi: \mathcal{B} \to \mathcal{S}$ a regular (orientation preserving) C^1 mapping. The *Piola transform* of $\boldsymbol{y}$ is given by

$$\boldsymbol{Y} = J\phi^* \boldsymbol{y}$$

where J is the Jacobian of ϕ. In coordinates, $Y^A = J(F^{-1})^A{}_b y^b$, where

$$J = \frac{\sqrt{\det g_{ab}}}{\sqrt{\det G_{AB}}} \frac{\partial(\phi^1, \ldots, \phi^n)}{\partial(X^1, \ldots, X^n)} \quad \text{and} \quad F^a{}_A = \frac{\partial \phi^a}{\partial X^A}.$$

We can phrase this in another useful way:

7.19 Proposition *Y is the Piola transform of y if and only if $\phi^*(i_y\, dv) = i_Y\, dV$.*

Proof Notice that $(n-1)$-forms and vector fields are in one-to-one correspondence by way of the mapping $Y \mapsto *Y$, where $*Y = i_Y\, dV$. (Above we calculated $i_Y\, dV$ in coordinates.) From Box 7.1, $\phi^*(i_y\, dv) = i_{\phi^* y}\phi^*\, dv = i_{\phi^* y} J\, dV = i_{J\phi^* y}\, dV$ so the assertion follows. ∎

Note. Y is an honest vector field, while $[\partial(\phi^1, \ldots, \phi^n)/\partial(X^1, \ldots, X^n)](F^{-1})^A{}_b y^b$ is *not*. The metric factors are important, even for curvilinear coordinates in $\mathbb{R}^3$, such as spherical coordinates.

7.20 Theorem (*Piola Identity*) *If Y is the Piola transform of y, then*

$$\boxed{\operatorname{DIV} Y = J\cdot(\operatorname{div} y)\circ\phi.}$$

We shall give two proofs of this important result.

First proof Let $\mathcal{U} \subset \mathcal{B}$ be a nice open set and $\partial\mathcal{U}$ its boundary. By the change of variables theorem and 7.19,

$$\int_{\partial\mathcal{U}} i_Y\, dV = \int_{\partial\phi_t(\mathcal{U})} i_y\, dv.$$

From 7.16 and 7.17,

$$\int_{\mathcal{U}} \operatorname{DIV} Y\, dV = \int_{\phi_t(\mathcal{U})} \operatorname{div} y\, dv = \int_{\mathcal{U}} J(\operatorname{div} y\circ\phi)\, dV.$$

Since $\mathcal{U}$ is arbitrary, the assertion follows. ∎

Second Proof We compute directly, using differential forms: $(\operatorname{DIV} Y)\, dV = \mathcal{L}_Y\, dV = d(i_Y\, dV)$ by the general formula $\mathcal{L}_Y\, \alpha = i_Y\, d\alpha + d i_Y \alpha$ and the fact that d of an n-form is zero. Thus $(\operatorname{DIV} Y)\, dV = d(i_Y\, dV) = d\phi^*(i_y\, dv) = \phi^* d(i_y\, dv)$ (the operations of pull-back and d commute) $= \phi^*(\operatorname{div} y\, dv) = J(\operatorname{div} y\circ\phi)\, dV$ (definition of J), and so $\operatorname{DIV} Y = J(\operatorname{div} y\circ\phi)$. ∎

From 7.16 we get another important way of expressing the Piola transformation: $Y\cdot N\, dA = y\cdot n\, da$, where da is the area element on $\phi(\mathcal{U})$; it is related to dA according to $(da)_b = J(F^{-1})^A{}_b(dA)_A$. See Figure 1.7.1. This equation shows how the area elements on $\partial\mathcal{U}$ and $\partial\phi(\mathcal{U})$ are related.

Since $Y^A = J(F^{-1})^A{}_a y^a$, the Piola identity may be read this way:

$$\operatorname{DIV}(JF^{-1}) = 0,$$

where

$$\begin{aligned}(\operatorname{DIV}(JF^{-1}))_a &= (J(F^{-1})^A{}_a)_{|A} \\ &= \frac{1}{\sqrt{\det G_{BC}}} \frac{\partial}{\partial X^A}\left(\sqrt{\det G_{BC}}\, J(F^{-1})^A{}_a\right).\end{aligned}$$

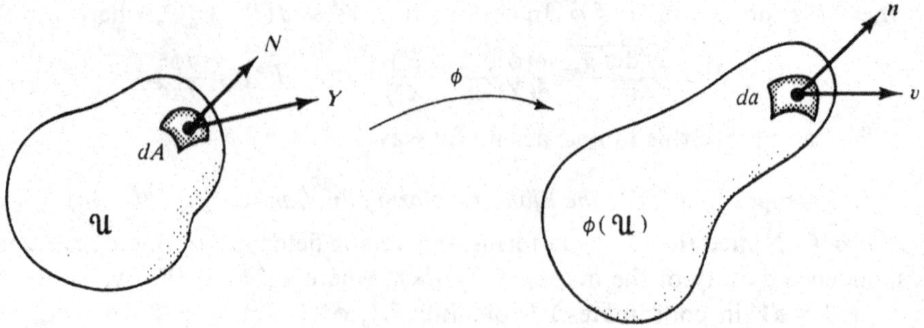

Figure 1.7.1

We can also make a Piola transformation on any index of a tensor. For example, let σ^{ab} be a given two tensor. If we make a Piola transformation on the last index, we get a two-point tensor P with components $P^{aB} = J(F^{-1})^B{}_b\sigma^{ab}$. The Piola identity, then, tells us that DIV $\boldsymbol{P} = J(\text{div}\,\boldsymbol{\sigma})\circ\phi$; that is, $P^{aB}{}_{|B} = J\sigma^{ab}{}_{|b}$.

Problem 7.7 If $\boldsymbol{y}\cdot\boldsymbol{n}$ is interpreted as a flux per unit area on $\phi(\mathfrak{U})$, show that one can interpret $\boldsymbol{Y}\cdot\boldsymbol{N}$ as the corresponding flux per unit of underformed area. Use this to give a physical interpretation of the Piola identity.

Box 7.4 ***Summary of Important Formulas for Section 1.7***

(See Box 7.1 for the key identities for differential forms.)

Definition of a k-form on $\mathfrak{M}$

$\boldsymbol{\alpha}$ is a $\binom{0}{k}$ tensor that is skew symmetric.

$$\boldsymbol{\alpha}(\boldsymbol{W}_{\pi(1)}, \ldots, \boldsymbol{W}_{\pi(k)}) = \text{sgn}\,\pi\boldsymbol{\alpha}(\boldsymbol{W}_1, \ldots, \boldsymbol{W}_k)$$

$\alpha_{A_1\cdots A_k}$ changes sign if any two indices are switched.

Wedge Product

$$(\boldsymbol{\alpha} \wedge \boldsymbol{\beta})(\boldsymbol{W}_1, \ldots, \boldsymbol{W}_{k+l}) = \frac{1}{k!\,l!}\sum_{\pi}(\text{sgn}\,\pi)\boldsymbol{\alpha} \otimes \boldsymbol{\beta} \cdot(\boldsymbol{W}_{\pi(1)}, \ldots, \boldsymbol{W}_{\pi(k+l)})$$

$(\boldsymbol{\alpha} \wedge \boldsymbol{\beta})_{A_1\cdots A_k B_1\cdots B_l}$ is the complete anti-symmetrization of $\alpha_{A_1\cdots A_k}\beta_{B_1\cdots B_l}$.

Interior Product

$$(\boldsymbol{i_W\alpha})(\boldsymbol{W}_2, \ldots, \boldsymbol{W}_k) = \boldsymbol{\alpha}(\boldsymbol{W}, \boldsymbol{W}_2, \ldots, \boldsymbol{W}_k)$$

$$(\boldsymbol{i_W\alpha})_{A_1\cdots A_k} = W^A\alpha_{AA_2\cdots A_k}$$

Exterior Derivative $\boldsymbol{d}$

Characterized by:

(i) $\boldsymbol{dd\alpha} = \boldsymbol{0}$

(ii) $\boldsymbol{d}(\boldsymbol{\alpha} \wedge \boldsymbol{\beta}) = \boldsymbol{d\alpha} \wedge \boldsymbol{\beta} + (-1)^k\boldsymbol{\alpha} \wedge \boldsymbol{d\beta}$

If $\boldsymbol{\alpha} = \alpha_{A_1\ldots A_k}\boldsymbol{d}X^{A_1} \wedge \cdots \wedge \boldsymbol{d}X^{A_k}$, then $\boldsymbol{d\alpha} = \frac{\partial\alpha_{A_1\cdots A_k}}{\partial X^A}\boldsymbol{d}X^A \wedge \boldsymbol{d}X^{A_1} \wedge \cdots \wedge \boldsymbol{d}X^{A_k}$

(iii) df = differential of the function f

(iv) d is local

Poincaré Lemma

If $d\boldsymbol{\alpha} = 0$, then locally there is a $\boldsymbol{\beta}$ such that $\boldsymbol{\alpha} = d\boldsymbol{\beta}$.

$$\beta_{A_1 \cdots A_{k-1}}(X) = \int_0^1 t^{k-1} X^A \alpha_{AA_1 \cdots A_{k-1}}(tX)\, dt$$

Euler–Lagrange Equations of a Variational Principle

$q(t) \in \mathbb{R}^n$ satisfies $(d/dt)(\partial \mathcal{L}/\partial \dot{q}^i) - \partial \mathcal{L}/\partial q^i = 0$ $(i = 1, \ldots, n)$ if and only if $q(\cdot)$ is a critical point of $L(q) = \int_0^1 \mathcal{L}(q, \dot{q})\, dt$ subject to the condition that $q(0)$, $q(1)$ be fixed.

Inverse Problem for a Variational Principle

A nonlinear operator A on a Banach space $\mathfrak{X}$ is a potential operator; that is, $\langle A(x), v\rangle = dL(x)\cdot v$ for some $L(x)$ if and only if $\langle DA(x)\cdot v_1, v_2\rangle$ is symmetric in v_1 and v_2 (i.e., $DA(x)$ is a symmetric operator).

Hamilton's Equations

Symplectic form: $\boldsymbol{\omega} = dq^i \wedge dp_i$

Hamiltonian vector field X_H:

$$i_{X_H}\boldsymbol{\omega} = dH \qquad X_H = \left(\frac{\partial H}{\partial p_i}, -\frac{\partial H}{\partial q^i}\right)$$

Change of Variables

$$\int_{\mathfrak{M}} \phi^*\boldsymbol{\mu} = \int_{\phi(\mathfrak{M})} \boldsymbol{\mu} \qquad \int_{\mathfrak{M}} (f\circ\phi) J\, dV = \int_{\phi(\mathfrak{M})} f\, dv$$

Stokes' Theorem

$$\int_{\mathfrak{M}} d\boldsymbol{\alpha} = \int_{\partial\mathfrak{M}} \boldsymbol{\alpha}$$

$$\int_{\mathfrak{M}} \frac{\partial \alpha_{A_1 \cdots A_{n-1}}}{\partial X^B}\, dX^B \wedge dX^{A_1} \wedge \cdots \wedge dX^{A_{n-1}} = \int_{\partial\mathfrak{M}} \alpha_{A_1 \cdots A_{n-1}}\, dX^{A_1} \wedge \cdots \wedge dX^{A_{n-1}}$$

Divergence Theorem

$$\int_{\mathfrak{M}} (\text{DIV}\, \boldsymbol{W})\, dV = \int_{\partial\mathfrak{M}} \boldsymbol{W}\cdot\boldsymbol{N}\, dA \qquad \int_{\mathfrak{M}} W^A{}_{|A}\, dV = \int_{\partial\mathfrak{M}} W^A N_A\, dA$$

Piola Transformation

$\boldsymbol{Y} = J\phi^*\boldsymbol{y}$ or $\qquad Y^A = J(F^{-1})^A{}_a y^a$

$i_{\boldsymbol{Y}}\, dV = \phi^*(i_{\boldsymbol{y}}\, dv)$

Piola Identity

$$\text{DIV}\, \boldsymbol{Y} = J(\text{div}\, \boldsymbol{y}\circ\phi) \qquad Y^A{}_{|A} = J y^a{}_{|a}$$

CHAPTER 2

BALANCE PRINCIPLES

This chapter presents the basic dynamical equations for continuum mechanics and some key inequalities from thermodynamics. The latter may be used to give functional form to the second Piola–Kirchhoff stress tensor—a fundamental ingredient in the dynamical equations. The study of this functional form is the main goal of Chapter 3 on constitutive theory.

We shall set up the basic equations (or inequalities) as integral balance conditions. Therefore, the first section of this chapter is devoted to their general study. For the dynamical equations, the basic postulate is the existence of a stress tensor and a momentum balance principle. In particle mechanics these are analogous to the postulate of Hamilton's equations, or Newton's second law. We shall present a detailed study of these ideas in both the material and spatial pictures by using the Piola transformation that was developed in Section 1.7.

The integral form of momentum balance is subject to an important criticism: it is not form invariant under *general* coordinate transformations, although the dynamical equations themselves are. We shall examine this and a number of related covariance questions. Covariance ideas are also useful in the discussion of constitutive theory, as we shall see in Chapter 3.

2.1 THE MASTER BALANCE LAW

Let $\mathcal{B}$ and $\mathcal{S}$ be Riemannian manifolds with metric tensors $\boldsymbol{G}$ and $\boldsymbol{g}$, respectively. Suppose $x = \phi(X, t)$ is a C^1 regular motion of $\mathcal{B}$ in $\mathcal{S}$, $\boldsymbol{V}(X, t)$ is the material velocity, and $\boldsymbol{v}(t, x)$ is the spatial velocity. For the moment we shall assume that $\mathcal{B}$ and $\mathcal{S}$ are the same dimension; for example, $\mathcal{B} \subset \mathcal{S}$ is open and ϕ is a regular C^1 motion. (For shells and related situations the results need to be modified along the lines indicated in Box 5.1, Chapter 1 and Box 1.2, Chapter 2.) Let dV be the volume element on $\mathcal{B}$ and dv that in $\mathcal{S}$, and let $J(X, t)$ be the Jacobian of $\phi_t(X)$. (See Section 1.5.) Recall that an open subset $\mathcal{U} \subset \mathcal{B}$ is called "nice" if it has a piecewise C^1 boundary $\partial\mathcal{U}$.

1.1 Transport Theorem *Let $f(x, t)$ be a given C^1 real-valued function of time t and position $x \in \phi_t(\mathcal{B})$, and let $\mathcal{U}$ be a nice open set in $\mathcal{B}$. Then (suppressing the arguments (x, t)),*

$$\boxed{\frac{d}{dt}\int_{\phi_t(\mathcal{U})} f\,dv = \int_{\phi_t(\mathcal{U})} (\dot{f} + f \operatorname{div} \boldsymbol{v})\,dv = \int_{\phi_t(\mathcal{U})} \left(\frac{\partial f}{\partial t} + \operatorname{div}(f\boldsymbol{v})\right) dv}$$

where $\dot{f} = \dfrac{\partial f}{\partial t} + df \cdot \boldsymbol{v} = \dfrac{\partial f}{\partial t} + \dfrac{\partial f}{\partial x^a} v^a$ is the material derivative of f and $\operatorname{div} \boldsymbol{v} =$ $v^a{}_{|a}$ *is the divergence of $\boldsymbol{v}$.*

Proof By change of variables, and differentiating under the integral sign,

$$\frac{d}{dt}\int_{\phi_t(\mathcal{U})} f\,dv = \frac{d}{dt}\int_{\mathcal{U}} f(\phi(X, t), t)J(X, t)\,dV$$

$$= \int_{\mathcal{U}} \left\{\dot{f}(\phi(X, t), t)J(X, t) + f(\phi(X, t), t)\frac{\partial J(X, t)}{\partial t}\right\} dV.$$

From 5.4, Chapter 1, $(\partial J/\partial t)(X, t) = (\operatorname{div} \boldsymbol{v})J$. Inserting this in the preceding expression and changing variables back to x gives the result. ∎

Problem 1.1 Generalize the transport theorem as follows. Let $\phi_t : \mathcal{B} \to \mathcal{S}$ be a regular motion of $\mathcal{B}$ in $\mathcal{S}$ and let $\mathcal{P} \subset \mathcal{B}$ be a k-dimensional submanifold. Let α be a k-form on $\mathcal{S}$ and let $\int_{\phi_t(\mathcal{P})}\alpha$ denote the integral of α over $\phi_t(\mathcal{P})$. Show that

$$\frac{d}{dt}\int_{\phi_t(\mathcal{P})} \boldsymbol{\alpha} = \int_{\phi_t(\mathcal{P})} \boldsymbol{L_v\alpha},$$

where $\boldsymbol{v}$ is the spatial velocity of the motion. Show that this includes the transport theorem as a special case by choosing $\mathcal{P} = \mathcal{U}$ and $\boldsymbol{\alpha} = \boldsymbol{f\,dv}$.

Many of the basic laws of continuum mechanics can be expressed in terms of integral equalities or inequalities. We abstract these by making the following.

1.2 Definition Let $a(x, t)$, $b(x, t)$ be given scalar functions defined for $t \in \mathbb{R}$ (or an open interval in $\mathbb{R}$), $x \in \phi_t(\mathcal{B})$, and $\boldsymbol{c}(x, t)$ a given vector field on $\phi_t(\mathcal{B})$. We say that a, b, and $\boldsymbol{c}$ satisfy the *master balance law* if, for any nice open set $\mathcal{U} \subset \mathcal{B}$, the integrals that appear in the following equation exist, $\int_{\phi_t(\mathcal{U})} a\,dv$ is t-differentiable and

$$\boxed{\frac{d}{dt}\int_{\phi_t(\mathcal{U})} a\,dv = \int_{\phi_t(\mathcal{U})} b\,dv + \int_{\partial\phi_t(\mathcal{U})} \langle \boldsymbol{c}, \boldsymbol{n}\rangle\,da}$$

where $\boldsymbol{n}$ is the unit outward normal to $\partial\phi_t(\mathcal{U})$ and da is the area element on this surface. If the above equality is replaced by the inequality

$$\boxed{\frac{d}{dt}\int_{\phi_t(U)} a\,dv \geq \int_{\phi_t(\mathcal{U})} b\,dv + \int_{\partial\phi_t(\mathcal{U})} \langle \boldsymbol{c}, \boldsymbol{n}\rangle\,da}$$

we say that a, b, $\boldsymbol{c}$ satisfy the *master balance inequality*.

If there is danger of confusion, we may refer to these as the *spatial master balance law* (or *inequality*, respectively).

The next theorem shows how to cast such a relationship into local form.

1.3 Localization Theorem (*Spatial*) *Let a and $\boldsymbol{c}$ be C^1 and b be C^0. They satisfy the master balance law if and only if*

$$\boxed{\frac{\partial a}{\partial t} + \operatorname{div}(a\boldsymbol{v}) = b + \operatorname{div}\boldsymbol{c}.} \tag{1}$$

They satisfy the master inequality if and only if

$$\boxed{\frac{\partial a}{\partial t} + \operatorname{div}(a\boldsymbol{v}) \geq b + \operatorname{div}\boldsymbol{c}.} \tag{1'}$$

Proof By the transport and divergence theorems, the master balance law is equivalent to

$$\int_{\phi_t(\mathcal{U})}\left(\frac{\partial a}{\partial t} + \operatorname{div}(a\boldsymbol{v})\right)dv = \int_{\phi_t(\mathcal{U})} b\,dv + \int_{\phi_t(\mathcal{U})} (\operatorname{div}\boldsymbol{c})\,dv$$

for any nice open set $\mathcal{U} \subset \mathcal{B}$. Thus if the identity (1) or inequality (1)′ hold, the corresponding integral relation holds. Conversely, we may use the following elementary calculus lemma: if $g(x)$ is continuous and $\int_{\mathcal{V}} g(x)\,dv = 0$ (resp. ≥ 0) on a family $\{\mathcal{V}\}$ of open sets of arbitrarily small diameter about each point, then $g(x) = 0$ (resp. is ≥ 0). The result therefore holds. ∎

The passage from the integral form to the localized form requires more differentiability than is needed to make sense of the integral form. Thus the integral form is more general and is crucial for the discussion of shocks. See Box 1.1, below.

Next we shall consider the material master balance law and inequality.

1.4 Definition Let $A(X, t)$ and $B(X, t)$ be given functions on $\mathcal{B}$ and $\boldsymbol{C}(X, t)$ a vector field. We say that A, B, and $\boldsymbol{C}$ satisfy the (*material*) *master balance law* if, for all nice open sets $\mathcal{U} \subset \mathcal{B}$, the integrals appearing in the following equation exist, $\int_{\mathcal{U}} AJ\, dV$ is t-differentiable and

$$\boxed{\frac{d}{dt}\int_{\mathcal{U}} AJ\, dV = \int_{\mathcal{U}} BJ\, dV + \int_{\partial U} \langle C, N\rangle\, dA}$$

where N is the unit outward normal to $\partial\mathcal{U}$ and dA is the area element on $\partial\mathcal{U}$. Similarly, we say that A, B, C obey the *master balance inequality* if

$$\boxed{\frac{d}{dt}\int_{\mathcal{U}} AJ\, dV \geq \int_{\mathcal{U}} BJ\, dV + \int_{\partial\mathcal{U}} \langle C, N\rangle\, dA}$$

for all nice open sets $\mathcal{U} \subset \mathcal{B}$.

Again we can localize, but the procedure is now simpler. We need only differentiate under the integral sign and equate integrands. Thus, we get the following:

1.5 Localization Theorem (*Material*) *Let AJ, BJ be C^0, $\partial(AJ)/\partial t$ exist and be C^0, and let $\boldsymbol{C}$ be C^1. Then A, B, C, satisfy the master balance law if and only if*

$$\boxed{\frac{\partial}{\partial t}(AJ) = BJ + \text{DIV}\, \boldsymbol{C},} \tag{2}$$

and the master balance inequality if and only if

$$\boxed{\frac{\partial}{\partial t}(AJ) \geq BJ + \text{DIV}\, \boldsymbol{C}.} \tag{2)'$$

Theorems 1.3 and 1.5 are connected as follows:

1.6 Proposition *Suppose A, B, $\boldsymbol{C}$, and a, b, $\boldsymbol{c}$, are related by*

$$A(X, t) = a(x, t) \quad (\text{where } x = \phi(X, t)),$$
$$B(X, t) = b(x, t),$$
$$\boldsymbol{C}(X, t) = J(X, t)\boldsymbol{F}^{-1}(X, t)\cdot \boldsymbol{c}(x, t)$$

(i.e., $\boldsymbol{C}$ is the Piola transform of $\boldsymbol{c}$ so that $C^A = J(F^{-1})^A{}_a c^a$). Then A, B, $\boldsymbol{C}$ satisfy the material master balance law (resp. inequality) if and only if a, b, $\boldsymbol{c}$ satisfy the spatial master balance law (resp. inequality).

Proof This follows from the integral forms in the definitions by changing variables and using the identity $\langle \boldsymbol{C}, \boldsymbol{N}\rangle\, dA = \langle \boldsymbol{c}, \boldsymbol{n}\rangle\, da$ derived in Section 1.7. ▮

Problem 1.2 Directly verify the equivalence of the localized forms of the balance laws (1) and (2).

Problem 1.3 Take $a = \rho$ (the mass density), $b = 0$, $\mathbf{c} = \mathbf{0}$, and show that the master balance law reduces to conservation of mass. Localize using 1.3, and convert to material form.

If we incorporate the mass density into the balance laws, some simplification results. Let $\rho(x, t)$ be the mass density and assume that conservation of mass holds; that is, $\partial\rho/\partial t + \operatorname{div}(\rho \boldsymbol{v}) = 0$. If f is replaced by ρf and conservation of mass is assumed, the transport theorem takes the following form:

$$\frac{d}{dt}\int_{\phi_t(\mathcal{U})} f\rho \, dv = \int_{\phi_t(\mathcal{U})} \dot{f}\rho \, dv. \tag{3}$$

If $a = \rho\tilde{a}$ and $b = \rho\tilde{b}$ in the master balance law, the localization (1) becomes

$$\rho\dot{\tilde{a}} = \rho\tilde{b} + \operatorname{div} \mathbf{c}. \tag{4}$$

Correspondingly, if we take $A = \rho\tilde{A}$ and $B = \rho\tilde{B}$ in the material form, the localized equations (2) become

$$\rho_{\text{Ref}} \frac{\partial\tilde{A}}{\partial t} = \rho_{\text{Ref}}\tilde{B} + \text{DIV } \boldsymbol{C}. \tag{5}$$

Similarly for the master balance inequalities.

Problem 1.4 Prove Equations (3), (4), and (5).

The material forms of the balance principles are convenient, since their localization does not involve the convective term[1] $da \cdot \boldsymbol{v} = (\partial a/\partial x^a)v^a$. Furthermore, the material form is valid for shells as it stands, but the spatial form requires modification; see Box 1.2 below.

Box 1.1 ***The Transport Theorem and Discontinuity Surfaces***

The transport theorem may be supplemented by the following version.

1.7 Proposition *Let $f(x, t)$ be a given bounded C^1 function on a moving open set $\mathcal{U}_t \subset \mathcal{S}$ and suppose f is continuous on $\partial\mathcal{U}_t$. Suppose the boundary of $\mathcal{U}_t$ is moving with velocity $\boldsymbol{w}$, and w_n is its (outward) normal component. Then*

$$\frac{d}{dt}\int_{\mathcal{U}_t} f \, dv = \int_{\mathcal{U}_t} \frac{\partial f}{\partial t} \, dv + \int_{\partial\mathcal{U}_t} f w_n \, da$$

[1]For instance, in fluid mechanics this convective term causes technical difficulties. Some existence and uniqueness proofs are considerably simplified if material coordinates are used. See Ebin and Marsden [1970] and Chapter 6.

Proof Extend w to a vector field and let ψ_t be its flow. Then we have $\mathcal{U}_t = \psi_t(\mathcal{U}_0)$. By the transport and divergence theorems,

$$\frac{d}{dt}\int_{\mathcal{U}_t} f\,dv = \int_{\mathcal{U}_t}\left(\frac{\partial f}{\partial t} + \operatorname{div}(fw)\right)dv = \int_{\mathcal{U}_t}\frac{\partial f}{\partial t}\,dv + \int_{\partial\mathcal{U}_t} fw_n\,da. \quad \blacksquare$$

1.8 Theorem *Let $f(x, t)$ be given, and let $\phi(X, t)$ be a regular C^1 motion. Suppose f has a jump discontinuity across a surface σ_t but is C^1 elsewhere. Assume that $\operatorname{div}(fv)$ is integrable on $\phi_t(\mathcal{U})$, as is $\partial f/\partial t$. Then for a nice open set $\mathcal{U} \subset \mathcal{S}$,*

$$\frac{d}{dt}\int_{\phi_t(\mathcal{U})} f\,dv = \int_{\phi_t(\mathcal{U})}\left(\frac{\partial f}{\partial t} + \operatorname{div}(fv)\right)dv + \int_{\sigma_t\cap\mathcal{U}_t}[f](v_n - w_n)\,da,$$

where w_n is the normal velocity of σ_t, v_n is the normal component of v, and $[f]$ denotes the jump in f across σ_t.[2]

Proof Write $\mathcal{U}_t = \phi_t(\mathcal{U}) = \mathcal{U}_t^+ \cup \mathcal{U}_t^-$, where $\mathcal{U}_t^+$ is on the "forward moving" side of σ_t, or the "+ side". Write $\partial\mathcal{U}_t^+ = \sigma_t \cup (\partial\mathcal{U}_t \cap (\phi_t(\mathcal{B}))^+)$ and similarly for $\partial\mathcal{U}_t^-$. See Figure 2.1.1.

By 1.7, and noting that $-w_n$ is the outward normal component of w on σ_t with respect to $\mathcal{U}_t^+$,

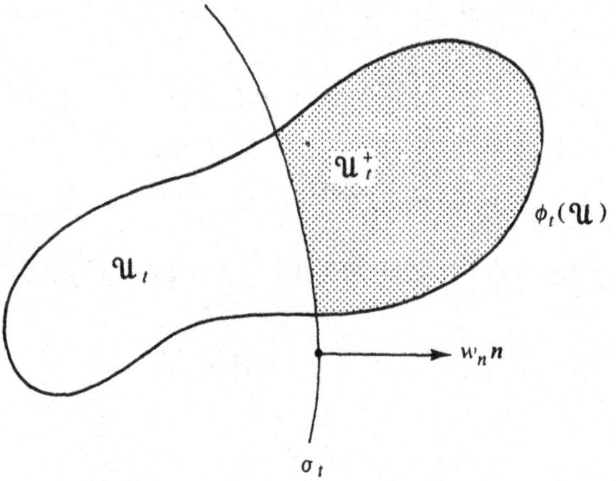

Figure 2.1.1

[2] We assume σ_t divides $\phi_t(\mathcal{B})$ into two pieces. One piece is denoted +, the other −. The jump in f is from the − side to the + side; that is, $[f] = f^+ - f^-$, where f^+ is the limiting value of f from the + side and f^- that from the − side.

$$\frac{d}{dt}\int_{\mathcal{U}_t^+} f\,dv = \int_{\mathcal{U}_t^+} \frac{\partial f}{\partial t}\,dv - \int_{\sigma_t \cap \mathcal{U}_t} f^+ w_n\,da + \int_{\partial\mathcal{U}_t \cap \phi_t(\mathcal{B})^+} f v_n\,da$$

$$= \int_{\mathcal{U}_t^+} \frac{\partial f}{\partial t}\,dv - \int_{\sigma_t \cap \mathcal{U}_t} f^+ w_n\,da + \int_{\mathcal{U}_t^+} \operatorname{div}(fv)\,dv$$

$$+ \int_{\sigma_t \cap \mathcal{U}_t} f^+ v_n\,da.$$

Adding this to the similar expression for $\mathcal{U}_t^-$ gives

$$\frac{d}{dt}\int_{\mathcal{U}_t} f\,dv = \int_{\mathcal{U}_t} \frac{\partial f}{\partial t}\,dv - \int_{\sigma_t \cap \mathcal{U}_t} (f^+ - f^-) w_n\,da + \int_{\mathcal{U}_t} \operatorname{div}(fv)\,dv$$

$$+ \int_{\sigma_t \cap \mathcal{U}_t} [f] v_n\,da. \quad \blacksquare$$

This result can be used to generalize the master balance laws to the case in which the arguments experience jump discontinuities across a smooth surface. Let the hypotheses be as above, but assume that a, b, $\mathbf{c}$ and A, B, C may be discontinuous across the surfaces σ_t and Σ_t, respectively, where $\sigma_t = \phi_t(\Sigma_t)$. From 1.8 and the divergence theorem, we have the following identities:

$$\frac{d}{dt}\int_{\phi_t(\mathcal{U})} a\,dv - \int_{\phi_t(\mathcal{U})} b\,dv - \int_{\partial\phi_t(\mathcal{U})} \langle \mathbf{c}, \boldsymbol{n}\rangle\,da$$

$$= \int_{\phi_t(\mathcal{U})} \left(\frac{\partial a}{\partial t} + \operatorname{div}(av) - b - \operatorname{div}\mathbf{c}\right) dv$$

$$+ \int_{\sigma_t \cap \mathcal{U}_t} [a(\langle v, \boldsymbol{n}\rangle - w_n) - \langle \mathbf{c}, \boldsymbol{n}\rangle]\,da$$

and

$$\frac{d}{dt}\int_{\mathcal{U}} AJ\,dV - \int_{\mathcal{U}} BJ\,dV - \int_{\partial\mathcal{U}} \langle C, N\rangle\,dA$$

$$= \left(\frac{\partial(AJ)}{\partial t} - BJ - \operatorname{DIV} C\right) dV$$

$$+ \int_{\Sigma_t} [AJ(\langle V, N\rangle - W_N) - \langle C, N\rangle]\,dA$$

where W_N is the normal velocity of the surface Σ_t. The integral balance laws imply the local form given previously as well as the following *jump discontinuity conditions*:

$$[a(\langle v, \boldsymbol{n}\rangle - w_n) - \langle \mathbf{c}, \mathbf{n}\rangle] = 0 \qquad \text{on } \sigma_t \text{ (spatial form)}$$

and $$[AJ(\langle \mathbf{V}, \mathbf{N}\rangle - W_N) - \langle \mathbf{C}, \mathbf{N}\rangle] = 0 \qquad \text{on } \Sigma_t \text{ (material form).}$$

These results are fundamental in wave-propagation theory and in shock waves. The surfaces σ_t and Σ_t are called *propagating singular surfaces*—the "singular" referring to the fact that functions of interest experience discontinuities across them.[3]

An important result, known as Cauchy's theorem, deals with the form of the integrand $\langle c, n\rangle$ in the surface integral term of the master balance law.

1.9 Cauchy's Theorem *Let $a(x, t)$ be C^1, and $b(x, t)$, $c(x, t, n)$ be C^0 scalar functions defined for all $t \in \mathbb{R}$ (or an open interval), all $x \in \phi_t(\mathcal{B})$, and all unit vectors n at x. Assume that a, b, and c satisfy the master balance law in the sense that, for any nice open set $\mathcal{U} \subset \mathcal{B}$, we have*

$$\frac{d}{dt}\int_{\phi_t(\mathcal{U})} a(x, t)\, dv = \int_{\phi_t(\mathcal{U})} b(x, t)\, dv + \int_{\partial\phi_t(\mathcal{U})} c(x, t, n)\, da,$$

where n is the unit outward normal to $\partial\phi_t(\mathcal{U})$. Then there exists a unique vector field $\mathbf{c}(x, t)$ on $\phi_t(\mathcal{B})$ such that $c(x, t, n) = \langle \mathbf{c}(x, t), n\rangle$.

Proof By the transport theorem, the hypothesis is

$$\int_{\phi_t(\mathcal{U})} (\dot{a} + a \operatorname{div} v - b)\, dv = \int_{\partial\phi_t(\mathcal{U})} c(x, t, n)\, da \tag{1}$$

for all nice regions $\mathcal{U}$.

Now fix the variable t (it will be omitted in what follows). Work in a neighborhood of a point $x_0 \in \mathcal{S}$ and choose an oriented coordinate system $\{x^a\}$ that is orthonormal at x_0.

We shall work in three dimensions for simplicity; see Figure 2.1.2. In the first quadrant of the chosen coordinates system, draw the tetrahedron $\mathcal{W}$ with one edge of length l, and the others in fixed proportions to it. Let n be the unit normal to the skew face. By construction,

$$\lim_{l\to 0} \frac{\text{volume } \mathcal{W}}{\text{area } \partial\mathcal{W}} = 0,$$

and so from (1),

$$\lim_{l\to 0} \frac{1}{\text{area } \partial\mathcal{W}} \int_{\partial\mathcal{W}} c(t, x, n)\, da = 0,$$

Let the coordinate faces be denoted Σ_i ($i = 1, 2, 3$), and the skew face be Σ. By

[3]For results on wave propagation theory, see Chen [1972] and Eringen and Suhubi [1974]. For results on shocks, see Courant and Friedrichs [1976], Hughes and Marsden [1976], and Chorin and Marsden [1979].

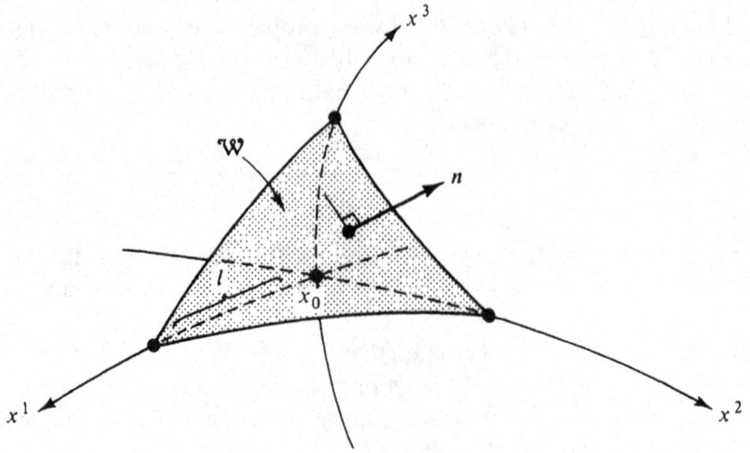

Figure 2.1.2

the mean value theorem for integrals, there is a $z \in \Sigma$ and points $z_i \in \Sigma_i$ such that

$$\frac{1}{\text{area } \partial\mathcal{W}} \int_{\partial\mathcal{W}} c\, da = \frac{1}{\text{area } \partial\mathcal{W}} \left\{ c(z, \boldsymbol{n})(\text{area } \Sigma) + \sum_{i=1}^{3} c(z_i, -\boldsymbol{n}_i)(\text{area } \Sigma_i) \right\}$$

where $-\boldsymbol{n}_i$ is the outward unit normal to Σ_i at z_i. As $l \longrightarrow 0$, (area Σ)/(area $\partial\mathcal{W}$) $\longrightarrow \alpha$, a constant, and (area Σ_i)/(area $\partial\mathcal{W}$) $\longrightarrow n^i\alpha$, where $\boldsymbol{n} = n^i\boldsymbol{e}_i$ is the limiting value of $\boldsymbol{n}$ as $l \longrightarrow 0$. (As the proportions of the sides vary, we get all possible vectors $\boldsymbol{n}$ in the first quadrant.) Since $\boldsymbol{n}_i \longrightarrow \boldsymbol{e}_i$ as $l \longrightarrow 0$, we get the identity

$$0 = c(x_0, \boldsymbol{n}) + \sum_{i=1}^{3} n^i c(x_0, -\boldsymbol{e}_i). \tag{2}$$

Taking the limit $\boldsymbol{n} \longrightarrow \boldsymbol{e}_i$, we get

$$c(x_0, \boldsymbol{e}_i) = -c(x_0, -\boldsymbol{e}_i). \tag{3}$$

Substituting back into (2) gives $c(x_0, \boldsymbol{n}) = \sum_{i=1}^{3} n^i c(x_0, \boldsymbol{e}_i)$. Similar reasoning holds for the other quadrants. ▮

Problem 1.5 Formulate Cauchy's theorem for material quantities.

Problem 1.6 This problem is the first step in generalizing Cauchy's theorem to allow $c(x, t, \boldsymbol{n})$'s that are not necessarily continuous. (See Gurtin and Martins [1976] for more information.) Define a *Cauchy flux* to be a map F that assigns to each oriented piecewise smooth surface (possibly with boundary) $\mathfrak{S} \subset \mathcal{S} = \mathbb{R}^3$ a vector $F(\mathfrak{S}) \in \mathbb{R}^3$ such that

(a) If $\mathfrak{S} = \bigcup_{i=1}^{n} \mathfrak{S}_i$, a disjoint union (except on boundaries) with compatible orientations, then $F(\mathfrak{S}) = \sum_{i=1}^{n} F(\mathfrak{S}_i)$, and

(b) F is area-continuous; that is, if area $(\mathfrak{S}_n) \longrightarrow 0$, then $F(\mathfrak{S}_n) \longrightarrow 0$.

Prove that $F(-\mathfrak{S}) = -F(\mathfrak{S})$, where $-\mathfrak{S}$ denotes $\mathfrak{S}$ with the opposite orientation. Compare Equation (3) above. [*Hint:* Consider a pill-box of thickness $\epsilon > 0$].

Box 1.2 ***Balance Laws for Shells***

The transport theorem for shells follows.

1.10 Theorem *Let* $\dim \mathcal{S} = \dim \mathcal{B} + 1$ *and let* $\phi_t: \mathcal{B} \longrightarrow \mathcal{S}$ *be a motion (of embeddings) of* $\mathcal{B}$ *into* $\mathcal{S}$. *Let* f *be a given* C^1 *function of* $t \in \mathbb{R}$ *and* $x \in \phi_t(\mathcal{B})$. *Let* $\mathcal{U} \subset \mathcal{B}$ *be a nice open set. Then*

$$\frac{d}{dt}\int_{\phi_t(\mathcal{U})} f\, dv = \int_{\phi_t(\mathcal{U})} \{\dot{f} + f(\overline{\operatorname{div}}\, \boldsymbol{v}_\parallel + v_n \operatorname{tr} \boldsymbol{k})\}\, dv$$

where $\boldsymbol{v}_\parallel$ *is the component of the spatial velocity field parallel to* $\phi_t(\mathcal{B})$, v_n *is its normal component, and* tr $\boldsymbol{k}$ *is the mean curvature of the hypersurface* $\phi_t(\mathcal{B})$. *(See Box 5.1, Chapter 1.)*

Proof

$$\frac{d}{dt}\int_{\phi_t(\mathcal{U})} f\, dv = \frac{d}{dt}\int_{\mathcal{U}} f(\phi(X, t), t) J(X, t)\, dV(X)$$

by the change of variables theorem. By 5.13 and 5.14 of Chapter 1, we have

$$\frac{\partial J}{\partial t} = ((\overline{\operatorname{div}}\, \boldsymbol{v}_\parallel)\circ\phi_t + v_n(\operatorname{tr} \boldsymbol{k})\circ\phi_t)J.$$

After differentiating under the integral sign, one may complete the proof as in 1.1. ∎

Given scalar functions a, b on $\phi_t(\mathcal{B})$ and a vector field $\boldsymbol{c}(x, t)$, we define the master balance law as before:

$$\frac{d}{dt}\int_{\phi_t(\mathcal{U})} a\, dv = \int_{\phi_t(\mathcal{U})} b\, dv + \int_{\partial\phi_t(\mathcal{U})} \langle \boldsymbol{c}, \boldsymbol{n} \rangle\, da.$$

The localized equations are

$$\frac{\partial a}{\partial t} + \overline{\operatorname{div}}(a\boldsymbol{v}_\parallel) + a v_n \operatorname{tr} \boldsymbol{k} = b + \overline{\operatorname{div}}\, \boldsymbol{c}$$

where $\overline{\operatorname{div}}$ means the intrinsic (Riemannian) divergence for vector fields on the surface $\phi_t(\mathcal{B})$.

The material picture for shells is the same as that in 1.4 and 1.5. The Piola transform still makes sense as a relation between vector fields $\boldsymbol{C}$ on $\mathcal{B}$ and vector fields $\boldsymbol{c}$ on $\phi_t(\mathcal{B})$.

Problem 1.7 Prove a Cauchy theorem for shells.

Problem 1.8 Prove a transport theorem for discontinuity surfaces in shells.

Box 1.3 ***Transport Theorem for a Distensible Tube***

Balance laws in mechanics involve terms of the form $\dot{f} + f \operatorname{div} \boldsymbol{v}$. Most of the difficulty in deriving one-dimensional theories is computing the correct one-dimensional counterpart of $\dot{f} + f \operatorname{div} \boldsymbol{v}$. The procedure given here explains how to do this.

Consider a section $\mathcal{S}$ of the tube illustrated in Figure 2.1.3. Let $\mathcal{S}$ be chosen normal to the z^3-axis. The *luminal area* of the tube is the area of $\mathcal{S}$ and is denoted $A(z^3, t) > 0$; $\mathcal{S}$ need not be circular. Let $\mathcal{C}$ denote $\partial\mathcal{S}$, the *luminal boundary* with line element dl. Let $f(z^1, z^2, z^3, t)$ be a C^1 real-valued function, and define its *area mean* to be $\bar{f} = (\int_{\mathcal{S}} f\, da)/A$: $\mathbb{R}^2 \to \mathbb{R}$, where da is the area element on $\mathcal{S}$. Let $\boldsymbol{v}$ denote the spatial velocity vector of material contained in the tube. The z^3-component of $\boldsymbol{v}$ is denoted v^3. Let $\boldsymbol{n}$ be the *unit outward normal vector* to the tube defined on the boundary of the tube. The tube in general will have a

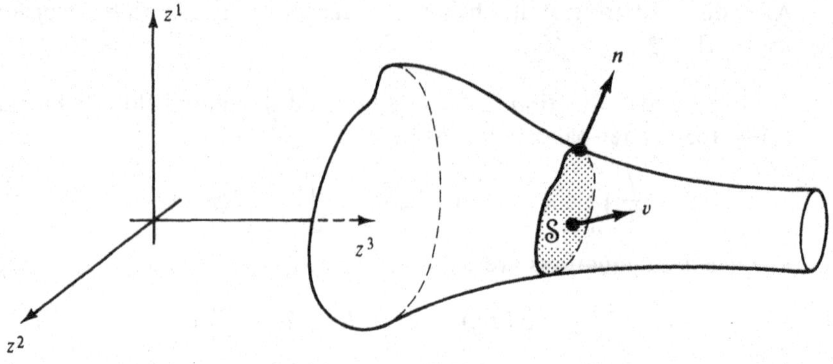

Figure 2.1.3

time-dependent taper so $\boldsymbol{n}$ does not necessarily lie in the z^1z^2-plane—that is, in the plane of $\mathcal{S}$. Let v_n represent the component of $\boldsymbol{v}$ in the direction $\boldsymbol{n}$; that is, $v_n = \boldsymbol{v} \cdot \boldsymbol{n}$. If u_n denotes the *normal velocity of the luminal surface*, then w_n, the *relative normal velocity of the lumen*, is defined by $u_n = v_n + w_n$. Thus the amount of fluid leaving the tube through the luminal surface is measured by $-w_n$. The preceding definitions and the transport theorem yield the following identity:

$$\frac{\partial}{\partial t}(A\bar{f}) + \frac{\partial}{\partial z^3}[A(\overline{fv^3})] = \int_{\mathcal{S}} (\dot{f} + f \operatorname{div} \boldsymbol{v})\, da + \int_{\mathcal{C}} f w_n\, dl.$$

This relation is useful in establishing one-dimensional theories of flow through distensible tubes.[4] The last term allows for outflow due to transverse branching.

Box 1.4 ***Summary of Important Formulas for Section 2.1***

Transport Theorem

$$\frac{d}{dt}\int_{\phi_t(\mathcal{U})} f\, dv = \int_{\phi_t(\mathcal{U})} (\dot{f} + f \operatorname{div} \boldsymbol{v})\, dv \qquad \dot{f} + f \operatorname{div} \boldsymbol{v} = \frac{\partial f}{\partial t} + (fv^a)_{|a}$$

$$= \int_{\phi_t(\mathcal{U})} \left(\frac{\partial f}{\partial t} + \operatorname{div}(f\boldsymbol{v})\right) dv$$

Master Balance Law

(i) For spatial quantities a, b, $\boldsymbol{c}$:

(Integral Form)

$$\frac{d}{dt}\int_{\phi_t(\mathcal{U})} a\, dv = \int_{\phi_t(\mathcal{U})} b\, dv + \int_{\partial\phi_t(\mathcal{U})} \langle \boldsymbol{c}, \boldsymbol{n}\rangle\, da \qquad \langle \boldsymbol{c}, \boldsymbol{n}\rangle = c^a n^b g_{ab}$$

(Local Form)

$$\frac{\partial a}{\partial t} + \operatorname{div}(a\boldsymbol{v}) = b + \operatorname{div} \boldsymbol{c} \qquad \frac{\partial a}{\partial t} + (av^b)_{|b} = b + c^b{}_{|b}$$

(ii) For material quantities A, B, $\mathbf{C}$:

(Integral Form)

$$\frac{d}{dt}\int_{\mathcal{U}} AJ\, dV = \int_{\mathcal{U}} BJ\, dV + \int_{\mathcal{U}} \langle C, N\rangle\, dA \qquad \langle C, N\rangle = C^A N^B G_{AB}$$

(Local Form)

$$\frac{\partial}{\partial t}(AJ) = BJ + \operatorname{DIV} \mathbf{C} \qquad \frac{\partial}{\partial t}(AJ) = BJ + C^A{}_{|A}$$

[4]See Hughes and Lubliner [1973].

Material vs. Spatial

$$A(X, t) = a(x, t) \qquad A(X^A, t) = a(x^a, t)$$
$$B(X, t) = b(x, t) \qquad B(X^A, t) = b(x^a, t)$$
$$C(X, t) = JF^{-1} c \qquad C^A(X^B, t) = J(F^{-1})^A{}_a C^a$$
$$x = \phi(X, t) \qquad x^a = \phi^a(X^A, t)$$

Balance Laws Using Mass Density $\rho(x, t)$

Transport: $\dfrac{d}{dt}\displaystyle\int_{\phi_t(\mathcal{U})} f\rho\, dv = \int_{\phi_t(\mathcal{U})} \dot{f}\rho\, dv$

Master Balance Law

$$\rho\dot{\tilde{a}} = \rho\tilde{b} + \operatorname{div} c \qquad a = \rho\tilde{a}, \qquad b = \rho\tilde{b}$$
$$\rho_{\text{Ref}}\frac{\partial \tilde{A}}{\partial t} = \rho_{\text{Ref}}\tilde{B} + \text{DIV}\, C \qquad A = \rho\tilde{A}, \qquad B = \rho\tilde{B}$$

Cauchy's Theorem

If $a(x, t)$, $b(x, t)$, and $c(x, t, \boldsymbol{n})$ satisfy the master balance law, then $c(x, t, \boldsymbol{n}) = \langle \mathbf{c}(x, t), \boldsymbol{n}\rangle$ for some vector field $\boldsymbol{c}$. Materially, if $A(X, t)$, $B(X, t)$, and $C(X, t, N)$ satisfy the master balance law, then $C(X, t, N) = \langle \boldsymbol{C}(X, t), N\rangle$.

Transport Theorem for Discontinuity Surfaces

$$\frac{d}{dt}\int_{\phi_t(\mathcal{U})} f\, dv = \int_{\phi_t(\mathcal{U})} (\dot{f} + f \operatorname{div} \boldsymbol{v})\, dv + \int_{\sigma_t \cap \mathcal{U}_t} [f](v_n - w_n)\, da,$$

where σ_t = discontinuity surface, w_n is its normal velocity, v_n is the normal component of $\boldsymbol{v}$, and $[f]$ is the jump of f across σ_t.

Jump Discontinuity Conditions for the Master Balance Law

$[a\langle \boldsymbol{v}, \boldsymbol{n}\rangle - w_n - \langle \boldsymbol{c}, \boldsymbol{n}\rangle] = 0$ on σ_t

$[AJ\langle V, N\rangle - W_N - \langle C, N\rangle] = 0$ on Σ_t, where $\sigma_t = \phi_t(\Sigma_t)$

Transport Theorem for Shells

$$\frac{d}{dt}\int_{\phi_t(\mathcal{U})} f\, dv = \int_{\phi_t(\mathcal{U})} [\dot{f} + f(\overline{\operatorname{div}}\, \boldsymbol{v}_{||} + v_n \operatorname{tr} \boldsymbol{k})]\, dv$$

2.2 THE STRESS TENSOR AND BALANCE OF MOMENTUM

This section develops the basic equations of continuum mechanics. The concept which sets continuum mechanics apart from particle mechanics is that of stress. The idea, introduced by Cauchy a century and a half after Newton, expresses the interaction of a material with surrounding material in terms of surface contact forces.

We quote Truesdell [1968], p. 186 concerning the *Stress Principle of Cauchy*: "Upon any smooth, closed, orientable surface $\mathfrak{S}$, be it an imagined surface within the body or the bounding surface of the body itself, there exists an integrable field of traction vectors $t_{\mathfrak{S}}$ equipollent (same resultant and moment) to the action exerted by the matter exterior to $\mathfrak{S}$ and contiguous to it on that interior to $\mathfrak{S}$." See Figure 2.2.1.

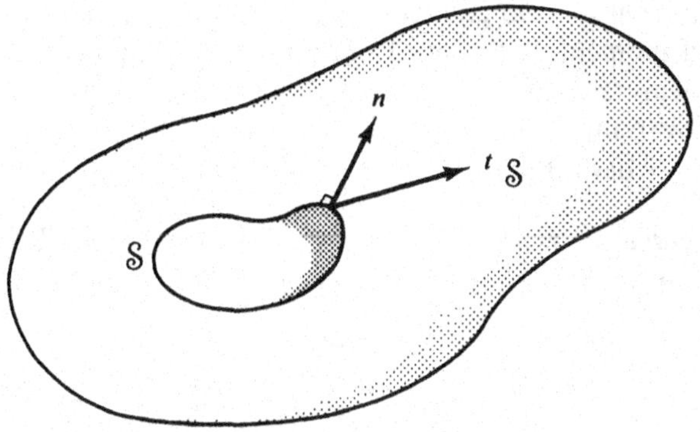

Figure 2.2.1

We postualte the existence of a vector field $t(x, t, n)$ depending on time t, the spatial point x, a unit vector n, and, implicitly, the motion $\phi(x, t)$ itself. Physically, $t(x, t, n)$ *represents the force per unit area exerted on a surface element oriented with normal* n. We shall call t the *Cauchy stress vector*.

Throughout this section we shall assume that $\mathcal{B}$ is a simple body moving in $\mathcal{S} = \mathbb{R}^n$. We shall let ϕ_t be a regular C^1 motion of $\mathcal{B}$ in $\mathcal{S}$, and, as usual, let $v(x, t)$ and $V(X, t)$ be its spatial and material velocities. Let $\rho(x, t)$ be a mass density function. We shall explicitly state when we are assuming conservation of mass. Finally, we assume that forces other than surface contact forces (tractions) arise from an external force field b, an example of which is the gravitational force.

The reader who has read Box 5.1, Chapter 1, and Box 1.2 in the preceding section may generalize the results below to shells. In Box 2.2 we shall indicate another possible generalization of the results given here.

2.1 Definition Given $\phi(X, t)$, $\rho(x, t)$, $t(x, t, n)$, and $b(x, t)$ we say that *balance of momentum* is satisfied provided that for every nice open set $\mathcal{U} \subset \mathcal{B}$,

$$\boxed{\frac{d}{dt}\int_{\phi_t(\mathcal{U})} \rho v \, dv = \int_{\phi_t(\mathcal{U})} \rho b \, dv + \int_{\partial\phi_t(\mathcal{U})} t \, da}$$

where t is evaluated on the unit outward normal n to $\partial\phi_t(\mathcal{U})$ at a point x.

The above equation states that the rate of change of linear momentum of $\phi_t(\mathcal{U})$ equals the net force acting on it.

Warning: Balance of momentum explicitly uses the linear structure of $\mathbb{R}^n$ since vector functions are integrated. It is correct to interpret this equation componentwise in Cartesian coordinates z_i but *not* in a general coordinate system. In this sense, *balance of momentum is not a tensorial postulate*. We shall discuss this objection in greater detail in Section 2.3. There we shall show how balance of momentum follows from an energy principle that does not require $\mathcal{S}$ to be linear. (For relativisitic elasticity it is essential to have a basic postulate that is covariant—such as balance of four momentum—and for classical elasticity it is desirable.)

2.2 Theorem *Assume that balance of momentum holds, that $\phi(X, t)$ is C^1, and $\boldsymbol{t}(x, t, \boldsymbol{n})$ is a continuous function of its arguments. Then there is a unique $\binom{2}{0}$ tensor field, denoted $\boldsymbol{\sigma}$, depending only on x and t such that*

$$\boxed{\boldsymbol{t}(x, t, \boldsymbol{n}) = \langle \boldsymbol{\sigma}(x, t), \boldsymbol{n} \rangle.}$$

In coordinates $\{x^a\}$ on $\mathcal{S}$, the preceding equation reads

$$\boxed{t^a(x, t, \boldsymbol{n}) = \sigma^{ac}(x, t) g_{bc} n^b = \sigma^a{}_b n^b.}$$

Proof Let $\boldsymbol{u}_0$ be a vector in $\mathbb{R}^n$. Then

$$\frac{d}{dt} \int_{\phi_t(\mathcal{U})} \rho \langle \boldsymbol{v}, \boldsymbol{u}_0 \rangle \, dv = \int_{\phi_t(\mathcal{U})} \rho \langle \boldsymbol{b}, \boldsymbol{u}_0 \rangle \, dv + \int_{\partial \phi_t(\mathcal{U})} \langle \boldsymbol{t}, \boldsymbol{u}_0 \rangle \, dv.$$

By Cauchy's theorem, 1.7, there is a vector field $\boldsymbol{c}$ such that $\langle \boldsymbol{t}, \boldsymbol{u}_0 \rangle = \langle \boldsymbol{c}, \boldsymbol{n} \rangle$. Since $\langle \boldsymbol{t}, \boldsymbol{u}_0 \rangle$ depends linearly on $\boldsymbol{u}_0$, $\boldsymbol{c}$ must also be a linear function of $\boldsymbol{u}_0$, and so defines the required tensor $\boldsymbol{\sigma}$. (In Cartesian coordinates, picking $\boldsymbol{u}_0 = \hat{\boldsymbol{\imath}}_i$ gives a vector field $\boldsymbol{c}^i$ such that $t^i = \langle \boldsymbol{c}^i, \boldsymbol{n} \rangle = n^j \sigma^{ik} \delta_{jk}$, where σ^{ik} are the components of $\boldsymbol{c}^i$.) Obviously, $\boldsymbol{\sigma}$ is uniquely determined. ∎

2.3 Definition The tensor σ^{ab} is called the *Cauchy stress tensor*. The associated tensor $\boldsymbol{\sigma}$ with components $\sigma^a{}_b$ is also called the Cauchy stress tensor.

2.4 Theorem *Assume that balance of momentum and conservation of mass hold. Then*

$$\boxed{\rho \dot{v} = \rho b + \operatorname{div} \boldsymbol{\sigma}, \qquad \text{where} \quad (\operatorname{div} \boldsymbol{\sigma})^a = \sigma^{ab}{}_{|b}.}$$

Proof The validity of this equation in Cartesian coordinates results from the spatial form of the master balance law (see Equation (4), Section 2.1). ∎

In general coordinates $\{x^a\}$, this basic set of dynamic equations in 2.4 reads as follows:

$$\rho\left(\frac{\partial v^a}{\partial t} + v^b \frac{\partial v^a}{\partial x^b} + \gamma^a_{bc} v^b v^c\right) = \rho b^a + \frac{\partial \sigma^{ab}}{\partial x^b} + \sigma^{ac} \gamma^b_{cb} + \sigma^{cb} \gamma^a_{cb}.$$

Notice that $\sigma \cdot n$ is the force per unit of deformed area. To find the force per unit of undeformed area, we perform a Piola transformation (see Section 1.7).

2.5 Definition The *first Piola–Kirchhoff stress tensor*[5] $\boldsymbol{P}$ is the two-point tensor obtained by performing a Piola transformation on the second index of $\boldsymbol{\sigma}$. In coordinates,

$$\boxed{P^{aA} = J(F^{-1})^A{}_b\,\sigma^{ab}.}$$

Here P^{aA} is a function of (X, t) and σ^{ab} is evaluated at (X, t), where $x = \phi(X, t)$.

From results in Section 2.1 we can read off the material version of balance of momentum.

2.6 Theorem *Assume conservation of mass holds. Then balance of momentum is equivalent to*

$$\frac{d}{dt}\int_{\mathcal{U}} \rho_{\text{Ref}} V\,dV = \int_{\mathcal{U}} \rho_{\text{Ref}} \boldsymbol{B}\,dV + \int_{\partial\mathcal{U}} \langle \boldsymbol{P}, \boldsymbol{N}\rangle\,dA$$

for any nice open set $\mathcal{U} \subset \mathcal{B}$. *Here* $\boldsymbol{B}(X, t) = b(x, t)$ *and* $\langle \boldsymbol{P}, \boldsymbol{N}\rangle = P^{aA}N_A$. *This, in turn, is equivalent to*

$$\boxed{\rho_{\text{Ref}}\boldsymbol{A} = \rho_{\text{Ref}}\boldsymbol{B} + \text{DIV}\,\boldsymbol{P}}$$

where $(\text{DIV}\,\boldsymbol{P})^a = P^{aA}{}_{|A}$.

Problem 2.1 Let $\boldsymbol{A}$ be an $n \times n$ matrix such that for any unit vector $\boldsymbol{n}$, $\boldsymbol{A} \cdot \boldsymbol{n}$ is parallel to $\boldsymbol{n}$. Show that $\boldsymbol{A} = p\boldsymbol{I}$, where p is a real number and $\boldsymbol{I}$ is the $n \times n$ identity matrix.

2.7 Example A *perfect fluid* is characterized by the fact that no shear forces are possible; that is, $\boldsymbol{\sigma} \cdot \boldsymbol{n}$ is parallel to $\boldsymbol{n}$. Problem 2.1 shows that $\boldsymbol{\sigma} = -p\boldsymbol{I}$ for some scalar function p called the *pressure*, and $\boldsymbol{I}$ the identity (equivalently, $\sigma^{ba} = -pg^{ba}$). The equations of motion then read

$$\rho\dot{v} = \rho b - \nabla p \quad \text{(Euler's equations)}$$

since $\text{div}\,(-p\boldsymbol{I}) = -\nabla p$. The reader should write out the equations in material form as an exercise.

If we pull the first leg of $\boldsymbol{P}$ back to $\mathcal{B}$ we get a $\binom{2}{0}$ tensor on $\mathcal{B}$. This tensor will turn out to play an important role in constitutive theory. The formal definition follows.

---

[5] See Piola [1845] and Kirchhoff [1852].

2.8 Definition The *second Piola–Kirchhoff stress tensor* $\boldsymbol{S}$ is obtained by pulling the first leg of $\boldsymbol{P}$ back by ϕ_t. In coordinates,

$$\boxed{S^{AB} = (F^{-1})^A{}_a P^{aB} = J(F^{-1})^A{}_a(F^{-1})^B{}_b\sigma^{ab}.}$$

Problem 2.2 Show that balance of momentum can be written as

$$\rho_{\text{Ref}}\boldsymbol{\alpha} = \rho_{\text{Ref}}\phi_t^*(\boldsymbol{B}) + \text{DIV}_C\boldsymbol{S}$$

where $\boldsymbol{\alpha}$ is the convected acceleration and $\text{DIV}_C\boldsymbol{S}$ is the divergence of $\boldsymbol{S}$ with respect to the metric $\boldsymbol{C}^\flat = \phi_t^*(g)$.

Box 2.1 ***The Unit Normal as a One-Form***

In retrospect, the position of the indices on the Cauchy stress tensor seem most natural up—that is, σ^{ab}. However, to contract with $\boldsymbol{n}$, $\sigma^a{}_b$ seems more natural. We can encompass both properties if we think of unit normals as one-forms rather than vectors; then $\sigma^{ab}n_b$ is the natural contraction.

To motivate this idea, consider a real-valued function f on $\mathbb{R}^3$. The level sets $f(x, y, z) = c$ are surfaces. In vector calculus we learn that the gradient vector $\nabla f(x, y, z)$ is a unit normal to the surface through the point (x, y, z). However, the one-form $\boldsymbol{d}f(x, y, z)$ carries similar information: if $\boldsymbol{u}$ is a unit vector, $\boldsymbol{d}f(x, y, z)\cdot\boldsymbol{u}$ is the rate of change of f in the direction of $\boldsymbol{u}$. We can think of $\boldsymbol{d}f(x, y, z)$ as a measure of how closely stacked the level sets "$f =$ constant" are near (x, y, z). It is as legitimate a way of thinking of the normal direction to the level set as that provided by the gradient. See Figure 2.2.2.

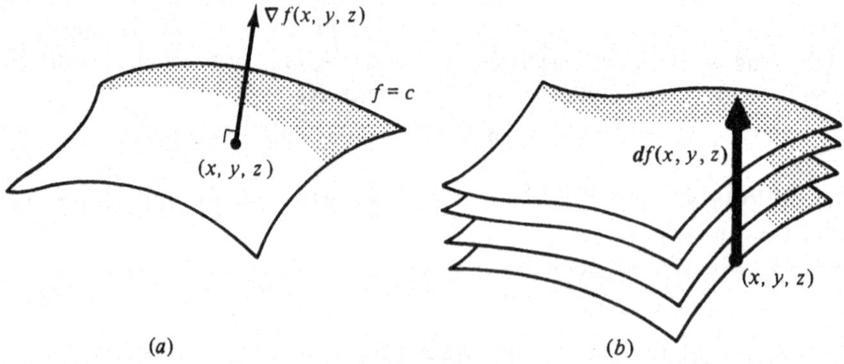

Figure 2.2.2

This idea fits naturally with the Cauchy stress postulate. Indeed, $\boldsymbol{t}(x, t, \boldsymbol{n})$ was supposed to be the force across a two-surface element at x with unit normal $\boldsymbol{n}$. But a two-surface element can be thought of as a two-form at x, say da. The one-form $\boldsymbol{n}^\flat$ such that $\boldsymbol{n}^\flat \wedge da = dv$ is the "unit normal" to da. The same unit normal thought of as a vector $\boldsymbol{n}$ satisfies $da = \boldsymbol{i}_{\boldsymbol{n}}\, dv$. (See also Misner, Thorne and Wheeler [1973] pp. 53–60.)

Box 2.2 ***The Convected and Co-rotational Stress Tensors***

In addition to the Cauchy and the Piola–Kirchhoff stress tensors, two other representations of the stress are of some importance since they take into account the deformational and rotational parts of the motion (see Section 1.3).[6]

The *convected stress tensor* $\boldsymbol{\Sigma}$ is obtained by pulling back the contravariant Cauchy stress tensor $\boldsymbol{\sigma}^\flat$ to $\mathcal{B}$. In components,

$$\Sigma_{AB} = F^a{}_A \sigma_{ab} F^b{}_B.$$

The *co-rotational* (*or rotated*) *stress tensor* $\boldsymbol{\mathfrak{S}}$ is obtained by rotating the Cauchy stress tensor $\boldsymbol{\sigma}$ back to $\mathcal{B}$. Recall from 3.12, Chapter 1, that the rotation matrix $\boldsymbol{R}\colon T_X\mathcal{B} \longrightarrow T_x\mathcal{S}$ is the orthogonal part of $\boldsymbol{F}$. In components,

$$\mathfrak{S}^{AB} = (R^{-1})^A{}_a \sigma^{ab} (R^{-1})^B{}_b.$$

Recall that $\boldsymbol{R}$ is orthogonal: $\boldsymbol{R}^{-1} = \boldsymbol{R}^T$; that is, $(R^{-1})^A{}_a = R^b{}_B g_{ab} G^{AB}$. Thus,

$$\mathfrak{S}_{AB} = R^a{}_A \sigma_{ab} R^b{}_B$$

That is, we can pull back either the covariant or contravariant form of $\boldsymbol{\sigma}$ to the same end.

These forms of the stress tensor are convenient in some numerical applications (see, for example, Oden [1972]).

Next we turn to balance of moment of momentum, which corresponds to the statement in particle mechanics that the rate of change of angular momentum equals torque. In particle mechanics, balance of moment of momentum is a corollary of balance of linear momentum, whereas in continuum mechanics it is an independent hypothesis.

[6]See, for example, Chapter 4, of Eringen [1975].

Assume we are in $\mathbb{R}^3$. The vector from the origin to the point x is denoted $\boldsymbol{x}$, and $\boldsymbol{x} \times \boldsymbol{y}$ is the usual cross product.

2.9 Definition We say that *balance of moment of momentum* is satisfied if, for every nice $\mathcal{U} \subset \mathcal{B}$,

$$\frac{d}{dt}\int_{\phi_t(\mathcal{U})} \rho(\boldsymbol{x} \times \boldsymbol{v})\, dv = \int_{\phi_t(\mathcal{U})} \rho(\boldsymbol{x} \times \boldsymbol{b})\, dv + \int_{\partial\phi_t(\mathcal{U})} \boldsymbol{x} \times \langle \boldsymbol{\sigma}, \boldsymbol{n}\rangle\, da.$$

This postulate leads to the important conclusion that the Cauchy stress tensor is symmetric.

2.10 Theorem *Let conservation of mass and balance of momentum hold. Then balance of moment of momentum holds if and only if $\boldsymbol{\sigma}$ is symmetric; that is,*

$$\boxed{\sigma^{ab} = \sigma^{ba}.}$$

Proof By the master balance law (Equation (4), Section 2.1), balance of moment of momentum is equivalent to $\rho(\boldsymbol{x} \times \dot{\boldsymbol{v}}) = \rho(\boldsymbol{x} \times \boldsymbol{b}) + \operatorname{div}(\boldsymbol{x} \times \boldsymbol{\sigma})$, where $\boldsymbol{x} \times \boldsymbol{\sigma}$ is the two-tensor given by $(\boldsymbol{x} \times \boldsymbol{\sigma})^{ij} = \epsilon^i{}_{lk} z^l \sigma^{kj}$, where $\epsilon^i{}_{lk}$ is the permutation symbol ($+1$ if ilk is an even permutation of (1, 2, 3), -1 if odd, and 0 if otherwise), and z^l are the Cartesian coordinates of $\boldsymbol{x}$. Thus

$$(\operatorname{div}(\boldsymbol{x} \times \boldsymbol{\sigma}))^i = \frac{\partial}{\partial z^j}(\epsilon^i{}_{lk} z^l \sigma^{kj}) = (\boldsymbol{x} \times \operatorname{div} \boldsymbol{\sigma})^i + \epsilon^i{}_{lk}\delta^l{}_j \sigma^{kj}.$$

Substituting this into $\rho(\boldsymbol{x} \times \dot{\boldsymbol{v}}) = \rho(\boldsymbol{x} \times \boldsymbol{b}) + \operatorname{div}(\boldsymbol{x} \times \boldsymbol{\sigma})$ and using balance of momentum leaves $0 = \epsilon^i{}_{jk}\sigma^{kj}$ $(j = 1, 2, 3)$, which is equivalent to symmetry of σ^{ik}. ∎

Symmetry of $\boldsymbol{\sigma}$ is a coordinate-independent statement. Also, note that symmetry of $\boldsymbol{\sigma}$ is equivalent to symmetry of the second Piola–Kirchhoff stress tensor: $S^{AB} = S^{BA}$. The first Piola–Kirchhoff stress tensor then has the following symmetry: $P^{aA}F^b{}_A = P^{bA}F^a{}_A$.

Box 2.3 ***Some Remarks on Cosserat Continua***[7]

Some materials, called Cosserat continua, or directed media, have an internal structure that influences their behavior when viewed as a continuous medium. For these materials it is useful to generalize the notion

[7]The earliest concepts of this kind were apparently espoused by Duhem [1906] and developed extensively by Cosserat and Cosserat [1909] with rods and shells in mind (cf. Truesdell and Toupin [1960], Toupin [1962], [1964], Antman [1972a], Naghdi [1972], and Box 2.1, Chapter 7). The history is thoroughly described in Truesdell and Noll [1965]. Recent important contributions to the theory of liquid crystals have been made by Ericksen and Leslie; see, for example, their review articles in Brown [1976].

of a simple body to take into account this structure. Liquid crystals are an example. This class of substances have rodlike molecules whose alignment influences their material behavior. We modify the containing space $\mathcal{S} = \mathbb{R}^3$ to allow for these extra (microscopic) degrees of freedom. There are at least three choices for liquid crystals:

(i) $\mathcal{S} = \mathbb{R}^3 \times \mathbb{R}^3$ (extensible directed rods),
(ii) $\mathcal{S} = \mathbb{R}^3 \times S^2$ (inextensible directed rods; "cholestric" case), and
(iii) $\mathcal{S} = \mathbb{R}^3 \times \mathbb{P}^2$ (inextensible undirected rods; "nematic" case).

Here S^2 is the unit two-sphere in $\mathbb{R}^3$ and is used to model vectors that are free to point in any direction, but which are inextensible. $\mathbb{P}^2$ denotes real projective two-space, defined to be S^2 with antipodal points identified. It is used to model inextensible rods that have indistinguishable ends.

We may summarize the internal variables by choosing the ambient space to be $\mathcal{S} = \mathbb{R}^3 \times \mathcal{R}$, where $\mathcal{R}$ is some manifold. The body should consist of an open set $\mathcal{P} \subset \mathbb{R}^3$ (points in physical space) together with a representative of $\mathcal{R}$ (rod variables) attached to each point. Picking such a representative to give us our "reference configuration" amounts to choosing a map $f_0 : \mathcal{P} \longrightarrow \mathcal{R}$. Let $\mathcal{B} \subset \mathcal{S}$ be the graph of f_0:

$$\mathcal{B} = \{(X_{\mathcal{P}}, f_0(X_{\mathcal{P}})) \mid X_{\mathcal{P}} \in \mathcal{P}\}.$$

A *configuration* of $\mathcal{B}$ is, as usual, a map $\phi : \mathcal{B} \longrightarrow \mathcal{S}$. We define $\phi_{\mathcal{P}} : \mathcal{P} \longrightarrow \mathbb{R}^3$ and $\phi_{\mathcal{R}} : \mathcal{P} \longrightarrow \mathcal{R}$ by writing $\phi(X_{\mathcal{P}}, f_0(X_{\mathcal{P}})) = (\phi_{\mathcal{P}}(X_{\mathcal{P}}), \phi_{\mathcal{R}}(X_{\mathcal{P}}))$. The velocity field $\boldsymbol{v}$ of a motion $\phi(X, t)$ likewise has two components, which we can write $\boldsymbol{v} = (\boldsymbol{v}_{\mathcal{P}}, \boldsymbol{v}_{\mathcal{R}})$. Let us also write $x = \phi(X) = (x_{\mathcal{P}}, x_{\mathcal{R}}) \in \mathbb{R}^3 \times \mathcal{R}$. See Figure 2.2.3 in which $\mathbb{R}^3$ has been replaced by $\mathbb{R}^2$ for visualization.

Let $\mathcal{U}_{\mathcal{P}}$ be an open set in $\mathcal{P}$ and $\mathcal{U}$ its graph in $\mathcal{B}$. Then $\phi(\mathcal{U})$ is a subset of $\mathcal{S}$, but is not open. Nevertheless, its component in $\mathbb{R}^3$—namely, $\phi_{\mathcal{P}}(\mathcal{U}_{\mathcal{P}})$—will be an open set with volume element $dv_{\mathcal{P}}$, if ϕ is regular.[8] We let J stand for the Jacobian of $\phi_{\mathcal{P}}$. The mass density $\rho(x, t)$ is assumed to depend only on $x_{\mathcal{P}}$. Then we get, as in Section 1.5,

$$\frac{\partial \rho}{\partial t} + d\rho \cdot \boldsymbol{v}_{\mathcal{P}} + \rho \operatorname{div} \boldsymbol{v}_{\mathcal{P}} = 0.$$

(If one assumed ρ depended on $x_{\mathcal{R}}$ as well, then an analysis like that for shells would be required.)

A formulation accounting for a *single director* may be given as follows. We assume the existence of a traction vector $\boldsymbol{t}(x, t, \boldsymbol{n}_{\mathcal{P}})$ for the

[8]Even if ϕ is regular, it could happen that $\phi(\mathcal{B})$ is not a graph; we postulate that $\phi(\mathcal{B})$ is a graph over $\phi_{\mathcal{P}}(\mathcal{B})$ as part of the definition of regularity. It is equivalent to $\phi_{\mathcal{P}}$ being regular.

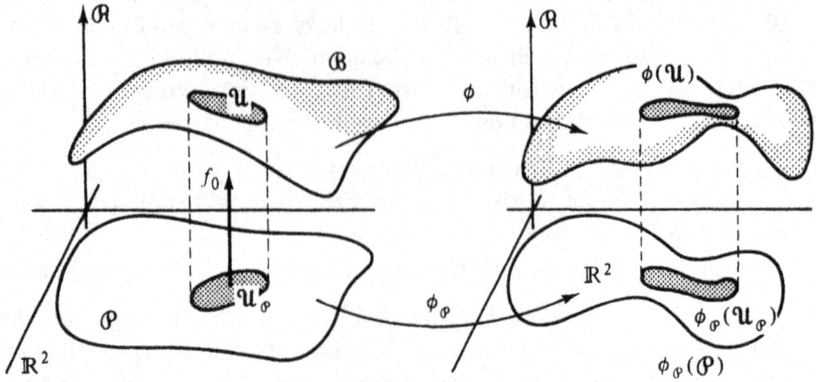

Figure 2.2.3

motion of $\mathcal{B}$ in $\mathcal{S}$, where $\boldsymbol{n}_{\mathcal{P}}$ denotes a unit normal in $\mathcal{P}$. Balance of momentum is now postulated in the form

$$\frac{d}{dt}\int_{\phi_{t,\mathcal{P}}(\mathcal{U}_{\mathcal{P}})} \rho v \, dv_{\mathcal{P}} = \int_{\phi_{t,\mathcal{P}}(\mathcal{U}_{\mathcal{P}})} f \, dv_{\mathcal{P}} + \int_{\partial\phi_{t,\mathcal{P}}(\mathcal{U}_{\mathcal{P}})} \boldsymbol{t}(x, t, \boldsymbol{n}_{\mathcal{P}}) \, da_{\mathcal{P}}.$$

where $\boldsymbol{\phi}_{t,\mathcal{P}}$ is the map $\boldsymbol{\phi}_{\mathcal{P}}$ at time t. This equation has an $\mathcal{R}$ component as well as the usual $\mathcal{P}$ component. For example,

$$\boldsymbol{f} = (\rho \boldsymbol{b}_{\mathcal{P}}, \rho \boldsymbol{b}_{\mathcal{R}} + \boldsymbol{\pi})$$

where $\boldsymbol{b}_{\mathcal{P}}$ is the usual *body force per unit mass*, $\boldsymbol{b}_{\mathcal{R}}$ is the *external director body force per unit mass* and $\boldsymbol{\pi}$ is the *intrisic director body force per unit volume*. By an argument like that in Cauchy's theorem, we find that $\boldsymbol{t}$ is a linear function of $\boldsymbol{n}_{\mathcal{P}}$. Thus we can write

$$\boldsymbol{t}(x, t, \boldsymbol{n}_{\mathcal{P}}) = (\sigma_{\mathcal{P}}^{ab}(x_{\mathcal{P}}, x_{\mathcal{R}}, t)n_b, \sigma_{\mathcal{R}}^{cb}(x_{\mathcal{B}}, x_{\mathcal{R}}, t)n_b)$$

where the n_b's are components of $\boldsymbol{n}_{\mathcal{P}}$, a refers to coordinates in $\mathbb{R}^3$ and c refers to coordinates in $\mathcal{R}$. One calls $\sigma_{\mathcal{R}}^{cb}$ the *director stress.*

The equations of motion then become the pair of vector equations:

$$\begin{cases} \rho a_{\mathcal{P}}^{ab} = \rho b_{\mathcal{P}}^{a} + \sigma_{\mathcal{P}}^{ab}{}_{|b} \\ \rho a_{\mathcal{R}}^{c} = \rho b_{\mathcal{R}}^{c} + \sigma_{\mathcal{R}}^{cb}{}_{|b} + \pi^c. \end{cases}$$

These equations are coupled because $\sigma_{\mathcal{P}}^{ab}$, $\sigma_{\mathcal{R}}^{cb}$ and π^c depend on $x_{\mathcal{P}}$ and $x_{\mathcal{R}}$.

In general, the stress tensor $\sigma_{\mathcal{P}}$ is not symmetric. However, in any specific situation, information can be obtained from balance of moment of momentum. This leads us to postulate balance of moment of momentum as follows:

$$\int_{\phi_t,\mathcal{P}(\mathcal{U})} \rho(\boldsymbol{x}_{\mathcal{P}} \times \boldsymbol{a}_{\mathcal{P}} + \boldsymbol{x}_{\mathcal{R}} \times \boldsymbol{a}_{\mathcal{R}})\, dv_{\mathcal{P}}$$

$$= \int_{\phi_t,\mathcal{P}(\mathcal{U})} \rho(\boldsymbol{x}_{\mathcal{P}} \times \boldsymbol{b}_{\mathcal{P}} + \boldsymbol{x}_{\mathcal{R}} \times \boldsymbol{b}_{\mathcal{R}})\, dv_{\mathcal{P}}$$

$$+ \int_{\partial\phi_t,\mathcal{P}(\mathcal{U})} (\boldsymbol{x}_{\mathcal{P}} \times \langle \boldsymbol{\sigma}_{\mathcal{P}}, \boldsymbol{n}_{\mathcal{P}}\rangle + \boldsymbol{x}_{\mathcal{R}} \times \langle \boldsymbol{\sigma}_{\mathcal{R}}, \boldsymbol{n}_{\mathcal{P}}\rangle)\, da_{\mathcal{P}}.$$

The argument in 2.10, shows that this assumption is equivalent to symmetry of the tensor k^{ab}, where

$$k^{ab} = \sigma_{\mathcal{P}}^{ab} - \pi^a x_{\mathcal{R}}^b + x_{\mathcal{R}|c}^b \sigma_{\mathcal{R}}^{ac}$$

There are analogous results when $\mathcal{R}$ is chosen in a more general manner such as in the case when there are several directors.

Oriented media concepts are often used as alternatives to the three-dimensional theory for the direct construction of rod, plate, and shell theories (see the references cited in footnote 7). Further properties of oriented media such as Euclidean invariance and balance of energy are considered in problems below.

Box 2.4 ***Summary of Important Formulas for Section 2.2***

Cauchy Stress Tensor

$\boldsymbol{\sigma}$ $\qquad\qquad$ σ^{ab}

Cauchy Traction Vector

$\boldsymbol{t} = \langle \boldsymbol{\sigma}, \boldsymbol{n}\rangle = \boldsymbol{\sigma}\cdot\boldsymbol{n}$ $\qquad\qquad$ $t^a = \sigma^{ab} n_b = \sigma^a{}_b n^b$

Balance of Momentum: Spatial

$$\frac{d}{dt}\int_{\phi_t(\mathcal{U})} \rho \boldsymbol{v}\, dv = \int_{\phi_t(\mathcal{U})} \rho \boldsymbol{b}\, dv + \int_{\partial\phi_t(\mathcal{U})} \langle \boldsymbol{\sigma}, \boldsymbol{n}\rangle\, da$$

$$\frac{d}{dt}\int_{\phi_t(\mathcal{U})} \rho v^i\, dv = \int_{\phi_t(\mathcal{U})} \rho b^i\, dv + \int_{\partial\phi_t(\mathcal{U})} \sigma^{ij} n_j\, da$$

(Euclidean coordinates only)

Balance of Momentum: Localization

$\rho\dot{\boldsymbol{v}} = \rho\boldsymbol{b} + \operatorname{div}\boldsymbol{\sigma}$ $\qquad\qquad$ $\rho\dfrac{\partial v^a}{\partial t} + v^b v^a{}_{|b} = \rho b^a + \sigma^{ab}{}_{|b}$

Piola–Kirchhoff Stress Tensors

First: $\boldsymbol{P} = J\boldsymbol{\sigma}\boldsymbol{F}^{-T} = \boldsymbol{F}\boldsymbol{S}$ $\qquad\qquad$ $P^{aA} = J(F^{-1})^A{}_b \sigma^{ab} = S^{BA} F^a{}_B$

Second: $\boldsymbol{S} = \boldsymbol{F}^{-1}\boldsymbol{P}$ $\qquad\qquad$ $S^{BA} = P^{aA}(F^{-1})^B{}_a$

Balance of Momentum: Material

$$\frac{d}{dt}\int_{\mathcal{U}} \rho_{\text{Ref}} \boldsymbol{V}\,dV = \int_{\mathcal{U}} \rho_{\text{Ref}} \boldsymbol{B}\,dV + \int_{\partial\mathcal{U}} \langle \boldsymbol{P}, \boldsymbol{N}\rangle\,dA$$

$$\frac{d}{dt}\int_{\mathcal{U}} \rho_{\text{Ref}} V^i\,dV = \int_{\mathcal{U}} \rho_{\text{Ref}} B^i\,dV + \int_{\partial\mathcal{U}} P^{iI} N_I\,dA$$

(Euclidean coordinates only)

Balance of Momentum: Localization

$$\rho_{\text{Ref}} \boldsymbol{A} = \rho_{\text{Ref}} \boldsymbol{B} + \text{DIV}\,\boldsymbol{P}$$

$$\rho_{\text{Ref}}\left(\frac{\partial V^a}{\partial t} + \gamma^a_{bc} V^b V^c\right) = \rho_{\text{Ref}} B^a + P^{aA}{}_{|A}$$

Euler Equations (Perfect Fluid)

$$\rho \dot{\boldsymbol{v}} = \rho \boldsymbol{b} - \nabla p$$

$$\rho\left(\frac{\partial v^a}{\partial t} + v^b v^a{}_{|b}\right) = \rho b^a - g^{ab}\frac{\partial p}{\partial x^b}$$

Symmetry of Stress Tensor (Balance of Moment of Momentum)

$\boldsymbol{\sigma} = \boldsymbol{\sigma}^T$ $\qquad \sigma^{ab} = \sigma^{ba}$

$\boldsymbol{S} = \boldsymbol{S}^T$ $\qquad S^{AB} = S^{BA}$

$\boldsymbol{F}\boldsymbol{P}^T = \boldsymbol{P}\boldsymbol{F}^T$ $\qquad F^b{}_A P^{aA} = P^{bA} F^a{}_A$

Convected Stress Tensor

$\boldsymbol{\Sigma} = \boldsymbol{F}^T \boldsymbol{\sigma}^\flat \boldsymbol{F}$ = pull-back of $\boldsymbol{\sigma}^\flat$ $\qquad \Sigma_{AB} = F^a{}_A \sigma_{ab} F^b{}_B$

Co-rotational Stress Tensor

$\boldsymbol{\mathfrak{S}}^\sharp = \boldsymbol{R}^{-1}\boldsymbol{\sigma}^\sharp \boldsymbol{R}^{-T}$ $\qquad \mathfrak{S}^{AB} = (R^{-1})^A{}_a \sigma^{ab} (R^{-1})^B{}_b$

$\boldsymbol{\mathfrak{S}}^\flat = \boldsymbol{R}^T \boldsymbol{\sigma}^\flat \boldsymbol{R}$ $\qquad \mathfrak{S}_{AB} = R^a{}_A \sigma_{ab} R^b{}_B$

Cosserat Continua and Liquid Crystals

$\mathcal{S} = \mathbb{R}^3 \times \mathcal{R}$ = body $\times$ director variables

$\mathcal{B} = \{(X_{\mathcal{P}}, f_0(X_{\mathcal{P}})) \mid X_{\mathcal{P}} \in \mathcal{P}\}$, where $f_0 : \mathcal{P} \to \mathcal{R}$ is given

$\boldsymbol{\sigma} = (\boldsymbol{\sigma}_{\mathcal{P}}, \boldsymbol{\sigma}_{\mathcal{R}})$ $\qquad \boldsymbol{\sigma} = (\sigma^{ab}_{\mathcal{P}}, \sigma^{cb}_{\mathcal{R}})$ a for $\mathbb{R}^3$ components, c for $\mathcal{R}$ components

$$\begin{cases} \rho \boldsymbol{a}_{\mathcal{P}} = \rho \boldsymbol{b}_{\mathcal{P}} + \text{div}(\boldsymbol{\sigma}_{\mathcal{P}}) \\ \rho \boldsymbol{a}_{\mathcal{R}} = \rho \boldsymbol{b}_{\mathcal{R}} + \text{div}(\boldsymbol{\sigma}_{\mathcal{R}}) + \boldsymbol{\pi} \end{cases} \qquad \begin{cases} \rho a^a_{\mathcal{P}} = \rho b^a_{\mathcal{P}} + \sigma^{ab}{}_{|b} \\ \rho a^c_{\mathcal{R}} = \rho b^c_{\mathcal{R}} + \sigma^{cb}{}_{|b} + \pi^c \end{cases}$$

$\boldsymbol{k}^\sharp$ is symmetric $\qquad k^{ab} = k^{ba} = \sigma^{ab}_{\mathcal{P}} - \pi^a x^b_{\mathcal{R}} + x^b_{\mathcal{R}|c}\sigma^{ac}_{\mathcal{R}}$

2.3 BALANCE OF ENERGY

The three sections that follow deal with thermomechanics, or the thermodynamics of continuous media. This section and the next deal with the balance of energy, or the "first law of thermodynamics," while Section 2.5 deals with the entropy production inequality, or the "second law of thermodynamics."

Balance of energy gives important insight into the other balance laws. Box 3.1 and Section 2.4 show how balance of energy can be used as a basis for the derivation of all the other balance equations.

Let $\mathcal{B}$ be a simple body in $\mathcal{S} = \mathbb{R}^3$, $\phi(X, t)$ a regular C^1 motion of $\mathcal{B}$ in $\mathcal{S}$, and $\boldsymbol{v}(x, t)$ its spatial velocity field. Let $\rho(x, t)$ (the mass density) and $\boldsymbol{t}(x, t, \boldsymbol{n})$ (the Cauchy traction vector) be defined and let $\boldsymbol{b}(x, t)$ denote the external force per unit mass. In addition to this structure, assume there exist functions with the stated physical interpretations:

$e(x, t)$, the internal energy function per unit mass;
$r(x, t)$, the heat supply per unit mass;
$h(x, t, \boldsymbol{n})$, the heat flux (across a surface with normal $\boldsymbol{n}$).

The function e represents energy stored internally in the body, which is a macroscopic reflection of things like chemical binding energy, intermolecular energy, and energy of molecular vibrations. Directly, it is exemplified by the mechanical energy stored in a coiled spring. In Chapter 3 we use the results of this and the next section to establish a fundamental link between the internal energy and the stress tensor. The function $r(x, t)$ represents incoming heat energy, such as radiation. Like $\boldsymbol{b}$, r is usually regarded as "external" or "given." The function $h(x, t, \boldsymbol{n})$ represents the rate of heat conduction across a surface with unit normal $\boldsymbol{n}$. It is analogous to $\boldsymbol{t}(x, t, \boldsymbol{n})$ and reflects the influence of one part of the material on its contiguous part.

3.1 Definition We say that *balance of energy* holds provided that for every nice open set $\mathcal{U} \subset \mathcal{B}$,

$$\frac{d}{dt}\int_{\phi_t(\mathcal{U})} \rho\left(e + \tfrac{1}{2}\langle \boldsymbol{v}, \boldsymbol{v}\rangle\right) dv = \int_{\phi_t(\mathcal{U})} \rho(\langle \boldsymbol{b}, \boldsymbol{v}\rangle + r)\, dv + \int_{\partial\phi_t(\mathcal{U})} (\langle \boldsymbol{t}, \boldsymbol{v}\rangle + h)\, da.$$

This states that the rate of increase of the total energy (internal and kinetic) of any portion of the body equals the rate of work done on that portion (from the body forces and surface tractions) plus the rate of increase of heat energy (from the heat supply r plus heat flow h across the boundary).

3.2 Proposition *Assume that* $\boldsymbol{t}(x, t, \boldsymbol{n}) = \boldsymbol{\sigma}(x, t)\cdot\boldsymbol{n}$ *for a two-tensor* $\boldsymbol{\sigma}(x, t)$ *(see 2.2). Balance of energy implies the existence of a unique vector field* $\boldsymbol{q}(x, t)$ *such that for all* $\boldsymbol{n}$, $h(x, t, \boldsymbol{n}) = -\langle \boldsymbol{q}(x, t), \boldsymbol{n}\rangle$. *We call* $\boldsymbol{q}$ *the* heat flux vector.

Proof Apply the Cauchy theorem 1.7 with $c = \langle \boldsymbol{t}, \boldsymbol{v}\rangle + h = \boldsymbol{v}\cdot\boldsymbol{\sigma}\cdot\boldsymbol{n} + h$ to establish the existence of a vector field $\boldsymbol{c}$ such that $c = \langle \boldsymbol{c}, \boldsymbol{n}\rangle$, and define $-\boldsymbol{q} = \boldsymbol{c} - \boldsymbol{v}\cdot\boldsymbol{\sigma}$. ∎

To reinforce the physical interpretation of balance of energy, we introduce the following terminologies with reference to a moving portion $\phi_t(\mathcal{U})$ of the body:

$$\mathfrak{K} = \int_{\phi_t(\mathfrak{U})} \tfrac{1}{2}\rho\langle v, v\rangle\, dv = \text{kinetic energy},$$

$$\mathfrak{E} = \int_{\phi_t(\mathfrak{U})} \rho e\, dv = \text{internal energy},$$

$$\mathfrak{W} = \int_{\phi_t(\mathfrak{U})} \rho\langle b, v\rangle\, dv + \int_{\partial\phi_t(\mathfrak{U})} \langle t, v\rangle\, da = \text{mechanical power},$$

$$\mathfrak{Q} = \int_{\phi_t(\mathfrak{U})} \rho r\, dv + \int_{\partial\phi_t(\mathfrak{U})} h\, da = \text{nonmechanical power}.$$

The first law then reads:

$$\frac{d}{dt}(\mathfrak{K} + \mathfrak{E}) = \mathfrak{Q} + \mathfrak{W}.$$

Now we shall obtain the local form of energy balance.

3.4 Theorem *Assume ϕ_t is a C^1 regular motion and the following balance principles hold: conservation of mass, balance of momentum, balance of moment of momentum, and balance of energy.*

Then

$$\rho\dot{e} + \operatorname{div} \boldsymbol{q} = \boldsymbol{\sigma} : \boldsymbol{d} + \rho r,$$

where $\dot{e} = \partial e/\partial t + v[e] = \partial e/\partial t + (\partial e/\partial x^a)v^a$, $\operatorname{div} \boldsymbol{q} = q^a{}_{|a}$, $\boldsymbol{d} = \frac{1}{2}\mathfrak{L}_v \boldsymbol{g}$ (*i.e.*, $d_{ab} = \frac{1}{2}(v_{a|b} + v_{b|a})$), *and* $\boldsymbol{\sigma} : \boldsymbol{d} = \sigma^{ab}d_{ab}$.

Proof By localization of the spatial master balance law (see Equation (4), Section 2.1), balance of energy is equivalent to the following:

$$\rho(e + \tfrac{1}{2}\langle v, v\rangle)^{\cdot} = \rho\langle b, v\rangle + \rho r + \operatorname{div}(v\cdot\boldsymbol{\sigma} - \boldsymbol{q}).$$

Notice that $\operatorname{div}(v\cdot\boldsymbol{\sigma} - \boldsymbol{q}) = (v_a\sigma^{ab} - q^b)_{|b} = v_{a|b}\sigma^{ab} + v_a\sigma^{ab}{}_{|b} - q^b{}_{|b} = \boldsymbol{\sigma} : \boldsymbol{d} + \langle(\operatorname{div}\boldsymbol{\sigma}), v\rangle - \operatorname{div}\boldsymbol{q}$, since $\boldsymbol{\sigma}$ is symmetric. Using this, $\langle v, v\rangle^{\cdot} = 2\langle\dot{v}, v\rangle$ (see Problem 4.7, Chapter 1) and balance of momentum $\rho\dot{v} = \rho b + \operatorname{div}\boldsymbol{\sigma}$, we get the result. ∎

It is useful to put balance of energy into material form. Let $x = \phi(X, t)$, $E(X, t) = e(x, t)$, $\boldsymbol{Q}(X, t) = J\boldsymbol{F}^{-1}\boldsymbol{q}(x, t)$ (Piola transform), $R(X, t) = r(x, t)$,

$$P^{aA} = J(F^{-1})^A{}_b\sigma^{ab} \quad \text{(first Piola–Kirchhoff stress tensor)},$$

and

$$S^{BA} = P^{aA}(F^{-1})^B{}_a \quad \text{(second Piola–Kirchhoff stress tensor)}.$$

If we multiply the localized balance of energy equation by J, use $\rho_{\text{Ref}}(X, t) = J(X, t)\rho(x, t)$, and the Piola identity $\text{DIV}\,\boldsymbol{Q} = J\operatorname{div}\boldsymbol{q}$, we get

$$\rho_{\text{Ref}}\frac{\partial E}{\partial t} + \text{DIV}\,\boldsymbol{Q} = P^{aA}F^b{}_A d_{ab} + \rho_{\text{Ref}}R.$$

From Section 1.4 and symmetry of $J\sigma^{ab} = P^{aA}F^b{}_A$, we get $F^a{}_A P^{bA} d_{ab} = P^{bA}v_{b|a}F^a{}_A = P^{bA}V_{b|A}$. Now write $P^{bA}V_{b|A} = F^b{}_B S^{BA} V_{b|A}$. Using symmetry of $\boldsymbol{S}$, we get $P^{bA}V_{b|A} = S^{AB}D_{AB}$, where $\boldsymbol{D}$ is the material rate of deformation tensor discussed in Section 1.3: $D_{AB} = \frac{1}{2}(F^b{}_B V_{b|A} + F^b{}_A V_{b|B})$. Thus we obtain:

3.5 Corollary *Balance of energy in localized material form is*

$$\rho_{\text{Ref}} \frac{\partial E}{\partial t} + \text{DIV}\, \boldsymbol{Q} = \boldsymbol{S} : \boldsymbol{D} + \rho_{\text{Ref}} R.$$

Problem 3.1 Derive the material form of balance of energy directly from the balance principle in integrated form:

$$\frac{d}{dt} \int_{\mathfrak{U}} \rho_{\text{Ref}}(E + \tfrac{1}{2}\langle V, V \rangle) dV = \int_{\mathfrak{U}} \rho_{\text{Ref}}(\langle \boldsymbol{B}, V \rangle + R)\, dV + \int_{\partial \mathfrak{U}} (\boldsymbol{V} \cdot \boldsymbol{P} \cdot \boldsymbol{N} - \boldsymbol{Q} \cdot \boldsymbol{N})\, dA.$$

3.6 Example (The Rigid Heat Conductor) A motion ϕ_t is called *rigid* if $\boldsymbol{F}_t^T \boldsymbol{F}_t =$ Identity and $\boldsymbol{F}_t$ is independent of $X \in \mathcal{B}$. Assume the existence of a positive function $\Theta(X, t)$, called the *absolute temperature*, and numbers c (=specific heat) and k (= conductivity) such that $E = c\Theta$ and $\mathbf{Q} = -k\, \text{GRAD}\, \Theta$ (Fourier's law). Let R be a given function of (X, t). In this case $\boldsymbol{D} = \mathbf{0}$ (since the motion is rigid), so by 3.5,

$$\rho_{\text{Ref}} c \frac{\partial \Theta}{\partial t} = \rho_{\text{Ref}} R + k \Delta \Theta,$$

where

$$\Delta \Theta = \text{DIV GRAD}\, \Theta = \left(\frac{\partial \Theta}{\partial X^B} G^{BA} \right)_{|A}$$

is the Laplacian of Θ. This is the classical linear heat equation. Exploration of the assumptions made in this example and its nonlinear generalizations are given in Section 3.4. (The equation for $\partial \Theta / \partial t$ for nonrigid heat conductors will generally be coupled with the equation for $\partial V / \partial t$.)

The following boxes investigate balance of energy a little more deeply. The first box shows that balance of energy, plus invariance under rigid body motions in $\mathbb{R}^3$, gives all the earlier balance principles. This is due to Noll–Truesdell or Green–Naghdi–Rivlin, depending on the interpretation. The second box shows that the two approaches are mathematically equivalent. The next section modifies the arguments so they are covariant.

Box 3.1 ***Energy Balance and Invariance Under Rigid Motions (Green–Naghdi–Rivlin Theorem)***

Let $\mathcal{B}$ be a simple body in $\mathbb{R}^3$ and $\phi(X, t)$ a fixed regular motion of $\mathcal{B}$ in $\mathbb{R}^3$. Assume that functions $e(x, t)$, $\rho(x, t)$, $\boldsymbol{b}(x, t)$, $\boldsymbol{t}(x, t, \boldsymbol{n})$, $r(x, t)$, and $h(x, t, \boldsymbol{n})$ are given and satisfy balance of energy: for all nice

$\mathcal{U} \subset \mathcal{B}$,

$$\frac{d}{dt}\int_{\phi_t(\mathcal{U})} \rho(e + \tfrac{1}{2}\langle v, v\rangle)dv = \int_{\phi_t(\mathcal{U})} \rho(\langle b, v\rangle + r)\,dv + \int_{\partial\phi_t(\mathcal{U})} (\langle t, v\rangle + h)\,da. \tag{1}$$

Let $\xi_t : \mathbb{R}^3 \to \mathbb{R}^3$ denote a motion of $\mathbb{R}^3$ and let $x' = \xi_t(x)$. In this box we shall assume that ξ_t is either a translational motion,

$$\xi_t(x) = x + (t - t_0)c,$$

where c is a constant vector, or a rotational motion,

$$\xi_t(x) = e^{(t-t_0)\Omega}x,$$

where Ω is a 3×3 skew matrix. (The later is a steady rotation about a fixed axis in space—infinitesimal rotations are represented by skew matrices and one exponentiates these to get finite rotations; see Problem 4.6 below).

The composition $\phi'_t = \xi_t \circ \phi_t$ defines a new motion of $\mathcal{B}$ in $\mathbb{R}^3$ called the *superposed motion*.[9] The spatial velocity of the new motion is given by

$$v'_t(x') = w(x') + \xi_{t*}v_t(x'),$$

where w is the velocity field of ξ_t. Explicitly,

$$v'_t(x') = c + v_t(x) = c + v_t(x' - (t - t_0)c)$$

if ξ_t is translational and

$$v'_t(x') = \Omega x' + e^{(t-t_0)\Omega}v_t(e^{-(t-t_0)\Omega}x')$$

if ξ_t is rotational. Let $a'(x', t)$ denote the spatial acceleration of the new motion ϕ'. Computing a', we have the equation

$$a'_t = a^\xi_t + \xi_{t*}a_t + 2\nabla_{\xi_{t*}v_t}w_t,$$

where a^ξ_t is the acceleration of the superposed motion ξ_t. For translational motion this reduces to $a'_t(x') = a_t(x)$, and for rotational motion, a^ξ_t corresponds to "centripetal force" and $2\nabla_{\xi_{t*}v_t}w_t$ corresponds to "Coriolis force" as is described in elementary mechanics texts, such as Goldstein [1980]. (This formula is derived in a more general context in the next section.)

---

[9]The following example may help fix ideas. Consider in $\mathbb{R}^3$, an elastic body in a (natural) unstressed state, so $t = 0$. The body thus satisfies the equilibrium equations with zero body force and zero traction. If the body is subjected to a uniform rotation, particles in the body are accelerated and the resulting motion satisfies balance of momentum with a non-zero body force (centrifugal and Coriolis forces).

Associated with the motion ϕ', we define new functions as follows:

$$\rho'(x', t) = \rho(x, t), \qquad e'(x', t) = e(x, t), \qquad r'(x', t) = r(x, t),$$
$$t'(x', t, n') = T\xi_t \cdot t(x, t, n), \qquad v' \text{ and } a' \text{ as explained above,}$$
$$b' = b + a_t^{\xi} + 2\nabla_{\xi_{t*}v_t} w_t.$$

Here n' is related to n as follows. Let Σ be a plane in $\mathbb{R}^3$ with unit normal n. Then $\Sigma' = T\xi_t(x) \cdot \Sigma$ is the plane transformed by the deformation gradient of ξ_t at x. We let n' be the unit normal to Σ'. If ξ_t is an isometry, then $n' = T\xi_t(x) \cdot n$, but in general, this relationship need not hold.

Problem 3.2 If the unit normal n is described in terms of a one-form α, show that n' is described in terms of $\alpha' = \xi_{t*}\alpha$. (*Note:* α and α' need not be normalized; see Box 2.1 in Section 2.2.)

These transformation formulas are all "natural" except possibly for the formula for the apparent body force b'. Indeed there is some controversy over how forces should be treated in continuum mechanics. However, in the present context there is little doubt that when one modifies a motion by superposing a steady rotational motion, then one should add to the forces given by Newtonian physics (in an intertial $\mathbb{R}^3$ background), the "fictitious" centrifugal and Coriolis forces. This is the content of the transformation law for b. Notice that b' is defined so that the identity

$$b' - a' = \xi_{t*}(b - a)$$

holds. Further information on transformation of forces is found in the next box and in the next section.

3.8 Theorem *Let ϕ_t be a given motion of $\mathcal{B}$ in $\mathbb{R}^3$ and let balance of energy hold. Assume that for any superposed translational or rotational motion, balance of energy still holds. Then there exist σ and q such that $t = \sigma \cdot n$ and $h = -q \cdot n$, and conservation of mass, balance of momentum, and balance of moment of momentum all hold.*

Conversely, if these balance principles all hold, then balance of energy remains valid under the superposition of rigid motions of $\mathbb{R}^3$, as defined above.

The original derivation of Green and Rivlin [1964a] proceeds slightly differently from ours. They deal with translations first, then simplify the balance of energy stement by eliminating b, and then assume a transformation property under rotations. This avoids the problem of transformation of forces mentioned above but has the conceptual problem of necessitating a hypothesis on a reduced energy equation in the middle of the derivation. This point has also been emphasized by Serrin [1975]. The proof, however, is not substantially changed by this modification.

Proof of Theorem 3.8 First choose $\xi_t(x) = x + (t - t_0)c$ for a constant vector c. Thus $\rho' = \rho$, $v' = v + c$, $a' = a$, and $b' = b$ (the primed quantities are evaluated at (x', t), while the unprimed ones are evaluated at (x, t)). By assumption,

$$\frac{d}{dt}\int_{\phi_t'(\mathfrak{U})} \rho'(e' + \tfrac{1}{2}\langle v', v'\rangle)\, dv$$
$$= \int_{\phi_t'(\mathfrak{U})} \rho'(\langle b', v'\rangle + r')\, dv + \int_{\partial\phi_t'(\mathfrak{U})} (t'\cdot v' + h')\, da.$$

Now use the transport theorem, the identity $(\dot{f})' = (f')^{\cdot}$ for scalar fields f, and the identity $\langle v, w\rangle^{\cdot} = \langle v, \dot{w}\rangle + \langle \dot{v}, w\rangle$ for vector fields (from Problem 4.7, Chapter 1) to give

$$\int_{\phi_t'(\mathfrak{U})} (\dot{\rho}' + \rho' \operatorname{div} v')(e' + \tfrac{1}{2}\langle v', v'\rangle) dv + \int_{\phi_t'(\mathfrak{U})} \rho'(\dot{e} + \langle a', v'\rangle)\, dv$$
$$= \int_{\phi_t'(\mathfrak{U})} \rho'(\langle b', v'\rangle + r')\, dv + \int_{\phi_t'(\mathfrak{U})} (t'\cdot v' + h')\, da. \tag{2}$$

Next, set $t = t_0$ and subtract identity (2) and the unprimed equation (1) for balance of energy for ϕ_t to give

$$\int_{\phi_t(\mathfrak{U})} (\dot{\rho} + \rho \operatorname{div} v)(\langle v, c\rangle + \tfrac{1}{2}\langle c, c\rangle) dv = \int_{\phi_t(\mathfrak{U})} \rho\langle (b - a), c\rangle\, dv$$
$$+ \int_{\partial\phi_t(\mathfrak{U})} t\cdot c\, da. \tag{3}$$

Cauchy's theorem applied to (3) gives the existence of a two tensor σ such that $t = \sigma\cdot n$. Applying Cauchy's theorem again to (1) then gives $h = -q\cdot n$ for a vector field q. Substituting $t = \sigma\cdot n$ in (3) and using the divergence theorem gives

$$\int_{\phi_t(\mathfrak{U})} (\dot{\rho} + \rho \operatorname{div} v)(\langle v, c\rangle + \tfrac{1}{2}\langle c, c\rangle) dv$$
$$= \int_{\phi_t(\mathfrak{U})} \langle \rho(b - a) + \operatorname{div} \sigma, c\rangle\, dv. \tag{4}$$

Substitute $c = \lambda u$ in (4), where u is a unit vector, and differentiate twice with respect to λ to give $\int_{\phi_t(\mathfrak{U})} (\dot{\rho} + \rho \operatorname{div} v)\, dv = 0$—that is, conservation of mass. The left-hand side of (4) thus vanishes identically, so the right-hand side gives balance of momentum.

Next we let $\xi_t(x) = e^{(t-t_0)\Omega}x$ for Ω a constant skew matrix, so that at $t = t_0$, $v' = \Omega x + v$. Using conservation of mass, (2) at $t = t_0$ becomes

$$\int_{\phi_t(\mathcal{U})} \rho(\dot{e} + \langle a' - b', v' \rangle - r)\, dv = \int_{\partial\phi_t(\mathcal{U})} (t \cdot v' + h)\, da. \tag{5}$$

Now using $a' - b' = a - b$ and subtracting (5) from the corresponding unprimed version yields

$$\int_{\phi_t(\mathcal{U})} \rho(\langle a - b, \Omega \cdot x \rangle\, dv + \int_{\partial\phi_t(\mathcal{U})} t \cdot (\Omega \cdot x)\, da = 0.$$

Thus $\displaystyle\int_{\phi_t(\mathcal{U})} \langle \rho(a - b - \operatorname{div} \sigma), \Omega \cdot x \rangle\, dv + \int_{\phi_t(\mathcal{U})} \sigma : \Omega\, dv = 0.$

Applying balance of momentum gives $\sigma : \Omega = 0$; so as Ω is an arbitrary skew matrix, σ is symmetric.

The converse assertion may be proved by substitution of the primed quantities into Equation (2). ∎

The same theorem is true if we replace translational and rotational motions by a general time-dependent rigid motion of the form $\xi_t(x) = x + c(t) + Q(t) \cdot x$, where $c(t) \in \mathbb{R}^3$ and $Q(t)$ is a time-dependent proper orthogonal matrix.

Problem 3.2 Prove a version of Theorem 3.8 suitable for Cosserat continua (see Toupin [1964]).

Box 3.2 ***Energy Balance and Frame Indifference (Noll's Theorem)***

A different formulation of Theorem 3.8 may be given by viewing the superposed motion ξ_t passively rather than actively. Instead of regarding ξ_t as a motion of space, regard it as a time-dependent coordinate chart. (This is analogous to regarding a matrix as either a linear transformation or a change of basis.[10]) In the passive context, the theorem was first given by Noll [1963].

To reformulate the result, we need some terminology. Let $\mathcal{V} = \mathcal{S} \times \mathbb{R} = \mathbb{R}^4$ denote *standard Newtonian spacetime.* A *framing or slicing* of $\mathcal{V}$ is a one-parameter family of embeddings $i_\lambda : \mathcal{S} \to \mathcal{V}$, where λ is a

[10] The reader may recall that a similar point was made by Dirac in 1928, who showed the equivalence between the Heisenberg and Schrödinger pictures of quantum mechanics. There is a similar relationship between the Green–Naghdi–Rivlin theorem and the Noll theorem.

real parameter, such that the map $i: \mathcal{S} \times \mathbb{R} \longrightarrow \mathcal{V}$, $i(p, t) = i_t(p)$, is a diffeomorphism. Let $\mathcal{S}_\lambda = i_\lambda(\mathcal{S})$; see Figure 2.3.1. We think of the point $p \in \mathcal{S}$ and $\lambda \in \mathbb{R}$ as giving space–time coordinates to the point $i_\lambda(p)$ in $\mathcal{V}$. A slicing will be called *Newtonian* if $\pi_2(i_\lambda(p)) = \lambda$ for all λ and p, where $\pi_2: \mathcal{S} \times \mathbb{R} \longrightarrow \mathbb{R}$ is the projection onto the second factor. In other words, the whole slice $i_\lambda(\mathcal{S})$ lies in $\mathcal{V}$ at "time" λ. A Newtonian slicing will be called *rigid* if for each fixed λ, $\pi_1 \circ i : \mathcal{S} \longrightarrow \mathcal{S}$ is an isometry, where $\pi_1 : \mathcal{V} \longrightarrow \mathcal{S}$ is the projection onto the first factor.

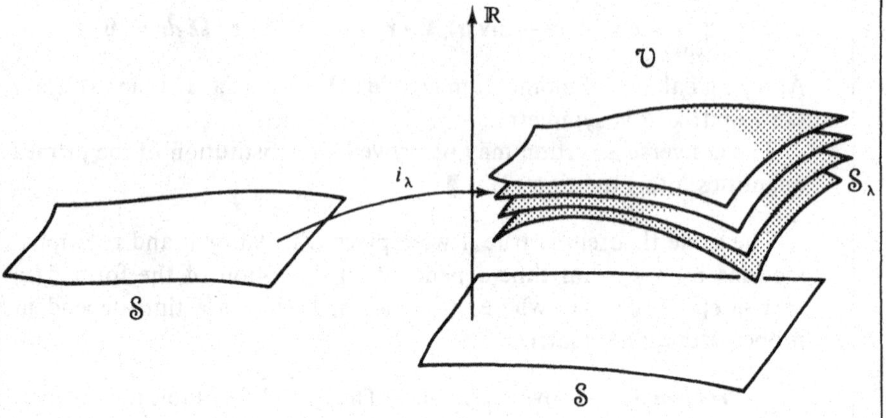

Figure 2.3.1

Now let ϕ_t be a given motion of $\mathcal{B}$ in $\mathcal{S}$. Associate with this motion the mapping $\Phi : \mathcal{B} \times \mathbb{R} \longrightarrow \mathcal{V}, \Phi(X, t) = (\phi(X, t), t)$. The *representation* of this motion in a slicing i_λ is the map $\Phi' : \mathcal{B} \times \mathbb{R} \longrightarrow \mathcal{S} \times \mathbb{R}$ given by $i^{-1} \circ \Phi$ and we write ϕ'_t for the map of $\mathcal{B}$ to $\mathcal{S}$ defined by $\phi'_t(X) = \pi_1(\Phi'(X, t))$, and also call ϕ'_t the *representation* of the motion ϕ_t in the slicing i_λ.[11]

Given functions e, $\boldsymbol{p}$, $\boldsymbol{b}$, $\boldsymbol{t}$, r, and h associated with the motion ϕ_t, we assume that balance of energy (Equation (1) of Box 3.1) holds. (Note that the representation of the motion ϕ_t in the identity slicing is just ϕ_t again.) Given a slicing we obtain the representation ϕ'_t and correspondingly transform the quantities e, $\boldsymbol{p}$, $\boldsymbol{b}$, $\boldsymbol{t}$, r, and h to obtain new ones. For example, define $e' : \mathcal{S} \times \mathbb{R} \longrightarrow \mathbb{R}$, $e' = e \circ i$ (actually e' may only be defined on the image of $\mathcal{B} \times \mathbb{R}$ by Φ, but we need not fuss with this point). Similarly, appropriately transform vectors tangent to the

[11] Φ is sometimes called the *world tube* and $\pi_1 \circ i^{-1} \circ \Phi : \mathcal{B} \times \mathbb{R} \longrightarrow \mathcal{S}$ is the *motion* associated to Φ by the framing. This terminology will all be expanded on in the next section.

images $\mathcal{S}_\lambda = i_\lambda(\mathcal{S})$ such as $\boldsymbol{n}$ (a unit normal to $\phi_t(\mathcal{U})$ in $\mathcal{S}_t$) and transform $\boldsymbol{t}$ by Ti_λ and transform $\boldsymbol{b}$ such that $\boldsymbol{b} - \boldsymbol{a}$ transforms as a vector, as in Box 3.1. The velocity $\boldsymbol{v}'$ and acceleration $\boldsymbol{a}'$ of the motion in a given slicing are just the usual velocity and acceleration of the representative ϕ'_t of the motion.

The energy-invariance theorem of Noll can now be stated as follows:

3.9 Theorem *Let ϕ_t be a given motion of $\mathcal{B}$ in $\mathbb{R}^3$ and let balance of energy hold. Assume that for any rigid Newtonian slicing of $\mathcal{V} = \mathcal{S} \times \mathbb{R}$ the representative ϕ'_t of ϕ_t still satisfies balance of energy. Then one can write $\boldsymbol{t} = \boldsymbol{\sigma} \cdot \boldsymbol{n}$, $h = -\boldsymbol{q} \cdot \boldsymbol{n}$, and conservation of mass, balance of momentum, and moment of momentum all hold.*

Conversely, if the motion satisfies these balance principles, then the representatives in any rigid Newtonian slicing satisfy balance of energy as well.

Proof This follows from 3.8 by choosing $\xi_t = (\pi_1 \cdot i_t)^{-1}$. ∎

We have formulated Noll's theorem this way to clarify the passive and active interpretations of ξ_t. The actual history is complicated by the fact that Noll views forces differently than Green, Naghdi, and Rivlin. Our view of classical nonrelativistic forces (expanded upon in the next section) is just this: *they are vector fields on spacetime given by physical laws in inertial frames*; guided by the way accelerations are represented in different slicings, we obtain transformation laws for forces. The Noll scheme, which takes a different point of view, is sketched below (see Truesdell's text [1977] for further details).

Box 3.3 ***Historical Remarks on Noll's Theorem***

To help understand Noll's approach to forces, we will use the method of representing the kinematics of a body in Newtonian spacetime relative to a given slicing, as explained in the preceding box. To this kinematic structure, Noll appends axioms on the specification of force as follows.

3.10 Noll's Force Axioms *Given a framing of (flat) spacetime, let ϕ_t be the representative of the motion*[12] *of $\mathcal{B}$ and let $\mathcal{P}$ be a part (subset) of $\mathcal{B}$.*

[12]Noll's terminology would be: ϕ_t is the motion of $\mathcal{B}$ relative to the framing; that is, you get a motion only after you pick a framing. We have made the notation fit with that in the rest of this book as far as possible. Our choice of terminology is influenced by standard practices in general relativity.

(a) The (total) force on $\mathcal{P}$ due to its external environment is specified by:

(i) *a vector field $\mathfrak{B}$ on $\mathcal{P}$ along ϕ_t, that is, $\mathfrak{B}(X, t) \in T_x\mathbb{R}^3 = \mathbb{R}^3$, where $x = \phi_t(X)$, and $X \in \mathcal{P}$;*

(ii) *a vector field $\mathfrak{T}$ on $\partial\mathcal{P}$ along ϕ_t, that is, $\mathfrak{T}(X, N(X), t) \in T_x\mathbb{R}^3 = \mathbb{R}^3$, where $x = \phi_t(X)$, $X \in \mathcal{P}$, and $N(X) =$ outward unit normal to $\partial\mathcal{P}$ at X.*

One calls $\mathfrak{B}$ and $\mathfrak{T}$ the total body force density *and the* contact force density, *respectively, relative to the framing.*

(b) Let $\bar{\phi}_t$ be the representative of the motion relative to a different framing. Let $\bar{\mathfrak{B}}$ and $\bar{\mathfrak{T}}$ be the body force and contact force densities relative to the second framing. If the change of frame is rigid that is, $\bar{x} = \xi_t(x) = Q(t)x + c(t)$, where Q is orthogonal, then $\bar{\mathfrak{B}}(X, t) = Q(t)\cdot \mathfrak{B}(X, t)$, and $\bar{\mathfrak{T}}(X, N, t) = Q(t)\cdot\mathfrak{T}(X, N, t)$.

Notice that $\mathfrak{B}$ and $\mathfrak{T}$ transform as vectors under a change of frame. Noll calls such vector fields *materially frame indifferent.* This transformation law is apparently in contradiction to what we had before and seems to ignore centrifugal and Coriolis forces. Noll successfully reconciles this, as we shall see. (Secretly, Noll's forces correspond to the quantity we called $\boldsymbol{B} - \boldsymbol{A}$ above, which *does* transform as a vector.)

The *power* or *working* of the forces on a part $\mathcal{P}$ of the body $\mathcal{B}$ relative to a framing is defined by

$$W(\phi_t) = \int_{\mathcal{P}} \dot{\phi}_t(X)\cdot\mathfrak{B}(X, t)\, dV + \int_{\partial\mathcal{P}} \dot{\phi}_t(X)\cdot\mathfrak{T}(X, N(X), t)\, dA$$

3.11 Noll's Axiom of Working *The working is frame indifferent; that is, $\bar{W}(\bar{\phi}_t) = W(\phi_t)$ for a rigid change of frame.*

3.12 Noll's Theorem *If the two axioms above hold, then the force system is balanced:*

$$\int_{\mathcal{P}} \mathfrak{B}(X, t)\, dV + \int_{\partial\mathcal{P}} \mathfrak{T}(X, N(X), t)\, dA = \mathbf{0}.$$

(This holds relative to any framing.)

The proof is easily given using the techniques we have developed in Theorem 3.8. A useful addition to Noll's theorem was given by Beatty [1967a], which parallels the version given in 3.9.

3.13 Beatty's Axiom of Energy *Let a motion of $\mathcal{B}$ in $\mathcal{S}$ be given. Assume that relative to any (rigid Newtonian) framing there are scalar functions E and H representing the internal energy and the nonmechanical working. Assume:*

(*a*) *E and H are materially frame-indifferent scalar quantities (transform as scalars);*
(*b*) *Noll's axiom of force holds and W is defined in terms of $\mathfrak{B}$ and $\mathfrak{T}$ as above;*
(*c*) *$\dot{E} = W + H$ holds and is invariant under changes of rigid Newtonian frame.*

3.14 Theorem *Beatty's axiom of energy implies Noll's axiom of working and hence (by Noll's theorem) the force system is balanced.*

Again the proof follows the usual pattern. The reader should write out the details.

There remains the job of reconciling the Noll version of force with the version employed by Green, Naghdi, and Rivlin. This, Noll does as follows. Forces are divided into a sum; the parts determined by a designated "great system" and the remainder. Each part is to be materially frame indifferent. The body force becomes of the form $\mathfrak{B} = \boldsymbol{\beta} - \mathfrak{A}$ and the traction $\mathfrak{T} = \boldsymbol{t} + \boldsymbol{t}_{\text{external}}$. Assuming $\boldsymbol{t}$ and $\boldsymbol{t}_{\text{external}}$ are contact-type forces, zero unless boundary contact is made, a theorem of Gurtin and Williams shows that $\boldsymbol{t}_{\text{external}}$ is zero. Let $\boldsymbol{A}$ denote the acceleration relative to a rigid framing, let $\boldsymbol{i} = \boldsymbol{A} - \mathfrak{A}$, the "coordinate body force density," and let $\boldsymbol{b} = \boldsymbol{\beta} + \boldsymbol{i}$ denote the "apparent body force density." Since $\boldsymbol{A}$ is frame dependent, so is $\boldsymbol{b}$ and $\boldsymbol{b} - \boldsymbol{A} = \mathfrak{B} - \mathfrak{A}$. Substituting this into the balance of forces now gives the standard balance of momentum equation. One also concludes from 3.10 that the traction $\boldsymbol{T}$ transforms as a vector. Noll now defines an *inertial frame* as one in which $\boldsymbol{i} = \boldsymbol{0}$ and assumes as an inertia axiom that one exists.[13] For further details on this approach, see Noll [1973], Wang and Truesdell [1973], and Truesdell [1977].

[13]These definitions seem to depend on the initial choice of the "great system." It is not clear to the authors how this works in practice. For example, what if you are ignorant of electromagnetic forces and don't include them in your great system? Surely strange inertial systems can result! Another approach to inertial systems, based on the classic work of Cartan [1923] is explained in the next section.

Box 3.4 ***Summary of Important Formulas for Section 2.3***

Balance of Energy (Spatial)

$$\frac{d}{dt}\int_{\phi_t(\mathfrak{U})} \rho(e + \tfrac{1}{2}\langle v, v\rangle)\, dv = \int_{\phi_t(\mathfrak{U})} \rho(\langle b, v\rangle + r)\, dv + \int_{\partial\phi_t(\mathfrak{U})} (\langle t, v\rangle + h)\, da$$

$$\frac{d}{dt}\int_{\phi_t(\mathfrak{U})} \rho(e + \tfrac{1}{2}g_{ab}v^a v^b)\, dv = \int_{\phi_t(\mathfrak{U})} \rho(g_{ab}b^a v^b + r)\, dv + \int_{\partial\phi_t(\mathfrak{U})} (g_{ab}v^a t^b + r)\, da$$

Heat Flux Vector

$$h = -\langle q, n\rangle \qquad h = -g_{ab}q^a n^b$$

Localization

$$\rho\dot{e} + \operatorname{div} q = \sigma : d + \rho r \qquad \rho\dot{e} + q^a{}_{|a} = \sigma^{ab}d_{ab} + \rho r$$

Balance of Energy (Material)

$$\frac{d}{dt}\int_{\mathfrak{U}} \rho_{\text{Ref}}(E + \tfrac{1}{2}\langle V, V\rangle)\, dV = \int_{\mathfrak{U}} \rho_{\text{Ref}}(\langle B, V\rangle + R)\, dV + \int_{\partial\mathfrak{U}} (\langle V\cdot P\cdot N\rangle - \langle Q, N\rangle)\, dA$$

$$\frac{d}{dt}\int_{\mathfrak{U}} \rho_{\text{Ref}}(E + \tfrac{1}{2}g_{ab}V^a V^b)\, dV = \int_{\mathfrak{U}} \rho_{\text{Ref}}(g_{ab}B^a V^b + R)\, dV + \int_{\partial\mathfrak{U}} (G_{AB}N^A P^{aB} V^b g_{ab} - G_{AB}Q^A N^B)\, dA$$

Piola Transformations of Heat Flux Vector

$$Q = JF^{-1}q \qquad Q^A = J(F^{-1})^A{}_a q^a$$

Localization

$$\rho_{\text{Ref}}\frac{\partial E}{\partial t} + \operatorname{DIV} Q = P : \frac{\partial F}{\partial t} + \rho_{\text{Ref}}R = S : D + \rho_{\text{Ref}}R$$

$$\rho_{\text{Ref}}\frac{\partial E}{\partial t} + Q^A{}_{|A} = P_a{}^A\frac{\partial F^a{}_A}{\partial t} + \rho_{\text{Ref}}R = S^{AB}D_{AB} + \rho_{\text{Ref}}R$$

Rigid Heat Conductor (Classical Fourier Theory)

$$\rho_{\text{Ref}}C\frac{\partial\Theta}{\partial t} = \rho_{\text{Ref}}R + k\Delta\Theta \qquad \rho_{\text{Ref}}C\frac{\partial\Theta}{\partial t} = \rho_{\text{Ref}}R + k\left(G^{AB}\frac{\partial\Theta}{\partial X^B}\right)_{|A}$$

2.4 CLASSICAL SPACETIMES, COVARIANT BALANCE OF ENERGY, AND THE PRINCIPLE OF VIRTUAL WORK*

Any theory (relativistic or nonrelativistic) that purports to be fundamental ought to be generalizable so the underlying physical space is a manifold and not just Euclidean (or Newtonian) space. This is a basic message we receive

*We thank Mort Gurtin and John Pierce for helpful discussions on this material.

from Einstein. In fact, he says more: when regarded as a relativistic theory, it should couple to general relativity and be generally "covariant" under all transformations, not just rigid body motions.

Dyson's comments, which follow, apply just as well to a large body of continuum mechanics literature as to quantum mechanics:

"The most glaring incompatibility of concepts in contemporary physics is that between Einstein's principle of general coordinate invariance and all the modern schemes for a quantum-mechanical description of nature. Einstein based his theory of general relativity on the principle that God did not attach any preferred labels to the points of space–time. This principle requires that the laws of physics should be invariant under the Einstein group $\mathcal{E}$, which consists of all one-to-one and twice-differentiable transformations of the coordinates. By making full use of the invariance under $\mathcal{E}$, Einstein was able to deduce the precise form of his law of gravitation from general requirements of mathematical simplicity without any arbitrariness. He was also able to reformulate the whole of classical physics (electromagnetism and hydrodynamics) in $\mathcal{E}$-invariant fashion, and so determine unambiguously the mutual interactions of matter, radiation and gravitation within the classical domain. There is no part of physics more coherent mathematically and more satisfying aesthetically than this classical theory of Einstein based upon $\mathcal{E}$-invariance.

On the other hand, all the currently viable formalisms for describing nature quantum-mechanically use a much smaller invariance group. The analysis of Bacry and Lévy–Leblond indicates the extreme range of quantum-mechanical kinematical groups that have been contemplated. In practice all serious quantum-mechanical theories are based either on the Poincaré group $\mathcal{P}$ or the Galilei group $\mathcal{G}$. This means that a class of preferred inertial coordinate-systems is postulated a priori, in flat contradiction to Einstein's principle. The contradiction is particularly uncomfortable, because Einstein's principle of general coordinate invariance has such an attractive quality of absoluteness. A physicist's intuition tells him that, if Einstein's principle is valid at all, it ought to be valid for the whole of physics, quantum-mechanical as well as classical. If the principle were not universally valid, it is difficult to understand why Einstein achieved such deeply coherent insights into nature by assuming it to be so." (From Dyson [1972]).

A number of prevalent concepts in mechanics cannot be fundamental in this sense. (This does *not* mean they are not useful in the context of $\mathbb{R}^3$). For example, the total momentum or total force acting on a body do not directly make sense if the containing space $\mathcal{S}$ is curved. However, energy balance does make sense on manifolds and can be used as a covariant basis for elasticity.

In this section we begin by giving the basic properties of classical spacetimes. Secondly, we present a covariant treatment of energy balance; the generalization from rigid motions to arbitrary ones gives a richer structure to the theory.[14] As an application, we show in Box 4.1 how to obtain the principle of virtual work *directly* from energy balance in its covariant form. Box 4.2 discusses another approach to stress and equilibrium which is also covariant, based on the theory of manifolds of maps.

We begin then with a discussion of classical spacetimes. (Relativistic spacetimes are discussed in Section 5.7.) The main features of classical spacetimes were first abstracted in the form we will use by Cartan [1923], and were later extended by Trautman [1965] and Misner, Thorne, and Wheeler [1973]. Classical spacetimes that possess affine structures have been used in continuum mechanics by Noll [1973], Wang and Truesdell [1973], and Truesdell [1977]. However, for our purposes here, this affine structure, while allowable, only confuses the issues under consideration. It is for this reason we go back and base our development on the older classic work of Cartan.

4.1 Definitions Let $\mathcal{V}$ be a four-dimensional manifold and $\mathcal{S}$ a three-dimensional manifold. A *slicing* of $\mathcal{V}$ by $\mathcal{S}$ is a diffeomorphism

$$i : \mathcal{S} \times \mathbb{R} \longrightarrow \mathcal{V}.$$

We write $i_t : \mathcal{S} \longrightarrow \mathcal{V}$ for the map $i_t(x) = i(x, t)$. Inversely, a diffeomorphism $f : \mathcal{V} \longrightarrow \mathcal{S} \times \mathbb{R}$ is called a *framing* of $\mathcal{V}$. See Figure 2.4.1.

We call $\mathcal{V}$ a *classical spacetime* (based on $\mathcal{S}$) if it has the following properties:

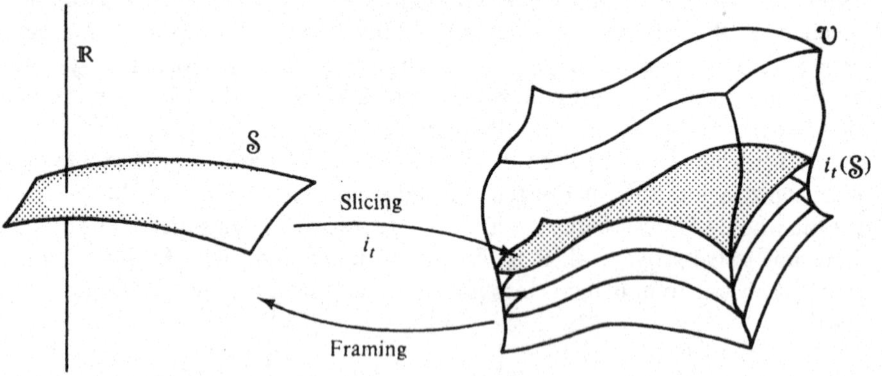

Figure 2.4.1

---

[14]Covariance may be explained in general terms as follows. Suppose we have a theory described by a number of (tensor) fields $\boldsymbol{u}, \boldsymbol{v}, \ldots$ on space (or spacetime) $\mathcal{S}$ and the equations of our theory take the form $F(\boldsymbol{u}, \boldsymbol{v}, \ldots) = 0$. These may be partial differential equations, integral equations, etc.) The equations are called covariant (or form invariant) when for any diffeomorphism $\phi : \mathcal{S} \longrightarrow \mathcal{S}$, $\phi^*(F(\boldsymbol{u}, \boldsymbol{v}, \ldots)) = F(\phi^*\boldsymbol{u}, \phi^*\boldsymbol{v} \ldots)$.

1. There is a surjective smooth map (called a universal time function) $\tau: \mathcal{V} \longrightarrow \mathbb{R}$ with nowhere-zero derivative such that for each $t \in \mathbb{R}$, $\mathcal{S}_t = \tau^{-1}(t)$ is diffeomorphic to $\mathcal{S}$.
2. There is a slicing $i: \mathcal{S} \times \mathbb{R} \longrightarrow \mathcal{V}$ such that $\tau(i(x, t)) = t$ for all $x \in \mathcal{S}$ and $t \in \mathbb{R}$. (Such a slicing will be called *Newtonian*.)
3. There is a symmetric connection ∇ on $\mathcal{V}$ such that $\nabla(d\tau) = 0$.
4. For each $t \in \mathbb{R}$, ∇ *restricts* to a connection ∇_t on $\mathcal{S}_t$; that is, if v_1 and v_2 are tangent to $\mathcal{S}_t$, so is $\nabla_{v_1} v_2$. Moreover, each $\mathcal{S}_t$ carries a Riemannian metric $\mathbf{g}_t$ whose Levi-Civita connection is ∇_t.

A classical spacetime should be denoted by the quintuple $(\mathcal{V}, \mathcal{S}, \tau, \nabla, g_t)$, although when no confusion will arise, we will just denote it by $\mathcal{V}$.

The basic example of a classical spacetime is $\mathcal{V} = \mathbb{R}^3 \times \mathbb{R}$ with τ the projection on the last factor, ∇ the trivial connection associated with the vector space structure of $\mathcal{V}$, (so $\nabla_v w = Dw \cdot v$), and with g_t the Euclidean metric on $\mathcal{S}_t = \mathbb{R}^3 \times \{t\}$ which is identified as $\mathbb{R}^3$. We will call this the *standard* classical spacetime.

Note that ∇ is also the connection of the Minkowski metric on $\mathcal{V}$; the thing that distinguishes classical from relativistic spacetimes is that in the former we select a time function and base constructions on it. In Minkowski space a basic point is that there is nothing special about the coordinate time τ.

Problem 4.1 Show that condition 3 in Definition 4.1 implies that ∇ is restrictable.

We will usually think of a slicing $i: \mathcal{S} \times \mathbb{R} \longrightarrow \mathcal{V}$ as being an "observer." Different "observers" will measure spacetime quantities, such as accelerations, in different ways. Two of the basic quantities associated with observers are the metric and connection. This idea leads us to make the following:

4.2 Definition Let $\mathcal{V}$ be a classical spacetime and let $i: \mathcal{S} \times \mathbb{R} \longrightarrow \mathcal{V}$ be a Newtonian slicing. The *metric observed by i on* $\mathcal{S}$ *at time t* is $g_t^i = i_t^* g_t$. Likewise, the *connection observed by i at time t* is $\nabla_t^i = i_t^* \nabla_t$ (defined by $(i_t^* \nabla_t)_X Y = (\nabla_t)_{(i_t \cdot X)}(i_t \cdot Y)$).

Problem 4.2 Show that ∇_t^i is the Levi-Civita connection of g_t^i.

Warning on Notation: The superscript ^{i} does not refer to tensorial components, but rather to the representation of the given object in the slicing i.

If $\mathcal{S}$ has a fixed Rimannian structure g, then a Newtonian slicing i is called *rigid* if $g_t^i = g$ for all $t \in \mathbb{R}$.

Remarks In the literature, spacetimes are often restricted to be affine and slicings restricted to be rigid. To understand elasticity in the covariant context, such a point of view is abandoned.

If one wishes to construct classical analogues of Fermi–Walker transport for gyroscopes, then additional restrictions on the curvature of the connection ∇ must be made. See Misner, Thorne, and Wheeler [1973] for details, as well as Section 5.7. For this section, however, this extra structure is not needed.

Now we turn to the kinematics of a body moving in spacetime. The spacetime version of a motion is called a world tube, defined as follows:

4.3 Definition Let $\mathcal{V}$ be a classical spacetime and $\mathcal{B}$ a reference manifold. A *world tube* for $\mathcal{B}$ is a mapping $\Phi : \mathcal{B} \times \mathbb{R} \to \mathcal{V}$ such that

(1) for each $t \in \mathbb{R}$, and $X \in \mathcal{B}$, $\tau(\Phi(x, t)) = t$, and
(2) for each $t \in \mathbb{R}$, the map $\Phi_t : \mathcal{B} \to \mathcal{S}_t$ given by $\Phi_t(X) = \Phi(X, t)$ is an embedding.

For fixed $X \in \mathcal{B}$ (a *material point*), the curve $t \mapsto \Phi(X, t)$ is called the *world line* of X. The velocity and acceleration of these curves define vector fields over the map Φ as follows: (a) *four velocity*,

$$V^\Phi : \mathcal{B} \times \mathbb{R} \to T\mathcal{V}. \qquad V^\Phi(X, t) = \frac{\partial}{\partial t}\Phi(X, t);$$

(b) *four acceleration*,

$$A^\Phi : \mathcal{B} \times \mathbb{R} \to T\mathcal{V}, \qquad A^\Phi(X, t) = \frac{DV^\Phi}{Dt}(X, t),$$

where D/Dt is the covariant derivative for the spacetime connection ∇. See Figure 2.4.2. (These are "absolute" concepts, independent of any framing.) The *spatial* velocity and acceleration fields are defined by $v^\Phi = V^\Phi \circ \Phi^{-1}$ and $a^\Phi = A^\Phi \circ \Phi^{-1}$; these are vector fields on $\Phi(\mathcal{B} \times \mathbb{R})$.

If $i : \mathcal{S} \times \mathbb{R} \to \mathcal{V}$ is a Newtonian slicing, we can view i as a world tube for $\mathcal{B} = \mathcal{S}$. Its *frame velocity* and *frame accelerations* are the associated four

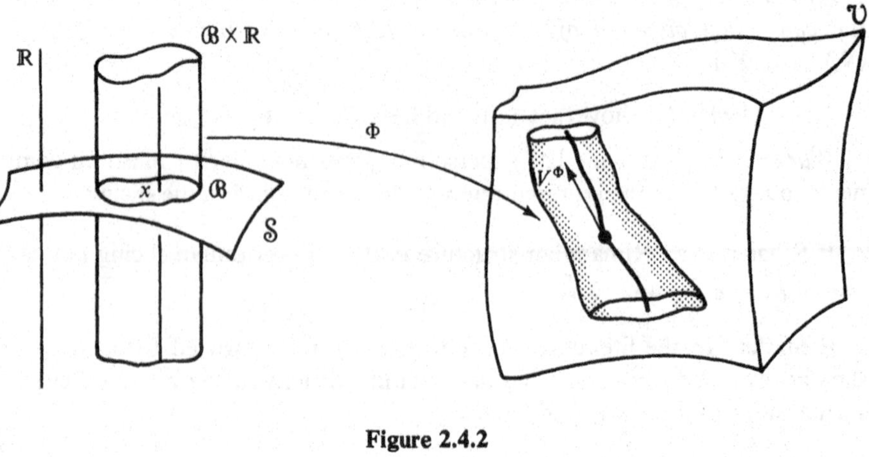

Figure 2.4.2

velocities and accelerations; these are denoted V^i, A^i materially and v^i, a^i spatially, so that

$$v^i = V^i \circ i^{-1} \quad \text{and} \quad a^i = A^i \circ i^{-1}.$$

The concept of world tube is the spacetime way of viewing a motion. In fact, if $\phi_t : \mathcal{B} \to \mathcal{S}$ is a motion of $\mathcal{B}$ in $\mathcal{S}$, the map $\Phi : \mathcal{B} \times \mathbb{R} \to \mathcal{S} \times \mathbb{R}$, $\Phi(X, t) = (\phi(X, t), t)$ is an associated world tube if we choose $\mathcal{V} = \mathcal{S} \times \mathbb{R}$.

Problem 4.3 Show that the frame velocity of a Newtonian slicing satisfies $d\tau \cdot v^i = 1$. Use property 3 of Definition 4.1 to deduce that $d\tau \cdot a^i = 0$.

Problem 4.4 Let a coordinate chart on $\mathcal{S}$ be given and let $\mathcal{V}$ inherit a chart via $i : \mathcal{S} \times \mathbb{R} \to \mathcal{V}$. If e_j ($j = 1, 2, 3$) are coordinate tangent vectors on $\mathcal{V}$ corresponding to the chart on $\mathcal{S}$, show that they are tangent to $\mathcal{S}_t$ and that the fourth coordinate tangent vector is $e_0 = v^i$. Let $\alpha, \beta, \ldots$ range from 0 to 3 and let $\Gamma^\alpha_{\beta\gamma}$ be the Christoffel symbols for ∇, that is, $\nabla_{e_\alpha} e_\beta = \Gamma^\sigma_{\alpha\beta} e_\sigma$. Show that (1) $\Gamma^l_{jk} = g_t^{lm} \Gamma_{mjk}$ are the Christoffel symbols for g_t, (2) $\Gamma^0_{jk} e_j = 0$, and (3) $\Gamma^l_{00} e_l$ is a^i, the frame acceleration. (*Note:* The interpretation of Γ^l_{0k} is related to Fermi–Walker transport. See Misner, Thorne and Wheeler [1973].)

4.4 Definitions Let $\mathcal{V}$ be a classical spacetime, $\Phi : \mathcal{B} \times \mathbb{R} \to \mathcal{V}$ a world tube, and $i : \mathcal{S} \times \mathbb{R} \to \mathcal{V}$ a (Newtonian) slicing.

The *motion of* $\mathcal{B}$ *in* $\mathcal{S}$ *relative to* i is the motion $\phi^i : \mathcal{B} \times \mathbb{R} \to \mathcal{S}$ defined by $\Phi(X, t) = i(\phi^i(X, t), t)$. The velocity field of ϕ^i will be called the *apparent velocity* of the world tube Φ relative to the slicing i and will be denoted $V^{\Phi,i}$ materially and $v^{\Phi,i}$ spatially.

We now relate the velocity of the world tube v^Φ, the apparent velocity of the world tube $v^{\Phi,i}$ and the frame velocity v^i (see Figure 2.4.3).

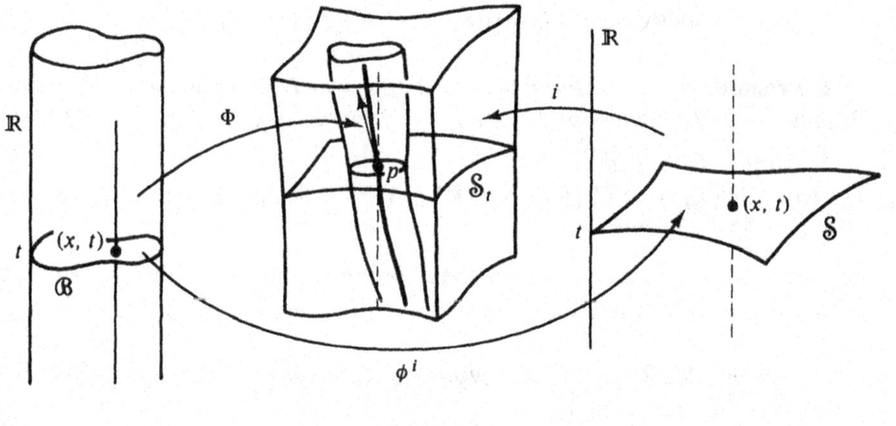

Figure 2.4.3

4.5 Proposition *Let $X \in \mathcal{B}$, $t \in \mathbb{R}$, and let $p \in \mathcal{V}$ be given by $p = \Phi(X, t)$. Let $p = i(x, t)$ so that $\phi^i(X, t) = x$. Then*

(a) $\boldsymbol{v}^\Phi(p) - \boldsymbol{v}^i(p)$ is tangent to $\mathcal{S}_t$ at p

and

$$\text{(b)} \qquad \boxed{Ti_t(x)\cdot \boldsymbol{v}^{\Phi,i}(x, t) = \boldsymbol{v}^\Phi(p) - \boldsymbol{v}^i(p).}$$

Proof (*a*) Since $T_p\mathcal{S}_t = \ker d\tau(p)$, we need to check that $d\tau(p)\cdot[\boldsymbol{v}^\Phi(p) - \boldsymbol{v}^i(p)] = 0$. But $d\tau(p)\cdot \boldsymbol{v}^\Phi(p) = 1 = d\tau(p)\cdot \boldsymbol{v}^i(p)$ by Problem 4.3.

(*b*) If we differentiate the defining relationship $\Phi(X, t) = i(\phi^i((X, t), t)$ in t we get $\boldsymbol{V}^\Phi(x, t) = Ti_t(x, t)\cdot \boldsymbol{V}^{\Phi,i}(X, t) + \boldsymbol{V}^i(X, t)$ by the chain rule. But $\boldsymbol{V}^\Phi(X, t) = \boldsymbol{v}^\Phi(p)$, $\boldsymbol{V}^{\Phi,i}(X, t) = \boldsymbol{v}^{\Phi,i}(x, t)$, and $\boldsymbol{V}^i(x, t) = \boldsymbol{v}^i(p)$, so the result follows. ∎

In words, the result 4.5(b) says that:

$$\begin{pmatrix}\text{Apparent velocity of}\\ \text{body relative to}\\ \text{observer } i\end{pmatrix} = \begin{pmatrix}\text{Absolute velocity}\\ \text{of body in spacetime}\end{pmatrix} - \begin{pmatrix}\text{Absolute velocity}\\ \text{of observer in}\\ \text{spacetime}\end{pmatrix}$$

We can alternatively interpret this as follows: for $p \in \mathcal{V}$, we can decompose $T_p\mathcal{V} = T_p\mathcal{S}_t \oplus \operatorname{span} \boldsymbol{v}^i(p)$. The component of $\boldsymbol{v}^\Phi(p)$ tangent to $\mathcal{S}_t$ at p is thus just $v^\Phi(p) - v^i(p)$. We can identify $T_p\mathcal{S}_t$ with $T_x\mathcal{S}$, where $i(x, t) = p$ via Ti_t. Decomposing $\boldsymbol{v}^\Phi(p)$ into its spatial and temporal parts then produces a vector tangent to $\mathcal{S}$ at x. Thus 4.5 says that this vector is just the material velocity of the motion relative to the slicing i. (In general, tensor fields on $\mathcal{V}$ can be decomposed into spatial, temporal, and mixed parts relative to any slicing. The various pieces may be regarded as tensor fields on $\mathcal{S}$ by way of pull-back by the slicing.)

Given a world tube Φ, its four velocity $\boldsymbol{V}^\Phi$ is well defined, independent of any (Newtonian) slicing. If i_1 and i_2 are two Newtonian slicings, we can compare the velocities relative to them in terms of the *change of framing*: $\xi : \mathcal{S} \times \mathbb{R} \to \mathcal{S}$ defined by $i_2^{-1}\circ i_1(x, t) = (\xi(x, t), t)$. Regarding ξ as a motion of $\mathcal{S}$ in $\mathcal{S}$, let $\boldsymbol{w}$ be its spatial velocity field and write $\xi_t : \mathcal{S} \to \mathcal{S}$, $\xi_t(x) = \xi(x, t)$.

4.6 Proposition (*Transformation of motions and velocities under a change of framing.*) *Using the notation in the previous paragraph,*

(a) $\phi_t^{i_2} = \xi_t \circ \phi_t^{i_1}$ *and*

(b) $\boldsymbol{V}^{\Phi,i_2}(X, t) = T\xi_t(\phi^{i_1}(X, t))\cdot \boldsymbol{V}^{\Phi,i_1}(X, t) + \boldsymbol{w}_t(\xi_t(\phi^{i_1}(X, t))$. (*Or, briefly,* $\boldsymbol{v}^{\Phi,i_2} = \xi_{t*}\boldsymbol{v}^{\Phi,i_1} + \boldsymbol{w}$.

Proof (a) follows directly from the definitions and (b) follows by differentiating (a) in t. One may alternatively derive (b) from 4.5(b). ∎

Problem 4.5 Regard i_1 as a world tube. Show that $\boldsymbol{w}$ is its velocity field relative to the slicing i_2.

Thus, under a change of framing, the observed velocities in the two framings "differ" by the relative velocities of the two frames.

Now we turn to acceleration.

4.7 Definitions Let $\mathcal{V}$ be a classical spacetime, Φ a world tube, and i a Newtonian slicing. The *apparent acceleration* of Φ relative to i is the acceleration of the motion ϕ^i of Φ relative to i, written $A^{\Phi,i}$ materially and $\boldsymbol{a}^{\Phi,i}$ spatially. This acceleration is computed at the same time using the connection observed by i. (Thus, the coordinate formulas are the same as those found in Chapter 1, Section 1.2, except now the Christoffel symbols might be time dependent.[15])

Again we wish to compute the relationship between the absolute accelerations $\boldsymbol{a}^\Phi$, $\boldsymbol{a}^i$ and the apparent acceleration $\boldsymbol{a}^{\Phi,i}$. The result will be of the following form:

$$\begin{pmatrix}\text{Apparent acceleration}\\ \text{of body relative to}\\ \text{observer } i\end{pmatrix} = \begin{pmatrix}\text{Absolute acceleration}\\ \text{of body in spacetime}\\ \end{pmatrix} - \begin{pmatrix}\text{Absolute acceleration}\\ \text{of observer in space-}\\ \text{time}\end{pmatrix} + \begin{pmatrix}\text{Ficticious}\\ \text{accelerations}\end{pmatrix}$$

The "ficticious" accelerations include the usual centrifugal and Coriolis accelerations one "sees" from a rotating frame (see Problem 4.6 below and Section 7 of Chapter 5 for more information.)

4.8 Proposition *Using the notation of 4.5, we have*

$$\boxed{Ti_t(x)\cdot \boldsymbol{a}^{\Phi,i}(x, t) = \boldsymbol{a}^\Phi(p) - \boldsymbol{a}^i(p) - 2(\nabla_{v^\Phi - v^i}\boldsymbol{v}^i)(p).}$$

Proof By definition we have

$$\boldsymbol{a}^{\Phi,i}(x, t) = \frac{\partial \boldsymbol{v}^{\Phi,i}}{\partial t}(x, t) + (\nabla^i_{v^{\Phi,i}}\boldsymbol{v}^{\Phi,i})(x, t). \tag{1}$$

Thus, by definition of the observed connection and 4.5(b), (1) gives

$$Ti_t(x, t)\cdot \boldsymbol{a}^{\Phi,i}(x, t) = Ti_t(x, t)\frac{\partial \boldsymbol{v}^{\Phi,i}}{\partial t}(x, t) + (\nabla_{(v^\Phi - v^i)}(\boldsymbol{v}^\Phi - \boldsymbol{v}^i))(p). \tag{2}$$

However, we claim that

$$Ti_t(x, t)\frac{\partial \boldsymbol{v}^{\Phi,i}}{\partial t}(x, t) = [\boldsymbol{v}^i, \boldsymbol{v}^\Phi - \boldsymbol{v}^i](p). \tag{3}$$

Indeed, each side equals

$$\frac{d}{ds}F^*_{s,t}(\boldsymbol{v}^\Phi - \boldsymbol{v}^i)\bigg|_{s=t},$$

[15]They will be time independent if the slicing is rigid.

where $F_s: \mathcal{S}_t \longrightarrow \mathcal{S}_s$ is given by $F_{s,t} = i_s \circ i_t^{-1}$ (the reader should fill in the details of this step).

Since ∇ on $\mathcal{V}$ is symmetric (torsion-free) we have

$$[\boldsymbol{v}^i, \boldsymbol{v}^\Phi - \boldsymbol{v}^i] = \nabla_{v^i}(\boldsymbol{v}^\Phi - \boldsymbol{v}^i) - \nabla_{v^\Phi - v^i}\boldsymbol{v}^i. \tag{4}$$

Also, we have

$$\boldsymbol{a}^i = \nabla_{v^i}\boldsymbol{v}^i \quad \text{and} \quad \boldsymbol{a}^\Phi = \nabla_{v^\Phi}\boldsymbol{v}^\Phi. \tag{5}$$

Substituting (4), (5), and (3) into (2) gives the proposition (observe that a cancellation of $\nabla_{v^i}\boldsymbol{v}^\Phi$ occurs). ∎

Problem 4.6 Let $\mathcal{V} = \mathbb{R}^3 \times \mathbb{R}$ be the standard classical spacetime and i is the rigid slicing $i(x, t) = (\boldsymbol{Q}(t)(x - \boldsymbol{x}_0), t)$, where $\boldsymbol{Q}(t)$ is a rotation, $\boldsymbol{Q}(t) = e^{t\boldsymbol{\Omega}}$, where $\boldsymbol{\Omega}$ is a constant skew matrix. Let $\boldsymbol{\omega}(t)$ be the angular velocity; that is, $\boldsymbol{\Omega v} = \boldsymbol{\omega} \times \boldsymbol{v}$. Show that 4.5(b) and 4.8 reduce to the classical formulas for the velocity and acceleration in a rotating frame (Goldstein [1980], p. 177), namely

$$\boldsymbol{v}_r = \boldsymbol{v}_s - \boldsymbol{\omega} \times \mathbf{r} \quad \text{and} \quad \boldsymbol{a}_r = \boldsymbol{a}_s - \boldsymbol{\omega} \times (\boldsymbol{\omega} \times \boldsymbol{r}) - 2\boldsymbol{\omega} \times \boldsymbol{v}_r.$$

($\boldsymbol{v}_r$ = velocity in the rotating frame, $\boldsymbol{v}_s$ = velocity in space, etc.) Modify these formulas for $i(x, t) = (\boldsymbol{Q}(t)(x - x_0) + \boldsymbol{c}(t), t)$, where $\boldsymbol{Q}(t)$ is an arbitrary curve in SO(3) and $\boldsymbol{c}(t)$ is a curve in $\mathbb{R}^3$.

Just as with velocities, we can compute the transformation of accelerations under a change of framing.

4.9 Proposition *Let i_1 and i_2 be two framings of $\mathcal{V}$ and let $\xi: \mathcal{S} \times \mathbb{R} \longrightarrow \mathcal{S}$ be the change of framing, as defined preceding 4.6. Then at $(x, t) \in \mathcal{S} \times \mathbb{R}$ we have*

$$\boldsymbol{a}^{\Phi, i_2} = \xi_{t*}\boldsymbol{a}^{\Phi, i_1} + \boldsymbol{a}^{\xi, i_2} + 2\nabla^{i_2}_{(\xi_{t*}v^{\Phi, i_1})}(\boldsymbol{w}),$$

where $\boldsymbol{a}^{\xi, i_2}$ is the acceleration of ξ computed using ∇^{i_2}.

The proof proceeds as in 4.8, so is left for the reader.

Problem 4.7 (a) Prove 4.9. (b) For $\mathcal{V} = \mathbb{R}^3 \times \mathbb{R}$ and i_1 the identity slicing and $i_2 = i$ an arbitrary one, show that 4.9 and 4.8 coincide.

Next we consider *inertial* slicings.

4.10 Definition A (Newtonian) slicing i of a classical spacetime $\mathcal{V}$ is called *inertial* if the spatial velocity field of i is covariant constant: $\nabla \boldsymbol{v}^i = \boldsymbol{0}$.

The acceleration of an inertial slicing vanishes since $\boldsymbol{a}^i = \nabla_{v^i}\boldsymbol{v}^i = 0$. Thus for each $x \in \mathcal{S}$, the curve $t \mapsto i(x, t)$ is a geodesic in $\mathcal{V}$. From 4.8 we see at once that for an inertial slicing, $Ti_t(x)\cdot\boldsymbol{a}^{\Phi, i}(x, t) = \boldsymbol{a}^\Phi(p)$. That is, we say the apparent acceleration equals the absolute acceleration. If this is to hold for any world tube, choosing $\Phi = i$ gives $\boldsymbol{a}^i = \boldsymbol{0}$, and then in 4.8 $\boldsymbol{v}^\Phi - \boldsymbol{v}^i$ is arbitrary, so $\nabla \boldsymbol{v}^i = \boldsymbol{0}$. Thus we have proved:

4.11 Proposition *A (Newtonian) slicing is inertial if and only if the apparent and absolute accelerations of any world tube coincide.*

The *axiom of inertia* states that an inertial slicing exists. We view *forces* as vector fields on $\mathcal{S}$ given relative to an inertial slicing. Viewing Newton's law for *test particles* in different slicings then allows us to deduce the transformation properties of forces. These will be stated and used below.

Now we turn to our covariant treatment of balance of energy. Let us assume that a rigid inertial frame is specified $i_0: \mathcal{S} \times \mathbb{R} \longrightarrow \mathcal{V}$ and that g is the corresponding fixed Riemannian metric on $\mathcal{S}$. Let ϕ_t be a regular motion of $\mathcal{B}$ in $\mathcal{S}$. If we wish, we can imagine that a world tube Φ is given and that ϕ_t is the motion of $\mathcal{B}$ relative to our inertial slicing i_0. The spatial velocity of ϕ_t is denoted $\boldsymbol{v}_t$ and the spatial acceleration $\boldsymbol{a}_t$, as usual.

Let $\rho(x, t)$, $\boldsymbol{b}(x, t)$, $h(x, t, \boldsymbol{n})$, $\boldsymbol{t}(x, t, \boldsymbol{n})$, $e(x, t)$, and $r(x, t)$ be given functions defined for $x \in \phi_t(\mathcal{B})$ and $\boldsymbol{n} \in T_x\mathcal{S}$.

As in the previous section, these functions are said to satisfy the *balance of energy principle* if, for all nice $\mathcal{U} \subset \mathcal{B}$,

$$\frac{d}{dt}\int_{\phi_t(\mathcal{U})} \rho(e + \tfrac{1}{2}\langle \boldsymbol{v}, \boldsymbol{v}\rangle)\, dv = \int_{\phi_t(\mathcal{U})} \rho(\langle \boldsymbol{b}, \boldsymbol{v}\rangle + r)\, dv + \int_{\partial\phi_t(\mathcal{U})} (\langle \boldsymbol{t}, \boldsymbol{v}\rangle + h)\, da,$$

where $\boldsymbol{t}$ and h are evaluated on the unit outward normal $\boldsymbol{n}$ of $\partial\phi_t(\mathcal{U})$.

4.12 Definition Let $\xi_t: \mathcal{S} \longrightarrow \mathcal{S}$ be a one-parameter family of regular mappings of $\mathcal{S}$ (interpreted either as a superposed motion or as a change of Newtonian slicing). Define the *primed quantities* as follows:

$$x' = \xi_t(x), \qquad dv' = \xi_{t*}\, dv \quad (\text{i.e., } J(\xi_t)\, dv' = dv),$$

$$\rho'(x', t) = \rho(x, t), \qquad r'(x', t) = r(x, t), \qquad e'(x', t) = e(x, t),$$

where $\boldsymbol{n} \in T_x\mathcal{S}$, and $\boldsymbol{n}'$ is the unit normal to the range of the orthogonal complement of $\boldsymbol{n}$ under $T\xi_t(x)$,

$$h'(x', t, \boldsymbol{n}') = h(x, t, \boldsymbol{n}), \qquad \boldsymbol{t}'(x', t, \boldsymbol{n}') = T\xi_t(x)\cdot \boldsymbol{t}(x, t, \boldsymbol{n}),$$

$$\phi'_t = \xi_t \circ \phi_t, \qquad \boldsymbol{v}'_t = \text{spatial velocity of } \phi'_t, \qquad \boldsymbol{a}'_t = \text{spatial acceleration of } \phi'_t,$$

$$\boldsymbol{b}' = \boldsymbol{b} + \boldsymbol{a}^{\xi}_t + 2\nabla_{\xi_{t*}\boldsymbol{v}_t}\boldsymbol{w}_t \text{ ("the apparent body force")},$$

where $\boldsymbol{w}_t$ and $\boldsymbol{a}^{\xi}_t$ are the velocity and acceleration of ξ_t.

Our transformation of forces has been arranged so that $\boldsymbol{b}' - \boldsymbol{a}' = \xi_{t*}(\boldsymbol{b} - \boldsymbol{a})$ by 4.9.

At this stage there has been no physical justification for the transformation law for the Cauchy stress vector: $\boldsymbol{t}' = \xi_{t*}\boldsymbol{t}$. If ξ_t is rigid, it is "clear" that this is appropriate for all continuum theories. However, for nonrigid slicings, it is less apparent. In fact, a few examples involving shearing will convince the reader that it is reasonable for elasticity, but *not necessarily* for other continuum theories. Viscous or plastic materials might be characterized by different transformation laws for the stress. Thus, we are only proposing the following covariant treatment for pure elasticity. It needs to be modified for other theories.[16] This point will be explored further in Section 3.3 on covariant constitutive theory.

One may be tempted to demand that the primed quantities satisfy balance of energy for every superposed motion. However, it will follow from the arguments used to prove Theorem 4.13 below that his can occur if and only if the Cauchy traction vector is identically zero! The additional object that must be transformed is the metric tensor.

The first assumptions of the covariant theory are as follows.

Assumption 1 Let $(\mathcal{S}, \mathbf{g})$ be a given Riemannian manifold and $\phi_t: \mathcal{B} \to \mathcal{S}$ a motion of $\mathcal{B}$ in $\mathcal{S}$. Let $\rho(x, t)$, $e(x, t)$, . . . be given functions which satisfy balance of energy.

Given a change of Newtonian slicings, we get maps $\xi_t: \mathcal{S} \to \mathcal{S}$. The representation of the old metric $\mathbf{g}$ on $\mathcal{S}$ in the new slicing is $\xi_{t*}\mathbf{g} = \mathbf{g}'$ and our "observer" does the calculations using the new metric $\mathbf{g}'$. See Figure 2.4.4.

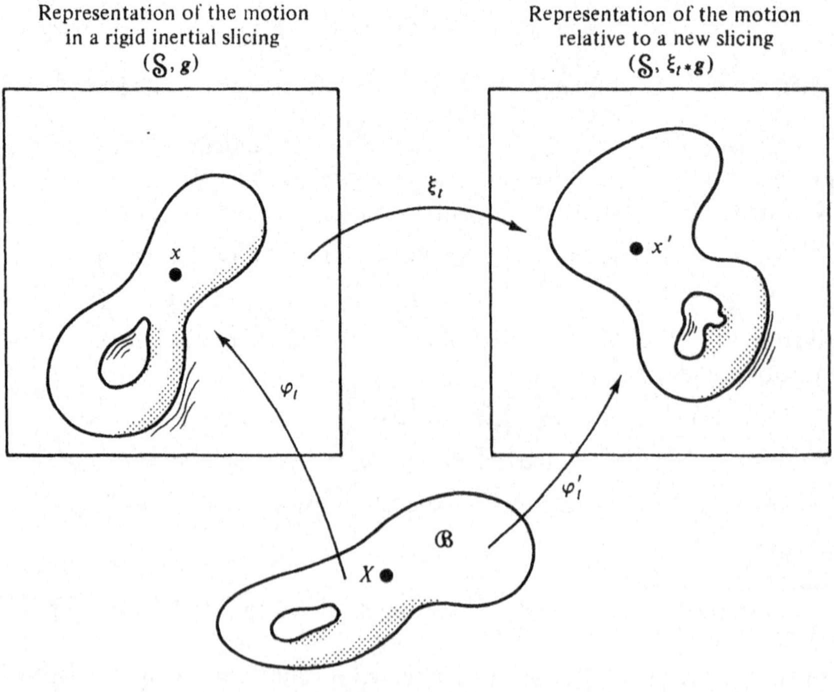

Figure 2.4.4

[16]It is possible, however, that the use of "internal variables" will enable a single theory to be developed, valid for a larger class of continua.

Since changes of metrics on $\mathcal{S}$ from g to say $\bar{g}$ affect the accelerations of particles, the internal energy must depend parametrically on the metric g. Write $e(x, t, g)$ for the dependence. The observer ξ_t also wishes to use the correct internal energy associated with the "true" metric g on space; that is, $e'(x', t, g)$. It is natural to suppose that e transforms tensorally:

$$e(x, t, \bar{g}) = e'(x', t, \xi_{t*}\bar{g}).$$

Thus we must choose

$$e'(x', t, g) = e(x, t, \xi_t^* g).$$

The metric $\xi_t^* g$ is "equivalent" to g in the sense that their Riemannian geometries are naturally related by ξ_t (i.e., curvature, geodesics, etc.). Let the collection of all metrics of the form $\eta^* g$ on $\mathcal{S}$, where $\eta: \mathcal{S} \longrightarrow \mathcal{S}$ is a diffeomorphism, be denoted $\mathcal{O}_g$ and be called the *orbit* of g; thus[17]

$$\mathcal{O}_g = \{\eta^* g \mid \eta: \mathcal{S} \longrightarrow \mathcal{S} \text{ is a diffeomorphism}\}.$$

This discussion leads to our second assumption:

Assumption 2 Assume there are functions $e, \rho, \ldots$ depending on the variables $x, t, \bar{g} \in \mathcal{O}_g$ coinciding with those in Assumption 1 when $\bar{g} = g$. Given a Newtonian slicing and associated maps $\xi_t: \mathcal{S} \longrightarrow \mathcal{S}$, define the representatives of $e, \rho, \ldots$ in that slicing by

$$e'(x', t) = e(x, t, \xi_t^* g),$$

etc. Assume that for each such slicing, balance of energy holds.

Using the methods of Chapter 3 one can prove that if e is a differentiable function of $\bar{g} \in \mathcal{O}_g$, then e is a function only of the point values of $\bar{g}$. That is to say, there is a function also denoted e, of x, t and symmetric positive-definite two-tensors at x, such that

$$e(x, t, \bar{g}) = e(x, t, \bar{g}(x)) \quad \text{for each } \bar{g} \in \mathcal{O}_g.$$

Let us take this for granted in what follows (or make it part of Assumption 2 if this is preferred).[18]

4.13 Theorem *If Assumptions 1 and 2 hold, then the following hold: there exist* $\boldsymbol{\sigma}$ *and* $\boldsymbol{q}$ *such that* $\boldsymbol{t} = \boldsymbol{\sigma} \cdot \boldsymbol{n}$ *and* $h = -\boldsymbol{q} \cdot \boldsymbol{n}$, *and*

(i) *conservation of mass,*
(ii) *balance of momentum,*
(iii) *balance of moment of momentum,*
(iv) *balance of energy, and*
(v) $\boldsymbol{\sigma} = 2\rho(\partial e / \partial \boldsymbol{g})$.

---

[17]One can show that $\mathcal{O}_g$ is a smooth infinite-dimensional manifold, although we shall not go into these technical points; see Box 4.2 below and Ebin [1970].

[18]The dependence of e only on point values of g (and not on derivatives) corresponds in general relativity to what is called *minimal coupling* of a field theory to gravity. In this more general context, g will be time dependent and e defined for all metrics, not just those in the orbit of a fixed one. (See Section 5.7).

Conversely, if (i)–(v) *hold, then balance of energy is covariant; that is, Assumptions 1 and 2 hold.*

Equation (v) relating the internal energy and stress is the additional condition needed to relax the rigidity assumptions of the slicing. Some additional benefits of this point of view are described in the following box and in Box 3.3.

As we shall see in Chapter 3, formula (v) is equivalent to formulas derived by completely different methods. As far as we know, a formula like (v) first appeared in the elasticity literature in Doyle and Ericksen [1956], and subsequently was overlooked. Doyle and Ericksen took a passive point of view, regarding points of $\mathcal{O}_g$ as different representations of the same metric. (This situation is similar to the distinction between the Green–Naghdi–Rivlin and Noll points of view.) In general relativity this formula is commonly associated to variational principles and may be found in standard textbooks such as Hawking and Ellis [1973], p. 66.

Proof of 4.13 Balance of energy for the primed and unprimed motions are:

$$\frac{d}{dt}\int_{\phi_t(\mathcal{U})} \rho(e + \tfrac{1}{2}\langle v, v\rangle)\, dv = \int_{\phi_t(\mathcal{U})} \rho(\langle b, v\rangle + r)\, dv + \int_{\partial\phi_t(\mathcal{U})} (t \cdot v + h)\, da, \tag{1}$$

$$\frac{d}{dt}\int_{\phi_t'(\mathcal{U})} \rho'(e' + \tfrac{1}{2}\langle v', v'\rangle) dv' = \int_{\phi_t(\mathcal{U})} \rho'(\langle b', v'\rangle + r')\, dv' + \int_{\partial\phi_t'(\mathcal{U})} (t' \cdot v' + h')\, da'. \tag{2}$$

We shall evaluate (2) at $t = t_0$ when $\xi_{t_0} =$ identity. The definition of material time derivative and the Lie derivative gives

$$(e')^{\cdot} = \dot{e} + \frac{\partial e}{\partial g} : \mathfrak{L}_w g$$

where w is the velocity of ξ_t (at time t_0). We also have (at $t = t_0$) $v' = w + v$. Using the transport theorem,[19]

$$\frac{d}{dt}\int_{\phi_t'(\mathcal{U})} f\, dv = \int_{\phi_t'(\mathcal{U})} (\dot{f} + f \operatorname{div} v)\, dv'$$

and evaluating at $t = t_0$, (2) becomes

$$\begin{aligned} &\int_{\phi_t(\mathcal{U})} (\dot{\rho} + \rho \operatorname{div} v)(e + \tfrac{1}{2}\langle v + w, v + w\rangle)\, dv \\ &\quad + \int_{\phi_t(\mathcal{U})} \rho\left(\dot{e} + \frac{\partial e}{\partial g} : \mathfrak{L}_w g + \langle a', v + w\rangle\right) dv \\ &= \int_{\phi_t(\mathcal{U})} \rho(\langle b', v + w\rangle + r)\, dv + \int_{\partial\phi_t(\mathcal{U})} (t \cdot (v + w) + h)\, da. \end{aligned} \tag{3}$$

---

[19] There is no div w term because dv', and not dv, is used.

Since $\xi_{t_0} =$ Identity, we have $a' - b' = a - b$. Substituting this in (3) and arranging terms yields

$$\int_{\phi_t(\mathcal{U})} \{(\dot{\rho} + \rho \operatorname{div} v)(e + \tfrac{1}{2}\langle v, v\rangle) + \rho(\dot{e} + \langle a - b, v\rangle)\}\, dv$$
$$+ \int_{\phi_t(\mathcal{U})} (\dot{\rho} + \rho \operatorname{div} v)\cdot\tfrac{1}{2}\langle w, w\rangle\, dv + \int_{\phi_t(\mathcal{U})} (\dot{\rho} + \rho \operatorname{div} v)(\langle v, w\rangle)\, dv$$
$$+ \int_{\phi_t(\mathcal{U})} \rho\left\{\frac{\partial e}{\partial g}\mathfrak{L}_w g + \langle a - b, w\rangle\right\} dv = \int_{\phi_t(\mathcal{U})} \rho r\, dv + \int_{\partial\phi_t(\mathcal{U})} (t\cdot v + h)\, da$$
$$+ \int_{\partial\phi_t(\mathcal{U})} t\cdot w\, da. \tag{4}$$

Subtracting (1) and (4) gives the identity

$$0 = \int_{\phi_t(\mathcal{U})} (\dot{\rho} + \rho \operatorname{div} v)\cdot\tfrac{1}{2}\langle w, w\rangle\, dv + \int_{\phi_t(\mathcal{U})} (\dot{\rho} + \rho \operatorname{div} v)(\langle v, w\rangle)\, dv$$
$$+ \int_{\phi_t(\mathcal{U})} \rho\left(\frac{\partial e}{\partial g} : \mathfrak{L}_w g + \langle a - b, w\rangle\right) dv - \int_{\partial\phi_t(\mathcal{U})} t\cdot w\, da. \tag{5}$$

Cauchy's theorem applied to (5) shows that $t\cdot w = w\cdot\sigma\cdot n$ for a two-tensor σ. Cauchy's theorem applied to (1) then shows $h = -q\cdot n$ for a vector field q.

Since w is arbitrary, (5) is equivalent to the two equations $\dot{\rho} + \rho \operatorname{div} v = 0$ and

$$0 = -\int_{\phi_t(\mathcal{U})} \rho\left(\frac{\partial e}{\partial g} : \mathfrak{L}_w g + \langle a - b, w\rangle\right) dv + \int_{\partial\phi_t(\mathcal{U})} t\cdot w\, da. \tag{6}$$

Applying the divergence theorem to the last term using

$$\operatorname{div}(w\cdot\sigma) = (\operatorname{div}\sigma)\cdot w + \sigma : \omega + \sigma : k,$$

where $\quad \omega_{ab} = \tfrac{1}{2}(w_{a|b} - w_{b|a}) \quad$ (the spin),

and $\quad k_{ab} = \tfrac{1}{2}(w_{a|b} + w_{b|a}) \quad (k = \tfrac{1}{2}\mathfrak{L}_w g)$,

equation (6) becomes

$$0 = -\int_{\phi_t(\mathcal{U})} \left[\rho\left(\frac{\partial e}{\partial g} - \tfrac{1}{2}\sigma\right) : \mathfrak{L}_w g - \sigma : \omega\right] dv - \int_{\phi_t(\mathcal{U})} \langle \rho a - \rho b - \sigma, \operatorname{div} w\rangle\, dv.$$

Since $\mathcal{U}$ is arbitrary and at any point $\mathfrak{L}_w g$, ω, and w can be chosen independently, we have separately,

$$2\rho\frac{\partial e}{\partial g} = \sigma, \quad \sigma \text{ is symmetric}, \quad \text{and} \quad \rho a - \rho b - \operatorname{div}\sigma = 0.$$

This proves the first part of the theorem. The reader can prove the converse by similar methods. ▮

Problem 4.8 Redo these calculations in material coordinates.

Box 4.1 ***The Principle of Virtual Work***

The principle of virtual work is often derived in the following way.

(1) Start with the integral balance law of momentum (see Section 2.1).

(2) Localize to give the differential equations in, say, material form:

$$\rho_{\mathrm{Ref}} \boldsymbol{A} = \rho_{\mathrm{Ref}} \boldsymbol{B} + \mathrm{DIV}\, \boldsymbol{P}.$$

(3) Contract these equations with a vector function $\boldsymbol{W}$ (over the motion), and integrate over the body:

$$\int_{\mathcal{B}} \rho_{\mathrm{Ref}} \boldsymbol{A} \cdot \boldsymbol{W}\, dV = \int_{\mathcal{B}} \rho_{\mathrm{Ref}} \boldsymbol{B} \cdot \boldsymbol{W}\, dV + \int_{\mathcal{B}} \boldsymbol{W} \cdot \mathrm{DIV}\, \boldsymbol{P}\, dV.$$

In the classical literature, $\boldsymbol{W}$ is often called a *virtual displacement* and denoted $\delta\phi$.

(4) Integrate the last term by parts:

$$\boxed{\int_{\mathcal{B}} \rho_{\mathrm{Ref}} \boldsymbol{A} \cdot \boldsymbol{W}\, dV = \int_{\mathcal{B}} \rho_{\mathrm{Ref}} \boldsymbol{B} \cdot \boldsymbol{W}\, dV - \int_{\mathcal{B}} \boldsymbol{P} : \nabla \boldsymbol{W}\, dV + \int_{\partial\mathcal{B}} \boldsymbol{W} \cdot \boldsymbol{P} \cdot \boldsymbol{N}\, dA.} \quad (1)$$

If $\boldsymbol{w}$ is the spatial form of $\boldsymbol{W}$, and $\tau^{ab} = P^{aB} F^b{}_B$ is the *Kirchhoff stress*, note that $\boldsymbol{P} : \nabla \boldsymbol{W} = \boldsymbol{\tau} : \nabla \boldsymbol{w} = \frac{1}{2} \boldsymbol{\tau} : \mathcal{L}_{\boldsymbol{w}} \boldsymbol{g}$, by symmetry of $\boldsymbol{\tau}$.

The validity of (1) for all variations $\boldsymbol{W} = \delta\phi$ is called the *principle of virtual work*. This principle must be modified, as we shall see in Chapter 5, if boundary conditions are prescribed.

The importance of the principle of virtual work rests on three facts.

(a) It is very useful numerically. (cf. Oden and Reddy [1976a].)

(b) It is believed to remain valid under conditions for which differential equations do not necessarily make sense.[20]

(c) Equation (1) coincides with the *weak form* of the differential equations for which there are many relevant mathematical theorems.

The paper of Antman and Osborn [1979] shows how to by-pass steps (2) and (3) above and thus show the equivalence of the integral balance laws and the principle of virtual work under conditions for which both are meaningful.

In the remainder of this box we shall show how to *directly* obtain the

[20]For this purpose the material version is a bit simpler, as there is no convective term $\boldsymbol{v} \cdot \nabla \boldsymbol{v}$ in the acceleration.

principle of virtual work from balance of energy. In doing so it is crucial to use the covariant version. Indeed it has often been noted as a defect of the "rigid" version of balance of energy, that one cannot directly get the principle of virtual work (see, e.g., Sewell [1966a]).

First of all, one should carry out the program of Theorem 4.13 in material coordinates (see Problem 4.8). Conservation of mass now merely reads ρ_{Ref} = constant in time, and is derivable from Assumptions 1 and 2. As in the proof of 4.13 (see Equation (6)), the condition of covariance leads directly to the following:

$$0 = -\int_{\mathcal{U}} \rho_{\text{Ref}}\left(\frac{\partial E}{\partial g}: \mathfrak{L}_{\boldsymbol{W}} g + \langle \boldsymbol{A} - \boldsymbol{B}, \boldsymbol{W}\rangle\right) dV + \int_{\partial\mathcal{U}} \boldsymbol{W}\cdot\boldsymbol{P}\cdot\boldsymbol{N}\, dA. \tag{2}$$

Arbitrariness of $\mathcal{U}$ gives the identity $\boldsymbol{\tau} = 2\rho_{\text{Ref}}\, \partial E/\partial g$. Substituting this into (2) and choosing $\mathcal{U} = \mathcal{B}$, we obtain exactly (1), written in terms of the Kirchhoff stress.

With an additional constitutive assumption (see Chapter 3) one can directly relate E and $\boldsymbol{P}$, avoiding the necessity of using $\boldsymbol{\tau}$. The covariant approach is compatible with such assumptions, as we shall see in Section 3.3.

Box 4.2 ***An Invariant Theory of Stress and Equilibrium***[21]

In this box we present an invariant formulation of some basic notions of continuum mechanics, including stress, equilibrium, and the principle of virtual work using the theory of manifolds of maps, which is summarized below. Manifolds of maps will also be useful in our discussion of Hamiltonian structures in Chapter 5.

A basic concept in any physical theory is that of its configuration space: i.e., the set of all possible states of the system. If the configuration space is a differentiable manifold, a *virtual displacement* may be defined to be an element of its tangent bundle. A *generalized force* is an element of the cotangent bundle, and its evaluation on a virtual displacement has the meaning of virtual work.

The placements of bodies in space are described by introducing two different mathematical models, the global model and the local model, and the fact that both ultimately serve to represent the same physical

[21]The material for this box was kindly communicated by Reuven Segev and Marcelo Epstein and is based on the ideas of Epstein and Segev [1980] and Segev [1981].

phenomenon is accounted for by imposing compatibility conditions. The global model deals with configurations $\phi: \mathcal{B} \longrightarrow \mathcal{S}$ from the point of view of analyzing the space $\mathcal{C}$ of configuration as a whole. In the local model a fiber space is attached to each point of the body $\mathcal{B}$ and the space $\mathcal{S}$. These spaces provide a framework for the representation of the internal structure of the body. If these spaces are the tangent spaces, we recover the usual theory: the Piola-Kirchhoff stress represents the local force, and compatibility with the global force yields the equilibrium and boundary conditions. Higher order and Cosserat theories (see Box 2.3, Chapter 2) may be dealt with by suitable choices of fiber spaces.

For each model we have a configuration space and corresponding sets of virtual displacements and forces. The local force is a generalization of the classical notion of stress field. The compatibility between the models is specified in terms of a map from the global to the local configuration space. This compatibility condition between local and global forces is a generalization of the principle of virtual work.

We now summarize some results from the theory of manifolds of maps that are needed. Some useful references for this subject are Palais [1968] Eliasson [1967] and Ebin and Marsden [1970]. For this box the technical proofs of the following results are not needed, so are omitted.

(i) Let $\mathfrak{M}$ be a compact C^∞ manifold (possibly with boundary) and $\mathfrak{N}$ a C^∞ manifold without a boundary. Then, the set $C^k(\mathfrak{M}, \mathfrak{N})$ of C^k mappings $\mathfrak{M} \longrightarrow \mathfrak{N}$, $0 \leq k < \infty$, can be given a structure of a C^∞ manifold modelled on a Banach space. (Similar results hold for other function spaces discussed in Chapter 6).

(ii) Let $\phi \in C^k(\mathfrak{M}, \mathfrak{N})$; then the tangent space to $C^k(\mathfrak{M}, \mathfrak{N})$ at ϕ is given by $T_\phi C^k(\mathfrak{M}, \mathfrak{N}) = \{V \in C^k(\mathfrak{M}, T\mathfrak{N}) \mid \tau \circ V = \phi\}$, where $\tau_{\mathfrak{N}}: T\mathfrak{N} \longrightarrow \mathfrak{N}$ is the projection.

This result should be plausible from our discussions of material velocity.

(iii) The set $\mathcal{C}$ of C^k embeddings is open in $C^k(\mathfrak{M}, \mathfrak{N})$ for $k \geq 1$.

This follows directly from the implicit function theorem.

(iv) Let $\pi: \mathcal{E} \longrightarrow \mathfrak{M}$ and $\rho: \mathcal{F} \longrightarrow \mathfrak{N}$ be C^∞ vector bundles where $\mathfrak{M}$ and $\mathfrak{N}$ are as in (i) and $r \leq k$. Then the set $C^r_k(\mathcal{E}, \mathcal{F})$ of C^r vector bundle maps of $\mathcal{E}$ to $\mathcal{F}$ covering C^k base maps has a C^∞ vector bundle structure $C^r_k(E, F) \longrightarrow C^k(\mathfrak{M}, \mathfrak{N})$ and the projection map assigns to each vector bundle morphism its base map. (See Verona [1970]).

(v) For $f \in C^r_k(\mathcal{E}, \mathcal{F})$, $T_f C^r_k(\mathcal{E}, \mathcal{F}) = \{g \in C^r_k(\mathcal{E}, T\mathcal{F}) \mid \tau_{\mathcal{F}} \circ g = f\}$, where $\tau_{\mathcal{F}}: T\mathcal{F} \longrightarrow \mathcal{F}$ and $T\rho: T\mathcal{F} \longrightarrow T\mathfrak{N}$ is the vector bundle structure used on $T\mathcal{F}$. The tangent bundle projection is given by $h \mapsto \tau_{\mathcal{F}} \circ h$ for $h \in TC_k^r(\mathcal{E}, \mathcal{F})$. In particular, the set of all C^{k-1} vector bundle morphisms of the tangent bundles is a vector bundle $C_k^{k-1}(T\mathfrak{M}, T\mathfrak{N}) \longrightarrow C^k(\mathfrak{M}, \mathfrak{N})$, and each $h \in TC_k^{k-1}(T\mathfrak{M}, T\mathfrak{N})$ is identified with a vector bundle map from the bundle $\tau_{\mathfrak{M}}: T\mathfrak{M} \longrightarrow \mathfrak{M}$ to the bundle $T\tau_{\mathfrak{N}}: TT\mathfrak{N} \longrightarrow T\mathfrak{N}$.

(iv) Let $\lambda_T: C^k(\mathfrak{M}, \mathfrak{N}) \longrightarrow C_k^{k-1}(T\mathfrak{M}, T\mathfrak{N})$ be the operation of taking the tangent on elements of the manifold of maps; i.e., $\lambda_T(f) = Tf$. Then λ_T is a C^∞ section of the vector bundle $C_k^{k-1}(T\mathfrak{M}, T\mathfrak{N}) \longrightarrow C^k(\mathfrak{M}, \mathfrak{N})$. The tangent map $T\lambda_T: TC^k(\mathfrak{M}, \mathfrak{N}) \longrightarrow TC_k^{k-1}(T\mathfrak{M}, T\mathfrak{N})$ is given by $T\lambda_T(v) = \omega \circ Tv$, where Tv, is the tangent to the map $\mathfrak{M} \longrightarrow T\mathfrak{N}$ with which v is identified and where ω is the canonical involution $T^2\mathfrak{N} \longrightarrow T^2\mathfrak{N}$: if (x, a, b, c) is the representative of an element in $T^2\mathfrak{N}$ with respect to a natural chart, then the local representative of ω with respect to this chart is given by $(x, a, b, c) \mapsto (x, b, a, c)$.

Since we are presently concerned with the geometric aspects only, we shall work with smooth maps for simplicity (see Chapter 6 for the function space setting needed for analysis).

The *global model* consists of the following:

GM1 *Physical space* is an m-dimensional differentiable manifold $\mathcal{S}$ without a boundary.

GM2 A *body* is an n-dimensional compact differentiable manifold $\mathcal{B}$ which may have a boundary, with $n \leq m$.

GM3 A *global configuration* is a (smooth) embedding $\phi: \mathcal{B} \longrightarrow \mathcal{S}$. By the results above, the configuration space $\mathcal{C}$ is an infinite dimensional differentiable manifold.

GM4 Let $\tau_{\mathcal{C}}: T\mathcal{C} \longrightarrow \mathcal{C}$ be the tangent bundle of the configuration space. A *global virtual displacement* is an element $\delta\phi \in T\mathcal{C}$.

GM5 A *global force* is an element $f \in T^*\mathcal{C}$. The evaluation $f(\delta\phi)$ for $f \in T_\phi^*\mathcal{C}$, $\delta\phi \in T_\phi\mathcal{C}$ is called the *virtual work* of the force f on the virtual displacement $\delta\phi$.

By GM5, forces are linear functionals on the vector space of vector fields (virtual displacements) over the image of the body in space. For linear functionals continuous in the C^0 topology, a global force can be represented by a vector measure over the image of the body (as in the definition of forces in general given in Truesdell and Toupin [1960] and Truesdell [1977].) In the more general case of the C^k topology, a global force will consist of distributions.

In the *local model*, configurations, virtual displacements, and forces are defined independent of the global model.

LM1 In the local model the *space* is vector bundle $\rho: \mathcal{F} \longrightarrow \mathcal{S}$.

LM2 The *body* is a vector bundle $\pi: \mathcal{E} \longrightarrow \mathcal{B}$.[22]

LM3 A *local configuration* is a (smooth) vector bundle map $F: \mathcal{E} \longrightarrow \mathcal{F}$ such that the induced base map $\phi: \mathcal{B} \longrightarrow \mathcal{S}$, is an embedding. By the properties of manifolds of maps, the local configuration space $\mathcal{R}$, consisting of all such embeddings, has a structure of a vector bundle $\pi_{\mathcal{C}}: \mathcal{R} \longrightarrow \mathcal{C}$ over the global configuration space.

LM4 Let $\tau_{\mathcal{R}}: T\mathcal{R} \longrightarrow \mathcal{R}$ be the tangent bundle of the local configuration space. The *local virtual displacement* is an element δF of the tangent bundle: $\delta F \in T\mathcal{R}$.
The relation between the configuration spaces and their tangent bundles can be represented by the following commutative diagram

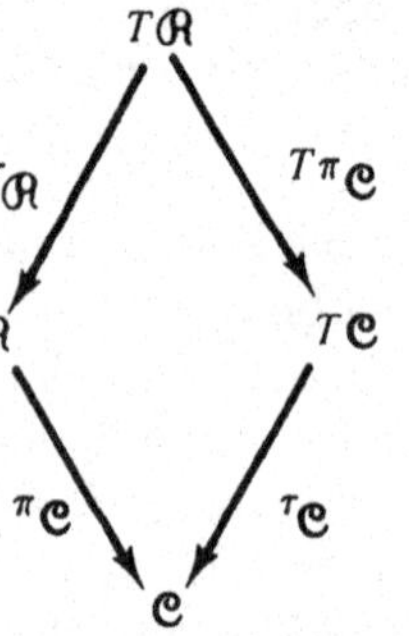

and when we make the identification of δF with a vector bundle morphism from $\pi: \mathcal{E} \longrightarrow \mathcal{B}$ to $T\rho: T\mathcal{F} \longrightarrow T\mathcal{N}$, we have

$$\tau_{\mathcal{R}}(\delta F) = F \qquad \pi_{\mathcal{C}}(F) = \phi$$

$$T\pi_{\mathcal{C}}(\delta F) = \delta\phi \qquad \tau_{\mathcal{C}}(\delta F) = \delta\phi$$

such that F, $\delta\phi$ and ϕ make the following diagram commutative.

[22]For many applications, such as Cosserat continua, these should be fiber bundles. However we use vector bundles for simplicity. The generalization is not difficult.

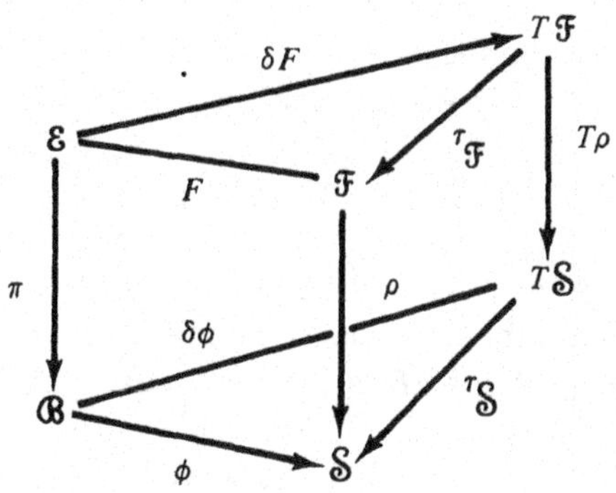

LM5 A *local force* is an element $\sigma \in T^*\mathcal{R}$. The evaluation $\sigma(\delta F)$ for $\sigma \in T^*_F\mathcal{R}$, $\delta F \in T_F\mathcal{R}$ is called the *virtual work* of the local force σ on the local virtual displacement δF.

The link between the global and the local models is provided by the *compatibility conditions.*

CC1 Let λ associate to each manifold $\mathfrak{M}$ a vector bundle $\lambda(\mathfrak{M})$ over it and to each map between manifolds a corresponding vector bundle map. We write $\lambda(\mathcal{B}) = \mathcal{E}$, $\lambda(\mathcal{S}) = \mathfrak{F}$, and if f is a map of $\mathcal{B}$ to $\mathcal{S}$, $\lambda(f)$ is a vector bundle map covering f. We may thus regard $\lambda: \mathcal{C} \longrightarrow \mathcal{R}$ as a section of $\pi_\mathcal{C}$ (assumed to be differentiable). We call λ a *compatibility operation.*

CC2 A local configuration F is called *compatible* with the global configuration ϕ if $F = \lambda(\phi)$.

CC3 A local virtual displacement δF is called *compatible* with the global virtual displacement $\delta\phi$ if $\delta F = T\lambda(\delta\phi)$.

CC4 A global force f is *compatible* with the local force σ if $f = T^*\lambda(\sigma)$, where $T^*\lambda: T^*\mathcal{R} \,|\, \lambda(\mathcal{C}) \longrightarrow T^*\mathcal{C}$ is the adjoint map of $T\lambda$.

The last compatibility condition is a generalization of the equilibrium equations of classical continuum mechanics, as we shall see later. The compatibility of the global force f with the local force σ implies $f(\delta\phi) = (T^*\lambda(\sigma))(\delta\phi)$ for all $\delta\phi \in T_\phi\mathcal{C}$. However, by the definition of the adjoint map $(T^*\lambda(\sigma))(\delta\phi) = \sigma(T\lambda(\delta\phi))$ and, using CC2, we obtain the equivalent form $f(\delta\phi) = \sigma(\delta F)$ for all compatible pairs $\delta\phi$, δF, which, as we shall see below, is the statement of the principle of virtual work.

Vector bundles are now constructed that allow us to relate the local configuration, the local virtual displacement, and the local force to the classical concepts of deformation gradient tensor field, its variation, and the Piola-Kirchhoff stress.

Given a configuration ϕ, consider the vector bundle $\mathcal{L}(\mathcal{E}, \phi^*\mathcal{F}) \to \mathcal{B}$, whose fiber over $X \in \mathcal{B}$ is $\mathcal{L}(\mathcal{E}, \phi^*\mathcal{F})_X = \mathcal{L}(\mathcal{E}_X, \mathcal{F}_{\phi(X)})$, the set of linear maps of $\mathcal{E}_X$ to $\mathcal{F}_{\phi(X)}$. The space $C^\infty(\mathcal{L}(\mathcal{E}, \phi^*\mathcal{F}))$ of C^∞ sections of $\mathcal{L}(\mathcal{E}, \phi^*\mathcal{F})$ can be identified with $\mathcal{R}_\phi$. An element of $C^\infty(\mathcal{L}(\mathcal{E}, \phi^*\mathcal{F}))$ has the nature of a two-point field $\boldsymbol{F}$, just as in Section 1.4.

Recall that $T_{\boldsymbol{F}}\mathcal{R}$ is the set of vector bundle maps $\delta\mathcal{F}: \mathcal{E} \to T\mathcal{F}$ satisfying $\tau_{\boldsymbol{F}} \circ \delta\boldsymbol{F} = \boldsymbol{F}$. The elements $\delta\boldsymbol{F}$ satisfying $Tp \circ \delta\boldsymbol{F} = 0$ are called *vertical*, and this set is denoted $T_{\boldsymbol{F},\mathrm{vert}}\mathcal{R}$. We can identify $\mathcal{R}_\phi$ with $T_{\boldsymbol{F},\mathrm{vert}}\mathcal{R}$ by mapping $\tilde{\boldsymbol{F}} \in \mathcal{R}_\phi$ to the tangent to the curve $t \mapsto \boldsymbol{F} + t\tilde{\boldsymbol{F}}$ at $t = 0$. Thus, vertical vectors are variations $\delta\boldsymbol{F}$ of the *values* of $\boldsymbol{F}$ only, keeping the base map ϕ fixed. If the bundle $\mathcal{F}$ has a symmetric connection, or we choose local charts, any variation $\delta\boldsymbol{F} \in T_{\boldsymbol{F}}\mathcal{R}$ can be split into the element $Tp \circ \delta\boldsymbol{F}$ of $C^\infty(T\mathcal{B}, \phi^*T\mathcal{S})$ and a vertical variation, which we denote $(\delta\boldsymbol{F})_{\mathrm{vert}}$.

Elements $\boldsymbol{P}$ of $C^\infty(\mathcal{L}(\phi^*\mathcal{F}, \mathcal{E}))$ may be regarded as elements of $T^*_{\boldsymbol{F},\mathrm{vert}}\mathcal{R}$ by

$$(\delta\boldsymbol{F})_{\mathrm{vert}} \mapsto \int_{\mathcal{B}} \operatorname{trace}\left[\boldsymbol{P} \circ (\delta\boldsymbol{F})_{\mathrm{vert}}\right] dV \tag{1}$$

The two-point field $\boldsymbol{P}$ is a generalization to bundles of the first Piola-Kirchhoff stress.

Let us turn now to the case of tangent bundles and assume $\mathcal{B}$ and $\mathcal{S}$ carry metrics $\boldsymbol{G}$ and $\boldsymbol{g}$ respectively (and so have corresponding connections).

By the *local model of tangent bundles* we mean:

LMT1 The vector bundle which models the space is the tangent bundle $\tau_{\mathcal{S}}: T\mathcal{S} \to \mathcal{S}$.

LMT2 The *body* is modelled by $\tau_{\mathcal{B}}: T\mathcal{B} \to \mathcal{B}$.

LMT3 The *local configuration* space $\mathcal{R}$ consists of vector bundle maps $\boldsymbol{F}: T\mathcal{B} \to T\mathcal{S}$.

Choose $\lambda = \lambda_T$ by $\lambda_T(\mathfrak{M}) = T\mathfrak{M}$ and $\lambda_T(\phi) = T\phi$. Thus the compatibility conditions become

CCT1 *The local configuration $\boldsymbol{F}$ is compatible with the global configuration ϕ if $\boldsymbol{F} = \lambda_T(\phi) = T\phi$*: i.e., *if $\boldsymbol{F}$ is the deformation gradient of ϕ.*

CCT2 *The local virtual displacement $\delta\boldsymbol{F}$ is compatible with the global virtual displacement δ if $\delta\boldsymbol{F} = T\lambda_T(\delta\phi)$: i.e., $\delta\boldsymbol{F} = \omega \circ T\delta\phi$, where ω is the canonical involution.*

CCT3 *The local force σ is compatible with the global force f if $f(\delta\phi) = \sigma(\delta F)$ for compatible pairs $\delta\phi$ and δF.*

Let us write $V = \delta\phi$ for the variation of ϕ, so V is a material vector field over ϕ. Then one can check that (CCT2) is the same as saying

and
$$\left.\begin{aligned}(\delta F)_{\text{vert}} &= \nabla V\\ (\delta F)_{\text{hor}} &= F\end{aligned}\right\} \tag{2}$$

The second of these equations says we are not free to vary the horizontal part. Thus, vertical variations are the important ones. Consistent with this, suppose the local force is given by the element P in (1). Then CCT3 is the same as saying

$$f(V) = \int_{\mathcal{B}} \operatorname{trace}[P(\nabla V)]\, dV \tag{3}$$

If $f(V) = \int_{\mathcal{B}} B \cdot V\, dV + \int_{\partial\mathcal{B}} \tau \cdot V\, dV$ is the usual form for a given global force, then (3) is just the statement of the principle of virtual work for equilibrium solutions, which is equivalent to the balance laws, and boundary conditions if solutions are smooth enough (see the previous box and Chapter 5).

If we work on spacetime rather than space, a similar construction yields the dynamic equations. If we chose different local bundles, one may obtain higher order or Cosserat theories.

Box 4.3 ***Summary of Important Formulas for Section 2.4***

Slicing of a Classical Spacetime $\mathcal{V}$ by Space $\mathcal{S}$

$$i: \mathcal{S} \times \mathbb{R} \longrightarrow \mathcal{V}, \qquad i(x, t) = p$$
$$\textit{Velocity:} \quad v^i(p) = \frac{\partial}{\partial t} i(x, t)$$
$$\textit{Acceleration:} \quad a^i(p) = (\nabla_{v^i} v^i)(p)$$

The inverse of a slicing is a *framing*.
A *world tube* is a map $\Phi: \mathcal{B} \times \mathbb{R} \longrightarrow \mathcal{V}$, $\Phi(x, t) = p$.

$$\textit{Velocity:} \quad v^\Phi(p) = \frac{\partial}{\partial t} \Phi(x, t)$$
$$\textit{Acceleration:} \quad a^\Phi(p) = (\nabla_{v^\Phi} v^\Phi)(p)$$

Motion of Φ relative to i: ϕ^i = first component of $i^{-1} \circ \Phi$

(velocity of Φ^i) $= v^{\Phi, i}$ = *apparent velocity*
(acceleration of ϕ^i) $= a^{\Phi, i}$ = *apparent acceleration*

Apparent vs. Absolute Velocities and Accelerations

$$i_{t*}\boldsymbol{v}^{\Phi,i} = \boldsymbol{v}^{\Phi} - \boldsymbol{v}^{i}$$
$$i_{t*}\boldsymbol{a}^{\Phi,i} = \boldsymbol{a}^{\Phi} - \boldsymbol{a}^{i} - 2\nabla_{v^{\Phi}-v^{i}}\boldsymbol{v}^{i}$$

Transformation under a Change of Framing $\xi_t: \mathcal{S} \to \mathcal{S}$

$$\boldsymbol{v}^{\Phi,i_2} = \xi_{t*}\boldsymbol{v}^{\Phi,i_1} + \boldsymbol{w}$$
$$\boldsymbol{a}^{\Phi,i_2} = \xi_{t*}\boldsymbol{a}^{\Phi,i_1} + \boldsymbol{a}^{\xi,i_2} + 2\nabla^{i_2}_{(\xi_{t*}v^{\Phi,i_1})}\boldsymbol{w}$$

where $\boldsymbol{w}$ = velocity field of ξ and $\boldsymbol{a}^{\xi,i_2}$ is its acceleration field.

Inertial Frame

$\nabla \boldsymbol{v}^{i} = \boldsymbol{0}$, or, equivalently, $i_{t*}\boldsymbol{a}^{\Phi,i} = \boldsymbol{a}^{\Phi}$ for all world tubes Φ.

Covariance and Balance of Energy

If balance of energy holds under all changes of slicings (or equivalently under all superposed motions, not necessarily rigid), then in addition to the usual consequences of the Green–Naghdi–Rivlin–Noll results we have the Doyle–Ericksen formula

$$\boldsymbol{\sigma} = 2\rho\frac{\partial e}{\partial \boldsymbol{g}}, \qquad \boldsymbol{\tau} = 2\rho_{\text{Ref}}\frac{\partial E}{\partial \boldsymbol{g}} \qquad \sigma^{ab} = 2\rho\frac{\partial e}{\partial g_{ab}}, \qquad \tau^{ab} = 2\rho_{\text{Ref}}\frac{\partial E}{\partial g_{ab}}$$

Principle of Virtual Work

$$\int_{\mathcal{B}} \rho_{\text{Ref}}\boldsymbol{A}\cdot\boldsymbol{W}\,dV = \int_{\mathcal{B}} \rho_{\text{Ref}}\boldsymbol{B}\cdot\boldsymbol{W}\,dV - \int_{\mathcal{B}} \boldsymbol{P}\cdot\nabla\boldsymbol{W}\,dV + \int_{\partial\mathcal{B}} \boldsymbol{W}\cdot\boldsymbol{P}\cdot\boldsymbol{N}\,dA$$

$$\int_{\mathcal{B}} \rho_{\text{Ref}}A^{a}W_{a}\,dV = \int_{\mathcal{B}} \rho_{\text{Ref}}B^{a}W_{a}\,dV - \int_{\mathcal{B}} P^{aA}W_{a|A}\,dV + \int_{\partial\mathcal{B}} W_{a}P^{aA}N_{A}\,dA$$

for all variations $\boldsymbol{W}$.

2.5 THERMODYNAMICS II; THE SECOND LAW

The second law of thermodynamics is frequently shrouded in mysterious physical jargon and less than adequate mathematical treatment. The authors' education was no exception. Our own frustration is consistent with the following quotation from Truesdell [1969], pp. 1–23: "The difference [between mechanics and thermodynamics] is that thermodynamics never grew up."

The results we shall need are given a concise mathematical treatment below. We make no attempt at detailed physical justification. In addition to Truesdell's work, the reader can profitably consult the texts Malvern [1969] and Sommerfeld [1964b] in which entropy is regarded as a measure of disorder and its origins in statistical mechanics via the Boltzmann's equation are described.[23]

[23]For more advanced mathematical treatments, consult, for example Ruelle [1969, 1978], Arnold and Avez [1971], Hermann [1975], and Souriau [1975].

Let $\phi(X, t)$ be a C^1 regular motion of a simple body $\mathcal{B} \subset \mathcal{S}$. In addition to the functions $\rho(x, t)$, $v(x, t)$, $r(x, t)$, and $h(x, t)$ introduced earlier, assume there are functions $\eta(x, t)$, the specific (per unit mass) *entropy* and $\theta(x, t) > 0$, the *absolute temperature.*

5.1 Definition The above functions are said to obey the *entropy production inequality* or the *Clausius–Duhen inequality*[24] if, for all nice $\mathcal{U} \subset \mathcal{B}$, we have

$$\boxed{\frac{d}{dt}\int_{\phi_t(\mathcal{U})} \rho\eta\, dv \geq \int_{\phi_t(\mathcal{U})} \frac{\rho r}{\theta}\, dv + \int_{\partial\phi_t(\mathcal{U})} \frac{h}{\theta}\, da.}$$

This inequality is often called the *second law of thermodynamics,* although there is some controversy over this nomenclature.

If $r = 0$ (no heat sources) and $h = 0$ (no heat flux), we get

$$\frac{d}{dt}\int_{\phi_t(\mathcal{U})} \rho\eta\, dv \geq 0,$$

that is, total entropy is nondecreasing.

5.2 Theorem *Assume ϕ_t is a C^1 regular motion, r is continuous, $h(x, t, \boldsymbol{n}) = -\langle \boldsymbol{q}(x, t), \boldsymbol{n}\rangle$, where $\boldsymbol{q}$ is a C^1 vector field, θ is C^1, and that conservation of mass holds. Then*

$$\boxed{\rho\dot{\eta} \geq \frac{\rho r}{\theta} - \operatorname{div}\left(\frac{\boldsymbol{q}}{\theta}\right) = \frac{\rho r}{\theta} - \frac{1}{\theta}\operatorname{div}\boldsymbol{q} + \frac{1}{\theta^2}\langle \boldsymbol{q}, \nabla\theta\rangle.}$$

Proof This is a direct consequence of the master balance inequality; see Equation (4) of Section 2.1. ∎

We call $\gamma = \rho\dot{\eta} - \rho r/\theta + \operatorname{div}(\boldsymbol{q}/\theta)$ the *rate of entropy production.* Thus the Clausius–Duhem inequality asserts that $\gamma \geq 0$.

To obtain the material form, let

$$x = \phi(X, t), \quad \Theta(X, t) = \theta(x, t), \quad N(X, t) = \eta(x, t), \quad R(X, t) = r(x, t),$$
$$\boldsymbol{Q}(X, t) = J\boldsymbol{F}^{-1}\cdot\boldsymbol{q}(x, t), \text{ and } \rho_{\text{Ref}}(X) = \rho(X, O).$$

5.3 Theorem *The material form of the entropy production inequality is*

$$\boxed{\frac{d}{dt}\int_{\mathcal{U}} \rho_{\text{Ref}} N\, dV \geq \int_{\mathcal{U}} \rho_{\text{Ref}}\frac{R}{\Theta}\, dV + \int_{\partial\mathcal{U}} \frac{\boldsymbol{Q}\cdot\boldsymbol{N}}{\Theta}\, dA,}$$

---

[24]There has been much controversy over how one should set the basic principles of thermodynamics. In fact, several criticisms have been lodged against the general use of the Clausius–Duhem inequality (see Rivlin [1977], Day and Silhavy [1977], and Green and Naghdi [1977, 1978a]). For accounts of alternative theories not subject to these criticisms, and recent developments, see Green and Naghdi [1977, 1978a, 1978b], Serrin [1979] and Naghdi [1981].

or, in localized form,

$$\boxed{\rho_{\text{Ref}} \frac{\partial N}{\partial t} \geq \rho_{\text{Ref}} \frac{R}{\Theta} - \text{DIV}\left(\frac{\boldsymbol{Q}}{\Theta}\right).}$$

The reader can supply the proof in routine fashion by employing the material master inequality of Section 2.1.

It is convenient to recast the entropy production inequality into a slightly different form by using the internal energy $e(x, t)$ introduced in the previous section.

5.4 Definition The *free energy* is defined by $\psi = e - \theta\eta$, or, materially, $\Psi = E - \Theta N$.

Later we shall see how to interpret the transformation of e to ψ as a Legendre transformation and how to regard θ and η as conjugate variables.

5.5 Theorem *Assume conservation of mass, balance of momentum, moment of momentum, energy and the entropy production inequality hold. Then*

$$\boxed{\rho(\eta\dot{\theta} + \dot{\psi}) - \boldsymbol{\sigma}:\boldsymbol{d} + \frac{1}{\theta}\langle \boldsymbol{q}, \nabla\theta\rangle \leq 0,}$$

or, materially,

$$\boxed{\rho_{\text{Ref}}\left(N\frac{\partial \Theta}{\partial t} + \frac{\partial \Psi}{\partial t}\right) - \mathbf{P}:\frac{\partial \mathbf{F}}{\partial t} + \frac{1}{\Theta}\langle \boldsymbol{Q}, \text{GRAD}\,\Theta\rangle \leq 0.}$$

These inequalities are called the reduced dissipation inequalities.

Proof From 5.4,

$$\dot{\psi} = \dot{e} - \theta\dot{\eta} - \dot{\theta}\eta; \qquad \text{that is,} \quad \theta\dot{\eta} = \dot{e} - \dot{\theta}\eta - \dot{\psi}.$$

Combining this with 5.2 gives

$$\rho(\dot{e} - \dot{\theta}\eta - \dot{\psi}) \geq \rho r - \text{div}\,\boldsymbol{q} + \frac{1}{\theta}\langle \boldsymbol{q}, \nabla\theta\rangle.$$

From 3.4, $\rho\dot{e} = \rho r - \text{div}\,\boldsymbol{q} + \boldsymbol{\sigma}:\boldsymbol{d}$, so substitution gives the spatial result.

The material form is proved in exactly the same way, using 5.3 in place of 5.2. ∎

Problem 5.1 Prove the identity

$$\frac{J}{\theta}\langle \boldsymbol{q}, \text{grad}\,\theta\rangle = \frac{1}{\Theta}\langle \boldsymbol{Q}, \text{GRAD}\,\Theta\rangle$$

directly in coordinates. Use it to prove the material reduced dissipation inequality by multiplying the spatial form by J.

5.6 Example Assume we have a system in equilibrium; that is, $\partial\Theta/\partial t = 0$, $\partial\Psi/\partial t = 0$, and $\partial F/\partial t = 0$. Since $\Theta > 0$, the reduced dissipation inequality becomes $\langle \boldsymbol{Q}, \text{GRAD}\,\Theta\rangle \leq 0$; that is, heat flows from hot to cold.

5.7 Terminology Processes for which the following hold are given the accompanying name.

(a) $\boldsymbol{q} = \boldsymbol{0}$; adiabatic; (c) GRAD $N = \boldsymbol{0}$; homentropic;
(b) $\dot{\eta} = 0$: isentropic; (d) θ = constant: isothermal.

Problem 5.2[25] (Covariance and the Second Law) Suppose that η, r, θ, and so on can depend on the metric $\bar{g} \in \mathfrak{S}_g$, as in Box 3.3. Assume that the rate of entropy production γ transforms as a scalar under slicings of spacetime. Prove that $\partial\eta/\partial \boldsymbol{g} = \boldsymbol{0}$; that is, η is independent of $\boldsymbol{g}$. (Further discussions of covariance and thermodynamics are given in Section 3.3.)

Box 5.1 ***Summary of Important Formulas for Section 2.5***

Entropy Production Inequality (Clausius–Duhem Inequality)

(a) Spatial: $$\frac{d}{dt}\int_{\phi_t(\mathfrak{U})} \rho\eta\, dv \geq \int_{\phi_t(\mathfrak{U})} \frac{\rho r}{\theta}\, dv + \int_{\partial\phi_t(\mathfrak{U})} \frac{\langle \boldsymbol{q}, \boldsymbol{n}\rangle}{\theta}\, da$$

(b) Material: $$\frac{d}{dt}\int_{\mathfrak{U}} \rho_{\text{Ref}} N\, dV \geq \int_{\mathfrak{U}} \rho_{\text{Ref}} \frac{R}{\Theta}\, dV + \int_{\partial\mathfrak{U}} \frac{\langle \boldsymbol{Q}, \boldsymbol{N}\rangle}{\Theta}\, dA$$

Localized Entropy Production Inequality

(a) Spatial: $$\rho\dot{\eta} \geq \frac{\rho r}{\theta} - \frac{1}{\theta}\operatorname{div}\boldsymbol{q} + \frac{1}{\theta^2}\langle \boldsymbol{q}, \text{grad}\,\theta\rangle$$

(b) Material: $$\rho_{\text{Ref}}\frac{\partial N}{\partial t} \geq \rho_{\text{Ref}}\frac{R}{\Theta} - \frac{1}{\Theta}\text{DIV}\,\boldsymbol{Q} + \frac{1}{\Theta^2}\langle \boldsymbol{Q}, \text{GRAD}\,\Theta\rangle$$

Free Energy

(a) Spatial: $\psi(x, t) = e(x, t) - \theta(x, t)\eta(x, t)$
(b) Material: $\Psi(X, t) = E(X, t) - \Theta(X, t)\, N(X, t)$

Reduced Dissipation Inequality

(a) Spatial: $$\rho(\eta\dot{\theta} + \dot{\psi}) - \boldsymbol{\sigma} : \boldsymbol{d} + \frac{1}{\theta}\langle \boldsymbol{q}, \text{GRAD}\,\theta\rangle \leq 0$$

(b) Material: $$\rho_{\text{Ref}}\left(N\frac{\partial\Theta}{\partial t} + \frac{\partial\Psi}{\partial t}\right) - \boldsymbol{P} : \frac{\partial \boldsymbol{F}}{\partial t} + \frac{1}{\Theta}\langle \boldsymbol{Q}, \text{GRAD}\,\Theta\rangle \leq 0, \quad \boldsymbol{P} : \frac{\partial \boldsymbol{F}}{\partial t} = \boldsymbol{S} : \boldsymbol{D}$$

[25]Suggested by M. Gurtin.

CHAPTER 3

CONSTITUTIVE THEORY

Constitutive theory gives functional form to the stress tensor, free energy, and heat flux vector in terms of the motion and temperature. This accomplishes two things. First, different general functional forms distinguish various broad classes of materials (elastic, fluid, materials with memory, etc.) and specific functional forms determine specific materials. Secondly, these additional relationships make the equations formally well-posed; that is, there are as many equations as unknowns.

In Section 3.1, we describe the features of constitutive theory in rather general terms. At this stage the functional form is very broad; the key objective is to isolate what should be a function of what. We can be more specific by making more hypotheses. Proceeding towards the traditional theories of elasticity, we specialize further in Section 3.2. We see that the hypothesis of locality restricts us to functions of the point values of the deformation gradient and temperature and their derivatives. The dependence on higher derivatives is then eliminated by way of the entropy production inequality. Finally, the assumption of invariance under rigid body motions—that is, material frame indifference—allows us to reduce the dependence on the deformation gradient $\boldsymbol{F}$ to dependence on the deformation tensor $\boldsymbol{C}$ (i.e., on the strain).

Section 3.3 treats constitutive theory using ideas of covariance and links this with our covariant treatment of energy balance in Section 2.3. The formula for the Cauchy stress $\boldsymbol{\sigma} = 2\rho(\partial\psi/\partial \boldsymbol{g})$ is shown to be equivalent to the standard formula $\boldsymbol{S} = 2\rho_{\mathrm{Ref}}(\partial\Psi/\partial \boldsymbol{C})$. An application to the identification problem for thermoelastic and inelastic materials is given as well. Additional links with the Hamiltonian structure and conservation laws will be made in Chapter 5. The constitutive information is inserted into the basic balance laws in Section 3.4;

there we summarize the equations of a thermoelastic solid. Section 3.5 makes a further specialization by considering the restrictions that arise when the material possesses some symmetry. Special attention is given to isotropic solids.

3.1 THE CONSTITUTIVE HYPOTHESIS

From the previous chapter, we have a number of basic equations expressed in the material picture as follows:

(i) $\rho_{\text{Ref}} = \rho J$ (conservation of mass);

(ii) $\rho_{\text{Ref}} \dfrac{\partial V}{\partial t} = \text{DIV}\, \boldsymbol{P} + \rho_{\text{Ref}} \boldsymbol{B}$ (balance of momentum);

(iii) $\boldsymbol{S} = \boldsymbol{S}^T$ (balance of moment of momentum);

(iv) $\rho_{\text{Ref}} \dfrac{\partial E}{\partial t} + \text{DIV}\, \boldsymbol{Q} = \rho_{\text{Ref}} R + \boldsymbol{S} : \boldsymbol{D}$ (balance of energy);

(v) $\rho_{\text{Ref}} N \dfrac{\partial \Theta}{\partial t} + \dfrac{\partial \Psi}{\partial t} - \boldsymbol{S} : \boldsymbol{D} + \dfrac{1}{\Theta} \langle \mathrm{Q}, \text{GRAD}\, \Theta \rangle \leq 0$
(reduced dissipation inequality);

(vi) $E = \Psi + N\Theta$ (relation between internal energy and free energy).

These equations are formally ill-posed in the sense that there are not enough equations to determine the evolution of the system. The situation is analogous to Newton's second law $m\ddot{x} = \boldsymbol{F}$; one cannot solve this equation without specifying how $\boldsymbol{F}$ depends on $\boldsymbol{x}$ and $\dot{\boldsymbol{x}}$.

Generally one treats the motion $\phi(X, t)$ and the temperature $\Theta(X, t)$ as the unknowns and attempts to solve for them from equations (ii) and (iv). Condition (i) determines ρ in terms of ρ_{Ref} (usually given) and ϕ, so we can eliminate condition (i). We often regard $\boldsymbol{B}$ and R as given externally. Thus, if we are going to determine ϕ and Θ, we must specify $\boldsymbol{S}$, $\boldsymbol{Q}$, and E as functions of ϕ and Θ. Since E is related to Ψ by (vi), we might also ask for N and Ψ to be functions of ϕ and Θ. These functions must be related in such a way that (iii) and (v) are satisfied. A particular functional form given to $\boldsymbol{S}$, $\boldsymbol{Q}$, N, and Ψ is supposed to characterize a particular material; these same functional forms are associated with any body made from that same material. As we shall see in the next section, the reduced dissipation inequality can be used in delineating the possible functional forms and their interrelationships. Section 3.3 shows how to derive the same basic relationships using covariance as an axiom in place of the second law of thermodynamics.

These ideas can be made more precise using the following terminology.

1.1 Definition Let r be a positive integer or $+\infty$. The set of all *past motions* up to time T is

$$\mathfrak{M}_T = \{\phi : \mathcal{B} \times (-\infty, T] \longrightarrow \mathcal{S} \mid \phi \text{ is a } C^r \text{ regular motion of } \mathcal{B} \text{ in } \mathcal{S} \text{ for } -\infty < t \leq T\}$$

and the set of all *past temperature fields* up to time T is

$$\mathfrak{J}_T = \{\Theta: \mathcal{B} \times (-\infty, T] \longrightarrow (0, \infty) \mid \Theta \text{ is a } C^r \text{ function}\}.$$

Let $\mathcal{H} = \bigcup_{T \in \mathbb{R}} (\mathfrak{M}_T \times \mathfrak{J}_T \times \{T\})$, the set of *past histories.*

Here our notation is the same as in previous chapters; $\mathcal{B}$ is a reference configuration of the body and $\mathcal{S}$ is the containing space.

1.2 Definition By a *constitutive equation* for the second Piola–Kirchhoff stress tensor $\boldsymbol{S}$ we mean a mapping

$$\hat{\boldsymbol{S}}: \mathcal{H} \longrightarrow S_2(\mathcal{B}),$$

where $S_2(\mathcal{B})$ denotes the space of C^s symmetric contravariant two-tensor fields on $\mathcal{B}$ (for some suitable positive integer s). The second Piola–Kirchhoff stress tensor *associated* with $\hat{\boldsymbol{S}}$, a motion $\phi(X, t)$, and a temperature field $\Theta(X, t)$ is

$$\boldsymbol{S}(X, t) = \hat{\boldsymbol{S}}(\phi_{[t]}, \Theta_{[t]}, t)(X),$$

where $\phi_{[t]} \in \mathfrak{M}_t$ and $\Theta_{[t]} \in \mathfrak{J}_t$ are defined to be ϕ and Θ restricted to $(-\infty, t]$.

Similarly, constitutive equations for $\boldsymbol{Q}$, Ψ, and N are defined to be mappings:

$$\hat{\boldsymbol{Q}}: \mathcal{H} \longrightarrow \mathfrak{X}(\mathcal{B}) \quad (\mathfrak{X} \text{ denotes the } C^s \text{ vector fields on } \mathcal{B}),$$

$$\hat{\Psi}: \mathcal{H} \longrightarrow \mathfrak{F}(\mathcal{B}) \quad (\mathfrak{F} \text{ denotes the } C^s \text{ scalar fields on } \mathcal{B}),$$

$$\hat{N}: \mathcal{H} \longrightarrow \mathfrak{F}(\mathcal{B}).$$

One sometimes sees the relationship between $\boldsymbol{S}$ and $\hat{\boldsymbol{S}}$ written as

$$\boldsymbol{S}(X, t) = \hat{\boldsymbol{S}}(\phi(X', t'), \Theta(X', t'), X, t).$$

This is a way of trying to say that $\hat{\boldsymbol{S}}$ is a function of the functions ϕ and Θ as a whole. For instance, $\hat{\boldsymbol{S}}$ can depend on spatial or temporal derivatives of ϕ and Θ.

So far the definition of constitutive equation is general enough to include rate effects, that is, dependence on velocities, and memory effects, that is, dependence on past histories. Such generality is important for the study of fluids, viscoelastic, or plastic materials. Our emphasis in this book is on the case of pure elasticity. To make this specialization, we eliminate rate and memory effects as follows.

1.3 Definition By a *thermoelastic constitutive equation* for $\boldsymbol{S}$ we mean a mapping

$$\hat{\boldsymbol{S}}: \mathcal{C} \times \mathfrak{J} \longrightarrow S_2(\mathcal{B}),$$

where $\mathcal{C} = \{\phi: \mathcal{B} \longrightarrow \mathcal{S} \mid \phi \text{ is a regular } C^r \text{ configuration}\}$, the configuration space, and $\mathfrak{J} = \{\Theta: \mathcal{B} \longrightarrow (0, \infty) \mid \Theta \text{ is a } C^r \text{ function}\}$. We assume that $\hat{\boldsymbol{S}}$ is differentiable (in the Fréchet sense; see Box 1.1 below).

The second Piola–Kirchhoff stress tensor *associated* with $\hat{\boldsymbol{S}}$, a configuration ϕ, and a temperature field Θ is given by

$$\boldsymbol{S}(X) = \hat{\boldsymbol{S}}(\phi, \Theta)(X).$$

If ϕ and Θ depend on time, we write

$$S(X, t) = \hat{S}(\phi_t, \Theta_t)(X),$$

where, as usual, $\phi_t(X) = \phi(X, t)$ and $\Theta_t(X) = \Theta(X, t)$.

Thermoelastic constitutive equations for Q, Ψ, and N are defined similarly.

In the next section we shall impose requirements on our constitutive functions $\hat{S}$, $\hat{Q}$, $\hat{N}$, and $\hat{\Psi}$ that will drastically simplify how they can depend on ϕ and Θ and that will establish key relationships between them.

Constitutive theory is given most conveniently in the material picture because the domain $\mathcal{B}$ of the functions remains fixed. However, it can all be done spatially as well. To do so properly, one needs considerations of covariance and dependence on the metric tensor g of $\mathcal{S}$. We shall take up this issue in Section 3.3.

Box. 1.1 ***The Fréchet Derivative***

In Section 1.1 we reviewed a bit of differential calculus in $\mathbb{R}^n$; the main point was familiarization with the idea that the derivative of a map $f: \mathcal{U} \subset \mathbb{R}^n \to \mathbb{R}^m$ at a point $x \in \mathcal{U}$ is a linear map $Df(x): \mathbb{R}^n \to \mathbb{R}^m$. This box will generalize those ideas to Banach spaces. It will be in the nature of a review with a number of quite easy proofs omitted. (Consult standard texts such as Dieundonné [1960] or Lang [1971] for proofs.) However, in Section 4.1 we shall present a complete proof of the inverse mapping theorem for mappings between Banach spaces.

If $\mathfrak{X}$ and $\mathcal{Y}$ are Banach spaces, a linear mapping $A: \mathfrak{X} \to \mathcal{Y}$ is called *bounded* if there is a constant $M > 0$ such that $\|Ax\|_{\mathcal{Y}} \leq M\|x\|_{\mathfrak{X}}$ for all $x \in \mathfrak{X}$. Here $\|\cdot\|_{\mathcal{Y}}$ denotes the norm on $\mathcal{Y}$ and $\|\cdot\|_{\mathfrak{X}}$ that on $\mathfrak{X}$. (The subscripts are dropped if there is no danger of confusion.) A linear operator is bounded if and only if it is continuous. (The proof is not hard: if A is continuous at 0, then for any $\epsilon > 0$ there is a $\delta > 0$ such that $\|Ax\|_{\mathcal{Y}} < \epsilon$ if $\|x\|_{\mathfrak{X}} < \delta$. Let $M = \epsilon/\delta$.) Let $\mathcal{B}(\mathfrak{X}, \mathcal{Y})$ denote the set of all bounded linear operators of $\mathfrak{X}$ to $\mathcal{Y}$. Define

$$\|A\| \equiv \|A\|_{\mathfrak{X},\mathcal{Y}} = \inf_{x \neq 0} \frac{\|Ax\|}{\|x\|}.$$

A straightforward check shows that this makes $\mathcal{B}(\mathfrak{X}, \mathcal{Y})$ into a Banach space; we call the topology associated with the preceding norm the *norm topology*. If $\mathfrak{X}$ and $\mathcal{Y}$ are finite dimensional, $\mathcal{B}(\mathfrak{X}, \mathcal{Y})$ coincides with $\mathcal{L}(\mathfrak{X}, \mathcal{Y})$, the space of all linear mappings of $\mathfrak{X}$ to $\mathcal{Y}$.

> ***Problem 1.1*** For a linear map $A: \mathbb{R}^n \to \mathbb{R}^m$, show that $\|A\|$ is the largest eigenvalue of the symmetric operator $A^T A$.

1.4 Definition Let $\mathfrak{X}$ and $\mathcal{Y}$ be Banach spaces, $\mathcal{U} \subset \mathfrak{X}$ be open, and $f: \mathcal{U} \subset \mathfrak{X} \to \mathcal{Y}$. We say f is *differentiable* at $x_0 \in \mathcal{U}$ if there is a bounded linear operator $Df(x_0) \in \mathcal{B}(\mathfrak{X}, \mathcal{Y})$ such that for each $\epsilon > 0$ there is a $\delta > 0$ such that $\|h\|_{\mathfrak{X}} < \delta$ implies

$$\|f(x_0 + h) - f(x_0) - Df(x_0)\cdot h\|_{\mathcal{Y}} \leq \epsilon \|h\|_{\mathfrak{X}}.$$

(This uniquely determines $Df(x_0)$.)

We say that f is C^1 if it is differentiable at each point of $\mathcal{U}$ and if $x \mapsto Df(x)$ is continuous from $\mathcal{U}$ to $\mathcal{B}(\mathfrak{X}, \mathcal{Y})$, the latter with the *norm* topology. (Recall that in Euclidean spaces, $Df(x_0)$ is the linear map whose matrix in the standard bases is the matrix of partial derivatives of f.) Also note that if $A \in \mathcal{B}(\mathfrak{X}, \mathcal{Y})$ then $DA(x) = A$.

The concept "f is of class C^r" ($0 \leq r \leq \infty$) is defined inductively. For example, f is C^2 if it is C^1 and $x \mapsto D^2f(x) \in \mathcal{B}(\mathfrak{X}, \mathcal{B}(\mathfrak{X}, \mathcal{Y}))$, the derivative of $x \mapsto Df(x)$, is norm continuous. The space $\mathcal{B}(\mathfrak{X}, \mathcal{B}(\mathfrak{X}, \mathcal{Y}))$ is isomorphic to $\mathcal{B}^2(\mathfrak{X}, \mathcal{Y})$, the space of all continuous bilinear maps $b: \mathfrak{X} \times \mathfrak{X} \to \mathcal{Y}$. An isomorphism between these spaces is $b \mapsto \tilde{b}$, $\tilde{b}(x_1)\cdot x_2 = b(x_1, x_2)$. Thus $D^2f(x)$ is usually regarded as a bilinear map of $\mathfrak{X} \times \mathfrak{X}$ to $\mathcal{Y}$. Its value at $(u, v) \in \mathfrak{X} \times \mathfrak{X}$ will be denoted $D^2f(x)\cdot(u, v)$.

1.5 Proposition *If f is C^2, then $D^2f(x)$ is symmetric; that is,*

$$D^2f(x)\cdot(u, v) = D^2f(x)\cdot(v, u).$$

For $f: \mathbb{R}^n \to \mathbb{R}$, the matrix of $D^2f(x)$ is the matrix of second derivatives of f, so 1.5 generalizes the usual notion of symmetry of the second partial derivatives.

If f is a function of two (or more) variables, say $f: \mathcal{U} \subset \mathfrak{X}_1 \times \mathfrak{X}_2 \to \mathcal{Y}$, then the partial derivatives are denoted by $D_1 f$ and $D_2 f$ (or sometimes $D_{x_1} f$, etc.). If we identify $\mathfrak{X}_1 \times \mathfrak{X}_2$ with $\mathfrak{X}_1 \oplus \mathfrak{X}_2$, $(u, 0)$ with u, and $(0, v)$ with v, then we can write Df as the sum of its partial derivatives: $Df(x) = D_1 f(x) + D_2 f(x)$.

1.6 Definition Suppose $f: \mathcal{U} \subset \mathcal{Y} \to \mathcal{Y}$ is differentiable. Define the *tangent* of f to be the map

$$Tf: \mathcal{U} \times \mathfrak{X} \to \mathcal{Y} \times \mathcal{Y} \qquad \text{given by} \qquad Tf(x, u) = (f(x), Df(x)\cdot u),$$

where $Df(x)\cdot u$ means $Df(x)$ applied to $u \in \mathfrak{X}$ as a linear map.

From the geometric point of view developed in Chapter 1, T is more natural than D. If we think of (x, u) as a vector with base point x, then $(f(x), Df(x)\cdot u)$ is the image vector with its base point. See Figure 3.1.1. Another reason for favoring T is its behavior under composition, as given in the next theorem.

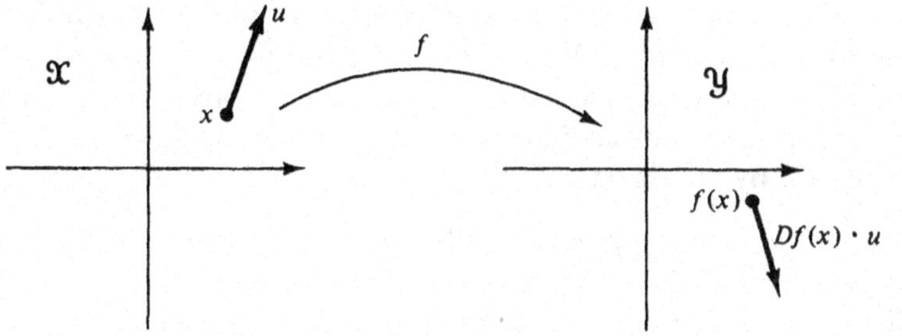

Figure 3.1.1

1.7 Proposition (*Chain Rule*) *Suppose* $f\colon \mathfrak{U} \subset \mathfrak{X} \to \mathfrak{V} \subset \mathfrak{Y}$ and g: $\mathfrak{V} \subset \mathfrak{Y} \to \mathfrak{Z}$ *are* C^1 *maps. Then the composite* $g \circ f\colon \mathfrak{U} \subset \mathfrak{X} \to \mathfrak{Z}$ *is also* C^1 *and*

$$\boxed{T(g \circ f) = Tg \circ Tf.}$$

In terms of D, this formula is equivalent to the usual chain rule

$$\boxed{D(g \circ f)(x) \cdot u = Dg(f(x)) \cdot (Df(x) \cdot u).}$$

For a proof, see Dieudonné [1960], p. 145, or Marsden [1974a], p. 168. For the validity of this chain rule, f and g need only be differentiable.

We will now show how the derivative Df is related to the usual directional derivative. A C^1 curve in $\mathfrak{X}$ is a C^1 map from $\mathfrak{I}$ into $\mathfrak{X}$, $c\colon \mathfrak{I} \to \mathfrak{X}$, where $\mathfrak{I}$ is an open interval of $\mathbb{R}$, Thus, for $t \in \mathfrak{I}$ we have $Dc(t) \in \mathfrak{B}(\mathbb{R}, \mathfrak{X})$, by definition. We identify $\mathfrak{B}(\mathbb{R}, \mathfrak{X})$ with $\mathfrak{X}$ by associating $Dc(t)$ with $Dc(t) \cdot 1$ (1 is the real number "one"). Let

$$\frac{dc}{dt}(t) = Dc(t) \cdot 1.$$

For $f\colon \mathfrak{U} \subset \mathfrak{X} \to \mathfrak{Y}$ of class C^1 we consider $f \circ c$, where $c\colon \mathfrak{I} \to \mathfrak{U}$. It follows from the chain rule that we have the very useful formula

$$\boxed{Df(x) \cdot u = \frac{d}{dt}\{f(x + tu)\}\Big|_{t=0}.}$$

1.8 Definition We call $Df(x) \cdot u$ the *directional derivative* of f in the direction of u. A map for which all the directional derivatives (defined by the preceding formula) exist at x is called *Gâteaux differentiable* at x.

On Euclidean space, d/dt defined this way coincides with the usual directional derivative. More specifically, suppose we have $f: \mathcal{U} \subset \mathbb{R}^m \longrightarrow \mathbb{R}^n$ of class C^1. Now $Df(x)$ is a linear map from $\mathbb{R}^m \longrightarrow \mathbb{R}^n$ and so it is represented by its components relative to the standard basis $e_1, \ldots, e_m$ of $\mathbb{R}^m$. By the above formula we see that

$$Df(u)\cdot e_i = \left(\frac{\partial f^1}{\partial x^i}(u), \ldots, \frac{\partial f^n}{\partial x^i}(u)\right)\cdot$$

Thus $Df(u)$ is represented by the usual Jacobian matrix.

If we fix $x, y \in \mathfrak{X}$ and apply the fundamental theorem of calculus to the map $t \mapsto f(tx + (1-t)y)$, assume f is C^1, and

$$\| Df(tx + (1-t)y) \| \leq M,$$

we obtain the *mean value inequality*:

1.9 Proposition *If $f: \mathcal{U} \subset \mathfrak{X} \longrightarrow \mathcal{Y}$ is C^1, $x, y \in \mathcal{U}$, the line joining x to y lies in $\mathcal{U}$ and $\| Df(z) \|_{\mathfrak{X},\mathcal{Y}} \leq M$ on this line, then*

$$\| f(x) - f(y) \|_{\mathcal{Y}} \leq M \| x - y \|_{\mathfrak{X}}.$$

The tangent is very convenient for dealing with higher derivatives. For example, the rth-order chain rule is obtained inductively to be $T^r(g \circ f) = T^r g \circ T^r f$, while a corresponding statement in terms of D is a good deal more complicated. Higher derivatives also occur in Taylor's theorem:

1.10 Proposition (*Taylor's Theorem*) *Suppose $f: \mathcal{U} \subset \mathfrak{X} \longrightarrow \mathcal{Y}$ is C^{r+1}. Then (for h sufficiently small)*

$$\boxed{\begin{aligned} f(x+h) &= f(x) + Df(x)\cdot h + \tfrac{1}{2}D^2 f(x)(h, h) + \cdots \\ &\quad + \frac{1}{r!}D^r f(x)\cdot(h, \ldots, h) + R_r(x, h), \end{aligned}}$$

where

$$R_r(x, h) = \frac{1}{r!}\int_0^1 (1-t)^r D^{r+1} f(x + th)\cdot(h, \ldots, h)\, dt$$

satisfies $\| R_r(x, h) \|_{\mathcal{Y}} \leq C \| h \|_{\mathfrak{X}}^{r+1}$ for $\| h \|_{\mathfrak{X}}$ small enough.

The proof proceeds in the same manner as in elementary calculus.

1.11 Proposition *Suppose $f: \mathfrak{X} \times \mathcal{Y} \longrightarrow \mathcal{Z}$ is a continuous bilinear mapping (i.e., $\| f(x, y) \|_{\mathcal{Z}} \leq M \| x \|_{\mathfrak{X}} \| y \|_{\mathcal{Y}}$ for some constant $M > 0$). Then f is C^∞ and*

$$Df(x, y)(u, v) = f(u, y) + f(x, v).$$

The verification of this *Leibniz rule* is straightforward.

Constitutive functions define mappings between Banach spaces of functions. To be able to use differential calculus on them we need to know how to differentiate them. The rest of this box will develop such skills. We begin with a simple example.

1.12 Example Let $\mathfrak{X}$ be the Banach space of continuous functions $\varphi: [0, 1] \longrightarrow \mathbb{R}$ with $\|\varphi\| = \sup_{x \in [0,1]} |\varphi(x)|$. Define

$$f: \mathfrak{X} \longrightarrow \mathfrak{X} \quad \text{by} \quad f(\varphi)(x) = [\varphi(x)]^2,$$

that is, $f(\varphi) = \varphi^2$. Then f is C^∞ and $Df(\varphi)\cdot\psi = 2\varphi\psi$.

To see this, note that the map $f_1: \varphi \mapsto (\varphi, \varphi)$ from $\mathfrak{X}$ to $\mathfrak{X} \times \mathfrak{X}$ is continuous linear and $f_2: \mathfrak{X} \times \mathfrak{X} \longrightarrow \mathfrak{X}$; $(\varphi, \psi) \mapsto \varphi \cdot \psi$ is continuous bilinear. Now apply 1.11 and the chain rule to $f = f_2 \circ f_1$.

Problem 1.2 Let $\mathfrak{X}$ be the set of C^1 functions on [0, 1] and $\mathcal{Y}$ be the set of C^0 functions. Define $f: \mathfrak{X} \longrightarrow \mathcal{Y}$ by $f(\varphi) = (\varphi')^3$. Prove that f is C^∞ and find Df and D^2f.

What about examples like $f(\varphi) = e^{\varphi'}$? The next theorem shows us how to differentiate such maps. Results of this type go back at least to Sobolev in the 1930s. (See also Abraham and Smale [1960].) For simplicity, we work in C^k spaces, but the same thing works in a variety of function spaces (such as Sobolev spaces $W^{s,p}$ that will be developed in Chapter 6). We stick to a special case that is relevant for elasticity.

1.13 Theorem (ω-*lemma*) *Let $\bar{\Omega} \subset \mathbb{R}^n$ be a region with smooth boundary; let $\mathfrak{X}$ be the Banach space of C^k maps*[1] *$u: \bar{\Omega} \longrightarrow \mathbb{R}^m$; let $\mathcal{Y}$ be the Banach space of C^{k-1} maps $g: \bar{\Omega} \longrightarrow \mathbb{R}^p$ $(1 \leq k < \infty)$. Let*

$$W: \bar{\Omega} \times \mathcal{L}(\mathbb{R}^n, \mathbb{R}^m) \longrightarrow \mathbb{R}^p$$

be C^r $(r \geq k - 1 - l)$, and define

$$f: \mathfrak{X} \longrightarrow \mathcal{Y}, \qquad f(u)(x) = W(x, Du(x)).$$

Then f is of class C^l and

$$(Df(u)\cdot v)(x) = D_2W(x, Du(x))\cdot[Dv(x)];$$

that is, $$(Df(u)\cdot v)^J = \frac{\partial W^J}{\partial(\partial u^i/\partial x^I)} \cdot \frac{\partial v^i}{\partial x^I}.$$

---

[1]A map $u: \bar{\Omega} = \Omega \cup \partial\Omega \longrightarrow \mathbb{R}^m$ is called C^k when it has a C^k extension to an open set containing $\bar{\Omega}$. The norm of u is defined by

$$\|u\| = \sup_{\substack{x \in \bar{\Omega} \\ 0 \leq j \leq k}} \|D^j u(x)\|$$

and this norm can be shown to make the C^k maps into a Banach space. (See Abraham and Robbin [1967], for example.)

Proof Induction reduces the argument to the case $r = 1$, $k = 1$. The following computation and the finite-demensional chain rule shows that f is Gâteaux differentiable with derivative as stated in the theorem:

$$[Df(u)\cdot v](x) = \frac{d}{d\epsilon} f(u + \epsilon v)(x)\Big|_{\epsilon=0} = \frac{d}{d\epsilon} W(x, Du(x) + \epsilon Dv))\Big|_{\epsilon=0}.$$

A straightforward uniform continuity argument shows that $u \mapsto Df(u) \in \mathfrak{B}(\mathfrak{X}, \mathfrak{Y})$ is norm continuous. The proof is now completed using the following:

1.14 Lemma *Let $f: \mathfrak{U} \subset \mathfrak{X} \to \mathfrak{Y}$ be Gâteaux differentiable and assume $u \mapsto Df(u) \in \mathfrak{B}(\mathfrak{X}, \mathfrak{Y})$ is continuous. Then f is C^1.*

Proof By the fundamental theorem of calculus,

$$[f(u_0 + h) - f(u_0)] - Df(u_0)\cdot h = \left[\int_0^1 \frac{d}{d\lambda} f(u_0 + \lambda h)\, d\lambda\right] - Df(u_0)\cdot h$$

$$= \int_0^1 [Df(u_0 + \lambda h)\cdot h - Df(u_0)\cdot h]\, d\lambda.$$

By continuity of Df, for any $\epsilon > 0$ there is a $\delta > 0$ such that $\| Df(u) - Df(u_0)\| < \epsilon$ if $\| u - u_0 \| < \delta$. Then $\| h \| < \delta$ implies:

$$\| f(u_0 + h) - f(u_0) - Df(u_0)\cdot h \|$$

$$\leq \int_0^1 \| Df(u_0 + \lambda h) - Df(x_0) \| \cdot \| h \|\, d\lambda \leq \epsilon \| h \|. \quad \blacksquare$$

Since f is continuous and linear in W, this shows that in fact f is C^1 as a function of the pair (u, W).

Problem 1.3 Let f be defined by composition with W as above. Show that if we *assume* that f maps C^k to C^{k-1} and is C^l, then W is C^r, $r = k - 1 + l$ as well.

Box. 1.2 ***Summary of Important Formulas for Section 3.1***

Thermoelastic Constitutive Equation for $\boldsymbol{S}$

$\hat{\boldsymbol{S}}: \mathfrak{C} \times \mathfrak{I} \to S_2(\mathfrak{B})$; $\qquad S^{AB}(X, t) = \hat{S}^{AB}(\phi^a(X', t), \Theta(X', t), X)$

$\boldsymbol{S}(X, t) = \hat{\boldsymbol{S}}(\phi_t, \Theta_t)(X)$

for $\boldsymbol{Q}$:

$\hat{\boldsymbol{Q}}: \mathfrak{C} \times \mathfrak{I} \to \mathfrak{X}(\mathfrak{B})$; $\qquad Q^A(X, t) = \hat{Q}^A(\phi^a(X', t), \Theta(X', t), X)$

$\boldsymbol{Q}(X, t) = \hat{\boldsymbol{Q}}(\phi_t, \Theta_t)(X)$

for Ψ:

$\hat{\Psi}: \mathcal{C} \times \mathcal{J} \longrightarrow \mathcal{F}(\mathcal{B})$; $\quad \Psi(X, t) = \hat{\Psi}(\phi^a(X', t), \Theta(X', t), X)$

$\Psi(X, t) = \hat{\Psi}(\phi_t, \Theta_t)(X)$

for N:

$\hat{N}: \mathcal{C} \times \mathcal{J} \longrightarrow \mathcal{F}(\mathcal{B})$; $\quad N(X, t) = \hat{N}(\phi^a(X', t), \Theta(X', t), X)$

$N(X, t) = \hat{N}(\phi_t, \Theta_t t)(X)$

Derivative of a Function-Space Mapping Defined by Composition

$f(u)(x) = W(x, Du(x))$

$(Df(u)\cdot v)(x)$ $\quad Df(u)\cdot v = \dfrac{\partial W}{\partial(\partial u^i/\partial x^j)}\cdot\dfrac{\partial v^i}{\partial x^j}$

$= D_2W(x, Du(x))\cdot Dv(x)$

3.2 CONSEQUENCES OF THERMODYNAMICS, LOCALITY, AND MATERIAL FRAME INDIFFERENCE

We now simplify the constitutive functions $\hat{S}, \hat{Q}, \hat{N}, \hat{\Psi}$ as follows. From an axiom of locality and the entropy production inequality, we shall deduce that $\hat{S}$ and $\hat{N}$ can be expressed in terms of $\hat{\Psi}$ and that $\hat{\Psi}$ depends only on points in $\mathcal{B}$, point values of $\boldsymbol{F}$ and Θ and not on ϕ or higher derivatives of $\boldsymbol{F}$ or Θ. Then, we postulate an axiom of material frame indifference to deduce that the dependence of $\hat{\Psi}$ on $\boldsymbol{F}$ is only through the Cauchy–Green tensor $\boldsymbol{C}$.

2.1 Definition A constitutive function for thermoelasticity

$$\hat{\Psi}: \mathcal{C} \times \mathcal{J} \longrightarrow \mathcal{F}(\mathcal{B})$$

is called *local* if for any open set $\mathcal{U} \subset \mathcal{B}$ and $\phi_1, \phi_2 \in \mathcal{C}$, which agree on $\mathcal{U}$, and $\Theta_1, \Theta_2 \in \mathcal{J}$, which agree on $\mathcal{U}$, then $\hat{\Psi}(\phi_1, \Theta_1)$ and $\hat{\Psi}(\phi_2, \Theta_2)$ agree on $\mathcal{U}$.

The idea of using locality as a basic postulate is due to Noll [1958]. It must be emphasized, however, that one sometimes may wish to impose *nonlocal* constraints, such as incompressibility (see Chapter 5). If the value of $\hat{\Psi}$ at X depends only on the values of ϕ and Θ and their derivatives up to order, say, k, at X, then $\hat{\Psi}$ is local. This is because knowledge of a mapping on an open set entails a knowledge of all its derivatives on that set. A constitutive function $\hat{\Psi}$ defined by a composition of this form is called a (nonlinear) *differential operator*. A simple example of a nonlocal operator is given by letting $\mathfrak{X} = C^0$ functions on [0, 1] and defining $f: \mathfrak{X} \longrightarrow \mathfrak{X}$ by $f(\phi)(x) = \int_0^x \phi(s)\, ds$.

While it is trivial that a differential operator is local, the converse is not so elementary. For linear operators, this is true and is due to Peetre [1959]. For nonlinear operators (satisfying some technical conditions), this is also true but is a deeper fact; see Epstein and Thurston [1979] (and for earlier versions, Dombrowski [1966] and Palais and Terng [1977]). Fortunately, we can bypass this theory in the present context, but it may be useful for higher-order materials.

We can summarize the situation as follows:

2.2 Axiom of Locality *Constitutive functions for thermoelasticity are assumed to be local.*

We next investigate the consequences of assuming the entropy production inequality. We wish to assume that it holds for all regular motions of the body. The momentum balance and energy balance are not taken into account, because any motion is consistent with them for a suitable choice of body force $\boldsymbol{B}$ and heat source R; that is, balance of momentum and energy *define* what $\boldsymbol{B}$ and R have to be. This is not unreasonable since we are supposed to be able to allow any choice of $\boldsymbol{B}$ and R.

2.3 Axiom of Entropy Production *For any (regular) motion of $\mathcal{B}$, the constitutive functions for thermoelasticity are assumed to satisfy the entropy production inequality:*

$$\rho_{\text{Ref}}\left(\hat{N}\frac{\partial \Theta}{\partial t} + \frac{\partial \hat{\Psi}}{\partial t}\right) - \hat{\boldsymbol{P}}: \frac{\partial \boldsymbol{F}}{\partial t} + \frac{1}{\Theta}\langle \hat{\boldsymbol{Q}}, \text{GRAD}\, \Theta \rangle \leq 0.$$

2.4 Theorem (*Coleman and Noll* [*1963*]) *Suppose the axioms of locality and entropy production hold. Then $\hat{\Psi}$ depends only on the variables X, $\boldsymbol{F}$, and Θ. Moreover, we have*

$$\boxed{\hat{N} = -\frac{\partial \hat{\Psi}}{\partial \Theta} \quad \text{and} \quad \hat{\boldsymbol{P}} = \rho_{\text{Ref}}\, \boldsymbol{g}^{\flat} \frac{\partial \hat{\Psi}}{\partial \boldsymbol{F}}, \qquad \text{that is,} \quad \hat{P}_a{}^A = \rho_{\text{Ref}} \frac{\partial \hat{\Psi}}{\partial F^a{}_A},}$$

and the entropy production inequality reduces to

$$\langle \boldsymbol{Q}, \text{GRAD}\, \Theta \rangle \leq 0.$$

Remarks (1) One can allow the constitutive functions to depend on derivatives of higher order than the first derivative $\boldsymbol{F}$ and still be consistent with the entropy production inequality provided one postulates the existence of higher-order stresses. This is the "multipolar" or "higher-order" theory. (See Green and Rivlin [1964b].)

(2) The precise meaning of "$\hat{\Psi}$ depends only on the variables X, $\boldsymbol{F}$, and Θ" is that for any configuration $\phi: \mathcal{B} \to \mathcal{S}$, there is a mapping (also denoted $\hat{\Psi}$) $\hat{\Psi}: \mathcal{B} \times \mathcal{L}_\phi(T\mathcal{B}, T\mathcal{S}) \times \mathbb{R} \to \mathbb{R}$ such that $\hat{\Psi}(\phi, \Theta)(X) = \hat{\Psi}(X, \boldsymbol{F}(X), \Theta(X))$, where $\mathcal{L}_\phi(T\mathcal{B}, T\mathcal{S})$ denotes the bundle over $\mathcal{B}$ of (1, 1) two-point tensors; that is, the fiber of $\mathcal{L}_\phi(T\mathcal{B}, T\mathcal{S})$ at $X \in \mathcal{B}$ consists of all linear maps $\boldsymbol{F}: T_X\mathcal{B} \to T_{\phi(X)}\mathcal{S}$.

(3) If $\mathcal{S}$ is a linear space and if one is willing to identify all the tangent spaces T_x with $\mathcal{S}$, then one can identify $\mathcal{L}_\phi(T\mathcal{B}, T\mathcal{S})$ with $\mathcal{L}(T\mathcal{B}, \mathcal{S})$, the bundle of linear maps of $T_X\mathcal{B}$ to $\mathcal{S}$. In this way, the apparent dependence on ϕ itself disappears.

(4) If $\hat{\Psi}$ is independent of X, the body is called *homogeneous*. This is a special case of a material symmetry, a topic treated in Section 3.5.

Proof of 2.4 (See Gurtin [1972b].) Given $\hat{\Psi}: \mathcal{C} \times \mathfrak{I} \to \mathfrak{F}$, we consider a motion $\phi_t \in \mathcal{C}$, a temperature field $\Theta_t \in \mathfrak{I}$, and form the composition $f(t) = \hat{\Psi}(\varphi_t, \Theta_t) \in \mathfrak{F}$. By the chain rule,

$$\frac{df}{dt} = D_\phi \hat{\Psi} \cdot V + D_\Theta \hat{\Psi} \cdot \dot{\Theta},$$

where $D_\phi \hat{\Psi}$ is the partial derivative of $\hat{\Psi}$ in the Frechet sense (see Box 1.1 of this Chapter). Note that for an arbitrary function $k(X, t)$, $\dot{k} \equiv \partial k/\partial t$.

Substituting the preceding equation into the entropy production inequality gives

$$\rho_{\text{Ref}} \hat{N} \dot{\Theta} + \rho_{\text{Ref}}(D_\phi \hat{\Psi} \cdot V + D_\Theta \hat{\Psi} \cdot \dot{\Theta}) - \hat{P}: \frac{\partial F}{\partial t} + \frac{1}{\Theta} \langle Q, \text{GRAD}\, \Theta \rangle \leq 0.$$

By assumption, this holds for all processes (ϕ_t, Θ_t). First of all, choose ϕ_t independent of time t; then we must have

$$\rho_{\text{Ref}}(\hat{N} \dot{\Theta} + D_\Theta \hat{\Psi} \cdot \dot{\Theta}) + \frac{1}{\Theta} \langle Q, \text{GRAD}\, \Theta \rangle \leq 0.$$

Suppose $\rho_{\text{Ref}}(\hat{N} \dot{\Theta} + D_\Theta \hat{\Psi} \cdot \dot{\Theta})$ did not vanish for some ϕ and all Θ_t. Then we can alter Θ_t to a new one $\tilde{\Theta}_t$ so that $\tilde{\Theta}_{t_0} = \Theta_{t_0}$ and $\dot{\tilde{\Theta}}_{t_0} = \alpha \dot{\Theta}_{t_0}$, where α is any prescribed constant. We can then choose the constant α to violate the assumed inequality. Therefore, we deduce our first identity

$$\rho_{\text{Ref}}(\hat{N} \dot{\Theta} + D_\Theta \hat{\Psi} \cdot \dot{\Theta}) = 0. \tag{1}$$

Similarly, fixing Θ and altering ϕ_t we get a second identity

$$\rho_{\text{Ref}} D_\phi \hat{\Psi} \cdot V = \hat{P}: \dot{F}. \tag{2}$$

Consider now this second identity. Fix $X_0 \in \mathcal{B}$ and fix Θ. Let ϕ_0 and ϕ_1 be two configurations with $\phi_0(X_0) = \phi_1(X_0)$ and $F_0(X_0) = F_1(X_0)$. In a chart on $\mathcal{S}$, let $\phi(X, t) = \phi_0(X) + t(\phi_1(X) - \phi_0(X))$ define a motion ϕ for X near X_0 and small t. Extend ϕ outside this neighborhood in a regular but otherwise arbitrary way. (This can be done by first extending the governing velocity field, for example.) By the axiom of locality, the way in which this extension is done does not affect $\hat{\Psi}$ in the neighborhood. This motion has a velocity field V_t and a deformation gradient F_t that satisfies

$$V_t(X_0) = \phi_1(X_0) - \phi_0(X_0) = \mathbf{0} \quad \text{and} \quad \dot{F}_t(X_0) = \mathbf{0}.$$

[Note that we *cannot* conclude that $(D_\phi \hat{\Psi} \cdot V_t)(X_0) = 0$ from $V_t(X_0) = \mathbf{0}$ alone, since $D_\phi \hat{\Psi} \cdot V_t$ is a linear operator on V_t and could depend on higher derivatives of V_t, for example.] Thus, from the second identity (2) above,

$$(\rho_{\text{Ref}} D_\phi \hat{\Psi} \cdot V_t)(X_0) = 0.$$

Therefore,
$$\frac{d}{dt}[\rho_{\text{Ref}}(X_0) \hat{\Psi}(\phi_t, \Theta)](X_0) = 0$$

and so
$$\rho_{\text{Ref}}(X_0)[\hat{\Psi}(\phi_1, \Theta)](X_0) = \rho_{\text{Ref}}(X_0)[\hat{\Psi}(\phi_0, \Theta)](X_0).$$

Thus, $\hat{\Psi}$ depends on ϕ only through $\boldsymbol{F}$. In a similar way, the first identity implies that $\hat{\Psi}$ depends only on the point values of Θ. Substituting this information back into the two identities (and using the results of Box 1.1) yields

$$\rho_{\text{Ref}}\left(\hat{N}\dot{\Theta} + \frac{\partial\hat{\Psi}}{\partial\Theta}\dot{\Theta}\right) = 0 \quad \text{and} \quad \rho_{\text{Ref}}\frac{\partial\hat{\Psi}}{\partial\boldsymbol{F}} : \dot{\boldsymbol{F}} = \boldsymbol{P} : \dot{\boldsymbol{F}}.$$

Arbitrariness of $\dot{\Theta}$ and $\dot{\boldsymbol{F}}$ then yields the stated identities. ∎

The relationships

$$P_a{}^A = \rho_{\text{Ref}}\frac{\partial\hat{\Psi}}{\partial F^a{}_A} \quad \text{and} \quad N = -\frac{\partial\hat{\Psi}}{\partial\Theta}$$

can be used to simplify the first law (energy balance). Indeed, balance of energy reads

$$\rho_{\text{Ref}}\frac{\partial\hat{E}}{\partial t} + \text{DIV}\,\boldsymbol{Q} = \boldsymbol{P} : \boldsymbol{F} + \rho_{\text{Ref}}R.$$

Write $\hat{E} = \hat{\Psi} + \hat{N}\Theta$ so that

$$\frac{\partial E}{\partial t} = \frac{\partial\hat{\Psi}}{\partial\boldsymbol{F}} : \dot{\boldsymbol{F}} + \frac{\partial\hat{\Psi}}{\partial\Theta}\dot{\Theta} + \frac{\partial\hat{N}}{\partial t}\Theta + \hat{N}\dot{\Theta} = \frac{\partial\hat{\Psi}}{\partial\boldsymbol{F}} : \dot{\boldsymbol{F}} + \frac{\partial\hat{N}}{\partial t}\Theta$$

using $\partial\hat{\Psi}/\partial\Theta = -N$. Substituting this expression into balance of energy and using $\rho_{\text{Ref}}\,(\partial\hat{\Psi}/\partial\boldsymbol{F}) = \boldsymbol{P}$ yields $\rho_{\text{Ref}}\Theta(\partial N/\partial t) + \text{DIV}\,\boldsymbol{Q} = \rho_{\text{Ref}}R$. In summary, we have proved:

2.5 Proposition *Assuming the identities in 2.4, balance of energy reduces to*

$$\boxed{\rho_{\text{Ref}}\Theta\frac{\partial N}{\partial t} + \text{DIV}\,\boldsymbol{Q} = \rho_{\text{Ref}}R.}$$

For pure elasticity (ignoring all thermal effects), the above constitutive conclusions may be obtained from balance of energy alone. Indeed, if $\boldsymbol{Q}$ and R are omitted from balance of energy, then we must have $\rho_{\text{Ref}}(\partial E/\partial t) = \boldsymbol{P} : \dot{\boldsymbol{F}}$ for all processes. Thus $\partial E/\partial\boldsymbol{F}$ and $\boldsymbol{P}$ must balance. We list the axioms special to this situation as follows:

Axiom 0. $\hat{E} : \mathcal{C} \longrightarrow \mathfrak{F}$ is a given differentiable map.
Axiom 1. $\hat{E}$ is local.
Axiom 2. There is a map $\hat{\boldsymbol{P}} : \mathcal{C} \longrightarrow$ two-point (0, 2) tensors such that for all motions $\phi_t \in \mathcal{C}$, balance of energy holds:

$$\rho_{\text{Ref}}\frac{\partial\hat{E}}{\partial t} = \boldsymbol{P} : \dot{\boldsymbol{F}}.$$

2.6 Theorem *Under Axioms 0, 1, and 2, $\hat{E}$ depends only on the point values of X and $\boldsymbol{F}$ and we have the identity*

$$\boxed{\rho_{\text{Ref}}\boldsymbol{g}^\flat\frac{\partial\hat{E}}{\partial\boldsymbol{F}} = \hat{\boldsymbol{P}}, \qquad \textit{that is,} \quad P_a{}^A = \rho_{\text{Ref}}\frac{\partial\hat{E}}{\partial F^a{}_A}.}$$

This is proved by the same techniques that were used to prove 2.4.

We now return to the thermoelastic context. From 2.4 we can draw no conclusion about the dependence of $\boldsymbol{Q}$ on ϕ and Θ; it could conceivably depend on many derivatives. It is often *assumed* that $\hat{\boldsymbol{Q}}$ depends only on the point values of X, $\boldsymbol{C}$, Θ, and $\nabla\Theta$; in this case, we say we have a *grade (1, 1)* material. In any case, one can generally say the following:

2.7 Proposition *$\hat{\boldsymbol{Q}}$ vanishes when its argument* GRAD Θ *vanishes.*

Proof Fixing all other arguments, let $f(\alpha) = \langle \hat{\boldsymbol{Q}}(\alpha \text{ GRAD } \Theta), \text{GRAD } \Theta\rangle$. Thus $\alpha f(\alpha) \leq 0$, so f changes sign at $\alpha = 0$. Since f is continuous, $f(0) = 0$. Thus $\langle \hat{\boldsymbol{Q}}(0), \text{GRAD } \Theta\rangle = 0$, so $\hat{\boldsymbol{Q}}(0) = \boldsymbol{0}$. ▮

Taylor's theorem implies that for GRAD Θ small, $\boldsymbol{Q}$ is well approximated by a matrix that is negative semi-definite times GRAD Θ: fixing all arguments but GRAD Θ,

$$\hat{\boldsymbol{Q}}(\text{GRAD } \Theta) = \boldsymbol{K}\cdot\text{GRAD } \Theta,$$

where

$$\boldsymbol{K} = \int_0^1 \frac{\partial \hat{\boldsymbol{Q}}}{\partial(\text{GRAD } \Theta)}(s \text{ GRAD } \Theta)\, ds$$

by the fundamental theorem of calculus. The inequality $\langle \hat{\boldsymbol{Q}}, \text{GRAD } \Theta\rangle \leq 0$ means $\boldsymbol{K}$ is negative semi-definite—that is, dissipative. If $\boldsymbol{K}$ were assumed constant one would recover the Fourier law (see Example 3.6, Chapter 2).

2.8 Example (Rigid Heat Conductor) The rigid heat conductor (cf. 3.6, Chapter 2) makes the assumption that the motion is fixed, say ϕ = identity, for all time. We also assume that $\boldsymbol{Q}$ depends only on X, Θ, and GRAD Θ. Then the evolution of Θ in time is determined from balance of energy:

$$\rho_{\text{Ref}}\Theta\frac{\partial \hat{N}}{\partial t} + \text{DIV } \hat{\boldsymbol{Q}} = \rho_{\text{Ref}}R.$$

Since $\hat{N}$ depends only on X and Θ, we get

$$\begin{aligned}\left(\rho_{\text{Ref}}\Theta\frac{\partial \hat{N}}{\partial \Theta}\right)\frac{\partial \Theta}{\partial t} &= -(\hat{Q}^A)_{|A} + \rho_{\text{Ref}}R \\ &= -\left(\frac{\partial \hat{Q}^A}{\partial \Theta_{|B}}\right)\Theta_{|B|A} - \frac{\partial \hat{Q}^A}{\partial \Theta}\Theta_{|A} + \rho_{\text{Ref}}R - R_I.\end{aligned} \tag{3}$$

where $R_I = \text{DIV}_X\hat{\boldsymbol{Q}}$, the divergence of $\hat{\boldsymbol{Q}}$ with Θ and GRAD Θ fixed. As was observed above, the matrix $\partial\hat{Q}^A/\partial\Theta_{|B}$ is negative semi-definite. We also *assume* the scalar function $\partial N/\partial\Theta = -\partial^2\Psi/\partial\Theta^2$ (= specific heat at constant volume) is positive, so the equation is formally parabolic. (3) is the general form of a nonlinear heat equation. Notice, finally, that positivity of $\partial^2\hat{\Psi}/\partial\Theta^2$ means that $\hat{\Psi}$ is a convex function of Θ.[2]

[2]The theory of monotone operators is applicable here [if $\rho_{\text{Ref}}\Theta(\partial\hat{N}/\partial\Theta)$ is not a constant, one presumably uses a weighted norm] to yield an existence and uniqueness theorem for $\Theta(X, t)$ for all time, given Θ at time $t = 0$ and given $\hat{\boldsymbol{Q}}$, $\hat{N}$. The extra assumption of positivity of $\partial^2\Psi/\partial\Theta^2$ may be regarded as formally equivalent to well-posedness of the equations (see Crandall and Nohel [1978] for details).

Now we turn to our final constitutive axiom, material frame indifference.

2.9 Axiom of Material Frame Indifference *Let $\hat{\Psi}$ be a thermoelastic constitutive function satisfying the above axioms, so that $\hat{\Psi}$ is a function of X, $\boldsymbol{F}$, and Θ. Assume that if $\xi\colon \mathcal{S} \to \mathcal{S}$ is a regular, orientation-preserving map taking x to x' and $T\xi$ is an isometry from $T_x\mathcal{S}$ to $T_{x'}\mathcal{S}$, then*

$$\hat{\Psi}(X, \boldsymbol{F}, \Theta) = \hat{\Psi}(X, \boldsymbol{F}', \Theta), \qquad \text{that is,} \quad \hat{\Psi}(X, \boldsymbol{F}, \Theta) = \hat{\Psi}(X, \xi_* \boldsymbol{F}, \Theta),$$

where $\boldsymbol{F}\colon T_X\mathcal{B} \to T_x\mathcal{S}$, $\boldsymbol{F}'\colon T_X\mathcal{B} \to T_{x'}\mathcal{S}$, and $\boldsymbol{F}' = T\xi \cdot \boldsymbol{F} = \xi_* \boldsymbol{F}$. See Figure 3.2.1.

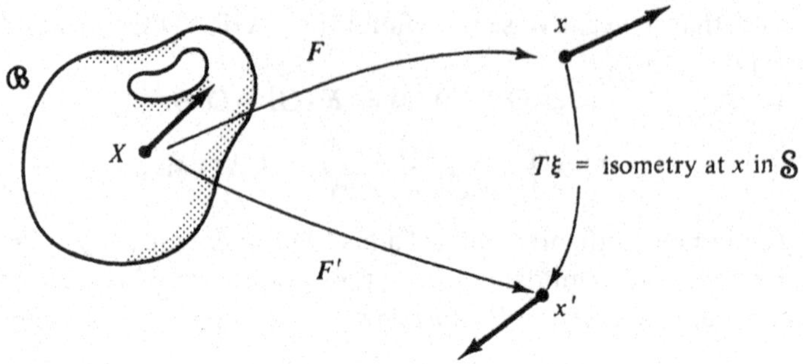

Figure 3.2.1

Stated loosely, this axiom means that our constitutive functions are invariant under rotations of the ambient space $\mathcal{S}$ in which our body moves. One sometimes says that $T\xi$ "rotates observer frames." One can interpret ξ either actively (superposed motion) or passively (change of framing) as explained in Boxes 3.1 and 3.2 of Chapter 1.

2.10 Theorem *Let $\hat{\Psi}$ satisfy the above axioms. Then $\hat{\Psi}$ depends only on X, $\boldsymbol{C}$, and Θ. By abuse of notation we shall write $\hat{\Psi}(X, \boldsymbol{C}, \Theta)$.*

Remarks (1) Equally well, we could say $\hat{\Psi}$ is a function only of X, Θ, and the right stretch tensor $\boldsymbol{U} = \boldsymbol{C}^{1/2}$ (see 3.11 and 3.13, Chapter 1).

(2) By abuse of notation we shall write $\boldsymbol{C}$ for both $\boldsymbol{C}$ and $\boldsymbol{C}^\flat$ when there is little danger of confusion.

(3) One can also state as an axiom the corresponding transformation law for the first Piola–Kirchhoff stress $\hat{\boldsymbol{P}}$; that is, for any ξ as above, $\xi_*\{\hat{\boldsymbol{P}}(X, \boldsymbol{F}, \Theta)\} = \hat{\boldsymbol{P}}(X, \xi_*\boldsymbol{F}, \Theta)$, where $\xi_*\hat{\boldsymbol{P}}$ means push-forward the spatial index of $\hat{\boldsymbol{P}}$. This axiom is a consequence of the axiom for $\hat{\Psi}$. In Euclidean space, material frame indifference for the stress reads:

$$\boxed{R^a{}_b \hat{P}^{bA}(X, F^c{}_C, \Theta) = \hat{P}^{aA}(X, R^c{}_b F^b{}_C, \Theta)}$$

for any proper orthogonal matrix $R^a{}_b$.

(4) Note that in $\mathbb{R}^3$, material frame indifference is the same as objectivity for orthogonal transformations (see Box 6.1, Chapter 1, for a discussion of objectivity).

Proof of 2.10 Suppose $\boldsymbol{F}_1: T_X\mathcal{B} \longrightarrow T_{x_1}\mathcal{S}$ and $\boldsymbol{F}_2: T_X\mathcal{B} \longrightarrow T_{x_2}\mathcal{S}$ and $\boldsymbol{F}_1^T\boldsymbol{F}_1 = \boldsymbol{F}_2^T\boldsymbol{F}_2$; that is, $\boldsymbol{F}_1$ and $\boldsymbol{F}_2$ give rise to the same $\boldsymbol{C}$ tensor. We have to show that $\hat{\Psi}(X, \boldsymbol{F}_1, \Theta) = \hat{\Psi}(X, \boldsymbol{F}_2, \Theta)$. Then $\hat{\Psi}(X, \boldsymbol{C}, \Theta)$ will be well defined if we let it be this common value.

Choose a regular map $\xi: \mathcal{S} \longrightarrow \mathcal{S}$ with $\xi(x_1) = x_2$ and such that $T\xi(x_1)\boldsymbol{F}_1 = \boldsymbol{F}_2$. This is possible since $\boldsymbol{F}_1$ and $\boldsymbol{F}_2$ are assumed to be invertible. The assumption $\boldsymbol{F}_1^T\boldsymbol{F}_1 = \boldsymbol{F}_2^T\boldsymbol{F}_2$ implies that $T\xi(x_1)$ is an isometry. Indeed,

$$\begin{aligned}\langle T\xi(x_1)\cdot \boldsymbol{F}_1\cdot \boldsymbol{V}_1, T\xi(x_1)\cdot \boldsymbol{F}_1\cdot \boldsymbol{V}_2\rangle &= \langle \boldsymbol{F}_2\cdot \boldsymbol{V}_1, \boldsymbol{F}_2\cdot \boldsymbol{V}_2\rangle \\ = \langle \boldsymbol{F}_2^T\boldsymbol{F}_2\boldsymbol{V}_1, \boldsymbol{V}_2\rangle &= \langle \boldsymbol{F}_1^T\boldsymbol{F}_1\boldsymbol{V}_1, \boldsymbol{V}_2\rangle = \langle \boldsymbol{F}_1\boldsymbol{V}_1, \boldsymbol{F}_1\boldsymbol{V}_2\rangle.\end{aligned}$$

Therefore, by our axiom, $\hat{\Psi}(X, \boldsymbol{F}_1, \Theta) = \hat{\Psi}(X, \boldsymbol{F}_2, \Theta)$. ∎

The last part of the proof can also be seen in coordinates as follows:

$$\begin{aligned} C_{AB} &= g_{ab}F^a_{1A}F^b_{1B} && (\boldsymbol{C} = \boldsymbol{F}_1^T\boldsymbol{F}_1) \\ &= g_{ab}F^a_{2A}F^b_{2B} && (\boldsymbol{F}_1^T\boldsymbol{F}_1 = \boldsymbol{F}_2^T\boldsymbol{F}_2) \\ &= g_{ab}\frac{\partial \xi^a}{\partial x^c}F^c_{1A}\frac{\partial \xi^b}{\partial x^d}F^d_{1B} && (T\xi\cdot \boldsymbol{F}_1 = \boldsymbol{F}_2). \end{aligned}$$

Thus, comparing the first and third lines, $g_{ab} = g_{cd}(\partial\xi^c/\partial x^a)(\partial\xi^d/\partial x^b)$ (evaluated at x_1) so ξ is an isometry—that is, leaves the metric tensor g_{ab} invariant.

Problem 2.1 What becomes of this proof for shells?

Problem 2.2 Show that even though $\hat{\boldsymbol{P}}$ is materially frame indifferent, $\partial\hat{\boldsymbol{P}}/\partial t$ need not be.

If we write the axiom of entropy production in terms of $\boldsymbol{S}$ and $\boldsymbol{C}$ by writing

$$\boldsymbol{P}: \frac{\partial \boldsymbol{F}}{\partial t} = \boldsymbol{S}: \boldsymbol{D} = \tfrac{1}{2}\boldsymbol{S}: \frac{\partial \boldsymbol{C}}{\partial t}$$

and repeat the argument of 2.4, Chapter 3 (or 2.6, Chapter 3 for pure elasticity), we find, instead of $\hat{\boldsymbol{P}} = \rho_{\text{Ref}}g^\sharp(\partial\hat{\Psi}/\partial \boldsymbol{F})$, the important identity

$$\boldsymbol{S} = 2\rho_{\text{Ref}}\frac{\partial\hat{\Psi}}{\partial \boldsymbol{C}}.$$

This identity may also be proved by the chain rule as follows: regarding $\boldsymbol{C}$ as a function of $\boldsymbol{F}$ through $\boldsymbol{C} = \boldsymbol{F}^T\boldsymbol{F}$—that is, $C_{AB} = g_{ab}F^a{}_AF^b{}_B$—the chain rule gives

$$\boldsymbol{S} = \boldsymbol{F}^{-1}\hat{\boldsymbol{P}} = \boldsymbol{F}^{-1}\rho_{\text{Ref}}g^\sharp\frac{\partial\hat{\Psi}}{\partial \boldsymbol{F}} = \boldsymbol{F}^{-1}\rho_{\text{Ref}}g^\sharp\frac{\partial\hat{\Psi}}{\partial \boldsymbol{C}}\frac{\partial \boldsymbol{C}}{\partial \boldsymbol{F}}.$$

Noting that

$$\frac{\partial C_{AB}}{\partial F^a{}_C} = g_{ab}\delta^C{}_A F^b{}_B + g_{ca}F^c{}_A\delta^C{}_B,$$

we get

$$\begin{aligned}
\hat{S}^{DC} &= \rho_{\mathrm{Ref}}(F^{-1})^D{}_d \frac{\partial\hat{\Psi}}{C_{AB}}\frac{\partial C_{AB}}{\partial F^a{}_C} g^{ad} \\
&= \rho_{\mathrm{Ref}}(F^{-1})^D{}_d \frac{\partial\hat{\Psi}}{C_{AB}}(g_{ab}\delta^C{}_A F^b{}_B + g_{ca}F^c{}_A\delta^C{}_B)g^{ad} \\
&= \rho_{\mathrm{Ref}}(F^{-1})^D{}_d \frac{\partial\hat{\Psi}}{\partial C_{AB}}(\delta^d{}_b\delta^C{}_A F^b{}_B + \delta^d{}_c\delta^C{}_B F^c{}_A) \\
&= \rho_{\mathrm{Ref}}(F^{-1})^D{}_b F^b{}_B \frac{\partial\hat{\Psi}}{C_{CB}} + \rho_{\mathrm{Ref}}(F^{-1})^D{}_c F^c{}_A \frac{\partial\hat{\Psi}}{\partial C_{AC}} \\
&= \rho_{\mathrm{Ref}}\delta^D{}_B \frac{\partial\hat{\Psi}}{\partial C_{CB}} + \rho_{\mathrm{Ref}}\delta^D{}_A \frac{\partial\hat{\Psi}}{\partial C_{AC}} \\
&= \rho_{\mathrm{Ref}}\left(\frac{\partial\hat{\Psi}}{\partial C_{CD}} + \frac{\partial\hat{\Psi}}{\partial C_{DC}}\right) = 2\rho_{\mathrm{Ref}}\frac{\partial\hat{\Psi}}{\partial C_{DC}}.
\end{aligned}$$

Thus, we have:

2.11 Proposition *Under the axioms for constitutive theory listed above, the second Piola–Kirchhoff stress is given by*

$$\boxed{S = 2\rho_{\mathrm{Ref}}\frac{\partial\hat{\Psi}}{\partial C}: \qquad \text{that is,} \quad S^{AB} = 2\rho_{\mathrm{Ref}}\frac{\partial\hat{\Psi}}{\partial C_{AB}}.}$$

Problem 2.3 (J. Ball; cf. Wang and Truesdell [1973]). Show that frame indifference of $\boldsymbol{P}$ and symmetry of $\boldsymbol{S}$ implies frame indifference of Ψ. Also, show that frame indifference of Ψ implies symmetry of $\boldsymbol{S}$ [assuming $\boldsymbol{P} = \rho_{\mathrm{Ref}}(\partial\hat{\Psi}/\partial\boldsymbol{F})$]. [*Hint*: Let $\boldsymbol{Q}$ be a proper orthogonal matrix; write $\boldsymbol{Q} = e^{\Omega}$, where Ω is skew, and join $\boldsymbol{F}$ to $\boldsymbol{QF}$ along the curve $\boldsymbol{F}(t) = e^{t\Omega}\boldsymbol{F}$ $(0 \le t \le 1)$. Consider $(d/dt)\hat{\Psi}(\boldsymbol{F}(t))$.]

So far we have formulas for the first and second Piola–Kirchhoff stress tensors in terms of the free energy $\hat{\Psi}$. What about the Cauchy stress tensor $\boldsymbol{\sigma}$? Since $\boldsymbol{C} = \phi^*\boldsymbol{g}$, we can regard $\boldsymbol{C}$ as a function of the point values of $\boldsymbol{F}$ and the spatial metric $\boldsymbol{g}$. Therefore, $\hat{\Psi}$ becomes a function of $X, \boldsymbol{F}(X), \boldsymbol{g}(x)$, and $\Theta(X)$. Set

$$\hat{\psi}(x, \boldsymbol{F}(X), \boldsymbol{g}(x), \Theta(X)) = \hat{\Psi}(X, \boldsymbol{C}(\boldsymbol{F}(X), \boldsymbol{g}(x)), \Theta(X)).$$

By the chain rule,

$$\frac{\partial\hat{\psi}}{\partial g_{ab}} = \frac{\partial\hat{\Psi}}{\partial C_{AB}}\frac{\partial C_{AB}}{\partial g_{ab}} = \frac{\partial\hat{\Psi}}{\partial C_{AB}}F^a{}_A F^b{}_B = \left(\phi_*\frac{\partial\hat{\Psi}}{\partial C}\right)^{ab}.$$

Therefore,

$$\sigma^{ab} = \frac{1}{J}(\phi_* S)^{ab} = \frac{1}{J}\left(\phi_*\left(2\rho_{\text{Ref}}\frac{\partial\hat{\Psi}}{\partial C}\right)\right)^{ab} = 2\rho\left(\phi_*\frac{\partial\hat{\Psi}}{\partial C}\right)^{ab} = 2\rho\frac{\partial\hat{\psi}}{\partial g_{ab}}.$$

Thus, we have proved:

2.12 Proposition *Defining the function* $\hat{\psi}$ *as above, the equation* $\boldsymbol{S} = 2\rho_{\text{Ref}}(\partial\hat{\Psi}/\partial C)$ *is equivalent to*

$$\boxed{\boldsymbol{\sigma} = 2\rho\frac{\partial\hat{\psi}}{\partial \boldsymbol{g}}.}$$

Thus, the equation for the Cauchy stress that we derived in Box 3.3, Chapter 2, is in agreement with and indeed is equivalent to the standard deduction 2.11 in constitutive theory. (See Box 2.1 below for the relation between $\partial\hat{\psi}/\partial\boldsymbol{g}$ and $\partial\hat{e}/\partial\boldsymbol{g}$.) As we noted in Box 3.3, Chapter 2, the formula in Proposition 2.12 was first given (in a slightly different from) by Doyle and Ericksen [1956].

Problem 2.4 Consider a stored energy function of the form $\hat{E} = h(J)$. Verify, using $\boldsymbol{P} = \rho_{\text{Ref}}\boldsymbol{g}^\flat(\partial\hat{E}/\partial\boldsymbol{F})$ and $\sigma = 2\rho(\partial\hat{E}/\partial\boldsymbol{g})$ that the same formula for the Cauchy stress results: $\sigma = ph'\boldsymbol{g}^\sharp$.

Box 2.1 ***Entropy and Temperature as Conjugate Variables***

Recall that given a Lagrangian $L(q^i, \dot{q}^j)$, one defines the momentum by $p_j = \partial L/\partial\dot{q}^j$ and the energy by $H(q^i, p_j) = p_j\dot{q}^j - L(q^j, \dot{q}^j)$, assuming $p_j = \partial L/\partial\dot{q}^j$ defines a legitimate change of variables $(q^i, \dot{q}^j) \mapsto (q^i, p_j)$. The relationship $N = -\partial\hat{\Psi}/\partial\Theta$ is analogous to $p_j = \partial L/\partial\dot{q}^j$.

Suppose $\hat{\Psi}$ is regarded as a function of ϕ and Θ, and the change of variables from (ϕ, Θ) to (ϕ, N) given by $\Theta \mapsto \Theta(\phi, N)$ is legitimate (invertible). Then the formula

$$\hat{E} = \Theta N + \hat{\Psi} \tag{1}$$

for the internal energy is a partial Legendre transform, so $\hat{E}$ is now a function of ϕ and N. (If $\hat{\Psi}$ is $-L$ above, then H is $\hat{E}$ so an overall sign is off; but this is convention.) We claim that our formula

$$\hat{P}_a{}^A = \rho_{\text{Ref}}\frac{\partial\hat{\Psi}}{\partial F^a{}_A} \quad \text{(variables } \phi, \Theta) \tag{2}$$

is equivalent to

$$\hat{P}_a{}^A = \rho_{\text{Ref}}\frac{\partial\hat{E}}{\partial F^a{}_A} \quad \text{(variables } \phi, N). \tag{3}$$

Indeed, from (1), with Θ a function of (ϕ, N), we get

$$\frac{\partial \hat{E}}{\partial F^a{}_A} = \frac{\partial \Theta}{\partial F^a{}_A} N + \left(\frac{\partial \hat{\Psi}}{\partial F^a{}_A} + \frac{\partial \hat{\Psi}}{\partial \Theta} \frac{\partial \Theta}{\partial F^a{}_A} \right).$$

The first and last terms cancel since $N = -\partial \hat{\Psi}/\partial \Theta$, so $\partial \hat{E}/\partial F^a{}_A = \partial \hat{\Psi}/\partial F^a{}_A$. Thus (2) and (3) are equivalent.

In the same way we have the spatial form, $\partial \hat{\psi}/\partial \mathbf{g} = \partial \hat{e}/\partial \mathbf{g}$, where $\hat{\psi}$ is a function of $\mathbf{g}$, ϕ and θ and $\hat{e}$ is a function of $\mathbf{g}$, ϕ and η. Thus, Proposition 2.12 is consistent with the results of Box 3.3, Chapter 2; in that box, if constitutive assumptions are added, the basic variables should be $\mathbf{g}$, ϕ and η.

Problem 2.5 Prove that $\Theta = \partial \hat{E}/\partial N$ and $\theta = \partial \hat{e}/\partial \eta$.

The fact that Θ and N are analogous to velocity and momentum indicates that results of symplectic geometry are relevant in thermodynamics. See Box 6.1, Chapters 1 and 5, and, for example, Oster and Perelson [1973].

Box 2.2 ***Summary of Important Formulas for Section 3.2***

If the constitutive function $\hat{\Psi}$ satisfies the axioms of locality, entropy production, and material frame indifference, then $\hat{\Psi}$ is a function only of X, $\mathbf{C}$, and Θ.

Constitutive Identities

$$\hat{N} = -\frac{\partial \hat{\Psi}}{\partial \Theta}$$

$$\hat{\mathbf{S}} = 2\rho_{\text{Ref}} \frac{\partial \hat{\Psi}}{\partial \mathbf{C}} \qquad \hat{S}^{AB} = 2\rho_{\text{Ref}} \frac{\partial \hat{\Psi}}{\partial C_{AB}}$$

$$\hat{\mathbf{P}} = \rho_{\text{Ref}} \mathbf{g}^{\sharp} \frac{\partial \hat{\Psi}}{\partial \mathbf{F}} = \mathbf{F}\hat{\mathbf{S}} \qquad \hat{P}^{aB} = \rho_{\text{Ref}} g^{ab} \frac{\partial \hat{\Psi}}{\partial F^b{}_A} = F^a{}_B \hat{S}^{BA}$$

$$\hat{\boldsymbol{\sigma}} = 2\rho \frac{\partial \hat{\psi}}{\partial \mathbf{g}} \qquad \hat{\sigma}^{ab} = 2\rho \frac{\partial \hat{\psi}}{\partial g_{ab}}$$

Entropy Production Inequality

$\langle \hat{\mathbf{Q}}, \text{GRAD}\, \Theta \rangle \leq 0 \qquad \hat{Q}^A \Theta_{|A} \leq 0$

$\hat{\mathbf{Q}}$ vanishes when its argument GRAD Θ does.

Balance of Energy $(\hat{E} = \hat{\Psi} + \hat{N}\Theta)$

$$\rho_{\text{Ref}} \Theta \frac{\partial \hat{N}}{\partial t} + \text{DIV}\, \hat{\mathbf{Q}} = \rho_{\text{Ref}} R \qquad \rho_{\text{Ref}} \Theta \frac{\partial \hat{N}}{\partial t} + \hat{Q}^A{}_{|A} = \rho_{\text{Ref}} R$$

Rigid Heat Conductor ($C \equiv 0$)

$$\left(\rho_{\mathrm{Ref}}\Theta \frac{\partial \hat{N}}{\partial \Theta}\right)\frac{\partial \Theta}{\partial t} + \mathrm{DIV}\,\hat{Q} = \rho_{\mathrm{Ref}} R$$

$$\left(\rho_{\mathrm{Ref}}\Theta \frac{\partial \hat{N}}{\partial \Theta}\right)\frac{\partial \Theta}{\partial t} = -\left(\frac{\partial \hat{Q}^A}{\partial \Theta_{|B}}\right)\Theta_{|B|A} - \frac{\partial \hat{Q}^A}{\partial \Theta}\Theta_{|A} + \rho_{\mathrm{Ref}} R - R_I$$

Material Frame Indifference

If ξ is an isometry, then

$$\xi_* \hat{P}(X, F, \Theta) = \hat{P}(X, \xi_* F, \Theta).$$

This implies $\hat{P}$ depends only on X, C, and Θ.

For any proper orthogonal matrix $R^a{}_b$,

$$R^a{}_b \hat{P}^{bA}(X, F^c{}_G, \Theta) = \hat{P}^{aA}(X, R^c{}_b F^b{}_G, \Theta).$$

Legendre Transformation: $\Theta \mapsto \Theta(\phi, N)$

$$\frac{\partial \hat{\Psi}}{\partial F} = \frac{\partial \hat{E}}{\partial F}, \quad \frac{\partial \hat{\Psi}}{\partial C} = \frac{\partial \hat{E}}{\partial C}, \quad \frac{\partial \hat{\psi}}{\partial g} = \frac{\partial \hat{e}}{\partial g}$$

$$\frac{\partial \hat{\Psi}}{\partial F^a{}_A} = \frac{\partial \hat{E}}{\partial F^a{}_A}, \quad \frac{\partial \hat{\Psi}}{\partial C_{AB}} = \frac{\partial \hat{E}}{\partial C_{AB}}, \quad \frac{\partial \hat{\psi}}{\partial g_{ab}} = \frac{\partial \hat{e}}{\partial g_{ab}}$$

3.3 COVARIANT CONSTITUTIVE THEORY

We shall now obtain the principal theorems of constitutive theory by using a different set of axioms. The idea is to reproduce the theorems of Section 3.2 from a point of view that is covariant, not relying on the rigid Euclidean structure of $\mathbb{R}^3$. (The results therefore directly generalize to manifolds, but this is not our main motivation.) In Section 3.2 and in Box 3.3, Chapter 2, we derived the basic relationship between stress and free energy: $\boldsymbol{\sigma} = 2\rho(\partial\hat{\psi}/\partial \boldsymbol{g})$. This relationship enables us to obtain a covariant description of constitutive theory, as well as of balance of energy. As in the previous section, the hypotheses here are intended to be relevant for thermoelasticity. For other continuum theories they must be modified using more general assumptions as indicated in Section 3.1.

We shall begin by discussing pure elasticity. This will involve combining Theorem 2.6 with a covariant version of material frame indifference. Secondly, we discuss thermoelasticity. Covariance assumptions allow one to make the conclusions of the Coleman–Noll Theorem (2.4) from balance of energy alone, *without* invoking the Clausius–Duhem inequality.

3.1 Covariant Constitutive Axioms (*Elasticity*) *Let* $\hat{E}\colon \mathfrak{C} \longrightarrow \mathfrak{F}$ *be a given differentiable map. Assume:*

1. *$\hat{E}$ is local.*
2. *There is a map* $\hat{\boldsymbol{P}}\colon \mathfrak{C} \longrightarrow$ *two-point tensors, such that for all motions* ϕ_t,

balance of energy holds (see 3.5 of Chapter 2):

$$\rho_{\text{Ref}} \frac{\partial \hat{E}}{\partial t} = \hat{P} : \dot{F}.$$

3. *Let g be a given metric on $\mathcal{S}$ and let $\mathcal{O}_g$ denote the set of metrics of the form $\xi_* g$, where $\xi: \mathcal{S} \to \mathcal{S}$ is a diffeomorphism. Assume there is a map $\hat{\hat{E}}: \mathcal{C} \times \mathcal{O}_g \to \mathcal{F}$ such that*

$$\hat{E}(\phi) = \hat{\hat{E}}(\xi \circ \phi, \xi_* g)$$

for all diffeomorphisms $\xi: \mathcal{S} \to \mathcal{S}$. (Taking ξ = identity, note that $\hat{E}(\phi) = \hat{\hat{E}}(\phi, g)$.)

Axioms 1 and 2 were discussed in the previous section. Axiom 3 may be interpreted as follows: think of ξ as being either a coordinate chart for $\mathcal{S}$, or a framing at one instant (see Box 4.1, Chapter 1). Axiom 3 states that one can compute $\hat{E}(\phi)$ in terms of the representative of ϕ and the representative of the metric in the framing ξ.

The following remarks are intended to make Axiom 3 plausible:

3.2 Thought Experiment to Justify Axiom 3 Imagine that ξ is a (nonlinear) coordinate chart. The representation of a configuration ϕ is $\xi \circ \phi$. (For example, if ϕ is given in Cartesian coordinates, ξ might be the coordinate change to spherical coordinates and $\xi \circ \phi$ becomes ϕ written in spherical coordinates.) How can the observer ξ determine the internal energy in the configuration he sees to be $\xi \circ \phi$? He could, for example, perform some experiment to see how much work this configuration can do, such as in Figure 3.3.1. The measurement of the magnitude of velocity uses the observers metric—namely, $\xi_* g$. Thus, the internal energy also should require a knowledge of $\xi \circ \phi$ and $\xi_* g$ for its measurement.

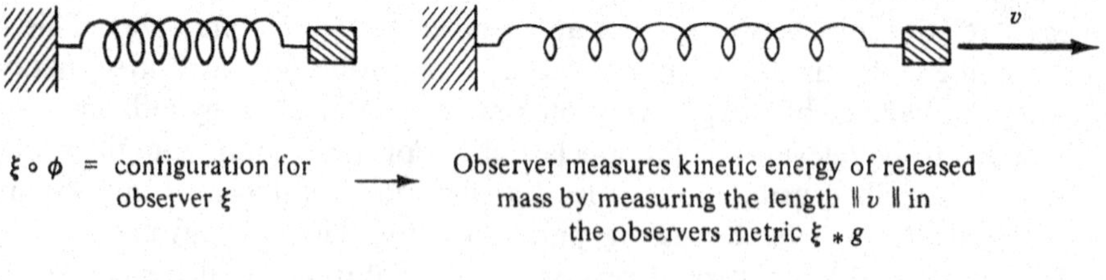
v
ξ ∘ φ = configuration for observer ξ
Observer measures kinetic energy of released mass by measuring the length ‖ v ‖ in the observers metric ξ * g

Figure 3.3.1 An observer ξ measuring the energy in a configuration.

3.3 Theorem *Assume Axioms* 1, 2, *and* 3. *Then*

(i) *$\hat{E}$ depends on ϕ only through the point values of $C = \phi^* g$ and*

(ii) $2\rho_{\text{Ref}}(\partial \hat{E}/\partial C) = S$; *in paticular, S is symmetric.*

Proof Suppose that $\phi_1^* g = \phi_2^* g$. Letting $\xi = \phi_2 \circ \phi_1^{-1}$, we see that $\xi_* g = g$ and so

$$\hat{E}(\phi_1) = \hat{\hat{E}}(\xi \circ \phi_1, \xi_* g) = \hat{\hat{E}}(\phi_2, g) = \hat{E}(\phi_2).$$

Thus $\hat{E}$ depends on ϕ only through C. Axiom 2 shows that $\hat{E}$ depends only on the point values of C and that $2\rho_{\text{Ref}}(\partial\hat{E}/\partial C) = S$ holds, by the same argument as in Theorem 2.4 (cf. Theorem 2.6). ∎

Axiom 3 is equivalent to the usual form of material frame indifference. Indeed, if $\hat{E}$ depends only on $\phi^* g = C$, then $\hat{\hat{E}}(\xi\circ\phi, \xi_* g) = \hat{E}(\phi)$ is well defined, for $(\xi\circ\phi)^*(\xi_* g) = \phi^* g = C$; so from $\xi\circ\phi$ and $\xi_* g$ we can construct C.

Once $\hat{E}$ is a function of C, it is reasonable to ask how it depends on the representation of the reference configuration. In order to form a scalar from C, the metric G on $\mathcal{B}$ is needed. Relative to various representations $\Xi: \mathcal{B} \longrightarrow \mathcal{B}$ of $\mathcal{B}$, G will look different. In fact, under such a change of reference configuration, G changes to $\Xi_* G$ and ϕ changes to $\phi\circ\Xi^{-1}$, so C changes to $\Xi_* C$. The assertion that $\hat{E}$ changes correspondingly is the content of the next concept.

3.4 Definition We say that $\hat{E}$ is *materially covariant* if there is a function

$$\check{E}: \mathcal{O}_G \times \mathcal{C} \longrightarrow \mathcal{F}$$

such that $$\hat{E}(\phi) = \check{E}(G, \phi)$$

and $$\check{E}(\Xi^* G, \phi\circ\Xi) = \check{E}(G, \phi)\circ\Xi$$

for every diffeomorphism $\Xi: \mathcal{B} \longrightarrow \mathcal{B}$.

When these axioms hold, we say that $\hat{E}$ is a *tensorial function* of G and C. Material covariance is essentially equivalent to isotropy of the material. The precise situation is given in Section 5.

The overall situation is indicated in Figure 3.3.2.

Problem 3.1 Define the *neo-hookean* constitutive function by

$$E(G, C) = \lambda_1^2 + \cdots + \lambda_n^2 - n,$$

where $n = \dim \mathcal{B} = \dim \mathcal{S}$ and $\lambda_1, \ldots, \lambda_n$ are the principal stretches—that is, eigenvalues of $\sqrt{C}$ using the metric G. Show that this is a materially covariant energy function satisfying Axioms 1–3.

Next we discuss the role of covariance in thermoelasticity. There are numerous ways of modifying the axioms in Section 3.2 to take into account the extra information covariance gives (see Problems 3.3, 3.4, and 3.5). We shall bypass the Clausius–Duhem inequality and derive all the relations in 2.4 by making reasonable covariance assumptions and assuming balance of energy. *As in the covariant approach in Box* 3.3, *Chapter* 2, *we focus not on all processes, but rather on all transformations of a given process.* Comparing the original and transformed process enables us to cancel out the heat source term R in balance of energy. This was the principal obstacle previously and the reason for the success of the entropy production inequality in 2.4.

If one has obtained the identities $\hat{N} = -\partial\hat{\Psi}/\partial\Theta$ and $P_a{}^A = \rho_{\text{Ref}}(\partial\hat{\Psi}/\partial F^a{}_A)$ by other means, then the entropy production inequality is *equivalent* to $\langle \boldsymbol{Q}, \text{GRAD}\,\Theta\rangle \leq 0$.

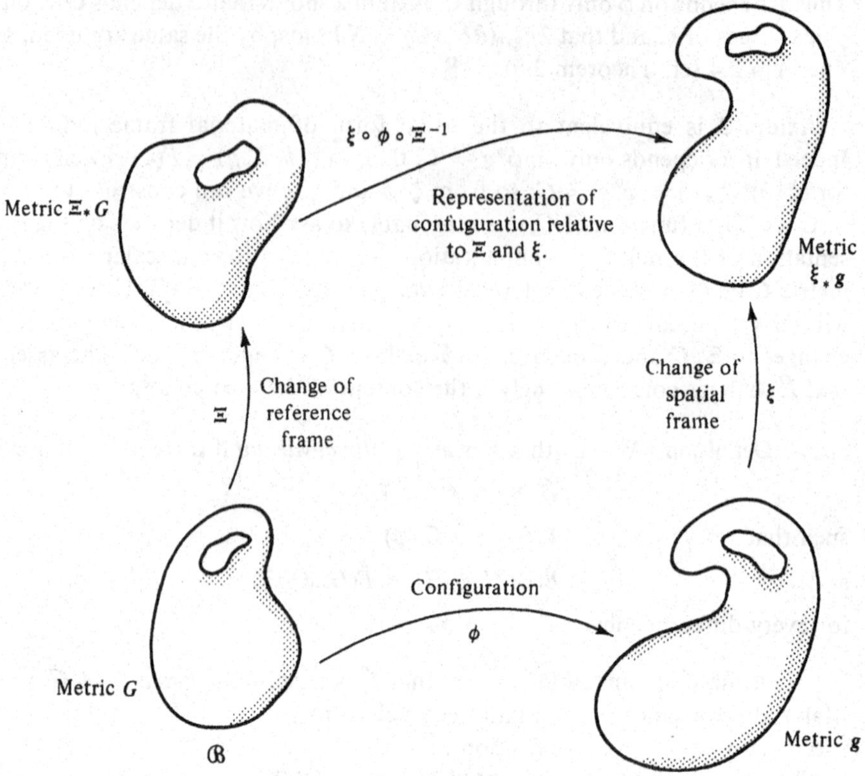

Figure 3.3.2

Our axioms involve not only coordinate changes $\xi: \mathcal{S} \longrightarrow \mathcal{S}$ of space, but also temperature rescalings (i.e., representations of the temperature in various units). This is expressed in terms of the space of monotone increasing diffeomorphisms $r: \mathbb{R}^+ \longrightarrow \mathbb{R}^+$. For our purposes it is enough to consider linear rescalings.[3]

3.5 Covariant Constitutive Axioms (*Thermoelasticity*) *Assume* $\hat{\Psi}$, $\hat{P}$, $\hat{Q}$, *and* $\hat{N}$ *are given thermoelastic constitutive functions satisfying:*

(1) *The axiom of locality.*

---

[3]We could have expressed our axioms in terms of a metric on $\mathcal{S} \times \mathbb{R}^+$ and transformations of the whole space $\mathcal{S} \times \mathbb{R}^+$. This may be important for theories with internal variables, for example, but for thermoelasticity, direct temperature rescalings are easier to understand. The relationship to metrics is as follows: In one dimension a diffeomorphism $r: \mathbb{R}^+ \longrightarrow \mathbb{R}^+$ can be reconstructed from $r_*\delta$, where δ is the standard metric on $\mathbb{R}$. If we identify metrics with positive functions, then $\delta = 1$ and $r_*\delta = (dr/dx)^2$, so $r(x) = \int^x \sqrt{r_*\delta}\, dx$ + constant. Here $r_*\delta$ plays the same role as ξ_*g.

(2) *For a given process* ϕ_t, Θ_t, *balance of energy*[4] *holds:*

$$\rho_{\text{Ref}} \frac{\partial E}{\partial t} + \text{DIV}\, Q = P : \dot{F} + \rho_{\text{Ref}} R,$$

where $E = \Psi + N\Theta$.

(3) *There is a map* $\hat{\Psi}: \mathcal{C} \times \mathcal{O}_g \times \mathfrak{I} \times \mathbb{R}^+ \to \mathfrak{F}$ *such that for any diffeomorphism* $\xi: \mathcal{S} \to \mathcal{S}$ *and any* $r: \mathbb{R}^+ \to \mathbb{R}^+$,

$$\hat{\Psi}(\phi, \Theta) = \hat{\Psi}(\xi \circ \phi, \xi_* g, r\Theta, r)$$

(4) *For curves* $\xi_t: \mathcal{S} \to \mathcal{S}$ *and* $r_t(x) \in \mathbb{R}^+$, *assume that* $\phi'_t = \xi_t \circ \phi_t$ *and* $\Theta'_t = r_t\Theta_t$ *satisfy balance of energy where we demand the transformation properties in 4.12, Chapter 2, except that* Ψ *and* N *should transform as scalars,* $Q'_t = r_t \xi_{t*} Q_t$ *and* $(R'_t - \Theta'_t(\partial N'_t/\partial t)) = r_t \cdot (R_t - \Theta_t(\partial N_t/\partial t))$. *(This latter transformation formula accounts for the "apparent heat supply" due to entropy production and is analogous to the transformation formula for body forces in 4.12, Chapter 2, in which there are "apparent body forces" due to the velocity and acceleration of* ξ_t.*)*

3.6 Theorem *Under these assumptions,* $\hat{\Psi}$ *depends only on the point values of* C *and* Θ *and we have*

$$\boxed{\hat{N} = -\frac{\partial \hat{\Psi}}{\partial \Theta}, \qquad S = 2\rho_{\text{Ref}} \frac{\partial \hat{\Psi}}{\partial C} \quad \left(\text{or equivalently, } P_a{}^A = \rho_{\text{Ref}} \frac{\partial \hat{\Psi}}{\partial F^a{}_A}\right).}$$

Proof Using the notation of 2.4, balance of energy reads:

$$\rho_{\text{Ref}}\left(D_\phi \hat{\Psi} \cdot V + D_\Theta \hat{\Psi} \cdot \dot{\Theta} + \frac{\partial \hat{N}}{\partial t}\Theta + \hat{N}\dot{\Theta}\right) + \text{DIV}\, Q = P : \dot{F} + \rho_{\text{Ref}} R.$$

If we write down the corresponding expression for the primed quantities at $\xi =$ identity, $\dot{\xi} = w$, $r = 1$, $\dot{r} = u$, and subtract we are led to the identity

$$\rho_{\text{Ref}}(D_\phi \hat{\Psi} \cdot w + D_\Theta \hat{\Psi} \cdot (u\Theta) + \hat{N} \cdot (u\Theta)) = P : \nabla w.$$

Thus we get the two identities

$$\rho_{\text{Ref}}(D_\phi \hat{\Psi} \cdot w) = P : \nabla w \tag{1}$$

and

$$D_\Theta \hat{\Psi} \cdot (u\Theta) + \hat{N} \cdot (u\Theta) = 0 \tag{2}$$

Since ∇w and $u = \dot{r}$ are arbitrary, the identities stated in the Theorem and the dependence on point values now follows by the same argument used to analyze (1) and (2) in the proof of Theorem 2.4. Axiom 3 implies material frame indifference; so, as usual, $\hat{\Psi}$ depends only on C. ∎

Problem 3.2 (J. Ball) Show that the entropy production inequality for all subregions of $\mathcal{B}$ follows from that for all of $\mathcal{B}$ and covariance under all superposed motions ξ_t.

---

[4] We have written balance of energy in a simplified form in which balance of momentum has been used. Of course one could go back to the primitive form of balance of energy and *derive* balance of momentum using Box 3.3, Chapter 2.

Problem 3.3 (M. Gurtin) Derive the constitutive identities by assuming an appropriate transformation property of the rate of entropy production

$$\Gamma = \rho_{\mathrm{Ref}}(N\dot{\Theta} + \dot{\Psi}) - \boldsymbol{S}:\boldsymbol{D} + \frac{1}{\Theta}\langle \boldsymbol{Q}, \nabla\Theta\rangle.$$

Problem 3.4 Show that preservation of the Clausius–Duhem inequality under superposed motions $\xi_t(x)$ and temperature rescalings $r_t(x)$ gives a monotonicity condition on $\boldsymbol{Q}$. Is this new information?

Box 3.1 ***The Duhamel–Neumann Hypothesis***[5]

The purpose of this box is to prove a decomposition of the rate of deformation tensor, which generalizes the "Duhamel–Neumann hypothesis" (Sokolnikoff [1956], p. 359). This decomposition, often made on an ad hoc basis, has proven to be very useful in the identification of constitutive functions. We shall confine ourselves to thermoelasticity, although more general theories using internal variables can be used as well. We assume the existence of a free energy Ψ, a function of the Cauchy–Green tensor C_{AB}, and the temperature Θ; that is, i.e. $\Psi = \hat{\Psi}(C_{AB}, \Theta)$. We also assume the usual relationship between the second (symmetric) Piola–Kirchhoff stress tensor S^{AB} and $\hat{\Psi}$, derived in this and the previous sections, namely,

$$S^{AB} = 2\rho_{\mathrm{Ref}} \frac{\partial \hat{\Psi}}{\partial C_{AB}} \tag{1}$$

Our derivation of the Duhamel–Neumann hypothesis depends on recasting this into an *equivalent* spatial form, and relating the free energy $\hat{\psi}$ in spatial coordinates and the *Kirchhoff stress* $\boldsymbol{\tau} = J\boldsymbol{\sigma}$ (= Jacobian of the deformation $\times$ Cauchy stress), namely,

$$\boldsymbol{\tau} = 2\rho_{\mathrm{Ref}} \frac{\partial \hat{\psi}}{\partial \boldsymbol{g}}, \qquad \text{that is,} \quad \tau^{ab} = 2\rho_{\mathrm{Ref}} \frac{\partial \hat{\psi}}{\partial g_{ab}}, \tag{2}$$

where we take $\hat{\psi}$ to be a function of the spatial metric $\boldsymbol{g}$, the (left) Cauchy–Green tensor $\boldsymbol{c}$ and the temperature θ. We have derived (2) in the first section of this chapter and showed its equivalence to (1). The idea now is to manipulate (2) using the Lie derivative and a Legendre transform. In particular, we recall that the rate of deformation tensor $d_{ab} = \frac{1}{2}(v_{a|b} + v_{b|a})$ is given in terms of the Lie derivative by

$$\boldsymbol{d} = \tfrac{1}{2}\boldsymbol{L}_{\boldsymbol{v}}\boldsymbol{g}, \tag{3}$$

where $\boldsymbol{v}$ denotes the spatial velocity field of a given motion and $\boldsymbol{L}_{\boldsymbol{v}}$ denotes Lie differentiation with respect to $\boldsymbol{v}$ (see Section 1.6). For a

[5]The results of this box were obtained in collaboration with K. Pister.

scalar function f of position and time, $L_v f = \dot{f} = \partial f/\partial t + v \cdot \nabla f$ is the material derivative. By the chain rule,

$$L_v \frac{\partial \hat{\psi}}{\partial g} = \frac{\partial^2 \hat{\psi}}{\partial g^2} L_v g + \frac{\partial^2 \hat{\psi}}{\partial g\, \partial c} L_v c + \frac{\partial^2 \hat{\psi}}{\partial g\, \partial \theta} \dot{\theta}.$$

By definition (3) and the simple identity $L_v c = \mathbf{0}$, we get

$$L_v \frac{\partial \hat{\psi}}{\partial g} = 2 \frac{\partial^2 \hat{\psi}}{\partial g^2} d + \frac{\partial^2 \hat{\psi}}{\partial g\, \partial \theta} \dot{\theta}. \tag{4}$$

Since $\dot{\rho}_{\mathrm{Ref}} = 0$, combining (4) and (3) yields the following:

3.7 Proposition

$$\boxed{L_v \tau = \rho_{\mathrm{Ref}}(\mathsf{a} \cdot 2d + m\dot{\theta}),} \tag{5}$$

where

$$\mathsf{a} = 2 \frac{\partial^2 \hat{\psi}}{\partial g^2} \tag{6}$$

is the tangential mechanical stiffness *and*

$$m = 2 \frac{\partial^2 \hat{\psi}}{\partial g\, \partial \theta} \tag{7}$$

are the thermal stress coefficients. *Using indices, Equations* (5), (6), *and* (7) *read:*

$$(L_v \tau)^{ab} = \rho_{\mathrm{Ref}}(2\mathsf{a}^{abcd}\, d_{cd} + m^{ab}\dot{\theta}), \tag{5$'$}$$

$$\mathsf{a}^{abcd} = 2 \frac{\partial^2 \hat{\psi}}{\partial g_{ab}\, \partial g_{cd}}, \tag{6$'$}$$

and

$$m^{ab} = 2 \frac{\partial^2 \hat{\psi}}{\partial g_{ab}\, \partial \theta}. \tag{7$'$}$$

For isothermal deformations we can regard θ as absent; so (5) reduces to $L_v \tau = \rho_{\mathrm{Ref}} \cdot \mathsf{a}(2d)$, an incremental form of the mechanical constitutive relation.

Now we perform a Legendre transformation to obtain the inverse form of (5). Define the *complementary free energy* by

$$\begin{aligned} \chi &= \hat{\chi}(\tau, c, \theta) \\ &= \frac{1}{\rho_{\mathrm{Ref}}} \operatorname{tr}(\tau \cdot g) - 2\hat{\psi}(g, c, \theta), \end{aligned} \tag{8}$$

We are assuming that the change of variables from g to τ defined by (2) is invertible.[6]

[6]This can be justified (locally at least) if we make the usual hypothesis that the material is strongly elliptic; the result then follows from the inverse function theorem. See Chapters 4, 5, and 6 for the necessary background.

In formula (8), $\mathrm{tr}(\boldsymbol{\tau}\cdot g) = \mathrm{trace}(\tau^{ab}g_{bc}) = \tau^{ab}g_{ab} = \tau : g$, the full contraction of $\boldsymbol{\tau}$ and g. We then have

$$\rho_{\mathrm{Ref}} \frac{\partial \hat{\chi}}{\partial \boldsymbol{\tau}} = g + \mathrm{tr}\left(\boldsymbol{\tau}\cdot\frac{\partial g}{\partial \boldsymbol{\tau}}\right) - 2\rho_{\mathrm{Ref}} \frac{\partial \hat{\psi}}{\partial g}\frac{\partial g}{\partial \boldsymbol{\tau}},$$

that is,
$$\rho_{\mathrm{Ref}} \frac{\partial \hat{\chi}}{\partial \tau^{ab}} = g_{ab} + \tau^{cd}\frac{\partial g_{cd}}{\partial \tau^{ab}} - 2\rho_{\mathrm{Ref}} \frac{\partial \hat{\psi}}{\partial g_{cd}}\frac{\partial g_{cd}}{\partial \tau^{ab}}.$$

By (2), the last two terms cancel, leaving

$$g = \rho_{\mathrm{Ref}} \frac{\partial \hat{\chi}}{\partial \boldsymbol{\tau}}. \tag{9}$$

Operating on (9) with the Lie derivative gives

$$2d = \rho_{\mathrm{Ref}} \frac{\partial^2 \hat{\chi}}{\partial \boldsymbol{\tau}^2}\cdot \boldsymbol{L}_v\boldsymbol{\tau} + \rho_{\mathrm{Ref}} \frac{\partial^2 \hat{\chi}}{\partial \boldsymbol{\tau}\,\partial \theta}\dot{\theta} \tag{10}$$

by (3) and $\boldsymbol{L}_v c = 0$. Define the *material compliance tensors*

$$\mathsf{r} = \frac{\partial^2 \hat{\chi}}{\partial \boldsymbol{\tau}^2} \quad \text{and} \quad s = \frac{\partial^2 \hat{\chi}}{\partial \boldsymbol{\tau}\,\partial \theta} \tag{11}$$

so that (10) yields:

3.8 Theorem

$$\boxed{2d = \rho_{\mathrm{Ref}}(\mathsf{r}\cdot \boldsymbol{L}_v\boldsymbol{\tau} + s\dot{\theta})} \tag{12}$$

In componential form these read

$$\mathsf{r}_{abcd} = \frac{\partial^2 \hat{\chi}}{\partial \tau^{ab}\,\partial \tau^{cd}} \quad \text{and} \quad s_{ab} = \frac{\partial^2 \hat{\chi}}{\partial \tau^{ab}\,\partial \theta} \tag{11$'$}$$

and

$$2d_{ab} = \rho_{\mathrm{Ref}}\mathsf{r}_{abcd}(\boldsymbol{L}_v\boldsymbol{\tau})^{cd} + \rho_{\mathrm{Ref}}s_{ab}\dot{\theta}, \tag{12$'$}$$

where

$$(\boldsymbol{L}_v\boldsymbol{\tau})^{cd} = \frac{\partial \tau^{cd}}{\partial t} - \tau^{ad}v^c{}_{|a} - \tau^{ca}v^d{}_{|a}.$$

Equation (12) has the following interpretation:
The total deformation rate = the mechanical rate + the thermal rate. This generalizes the Duhamel–Neumann hypothesis for infinitesimal (linearized) thermoelasticity.

3.9 Remarks (1) Differentiating (9) in g, we have

$$\mathsf{I} = \rho_{\mathrm{Ref}} \frac{\partial}{\partial g}\left(\frac{\partial \hat{\chi}}{\partial \boldsymbol{\tau}}\right) = \rho_{\mathrm{Ref}} \frac{\partial^2 \hat{\chi}}{\partial \boldsymbol{\tau}^2}\cdot\frac{\partial \boldsymbol{\tau}}{\partial g},$$

where I is the fourth-order Kroneker delta with components $\delta^{ef}_{ab} = 1$ if $(e, f) = (a, b)$ and 0 otherwise. Substituting from (3), (6), and (11), we get

$$\mathbf{I} = \left(\rho_{\mathrm{Ref}}\frac{\partial^2 \hat{\chi}}{\partial \boldsymbol{\tau}^2}\right)2\rho_{\mathrm{Ref}}\frac{\partial^2 \hat{\psi}}{\partial \boldsymbol{g}^2} = (\rho_{\mathrm{Ref}}\mathbf{r})\cdot(\rho_{\mathrm{Ref}}\mathbf{a}),$$

that is,

$$\delta^{ef}_{ab} = \rho^2_{\mathrm{Ref}}\mathsf{r}_{abcd}\mathsf{a}^{cdef}.$$

In other words, $\rho_{\mathrm{Ref}}\mathbf{r}$ and $\rho_{\mathrm{Ref}}\mathbf{a}$ are inverse tensors.

(2) As explained in Box 6.1, Chapter 1, various stress rates in common use can be related to the Lie derivative of tensors associated to $\boldsymbol{\tau}$ or $\boldsymbol{\sigma}$. The Lie derivative of $\boldsymbol{\tau}$ (the "Truesdell stress rate", see Box 6.1, Chapter 1) is the most useful for the present context.

Problem 3.5 Do all this materially.

Box 3.2 Summary of Important Formulas for Section 3.3

Internal Energy Under a Change of Spatial Frame

$$E(\phi) = \hat{E}(\xi\circ\phi, \xi_* g), \qquad \xi: \mathcal{S} \to \mathcal{S}$$

Material Covariance (Change of Reference Frame)

$$\check{E}(\Xi^* G, \phi\circ\Xi) = \check{E}(G, \phi)\circ\Xi$$

Covariance

Covariance of energy balance under superposed motions and temperature recalings implies

$$\hat{N} = -\frac{\partial \hat{\Psi}}{\partial \Theta} \quad \text{and} \quad \hat{S} = 2\rho_{\mathrm{Ref}}\frac{\partial \hat{\Psi}}{\partial C}.$$

Kirchhoff Stress

$$\boldsymbol{\tau} = J\boldsymbol{\sigma} \qquad \tau^{ab} = J\sigma^{ab}$$

Mechanical Stiffness and Thermal Stress Tensors

$$\mathbf{a} = 2\frac{\partial^2 \hat{\psi}}{\partial \boldsymbol{g}^2}, \quad \boldsymbol{m} = 2\frac{\partial^2 \hat{\psi}}{\partial \boldsymbol{g}\,\partial\theta} \qquad \mathsf{a}^{abcd} = 2\frac{\partial^2 \hat{\psi}}{\partial g_{ab}\,\partial g_{cd}}, \quad m^{ab} = 2\frac{\partial^2 \psi}{\partial g_{ab}\,\partial\theta}$$

Stress Rate

$$L_v\boldsymbol{\tau} = \rho_{\mathrm{Ref}}(\mathbf{a}\cdot 2\boldsymbol{d} + \boldsymbol{m}\dot{\theta}) \qquad (L_v\boldsymbol{\tau})^{ab} = \rho_{\mathrm{Ref}}(2\mathsf{a}^{abcd}d_{cd} + m^{ab}\dot{\theta})$$

Complementary Free Energy

$$\hat{\chi}(\boldsymbol{\tau}, c, \theta) = \frac{1}{\rho_{\mathrm{Ref}}}\boldsymbol{\tau}:\boldsymbol{g} - 2\hat{\psi}(\boldsymbol{g}, c, \theta) \qquad \hat{\chi} = \frac{1}{\rho_{\mathrm{Ref}}}\tau^{ab}g_{ab} - 2\hat{\psi}$$

Compliance Tensors

$$\mathbf{r} = \frac{\partial^2 \hat{\chi}}{\partial \boldsymbol{\tau}^2}, \quad \boldsymbol{s} = \frac{\partial^2 \hat{\chi}}{\partial \tau\,\partial\theta} \qquad \mathsf{r}_{abcd} = \frac{\partial^2 \hat{\chi}}{\partial \tau^{ab}\,\partial\tau^{cd}}, \quad s_{ab} = \frac{\partial^2 \hat{\chi}}{\partial \tau^{ab}\,\partial\theta}$$

Duhamel–Neumann Relation

$$2\boldsymbol{d} = \rho_{\mathrm{Ref}}(\mathbf{r}\cdot L_v\boldsymbol{\tau} + \boldsymbol{s}\dot{\theta}) \qquad 2d_{ab} = \rho_{\mathrm{Ref}}(r_{abcd}(L_v\boldsymbol{\tau})^{cd} + s_{ab}\dot{\theta})$$

3.4 THE ELASTICITY TENSOR AND THERMOELASTIC SOLIDS

The basic equations of motion for a continuum were derived in Section 2.2:

$$\rho_{\mathrm{Ref}}\boldsymbol{A} = \rho_{\mathrm{Ref}}\boldsymbol{B} + \mathrm{DIV}\,\boldsymbol{P}, \qquad \text{that is,} \quad \rho_{\mathrm{Ref}}\left(\frac{\partial V^a}{\partial t} + \gamma^a_{bc}V^bV^c\right) = \rho_{\mathrm{Ref}}B^a + P^{aA}{}_{|A}.$$

If we use the constitutive hypothesis for a thermoelastic material, $\boldsymbol{P}$ will be a function $\hat{\boldsymbol{P}}$ of X, $\boldsymbol{F}$, and Θ. Then we can compute DIV $\hat{\boldsymbol{P}}$ by the chain rule, as follows:

$$\mathrm{DIV}\,\hat{\boldsymbol{P}} = \frac{\partial \hat{\boldsymbol{P}}}{\partial \boldsymbol{F}}\cdot\nabla_X \boldsymbol{F} + \mathrm{DIV}_X\,\hat{\boldsymbol{P}} + \frac{\partial \hat{\boldsymbol{P}}}{\partial \Theta}\cdot\nabla_X\Theta,$$

where $\mathrm{DIV}_X\,\hat{\boldsymbol{P}}$ means the divergence of $\hat{\boldsymbol{P}}$ holding the variables $\boldsymbol{F}$ and Θ constant. In coordinates, this equation reads as follows:

$$(\mathrm{DIV}\,\hat{\boldsymbol{P}})^a = \frac{\partial \hat{P}^{aA}}{\partial F^b{}_B}F^b{}_{B|A} + \left(\frac{\partial \hat{P}^{aA}}{\partial X^A} + \hat{P}^{aA}\Gamma^B_{AB} + \hat{P}^{aA}\gamma^b_{bc}F^c{}_A\right) + \frac{\partial \hat{P}^{aA}}{\partial \Theta}\frac{\partial \Theta}{\partial X^A},$$

where we have written out $\mathrm{DIV}_X\,\hat{\boldsymbol{P}}$ explicitly using the formula for the covariant derivative of a two-point tensor from Section 1.4, Chapter 1.[7] The covariant derivative of the deformation gradient is

$$F^b{}_{B|A} = \frac{\partial^2 \phi^b}{\partial X^A\,\partial X^B} + \frac{\partial \phi^e}{\partial X^B}\gamma^b_{ec}\frac{\partial \phi^c}{\partial X^A} - \frac{\partial \phi^b}{\partial X^C}\Gamma^C_{AB}.$$

Thus the leading term in DIV $\hat{\boldsymbol{P}}$ containing second derivatives of ϕ is

$$\frac{\partial \hat{P}^{aA}}{\partial F^b{}_B}\frac{\partial^2 \phi^b}{\partial X^A\,\partial X^B} = \frac{1}{2}\left(\frac{\partial \hat{P}^{aA}}{\partial F^b{}_B} + \frac{\partial \hat{P}^{aB}}{\partial F^b{}_A}\right)\frac{\partial^2 \phi^b}{\partial X^A\,\partial X^B}.$$

The term $\partial \hat{P}^{aA}/\partial F^b{}_B$ involved in this calculation is closely related to one of the basic tensors in elasticity theory:

4.1 Definition Let $\hat{\boldsymbol{P}}$ be a constitutive function for the first Piola–Kirchhoff stress for thermoelasticity, depending on the point values of X, $\boldsymbol{F}$, and θ. The (*first*) *elasticity tensor* is the two-point tensor A defined by

$$\boxed{\mathsf{A} = \frac{\partial \hat{\boldsymbol{P}}}{\partial \boldsymbol{F}}, \qquad \text{that is,} \quad \mathsf{A}^{aA}{}_b{}^B = \frac{\partial \hat{P}^{aA}}{\partial F^b{}_B}.}$$

We shall also write $\mathsf{A}^\flat$ for A with its first index lowered—that is, $\mathsf{A}_a{}^A{}_b{}^B$—and shall write A_s for A symmetrized on its large indices; its components are written as follows:

---

[7] Actually there are technical subtleties involved here in understanding this calculation in non-Euclidean spaces. $\hat{\boldsymbol{P}}$ is a mapping of vector bundles: $\hat{\boldsymbol{P}}: \mathcal{E} \longrightarrow \mathcal{F}$, where $\mathcal{E}$ is the bundle over $\mathcal{B}$ whose fiber at $X \in \mathcal{B}$ is the direct sum of the space of linear maps of $T_X\mathcal{B}$ to $T_{\phi(X)}\mathcal{S}$ and the real numbers ($\mathbb{R}$), and $\mathcal{F}$ is the bundle of two-point tensors over ϕ. The fiber derivative $\partial\hat{\boldsymbol{P}}/\partial\boldsymbol{F}$ is well defined, but to write $\mathrm{DIV}_X\,\hat{\boldsymbol{P}} = \mathrm{tr}(\nabla_X\hat{\boldsymbol{P}})$, one has to put a connection on $\mathcal{E}$ and $\mathcal{F}$, compute $\nabla\boldsymbol{P}$, and take its horizontal part. This process is equivalent to the computation given in the text.

$$\mathsf{A}^{a(A}{}_b{}^{B)} = \tfrac{1}{2}(\mathsf{A}^{aA}{}_b{}^B + \mathsf{A}^{aB}{}_b{}^A).$$

The tensor $\mathbf{A}_s{}^\flat$ is defined similarly.

Problem 4.1 Prove that $\mathbf{A}^\flat = \partial\hat{\boldsymbol{P}}^\flat/\partial \boldsymbol{F}$, where $\hat{\boldsymbol{P}}^\flat$ is the tensor with components $\hat{P}_a{}^A$: that is, prove that

$$\mathsf{A}_a{}^A{}_b{}^B = \frac{\partial P_a{}^A}{\partial F^b{}_B}.$$

To exploit material frame indifference and balance of moment of momentum it is useful to work with the second Piola–Kirchhoff stress tensor. This leads to the following.

4.3 Definition Let S be a constitutive function for the second Piola–Kirchhoff stress depending on X, C, and Θ, as in Section 3.2. Then the tensor on the body $\mathcal{B}$ defined by

$$\boxed{\mathsf{C} = \frac{\partial S}{\partial C}, \qquad \text{that is,} \quad \mathsf{C}^{ABCD} = \frac{\partial \hat{S}^{AB}}{\partial C_{CD}},}$$

is called the (*second*) *elasticity tensor* or the *elasticities.*

Notice that $\mathbf{C}$ is a fourth-order tensor on $\mathcal{B}$; that is, it is not a two-point tensor. In the previous two sections, we saw that S and $\hat{P}$ are related to the free energy $\hat{\Psi}$ by $\hat{S}^{AB} = 2\rho_{\text{Ref}}(\partial\hat{\Psi}/\partial C_{AB})$ and $\hat{P}_a{}^A = \rho_{\text{Ref}}(\partial\hat{\Psi}/\partial F^a{}_A)$; so we get the following important formula:

4.4 Proposition

$$\text{(a)} \quad \mathsf{C}^{ABCD} = 2\rho_{\text{Ref}} \frac{\partial^2\hat{\Psi}}{\partial C_{AB}\,\partial C_{CD}},$$

and so we have the symmetries: $\mathsf{C}^{ABCD} = \mathsf{C}^{BACD} = \mathsf{C}^{ABDC} = \mathsf{C}^{CDAB}$.

$$\text{(b)} \quad \mathsf{A}_a{}^A{}_b{}^B = \rho_{\text{Ref}} \frac{\partial^2\hat{\Psi}}{\partial F^a{}_A\,\partial F^b{}_B},$$

so we have the symmetry $\mathsf{A}_a{}^A{}_b{}^B = \mathsf{A}_b{}^B{}_a{}^A$.

We can relate $\mathbf{A}$ and $\mathbf{C}$ using the formula $P = FS$, that is, $P^{aA} = F^a{}_B S^{BA}$. The following proposition contains the relevant computations.

4.5 Proposition *The following formulas hold in general coordinates:*

$$\boxed{\begin{aligned} &\text{(a)} \quad \mathsf{A}^{aA}{}_b{}^B = 2\mathsf{C}^{CADB} F^c{}_D F^a{}_C g_{cb} + \hat{S}^{AB}\delta^a_b \\ &\text{(b)} \quad \mathsf{A}_a{}^A{}_b{}^B = 2\mathsf{C}^{CADB} F^c{}_D F^d{}_C g_{cb} g_{da} + \hat{S}^{AB} g_{ab} \end{aligned}}$$

Proof The chain rule applied to $P^{aA} = F^a{}_B S^{BA}$ gives

$$\frac{\partial \hat{P}^{aA}}{\partial F^b{}_B} = \frac{\partial \hat{S}^{CA}}{\partial C_{DE}} \frac{\partial C_{DE}}{\partial F^b{}_B} F^a{}_C + \hat{S}^{CA} \frac{\partial F^a{}_C}{\partial F^b{}_B}$$

From $C_{DE} = F^d{}_D F^c{}_E g_{dc}$, we have

$$\frac{\partial C_{DE}}{\partial F^b{}_B} = \delta^B{}_D F^c{}_E g_{bc} + F^d{}_D \delta^B{}_E g_{db}.$$

Substituting,

$$\frac{\partial \hat{P}^{aA}}{\partial F^b{}_B} = \mathsf{C}^{CADE}(\delta^B{}_D F^c{}_E g_{bc} + F^d{}_D \delta^B{}_E g_{db}) F^a{}_C + \hat{S}^{CA} \delta^a{}_b \delta^B{}_C$$

$$= \mathsf{C}^{CABE} F^c{}_E F^a{}_C g_{bc} + \mathsf{C}^{CADB} F^d{}_D F^a{}_C g_{db} + \hat{S}^{BA} \delta^a{}_b.$$

Using the symmetries $\mathsf{C}^{CABE} = \mathsf{C}^{CAEB}$ and $S^{AB} = S^{BA}$, (a) follows. Part (b) follows by lowering the first index. ▮

The tensor $\mathbf{A}^\flat$ is not necessarily symmetric in each pair of indices AB and ab separately, but only when both pairs are *simultaneously* transposed. In three dimensions it is easy to see that the tensors with this symmetry form (pointwise) a space of dimension 45. However the dimension of the space of tensors with the symmetries of the $\mathbf{C}$ tensor is only 21. Thus in the second elasticity tensor there is less to keep track of, in principle.

Problem 4.2 Write out a set of 21 independent components of $\mathbf{C}$ explicitly.

Often $\hat{S}$ or $\hat{P}$ are taken as primitive objects and are not necessarily assumed to be derived from a free energy function $\hat{\Psi}$. Notice that even if we don't assume $\hat{\Psi}$ exists, C^{ABCD} is always symmetric in AB and CD separately and 4.5 holds. From the Poincaré lemma (Box 7.2, Chapter 1) or basic vector calculus, together with the observation that the set of variables C_{AB} is an open convex cone, we see that the symmetry condition $\mathsf{C}^{ABCD} = \mathsf{C}^{CDAB}$ is *equivalent* to the existence of a free energy function. Also notice that this symmetry is in turn equivalent to 4.4(b). We summarize:

4.6 Proposition *If the axiom of entropy production is (temporarily) dropped, the following assertions are equivalent:*

(i) *There is a function $\hat{\Psi}$ of X, C, and Θ such that*

$$\hat{S}^{AB} = 2\rho_{\mathrm{Ref}} \frac{\partial \hat{\Psi}}{\partial C_{AB}}.$$

(ii) $\mathsf{C}^{ABCD} = \mathsf{C}^{CDAB}$.

(iii) $\mathsf{A}_a{}^A{}_b{}^B = \mathsf{A}_b{}^B{}_a{}^A$.

One sometimes refers to either of these three conditions as defining *hyperelasticity*. Further discussion of these points occurs in Chapter 5.

4.7 Notation In case thermal effects are ignored—that is, if Θ is omitted—we are in the case of isothermal hyperelasticity. We shall mean this if we just say "elasticity" in the future. In this case the free energy $\hat{\Psi}$ coincides with the internal energy $\hat{E}$ and is sometimes denoted W and is called the *stored energy function.*

Thus W will be a function of (X, C) and so

$$\hat{P} = \rho_{\text{Ref}} g^{\sharp} \frac{\partial W}{\partial F}, \qquad \hat{S} = 2\rho_{\text{Ref}} \frac{\partial W}{\partial C}, \quad \text{etc.}$$

The elasticity tensor $\mathbf{C}$ measures how the stresses S are changing with the measures of strain C. In principle, these are measurable for a given material. However, in general, the fact that S is a nonlinear function of C makes this problem of *identification* difficult or impossible in practice. In specific situations, the Duhamel–Neumann hypothesis is sometimes used (see Box 3.1). In the linearized approximation, identification involves measuring 21 numbers—that is, the 21 independent components of $\mathbf{C}$. We shall discuss the process of linearization in Chapter 4.

Now we return to the equations of motion for a thermoelastic material, insert the first elasticity tensor, and formulate the basic boundary value problems. The following notation will be convenient.

4.8 Definition The vector

$$B_I = \text{DIV}_X \hat{P}, \qquad \text{that is,} \quad B_I{}^a = \frac{\partial \hat{P}^{aA}}{\partial X^A} + \hat{P}^{aA}\Gamma^B_{AB} + \hat{P}^{bA}\gamma^a_{bc}F^c{}_A,$$

is called the *resultant force due to inhomogeneities*. (B_I is a function of X, F, and Θ.)

The basic equation of motion reads

$$\rho_{\text{Ref}} \dot{V} = \rho_{\text{Ref}} B + \text{DIV}\, P = \rho_{\text{Ref}} B + \mathbf{A} \cdot \nabla_X F + B_I + \frac{\partial \hat{P}}{\partial \Theta} \cdot \text{GRAD}\, \Theta.$$

This equation is usually thught of as governing the evolution of the configuration ϕ and is coupled to the equation of energy balance that is usually thought of as governing the evolution of Θ. In addition to these evolution equations, some boundary conditions must be imposed. For each of ϕ and Θ there are three types in common use:

4.9 Definition

(I) **Boundary conditions for ϕ:**

(a) *displacement*—ϕ is prescribed on $\partial\mathcal{B}$, the boundary of $\mathcal{B}$;

(b) *traction*—the tractions $\langle P, N \rangle^a = P^{aA}N_A$ are prescribed on $\partial\mathcal{B}$; or

(c) *mixed*—ϕ is prescribed on a part ∂_d of $\partial\mathcal{B}$ and $\langle \hat{P}, N \rangle$ on part ∂_τ of $\partial\mathcal{B}$ where $\partial_d \cap \partial_\tau = \varnothing$ and $\overline{\partial_d \cup \partial_\tau} = \partial\mathcal{B}$.

(II) **Boundary conditions for Θ:**

(a) *prescribed temperature*—Θ is prescribed on $\partial\mathcal{B}$ (= Dirichlet boundary conditions);

(b) *prescribed flux*—$\langle Q, N \rangle$ is precribed on $\partial\mathcal{B}$ (= Neumann boundary conditions); or

(c) *mixed*—Θ is prescribed on a part ∂_Θ of $\partial\mathcal{B}$ and $\langle Q, N \rangle$ on another part ∂_f of $\partial\mathcal{B}$ where $\partial_\Theta \cap \partial_f = \varnothing$, $\overline{\partial_\Theta \cup \partial_f} = \partial\mathcal{B}$.

Notice that the conditions I(b) and II(b) are, in general, nonlinear boundary conditions because $\hat{\boldsymbol{P}}$ and $\hat{\boldsymbol{Q}}$ are nonlinear functions of $\boldsymbol{F}$ and Θ.

Prescribing the traction $\langle \boldsymbol{P}, \boldsymbol{N} \rangle$ to be constant is an example of *dead loading* (see Figure 3.4.1). The reason for this is that $\langle \boldsymbol{P}, \boldsymbol{N} \rangle$ is actually a traction vector that assigns to $X \in \mathcal{B}$ a vector attached to the new configuration point $\phi(X, t) = x$. This traction vector is assigned in advance, independent of x. Notice, therefore, that the dead loading boundary condition requires $\mathcal{S}$ to be, for example, a linear space for it to make sense, since it implicitly identifies vectors at different points.

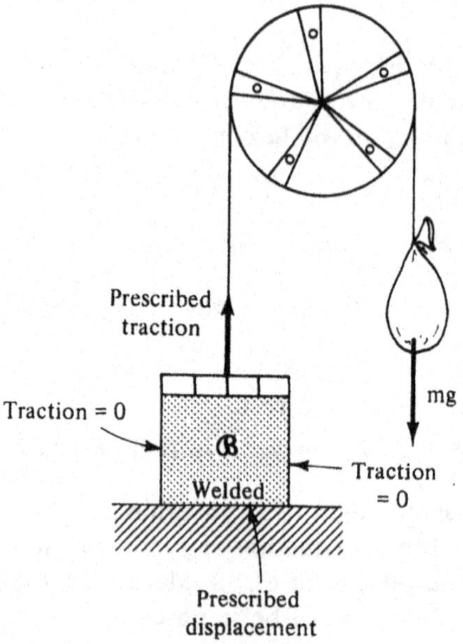

Figure 3.4.1

Problem 4.4 Formulate spatial versions of these boundary conditions.

4.10 Definition By the *initial boundary value problem for thermoelasticity,* we mean the problem of finding $\phi(X, t)$ and $\Theta(X, t)$ such that

(i) $\rho_{\text{Ref}} \dot{V} = \mathsf{A} \cdot \nabla_X F + \rho_{\text{Ref}} B + B_I + (\partial \hat{P}/\partial \Theta) \cdot \text{GRAD}\, \Theta$,
(ii) $\rho_{\text{Ref}} \Theta\, (\partial N/\partial t) + \text{DIV}\, \hat{Q} = \rho_{\text{Ref}} R$,
(iii) boundary conditions (I) and (II) hold, and
(iv) ϕ, V, and Θ are given at $t = 0$ (*initial conditions*),

where $\hat{\Psi}$ is a given constitutive function depending on X, C, and Θ and $\hat{N}$, $\hat{P}$, A, and B_I are given in terms of it as above, where B, ρ_{Ref}, and R are given and $\hat{Q}$ is a given function of X, C, Θ, and GRAD Θ satisfying $\langle \hat{Q}, \text{GRAD}\, \Theta \rangle \leq 0$.

If the motions are not sufficiently differentiable and shocks can develop, these conditions have to be supplemented by appropriate discontinuity conditions.

Notice that in (ii),

$$\rho_{\text{Ref}} \frac{\partial \hat{N}}{\partial t} = \rho_{\text{Ref}} \frac{\partial \hat{N}}{\partial C} \frac{\partial C}{\partial t} + \rho_{\text{Ref}} \frac{\partial \hat{N}}{\partial \Theta} \frac{\partial \Theta}{\partial t} = -\rho_{\text{Ref}} \frac{\partial}{\partial C} \frac{\partial \hat{\Psi}}{\partial \Theta} \cdot 2D - \rho_{\text{Ref}} \frac{\partial^2 \hat{\Psi}}{\partial \Theta^2} \frac{\partial \Theta}{\partial t}$$

$$= -\frac{\partial \boldsymbol{S}}{\partial \Theta} : \boldsymbol{D} - \rho_{\text{Ref}} \frac{\partial^2 \hat{\Psi}}{\partial \Theta^2} \frac{\partial \Theta}{\partial t},$$

where $\boldsymbol{D}$ is the rate of deformation tensor.

4.11 Definition If we omit Θ, the corresponding equation (ii), the term $(\partial \hat{P}/\partial \Theta) \cdot \text{GRAD}\, \Theta$ in (i), and the boundary and initial conditions for Θ, the resulting problem for determination of ϕ is called the *initial boundary value problem for elasticity* (or hyperelasticity).

In Chapter 6 we shall study the initial boundary value problem posed here in some detail, as well as the corresponding static problem.

4.12 Definition The *boundary value problem for thermoelastostatics* consists of finding ϕ and Θ as functions of X alone such that

(i) $\text{DIV}\, \hat{\boldsymbol{P}} + \rho_{\text{Ref}} \boldsymbol{B} = \mathbf{A} \cdot \nabla_X F + \rho_{\text{Ref}} \boldsymbol{B} + \boldsymbol{B}_I + (\partial \hat{P}/\partial \Theta) \cdot \text{GRAD}\, \Theta = \mathbf{0}$,
(ii) $\text{DIV}\, \hat{\boldsymbol{Q}} = \rho_{\text{Ref}} R$, and
(iii) boundary conditions (I) and (II) hold.

The *boundary value problem for elastostatics* consists of finding a (regular) deformation ϕ such that (i) holds (with the Θ term omitted) and boundary conditions (I) hold.

The static problem is, of course, obtained from the dynamic one by dropping time derivatives. There is a simple basic necessary condition the traction must satisfy in order that the static problem be soluble (assuming, as usual, regularity).

4.13 Proposition *The prescribed tractions $\langle \hat{\boldsymbol{P}}, \boldsymbol{N} \rangle$ in $I(b)$ on $\partial \mathcal{B}$ must satisfy the necessary condition*

$$\int_{\partial \mathcal{B}} \langle \hat{\boldsymbol{P}}, \boldsymbol{N} \rangle \, dA + \int_{\mathcal{B}} \rho_{\text{Ref}} \boldsymbol{B} \, dV = \mathbf{0} \quad (\textit{using Euclidean coordinates in } \mathbb{R}^n)$$

if the traction boundary value problem for (thermo-) elastostatics has a (regular) solution. Similarly the prescribed fluxes in $II(b)$ must satisfy

$$\int_{\partial \mathcal{B}} \langle \hat{\boldsymbol{Q}}, \boldsymbol{N} \rangle \, dA - \int_{\mathcal{B}} \rho_{\text{Ref}} R \, dV = 0.$$

Proof This follows from $\text{DIV}\, \hat{\boldsymbol{P}} + \rho_{\text{Ref}} \boldsymbol{B} = \mathbf{0}$ by integration over $\mathcal{B}$ and use of the divergence theorem, etc. ∎

There are more general boundary conditions that are compatible with the principle of virtual work (see Box 3.1, Chapter 2). The interested reader should consult Antman and Osborne [1979].

All of the above can be formulated in terms of the spatial picture and it is sometimes useful to do so. We give some of the key ideas in the following:

4.14 Definition Define the *spatial elasticity tensors* a and c by these coordinate formulas:

$$\mathsf{a}^{ac}{}_b{}^d = \frac{1}{J} F^c{}_A F^d{}_B \mathsf{A}^{aA}{}_b{}^B,$$

$$\mathsf{c}^{abcd} = \frac{2}{J} F^a{}_A F^b{}_B F^c{}_C F^d{}_D \mathsf{C}^{ABCD},$$

that is, a and c are push-forwards and Piola transforms of A and C on each large index.

The following relations then follow from the corresponding material results and the Piola identity.

4.15 Proposition *The following hold:*

(a) $\mathsf{a}^{ac}{}_b{}^d = \sigma^{cd}\delta^a{}_b + \mathsf{c}^{aced}g_{eb}$, *that is,* $\mathsf{a}^{acbd} = \sigma^{cd}g^{ab} + \mathsf{c}^{acbd}$.

(b) a *and* c *have these symmetries:*

$$\mathsf{a}^{acbd} = \mathsf{a}^{bdac},$$

$$\mathsf{c}^{abcd} = \mathsf{c}^{bacd} = \mathsf{c}^{abdc} = \mathsf{c}^{cdab}.$$

(c) *If* $U^a{}_B$ *is a two-point tensor field over* ϕ, *then*

$$(\mathsf{A}^{aA}{}_b{}^B U^b{}_B)_{|A} = J(\mathsf{a}^{ad}{}_b{}^e u^b{}_e)_{|d},$$

where $u^d{}_b = (F^{-1})^B{}_b U^d{}_B$ *is the push-forward of* U.

The tensors a and c play an important role in the theory of linearization (see Chapter 4).

The equations of motion in the spatial picture

$$\rho a = \operatorname{div} \boldsymbol{\sigma} + \rho b$$

become

$$\rho a^a = \mathsf{e}^{ab}{}_d{}^B F^d{}_{B|b} + (\operatorname{div}_x \boldsymbol{\sigma})^a + \rho b^a,$$

where e is the tensor given by

$$\mathsf{e}^{ab}{}_d{}^B = \frac{\partial \sigma^{ab}}{\partial F^d{}_B}$$

and div_x is the divergence holding F constant. The reader may wish to work out the relationship between e and a. The tensor c is *not* given by $\partial\sigma^{ab}/\partial c_{cd}$, and $\sigma^{ab} \neq 2\rho(\partial\psi/\partial c_{ab})$. Rather, the correct relations are

$$\mathsf{c}^{abcd} = \frac{\partial \sigma^{ab}}{\partial g_{cd}} \quad \text{and} \quad \sigma^{ab} = 2\rho\frac{\partial \hat{\psi}}{\partial g_{ab}} \quad \text{(see Section 3.3).}$$

We conclude this section with a brief remark on incompressible elasticity, Here one imposes the constraint that ϕ be volume preserving—that is, $J = 1$. For instance, such a condition is often imposed on rubber. (The Mooney–Rivlin–Ogden constitutive assumption for rubber is given in the next section.) The condition of incompressibility is perhaps best understood in terms of Hamiltonian systems with constraints, and we shall discuss this point of view in Chapter 5. For now we merely remark that this condition introduces a Lagrange multiplier into the equation as follows: we replace

$$\boldsymbol{\sigma} \text{ by } \boldsymbol{\sigma} - p\boldsymbol{g}^\sharp, \qquad \text{that is, } \sigma^{ab} \text{ by } \sigma^{ab} - pg^{ab},$$

where p is an unknown function, the *pressure*, to be determined by the condition of incompressibility. In terms of the first Piola–Kirchhoff tensor $\boldsymbol{P}$, we replace $\hat{\boldsymbol{P}}$ from our constitutive theory by $\hat{\boldsymbol{P}} - JPF^{-1}$, where P is a function of (X, t) to be determined by $J = 1$. We emphasize that in an initial boundary value problem, P becomes an unknown and it depends on ϕ in a non-local way, as in fluid mechanics (see Hughes and Marsden [1976] and Ebin and Marsden [1970]).

Box 4.1 ***Summary of Important Formulas for Section 3.4***

Elasticity Tensors

$$\mathsf{A} = \frac{\partial \hat{P}}{\partial F}, \quad \mathsf{C} = \frac{\partial \hat{S}}{\partial C} \qquad \mathsf{A}^{aA}{}_b{}^B = \frac{\partial \hat{P}^{aA}}{\partial F^b{}_B}, \quad \mathsf{C}^{ABCD} = \frac{\partial \hat{S}^{AB}}{\partial C_{CD}}$$

Symmetries of Elasticity Tensors

C is symmetric in its first and second slots, and in its third and fourth slots: $\mathsf{C}^{ABCD} = \mathsf{C}^{BACD} = \mathsf{C}^{ABDC}$

If $\hat{\boldsymbol{P}}, \hat{\boldsymbol{S}}$ are derived from an internal energy function (hyperelasticity): $\mathsf{C}^{ABCD} = \mathsf{C}^{CDAB}, \quad \mathsf{A}_a{}^A{}_b{}^B = \mathsf{A}_b{}^B{}_a{}^A$

Relationship Between A *and* C

$$\mathsf{A} = 2\mathsf{C}\cdot F\cdot F\cdot g + \hat{S}\otimes\boldsymbol{\delta} \qquad \mathsf{A}^{aA}{}_b{}^B = 2\mathsf{C}^{CADB}F^c{}_D F^a{}_C g_{cb} + \hat{S}^{AB}\delta^a{}_b$$

Stored Energy Function (for Isothermal Hyperelasticity)

If Θ is absent, $\hat{E} = \hat{\Psi}$ is often denoted by W; it depends on X, C:

$$\hat{P} = \rho_{\text{Ref}}\, g^\sharp \frac{\partial W}{\partial F} \qquad \hat{P}^{aA} = \rho_{\text{Ref}}\, g^{ab}\frac{\partial W}{\partial F^b{}_A}$$

$$\hat{S} = 2\rho_{\text{Ref}}\frac{\partial W}{\partial C} \qquad \hat{S}^{AB} = 2\rho_{\text{Ref}}\frac{\partial W}{\partial C_{AB}}$$

Resultant Force due to Inhomogeneities

$$B_I = \text{DIV}_X \hat{P} \qquad B_I^a = \frac{\partial \hat{P}^{aA}}{\partial X^A} + \hat{P}^{aA}\Gamma^B_{AB} + \hat{P}^{bA}\gamma^a_{bc}F^c{}_A$$

Boundary Conditions for ϕ:

(a) *displacement*—ϕ given on $\partial\mathcal{B}$;
(b) *traction*—$\langle P, N\rangle^a = P^{aA} N_A$ given on $\partial\mathcal{B}$; or
(c) *mixed*—ϕ given on ∂_d, traction on ∂_τ, where $\partial_d \cap \partial_\tau = \varnothing$ and $\overline{\partial_d \cup \partial_\tau} = \partial\mathcal{B}$.

Boundary Conditions for Θ:

(a) *temperature*—Θ given on $\partial\mathcal{B}$;
(b) *flux*—$\langle Q, N\rangle = Q^A N_A$ given on $\partial\mathcal{B}$; or
(c) *mixed*—Θ given on ∂_Θ, $\langle Q, N\rangle$ given on ∂_f, where $\partial_\Theta \cap \partial_f = \varnothing$, and $\overline{\partial_\Theta \cup \partial_f} = \partial\mathcal{B}$.

Evolution Equation for ϕ for a Thermoelastic Solid (remove Θ for elasticity)

$$\rho_{\text{Ref}}\dot{V} = \mathbf{A}\cdot\nabla_X F + \rho_{\text{Ref}} B + B_I + \frac{\partial \hat{P}}{\partial \Theta}\cdot \text{GRAD}\,\Theta$$

$$\rho_{\text{Ref}}\frac{\partial^2 \phi^a}{\partial t^2} + \rho_{\text{Ref}}\frac{\partial \phi^b}{\partial t}\frac{\partial \phi^c}{\partial t}\gamma^a_{bc} = \mathsf{A}^{aA}{}_b{}^B \phi^b{}_{|A|B} + \rho_{\text{Ref}} B^a + B^a_I + \frac{\partial \hat{P}^{aA}}{\partial \Theta}\frac{\partial \Theta}{\partial X^A}$$

Evolution Equation for Θ for a Thermoelastic Solid

$$-\rho_{\text{Ref}}\frac{\partial^2 \hat{\Psi}}{\partial \Theta^2}\frac{\partial \Theta}{\partial t} - \frac{\partial \hat{S}}{\partial \Theta} : D = \rho_{\text{Ref}} - \text{DIV}\,\hat{Q}$$

$$-\rho_{\text{Ref}}\frac{\partial^2 \hat{\Psi}}{\partial \Theta^2}\frac{\partial \Theta}{\partial t} - \frac{\partial \hat{S}^{AB}}{\partial \Theta}\cdot D_{AB} = \rho_{\text{Ref}} R - \hat{Q}^A{}_{|A}$$

Initial Boundary Value Problem for Thermoelastodynamics (ignore Θ for elastodynamics)

Find $\phi(X, t)$, $\Theta(X, t)$ satisfying their respective evolution equations, boundary conditions, and with ϕ, $\partial\phi/\partial t$ and Θ given at $t = 0$.

Boundary Value Problem for Thermoelastostatics (ignore Θ for elastostatics)

Find $\phi(X)$, $\Theta(X)$ such that the right-hand sides of the above evolution equations are zero, and the boundary conditions are satisfied.

Necessary Conditions on the Boundary Data for Thermoelastostatics (ignore Θ for elastostatics)

If the traction boundary condition for ϕ and the flux boundary condition for Θ are used, then in a Cartesian frame

$$\int_{\partial\mathcal{B}} \langle \hat{P}, N\rangle \, dA + \int_{\mathcal{B}} \rho_{\text{Ref}} B \, dV = \mathbf{0}$$

and

$$\int_{\partial\mathcal{B}} \langle \hat{Q}, N\rangle \, dA - \int_{\mathcal{B}} \rho_{\text{Ref}} R \, dV = 0$$

The Spatial Elasticity Tensors

$$\mathsf{a} = \frac{1}{J}\phi_* \mathsf{A} \qquad\qquad \mathsf{a}^{ac}{}_b{}^d = \frac{1}{J} F^c{}_A F^d{}_B \mathsf{A}^{aA}{}_b{}^B$$

$\mathbf{c} = 2\frac{1}{J}\phi_*\mathbf{C}$	$\mathsf{c}^{abcd} = 2\frac{1}{J}F^a{}_A F^b{}_B F^c{}_C F^d{}_D \mathsf{C}^{ABCD}$
$\mathbf{a} = \mathbf{c}\cdot g + \boldsymbol{\sigma} \otimes \delta$	$\mathsf{a}^{ac}{}_b{}^d = \mathsf{c}^{aced} g_{eb} + \sigma^{cd}\delta^a{}_b$
Symmetries	
	$\mathsf{a}^{acbd} = \mathsf{a}^{bdac}$,
	$\mathsf{c}^{abcd} = \mathsf{c}^{bacd} = \mathsf{c}^{abdc} = \mathsf{c}^{cdab}$
Piola Identity	
$\mathrm{DIV}(\mathbf{A}\cdot U) = J\,\mathrm{div}(\mathbf{a}\cdot u)$,	$(\mathsf{A}^{aA}{}_b{}^B U^b{}_B)_{\vert A} = J(\mathsf{a}^{ac}{}_b{}^d u^b{}_d)_{\vert c}$
where $u = \phi_* U$	

3.5 MATERIAL SYMMETRIES AND ISOTROPIC ELASTICITY

In Sections 3.2 and 3.3 we proved that various locality, thermodynamic, or covariance assumptions simplified the functional form of the constitutive functions. In this section we investigate further simplifications when the material has some symmetries. In particular, we are concerned with isotropic materials, and within that class shall give the St. Venant–Kirchhoff and Mooney–Rivlin–Ogden materials as examples. We work primarily with the free energy $\hat{\Psi}$, since the other constitutive functions and the elasticity tensors are derived from it.

5.1 Definition A *material symmetry* for $\hat{\Psi}$ at a point $X_0 \in \mathcal{B}$ is a linear isometry $\boldsymbol{\lambda}: T_{X_0}\mathcal{B} \longrightarrow T_{X_0}\mathcal{B}$ (i.e., $\boldsymbol{\lambda}$ preserves the inner product G_{X_0}) such that

$$\hat{\Psi}(X_0, C, \Theta) = \hat{\Psi}(X_0, \boldsymbol{\lambda}^*C, \Theta),$$

where C is an arbitrary symmetric positive-definite 2-tensor at X_0 and $\boldsymbol{\lambda}^*C$ is its transformation (pull-back) under $\boldsymbol{\lambda}$. The set of all material symmetries of $\hat{\Psi}$ at X_0 is denoted $\mathfrak{S}_{X_0}(\hat{\Psi})$ as is called the *material symmetry group* of $\hat{\Psi}$ at X_0. Similarly, $\boldsymbol{\lambda}$ is a *material symmetry for* $\hat{Q}$ at X_0 if

$$\hat{Q}(X_0, \boldsymbol{\lambda}^*C, \Theta, \boldsymbol{\lambda}^*\nabla\Theta) = \boldsymbol{\lambda}^*\hat{Q}(X_0, C, \Theta, \nabla\Theta).$$

($\hat{Q}$ is assumed to be grade (1, 1).) The set of such symmetries is denoted $\mathfrak{S}_{X_0}(\hat{Q})$. The *material symmetries for a thermoelastic material* are the simultaneous symmetries for $\hat{\Psi}$ and $\hat{Q}$. For pure elasticity, we look only at the symmetries of the internal energy $\hat{E}$ (i.e., $\hat{W}$). We shall write $\mathfrak{S}_{X_0} = \mathfrak{S}_{X_0}(\hat{\Psi}) \cap \mathfrak{S}_{X_0}(\hat{Q})$ for the material symmetries and understand $\mathfrak{S}_{X_0}$ to mean $\mathfrak{S}_{X_0}(\hat{E})$ for pure elasticity.

Relative to a coordinate system $\{X^A\}$ on $\mathcal{B}$ we write the components of $\boldsymbol{\lambda}$ as $\lambda^A{}_B$ so that if $V \in T_{X_0}\mathcal{B}$ has components V^A, then $(\boldsymbol{\lambda}\cdot V)^A = \lambda^A{}_B V^B$. The relationship $\hat{\Psi}(X_0, C, \Theta) = \hat{\Psi}(X_0, \boldsymbol{\lambda}^*C, \Theta)$ in coordinates reads

$$\hat{\Psi}(X_0, C_{AB}, \Theta) = \hat{\Psi}(X_0, C_{CD}\lambda^C{}_A\lambda^D{}_B, \Theta).$$

Similarly, the coordinate version for the symmetry condition on $\hat{Q}$ reads

$$\hat{Q}^E(X_0, C_{CD}\lambda^C{}_A\lambda^D{}_B, \Theta, \Theta_{|A}\lambda^A{}_B) = (\lambda^{-1})^E{}_F\hat{Q}^F(X_0, C_{AB}, \Theta, \Theta_{|B}).$$

It is easily checked that $\mathfrak{S}_{X_0}$ is a group. It is a subgroup of the group $GL(T_{X_0}\mathcal{B})$ of invertible linear transformations of $T_{X_0}\mathcal{B}$ to itself, and in fact a subgroup of $O(T_{X_0}\mathcal{B})$, the orthogonal linear transformations of $T_{X_0}\mathcal{B}$ to itself (orthogonal with respect to the inner product $\boldsymbol{G}_{X_0}$). Furthermore, one can show that $\mathfrak{S}_{X_0}$ is a Lie group; that is, it is a smooth manifold and group multiplication and inversion are smooth maps.[8]

5.2 Definition If $\boldsymbol{\lambda}(s)$ is a smooth curve in $\mathfrak{S}_{X_0}$ with $\boldsymbol{\lambda}(0) =$ identity, the linear transformation $\boldsymbol{\xi}: T_{X_0}\mathcal{B} \longrightarrow T_{X_0}\mathcal{B}$ defined by

$$\boldsymbol{\xi}(\boldsymbol{V}) = \frac{d}{ds}\boldsymbol{\lambda}(s)\cdot\boldsymbol{V}_{|s=0}$$

is called an *infinitesimal material symmetry*. The collection of all such $\boldsymbol{\xi}$ is called the *Lie algebra* of $\mathfrak{S}_{X_0}$ and is denoted $\mathfrak{s}_{X_0}$. The *Lie bracket* of $\boldsymbol{\xi} \in \mathfrak{s}_{X_0}$ and $\boldsymbol{\eta} \in \mathfrak{s}_{X_0}$ is given by

$$[\boldsymbol{\xi}, \boldsymbol{\eta}](\boldsymbol{V}) = \boldsymbol{\xi}(\boldsymbol{\eta}(\boldsymbol{V})) - \boldsymbol{\eta}(\boldsymbol{\xi}(\boldsymbol{V})).$$

5.3 Example Let $\mathcal{B} = \mathbb{R}^3$ and let X_0 be the origin. Suppose $\mathfrak{S}_{X_0}$ consists of rotations about the z-axis. Then $\mathfrak{S}_{X_0}$ is one-dimensional, parametrized by the angle of rotation. A curve in $\mathfrak{S}_{X_0}$ is

$$\boldsymbol{\lambda}(s) = \begin{pmatrix} \cos s & \sin s & 0 \\ -\sin s & \cos s & 0 \\ 0 & 0 & 1 \end{pmatrix}$$

and the corresponding infinitesimal symmetry is given by

$$\boldsymbol{\xi}(\boldsymbol{V}) = -\boldsymbol{k} \times \boldsymbol{V} \quad (\boldsymbol{k} = (0, 0, 1)).$$

If $\mathfrak{S}_{X_0}$ is the full rotation group $O(3)$ and if the infinitesimal generators $\boldsymbol{\xi}$ are identified with the axes $\boldsymbol{\omega}$ about which they are rotations, that is, if

$$\boldsymbol{\xi}(\boldsymbol{V}) = -\boldsymbol{\omega} \times \boldsymbol{V},$$

then the Lie algebra is identified with $\mathbb{R}^3$ and the Lie bracket with the cross product. (The reader should verify these statements.) ▮

Now we proceed to investigate the consequences of a material symmetry group on the constitutive functions.

5.4 Proposition *For* $\boldsymbol{\lambda} \in \mathfrak{S}_{X_0}(\hat{\Psi})$, *we have:*

(i) $\hat{N}(X_0, \boldsymbol{\lambda}^*\boldsymbol{C}, \Theta) = \hat{N}(X_0, \boldsymbol{C}, \Theta)$,

---

[8] We shall not require a background in Lie group theory for what follows, although such knowledge may be helpful. A concise summary of some of the important facts may be found in Abraham and Marsden [1978], Sect. 4-1.

(ii) $\hat{S}(X_0, \boldsymbol{\lambda}^*C, \Theta) = \boldsymbol{\lambda}^*\hat{S}(X_0, C, \Theta)$, and
(iii) $\mathbf{C}(X_0, \boldsymbol{\lambda}^*C, \Theta) = \boldsymbol{\lambda}^*\mathbf{C}(X_0, C, \Theta)$.

Proof This follows by differentiation of the relationship $\hat{\Psi}(X_0, \boldsymbol{\lambda}^*C, \Theta) = \hat{\Psi}(X_0, C, \Theta)$ with respect to Θ and to C, respectively. For instance, differentiation with respect to C and using $\hat{S} = 2\rho_{\text{Ref}}(\partial\hat{\Psi}/\partial C)$ gives

$$\hat{S}(X_0, \boldsymbol{\lambda}^*C, \Theta)\cdot\boldsymbol{\lambda}^* = \hat{S}(X_0, C, \Theta),$$

that is,

$$\hat{S}^{AB}(X_0, \boldsymbol{\lambda}^*C, \Theta)\lambda^C{}_A\lambda^D{}_B = \hat{S}^{CD}(X_0, C, \Theta).$$

Thus

$$\hat{S}^{AB}(X_0, \lambda^*C, \Theta) = (\lambda^{-1})^A{}_C(\lambda^{-1})^B{}_D\hat{S}^{CD}(X_0, C, \Theta).$$

This yields (ii) of our proposition. ∎

The infinitesimal versions of 5.4 lead to differential identities that are useful in studying a given symmetry. Some preparatory remarks are needed. Recall that if ϕ_t is a motion and $\mathbf{t}$ a time-independent tensor field, $(d/dt)\phi_t^*\mathbf{t} = \phi_t^*\mathfrak{L}_v\mathbf{t}$ defines the Lie derivative. If $\phi_t(X_0) = X_0$ for all t, then $\boldsymbol{v}(X_0) = \mathbf{0}$ and $\mathfrak{L}_v\mathbf{t}$ does not involve spatial derivatives of $\mathbf{t}$ and depends only on the first derivative of $\boldsymbol{v}$. If $\boldsymbol{\lambda}(s)$ is a curve in $GL(T_{X_0}\mathfrak{B})$, and $\mathbf{t}$ is a tensor at X_0, then $(d/ds)\boldsymbol{\lambda}(s)^*\mathbf{t}_{|s=0} = \mathfrak{L}_\xi\mathbf{t}$ is defined; if $\boldsymbol{\lambda}(s) = T\phi_s(X_0)$, then $\xi = D\boldsymbol{v}(X_0)$ and the two concepts agree. Now we are ready to linearize 5.4.

5.5 Theorem *For each $\xi \in \mathscr{s}_{X_0}$, the following identities must hold at X_0:*

(i) $\hat{S}(X_0, C, \Theta)\cdot\mathfrak{L}_\xi C = 0$,

(ii) $\dfrac{\partial\hat{N}}{\partial C}(X_0, C, \Theta)\cdot\mathfrak{L}_\xi C = 0$,

(iii) $\mathbf{C}(X_0, C, \Theta)\cdot\mathfrak{L}_\xi C = \mathfrak{L}_\xi\hat{S}(X_0, C, \Theta)$
(*the contraction in the left-hand side is in the last two slots of* $\mathbf{C}$), *and*

(iv) $$\frac{\partial\hat{Q}}{\partial C}(X_0, C, \Theta, \text{GRAD}\,\Theta)\cdot\mathfrak{L}_\xi C + \frac{\partial\hat{Q}}{\partial(\text{GRAD}\,\Theta)}(X_0, C, \Theta, \text{GRAD}\,\Theta)\cdot\mathfrak{L}_\xi(\text{GRAD}\,\Theta) = \mathfrak{L}_\xi\hat{Q}(X_0, C, \Theta, \text{GRAD}\,\Theta).$$

Proof Let ξ be tangent to the curve $\boldsymbol{\lambda}(s) \in \mathfrak{S}_{X_0}$ at $s = 0$. Then differentiation of

$$\hat{\Psi}(X_0, \boldsymbol{\lambda}(s)^*C, \Theta) = \hat{\Psi}(X_0, C, \Theta)$$

in s at $s = 0$ using the chain rule and the definition of the Lie derivative immediately yields (i). The others are similarly obtained. ∎

We shall not study the general implications of these identities, nor the classification of possible symmetry groups; the interested reader should consult Wang and Truesdell [1973]. Instead, we specialize immediately to the isotropic case.

5.6 Definition Let $\mathcal{B}$ be a simple thermoelastic body in $\mathbb{R}^3$. We say $\mathcal{B}$ is *isotropic* at $X_0 \in \mathcal{B}$ if $\mathfrak{S}_{X_0} \supset \mathrm{SO}(T_{X_0}\mathcal{B}) = \mathrm{SO}(3)$, the group of proper orthogonal 3×3 matrices. A material is *isotropic* if it is isotropic at every point.

In Section 3.3 we defined $\hat{\Psi}$ to be materially covariant if, for all diffeomorphisms $\Xi: \mathcal{B} \longrightarrow \mathcal{B}$, we have

$$\Xi^*\{\hat{\Psi}(G, C, \Theta)\} = \hat{\Psi}(\Xi^*G, \Xi^*C, \Xi^*\Theta)$$

(the base points X are suppressed since C, Θ are now fields, and G is included).

5.7 Proposition *Suppose $\hat{\Psi}$ is materially covariant. Then for each $X_0 \in \mathcal{B}$,*

$$\mathfrak{S}_{X_0}(\hat{\Psi}) \supset \mathrm{SO}(T_{X_0}\mathcal{B}).$$

In particular, any materially covariant body in $\mathbb{R}^3$ is isotropic.

Proof Let $\boldsymbol{\lambda} \in \mathrm{SO}(T_{X_0}\mathcal{B})$. We claim that there is a diffeomorphism $\Xi: \mathcal{B} \longrightarrow \mathcal{B}$ such that $\Xi(X_0) = X_0$ and $D\Xi(X_0) = \boldsymbol{\lambda}$. Indeed, we can write $\boldsymbol{\lambda} = \exp(\boldsymbol{\xi})$ for $\boldsymbol{\xi}$ a skew 3×3 matrix; find a vector field V on $\mathcal{B}$ such that $V(X_0) = \mathbf{0}$ and $DV(X_0) = \boldsymbol{\xi}$. We can do this in local coordinates and extend V to be arbitrary outside this neighborhood, say zero near $\partial\mathcal{B}$. Let Ξ_t be the flow of V and let $\Xi = \Xi_1$. By uniqueness of solutions to differential equations, $D\Xi_t(X_0) = \exp(t\boldsymbol{\xi})$, so Ξ has the required properties. At X_0, the relation $\Xi^*(\hat{\Psi}(G, C, \Theta)) = \hat{\Psi}(\Xi^*G, \Xi^*C, \Xi^*\Theta)$ then reduces to $\hat{\Psi}(X_0, C, \Theta) = \hat{\Psi}(X_0, \boldsymbol{\lambda}^*C, \Theta)$, so λ is a material symmetry. ∎

Roughly speaking, this proposition states that if one makes up a free energy function in a fully covariant or "tensorial" manner out of G, C, and Θ, then it necessarily will be isotropic. To describe non-isotropic materials then requires some "non-tensorial" constructions or the introduction of additional variables.

By definition, an isotropic free energy constitutive function $\hat{\Psi}$ is to be a "rotationally invariant" function of the argument C. Since C is symmetric, it can be brought to diagonal form by an orthogonal transformation, so $\hat{\Psi}$ is a function only of the eigenvalues of C; that is, $\hat{\Psi}$ depends only on the principal stretches. Since the eigenvalues are reasonably complicated functions of C, it is sometimes convenient to use the invariants of C.

5.8 Definition The *invariants* of a symmetrix matrix C (in an inner product space) are defined by

$$I_1(C) = \operatorname{tr} C, \qquad I_2(C) = \det C \operatorname{tr} C^{-1} \quad \text{and} \quad I_3(C) = \det C.$$

5.9 Proposition *The invariants of C are related to the coefficients in the characteristic polynomial $P(v)$ of C as follows:*

$$P(v) = v^3 - I_1(C)v^2 + I_2(C)v - I_3(C).$$

In terms of the (eigenvalues) v_1, v_2, v_3, *we have*

$$I_1(C) = v_1 + v_2 + v_3, \qquad I_2(C) = v_1v_2 + v_1v_3 + v_2v_3, \quad \text{and} \quad I_3(C) = v_1v_2v_3$$

(these are the elementary symmetric functions of ν_1, ν_2, ν_3). Moreover, the following formula for I_2 holds:

$$I_2(C) = \tfrac{1}{2}[(\operatorname{tr} C)^2 - \operatorname{tr}(C^2)]$$

This is verified by using an orthonormal basis in which C is diagonal, noting that I_1, I_2, I_3 are rotationally invariant functions.

Since ν_1, ν_2, ν_3 completely determine the characteristic polynomial and hence the invariants, and vice versa, we obtain the following:

5.10 Proposition *The following are equivalent:*

(a) *A scalar function f of C is invariant under orthogonal transformations.*
(b) *f is a function of the invariants of C.*
(c) *f is a symmetric function of the principal stretches.* ▮

Thus for isotropic materials we can regard $\hat{\Psi}$ as a function of X, I_1, I_2, I_3, and Θ. Note that the number of arguments in the C variable is thus reduced from 6 to 3. The invariants are covariant scalar functions of C, so we obtain from 5.7 and 5.10 the following:

5.11 Corollary *A body is materially covariant if and only if it is isotropic.*

Next we compute the second Piola–Kirchhoff stress tensor in terms of this data.

5.12 Theorem *For isotropic materials, the following constitutive relation holds:*

$$\boxed{\hat{S} = \alpha_0 G + \alpha_1 C + \alpha_2 C^2, \qquad \text{that is,} \quad \hat{S}_{AB} = \alpha_0 G_{AB} + \alpha_1 C_{AB} + \alpha_2 C_A{}^D C_{DB},}$$

where the α_i's are scalar functions of X, the invariants of C, and Θ.

Proof We have $\hat{S} = 2\rho_{\text{Ref}}(\partial\hat{\Psi}/\partial C)$. By the chain rule,

$$\frac{\partial\hat{\Psi}}{\partial C} = \frac{\partial\hat{\Psi}}{\partial I_1}\frac{\partial I_1}{\partial C} + \frac{\partial\hat{\Psi}}{\partial I_2}\frac{\partial I_2}{\partial C} + \frac{\partial\hat{\Psi}}{\partial I_3}\frac{\partial I_3}{\partial C}.$$

Now $I_1(C) = C_{AB}G^{AB}$, so $\partial I_1/\partial C = G^\sharp$, that is, $\partial I_1/C_{AB} = G^{AB}$. Next we calculate the derivative of I_3.

5.13 Lemma

$$\frac{\partial I_3}{\partial C} = (\det C)\cdot C^{-1}$$

Proof From the definition of the determinant,

$$\det C = \epsilon^{BCD} C_{A_1B} C_{A_2C} C_{A_3D},$$

where $\epsilon^{BCD} = \pm 1$ depending on whether (B, C, D) is an even or odd permutation of (1, 2, 3) and where (A_1, A_2, A_3) is a fixed even permutation of (1, 2, 3). Thus

$$\frac{\partial}{\partial C_{A_1B}}(\det C) = \epsilon^{BCD}C_{A_2C}C_{A_3D} = \epsilon^{BCD}C_{A_1E}C_{A_2C}C_{A_3D}(C^{-1})^{A_1E}$$
$$= (\det C)\delta^B{}_E C^{A_1E} = (\det C)(C^{-1})^{A_1B} \quad \blacksquare$$

The derivative of I_2 is given by

$$\frac{\partial I_2}{\partial C} = \left(\frac{\partial}{\partial C}\det C\right)\operatorname{tr} C^{-1} + \det C\frac{\partial}{\partial C}\operatorname{tr} C^{-1}$$
$$= (C^{-1}\det C \operatorname{tr} C^{-1}) - (\det C)\operatorname{tr}\frac{\partial C^{-1}}{\partial C}.$$

To carry on we need to compute $\partial(C^{-1})/\partial C$:

5.14 Lemma

$$\frac{\partial C^{-1}}{\partial C}\cdot H = -C^{-1}\cdot H\cdot C^{-1}, \qquad \text{that is,} \quad \frac{\partial(C^{-1})^{AB}}{\partial C_{CD}} = -(C^{-1})^{AC}(C^{-1})^{DB}.$$

Proof Differentiate the identity $C\cdot C^{-1} = Id$ in the direction H to get

$$H\cdot C^{-1} + C\cdot\frac{\partial C^{-1}}{\partial C}\cdot H = 0,$$

which gives the result. $\blacksquare$

We now note that

$$\operatorname{tr}\left(\frac{\partial C^{-1}}{\partial C}\right)^{CD} = -(C^{-1})^{AC}(C^{-1})^{D}{}_{A}, \qquad \text{that is,} \quad \operatorname{tr}\left(\frac{\partial C^{-1}}{\partial C}\right) = -C^{-2}.$$

Substitution of these formulas yields:

$$\hat{S} = 2\rho_{\mathrm{Ref}}\left[\frac{\partial\hat{\Psi}}{\partial I_1}G^{\flat} + \left(\frac{\partial\hat{\Psi}}{\partial I_2}I_2 + \frac{\partial\hat{\Psi}}{\partial I_3}I_3\right)C^{-1} - \frac{\partial\hat{\Psi}}{\partial I_2}I_3C^{-2}\right].$$

The Cayley–Hamilton theorem from linear algebra tells us that C satisfies its characteristic equation:

$$C^3 - I_1(C)C^2 + I_2(C)C - I_3(C) = 0.$$

Thus

$$C^{-1} = \frac{1}{I_3(C)}\{C^2 - I_1(C)C + I_2(C)\}$$

and

$$C^{-2} = \frac{1}{I_3(C)}\{C - I_1(C) + I_2(C)C^{-1}\}$$
$$= \frac{1}{I_3(C)}\left\{\frac{I_2(C)}{I_3(C)}C^2 + \left(Id - \frac{I_1(C)I_2(C)}{I_3(C)}\right)C + \left(\frac{I_2(C)}{I_3(C)} - I_1(C)\right)\right\}.$$

Inserting these expressions into the above formula for $\hat{S}$ yields the desired conclusion. $\blacksquare$

Recalling that $B = C^{-1}$, notice that we have also proved the following:

5.15 Corollary

$$\hat{S} = 2\rho_{\text{Ref}}\left\{\frac{\partial\hat{\Psi}}{\partial I_1}G^{\sharp} + \left(\frac{\partial\hat{\Psi}}{\partial I_2} + \frac{\partial\hat{\Psi}}{\partial I_3}I_3\right)B - \frac{\partial\hat{\Psi}}{\partial I_2}I_3B^2\right\}$$

In the spatial picture one can write $\boldsymbol{\sigma} = \beta_0 g + \beta_1 b + \beta_2(b^2)$, where b is the Finger deformation tensor (see Section 1.3).

5.16 Corollary *For isotropic materials, the (second) elasticity tensor* C *has the following component form.*

$$\begin{aligned}\mathsf{C}^{ABCD} &= \gamma_1\cdot G^{AB}G^{CD} + \gamma_2\cdot\{C^{AB}G^{CD} + G^{AB}C^{CD}\} + \gamma_3\{(C^2)^{AB}G^{CD} + G^{AB}(C^2)^{CD}\}\\ &\quad + \gamma_4\cdot C^{AB}C^{CD} + \gamma_5\cdot\{(C^2)^{AB}C^{CD} + C^{AB}(C^2)^{CD}\} + \gamma_6\cdot(C^2)^{AB}(C^2)^{CD}\\ &\quad + \gamma_7\{G^{AC}G^{BD} + G^{BC}G^{AD}\} + \gamma_8\{G^{AC}C^{BD} + G^{BC}C^{AD} + G^{AD}C^{BC}\\ &\quad + G^{BD}C^{AC}\},\end{aligned}$$

where $\gamma_1, \ldots, \gamma_8$ *are scalar functions of* X, *the invariants of* C *and, if the material is thermoelastic,* Θ.

Proof We differentiate the expression for $\hat{S}$ and proceed as in the proof of 5.12. The details, including the verification of the equality of the various coefficients in the expression for C using symmetry of the second derivatives of $\hat{\Psi}$, will be left to the reader. (It is straightforward, although tedious.) ∎

Isotropy reduces the number of pointwise independent components in S from 6 to 3 and for C, from 21 to 8.

Problem 5.1 In plane strain dim $\mathcal{B} = 2$ and dim $\mathcal{S} = 2$. How many independent components does C have? Compute C^{ABCD} and $\mathsf{A}^{A}{}_{a}{}^{B}{}_{b}$ in terms of $\hat{\Psi}$. Compare with Knowles and Sternberg [1977], Formula (1.17). (It will be necessary to translate the notation.)

We give, finally, two examples of constitutive relations for purely elastic isotropic materials. The first was proposed by St. Venant and Kirchhoff around 1860.

5.17 Example If $\hat{S}$ is a linear function of E and describes an isotropic material, then it has the form

$$\hat{S} = \lambda(\operatorname{tr} E)G^{\sharp} + 2\mu E^{\sharp}$$

for λ and μ functions of X. [*Warning*: This does *not* lead to linear equations of motion! Fosdick and Serrin [1979] show that $\hat{P}$ cannot be a linear function of F and be materially frame indifferent.] One calls $\lambda + \frac{2}{3}\mu$ the *bulk modulus*. The elasticity tensor is, as in our computations above,

$$\mathsf{C}^{ABCD} = \lambda G^{AB}G^{CD} + \mu(G^{AC}G^{BD} + G^{AD}G^{CB}).$$

Thus, in 5.15, $\lambda = \gamma_1$, $\mu = \gamma_7$, and the other γ_i are zero.

Problem 5.2 Find $W(F)$ for this example and verify material covariance.

5.18 Example The previous example is only appropriate in the small strain regime. It is used in the derivation of the von Karmen equations by Ciarlet [1983]. The following example, due to Mooney and Rivlin and Ogden, is often used to model rubber. Again, the material is isotropic. It is convenient to express the stored energy function W as a symmetric function of the principal stretches $\lambda_1, \lambda_2, \lambda_3$ (the eigenvalues of $C^{1/2}$). The form proposed by Ogden [1972] is as follows:

$$W = \sum_{i=1}^{M} a_i(\lambda_1^{\alpha_i} + \lambda_2^{\alpha_i} + \lambda_3^{\alpha_i} - 3)$$
$$+ \sum_{j=1}^{N} b_j((\lambda_2\lambda_3)^{\beta_j} + (\lambda_3\lambda_2)^{\beta_j} + (\lambda_1\lambda_2)^{\beta_j} - 3) + h(\lambda_1\lambda_2\lambda_3),$$

where a_i, b_j are positive constants, $\alpha_i \geq 1$, $\beta_j \geq 1$, and h is a convex function of one variable. The term "3" is a normalization constant such that the first two terms vanish when there is no deformation. The special case when $M = N = 1$, $\alpha_1 = \beta_1 = 2$, and $h = 0$ is called the *Mooney–Rivlin material*; W may then be written

$$\begin{aligned} W &= a_1(\lambda_1^2 + \lambda_2^2 + \lambda_3^2 - 3) + b_1((\lambda_2\lambda_3)^2 + (\lambda_3\lambda_1)^2 + (\lambda_1\lambda_2)^2 - 3) \\ &= a_1(I_1 - 3) + b_1(I_2 - 3), \end{aligned}$$

where I_i are the principal invariants of C. The further special case $W = a_1(I_1 - 3)$ is called a *neo-Hookean material.*

Box 5.1 ***Summary of Important Formulas for Section 5.1***

Material Symmetry at X_0

$\boldsymbol{\lambda}: T_{X_0}\mathcal{B} \longrightarrow T_{X_0}\mathcal{B}$ is orthogonal and

(i) $\hat{\Psi}(X_0, \boldsymbol{\lambda}^*C, \Theta) = \hat{\Psi}(X_0, C, \Theta)$ (i) $\hat{\Psi}(_0, \lambda^C{}_A C_{CD} \lambda^D{}_B, \Theta) = \hat{\Psi}(X_0, C_{AB}, \Theta)$

(ii) $\hat{Q}(X_0, \lambda^*C, \Theta, \lambda^*(\text{GRAD}\,\Theta)) = \boldsymbol{\lambda}^*\hat{Q}(X_0, C, \Theta, \text{GRAD}\,\Theta)$ (ii) $\hat{Q}^E(X_0, \lambda^C{}_A C_{CD} \lambda^D{}_B \Theta, (\boldsymbol{\lambda}^{-1})^A{}_B \Theta^{|B}) = (\boldsymbol{\lambda}^{-1})^E{}_F \hat{Q}^F(X_0, C_{AB}, \Theta, \Theta^{|A})$

Infinitesimal Material Symmetry at X_0

$\boldsymbol{\xi} = \dfrac{d}{ds}\boldsymbol{\lambda}_s$ $\xi^A{}_B = \dfrac{\partial}{\partial s}\lambda^A{}_B(s)_{|s=0}$

where $\boldsymbol{\lambda}_s$ is a curve of material symmetries at X_0, such that $\lambda_0 = Id$.

Transformation of $\hat{S}$ under a material symmetry $\boldsymbol{\lambda}$ at X_0

$\hat{S}(X_0, \boldsymbol{\lambda}^*C, \Theta) = \boldsymbol{\lambda}^*\hat{S}(X_0, \Theta)$ $\hat{S}^{AB}(X_0, \lambda^E{}_C C_{EF} \lambda^F{}_D, \Theta) = (\boldsymbol{\lambda}^{-1})^A{}_E (\boldsymbol{\lambda}^{-1})^B{}_F \cdot \hat{S}^{EF}(X_0, C_{CD}, \Theta)$

Infinitesimal Symmetry Identities

$$\hat{S}(X_0, C, \Theta)\cdot \mathfrak{L}_\xi C = 0 \qquad \hat{S}^{AB}(X_0, C_{CD}, \Theta)\cdot\{\xi^C{}_A(X_0)C_{CB} + \xi^C{}_B(X_0)C_{AC}\} = 0$$

$$\mathsf{C}(X_0, C, \Theta)\cdot \mathfrak{L}_\xi C = \mathfrak{L}_\xi \hat{S}(X_0, C, \Theta) \qquad \mathsf{C}^{ABCD}\cdot(\xi^E{}_C C_{ED} + \xi^E{}_D C_{EC}) = -\hat{S}^{EB}\xi^A{}_E - \hat{S}^{AE}\xi^B{}_E$$

Invariants of C

$$I_1 = \operatorname{tr} C \qquad I_1 = C^A{}_A$$

$$I_2 = \det C \operatorname{tr} C^{-1} \qquad I_2 = (\det(C_{AB}))(C^{-1})^A{}_A$$

$$I_3 = \det C \qquad I_3 = \det(C_{AB})$$

Second Piola–Kirchhoff Stress Tensor for an Isotropic Material

$$S^\flat = \alpha_0 G + \alpha_1 C + \alpha_2 C^2 \qquad S_{AB} = \alpha_0 G_{AB} + \alpha_1 C_{AB} + \alpha_2 C_A{}^D C_{DB}$$

where $\alpha_1, \alpha_2, \alpha_3$ are scalar functions of X, the invariants of C and Θ.

$$S = 2\rho_{\text{Ref}}\left\{\frac{\partial\hat{\Psi}}{\partial I_1}G^\sharp + \left(\frac{\partial\hat{\Psi}}{\partial I_2}I_2 + \frac{\partial\hat{\Psi}}{\partial I_3}I_3\right)B - \frac{\partial\hat{\Psi}}{\partial I_2}I_3 B^2\right\}$$

$$S^{AB} = 2\rho_{\text{Ref}}\{\hat{\Psi}_{,I_1}G^{AB} + (\hat{\Psi}_{,I_2}I_2 + \hat{\Psi}_{,I_3}I_3)B^{AB} - \hat{\Psi}_{,I_2}I_3 B^{AD}B_D{}^B\}$$

Elasticity Tensor for Isotropic Materials

$$\mathsf{C}^{ABCD} = \gamma_1 G^{AB}G^{CD} + \gamma_2\{C^{AB}G^{CD} + G^{AB}C^{CD}\} + \gamma_3\{C^{AE}C_E{}^B G^{CD} + G^{AB}C^{CE}C_E{}^D\} + \gamma_4 C^{AB}C^{CD} + \gamma_5\{C^{AE}C_E{}^B C^{CD} + C^{AB}C^{CE}C_E{}^D\} + \gamma_6 C^{AE}C_E{}^B C^{CF}C_F{}^D + \gamma_7\{G^{AC}G^{BD} + G^{BC}G^{AD}\} + \gamma_8\{G^{AC}C^{BD} + G^{BC}C^{AD} + G^{AD}C^{BC} + G^{BD}C^{AC}\}$$

Stress For St. Venant–Kirchhoff Material

$$\hat{S} = \lambda(\operatorname{tr} E)G^\sharp + 2\mu E^\sharp \qquad S^{AB} = \lambda E^D{}_D G^{AB} + 2\mu E^{AB}$$

CHAPTER **4**

LINEARIZATION

The process of linearization provides a key link between the linear and nonlinear theories of elasticity. After reviewing some differential calculus in Banach spaces in Section 4.1, we linearize the equations of nonlinear elasticity in Section 4.2. The procedure is systematic and applies to nonlinear theories in general. The linear theory can also be developed on a separate footing, as in Gurtin [1972a].

Often, linearization of the equations of continuum mechanics is done in Euclidean coordinates and then, at the end, partial derivatives are replaced by covariant derivatives. This is unsatisfactory. Surprisingly, it is not entirely trivial to give a covariant linearization procedure. One of our first goals is to do so.

Using the implicit function theorem proved in Section 4.1, we shall sketch in Section 4.2 how this can be used to establish local existence and uniqueness theorems in elastostatics. This will be detailed in Chapter 6.

The final section discusses in general terms when linearization is "valid." This will be applied in Section 7-3 to the traction problem in elastostatics; the linearization instabilities here go back to Signorini in the 1930s. It has recently been discovered that linearization instabilities occur rather generally at symmetric solutions of classical field theories. (See, for example, Arms, Marsden, and Moncrief [1981], and references therein.)

4.1 THE IMPLICIT FUNCTION THEOREM

In Box 1.1 of Chapter 3, the Fréchet derivative of a map between Banach spaces is defined and some of the basic properties are discussed. That box should be either reviewed or, if it was omitted, studied at this point.

Let us begin here with the concept of the linearization of a given set of equations written abstractly as

$$f(x) = 0,$$

where $f\colon \mathcal{U} \subset \mathfrak{X} \to \mathcal{Y}$ is a given map. If we write $x = x_0 + v$, where x_0 is fixed and expand $f(x_0 + v)$ in a Taylor series, the first two terms are just $f(x_0) + Df(x_0)\cdot v$.

1.1 Definition Let $\mathfrak{X}$ and $\mathcal{Y}$ be Banach spaces and let $\mathcal{U} \subset \mathfrak{X}$ be open. Let $f\colon \mathcal{U} \to \mathcal{Y}$ be a C^1 map and let $x_0 \in \mathcal{U}$ (not necessarily satisfying $f(x_0) = 0$). The *linearization* of the equations $f(x) = 0$ about x_0 are the equations

$$L_{x_0}(v) = 0,$$

where

$$L_{x_0}(v) = f(x_0) + Df(x_0)\cdot v.$$

If $L_{x_0}(v) = 0$, then $x_0 + \epsilon v$ satisfies $f(x_0 + \epsilon v) = 0$ to first order in ϵ. This is *not* the same as saying that, to first order, $x_0 + \epsilon v$ is a solution of $f(x) = 0$. For example, consider a function that arises in bifurcation theory—see Section 7.1. Let $f\colon \mathbb{R}^2 \to \mathbb{R}$ be defined by $f(x, \lambda) = x^3 - \lambda x$. The solution set $f(x, \lambda) = 0$ is the "pitchfork" consisting of the line $x = 0$ and the parabola $\lambda = x^2$. Consider the solution $(0, 0)$ and let $v = (h, k)$. Then since $(\partial f/\partial x) = (\partial f/\partial \lambda) = 0$ at $(0, 0)$, any (h, k) gives a solution to the linearized equations. However ϵv does not approximate a solution to $f(x, \lambda) = 0$ unless $h = 0$ or $k = 0$. The distinction is important and is discussed in Section 4.4.

A basic result that relates linearized and nonlinear theories is the implicit function theorem. It will be a basic tool in Chapter 6, so we give a complete proof of it here. We begin with the inverse mapping theorem.

1.2 Theorem (*Inverse Mapping Theorem*) *Let $f\colon \mathcal{U} \subset \mathfrak{X} \to \mathcal{Y}$ be of class C^r $(r \geq 1)$, $x_0 \in \mathcal{U}$, and suppose $Df(x_0)$ is a linear isomorphism. Then f is a C^r diffeomorphism of some neighborhood of x_0 onto some neighborhood of $f(x_0)$.*

It is essential to have Banach spaces in this result rather than more general spaces such as topological vector spaces or Fréchet spaces. The following problem shows the failure of Theorem 1.2 in Fréchet spaces.

Problem 1.1 (M. McCracken). Let $\mathcal{H}(\Delta)$ denote the set of all analytic functions on the unit disk with the topology of uniform convergence on compact subsets. Let $F\colon \mathcal{H}(\Delta) \to \mathcal{H}(\Delta)$ be defined by

$$\sum_{n=0}^{\infty} a_n z^n \mapsto \sum_{n=0}^{\infty} a_n^2 z^n.$$

Show that F is C^∞ and

$$DF\left(\sum_0^\infty a_n z^n\right)\cdot\left(\sum_0^\infty b_n z^n\right) = \sum_0^\infty 2a_n b_n z^n$$

(see Box 1.1, Chapter 3). If $a_0 = 0$ and $a_n = 1/n$ $(n \geq 1)$, then prove that $DF(\sum_1^\infty z^n/n)$ is a bounded, linear isomorphism. Since

$$F\left(z+\frac{z^2}{2}+\cdots+\frac{z^{k-1}}{k-1}-\frac{z^k}{k}+\frac{z^{k+1}}{k+1}+\cdots\right)=F\left(\sum_{n=1}^{\infty}\frac{z^n}{n}\right),$$

deduce that F is not locally one-to-one. (Consult Schwartz [1967] for more sophisticated versions of the inverse function theorem valid in Fréchet spaces.)

Relevant to the hypotheses of Theorem 1.2 is the *open mapping theorem*: if $T\colon \mathfrak{X} \to \mathfrak{Y}$ is a bijective and continuous linear mapping, then T^{-1} is continuous. (For the proof, see Choquet [1969], p. 322), although this theorem is not essential in what follows.

Proof of Theorem 1.2 We begin by assembling a few standard lemmas;

1.3 Lemma *Let $\mathfrak{M}$ be a complete metric space with distance function $d\colon \mathfrak{M} \times \mathfrak{M} \to \mathbb{R}$. Let $F\colon \mathfrak{M} \to \mathfrak{M}$ and assume there is a constant λ $(0 \le \lambda < 1)$ such that for all $x, y \in \mathfrak{M}$,*

$$d(F(x), F(y)) \le \lambda d(x, y).$$

Then F has a unique fixed point $x_0 \in \mathfrak{M}$; that is, $F(x_0) = x_0$.

This result (1.3) is usually called the *contraction mapping principle* and is the basis of many important existence theorems in analysis. [The other fundamental fixed-point theorem in analysis is the Schauder fixed-point theorem, which states that a continuous map of a compact convex set (in a Banach space, say) to itself, has a fixed point—not necessarily unique, however.]

The proof of Lemma 1.3 is as follows. Pick $x_1 \in \mathfrak{M}$ and define x_n inductively by $x_{n+1} = F(x_n)$. By induction we clearly have

$$d(x_{n+1}, x_n) \le \lambda^{n-1}\, d(F(x_1), x_1)$$

and so

$$d(x_n, x_{n+k}) \le \left(\sum_{j=n-1}^{n-1+k} \lambda^j\right) d(F(x_1), x_1).$$

Thus x_n is a Cauchy sequence. Since F is obviously uniformly continuous, $x_0 = \lim_{n\to\infty} x_n = \lim_{n\to\infty} x_{n+1} = \lim_{n\to\infty} F(x_n) = F(x_0)$. Since $\lambda < 1$ it follows that F has at most one fixed point. ∎

1.4 Lemma *Let $GL(\mathfrak{X}, \mathfrak{Y})$ denote the set of linear isomorphisms from $\mathfrak{X}$ onto $\mathfrak{Y}$. Then $GL(\mathfrak{X}, \mathfrak{Y}) \subset \mathcal{B}(\mathfrak{X}, \mathfrak{Y})$ is open.*

Proof Let

$$\|\alpha\| = \sup_{\substack{x\in\mathfrak{X}\\ \|x\|=1}} \|\alpha(x)\|$$

be the norm on $\mathcal{B}(\mathfrak{X}, \mathfrak{Y})$, the space of all bounded operators from $\mathfrak{X}$ to $\mathfrak{Y}$.

We can assume $\mathfrak{X} = \mathfrak{Y}$. Indeed if $\phi_0 \in GL(\mathfrak{X}, \mathfrak{Y})$, the map $\psi \mapsto \phi_0^{-1}\circ\psi$ from $\mathcal{B}(\mathfrak{X}, \mathfrak{Y})$ to $\mathcal{B}(\mathfrak{X}, \mathfrak{X})$ is continuous and $GL(\mathfrak{X}, \mathfrak{Y})$ is the inverse image of $GL(\mathfrak{X}, \mathfrak{X})$.

For $\phi \in GL(\mathfrak{X}, \mathfrak{X})$, we shall prove that any ψ sufficiently near ϕ is also invertible, which will give the result. More precisely, $\|\psi - \phi\| < \|\phi^{-1}\|^{-1}$ implies $\psi \in GL(\mathfrak{X}, \mathfrak{X})$. The key is that $\|\cdot\|$ is an algebra norm. That is, $\|\beta\circ\alpha\|$

$\leq \|\beta\|\|\alpha\|$ for $\alpha, \beta \in \mathcal{B}(\mathfrak{X}, \mathfrak{X})$. Since $\psi = \phi \circ (I - \phi^{-1} \circ (\phi - \psi))$, ϕ is invertible, and our norm assumption shows that $\|\phi^{-1} \circ (\phi - \psi)\| < 1$, it is sufficient to show that $I - \xi$ is invertible whenever $\|\xi\| < 1$. (I is the identity operator.) Consider the following sequence (called the Neumann series):

$$\xi_0 = I$$
$$\xi_1 = I + \xi$$
$$\xi_2 = I + \xi + \xi \circ \xi$$
$$\vdots$$
$$\xi_n = I + \xi + \xi \circ \xi + \cdots + (\xi \circ \xi \circ \cdots \circ \xi).$$

Using the triangle inequality and the norm inequality $\|\beta \circ \alpha\| \leq \|\beta\|\|\alpha\|$, we can compare this sequence with the sequence of real numbers, $1, 1 + \|\xi\|$, $1 + \|\xi\| + \|\xi\|^2, \ldots$, which we know is a Cauchy sequence since $\|\xi\| < 1$. Because $\mathcal{B}(\mathfrak{X}, \mathfrak{X})$ is complete, ξ_n must converge. The limit, say ρ, is the inverse of $I - \xi$. Indeed $(I - \xi)\xi_n = I - (\xi \circ \xi \circ \cdots \circ \xi) \longrightarrow I$ as $n \longrightarrow \infty$, so the result follows. ∎

1.5 Lemma *Let $\mathcal{I}: GL(\mathfrak{X}, \mathfrak{Y}) \longrightarrow GL(\mathfrak{Y}, \mathfrak{X})$, $\phi \mapsto \phi^{-1}$. Then $\mathcal{I}$ is of class C^∞ and $D\mathcal{I}(\phi) \cdot \psi = -\phi^{-1}\psi\phi^{-1}$.*

Proof We may assume $GL(\mathfrak{X}, \mathfrak{Y}) \neq \varnothing$. If we can show that $D\mathcal{I}(\phi) \cdot \psi = -\phi^{-1} \circ \phi \circ \phi^{-1}$, then it will follow from Leibniz' rule that $\mathcal{I}$ is of class C^∞. Since $\psi \mapsto -\phi^{-1}\psi\phi^{-1}$ is linear, we must show that

$$\lim_{\psi \to \phi} \frac{\|\psi^{-1} - (\phi^{-1}\psi\phi^{-1} + \phi^{-1}\phi\phi^{-1})\|}{\|(\psi - \phi\|} = 0.$$

Note that

$$\psi^{-1} - (\phi^{-1} - \phi^{-1}\psi\phi^{-1} + \phi^{-1}\phi\phi^{-1}) = \psi^{-1} - 2\phi^{-1} + \phi^{-1}\psi\phi^{-1}$$
$$= \psi^{-1}(\psi - \phi)\phi^{-1}(\psi - \phi)\phi^{-1}.$$

Again, using $\|\beta \circ \alpha\| \leq \|\alpha\|\|\beta\|$ for $\alpha \in \mathcal{B}(\mathfrak{X}, \mathfrak{Y})$ and $\beta \in \mathcal{B}(\mathfrak{Y}, \mathfrak{Z})$,

$$\|\psi^{-1}(\psi - \phi)\phi^{-1}(\psi - \phi)\phi^{-1}\| \leq \|\psi^{-1}\|\|\psi - \phi\|^2\|\phi^{-1}\|^2.$$

With this inequality, the preceding limit is clearly zero. ∎

To prove Theorem 1.2 it is useful to note that it is enough to prove it under the simplifying assumptions that $x_0 = 0$, $f(x_0) = 0$, $\mathfrak{X} = \mathfrak{Y}$, and $Df(0)$ is the identity. (Indeed, replace f by $h(x) = Df(x_0)^{-1} \circ [f(x + x_0) - f(x_0)]$.)

Now let $g(x) = x - f(x)$ so $Dg(0) = 0$. Choose $r > 0$ such that $\|x\| \leq r$ implies $\|Dg(x)\| \leq \frac{1}{2}$, which is possible by continuity of Dg. Thus by the mean value inequality, $\|x\| \leq r$ implies $\|g(x)\| \leq r/2$. Let $\bar{\mathcal{B}}_\epsilon(0) = \{x \in \mathfrak{X} \mid \|x\| \leq \epsilon\}$ the closed ball of radius ϵ. For $y \in \bar{\mathcal{B}}_{r/2}(0)$, let $g_y(x) = y + x - f(x)$. By the mean value inequality, if

$$y \in \bar{\mathcal{B}}_{r/2}(0) \quad \text{and} \quad x_1, x_2 \in \bar{\mathcal{B}}_r(0),$$

then

(a) $\|g_y(x)\| \le \|y\| + \|g(x)\| \le r$ and
(b) $\|g_y(x_1) - g_y(x_2)\| \le \frac{1}{2}\|x_1 - x_2\|$.

Thus by Lemma 1.3, $g_y(x)$ has a unique fixed point x in $\bar{\mathcal{B}}_r(0)$. This point x is the unique solution of $f(x) = y$. Thus f has an inverse

$$f^{-1}: \mathcal{V}_0 = \mathcal{B}_{r/2}(0) \to \mathcal{U}_0 = f^{-1}(\mathcal{B}_{r/2}(0)) \subset \mathcal{B}_r(0).$$

From (b) above, $\|f^{-1}(y_1) - f^{-1}(y_2)\| \le 2\|y_1 - y_2\|$, so f^{-1} is continuous.

From Lemma 1.4 we can choose r small enough so that $Df(x)^{-1}$ will exist for $x \in \mathcal{B}_r(0)$. Moreover, by continuity, $\|Df(x)^{-1}\| \le M$ for some M and all $x \in \mathcal{B}_r(0)$ can be assumed as well. If $y_1, y_2 \in \mathcal{B}_{r/2}(0)$, $x_1 = f^{-1}(y_1)$, and $x_2 = f^{-1}(y_2)$, then

$$\begin{aligned}\|f^{-1}(y_1) - f^{-1}(y_2) - Df(x_2)^{-1}(y_1 - y_2)\| &= \|x_1 - x_2 - Df(x_2)^{-1}[f(x_1) - f(x_2)]\| \\ &= \|Df(x_2)^{-1}\{Df(x_2)\cdot(x_1 - x_2) - f(x_1) - f(x_2)\}\| \\ &\le M\|f(x_1) - f(x_2) - Df(x_2)(x_1 - x_2)\|.\end{aligned}$$

This, together with (b) above, shows that f^{-1} is differentiable with derivative $Df(x)^{-1}$ at $f(x)$. By continuity of inversion (Lemma 1.5) we see that f^{-1} is C^1. Also from Lemma 1.5 and $Df^{-1}(y) = [Df(f^{-1}(y))]^{-1}$, we see that if f is C^2, then Df^{-1} is C^1 so f^{-1} is C^2. The general case follows by induction. ∎

In the study of manifolds and submanifolds, the argument used in the following is of central importance.

1.6 Theorem (*Implicit Function Theorem*) *Let $\mathcal{U} \subset \mathcal{X}$, $\mathcal{V} \subset \mathcal{Y}$ be open and $f: \mathcal{U} \times \mathcal{V} \to \mathcal{Z}$ be C^r ($r \ge 1$). Given $x_0 \in \mathcal{U}$, and $y_0 \in \mathcal{V}$, assume $D_2f(x_0, y_0): \mathcal{Y} \to \mathcal{Z}$ is an isomorphism. Then there are neighborhoods $\mathcal{U}_0$ of x_0, $\mathcal{V}_0$ of y_0 and $\mathcal{W}_0$ of $f(x_0, y_0)$ and a unique C^r map $g: \mathcal{U}_0 \times \mathcal{W}_0 \to \mathcal{V}_0$ such that for all $(x, w) \in \mathcal{U}_0 \times \mathcal{W}_0$,*

$$f(x, g(x, w)) = w.$$

Proof Consider the map $\Phi: \mathcal{U} \times \mathcal{V} \to \mathcal{X} \times \mathcal{Z}$, $(x, y) \mapsto (x, f(x, y))$. Then $D\Phi(x_0, y_0)$ is given by

$$D\Phi(x_0, y_0)\cdot(x_1, y_1) = \begin{pmatrix} I & 0 \\ D_1f(x_0, y_0) & D_2f(x_0, y_0)\end{pmatrix}\begin{pmatrix} x_1 \\ y_1\end{pmatrix},$$

which is easily seen to be an isomorphism of $\mathcal{X} \times \mathcal{Y}$ with $\mathcal{X} \times \mathcal{Z}$. Thus Φ has a unique C^r local inverse, say $\Phi^{-1}: \mathcal{U}_0 \times \mathcal{W}_0 \to \mathcal{U}_0 \times \mathcal{V}_0$, $(x, w) \mapsto (x, g(x, w))$. The g so defined is the desired map. ∎

In particular, setting $w = 0$, this theorem implies that $f(x, y) = 0$ is solvable for y as a function of x if $D_2f(x_0, y_0)$ is an isomorphism.

If $\mathfrak{X}$ is a Banach space and $\mathfrak{X}_1 \subset \mathfrak{X}$ is a closed subspace, then $\mathfrak{X}_1$ is said to *split* or be *complemented* when there is a closed subspace $\mathfrak{X}_2$ such that $\mathfrak{X} = \mathfrak{X}_1 \oplus \mathfrak{X}_2$. If $\mathfrak{X}$ is a Hilbert space, we can choose $\mathfrak{X}_2 = \mathfrak{X}_1^{\perp}$. We shall see in Chapter 6 that such decompositions of $\mathfrak{X}$ are naturally associated with elliptic operators.

1.7 Corollary *Let $\mathfrak{U} \subset \mathfrak{X}$ be open and $f\colon \mathfrak{U} \to \mathfrak{Y}$ be C^r $(r \geq 1)$. Suppose $Df(x_0)$ is surjective and* ker $Df(x_0)$ *is complemented. Then $f(\mathfrak{U})$ contains a neighborhood of $f(x_0)$.*[1]

Proof Let $\mathfrak{X}_1 = \ker Df(x_0)$ and $\mathfrak{X} = \mathfrak{X}_1 \oplus \mathfrak{X}_2$. Then $D_2 f(x_0)\colon \mathfrak{X}_2 \to \mathfrak{Y}$ is an isomorphism. Thus the hypotheses of Theorem 1.6 are satisfied and so $f(\mathfrak{U})$ contains $\mathfrak{W}_0$ provided by that theorem. ∎

We conclude with an example of the use of the implicit function theorem to prove an existence theorem for differential equations. For this and related examples we choose the spaces to be infinite dimensional. In fact, $\mathfrak{X}, \mathfrak{Y}, \mathfrak{Z}, \ldots$ will usually be spaces of functions and the map f will often be a nonlinear differential operator.

1.8 Example Let $\mathfrak{X}$ = all C^1 functions $f\colon [0, 1] \to \mathbb{R}$ with the norm

$$\|f\|_1 = \sup_{x \in [0,1]} |f(x)| + \sup_{x \in [0,1]} \left|\frac{df(x)}{dx}\right|$$

and $\mathfrak{Y}$ = all C^0 functions with $\|f\|_0 = \sup_{x \in [0,1]} |f(x)|$. These are Banach spaces. Let $F\colon \mathfrak{X} \to \mathfrak{Y}$, $F(f) = df/dx + f^3$. It is easy to check (see Box 1.1, Chapter 3) that F is C^∞ and $DF(0) = d/dx\colon \mathfrak{X} \to \mathfrak{Y}$. Clearly $DF(0)$ is surjective (by the fundamental theorem of calculus). Also, ker $DF(0) = \mathfrak{X}_1$ = all constant functions. This is complemented because it is finite dimensional; explicitly, a complement consists of functions with zero integral. Thus Corollary 1.7 yields the following statement:

There is an $\epsilon > 0$ such that if g is any continuous function, $g\colon [0, 1] \to \mathbb{R}$, $|g(x)| < \epsilon$, there is a C^1 function $f\colon [0, 1] \to \mathbb{R}$ such that

$$\frac{df}{dx} + f^3(x) = g(x).$$

The ω-lemma discussed in Box 1.1, Chapter 3 shows how to differentiate the basic operators that occur in nonlinear elasticity. This is sufficient for this chapter if one is willing to stick to Euclidean coordinates and then "handwave" while passing to general coordinates. In order to deal with linearization covariantly the following set-up is useful:[2]

---

[1]Actually, the hypothesis that ker $Df(x_0)$ split is not needed for the validity of 1.7. The idea is to work with the quotient space $\mathfrak{X}/\ker Df(x_0)$ on which $Df(x_0)$ induces an isomorphism. See Lusternik and Sobolev [1974], §8.8, and Luenberger [1969], §9.2.

[2]What follows is optional; those skipping it should restrict the results of the next section to Euclidean coordinates.

1.9 Notation and Definition Suppose $\mathcal{C}$ is a manifold, possibly infinite dimensional. Let $\pi: \mathcal{E} \to \mathcal{C}$ be a vector bundle over $\mathcal{C}$ and let $f: \mathcal{C} \to \mathcal{E}$ be a section of this bundle. Assume that $\mathcal{E}$ has a connection so that the covariant derivative of f at each point $\phi \in \mathcal{C}$,

$$\nabla f(\phi): T_\phi \mathcal{C} \longrightarrow \mathcal{E}_\phi = \pi^{-1}(\phi),$$

is defined. If α_t denotes parallel translation of elements of $\mathcal{E}_{\phi(t)}$ to $\mathcal{E}_\phi$ along a curve $\phi(t)$ tangent to V at ϕ, then

$$\nabla_V f(\phi) = \nabla f(\phi) \cdot V = \frac{d}{dt} \alpha_t \cdot f(\phi_t)\big|_{t=0}$$

just as in Riemannian geometry (see Section 1.4).

The *linearization* of the equations $f(\phi) = 0$ at a point $\phi_0 \in \mathcal{C}$ are the equations

$$L_{\phi_0}(V) = 0,$$

where $L_{\phi_0}(V) = f(\phi_0) + \nabla f(\phi_0) \cdot V$ and $V \in T_{\phi_0}\mathcal{C}$.

[If $f(\phi_0) = 0$, then $\nabla f(\phi_0)$ is independent of the connection and equals $Df(\phi_0)$ in charts.]

In elasticity, $\mathcal{E}_\phi$ will be chosen to be a space of tensors or two-point tensors over the configuration ϕ, and parallel translation of them will be defined by pointwise parallel translation defined in Section 1.4.

Box 1.1 ***Summary of Important Formulas for Section 4.1***

Linearization

If f is a mapping between linear spaces, the linearization of $f(x) = 0$ at x_0 is $L_{x_0}(u) = 0$, where $L_{x_0}(u) = f(x_0) + Df(x_0) \cdot u$. If f is a mapping from a manifold to a vector bundle, replace $Df(x_0)$ by a covariant derivative.

Inverse Mapping Theorem

If $f(x)$ is C^1 and $Df(x_0)$ is an isomorphism, $f(x) = y$ is locally uniquely solvable for x near x_0 as a C^1 function of y near $y_0 = f(x_0)$.

Implicit Function Theorem

If $f(x, y)$ is C^1, $f(x_0, y_0) = 0$ and $D_2 f(x_0, y_0)$ is an isomorphism, then $f(x, y) = 0$ is solvable for y as a C^1 function of x near (x_0, y_0).

4.2 LINEARIZATION OF NONLINEAR ELASTICITY

We now apply the ideas of the previous section to the equations of nonlinear elasticity. We shall use the set-up of 1.9 (or that of 1.1 if you wish to restrict to Euclidean structures at first).

2.1 Notation Let $\mathcal{C}$ denote the set of all regular C^k configurations ϕ: $\mathcal{B} \longrightarrow \mathcal{S}$ (impose boundary conditions of place if there are any). If there are no displacement boundary conditions *and* if $\mathcal{S} = \mathbb{R}^n$, then we can use the Euclidean structure to regard $\mathcal{C}$ as an open set in the Banach space $\mathfrak{X}$ of all C^k maps of $\mathcal{B}$ to $\mathbb{R}^n$. Even for general $\mathcal{S}$, one can show that $\mathcal{C}$ is a C^∞ infinite-dimensional manifold. (See Palais [1968], Ebin and Marsden [1970], and references therein.) A tangent vector to $\mathcal{C}$ at $\mathring{\phi} \in \mathcal{C}$ is the tangent to a curve $\phi_t \in \mathcal{C}$ with $\phi_0 = \mathring{\phi}$—that is, to a *motion.* Thus, from Section 1.2, (see also Box 4.2, Chapter 2) a tangent vector to $\mathcal{C}$ at $\mathring{\phi}$ is a vector field $\boldsymbol{U}$ covering $\mathring{\phi}$, zero on any portion of the boundary where displacement boundary conditions are imposed (see Figure 4.2.1). Sometimes $\boldsymbol{U}$ is spoken of as an *infinitestimal deformation imposed on the finite deformation* $\mathring{\phi}$, or as a *variation* of the configuration and one writes $\boldsymbol{U} = \delta\phi$.

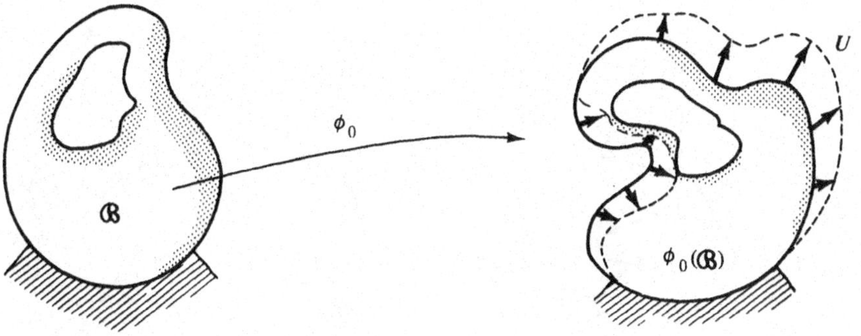

Figure 4.2.1

We shall build up the linearized equations in several steps. First, consider the association $\phi \mapsto \boldsymbol{F} = T\phi$. Here we let $\mathcal{E}$ be the vector bundle over $\mathcal{C}$ whose fiber at $\mathring{\phi}$ consists of all C^{k-1} maps $\boldsymbol{F}$: $T\mathcal{B} \longrightarrow T\mathcal{S}$ that cover $\mathring{\phi}$. Thus $\mathcal{E}$ is the space of sections of the bundle $T\mathcal{B} \otimes T^*\mathcal{S}$. There is a natural notion of parallel translation on ϕ obtained by pointwise parallel translation of two-point tensors $\boldsymbol{F}$ over curves in $\mathcal{S}$—see 4.28 of Chapter 1. (In Euclidean space this operation is just ordinary translation.)

2.2 Proposition *The linearization of the map f:* $\phi \mapsto \boldsymbol{F} = T\phi$ *at* $\mathring{\phi}$ *is given by*

$$L(\mathring{\phi}, \boldsymbol{U}) = \mathring{\boldsymbol{F}} + \nabla \boldsymbol{U},$$

where $\mathring{\boldsymbol{F}} = T\mathring{\phi}$. *In coordinates,*

$$(\nabla \boldsymbol{U})^a{}_A = U^a{}_{|A} = \gamma^a_{bc} U^c \mathring{F}^b{}_A + \frac{\partial U^a}{\partial X^A},$$

which is the covariant derivative of $\boldsymbol{U}$.

Proof By definition,

$$L(\mathring{\phi}, U) = \mathring{F} + \frac{d}{dt}\alpha_t \cdot F_t|_{t=0}$$

where $\phi_0 = \mathring{\phi}$, α_t denotes parallel translation, $F_t = T\phi_t$, and ϕ_t is tangent to U at $t = 0$. At $X \in \mathcal{B}$, $\alpha_t \cdot F_t(X)$ is, by definition of α_t, the parallel transport in $\mathcal{S}$ of $F_t(X)$ from $\phi_t(X)$, along the curve $t \mapsto \phi_t(X)$, to $\phi_0(X)$, that is, for $W \in T_X\mathcal{B}$, $(\alpha_t \cdot F_t)(X) \cdot W$ is the parallel transport in $\mathcal{S}$ of the vectors $F_t(X) \cdot W \in T_{\phi_t(X)}\mathcal{S}$ along the curve $c(t) = \phi_t(X) \in \mathcal{S}$. In coordinate charts we use the formula

$$\frac{d}{dt}(\alpha_t)^a{}_b|_{t=0} = \gamma^a{}_{cb}U^c$$

from Section 1.4. See Figure 4.2.2. Thus

$$\frac{d}{dt}(\alpha_t \cdot F_t)^a{}_A(X)|_{t=0} = \gamma^a{}_{cb}U^c\mathring{F}^b{}_A + \frac{d}{dt}\left(\frac{\partial \phi_t^a}{\partial X^A}\right)\bigg|_{t=0}$$

$$= \gamma^a{}_{cb}U^c\mathring{F}^b{}_A + \frac{\partial U^a}{\partial X^A} = U^a{}_{|A}. \quad \blacksquare$$

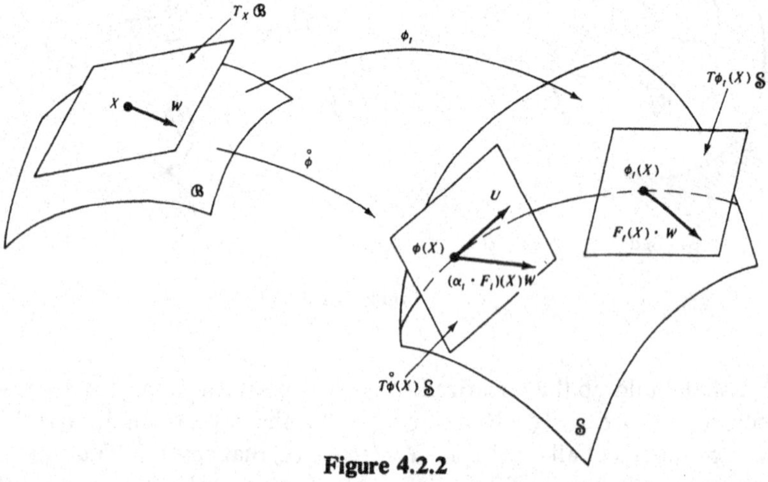

Figure 4.2.2

2.3 Proposition *The linearized equations at $\mathring{\phi}$ for the map $\phi \mapsto J =$ Jacobian of ϕ are*

$$L(\mathring{\phi}, V) = \mathring{J} + \mathring{J}[(\text{div } v)\circ\mathring{\phi}],$$

where $\mathring{J}$ is the Jacobian of $\mathring{\phi}$ and $v = V\circ\mathring{\phi}^{-1}$.

Proof Since the scalars over maps form a linear space, the definition gives

$$L(\mathring{\phi}, V) = \mathring{J} + \frac{d}{dt}J(\phi_t)|_{t=0},$$

where V is tangent to ϕ_t at $t = 0$. From 5.4, Chapter 1, this is $\mathring{J} + \mathring{J}[(\text{div } v)\circ\mathring{\phi}]$, where v is the corresponding spatial velocity. $\blacksquare$

2.4 Theorem *Let $\mathcal{C}$ denote the space of all C^k regular configurations ϕ: $\mathcal{B} \longrightarrow \mathcal{S}$, let $\mathcal{E}$ denote the bundle of C^{k-1} two-point tensors $\boldsymbol{F}$ over $\mathcal{C}$ as above, and let $\hat{\boldsymbol{P}}$ map two-point tensors $\boldsymbol{F}$ pointwise to two-point tensors $\hat{\boldsymbol{P}}(\boldsymbol{F})$ (perhaps of a different rank) and be C^r ($r \geq k$). Let $\mathcal{F}$ denote a bundle of C^{k-1} two-point tensor fields over $\mathcal{C}$ containing the range of $\hat{\boldsymbol{P}}$ and let*

$$f: \mathcal{C} \longrightarrow \mathcal{F}, \qquad \phi \mapsto \hat{\boldsymbol{P}} \circ \boldsymbol{F} \quad (\text{where } \boldsymbol{F} = T\phi).$$

Then f is C^r and the linearized equations of $f(\phi) = 0$ at $\mathring{\phi}$ are

$$L(\mathring{\phi}, V) = \mathring{\boldsymbol{P}} + \frac{\partial \mathring{\boldsymbol{P}}}{\partial \boldsymbol{F}} \cdot \nabla V,$$

where $\mathring{\boldsymbol{P}} = \hat{\boldsymbol{P}}(\mathring{\boldsymbol{F}})$, and so on.

This theorem is a covariant reformulation of the ω-lemma given in Box 1.1, Chapter 3. It is proved by combining the argument given there with that for the linearization of the map $f: \phi \mapsto T\phi = \boldsymbol{F}$ given above; we omit the details.

We now apply this result when $\hat{\boldsymbol{P}}$ is the first Piola–Kirchhoff stress tensor (with a given constitutive dependence on $\boldsymbol{F}$ assumed). Then we get

$$\boxed{L(\mathring{\phi}, V) = \mathring{\boldsymbol{P}} + \mathring{\mathbf{A}} \cdot \nabla V,}$$

where, from Section 3.3,

$$\mathring{\mathbf{A}} = \frac{\partial \mathring{\boldsymbol{P}}}{\partial \boldsymbol{F}}, \qquad \text{that is,} \quad \mathring{\mathrm{A}}^{aA}{}_b{}^B = \frac{\partial \hat{P}^{aA}}{\partial F^b{}_B}(\mathring{\boldsymbol{F}})$$

is the elasticity tensor evaluated at $\mathring{\phi}$. Also note that

$$(\mathring{\mathbf{A}} \cdot \nabla V)^{aA} = \mathring{\mathrm{A}}^{aA}{}_b{}^B V^b{}_{|B}.$$

2.5 Example The nonlinear equations of elastostatics are

$$\mathrm{DIV}\, \boldsymbol{P} + \rho_{\mathrm{Ref}} \boldsymbol{B} = \mathbf{0}.$$

The corresponding equations linearized at a configuration $\mathring{\phi}$ with $\boldsymbol{B}$ given[3] are

$$\boxed{\begin{aligned} &\mathrm{DIV}\, \mathring{\boldsymbol{P}} + \rho_{\mathrm{Ref}} \mathring{\boldsymbol{B}} + \mathrm{DIV}(\mathring{\mathbf{A}} \cdot \nabla V) = 0, \\ \text{that is,} \quad &\mathrm{DIV}[\mathring{\boldsymbol{P}} + (\mathring{\boldsymbol{S}} \otimes \boldsymbol{\delta} + 2\mathring{\mathbf{C}} \cdot \mathring{\boldsymbol{F}} \cdot \mathring{\boldsymbol{F}} \cdot \mathring{\boldsymbol{g}} \cdot \nabla V] + \rho_{\mathrm{Ref}} \mathring{\boldsymbol{B}} = \mathbf{0}. \end{aligned}}$$

From the Piola identity, $\mathrm{DIV}(\mathring{\mathbf{A}} \cdot \nabla V) = \mathring{J}\, \mathrm{div}(\mathring{\mathbf{a}} \cdot \nabla v) \circ \mathring{\phi}$, so the linearized equations may be written in spatial coordinates as

[3]Strictly speaking, $\boldsymbol{B}$ "given" requires $\mathcal{S}$ to be a linear space to make sense, and similarly for "prescribed" tractions $\boldsymbol{\tau}$ on ∂B. Given $\boldsymbol{B}$ or $\boldsymbol{\tau}$ as functions of X is usually called *dead loading* (see Section 3.3). If $\boldsymbol{b} = \boldsymbol{B} \circ \phi$ is regarded as given, then the extra term $\nabla \boldsymbol{b} \cdot V$ must be added to the linearized equations.

$$\rho\mathring{b} + \operatorname{div}(\mathring{\sigma} + \mathring{\mathbf{a}}\cdot\nabla v) = 0,$$
$$\text{that is,}\quad \rho\mathring{b} + \operatorname{div}(\mathring{\sigma} + [\mathring{\sigma}\otimes\boldsymbol{\delta} + \mathring{\mathbf{c}}\cdot g]\cdot\nabla v) = 0.$$

(The componential forms of **A** and **a** are given in Box 4.1 of Chapter 3.) If $\mathring{\phi}$ is a *stress-free state*, that is, $\mathring{\sigma} = \mathbf{0}$, then these reduce to the equations of *classical linear elastostatics*:

$$\rho b + \operatorname{div}(\mathbf{c}\cdot\nabla e) = \mathbf{0}, \qquad \text{that is,}\quad \rho b^a + (\mathrm{c}^{abcd}e_{cd})_{|b} = 0,$$

where $e_{cd} = \frac{1}{2}(v_{c|d} + v_{d|c}) = \frac{1}{2}\mathfrak{L}_v \mathbf{g}$, in view of the symmetries of **c**.

Problem 2.1 Show that the linearization of the symmetry $F^a{}_A P^{Ab} = F^b{}_A P^{Aa}$ (balance of moment of momentum) gives symmetry of $\mathring{J}[\mathring{\sigma}\cdot\nabla v + \mathring{\mathbf{a}}\cdot\nabla v]$.

If we consider the space of motions in place of the space of configurations, we can derive in the same way the linearization of the equations of motion.

2.6 Theorem *The linearization of the equations of motion*

$$\rho_{\mathrm{Ref}} A = \mathrm{DIV}\, P + \rho_{\mathrm{Ref}} B$$

at a motion $\mathring{\phi}_t$ *are*

$$\rho_{\mathrm{Ref}}(\mathring{A} + \ddot{U} - \mathring{B}) = \mathrm{DIV}(\mathring{P} + \mathring{\mathbf{A}}\cdot\nabla U),$$

where U *is a vector field over the motion* $\mathring{\phi}_t$.

In spatial coordinates, these read

$$\rho(\mathring{a} + \ddot{u} - \mathring{b}) = \operatorname{div}(\mathring{\sigma} + \mathring{\mathbf{a}}\cdot\nabla u).$$

The terms $-\rho_{\mathrm{Ref}}\mathring{A} + \rho_{\mathrm{Ref}}\mathring{B} + \mathrm{DIV}\,\mathring{P}$, which measure how close $\mathring{\phi}_t$ is to an actual solution, are called the *out-of-balance forces*. They arise naturally whenever a nonlinear problem is solved by iteration on a linearized problem.

The boundary conditions can be linearized by following the same procedures. We can state the results as follows:

2.7 Proposition *The linearization of the boundary condition of place, namely,* $\phi = \phi_d$ *given on* $\partial\mathcal{B}$ *at a configuration* $\mathring{\phi}$, *is*

$$L(\mathring{\phi}, U) = \mathring{\phi} - \phi_d + U = \mathbf{0} \quad \text{on} \quad \partial\mathcal{B}.$$

(For this to make sense, we assume $\mathcal{S} = \mathbb{R}^n$.*)*

The linearization of the boundary condition of traction, namely, $\langle P, N\rangle = \boldsymbol{\tau}$ *given on* $\partial\mathcal{B}$, *is*

$$\langle\mathring{\mathbf{A}}\cdot\nabla U, N\rangle = \boldsymbol{\tau} - \langle\mathring{P}, N\rangle \quad \text{on} \quad \partial\mathcal{B}.$$

Here, as usual, U *is a vector field over* $\mathring{\phi}$.

Similarly, the linearization of the initial conditions ($\phi_0 = d$ given and $V_0 = V$ given) yields the conditions

$$\mathring{\phi}_0 - d + U_0 = \mathbf{0} \quad \text{and} \quad \mathring{V}_0 - V + \dot{U}_0 = \mathbf{0}.$$

The process of linearization is subject to a number of variations. For example, suppose we consider the loads $(\boldsymbol{B},\boldsymbol{\tau})$ as variables in addition to the motion ϕ. Then the linearized equations involve the infinitesimal deformation $U = \delta\phi$ and load increments $B^* = \delta B$ and $\tau^* = \delta\tau$. As above, we get the following.

2.8 Proposition *The material form of the equations of elastodynamics linearized around a solution $\mathring{\phi}$, $\mathring{B}$, $\mathring{\tau}$ with incremental deformation $U = \delta\phi$ and incremental loads $B^* = \delta B$ and $\tau^* = \delta\tau$ are*

$$\rho_{\text{Ref}}(\mathring{A} - \mathring{B} + \ddot{U} - B^*) = \text{DIV}(\mathring{P} + \mathring{\mathbf{A}}\cdot\nabla U) \quad \text{in} \quad \mathcal{B},$$
$$\langle \mathring{\mathbf{A}}\cdot\nabla U, N\rangle = \mathring{\tau} - \langle \mathring{P}, N\rangle + \tau^* \quad \text{on} \quad \partial\mathcal{B}.$$

If $(\mathring{\phi}, \mathring{B}, \mathring{\tau})$ is a solution, these become

$$\rho_{\text{Ref}}(\ddot{U} - B^*) = \text{DIV}(\mathring{\mathbf{A}}\cdot\nabla U) \quad \text{in} \quad \mathcal{B},$$
$$\langle \mathring{\mathbf{A}}\cdot\nabla U, N\rangle = \tau^* \quad \text{on} \quad \partial\mathcal{B}.$$

Box 2.1 ***Summary of Important Formulas for Section 4.2***

Linearization of the map $\hat{\mathbf{P}}(F)$ at $\mathring{\phi}$

$$\hat{\mathbf{P}}(F) + \mathring{\mathbf{A}}\cdot\nabla V \qquad \mathring{P}^{aA} + \mathring{\mathsf{A}}^{aA}{}_b{}^B V^b{}_{|B}$$

Linearized Equations of Elastostatics: $\text{DIV}\, P + \rho_{\text{Ref}} B = \mathbf{0}$

Materially:

$$\text{DIV}\,\mathring{P} + \rho_{\text{Ref}}\mathring{B} + \text{DIV}(\mathring{\mathbf{A}}\cdot\nabla V) = 0 \qquad \mathring{P}^{aA}{}_{|A} + \rho_{\text{Ref}}\mathring{B}^a + (\mathring{\mathsf{A}}^{aA}{}_b{}^B V^b{}_{|B})_{|A} = 0$$

Spatially:

$$\text{div}\,\mathring{\boldsymbol{\sigma}} + \mathring{\rho}\mathring{\boldsymbol{b}} + \text{div}(\mathring{\mathbf{a}}\cdot\nabla\boldsymbol{v}) = \mathbf{0} \qquad \mathring{\sigma}^{ab}{}_{|b} + \mathring{\rho}\mathring{b}^a + (\mathring{\mathsf{a}}^{ac}{}_b{}^d v^b{}_{|d})_{|c} = 0$$

Classical Linear Elastostatics ($\mathring{\phi}$ stress free)

$$\rho\boldsymbol{b} + \text{div}(\mathbf{c}\cdot\nabla\boldsymbol{v}) = \mathbf{0} \qquad \rho b^a + (\mathsf{c}^{abcd} e_{cd})_{|b} = 0$$

Linearized Equations of Elastodynamics: $\rho_{\text{Ref}} A = \text{DIV}\, P + \rho_{\text{Ref}} B$

Materially:

$$\rho_{\text{Ref}}(\mathring{A} + \ddot{U} - \mathring{B}) = \text{DIV}(\mathring{P} + \mathring{\mathbf{A}}\cdot\nabla U) \qquad \rho_{\text{Ref}}(A^a + \ddot{U}^a - \mathring{B}^a) = (\mathring{P}^{aA} + \mathring{\mathsf{A}}^{aA}{}_b{}^B U^b{}_{|B})_{|A}$$

or, with incremental load B^*,

$$\rho_{\text{Ref}}(\mathring{A} + \ddot{U} - \mathring{B} - B^*) = \text{DIV}(\mathring{P} + \mathring{\mathbf{A}}\cdot\nabla U) \qquad \rho_{\text{Ref}}(\mathring{A}^a + \ddot{U}^a - \mathring{B}^a - B^{*a}) = (\mathring{P}^{aA} + \mathring{\mathsf{A}}^{aA}{}_b{}^B U^b{}_{|B})_{|A}$$

Spatially:

$$\rho(\mathring{a} + \ddot{u} - \mathring{b}) = \operatorname{div}(\mathring{\sigma} + \mathring{\mathbf{a}}\cdot\nabla u) \qquad \rho(\mathring{a}^a + \ddot{u}^a - \mathring{b}^a) = (\mathring{\sigma}^{ab} + \mathring{\mathbf{a}}^{ab}{}_c{}^d u^c{}_{|d})_{|b}$$

Linearized Boundary Conditions

Place: $\phi = \phi_d$ given on $\partial\mathcal{B}$: $\mathring{\phi} - \phi_d + U = 0$

Traction: $\langle P, N\rangle = \tau$ given on $\partial\mathcal{B}$:

$$\langle \mathring{\mathbf{A}}\cdot\nabla U, N\rangle = \mathring{\tau} - \langle \mathring{P}, N\rangle \text{ on } \partial\mathcal{B} \qquad \mathring{\mathbf{A}}^{aA}{}_b{}^B U^b{}_{|B} N_A = \mathring{\tau}^a - \mathring{P}^{aA} N_A \text{ on } \partial\mathcal{B}$$

or, with incremental traction τ^*:

$$\langle \mathring{\mathbf{A}}\cdot\nabla U, N\rangle = \mathring{\tau} - \langle \mathring{P}, N\rangle + \tau^* \text{ on } \partial\mathcal{B} \qquad \mathring{\mathbf{A}}^{aA}{}_b{}^B U^b{}_{|B} N_A = \mathring{\tau}^a - \mathring{P}^{aA} N_A + \tau^{*a} \text{ on } \partial\mathcal{B}$$

4.3 LINEAR ELASTICITY

The previous section derived the linearized equations of elasticity from the nonlinear ones. In this section we shall make a few remarks on the linear theory itself. In Chapter 6 further results are given in connection with the basic existence and uniqueness theorems for linear elasticity. Other than these topics, our discussions are not intended to be complete; the reader interested in further details of the theory should consult Gurtin [1972a]. In particular, this reference develops the theory on its own footing by way of linearity hypotheses and balance laws rather than as the linearization of the nonlinear theory.

Let us begin by summarizing what we found in the previous section; we simplify things slightly by linearizing about a given *stress-free, underformed* state (also called a *natural state*).

3.1 Equations of Linear (Classical) Elasticity Let $\mathcal{B}$ be open in $\mathcal{S}$. The equations for a vector field $u(x, t)$ representing an infinitesimal displacement of $\mathcal{B}$ are:

$$\rho\ddot{u} = \rho b + \operatorname{div}(\mathbf{c}\cdot\nabla u),$$

where $\rho(x)$ is the mass density, b is an (external) body force field, and $\mathbf{c}(x)$ is a given fourth-order tensor field (the elasticity tensor) on $\mathcal{B}$ with the symmetries

$$\mathbf{c}^{abcd} = \mathbf{c}^{bacd} = \mathbf{c}^{abdc} = \mathbf{c}^{cdab}.$$

(For linear elasto*statics*, drop the $\ddot{u}$ term.)

3.2 Boundary Conditions of Linear (Classical) Elasticity

(i) *Displacement:* $u = d$ on $\partial\mathcal{B}$.

(ii) *Traction:* $(\mathbf{c}\cdot\nabla \boldsymbol{u})\cdot\boldsymbol{n} = \boldsymbol{\tau}$ on $\partial\mathcal{B}$.
(iii) *Mixed:* (i) and (ii) hold on disjoint portions ∂_d and ∂_τ of $\partial\mathcal{B}$.

In coordinates the equations of motion read

$$\rho\ddot{u}^a = \rho b^a + (\mathsf{c}^{abcd}u_{c|d})_{|b}$$

and the traction boundary conditions read $\mathsf{c}^{abcd}u_{c|d}n_b = \tau^a$.

The following consequence of the divergence theorem is the linearized version of Proposition 4.13 in Chapter 3.

3.3 Proposition *Suppose* $\boldsymbol{u}$ *satisfies the equations of classical elastostatics with traction boundary conditions. Then*

$$\int_{\partial\mathcal{B}} \boldsymbol{\tau}\, dA + \int_{\mathcal{B}} \rho \boldsymbol{b}\, dV = \mathbf{0}.$$

Next we define the stress and strain tensors for the linear theory.

3.4 Definition The *strain tensor* is defined by $\boldsymbol{e} = \frac{1}{2}\mathcal{L}_{\boldsymbol{u}}\boldsymbol{g}$; that is,

$$e_{ab} = \tfrac{1}{2}(u_{a|b} + u_{b|a}).$$

The *stress tensor* is $\boldsymbol{s} = \mathbf{c}\cdot\nabla\boldsymbol{u}$; that is, $s^{ab} = \mathsf{c}^{abcd}u_{c|d}$.

The strain tensor is the linearization of the Lagrangian strain tensor

$$\boldsymbol{E} = \tfrac{1}{2}(\boldsymbol{C} - \boldsymbol{G}).$$

Note that $\boldsymbol{s}$ is symmetric, $s^{ab} = s^{ba}$, and that, by the symmetries of $\mathbf{c}$,

$$\boldsymbol{s} = \mathbf{c}\cdot\boldsymbol{e}.$$

The fact that $\boldsymbol{s}$ depends on $\boldsymbol{u}$ only through $\boldsymbol{e}$ is the infinitesimal version of material frame indifference.

3.5 Proposition *Let the* elastic stored energy function *be defined by*

$$\epsilon = \tfrac{1}{2}\boldsymbol{e}\cdot\mathbf{c}\cdot\boldsymbol{e} = \tfrac{1}{2}e_{ab}\mathsf{c}^{abcd}e_{cd}.$$

Then
$$\boldsymbol{s} = \frac{\partial\epsilon}{\partial\boldsymbol{e}}, \quad \text{that is,} \quad s^{ab} = \frac{\partial\epsilon}{\partial e_{ab}},$$

and
$$\mathbf{c} = \frac{\partial^2\epsilon}{\partial\boldsymbol{e}\,\partial\boldsymbol{e}}, \quad \text{that is,} \quad \mathsf{c}^{abcd} = \frac{\partial^2\epsilon}{\partial e_{ab}\,\partial e_{cd}}.$$

This is a straightforward verification. Note that the existence of a stored energy function is equivalent to the symmetry $\mathsf{c}^{abcd} = \mathsf{c}^{cdab}$.

It is perhaps reasonable to suppose that $\epsilon > 0$ when $\boldsymbol{e} \neq 0$. This corresponds to the necessity of work being done to deform $\mathcal{B}$ from its natural state.

3.6 Definition We say that the elasticity tensor $\mathbf{c}$ is *pointwise stable* if $\boldsymbol{e} \neq 0$ implies $\epsilon > 0$, for all symmetric two tensors e_{ab}.

In Chapter 6 on existence and uniqueness we shall study conditions of this sort in detail. For now we merely show that it implies that elastic waves have *real* wave speeds—that is, the *hyperbolicity* of the equations. To explain this, we require the following:

3.7 Definition Assume the mass density ρ is a positive constant, and that $\mathcal{S} = \mathbb{R}^3$ and $\mathbf{c}$ is a constant tensor (independent of x). By a *plane progressive wave* we mean a vector field $\boldsymbol{u}(x, t)$ on $\mathcal{B}$ of the form

$$\boldsymbol{u}(x, t) = \boldsymbol{\alpha}\phi(\boldsymbol{x}\cdot\boldsymbol{k} - ct),$$

where $\boldsymbol{\alpha}$, $\boldsymbol{k}$ are fixed vectors in $\mathbb{R}^3$, $\boldsymbol{x}$ denotes the vector from the origin to x, c is a positive real constant called the *speed of propagation*, and ϕ is a C^2 map of $\mathbb{R}$ to $\mathbb{R}$ with somewhere nonvanishing second derivative.

If $\rho\ddot{\boldsymbol{u}} = \operatorname{div}(\mathbf{c}\cdot\nabla\boldsymbol{u})$ holds, we call $\boldsymbol{u}$ an *elastic* plane progressive wave.

Notice that $\rho\ddot{\boldsymbol{u}} = \boldsymbol{\alpha}c^2\phi''$ (omitting the independent variables) and

$$\operatorname{div}(\mathbf{c}\cdot\nabla\boldsymbol{u}) = \operatorname{div}(\mathbf{c}\cdot[\phi'\boldsymbol{\alpha}\otimes\boldsymbol{k}]) = \phi''\mathbf{c}\cdot(\boldsymbol{k}\otimes\boldsymbol{\alpha}\otimes\boldsymbol{k}).$$

In components, this calculation reads

$$(\mathrm{c}^{abcd}u_{c|d})_{|b} = (\mathrm{c}^{abcd}\phi'\alpha_c k_d)_{|b} = \phi''\mathrm{c}^{abcd}k_b\alpha_c k_d.$$

Thus, for a plane progressive wave, $\rho\ddot{\boldsymbol{u}} = \operatorname{div}(\mathbf{c}\cdot\nabla\boldsymbol{u})$ reads $\boldsymbol{\alpha}c^2\phi'' = \phi''\mathbf{c}\cdot(\boldsymbol{k}\otimes\boldsymbol{\alpha}\otimes\boldsymbol{k})$. This simple calculation proves the following result of Fresnel and Hadamard:

3.8 Proposition *A plane progressive wave is elastic if and only if*

$$\mathbf{c}\cdot(\boldsymbol{k}\otimes\boldsymbol{\alpha}\otimes\boldsymbol{k}) = c^2\boldsymbol{\alpha} \qquad \textit{that is,} \quad \mathrm{c}^{abcd}k_b\alpha_c k_d = c^2\alpha^a.$$

The vector $\boldsymbol{k}$ represents the *direction of propagation* since for t constant, $\boldsymbol{u}$ is constant on the planes $\boldsymbol{x}\cdot\boldsymbol{k} - ct =$ constant, which are orthogonal to $\boldsymbol{k}$. For each $\boldsymbol{k}$ the preceding equation is an eigenvalue equation for the *polarization vector* $\boldsymbol{\alpha}$ with eigenvalue c^2, the square of the speed, and the matrix $\Lambda^{ac} = c^{abcd}k_b k_d$.

The tensor $\mathbf{A}(\boldsymbol{k})$ just defined is real and symmetric, so has real eigenvalues. Thus, if all possible wave speeds are to be real, $\mathbf{A}$ must have positive eigenvalues, so is positive-definite. This leads to the following *strong ellipticity conditions of Legendre and Hadamard*:

3.9 Proposition *The tensor* $\mathbf{c}$ *admits plane progressive elastic waves in all possible directions* $\mathbf{k}$ *if and only if for all non-zero vectors* $\mathbf{k}$ *and* $\boldsymbol{\alpha}$,

$$\mathrm{c}^{abcd}\alpha_a k_b \alpha_c k_d > 0.$$

In this case $\mathbf{c}$ *is called* strongly elliptic.

The relationship between 3.6 and 3.9 follows.

3.10 Proposition *If* $\mathbf{c}$ *is pointwise stable, then it is strongly elliptic.*

Proof Let $e = \frac{1}{2}(\boldsymbol{\alpha} \otimes \boldsymbol{k} + \boldsymbol{k} \otimes \boldsymbol{\alpha})$, that is, $e_{ab} = \frac{1}{2}(\alpha_a k_b + \alpha_b k_a)$. Stability implies that $\epsilon = \frac{1}{2}\boldsymbol{e}\cdot\mathbf{c}\cdot\boldsymbol{e} > 0$. Using symmetries of $\mathbf{c}$, we get $\epsilon = \frac{1}{2}\{\frac{1}{2}(\alpha_a k_b + \alpha_b k_a)\cdot \mathrm{c}^{abcd} e_{cd}\} = \frac{1}{2}\alpha_a k_b \mathrm{c}^{abcd} e_{cd} = \frac{1}{2}\alpha_a k_b \mathrm{c}^{abcd}\alpha_c k_d$, so for this choice of e, $\epsilon > 0$ is exactly the strong ellipticity condition. ∎

As we shall see shortly, strong ellipticity does *not* imply pointwise stability. Further investigation of the relationship between these two conditions occurs in Chapter 6. It turns out that these conditions play a crucial role in the problem of solvability of the linearized equations. This in turn affects the solvability of the nonlinear problem by way of the inverse function theorem.

Finally, we briefly discuss the case of isotropic linear elasticity. (For more general symmetry groups, consult Love [1927] and Gurtin [1972a].)

3.11 Definition Let $\mathbf{c}$ be an elasticity tensor, and $\epsilon(e) = \frac{1}{2}\boldsymbol{e}\cdot\mathbf{c}\cdot\boldsymbol{e}$ the corresponding energy function. We say that $\mathbf{c}$ is *isotropic* if, for all proper orthogonal linear transformations Q on $\mathbb{R}^3$, $\epsilon(Q^*e) = \epsilon(e)$, where $(Q^*e)_{ab} = Q^c{}_a e_{cd} Q^d{}_b$ and $Q^c{}_a$ is the matrix of Q in a given coordinate system.

The arguments presented in Section 3.5 can be used to prove the following:

3.12 Proposition *If* $\mathbf{c}$ *is isotropic and homogeneous (independent of* x*), then there are constants* λ *and* μ *called the* Lamé moduli *such that*

$$\boxed{\epsilon = \mu e\cdot e + \frac{\lambda}{2}(\operatorname{tr} e)^2,}$$

The corresponding stress is

$$\boxed{s^\flat = 2\mu e + \lambda(\operatorname{tr} e)g,}$$

and the components of the elasticity tensor $\mathbf{c}$ *are*

$$\boxed{\mathrm{c}^{abcd} = \mu(g^{ac}g^{bd} + g^{ad}g^{bc}) + \lambda g^{ab}g^{cd}.}$$

Problem 3.1 Give a direct proof of 3.12.

Problem 3.2 Show that the mean stress is $\frac{1}{3}\operatorname{tr}\mathbf{s} = k \operatorname{div}\mathbf{u}$, where $k = (3\lambda + 2\mu)/3$ is the *modulus of compression.* By consulting Gurtin [1972a] or other texts, give geometric interpretations of k, of *Young's modulus* $E = \mu(3\lambda + 2\mu)/(\lambda + \mu)$ and of *Poisson's ratio* $\nu = \lambda/2(\mu + \lambda)$.

3.13 Proposition *Let* $\mathbf{c}$ *be an isotropic and homogeneous elasticity tensor with Lamé moduli* λ *and* μ*. Then:*

(i) $\mathbf{c}$ *is pointwise stable if and only if* $\mu > 0$ *and* $k = (3\lambda + 2\mu)/3 > 0$;
(ii) $\mathbf{c}$ *is strongly elliptic if and only if* $\mu > 0$ *and* $\lambda + 2\mu > 0$.

Proof Writing

$$\epsilon = \mu e \cdot e + \frac{\lambda}{2}(\operatorname{tr} e)^2$$

$$= \mu(e - \tfrac{1}{3}(\operatorname{tr} e)g)\cdot(e - \tfrac{1}{3}(\operatorname{tr} e)g) + \frac{k}{2}(\operatorname{tr} e)^2,$$

we obtain (i) since $e - \frac{1}{3}(\operatorname{tr} e)g$ and $\operatorname{tr} e$ can be independently specified. Using 3.12, write

$$\mathsf{c}^{abcd}\alpha_a k_b \alpha_c k_d = \mu \boldsymbol{\alpha}\cdot\boldsymbol{\alpha} \boldsymbol{k}\cdot\boldsymbol{k} + (\lambda + \mu)(\boldsymbol{\alpha}\cdot\boldsymbol{k})^2$$
$$= \mu(\boldsymbol{k}_\perp \cdot \boldsymbol{k}_\perp)\boldsymbol{\alpha}\cdot\boldsymbol{\alpha} + (\lambda + 2\mu)(\boldsymbol{k}\cdot\boldsymbol{\alpha})^2$$

where $\boldsymbol{k}_\perp = \boldsymbol{k} - [(\boldsymbol{k}\cdot\boldsymbol{\alpha})/(\boldsymbol{\alpha}\cdot\boldsymbol{\alpha})]\boldsymbol{\alpha}$ is the component of $\boldsymbol{k}$ orthogonal to $\boldsymbol{\alpha}$. Since $\boldsymbol{k}_\perp$ and $\boldsymbol{k}\cdot\boldsymbol{\alpha}$ can be specified independently, we get (ii). ∎

Thus we see that pointwise stability implies strong ellipticity, but not conversely.

Problem 3.3 Show that an isotropic c is pointwise stable if and only if $E > 0$ and $-1 < \nu < \frac{1}{2}$ and is strongly elliptic if and only if $\mu > 0$ and $\nu < \frac{1}{2}$ or $\nu > 1$.

Problem 3.4 For an isotropic strongly elliptic c, show that there are two speeds of propagation of plane progressive elastic waves given by

$$c_1 = \sqrt{\frac{2\mu + \lambda}{\rho}} \quad \text{and} \quad c_2 = \sqrt{\frac{\mu}{\rho}}$$

(c^2 is a double eigenvalue).

Box 3.1 ***Summary of Important Formulas for Section 4.3***

Equations of Motion for Classical Elasticity

$\rho\ddot{u} = \rho b + \operatorname{div}(\mathsf{c}\cdot\nabla u)$ $\qquad$ $\rho\ddot{u}^a = \rho b^a + (\mathsf{c}^{abcd}u_{c|d})_{|b}$

Boundary Conditions

Displacement: $u = d$ given on $\partial\mathcal{B}$ $\qquad$ $u^a = d^a$ given on $\partial\mathcal{B}$

Traction: $s\cdot n = \tau$ given on $\partial\mathcal{B}$ $\qquad$ $s^{ab}n_b = \tau^a$ given on $\partial\mathcal{B}$

Strain Tensor

$e = \frac{1}{2}\mathfrak{L}_u g$ $\qquad$ $e_{ab} = \frac{1}{2}(u_{a|b} + u_{b|a})$

Stress Tensor

$s = \mathsf{c}\cdot\nabla u$ $\qquad$ $s^{ab} = \mathsf{c}^{abcd}u_{c|d}$

Stored Energy

$\epsilon = \frac{1}{2}e\cdot\mathsf{c}\cdot e = \frac{1}{2}\mathsf{c}\!:\!(e\otimes e)$ $\qquad$ $\epsilon = \frac{1}{2}\mathsf{c}^{abcd}e_{ab}e_{cd}$

$s = \dfrac{\partial\epsilon}{\partial e}, \quad \mathsf{c} = \dfrac{\partial^2\epsilon}{\partial e\,\partial e}$ $\qquad$ $s^{ab} = \dfrac{\partial\epsilon}{\partial e_{ab}}, \quad \mathsf{c}^{abcd} = \dfrac{\partial^2\epsilon}{\partial e_{ab}\,\partial e_{cd}}$

Pointwise Stability

$\epsilon(e) > 0 \quad \text{if} \quad e \neq 0$

Strong Ellipticity

$(\boldsymbol{\alpha} \otimes \boldsymbol{k}) \cdot \mathbf{c} \cdot (\boldsymbol{\alpha} \otimes \boldsymbol{k}) > 0$ if $\boldsymbol{\alpha} \neq \mathbf{0}, \boldsymbol{k} \neq \mathbf{0}$ $\qquad$ $\mathrm{c}^{abcd}\alpha_a k_b \alpha_c k_d > 0$ if $\boldsymbol{\alpha} \neq \mathbf{0}$ and $\boldsymbol{k} \neq \mathbf{0}$

Pointwise Stability $\Rightarrow$ *Strong Ellipticity*

Plane Progressive Wave

$\boldsymbol{u}(x, t) = \boldsymbol{\alpha}\phi(\boldsymbol{x} \cdot \boldsymbol{k} - ct)$ $\qquad$ $u^a(x, t) = \alpha^a \phi(x^b k_b - ct)$

Condition for Waves to be Elastic

$\mathbf{c}(\boldsymbol{k} \otimes \boldsymbol{\alpha} \otimes \boldsymbol{k}) = c^2 \boldsymbol{\alpha}$ $\qquad$ $\mathrm{c}^{abcd} k_b \alpha_c k_d = c^2 \alpha^a$

Reality of Wave Speeds $\Leftrightarrow$ *Strong Ellipticity*

Isotropic Linear Elasticity

(i) $\epsilon = \mu e \cdot e + \frac{\lambda}{2}(\operatorname{tr} e)^2$ $\qquad$ $\epsilon = \mu e^{ab} e_{ab} + \lambda (e^a{}_a)^2$

(ii) pointwise stability $\Leftrightarrow \mu > 0, \lambda > -\frac{2}{3}\mu$

(iii) strong ellipticity $\Leftrightarrow \mu > 0, \lambda > -\frac{1}{2}\mu$

(iv) wave speeds $\sqrt{\frac{2\lambda + \mu}{\rho}}$ and $\sqrt{\frac{\mu}{\rho}}$

4.4 LINEARIZATION STABILITY

In a number of important cases the linearization procedure "fails". This failure can be due to two reasons:

(1) The function spaces are chosen such that the operators of nonlinear elasticity are not differentiable.

(2) The linearized operator fails to be surjective.

Here we investigate abstractly what happens when (2) occurs. In elasticity, this can arise in several ways. First, strong ellipticity could fail, or, second, strong ellipticity can hold, but the linearized problem fails to be always solvable.

Some investigators believe that in some circumstances related to phase transitions, strong ellipticity will fail (see the recent works of Ericksen, Knowles, and Sternberg cited in the bibliography). On the other hand, strong ellipticity often holds, but stability is lost and the linearized operator fails to be invertible. Either situation is indicative that some kind of bifurcation is occurring. The latter situation will be studied in Chapter 7. Examples, including buckling and the traction problem of Signorini, will be presented. In Chapter 6 we shall discuss some function spaces that are important for the study of elastic stability but for which the possibility (1) above pertains.

Let $\mathfrak{X}$, $\mathfrak{Y}$ be Banach spaces and $f\colon \mathfrak{U} \subset \mathfrak{X} \to \mathfrak{Y}$ be a C^1 map. We are interested in solving $f(x) = 0$ for $x \in \mathfrak{X}$. Suppose $x_0 \in \mathfrak{X}$ is a given solution, that is, $f(x_0) = 0$. The linearized equations are simply $Df(x_0)\cdot h = 0$.

4.1 Definition[4] We say f is *linearization stable* at x_0 if for every solution h of the linearized equations there exists a C^1 curve $x(\epsilon) \in \mathfrak{X}$ defined for ϵ in some half-open interval $0 \leq \epsilon < \epsilon_0$ with $x(0) = x_0$, $f(x(\epsilon)) = 0$ and $x'(0) = h$.

Another way of putting this is as follows: f is linearization stable when for any solution h of the linearized equations, we can find a solution of $f(x) = 0$ as a perturbation expansion:

$$x(\epsilon) = x_0 + \epsilon h + \epsilon^2 h^{(2)} + \ldots \quad (\epsilon \geq 0),$$

where $h^{(2)}$ is obtained as in perturbation methods. Of course, this is not exactly the definition since we did not require $x(\epsilon)$ to be analytic in ϵ but only C^1. However, this is a technical point that can be adjusted to suit the situation. Also, depending on the situation, one may wish to demand $x(\epsilon)$ be defined for ϵ small, but both positive and negative (Example 4.3 will illustrate why).

We shall speak of h as an *infinitesimal deformation* of the equations $f(x) = 0$ and of a curve $x(\epsilon)$ of exact solutions through x_0 as an actual, or *finite deformation.* Thus linearization stability can be phrased this way: every infinitesimal deformation is tangent to a finite deformation. We also say that infinitesimal deformations that are tangent to finite deformations are *integrable*. If the conditions of Definition 4.1 fail, we say f is *linearization unstable* at x_0.

4.2 Theorem *Let $f\colon \mathfrak{U} \subset \mathfrak{X} \to \mathfrak{Y}$ be C^1 and $f(x_0) = 0$. Assume $Df(x_0)$ is surjective and its kernel splits. Then f is linearization stable at x_0.*

Proof Write $\mathfrak{X}_1 = \ker Df(x_0)$ and $\mathfrak{X} = \mathfrak{X}_1 \oplus \mathfrak{X}_2$. Then $D_2 f(x_0)\colon \mathfrak{X}_2 \to \mathfrak{Y}$ is an isomorphism. By the implicit function theorem the equation $f(x_1, x_2) = 0$ can be solved for a C^1 function $x_2 = g(x_1)$ for (x_1, x_2) near $x_0 = (x_{01}, x_{02})$ and satisfying $g(x_{01}) = x_{02}$. By implicit differentiation, $D_1 f(x_{01}, x_{02}) + D_2 f(x_{01}, x_{02}) \circ Dg(x_{01}) = 0$. Thus, $Dg(x_{01})$ is zero on $\mathfrak{X}_1$.

Let $x(\epsilon) = (x_{01} + \epsilon h, g(x_{01} + \epsilon h))$, which makes sense since $h \in \mathfrak{X}_1$. Clearly $x(\epsilon)$ is a C^1 curve and $x(0) = x_0$. Also,

$$x'(0) = (h, Dg(x_{01})\cdot h) = (h, 0) = h$$

since $Dg(x_{01})\cdot h = 0$. ∎

If f is C^r or analytic, so is $x(\epsilon)$. (See example (c) below.) While the conditions of this theorem are sufficient, they are not always necessary. A trivial example showing that the conditions are not always necessary is given by $f(x, y) =$

---

[4]This definition was first given in the context of perturbation theory of general relativity by Fischer and Marsden [1973].

$(x, 0), f\colon \mathbb{R}^2 \longrightarrow \mathbb{R}^2$. However, in some important cases they can be shown also to be necessary.[5]

Let us now consider three simple examples to clarify the sort of things that can happen (see Figure 4.4.1).

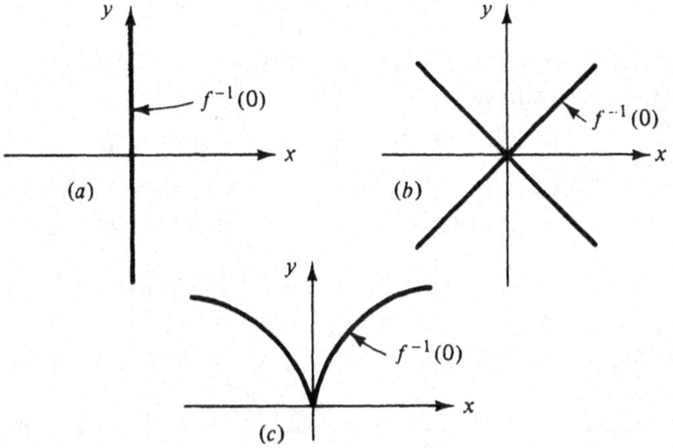

Figure 4.4.1

4.3 Examples (a) Let $f\colon \mathbb{R}^2 \longrightarrow \mathbb{R}, f(x, y) = x(x^2 + y^2)$. Then $f^{-1}(0) = \{(x, y) \mid f(x, y) = 0\}$ is the y-axis and $Df(0, 0) = 0$. We claim that f is linearization unstable at (0, 0). Indeed, a vector (h_1, h_2) is integrable if and only if $h_1 = 0$—that is, is tangent to the y-axis, although any (h_1, h_2) is an infinitesimal deformation of $f(x, y) = 0$.

(b) Let $f\colon \mathbb{R}^2 \longrightarrow \mathbb{R}, f(x, y) = x^2 - y^2$. Again inearization stability at (0, 0) fails. We have $f^{-1}(0) = \{(x, y) \mid x = \pm y\}$, which is not a manifold. An infinitesimal deformation (h_1, h_2) is integrable if and only if $h_1 = \pm h_2$.

(c) Let $f\colon \mathbb{R}^2 \longrightarrow \mathbb{R}, f(x, y) = x^2 - y^3$. The origin is linearization unstable and the integrable directions are along the *positive* y-axis. The curves of exact solutions in the direction of $(0, h_2)$ $(h_2 > 0)$ are given by

$$(x(\epsilon), y(\epsilon)) = (\pm\epsilon^{3/2}, \epsilon h) \quad (\epsilon > 0).$$

Notice that $(x(\epsilon), y(\epsilon))$ is C^1 and although not analytic in ϵ, is, however, analytic in $\sqrt{\epsilon}$. This sort of phenomena is an important feature of linearization instability (and occurs in the traction problem; see Chapter 7).

Next we shall derive some necessary second order conditions that must be satisfied if an infinitesimal deformation is integrable. (For the traction problem these become the "Signorini compatibility conditions.")

---

[5]For the equations of general relativity this is the case (cf. Arms and Marsden [1979]).

4.4 Theorem (Compatability Conditions) *Assume* $f: \mathcal{U} \subset \mathcal{X} \to \mathcal{Y}$ *is* C^2, $h \in \ker Df(x_0)$, *and that* h *is integrable to a* C^2 *finite deformation,* $x(\epsilon)$. *If* $l \in \mathcal{Y}^*$ *is orthogonal to the range of* $Df(x_0)$*—that is, if* $l(Df(x_0)\cdot u) = 0$ *for all* $u \in \mathcal{X}$*—then* h *must satisfy*

$$\boxed{l(D^2f(x_0)\cdot(h, h)) = 0.}$$

Proof Differentiation of $f(x(\epsilon)) = 0$ in ϵ gives $Df(x(\epsilon))\cdot x'(\epsilon) = 0$. Differentiating again, we have

$$D^2f(x(\epsilon))\cdot(x'(\epsilon), x'(\epsilon)) + Df(x(\epsilon))\cdot x''(\epsilon) = 0.$$

Setting $\epsilon = 0$ gives $D^2f(x_0)\cdot(h, h) + Df(x_0)\cdot x''(0) = 0$. Applying l then gives the result since l applied to the second term is zero. ∎

Likewise we can develop conditions of the third and higher order. For instance, if l is orthogonal to the ranges of $v \mapsto Df(x_0)\cdot v$ and $v \mapsto D^2f(x_0)(h, v)$, then we must have $l(D^3f(x_0)\cdot(h, h, h)) = 0$ if h is to be integrable to a C^3 finite deformation.

In examples (a) and (b) above, the third- and second-order conditions, respectively, pick out the directions of linearization stability. In example (c), if we rescale to make f homogeneous by considering instead

$$F(x, y, \lambda) = f(\sqrt{\lambda}\, x, y) = \lambda x^2 - y^3 \quad (\lambda > 0),$$

then the third-order condition on F yields the directions of linearization stability.

The results presented here are useful in understanding the local solutions of a nonlinear equation $f(x) = 0$ when the inverse function theorem fails. Conditions under which the compatability conditions give sufficient conditions for integrability are given using the techniques of bifurcation theory in Chapter 7.

Box 4.1 ***Summary of Important Formulas in Section 4.4***

Linearization Stability

The equation $f(x) = 0$ is linearization stable at a solution x_0 if every h satisfying $Df(x_0)\cdot h = 0$ is integrable; that is, there is a curve $x(\epsilon)$ satisfying $f(x(\epsilon)) = 0$, $x(0) = x_0$, and $x'(0) = h$.

Sufficient Conditions

If f is C^1, $Df(x_0)$ is surjective and its kernel splits, then f is linearization stable at x_0.

Compatibility Conditions

If $Df(x_0)$ is not surjective, l is a linear functional orthogonal to its range, and if $h \in \ker Df(x_0)$ is integrable, then $l(D^2f(x_0)\cdot(h, h)) = 0$.

CHAPTER 5

HAMILTONIAN AND VARIATIONAL PRINCIPLES

This chapter presents some of the basic facts about Hamiltonian systems and applies them to the special case of elasticity. This is a rich area and we do not intend this chapter to be exhaustive. We have chosen two applications to elasticity: conservation laws and reciprocity. These were selected primarily to demonstrate the unifying power of Hamiltonian concepts. In this context, these topics are simple and natural.

The basic principles in this chapter will be useful in the subsequent two chapters. Boxes 7.2 and 7.3 of Chapter 1 are prerequisites for this chapter, so they should be reviewed at this time.

5.1 THE FORMAL VARIATIONAL STRUCTURE OF ELASTICITY

This section presents the basic equations of elasticity in variational form. Subsequent sections will recast this same structure into the modern context of Hamiltonian mechanics. The variational (i.e., weak) form of the equations of elasticity is very useful for three reasons.

(1) It is convenient for numerical computation.

(2) The equations in weak form are believed to hold in situations (such as when shock waves are present) where the localized form doesn't make sense.

(3) It is mathematically helpful in the study of existence and uniqueness (see Chapter 6).

The reader who wishes additional information should consult Truesdell and

Toupin [1960], Vainberg [1964], Washizu [1982], Oden and Reddy [1976b], and Duvaut and Lions [1972].

We shall begin by considering statics. Recall from 4.12, Chapter 3 that the usual or strong form of the boundary value problem of nonlinear elastostatics is as follows:

$$\left.\begin{array}{l}\text{Find } \boldsymbol{\phi} \text{ such that DIV } \boldsymbol{P} + \rho_{\text{Ref}}\boldsymbol{B} = 0 \text{ in the body } \mathcal{B} \text{ and the following}\\ \text{boundary conditions hold: the configuration } \boldsymbol{\phi} \text{ is prescribed to be } \boldsymbol{\phi}_d\\ \text{on a portion } \partial_d \text{ of } \partial\mathcal{B} \text{ and the traction } \boldsymbol{P}\cdot\boldsymbol{N} = \boldsymbol{\tau} \text{ is prescribed on the}\\ \text{remainder } \partial_\tau.\end{array}\right\} \qquad \text{(S)}$$

For the moment we shall assume $\mathcal{B}$ is an open set in $\mathbb{R}^3$. Also we assume a constitutive relation on $\mathcal{B}$ so $\boldsymbol{P}$ is a function $\hat{\boldsymbol{P}}$ of $\boldsymbol{\phi}$, and for now we assume that the loads are *dead* (see Section 3.4). For the weak formulation, we recall some basic terminology:

1.1 Definitions Let the *space of configurations* be

$$\mathcal{C} = \{\boldsymbol{\phi} : \mathcal{B} \longrightarrow \mathbb{R}^3 \,|\, \boldsymbol{\phi} = \boldsymbol{\phi}_d \quad \text{on} \quad \partial_d\}$$

(the smoothness of $\boldsymbol{\phi}$ is discussed in Chapter 6; for now it is unspecified). The *space of variations* is

$$\mathcal{V} = \{\boldsymbol{\eta} : \mathcal{B} \longrightarrow \mathbb{R}^3 \,|\, \boldsymbol{\eta} = \boldsymbol{0} \quad \text{on} \quad \partial_d\}.$$

(So $\mathcal{V}$ is the tangent space to $\mathcal{C}$; see Chapter 4.)

We call $\boldsymbol{\phi} = \boldsymbol{\phi}_d$ on ∂_d the *essential boundary condition* and build it directly into the configuration space. The traction, or *natural boundary condition* $\boldsymbol{P}\cdot\boldsymbol{N} = \boldsymbol{\tau}$, will be built into our functional.

1.2 Definitions Let $G : \mathcal{C} \times \mathcal{V} \longrightarrow \mathbb{R}$ be given by

$$G(\boldsymbol{\phi}, \boldsymbol{\eta}) = \int_{\mathcal{B}} \boldsymbol{P} : \text{GRAD}\, \boldsymbol{\eta} - \rho_{\text{Ref}}\boldsymbol{B}\cdot\boldsymbol{\eta})\, dV - \int_{\partial_\tau} \boldsymbol{\tau}\cdot\boldsymbol{\eta}\, dA,$$

or, in coordinates,

$$G(\boldsymbol{\phi}, \boldsymbol{\eta}) = \int_{\mathcal{B}} (P_a{}^A \eta^a{}_{|A} - \rho_{\text{Ref}} B^a \eta_a)\, dV - \int_{\partial_\tau} \tau^a \eta_a\, dA.$$

The *weak form* of the boundary value problem of elastostatics is given by

$$\text{Find } \boldsymbol{\phi} \in \mathcal{C} \text{ such that } G(\boldsymbol{\phi}, \boldsymbol{\eta}) = 0 \text{ for all } \boldsymbol{\eta} \in \mathcal{V} \qquad \text{(W)}$$

1.3 Proposition *If $\boldsymbol{P}$ and $\boldsymbol{\eta}$ are assumed to be C^1, then the weak and strong forms of the boundary value problem are equivalent.*

Proof Observe that

$$\text{DIV}(\boldsymbol{\eta}\cdot\boldsymbol{P}) = (\eta^a P_a{}^A)_{|A} = \text{GRAD}\, \boldsymbol{\eta} : \boldsymbol{P} + \boldsymbol{\eta}\cdot\text{DIV}\, \boldsymbol{P},$$

so by the divergence theorem,

$$\int_{\partial\mathcal{B}} \boldsymbol{\eta}\cdot\boldsymbol{P}\cdot\boldsymbol{N}\, dA = \int_{\mathcal{B}} \boldsymbol{P} : \text{GRAD}\, \boldsymbol{\eta}\, dV + \int_{\mathcal{B}} \boldsymbol{\eta}\cdot\text{DIV}\, \boldsymbol{P}\, dV.$$

Therefore,

$$G(\phi, \eta) = -\int_{\mathcal{B}} \eta \cdot (\mathrm{DIV}\, \boldsymbol{P} + \rho_{\mathrm{Ref}} \boldsymbol{B})\, dV + \int_{\partial_\tau} \boldsymbol{\eta} \cdot (\boldsymbol{P} \cdot \boldsymbol{N} - \boldsymbol{\tau})\, dA.$$

If (S) holds, then clearly (W) holds. Conversely, if (W) holds and $\mathrm{DIV}\, \boldsymbol{P} + \rho_{\mathrm{Ref}} \boldsymbol{B}$ is continuous, choose $\boldsymbol{\eta}$ to be zero on ∂_d, but otherwise arbitrary, to conclude $\mathrm{DIV}\, \boldsymbol{P} + \rho_{\mathrm{Ref}} \boldsymbol{B} = 0$. Then choose $\boldsymbol{\eta}$ to be arbitrary on ∂_d to conclude that $\boldsymbol{P} \cdot \boldsymbol{N} = \boldsymbol{\tau}$ on ∂_τ. ∎

Next, assume there is a stored energy function (or a free energy) $\hat{\psi}(X, \boldsymbol{F})$ such that $P_a{}^A = \rho_{\mathrm{Ref}}(\partial \hat{\psi} / \partial F^a{}_A)$.

1.4 Definition The *energy functional* is given by $E: \mathcal{C} \to \mathbb{R}$,

$$E(\phi) = \int_{\mathcal{B}} \{\rho_{\mathrm{Ref}}(X) \hat{\psi}(X, \boldsymbol{F}) - \rho_{\mathrm{Ref}}(X) \boldsymbol{B}(X) \cdot \boldsymbol{\phi}(X)\}\, dV - \int_{\partial_\tau} \boldsymbol{\tau}(X) \cdot \boldsymbol{\phi}(X)\, dA,$$

where $\boldsymbol{\phi}(X)$ stands for the vector from the origin to the point $\phi(X)$ in $\mathbb{R}^3$.

1.5 Proposition *The derivative of E at ϕ in the direction η is given by*

$$DE(\phi) \cdot \boldsymbol{\eta} = G(\phi, \boldsymbol{\eta}).$$

Proof We have

$$DE(\phi) \cdot \boldsymbol{\eta} = \frac{d}{d\epsilon} E(\phi + \epsilon \boldsymbol{\eta})$$

$$= \frac{d}{d\epsilon} \int_{\mathcal{B}} [\rho_{\mathrm{Ref}}(X) \hat{\psi}(X, \boldsymbol{F} + \epsilon\, \mathrm{GRAD}\, \boldsymbol{\eta}) - \rho_{\mathrm{Ref}} B(X) \cdot (\boldsymbol{\phi} + \epsilon \boldsymbol{\eta})]\, dV$$

$$- \int_{\partial_\tau} \boldsymbol{\tau}(X) \cdot (\boldsymbol{\phi}(X) + \epsilon \boldsymbol{\eta}(X))\, dA \Big|_{\epsilon=0}$$

Assuming enough differentiability to pass $d/d\epsilon$ under the integral sign and to apply the chain rule, we get

$$\int_{\mathcal{B}} \left[\rho_{\mathrm{Ref}} \frac{\partial \hat{\psi}}{\partial \boldsymbol{F}} \cdot \mathrm{GRAD}\, \boldsymbol{\eta} - \rho_{\mathrm{Ref}} \boldsymbol{B} \cdot \boldsymbol{\eta}\right] dV - \int_{\partial_\tau} \boldsymbol{\tau} \cdot \boldsymbol{\eta}\, dA,$$

which is precisely $G(\phi, \boldsymbol{\eta})$. ∎

If the loads are not dead but depend on the point values of ϕ, and if they are conservative—that is, derivable from potentials in the sense that $B(\phi) \cdot \boldsymbol{\eta} = DV_B(\phi) \cdot \boldsymbol{\eta}$ and $\boldsymbol{\tau}(\phi) \cdot \boldsymbol{\eta} = DV_\tau(\phi) \cdot \boldsymbol{\eta}$—then the same result holds if in the definition of E we replace $\rho_{\mathrm{Ref}} \boldsymbol{B} \cdot \boldsymbol{\phi}$ by $\rho_{\mathrm{Ref}} V_B(\phi)$ and $\boldsymbol{\tau} \cdot \boldsymbol{\phi}$ by $V_\tau(\phi)$.

In classical terminology, $\boldsymbol{\eta}$ is called a *variation*, or a *virtual displacement*, and is denoted $\delta\phi$, and $DE(\phi) \cdot \boldsymbol{\eta}$ is denoted δE. We call ϕ a *critical point* for E if $DE(\phi) \cdot \boldsymbol{\eta} = 0$ for all $\boldsymbol{\eta}$—that is, if $\delta E = 0$ for any variation $\delta\phi$. Thus, we have proved the following:

1.6 Theorem.[1] *ϕ is a solution to the weak form of the boundary value problem of elastostatics if and only if ϕ is a critical point for the energy functional E.*

This formulation of the boundary value problem of elastostics is also called the *principle of virtual work* (cf. Box 4.1, Chapter 2). One of the reasons this formulation is computationally convenient is that the boundary condition $\boldsymbol{P}\cdot\boldsymbol{N} = \tau$ need not be imposed a priori, but comes out automatically in the search for a critical point. This is useful because $\boldsymbol{P}$ is generally a nonlinear function of ϕ, so directly imposing $\boldsymbol{P}\cdot\boldsymbol{N} = \boldsymbol{\tau}$ could be awkward.

Box 1.1 ***The Existence of Potentials***

Suppose $G: \mathcal{C} \times \mathcal{V} \to \mathbb{R}$ is defined by 1.2. When does there exist an $E: \mathcal{C} \to \mathbb{R}$ such that $G(\phi, \boldsymbol{\eta}) = DE(\phi)\cdot\boldsymbol{\eta}$? This is an *inverse problem* in the calculus of variations since it asks when a given set of equations can be written as the Euler–Lagrange equations for some energy functional.

The answer to this question is given by the infinite-dimensional version of Theorem 7.11 of Chapter 1. We summarize:

1.7 Proposition (*i*) *There is an $E: \mathcal{C} \to \mathbb{R}$ such that $G(\phi, \boldsymbol{\eta}) = DE(\phi)\cdot\boldsymbol{\eta}$ if and only if*

$$D_1 G(\phi, \boldsymbol{\eta})\cdot\boldsymbol{\xi} = D_1 G(\phi, \boldsymbol{\xi})\cdot\boldsymbol{\eta}$$

for all $\phi \in \mathcal{C}$ and $\boldsymbol{\eta}, \boldsymbol{\xi} \in \mathcal{V}$.

(*ii*) *If the condition in* (*i*) *holds, E is given by*

$$E(\phi) = \int_0^1 G(t\phi, \phi)\, dt.$$

The reader may supply a direct proof of this without difficulty. The symmetry condition in (i) is just the symmetry of $D^2E(\phi)(\boldsymbol{\eta}, \boldsymbol{\xi})$ in $\boldsymbol{\eta}$ and $\boldsymbol{\xi}$. The proposition as stated has nothing to do with elasticity. If we specialize G to be given by 1.2, then we have

$$D_1 G(\phi, \boldsymbol{\eta})\cdot\boldsymbol{\xi} = \int_{\mathcal{B}} \mathsf{A}_a{}^A{}_b{}^B \eta^a{}_{|A} \xi^b{}_{|B}\, dV$$
$$= \int_{\mathcal{B}} \mathbf{A}\cdot(\text{GRAD}\, \boldsymbol{\eta}, \text{GRAD}\, \boldsymbol{\xi})\, dV,$$

where $\mathbf{A}$ is the elasticity tensor. Thus the symmetry of $D_1 G$ holds if and only if we have the symmetry $\mathsf{A}_a{}^A{}_b{}^B = \mathsf{A}_b{}^B{}_a{}^A$. This is, in turn, equivalent

[1] Further aspects of the significance of the weak form are discussed in the next chapter in connection with the existence and uniqueness theory. In particular, we shall address the question: can E be minimized?

to the existence of $\hat{\psi}$ such that $P_a{}^A = \rho_{\text{Ref}}(\partial\hat{\psi}/\partial F^a{}_A)$, reproducing an observation we made in Section 3.4.

Problem 1.1 (J. Mesirov) Show that the equations $u'' = h(x, u, u')$, where $u' = du/dx$, $u'' = d^2u/dx^2$, are the Euler–Lagrange equations for some variational principle (with $u: [0, 1] \longrightarrow \mathbb{R}$, $u(0) = a$, $u(1) = b$) if and only if h is independent of u'. Do this (a) by directly identifying $u'' = h(x, u, u')$ with the Euler–Lagrange equations for $\int_0^1 F(x, u, u')\, dx$ and (b) by applying 1.7 to $G(u, \eta) = \int_0^1 [u'' - h(x, u, u')]\eta\, dx$.

There are several ways to formulate variational principles for elastodynamics. We shall go into these in detail in subsequent sections. For now we state the weak form as follows: let G map the pairs (motions satisfying the essential boundary condition, curves of virtual displacements) to functions of time by

$$G_{\text{dyn}}(\phi_t, \boldsymbol{\eta}_t) = \int_{\mathcal{B}} \rho_{\text{Ref}}(\boldsymbol{A} - \boldsymbol{B})\cdot\boldsymbol{\eta}_t\, dV + \int_{\mathcal{B}} (\boldsymbol{P}: \text{GRAD}\,\boldsymbol{\eta}_t)\, dV - \int_{\partial_\tau} \boldsymbol{\tau}\cdot\boldsymbol{\eta}_t\, dA,$$

where $\boldsymbol{A}$ is the material acceleration of the motion. Then the weak form is

$$\text{Find } \phi_t \text{ such that } G(\phi_t, \boldsymbol{\eta}_t) = 0 \text{ for all } \boldsymbol{\eta}_t \in \mathcal{V}. \qquad (W_{\text{dyn}})$$

The calculations above show that if $\boldsymbol{P}$ and $\boldsymbol{\eta}$ are C^1, then this is equivalent to the strong form: $\rho_{\text{Ref}}\boldsymbol{A} + \text{DIV}\ \boldsymbol{P} = \rho_{\text{Ref}}\boldsymbol{B}$ with boundary conditions $\phi = \phi_d$ on ∂_d and $\boldsymbol{P}\cdot\boldsymbol{N} = \boldsymbol{\tau}$ on ∂_τ. Also, if $\boldsymbol{P}$ is derived from a potential $\hat{\psi}$, then $G_{\text{dyn}}(\phi_t, \boldsymbol{\eta}_t) = 0$ are the Euler–Lagrange equations for the Lagrangian $L = \frac{1}{2}\int_{\mathcal{B}} \rho_{\text{Ref}}\, \|\boldsymbol{V}\|^2\, dV - E(\phi_t)$ in the sense that $\delta \int_{t_1}^{t_2} L\, dt = 0$ for all variations $\boldsymbol{\eta}_t$ that vanish at two specified times, say $t = 0$ and $t = T$.

Problem 1.2 Carry out this idea in detail using Proposition 7.9 in Box 7.2, Chapter 1.

Problem 1.3 Write $G(\phi, \boldsymbol{\eta}) = \int_{\mathcal{B}} \boldsymbol{f}(\phi)\cdot\boldsymbol{\eta}\, dV + \int_{\partial_\tau} \boldsymbol{g}(\phi)\cdot\boldsymbol{\eta}\, dA$ and observe that the elastostatic equations linearized about $\mathring{\phi}$, in strong form, are

$$\begin{cases} D\boldsymbol{f}(\mathring{\phi})\cdot\boldsymbol{u} + \boldsymbol{f}(\mathring{\phi}) = \boldsymbol{0} & \text{in} \quad \mathcal{B}, \\ D\boldsymbol{g}(\mathring{\phi})\cdot\boldsymbol{u} + \boldsymbol{g}(\mathring{\phi}) = \boldsymbol{0} & \text{on} \quad \partial_\tau. \end{cases}$$

Suppose $G(\phi, \boldsymbol{\eta}) = DE(\phi)\cdot\boldsymbol{\eta}$. Show that the equations linearized at $\mathring{\phi}$ are derived from the potential

$$E_{\text{lin}}(\boldsymbol{u}) = DE(\mathring{\phi})\cdot\boldsymbol{u} + \tfrac{1}{2}\, D^2E(\mathring{\phi})\cdot(\boldsymbol{u}, \boldsymbol{u}).$$

Work this out for linear elasticity and show that E_{lin} agrees with the energy function derived in Section 4.3.

Problem 1.4 (Hu–Washizu Theorem) In classical elastostatics, define

$$H(\boldsymbol{u}, \boldsymbol{e}, \boldsymbol{\sigma}, \boldsymbol{t}) = \int_{\mathcal{B}} \tfrac{1}{2}\, \mathsf{c}^{abcd}\, e_{ab} e_{cd}\, dV - \int_{\mathcal{B}} B^i u_i\, dV$$

$$- \int_{\mathcal{B}} \sigma^{ab}(e_{ab} - \tfrac{1}{2}(u_{a,b} + u_{b,a}))\, dV - \int_{\partial_\tau} \tau^i u_i\, dA - \int_{\partial_d} t^i(u_i - \bar{u}_i)\, dA.$$

Show that $DH(\boldsymbol{u}, \boldsymbol{e}, \boldsymbol{\sigma}, \boldsymbol{t}) = 0$ (D taken in all variables) reproduces six sets of equations in classical elastostatics (see Washizu [1982], p. 31).

Box 1.2 ***Summary of Important Formulas for Section 5.1***

Configuration Space

$\mathcal{C} = \{\phi\colon \mathcal{B} \to \mathbb{R}^3 \mid \phi$ satisfies the essential (displacement) boundary condition on $\partial_d \subset \partial\mathcal{B}\}$

Space of Variations

$\mathcal{V} = \{\boldsymbol{\eta}\colon \mathcal{B} \to \mathbb{R}^3 \mid \boldsymbol{\eta} = \boldsymbol{0} \text{ on } \partial_d\}$

Weak Form of the Boundary Value Problem of Elastostatics

$G(\phi, \boldsymbol{\eta}) = \boldsymbol{0}$ for all $\boldsymbol{\eta} \in \mathcal{V}$,

where

$$G(\phi, \boldsymbol{\eta}) = \int_{\mathcal{B}} (\boldsymbol{P}\colon \mathrm{GRAD}\, \boldsymbol{\eta} - \rho_{\mathrm{Ref}} \boldsymbol{B}\cdot\boldsymbol{\eta})\, dV - \int_{\partial_\tau} \boldsymbol{\tau}\cdot\boldsymbol{\eta}\, dA$$

$$= \int_{\mathcal{B}} (P_a{}^A \eta^a{}_{|A} - \rho_{\mathrm{Ref}}\, B^a \eta_a)\, dV - \int_{\partial_\tau} \tau^a \eta_a\, dA.$$

Energy Functional

$E\colon \mathcal{C} \to \mathbb{R}$

$$E(\phi) = \int_{\mathcal{B}} \rho_{\mathrm{Ref}}(\hat{\psi} - \boldsymbol{B}\cdot\boldsymbol{\phi})\, dV - \int_{\partial_\tau} \boldsymbol{\tau}\cdot\boldsymbol{\phi}\, dA$$

$$= \int_{\mathcal{B}} (\rho_{\mathrm{Ref}} \hat{\psi} - B_a \phi^a)\, dV - \int_{\partial_\tau} \tau_a \phi^a\, dA$$

If $P_a{}^A = \rho_{\mathrm{Ref}}(\partial\hat{\psi}/\partial F^a{}_A)$, then $DE(\phi)\cdot\boldsymbol{\eta} = G(\phi, \boldsymbol{\eta})$.

Potentials

$\hat{\psi}$ exists if and only if $D_1 G(\phi, \boldsymbol{\eta})\cdot\boldsymbol{\xi} = D_1 G(\phi, \boldsymbol{\xi})\cdot\boldsymbol{\eta}$ if and only if the elasticity tensor is symmetric: $\mathsf{A}_a{}^A{}_b{}^B = \mathsf{A}_b{}^B{}_a{}^A$.

5.2 LINEAR HAMILTONIAN SYSTEMS AND CLASSICAL ELASTICITY

There are a number of subleties involved in generalizing the ideas of Hamiltonian mechanics to infinite dimensions. For example, how are Poisson brackets and canonical transformations defined? In this and the next section we develop some of the basic ideas of this theory and apply them to elasticity. We assume the reader has read Boxes 7.2 and 7.3 in Chapter 1, but shall occasionally give the finite-dimensional coordinate analogues to help fix the concepts.

2.1 Definitions Let $\mathfrak{X}$ be a Banach space.

(a) A (weak) *symplectic form* on $\mathfrak{X}$ is a continuous bilinear map ω: $\mathfrak{X} \times \mathfrak{X} \longrightarrow \mathbb{R}$ that is (i) skew (i.e., $\omega(u, v) = -\omega(v, u)$) and (ii) (weakly) non-degenerate (i.e., $\omega(u, v) = 0$ for all $v \in \mathfrak{X}$ implies $u = 0$).[2] We often speak of $\mathfrak{X}$ with a symplectic structure ω as a *phase space*.

(b) A linear operator $A: \mathfrak{D}(A) \longrightarrow \mathfrak{X}$ with domain $\mathfrak{D}(A)$ a linear subspace of $\mathfrak{X}$ is called *Hamiltonian* if it is ω-skew; that is, $\omega(Ax, y) = -\omega(x, Ay)$ for all $x, y \in \mathfrak{D}(A)$.

(c) The *Hamiltonian or energy function* of A is defined by $H(u) = \frac{1}{2}\omega(Au, u)$, $u \in \mathfrak{D}(A)$.

(d) A bounded linear operator $T: \mathfrak{X} \longrightarrow \mathfrak{X}$ is called a *canonical transformation* if it preserves ω; that is, $\omega(Tu, Tv) = \omega(u, v)$ for all $u, v \in \mathfrak{X}$. (In the pull-back notation of Chapter 1, this reads: $T^*\omega = \omega$.)

(e) An operator $A = \mathfrak{D}(A) \longrightarrow \mathfrak{X}$ is called *closed* when its domain $\mathfrak{D}(A)$ is a Banach space using the norm

$$|||u|||^2 = \|Au\|^2 + \|u\|^2,$$

$\|\cdot\|$ being the $\mathfrak{X}$-norm. We shall assume in the following that A is closed and that this Banach space structure is given to $\mathfrak{D}(A)$.

In finite dimensions we often choose $\mathfrak{X} = \mathbb{R}^{2n}$ with coordinates $(q^1, \ldots, q^n, p_1, \ldots, p_n)$ and $\boldsymbol{\omega} = dq^i \wedge dp_i$. As a bilinear map, $\boldsymbol{\omega}$ is given by $\boldsymbol{\omega}((q, p), (\bar{q}, \bar{p})) = \bar{p}_i q^i - \bar{q}^i p_i$. The form $\boldsymbol{\omega}$ has the matrix

$$\mathbb{J} = \begin{bmatrix} 0 & I \\ -I & 0 \end{bmatrix}, \qquad I = \text{Identity}$$

in the sense that $\boldsymbol{\omega}(u, v) = u^T \cdot \mathbb{J} \cdot v$. A linear operator in this case may be represented in block form as $A = \begin{bmatrix} a & b \\ c & d \end{bmatrix}$. It is easily checked that A is ω-skew when b and c are symmetric and $a^T = -d$. The reader should also check that A and H are, in this case, related by Hamilton's equations: if $A(q, p) = (\dot{q}, \dot{p})$, then

$$\dot{q}^i = (aq + bp)^i = a^i{}_j q^j + b^{ij} p_j = \frac{\partial H}{\partial p_i},$$

$$\dot{p}_j = (cq + dp)_j = c_{ij} q^i + d^i{}_j p_i = -\frac{\partial H}{\partial q^j}.$$

This relation between A and H can also be written $Au = \mathbb{J}\, DH(u)$.

---

[2] We call ω strongly *non-degenerate* if for every continuous linear function $\alpha: X \longrightarrow \mathbb{R}$ (i.e., $\alpha \in X^*$), there exists a unique $u \in X$ such that $\omega(u, v) = \alpha(v)$ for all $v \in X$. In finite dimensions, strong and weak non-degeneracy are coincident concepts, but in infinite dimensions they are not; weak nondegeneracy is more important.

Problem 2.1 If $T = \begin{bmatrix} e & f \\ g & h \end{bmatrix}$, what conditions on e, f, g, and h make T a canonical transformation?

2.2 Proposition *In Definition 2.1 we have $DH(u)\cdot v = \omega(Au, v)$ for all $u, v \in \mathfrak{D}(A)$.*

Proof By definition, $H(u) = \frac{1}{2}\omega(Au, u)$. Differentiating, we get $DH(u)\cdot v = \frac{1}{2}\omega(Av, u) + \frac{1}{2}\omega(Au, v)$. Since A is ω-skew, we get the result. ∎

In Box 7.3, Chapter 1 we defined the Hamiltonian vector field X_H of an energy function H by the relationship $i_{X_H}\omega = dH$. Proposition 2.2 shows that A is the Hamiltonian vector field for its energy function. The next proposition takes this converse point of view:

2.3 Proposition *Let A be a closed operator and let $H: \mathfrak{D}(A) \to \mathbb{R}$ be a C^1 function such that $DH(u)\cdot v = \omega(Au, v)$ for all $u, v \in \mathfrak{D}(A)$. Then A is Hamiltonian with energy H, possibly after modifying H by the addition of a constant.*

Proof From the above formula, H is C^2 and $d^2H(0)(u, v) = \omega(Au, v)$. Since this is symmetric in (u, v), A is ω-skew; so A is Hamiltonian. Since $DH(u)\cdot v = D[\frac{1}{2}\omega(Au, u)]\cdot v$, H is the energy, up to a constant. ∎

2.4 Examples

1. (a) Let $\mathfrak{X}_1$ and $\mathfrak{Y}_1$ be Banach spaces with $\mathfrak{Y}_1$ continuously and densely included in $\mathfrak{X}_1$. Let $\mathfrak{X} = \mathfrak{Y}_1 \times \mathfrak{X}_1^*$ and define $\omega: \mathfrak{X} \times \mathfrak{X} \to \mathbb{R}$ by $\omega((y, \alpha), (\bar{y}, \bar{\alpha})) = \bar{\alpha}(y) - \alpha(\bar{y})$. It is straightforward to check that ω is a (weak) symplectic form. [The reader can check that if $\mathfrak{Y}_1 = \mathfrak{X}_1$, ω is non-degenerate in the strong sense if and only if $\mathfrak{X}_1$ is reflexive (i.e., the canonical inclusion $\mathfrak{X}_1 \subset \mathfrak{X}_1^{**}$ is onto).]

(b) If $\langle\ ,\ \rangle$ is a weakly non-degenerate symmetric bilinear form on $\mathfrak{X}_1$ and $\mathfrak{X} = \mathfrak{Y}_1 \times \mathfrak{X}_1$, then

$$\omega((y, x), (\bar{y}, \bar{x})) = \langle \bar{x}, y\rangle - \langle x, \bar{y}\rangle$$

is a (weak) symplectic form.

In (a) we call ω the *canonical symplectic form* on $\mathfrak{X}$ and in (b) we say ω is the symplectic form *induced* by $\langle\ ,\ \rangle$. In (a) note that $\omega((y, \alpha), (\bar{y}, \bar{\alpha})) = (\bar{y}, \bar{\alpha})\cdot \mathbb{J}\cdot \binom{y}{\alpha}$, where $\mathbb{J} = \begin{bmatrix} 0 & I \\ -I & 0 \end{bmatrix}$, and that A is related to H by $A(y, \alpha) = \mathbb{J}\cdot DH(y, \alpha)$. If $\mathfrak{X}_1 = \mathfrak{Y}_1 = \mathfrak{H}_1$ is a Hilbert space and if $\mathfrak{H}_1$ is identified with $\mathfrak{H}_1^*$ (by the Riesz representation theorem), then (a) and (b) are identical. However, the generality given is needed for the examples.

(c) If $\mathfrak{H}_1$ is a real Hilbert space and we let $\mathfrak{H} = \mathfrak{H}_1 \oplus \mathfrak{H}_1$ be its complexification, then the canonical symplectic form induced on $\mathfrak{H}$ in (b) (with $\mathfrak{Y}_1 = \mathfrak{H}_1$ and $\mathfrak{X}_1 = \mathfrak{H}_1$ can be written

$$\omega\colon \mathcal{H} \times \mathcal{H} \to \mathbb{R},\ \omega(\phi, \psi) = -\mathrm{Im}\,\langle \phi, \psi \rangle,$$

as is easily checked.

2. Let $\mathfrak{X} = \mathbb{R}^n \times (\mathbb{R}^n)^*$ and let g^{ij} and c_{ij} be symmetric matrices. Let points in $\mathfrak{X}$ be denoted by $(\boldsymbol{q}, \boldsymbol{p}) = (q^i, p_j)$ and set

$$H(q^i, p_j) = \tfrac{1}{2} g^{ij} p_i p_j + \tfrac{1}{2} c_{ij} q^i q^j,$$

and let $\boldsymbol{\omega}((\boldsymbol{q}, \boldsymbol{p}), (\bar{\boldsymbol{q}}, \bar{\boldsymbol{p}})) = \bar{p}_i q^i - p_i \bar{q}^i$. The corresponding linear Hamiltonian operator is

$$X_H(\boldsymbol{q}, \boldsymbol{p}) = A(\boldsymbol{q}, \boldsymbol{p}) = \left(\frac{\partial H}{\partial p_i}, -\frac{\partial H}{\partial q_i}\right) = (g^{ij} p_j,\ -c_{ij} q^j).$$

If $\boldsymbol{P}$ is an $n \times n$ orthogonal matrix diagonalizing $\boldsymbol{C} = (c_{ij})$—that is, if $\boldsymbol{P}^{-1}\boldsymbol{C}\boldsymbol{P} = \boldsymbol{\Lambda} = \mathrm{diag}\,(\lambda_1, \ldots, \lambda_n)$ is diagonal—then $\boldsymbol{T}\colon \mathfrak{X} \to \mathfrak{X}$, $T(\boldsymbol{q}, \boldsymbol{p}) = (\boldsymbol{P}^{-1}\boldsymbol{q}, \boldsymbol{p}\cdot\boldsymbol{P}^T)$ is a canonical transformation uncoupling the terms $c_{ij} q^i q^j$. If g^{ij} is positive-definite, we can simultaneously diagonalize g^{ij} and c_{ij} and thereby uncouple the equations into n harmonic oscillators.

3. The wave equation $\ddot{\phi} = \Delta\phi$ is Hamiltonian on $\mathfrak{X} = H^1(\mathbb{R}^n) \times L^2(\mathbb{R}^n)$ with

$$H(\phi, \dot{\phi}) = \tfrac{1}{2} \int_{\mathbb{R}^n} |\dot{\phi}|^2\, dx + \tfrac{1}{2} \int_{\mathbb{R}^n} |\nabla\phi|^2\, dx,$$

Here, $H^1(\mathbb{R}^n)$ denotes all functions $\phi\colon \mathbb{R}^n \to \mathbb{R}$ such that ϕ and $\nabla\phi$ are L^2—that is, such that $\int_{\mathbb{R}^n} |\phi|^2\, dx < \infty$ and $\int_{\mathbb{R}^n} |\nabla\phi|^2\, dx < \infty$. This is an example of a Sobolev space, some of whose properties are discussed in the next chapter. Thus, $\mathfrak{X}$ is designed so that $H(\phi, \dot{\phi}) < \infty$ for $(\phi, \dot{\phi}) \in \mathfrak{X}$. The symplectic form is

$$\omega\,((\phi, \dot{\phi}), (\psi, \dot{\psi})) = \int (\dot{\psi}\phi - \dot{\phi}\psi)\, dx,$$

which is the symplectic form of 1(b) with $\mathcal{Y}_1 = H^1$, $\mathfrak{X} = L^2$, and $\langle\, ,\rangle$ the L^2 inner product. The Hamiltonian operator is given by

$$A(\phi, \dot{\phi}) = (\dot{\phi}, \Delta\phi)$$

with $\mathfrak{D}(A) = H^2(\mathbb{R}^n) \times H^1(\mathbb{R}^n)$, where $H^2(\mathbb{R}^n)$ denotes those ϕ such that ϕ, $D\phi$, and $D^2\phi$ all lie in L^2.

4. If H_{op} is a symmetric operator on complex Hilbert space $\mathcal{H}$, the abstract Schrodinger equation,

$$\frac{\partial\psi}{\partial t} = iH_{\mathrm{op}}\psi,$$

is Hamiltonian with $H(\psi) = \tfrac{1}{2}\langle H_{\mathrm{op}}\psi, \psi\rangle$, $\omega(\psi, \phi) = -\mathrm{Im}\,\langle\psi, \phi\rangle$. (See Example 1(c)), and $A\psi = iH_{\mathrm{op}}\psi$.)

Now we turn to a few properties of the *dynamics* of linear Hamiltonian systems. Our results at this point will be preliminary; the necessary theorems for the full theory are presented in Section 6.3.

2.5 Definition Let $A: \mathfrak{D}(A) \to \mathfrak{X}$ be a closed linear operator in $\mathfrak{X}$, with $\mathfrak{D}(A) \subset \mathfrak{X}$ a dense linear subspace. We shall say *A generates a semigroup* $\{U_t \mid t \geq 0\}$ if each $U_t: \mathfrak{X} \to \mathfrak{X}$ is a bounded operator defined for $t \geq 0$ and:

(i) $U_0 =$ Identity and $\| U_t \|$ is uniformly bounded on bounded t-intervals;
(ii) $U_{t+s} = U_t \circ U_s$ for $s, t \geq 0$;
(iii) for each $x \in X$, $U_t x$ is continuous in t;
(iv) for each $x \in \mathfrak{D}(A)$, $U_t x$ lies in $\mathfrak{D}(A)$, is differentiable in t as a curve in X (differentiable from the right at $t = 0$) and

$$\frac{d}{dt} U_t x = A \cdot U_t x;$$

(v) if $x \in \mathfrak{X}$ and $U_t x$ is t-differentiable at $t = 0+$, then $x \in \mathfrak{D}(A)$.

If U_t is defined for all $t \in \mathbb{R}$, we say A generates the (one-parameter) *group* $U(t)$.

Conditions under which A generates a semigroup are given in Section 6.3 along with general properties of semigroups. For example, some of the conditions above are redundant. (Semi-) groups are almost synonymous with (semi-) flows in the sense of Section 1.6. However, semigroups are linear for each t, and A is only densely defined. The latter is an essential difference between ordinary and partial differential equations. In finite dimensions or if A is bounded, then $U_t = e^{tA}$, where e^{tA} is defined by, say, a power series, but when A is unbounded, this relationship is just formal.

The following shows that two of the most basic facts about Hamiltonian systems hold in this context.

2.6 Proposition *Let A be Hamiltonian with energy H and suppose A generates a semigroup* $\{U_t \mid t \geq 0\}$. *Then*

(*i*) *each* U_t *is a canonical transformation and*
(*ii*) $H(U_t x) = H(x)$ *for each* $x \in \mathfrak{D}(A)$ *and* $t \geq 0$ (*conservation of energy*).

Proof (*i*) If $u, v \in \mathfrak{D}(A)$ then

$$\frac{d}{dt}\omega(U_t u, U_t v) = \omega(AU_t u, U_t v) + \omega(U_t u, AU_t v) = 0,$$

since A is ω-skew. Hence, $\omega(U_t u, U_t v) = \omega(u, v)$ for $u, v \in \mathfrak{D}(A)$ and hence everywhere by continuity of U_t and ω, and denseness of $\mathfrak{D}(A)$.

(*ii*) Differentiating $U_{t+s} = U_t \cdot U_s$ in s at $s = 0$, we see that $AU_t = U_t A$ on $\mathfrak{D}(A)$. Thus, for $u \in \mathfrak{D}(A)$, $H(U_t u) = \frac{1}{2}\omega(AU_t u, \; U_t u) = \frac{1}{2}\omega(U_t Au, \; U_t u) = \frac{1}{2}\omega(Au, u) = H(u)$. ∎

The following definition and proposition is a basic criterion for the dynamic stability of linear Hamiltonian systems:

2.7 Definition We say that the semigroup $\{U_t \mid t \geq 0\}$ is (dynamically) *stable* if there is a constant $M > 0$ such that $\|U_t x\| \leq M\|x\|$ for all $x \in \mathfrak{X}$ and $t \geq 0$.

This condition implies that each orbit $t \mapsto U_t x$ remains in a bounded set for all time. The following is the *energy criterion* for stability for linear systems.

2.8 Proposition *Suppose A is a linear Hamiltonian system with energy H. Suppose there are positive constants c and C such that*

$$c\|x\|^2 \leq H(x) \leq C\|x\|^2$$

for all x in $\mathfrak{D}(A)$. Let A generate a semigroup $\{U_t \mid t \geq 0\}$. Then $\{U_t \mid t \geq 0\}$ is stable.

Proof By conservation of energy, $H(U_t x) = H(x)$. Thus

$$\|U_t x\|^2 \leq \frac{1}{c} H(U_t x) = \frac{1}{c} H(x) \leq \frac{C}{c}\|x\|^2$$

so $\{U_t \mid t \geq 0\}$ is stable with $M = C/c$. ∎

It turns out that the condition $H(x) \geq c\|x\|^2$ is also one of the important hypotheses in a theorem guaranteeing that A generates a semigroup. We shall study this issue in Section 6.3.

Although this proof does not use linearity of A, the hypotheses are not easy to satisfy for nonlinear elasticity. The difficulties involved will be apparent when we study potential wells and the energy criterion in Section 6-5.

Problem 2.2 Show that $\{U_t \mid t \geq 0\}$ is stable in $\mathfrak{D}(A)$ if $H(x) \geq c\|x\|_{\mathfrak{D}(A)}$.

2.9 Example (Abstract Wave Equation) Let $\mathfrak{H}$ be a real Hilbert space and $B\colon \mathfrak{D}(B) \longrightarrow \mathfrak{H}$ a symmetric operator (i.e., $\langle Bx, y\rangle = \langle x, By\rangle$ for all $x, y \in \mathfrak{D}(B)$) that is positive; that is, there is an $m > 0$ such that $\langle Bx, x\rangle \geq m^2\|x\|^2$. Let $\sqrt{B}$ be a (closed) positive square root for B with domain denoted $\mathfrak{D}(B^{1/2})$. Let $\mathfrak{X} = \mathfrak{D}(B^{1/2}) \times \mathfrak{H}$ and let

$$A = \begin{bmatrix} 0 & I \\ -B & 0 \end{bmatrix}$$

with $\mathfrak{D}(A) = \mathfrak{D}(B) \times \mathfrak{D}(B^{1/2})$, and let

$$\omega((x, y), (\bar{x}, \bar{y})) = \langle \bar{y}, x\rangle - \langle \bar{x}, y\rangle.$$

Then A is Hamiltonian with energy

$$H(x, y) - \tfrac{1}{2}\|y\|^2 + \tfrac{1}{2}\langle Bx, x\rangle.$$

Proposition 2.8 shows that A generates stable dynamics. The evolution is determined by the abstract wave equation $\ddot{x} = -Bx$. This example includes the wave equation $\ddot{u} = \Delta u - m^2 u$ for $m > 0$, with $\mathfrak{H}$ the L^2 functions. (A slight modification of the spaces is needed to include $m = 0$.)

In the next example, we study the case of classical elastodynamics with no external forcing. (This is needed so the system is linear.)

2.10 Example (Classical Elastodynamics) We now show that the homogeneous equations of classical linear elastodynamics,

$$\rho \ddot{\boldsymbol{u}} = \operatorname{div}(\mathbf{c}\cdot\nabla \boldsymbol{u}), \qquad \text{that is,} \quad \rho \ddot{u}^a = (\mathrm{c}^{abcd}u_{c|d})_{|b},$$

define a Hamiltonian system. We work in a region $\mathcal{B} \subset \mathbb{R}^3$ with displacement boundary conditions $\boldsymbol{u} = \mathbf{0}$ on ∂_d and/or traction boundary conditions, $(\mathbf{c}\cdot\nabla\boldsymbol{u})\cdot\boldsymbol{n} = \mathbf{0}$ on ∂_τ. Assume, as in Section 4.3, that c^{abcd} comes from a stored energy function; that is, $\mathrm{c}^{abcd} = \mathrm{c}^{cdab}$. We choose $\mathfrak{X} = H^1 \times L^2$, with displacement boundary conditions imposed. We set

$$\omega((\boldsymbol{u}, \dot{\boldsymbol{u}}), (\boldsymbol{v}, \dot{\boldsymbol{v}})) = \int_{\mathcal{B}} \rho(\dot{\boldsymbol{v}}\cdot\boldsymbol{u} - \dot{\boldsymbol{u}}\cdot\boldsymbol{v})\, dx,$$

$$A(\boldsymbol{u}, \dot{\boldsymbol{u}}) = \left(\dot{\boldsymbol{u}}, -\frac{1}{\rho}\operatorname{div}(\mathbf{c}\cdot\nabla\boldsymbol{u})\right)$$

$\mathfrak{D}(A) = H^2 \times H^1$, with displacement *and* traction boundary conditions imposed, and

$$H(\boldsymbol{u}, \dot{\boldsymbol{u}}) = \tfrac{1}{2}\int_{\mathcal{B}} \rho\|\dot{\boldsymbol{u}}\|^2\, dx + \tfrac{1}{2}\int_{\mathcal{B}} \mathrm{c}^{abcd}e_{ab}e_{cd}\, dx,$$

where $e_{ab} = \frac{1}{2}(u_{a|b} + u_{b|a})$. Notice that $A = \begin{bmatrix} 0 & I \\ -B & 0 \end{bmatrix}$, where $B\boldsymbol{u} = (1/\rho)\operatorname{div}(\mathbf{c}\cdot\nabla\boldsymbol{u})$. By symmetry of $\mathbf{c}$, B is a symmetric operator in the space of L^2 maps $\boldsymbol{u}\colon \mathcal{B} \longrightarrow \mathbb{R}^3$, with the inner product

$$\langle \boldsymbol{u}_1, \boldsymbol{u}_2 \rangle = \int_{\mathcal{B}} \rho(x)\boldsymbol{u}_1(x)\cdot\boldsymbol{u}_2(x)\, dx.$$

Thus, as A is ω-skew, we have a Hamiltonian system in $\mathfrak{X}$. It will generate a stable group $\{U(t)\}$ if the energy is positive-definite (and other technical conditions discussed in Section 6.3 hold). From the form of H, we are thus lead to the condition that there is an $\epsilon > 0$ such that for any *symmetric* tensor e_{ab},

$$\mathrm{c}^{abcd}e_{ab}e_{cd} \geq \epsilon\|e\|^2 \quad (\textit{stability condition}^3)$$

As we saw in Section 4.3, this condition implies (but is not implied by) the *strong ellipticity condition*

$$\mathrm{c}^{abcd}\xi_a\xi_c\eta_b\eta_d \geq \epsilon\|\boldsymbol{\xi}\|^2\,\|\boldsymbol{\eta}\|^2$$

(take $e_{ab} = \xi_a\eta_b + \xi_b\eta_a$). The latter condition was discussed in that section and was related to the reality of wave speeds. As we shall see in Section 6.4, the strong ellipticity condition is enough to guarantee that A generates a group

---

[3]That this implies positive-definiteness of the energy for the displacement problem relies on Korn's inequality. See Sections 6.2 and 6.4.

$\{U(t)\}$ and the stability condition guarantees $\{U(t)\}$ is stable for the displacement problem.

In Sections 5.3 and 5.4 we shall derive conservation laws for general Hamiltonian systems. They apply, in particular, to linear and nonlinear elasticity. In preparation for that, it is appropriate to make a few comments here on conservation laws in the framework of abstract linear Hamiltonian systems.

2.11 Definition Let $\mathfrak{X}$ be a Banach space with symplectic form ω and let A and B be two Hamiltonian operators in $\mathfrak{X}$ with corresponding energy functions

$$H_A(u) = \tfrac{1}{2}\omega(Au, u), \qquad H_B(u) = \tfrac{1}{2}\omega(Bu, u).$$

The *Poisson bracket* of H_A and H_B is defined for each $x \in \mathfrak{D}(A) \cap \mathfrak{D}(B)$ by

$$\{H_A, H_B\}(x) = \omega(Ax, Bx).$$

The reader can check that for finite-dimensional systems this coincides with the usual expression

$$\{f, g\} = \sum_{i=1}^{n} \left(\frac{\partial f}{\partial q^i}\frac{\partial g}{\partial p_i} - \frac{\partial f}{\partial p_i}\frac{\partial g}{\partial q^i}\right).$$

2.12 Proposition *At $x \in \mathfrak{D}([A, B])$, we have $\{H_A, H_B\} = H_{[A,B]}$, where $[A, B] = AB - BA$ is the operator commutator.*

Proof By definition,

$$\begin{aligned}\{H_A, H_B\}(x) &= \omega(Ax, Bx) = \tfrac{1}{2}\omega(Ax, Bx) - \tfrac{1}{2}\omega(Bx, Ax)\\ &= -\tfrac{1}{2}\omega(BAx, x) + \tfrac{1}{2}\omega(ABx, x) = \tfrac{1}{2}\omega([A, B]x, x) = H_{[A,B]}(x). \quad \blacksquare\end{aligned}$$

Suppose that A generates a semigroup U_t and B generates a group V_t. As we have seen, U_t and V_t are symplectic transformations.

2.13 Proposition *Suppose that V_t is a symmetry group of the function H_A in the following sense: each map V_t leaves $\mathfrak{D}(A)$ invariant, and $H_A \circ V_t = H_A$. Then H_B is a constant of the motion for A; that is, U_t leaves $\mathfrak{D}(B)$ invariant and $H_B \circ U_t = H_B$. Moreover, $\{U_t\}$ and $\{V_t\}$ commute; that is, $U_s V_t = V_t U_s$ for all $s, t \geq 0$.*

Proof Fix $t \geq 0$ and $x \in \mathfrak{D}(A)$. Since $H_A \circ V_t(x) = H_A(x)$ and V_t is symplectic,

$$\omega(Ax, x) = \omega(AV_t x, V_t x) = \omega(V_{-t}AV_t x, x).$$

Thus $\omega(Lx, x) = 0$, where $L = A - V_{-t}AV_t$. Now we invoke the following to show that $A = V_{-t}AV_t$.

2.14 Lemma *Suppose L is ω-skew and $\omega(Lx, x) = 0$ for all $x \in \mathfrak{D}(A)$, a dense subset. Then $Lx = 0$ for all $x \in \mathfrak{D}(L)$.*

Proof $0 = \omega(L(x+y), x+y) = \omega(Lx, x) + \omega(Ly, x) + \omega(Lx, y) + \omega(Ly, y) = 2\omega(Lx, y)$. Since this vanishes for all $x, y \in \mathfrak{D}(L)$ and ω is nondegenerate, $Lx = 0$. ▮

To complete the proof of 2.13, we shall also need to know that the semigroup generated by A is unique.

2.15 Lemma (*Uniqueness of Semigroups*) *Suppose A generates semigroups $\{U_t\}$ and $\{\bar{U}_t\}$. Then $U_t = \bar{U}_t$ for all $t \geq 0$.*

Proof In fact, we shall prove uniqueness of individual integral curves. Suppose $c(t)$ is t-differentiable and satisfies $c'(t) = Ac(t)$. We shall prove $c(t) = U_t c(0)$. Let $t_0 > 0$ be fixed and consider $h(t) = U_{t_0-t}c(t)$ for $t_0 - t \geq 0$. We have

$$\begin{aligned} \| h(t+\tau) - h(t) \| &= \| U_{t_0-t-\tau}c(t+\tau) - U_{t_0-t}c(t) \| \\ &= \| U_{t_0-t-\tau}c(t+\tau) - U_{t_0-t-\tau}U_\tau c(t) \| \\ &\leq C \| c(t+\tau) - U_\tau c(t) \|. \end{aligned}$$

Now $(1/\tau)[c(t+\tau) - U_\tau c(t)] = (1/\tau)[c(t+\tau) - c(t)] + (1/\tau)[c(t) - U_\tau c(t)] \longrightarrow Ac(t) - Ac(t) = 0$ as $\tau \longrightarrow 0+$. Thus, h is (right) differentiable with derivative zero, so h is constant. Hence $c(t_0) = U_{t_0-t}c(t)$; that is, $U_t c(t_0) = U_{t_0}c(t)$. Letting $t_0 \longrightarrow 0$ gives $c(t) = U_t c(0)$. ▮

Resuming the proof of 2.13, we have shown that $A = V_{-t}AV_t$. Now it is easily checked that $V_{-t}AV_t$ generates the semigroup $\{V_{-t}U_sV_t \mid s \geq 0\}$. By 2.15, we must have $U_s = V_{-t}U_sV_t$. That is, $U_sV_t = V_tU_s$. From (*v*) of 2.5, if follows that $U_s\mathfrak{D}(B) \subset \mathfrak{D}(B)$. Finally, we have the relation

$$U_sBx = \frac{d}{dt}U_sV_tx\Big|_{t=0} = \frac{d}{ds}V_tU_sx\Big|_{t=0} = BU_sx, \qquad x \in \mathfrak{D}(B).$$

Hence

$$H_B(U_sx) = \tfrac{1}{2}\omega(BU_sx, U_sx) = \tfrac{1}{2}\omega(U_sBx, U_sx) = \tfrac{1}{2}\omega(Bx, x) = H_B(x). \quad ▮$$

In order to conclude that the flows U_s and V_t commute, it is *not* enough to just have $\{H_A, H_B\} = 0$; that is, $[A, B] = 0$. (See Nelson [1959] for an example.) The infinite-dimensional case is much subtler than the finite-dimensional case, and it is well to be wary of reliance on formal calculations alone.

In section 5.5 we shall see how to use 2.13 for conservation laws. For example, we can conclude that if a one parameter group obtained from rotations leaves a Hamiltonian generator invariant, then the Hamiltonian flow commutes with rotations and angular momentum is conserved. (See section 6.2 and Chernoff and Marsden [1974] for the technical details needed to apply this to the wave equation).

Box 2.1 ***Summary of Important Formulas for Section 5.2***

Symplectic Form on a Banach Space $\mathfrak{X}$.

$\omega\colon X \times X \to \mathbb{R}$ is bilinear, skew ($\omega(u, v) = -\omega(v, u)$), and nondegenerate: $\omega(u, v) = 0$ for all v implies $u = 0$.

Hamiltonian Operator

$A\colon \mathfrak{D}(A) \to \mathfrak{X}$ is ω-skew: $\omega(Ax, y) = -\omega(x, Ay)$.

Energy of A

$H(x) = H_A(x) = \frac{1}{2}\omega(Ax, x), \quad x \in \mathfrak{D}(A)$

Canonical Transformation $T\colon \mathfrak{X} \to \mathfrak{X}$

$\omega(Tu, Tv) = \omega(u, v), \quad u, v \in \mathfrak{X}$

Semigroup of A. $\{U_t\}$ satisfies

$$U_0 = \mathrm{Id}, \quad U_{t+s} = U_t \circ U_s, \text{ and } \frac{d}{dt} U_t x = A U_t x \text{ for all } x \in \mathfrak{D}(A).$$

Properties of $\{U_t\}$

(i) $H(U_t x) = H(x)$

(ii) U_t is a canonical transformation.

Stability

$\{U_t\}$ is a stable semigroup when $\| U_t \| < M$ for all $t \geq 0$.

Energy Criterion

If $c\| x \|^2 \leq H(x) \leq C\| x \|^2$ for constants $C, c > 0$, then $\{U_t\}$ is stable.

Conservation Laws

If A generates U_t, B generates V_t; and if V_t leaves $\mathfrak{D}(A)$ invariant and $H_A \circ V_t = H_A$, then H_B is a constant of the motion for A: $H_B \circ U_t = H_B$.

Classical Elastodynamics with Homogeneous Boundary Conditions

$\rho \ddot{\boldsymbol{u}} = \operatorname{div}(\mathbf{c} \cdot \nabla \boldsymbol{u})$ on $\mathcal{B}$

$\boldsymbol{u} = 0$ on ∂_d, $\quad (\mathbf{c} \cdot \nabla \boldsymbol{u}) \cdot \boldsymbol{n} = 0$ on ∂_τ

$$\mathfrak{X} = \left\{ (\boldsymbol{u}, \dot{\boldsymbol{u}}) \,\middle|\, \int_{\mathcal{B}} |\boldsymbol{u}|^2\, dx < \infty, \int_{\mathcal{B}} |\nabla \boldsymbol{u}|^2\, dx < \infty, \int_{\mathcal{B}} |\dot{\boldsymbol{u}}|^2\, dx < \infty, \text{ and } \boldsymbol{u} = 0 \text{ on } \partial_d \right\}$$

$$\omega((\boldsymbol{u}, \dot{\boldsymbol{u}}), (\boldsymbol{v}, \dot{\boldsymbol{v}})) = \int_{\mathcal{B}} \rho(\dot{\boldsymbol{v}} \cdot \boldsymbol{u} - \dot{\boldsymbol{u}} \cdot \boldsymbol{v})\, dx$$

$$A(\boldsymbol{u}, \dot{\boldsymbol{u}}) = \left(\dot{\boldsymbol{u}}, \frac{1}{\rho} \operatorname{div}(\mathbf{c} \cdot \nabla \boldsymbol{u})\right)$$

$$\mathfrak{D}(A) = \left\{ (\boldsymbol{u}, \dot{\boldsymbol{u}}) \in \mathfrak{X} \,\middle|\, \int_{\mathcal{B}} |D^2 \boldsymbol{u}|^2\, dx < \infty, \int_{\mathcal{B}} |\nabla \dot{\boldsymbol{u}}|^2\, dx < \infty, \text{ and } (\mathbf{c} \cdot \nabla \boldsymbol{u}) \cdot \boldsymbol{n} = 0 \text{ on } \partial_\tau \right\}$$

$$H(\boldsymbol{u}, \dot{\boldsymbol{u}}) = \frac{1}{2} \int_{\mathcal{B}} \rho |\dot{\boldsymbol{u}}|^2\, dx + \frac{1}{2} \int_{\mathcal{B}} \mathrm{c}^{abcd} e_{ab} e_{cd}\, dx, \quad e_{ab} = \frac{1}{2}(u_{a|b} + u_{b|a})$$

5.3 ABSTRACT HAMILTONIAN AND LAGRANGIAN SYSTEMS

This section generalizes the notion of a Hamiltonian system from the linear to the nonlinear case. The next section shows how the equations of nonlinear elasticity fit into this scheme. For the beginning of the section, we shall assume the spaces are linear for simplicity; the reader is invited to make the generalization to Banach manifolds on a second reading.

3.1 Definitions Let $\mathfrak{X}$ and $\mathcal{Y}$ be Banach spaces, with $\mathcal{Y}$ continuously and densely included in $\mathfrak{X}$. Let $\omega: \mathfrak{X} \times \mathfrak{X} \longrightarrow \mathbb{R}$ be a weakly non-degenerate skew-symmetric bilinear form (as in the previous section). Let $\mathfrak{D} \subset \mathcal{Y}$ be an open set and let $G: \mathfrak{D} \longrightarrow \mathfrak{X}$ be a given mapping. We will say G is *Hamiltonian* if there is a C^1 function $H: \mathfrak{D} \longrightarrow \mathbb{R}$ such that

$$\boxed{\omega(Gx, y) = dH(x) \cdot y} \tag{1}$$

for all $x, y \in \mathcal{Y}$. As in the finite-dimensional case, we write $G = X_H$.

The criteria for an operator to be potential (see Box 1.1 in Section 5.1) yields the following:

3.2 Proposition *Let* $G: \mathfrak{D} \subset \mathcal{Y} \longrightarrow \mathfrak{X}$ *be* C^1. *Then* G *is Hamiltonian if and only if the linear operator* $DG(x): \mathcal{Y} \longrightarrow \mathfrak{X}$ *is* ω*-skew for each* $x \in \mathfrak{D}$; *that is,*

$$\omega(DG(x) \cdot u, v) = -\omega(u, DG(x) \cdot v)$$

for all $u, v \in \mathcal{Y}$. *If this is the case, we can choose*

$$H(x) = \int_0^1 \omega(G(sx), x)\, ds + \text{const.}$$

A principal goal is to study properties of the evolution equation

$$\frac{dx}{dt} = G(x).$$

This is the nonlinear analogue of the linear Hamiltonian evolution equations studied in the previous section. Later in this section we will write the equations of conservative, nonlinear elastodynamics in this form. To study this abstract evolution equation, we need to introduce its flow. Conditions under which there is a flow and its detailed properties will be studied in Chapter 6. For now, we just assume the properties we need.

3.3 Definition A (continuous local) flow (resp. semiflow) on $\mathcal{Y}$ is a continuous map $F: \mathfrak{D} \subset \mathcal{Y} \times \mathbb{R} \longrightarrow \mathcal{Y}$ (resp. $F: \mathfrak{D} \subset \mathcal{Y} \times \mathbb{R}^+ \longrightarrow \mathcal{Y}$), where $\mathfrak{D}$ is open, such that:

(i) $\mathcal{Y} \times \{0\} \subset \mathfrak{D}$ and $F(x, 0) = x$;

(ii) if $(x, t) \in \mathfrak{D}$, then $(x, t+s) \in \mathfrak{D}$ iff $(F(x, t), s) \in \mathfrak{D}$ and in this case $F(x, t+s) = F(F(x, t), s)$.

We shall usually write $F_t(x) = F(x, t)$, so (i) and (ii) respectively read $F_0 =$ identity and $F_{t+s} = F_s \circ F_t$ (where defined). For $x \in \mathfrak{D}$, the $t \geq 0$ for which $F_t(x)$ is defined is easily seen to be an interval, say $[0, T_x)$; we call T_x the *lifetime* of x, which may be $+\infty$.

We shall say that $G\colon \mathfrak{D} \subset \mathcal{Y} \to \mathfrak{X}$ *generates the flow* F_t if for each $x \in \mathfrak{D}$,

$$\frac{d}{dt} F_t(x) = G(F_t(x)) \tag{2}$$

(d/dt is a right derivative at $t = 0$ if F_t is only a semiflow).

Some theorems carried over from finite-dimensional Hamiltonian mechanics do not make sense unless $DF_t(x)$ exists. For example, F_t being a canonical transformation means $F_t^*\omega = \omega$, or explicitly,

$$\omega(DF_t(x)\cdot u, DF_t(x)\cdot v) = \omega(u, v) \tag{3}$$

for all $x \in \mathfrak{D}$ and $u, v \in \mathcal{Y}$. Conditions under which F_t is differentiable (in an appropriate sense) are given in the next chapter. In the following proposition, we assume what we need. Formally, by differentiating (2) with respect to x, we see that $DF_t(x)$ should satisfy the "equation of variation":

$$\frac{d}{dt} DF_t(x)\cdot u = DG(F_t(x))\cdot[DF_t(x)\cdot u]. \tag{4}$$

3.4 Proposition *Let G be C^1, be Hamiltonian, and generate a (semi) flow F_t on $\mathcal{Y}$. Assume there exist bounded linear operators $DF_t(x)\colon \mathcal{Y} \to \mathcal{Y}$ satisfying Equation* (4) *and $DF_0(x) =$ Identity. (We shall say in what sense DF_t is the derivative of F_t in Chapter 6; see Box 5.1.) Then F_t is symplectic (where defined); that is,* (3) *holds.*

Proof By (4),

$$\frac{d}{dt}\omega(DF_t(x)\cdot u, DF_t(x)\cdot v) = \omega(DG(F_t(x))\cdot[DF_t(x)\cdot u], DF_t(x)\cdot v)$$
$$+ \omega(DF_t(x)\cdot u, DG(F_t(x))\cdot[DF_t(x)\cdot v]).$$

Since $DG(F_t(x))$ is ω-skew, this vanishes, so the proposition follows. ∎

Next we turn to conservation of energy. In fact, we will prove a more general theorem involving Poisson brackets.

3.5 Definition Let X_H and X_K be two Hamiltonian operators with domains $\mathfrak{D} = \mathfrak{D}_H$ and $\mathfrak{D}_K$, respectively, where $\mathfrak{D}_H$ and $\mathfrak{D}_K$ are open in $\mathcal{Y} \subset \mathfrak{X}$. The *Poisson bracket* of H and K is defined on $\mathfrak{D}_H \cap \mathfrak{D}_K$ by

$$\boxed{\{H, K\}(x) = \omega(X_H(x), X_K(x)).}$$

For the following, we assume K and X_K are defined on a domain $\mathfrak{D}_K \supset \mathfrak{D} = \mathfrak{D}_H$, so $\{H, K\}$ is defined on $\mathfrak{D}$.

3.6 Proposition (*Chernoff–Marsden* [1974]) *Suppose $X_H: \mathfrak{D} \longrightarrow \mathfrak{X}$ is continuous, has energy $H: \mathfrak{D} \longrightarrow \mathbb{R}$, and generates a (continuous local semi) flow F_t. Let $X_K: \mathfrak{D} \longrightarrow \mathfrak{X}$ be Hamiltonian with energy $K: \mathfrak{D} \longrightarrow \mathbb{R}$. Then for $x \in \mathfrak{D}$, $K(F_t(x))$ is t-differentiable with derivative*

$$\frac{d}{dt} K(F_t(x)) = \{K, H\}(F_t(x))$$

for $0 \leq t < T_x$, the lifetime of x. In particular, if $\{K, H\} = 0$, then $K(F_t(x)) = K(x)$. Taking $K = H$, it follows that $H(F_t(x)) = H(x)$.

Proof The tricky point is that $K(F_t(x))$ is not obviously t-differentiable under these hypotheses; the chain rule does not apply since $t \mapsto F_t(x) \in \mathfrak{D}$ is *not* differentiable in the $\mathfrak{Y}$-topology.

It suffices to prove that for each $x_0 \in \mathfrak{D}$,

$$\left.\frac{d}{dt} K(F_t(x_0))\right|_{t=0} = \{K, H\}(x_0)$$

by the semiflow property of F_t. We can also take $x_0 = 0$ without loss of generality.

By 3.2, we can relate K to X_K by

$$K(x) = K(0) + \int_0^1 \omega(X_K(\tau x), x)\, d\tau.$$

Let $x_t = F_t(0)$. Then

$$\frac{1}{t}\{K(x_t) - K(0)\} = \int_0^1 \omega\left(X_K(\tau x_t), \frac{x_t}{t}\right) d\tau.$$

Now as $t \longrightarrow 0$, $x_t \longrightarrow 0$ in $\mathfrak{D}$ and as X_K is continuous, $X_K(\tau x_t) \longrightarrow X_K(0)$ uniformly for $0 \leq \tau \leq 1$. Also

$$\frac{x_t}{t} = \frac{x_t - x_0}{t} \longrightarrow X_H(0) \quad \text{as} \quad t \longrightarrow 0,$$

since X_H is the generator of F_t. Thus, we can pass to the limit under the integral sign to obtain

$$\lim_{t \to 0+} \frac{1}{t}\{K(x_t) - K(0)\} = \int_0^1 \omega(X_K(0), X_H(0))\, d\tau = \omega(X_K(0), X_H(0))$$
$$= \{K, H\}(0). \quad \blacksquare$$

For applications to elasticity, it will be necessary to use a number of results concerning Lagrangian systems. These applications also must allow $\mathfrak{X}$ and $\mathfrak{Y}$ to be manifolds. For example, the basic configuration space for the place problem of elastodynamics is a *nonlinear space $\mathfrak{C}$ of maps*; it is not even an open set in a linear space. For the traction problem it is an open set in a linear space, but this is lost if constraints such as incompressibility are imposed. Thus, if we are going to be honest about the configuration spaces, the use of nonlinear function spaces is necessary.

We shall now switch over to this manifold mode. The reader should take time to notice that much of what we did in Chapter 1 carries over to infinite-dimensional manifolds. At the appropriate points in what follows, the reader should go back to the relevant sections of that chapter and make the generalization (consult Abraham, Marsden and Ratiu [1982] or Lang [1972] if necessary). For example, let us formulate Definition 3.1(a) in this language:

3.7 Definition A *symplectic manifold* is a pair $(\mathcal{P}, \omega)$, where $\mathcal{P}$ is a manifold modeled on a Banach space $\mathfrak{X}$ and ω is a two-form such that

(i) $d\omega = 0$ and
(ii) ω is weakly non-degenerate, that is, for each $x \in \mathcal{P}$, $\omega(v_x, w_x) = 0$ for all $w_x \in T_x\mathcal{P}$ implies $v_x = 0$.

A basic symplectic manifold—the cotangent bundle—is constructed as follows

3.8 Definition Let $\mathbb{Q}$ be a manifold modeled on a Banach space $\mathfrak{X}$. Let $T^*\mathbb{Q}$ be its cotangent bundle, and $\pi^*: T^*\mathbb{Q} \longrightarrow \mathbb{Q}$ the projection. The *canonical one-form* θ on $T^*\mathbb{Q}$ is the one-form defined by

$$\theta(\alpha)\cdot w = \alpha \cdot T\pi^*(w),$$

where $\alpha \in T^*_x\mathbb{Q}$, $(x \in \mathbb{Q})$ and $w \in T_\alpha(T^*\mathbb{Q})$.

In a local coordinate chart $\mathfrak{U} \subset \mathfrak{X}$, the formula for θ reads

$$\theta(x, \alpha)\cdot(u, \beta) = \alpha(u),$$

where $(x, \alpha) \in \mathfrak{U} \times \mathfrak{X}^*$ and $(u, \beta) \in \mathfrak{X} \times \mathfrak{X}^*$. If $\mathbb{Q}$ is finite dimensional, θ can be written

$$\theta = p_i\, dq^i,$$

where $q^1, \ldots, q^n, p_1, \ldots, p_n$ are coordinates for $T^*\mathbb{Q}$ corresponding to coordinates $q^1, \ldots, q^n$ for $\mathbb{Q}$.

3.9 Definition The *canonical two-form* is defined by $\omega = -d\theta$, where d is the exterior derivative.

Problem 3.1 Using the local formula for d from Section 1.7 (see Formula 4(b) in Box 7.1, Chapter 1), show that in a local coordinate chart $\mathfrak{U} \subset \mathfrak{X}$ for $\mathbb{Q}$,

$$\omega(x, \alpha)\cdot((u_1, \alpha_1), (u_2, \alpha_2)) = \alpha_2(u_1) - \alpha_1(u_2),$$

which coincides with the canonical symplectic form in the linear case discussed in the previous section (see 2.4, Example 1). In the finite-dimensional case, show that this formula for $\boldsymbol{\omega}$ becomes $\boldsymbol{\omega} = \boldsymbol{dq^i \wedge dp_i}$ (as usual, there is a sum on repeated indices). Conclude that the canonical two-form ω on $T^*\mathcal{Q}$ is a *weak symplectic form*; that is, conditions (i) and (ii) of 3.7 hold for this ω on $\mathcal{P} = T^*\mathcal{Q}$.

Next we consider symplectic forms induced by metrics.

3.10 Definition Let $\langle\ ,\ \rangle$ be a (weak) Riemannian metric on $\mathcal{Q}$; that is, for each $x \in \mathcal{Q}$, $\langle\ ,\ \rangle_x$ is an inner product on $T_x\mathcal{Q}$. Then we have a smooth map $l: T\mathcal{Q} \to T^*\mathcal{Q}$ defined by $l(v_x)w_x = \langle v_x, w_x\rangle_x$, where $x \in \mathcal{Q}$ and $v_x, w_x \in T_x\mathcal{Q}$. [If $\langle\ ,\ \rangle$ is a (strong) Riemannian metric, it follows (from the implicit function theorem) that l is a diffeomorphism of $T\mathcal{Q}$ *onto* $T^*\mathcal{Q}$, but this is not the situation in most examples.] Set $\Omega = l^*(\omega)$, the pull-back of ω by l, where ω is the canonical form on $T^*\mathcal{Q}$. [Clearly Ω is exact since $\Omega = -d(l^*(\theta))$.]

Problem 3.2 (Polar Fluids) Let $\mathcal{Q} = \mathcal{C}$ be the configuration space of maps $\phi: \mathcal{B} \to \mathcal{S}$. Let C_{ab}^{AB} be a given tensor field and set

$$\langle V_1, V_2\rangle = \int_{\mathcal{B}} \rho g_{ab} V_1{}^a V_2{}^b \, dV + \int_{\mathcal{B}} \rho C_{ab}^{AB} V_1{}^a{}_{|A} V_2{}^b{}_{|B} \, dV$$

Show that $l(V)$ may be identified with the one form with components

$$\alpha_b = \rho[g_{ab}V^a - (C_{ab}^{AB} V^a{}_{|A})_{|B}]$$

Using the definition of pull-back one can readily verify the following:

3.11 Proposition

(a) If $\langle\ ,\ \rangle_x$ is a weak metric on $\mathcal{Q}$, then Ω is a weak symplectic form. In a chart $\mathcal{U} \subset \mathfrak{X}$ for $\mathcal{Q}$, we have

$$\begin{aligned} &\Omega(x, u)((v_1, w_1), (v_2, w_2)) \\ &= D_x\langle u, v_1\rangle_x v_2 - D_x\langle u, v_2\rangle_x v_1 + \langle w_2, v_1\rangle_x - \langle w_1, v_2\rangle_x, \end{aligned} \tag{5}$$

where D_x denotes the (Fréchet) derivative with respect to x.
(b) $\Omega = -d\Theta$, where, locally, $\Theta(x, e)(e_1, e_2) = \langle e, e_1\rangle_x$.

Problem 3.3 Prove 3.11, and, in the finite-dimensional case, show that Formula (5) for Ω becomes

$$\Omega = g_{ij} dq^i \wedge d\dot{q}^j + \frac{\partial g_{ij}}{\partial q^k} \dot{q}^i dq^j \wedge dq^k,$$

where $q^1, \ldots, q^n, \dot{q}^1, \ldots, \dot{q}^n$ are coordinates for $T\mathcal{Q}$.

Generalizing the case of $\mathcal{Y} \subset \mathfrak{X}$, two Banach spaces with the inclusion dense and continuous, let us call a *manifold domain* of $\mathcal{P}$, a subset $\mathcal{D} \subset \mathcal{P}$ such that $\mathcal{D}$ has its own manifold structure for which the inclusion $i: \mathcal{D} \to \mathcal{P}$ is C^∞ and such that its tangent $Ti: T\mathcal{D} \to T\mathcal{P}$ is also injective.

We shall also refer to generators as vector fields. If $(\mathcal{P}, \omega)$ is a weak symplectic manifold, the condition that a vector field $X_H = G: \mathfrak{D} \dashrightarrow T\mathcal{P}$ (with manifold domain $\mathfrak{D} \subset T\mathcal{P}$) be Hamiltonian with energy H is that

$$i_G\omega = dH.$$

That is, for all $x \in \mathcal{P}$ and $v \in T_x\mathcal{P}$ we have

$$\omega_x(G(x), v) = dH(x)\cdot v.$$

The reader will note that this is a generalization of the definition in the linear case. Also note that locally, G is Hamiltonian if and only if $d(i_G\omega) = 0$; in a chart in which ω is constant[4] this says that the linear operator $DG(x)$ is ω-skew, that is, just as in 3.2.

We often write $G = X_H$ because usually in examples H is given and then one constructs the Hamiltonian vector field X_H. In the finite-dimensional case, where $\boldsymbol{\omega} = dq^i \wedge dp_i$, the generator takes the familiar form of Hamilton's equations: $X_H = (\partial H/\partial p_i, -\partial H/\partial q^i)$ (see Box 7.1 of Chapter 1).

Because $\boldsymbol{\omega}$ is only weak, given $H: \mathfrak{D} \longrightarrow \mathbb{R}$, X_H need not exist. Also, even if H is smooth on all of $\mathcal{P}$, X_H will in general be defined only on a certain subset $\mathfrak{D}$ of $\mathcal{P}$, but where it is defined, it is unique. The linear wave equation discussed in the previous section is a case in point. Notice, however, that even if H is only defined on $\mathfrak{D}$, for $x \in \mathfrak{D}$, $dH(x)$ must extend to a bounded linear functional on $T_x\mathcal{P}$ because of the defining relation between $X_H = G$ and H.

The reader is invited to extend the proofs given in 3.4 and 3.6 on conservation of energy and the symplectic nature of the flow to the case in which the phase space is not necessarily linear and ω is not necessarily constant.

For elastodynamics we shall be especially interested in infinite-dimensional Lagrangian systems. Before developing this, we recall some of the key definitions in finite-dimensional coordinate language to help bridge the gap to the infinite-dimensional case.

Consider a finite dimensional configuration manifold $\mathbb{Q}$ with coordinates $(q^1, \ldots, q^n)$. The tangent bundle $T\mathbb{Q}$ has coordinates $(q^1, \ldots, q^n, \dot{q}^1, \ldots, \dot{q}^n)$ and a *Lagrangian* is a real-valued function $L(q^i, \dot{q}^i)$ that is, $L: T\mathbb{Q} \longrightarrow \mathbb{R}$. The *Legendre transformation* (or the *fiber derivative*) is the map $FL: T\mathbb{Q} \longrightarrow T^*\mathbb{Q}$ defined by $(q^i, \dot{q}^i) \mapsto (q^i, p_i)$, where $p_i = \partial L/\partial \dot{q}^i$. The *Lagrange two-form* $\boldsymbol{\omega}_L$ is the pull-back of the canonical two-form $\boldsymbol{\omega} = dq^i \wedge dp_i$ on $T^*\mathbb{Q}$ to $T\mathbb{Q}$, obtained by substituting $p_i = \partial L/\partial \dot{q}^i$ for p_i. By the chain rule, we get

$$\boldsymbol{\omega}_L = dq^i \wedge d\left(\frac{\partial L}{\partial \dot{q}^i}\right) = \frac{\partial^2 L}{\partial q^j\, \partial \dot{q}^i}\, dq^i \wedge dq^j + \frac{\partial^2 L}{\partial \dot{q}^i\, \partial \dot{q}^j}\, dq^i \wedge d\dot{q}^j.$$

If L is the kinetic energy of a metric, then $\boldsymbol{\omega}_L$ reduces to $\boldsymbol{\omega}$ defined in 3.10.

---

[4]Such charts always exist under some reasonable technical conditions. This result is called *Darboux's theorem*; for the finite-dimensional case, see Abraham and Marsden [1978], and for the infinite-dimensional case, see Marsden [1981].

The *action* is defined by $A(q^i, \dot{q}^i) = (\partial L/\partial \dot{q}^i)\dot{q}^i$ and the *energy* is $E = A - L$. Note that

$$dE = \frac{\partial L}{\partial \dot{q}^i}\,d\dot{q}^i + \frac{\partial^2 L}{\partial q^j\,\partial \dot{q}^i}\dot{q}^i\,dq^j + \frac{\partial^2 L}{\partial \dot{q}^i\,\partial \dot{q}^j}\dot{q}^i\,d\dot{q}^j - \frac{\partial L}{\partial q^i}\,dq^i - \frac{\partial L}{\partial \dot{q}^i}\,d\dot{q}^i$$

in which the term $(\partial L/\partial \dot{q}^i)\,d\dot{q}^i$ cancels. The *Euler–Lagrange equations* are (see Box 7.2, Chapter 1)

$$\frac{d}{dt}\frac{\partial L}{\partial \dot{q}^i} = \frac{\partial L}{\partial q^i}.$$

Let

$$X_E(q^i, \dot{q}^i) = \left[\dot{q}^i, \left(\frac{\partial^2 L}{\partial \dot{q}^i\,\partial \dot{q}^j}\right)^{-1}\left(\frac{\partial L}{\partial q^i} - \frac{\partial^2 L}{\partial \dot{q}^i\,\partial q^k}\dot{q}^k\right)\right],$$

assuming L is *regular*; that is, $\partial^2 L/\partial \dot{q}^i\,\partial \dot{q}^j$ is non-singular. Here $(\partial^2 L/\partial \dot{q}^i\,\partial \dot{q}^j)^{-1}$ stands for the matrix $[(\partial^2 L/\partial \dot{q}^k\,\partial \dot{q}^l)^{-1}]^{ij}$. Then Lagrange's equations in first-order form are

$$\frac{d}{dt}\begin{pmatrix} q^i \\ \dot{q}^i \end{pmatrix} = X_E(q^i, \dot{q}^i)$$

by the chain rule.

The basic connection with Hamiltonian systems is this fact: ***X_E** is the Hamiltonian vector field for the Hamiltonian E, using the symplectic form $\boldsymbol{\omega}_L$* ($\boldsymbol{\omega}_L$ is non-degenerate if L is regular). In other words, $i_{X_E}\boldsymbol{\omega}_L = \mathbf{d}E$. Thus, in this sense, the theory of Lagrangian systems is a special case of that of Hamiltonian systems. To prove this fact, we compute as follows: if we write $X_E = (G_1^i, G_2^i)$, then from the formula for $\boldsymbol{\omega}_L$, and $i_{G_E}(\mathbf{d}q^i \wedge \mathbf{d}q^j) = G_1^i\,\mathbf{d}q^j - \mathbf{d}q^i G_1^j$, and so on, we get

$$i_{X_E}\boldsymbol{\omega}_L = G_1^i \frac{\partial^2 L}{\partial q^j\,\partial \dot{q}^i}\,dq^j - G_2^j \frac{\partial^2 L}{\partial q^i\,\partial \dot{q}^j}\,dq^i + G_1^i \frac{\partial^2 L}{\partial \dot{q}^i\,\partial \dot{q}^j}\,d\dot{q}^j - G_2^j \frac{\partial^2 L}{\partial \dot{q}^i\,\partial \dot{q}^j}\,d\dot{q}^i.$$

Now substitute

$$G_1^i(q^j, q^j) = \dot{q}^i \quad \text{and} \quad G_2^i = \left(\frac{\partial^2 L}{\partial \dot{q}^i\,\partial \dot{q}^j}\right)^{-1}\left(\frac{\partial L}{\partial q^i} - \frac{\partial^2 L}{\partial \dot{q}^i\,\partial q^k}\dot{q}^k\right)$$

and the previous formula for $\mathbf{d}E$ results. (We will re-do these calculations in index-free notation below.) Here are two basic examples (also re-done below).

1. Let $\mathbb{Q}$ be a Riemannian manifold with metric tensor g_{ij}. Let $L\colon T\mathbb{Q} \to \mathbb{R}$ be given by $L(q^i, \dot{q}^i) = \frac{1}{2} g_{ij}\dot{q}^i\dot{q}^j$, the kinetic energy. *Then $(q^i(t), \dot{q}^i(t))$ satisfies Lagrange's equations if and only if $q^i(t)$ is a geodesic.*

Proof $d/dt\,(\partial L/\partial \dot{q}^j) = \partial L/\partial q^j$ reads

$$\frac{d}{dt}(g_{ij}(q(t))\dot{q}^i) = \frac{1}{2}\frac{\partial g_{kl}}{\partial q^j}\dot{q}^k\dot{q}^l, \text{ that is,}$$

$$g_{ij}\ddot{q}^i + \frac{\partial g_{ij}}{\partial q^k}\dot{q}^i\dot{q}^k = \frac{1}{2}\frac{\partial g_{kl}}{\partial q^j}\dot{q}^k\dot{q}^l,$$

$$g_{ij}\ddot{q}^i = g_{ij}\Gamma^i_{kl}\dot{q}^k\dot{q}^l; \quad \text{that is,} \quad \frac{D\dot{q}^i}{Dt} = 0.$$

which are the geodesic equations ∎

2. Let $\mathcal{Q}$ be as in 1, $V: \mathcal{Q} \to \mathbb{R}$ a given potential and let $L(q^i, \dot{q}^i) = \frac{1}{2} g_{ij}\dot{q}^i\dot{q}^j - V(q^i)$. This time, Lagranges equations read

$$\frac{D\dot{q}^i}{Dt} = -(\nabla V)^i \quad \text{(i.e., acceleration = force)},$$

where $(\nabla V)^i = (\text{grad } V)^i = g^{ij}(\partial V/\partial q^j)$ is the gradient of V.

Now we are ready to re-do these calculations in infinite dimension. One of the things that complicates matters is the domains of the various quantities. However, the linear case already shows this is quite basic.

3.12 Definitions Let $\mathfrak{M}$ be a manifold modeled on a Banach space $\mathfrak{X}_1$, and let $\mathcal{Q} \subset \mathfrak{M}$ be a manifold domain. Consider the following subset of the tangent bundle $T\mathfrak{M}$:

$$\mathcal{P} = \bigcup_{x \in \mathcal{Q}} T_x\mathfrak{M}.$$

We call $\mathcal{P}$ the *restriction of* $T\mathfrak{M}$ *to* $\mathcal{Q}$ and write $\mathcal{P} = T\mathfrak{M} \mid \mathcal{Q}$. If $\mathcal{Y}_1$ is the model space for $\mathcal{Q}$, we can endow $\mathcal{P}$ with a manifold domain structure by giving it the local product structure $\mathfrak{X} = \mathcal{Y}_1 \times \mathfrak{X}_1$.

By a *Lagrangian* on $\mathcal{P} \subset T\mathfrak{M}$ we mean a smooth function $L: \mathcal{P} \to \mathbb{R}$. In particular, for each $x \in \mathcal{Q}$, L restricts to a smooth function on $T_x\mathfrak{M}$. We form the *fiber derivative* $FL: \mathcal{P} \to T^*\mathfrak{M}$, defined as follows

$$FL(v)\cdot w = \frac{d}{dt} L(v + tw)\bigg|_{t=0} \quad \text{where} \quad v, w \in T_x\mathfrak{M}.$$

Define $\omega_L = (FL)^*\omega$, a two-form on $\mathcal{P}$, where ω is the canonical two-form on $T^*\mathfrak{M}$. Thus $\omega_L = -d\theta_L$, where $\theta_L = (FL)^*\theta$, θ being the canonical one-form on $T^*\mathfrak{M}$. We will call L *regular* if ω_L is weakly non-degenerate—that is, if $D_2D_2L(x, v)$ is weakly non-degenerate. This assumption will be made here.

The *action* A and *energy* E are defined by

$$A(v) = FL(v)\cdot v \quad \text{and} \quad E = A - L.$$

Let $\mathfrak{D}$ be the subset of $\mathcal{P}$ consisting of all points v such that $X_E(v) \in T\mathcal{P}$ is defined. Thus we regard X_E as a vector field on $\mathcal{P}$ with domain $\mathfrak{D}$. We call X_E the *Lagrangian vector field* determined by L.

Problem 3.4 Verify the following local formula for ω_L:

$$\begin{aligned}\omega_L(x, v)((x_1, v_1), (x_2, v_2)) = {} & D_1(D_2L(x, v)\cdot x_1)\cdot x_2 \\ & - D_1(D_2L(x, v)\cdot x_2)\cdot x_1 + D_2D_2L(x, v)\cdot v_2\cdot x_1 \\ & - D_2D_2L(x, v)\cdot v_1\cdot x_2. \end{aligned} \tag{6}$$

The classical Lagrange equations are *second-order equations.* There is a general notion of second-order fields on a tangent bundle $T\mathfrak{M}$, and we can extend this notion to the case of vector fields defined on $\mathcal{P} \subset T\mathfrak{M}$. Indeed the projection $\pi: T\mathfrak{M} \to \mathfrak{M}$ restricts to a map from $\mathcal{P}$ to $\mathcal{Q}$; we say that X_E is of *second order* when $T\pi(X_E(v)) = v$ for each $v \in \mathfrak{D}$. If, in a chart, $X_E(x, u) = (G_1(x, u), G_2(x, u))$ (dropping base points), then X_E is second order when $G_1(x, u) = u$. In this case, the first-order equations (2)

$$\dot{x} = G_1(x, u), \qquad \dot{u} = G_2(x, u)$$

are equivalent to the second-order equations

$$\ddot{x} = G_2(x, \dot{x}).$$

3.13 Theorem *Let L be a regular Lagrangian, with associated vector field X_E defined on $\mathfrak{D} \subset \mathcal{P}$, as above. Then the following hold:*

(*i*) $\mathfrak{D} \subset T\mathcal{Q}$,

(ii) *X_E is a second-order vector field,*

(iii) *In local coordinates, a point (x, v) of $T\mathcal{Q}$ belongs to $\mathfrak{D}$ if and only if*

$$\phi(x, v) = D_1L(x, v) - D_2(D_1L(x, v)\cdot v)$$

lies in the range of $D_2D_2(x, v)$, regarded as a map from $\mathfrak{X}_1$ into $\mathcal{Y}_1^*$. If this condition is met, we have the formula

$$X_E(x, v) = (v, [D_2D_2L(x, v)]^{-1}\cdot\phi(x, v)).$$

(iv) *A curve $(c(t), c'(t))$ in $T\mathcal{Q}$ is an integral curve of X_E if and only if* Lagrange's equations *hold:*

$$\frac{d}{dt} D_2L(c(t), c'(t)) = D_1L(c(t), c'(t)).$$

Proof Let $\mathcal{U} \subset \mathcal{Y}_1$ be a local chart for $\mathcal{Q}$, and consider $\mathcal{U} \times \mathfrak{X}_1$ the corresponding local chart for $\mathcal{P}$. If $(x, u) \in \mathcal{U} \times \mathfrak{X}_1$, we have $E(x, u) = D_2L(x, u)\cdot u - L(x, u)$ and hence

$$dE(x, u)\cdot(u_1, u_2) = D_1D_2L(x, u)\cdot u\cdot u_1 - D_1L(x, u)\cdot u_1 + D_2D_2L(x, u)\cdot u\cdot u_2,$$

where $u_1, u_2 \in \mathcal{Y}_1 \times \mathfrak{X}_1 \subset \mathfrak{X}_1 \times \mathfrak{X}_1$.

Locally, at points in $\mathfrak{D}$, let $X_E = (G_1, G_2)$ (base points are omitted). The defining condition on X_E is the relation

$$\omega_L(x, u)(X_E(x, u), (u_1, u_2)) = dE(x, u)\cdot(u_1, u_2).$$

Using the local formula (6) for ω_L, this becomes

$$\begin{aligned} D_1D_2L(x, u)\cdot u_1\cdot G_1 - D_1D_2L(x, u)\cdot G_1\cdot u_1 + D_2D_2L(x, u)\cdot u_2\cdot G_1 \\ - D_2D_2L(x, u)\cdot G_2\cdot u_1 \\ = D_1D_2L(x, u)\cdot u\cdot u_1 - D_1L(x, u)\cdot u_1 + D_2D_2L(x, u)\cdot u\cdot u_2. \end{aligned} \tag{7}$$

If we set u_1, and then u_2, equal to zero in relation (7), and make use of the weak non-degeneracy of $D_2D_2L(x, u)$, we obtain the following conditions:

$$G_1(x, u) = u, \tag{8}$$

$$D_2D_2L(x, u)\cdot G_2\cdot u_1 = D_1L(x, u)\cdot u_1 - D_1D_2L(x, u)\cdot u\cdot u_1. \tag{9}$$

From these local formulas all our conclusions follow. Indeed, because $G_1(x, u) \in \mathcal{Y}_1$, the relation $u = G_1(x, u)$ shows that $\mathfrak{D} \subset T\mathcal{Q}$. Conclusions (ii) and (iii) follow from (7) and (9). Finally, (iv) follows from the relation (9) satisfied by G_2, as one sees by using the chain rule to differentiate $D_2L(c(t), c'(t))$. ∎

Remarks (1) As we shall see below, this form of Lagrange's equations for field theories is a statement of the *weak form* of the equations.

(2) By a *base integral curve* we mean the projection to $\mathcal{Q}$ of an integral curve of X_E. Explicitly, let $\tilde{c}(t) \in \mathfrak{D}$ be an integral curve of X_E. (This means that $\tilde{c}'(t) = X_E(\tilde{c}(t))$, where $\tilde{c}'(t)$ is computed relative to the manifold structure of $\mathcal{P}$.) Define $c(t) = \pi(\tilde{c}(t))$, a curve in $\mathcal{Q}$. This is the base integral curve. Because X_E is a second-order vector field, we have $c'(t) = \tilde{c}(t)$, where the derivative is computed relative to the manifold structure of $\mathfrak{M}$. In components, if $X_E(x, u) = (u, G_2(x, u))$, then the integral curve $(x(t), u(t))$ satisfies $\dot{x} = u, \dot{u} = G_2(x, u)$. The curve $x(t)$ is thus the base integral curve.

(3) Because E is a smooth function on $\mathcal{P}$, it was not necessary to introduce a manifold structure on $\mathfrak{D}$ in order to construct X_E. (Although, in many instances, $\mathfrak{D}$ will have its own manifold structure)

3.14 Example To illustrate these constructions, consider the wave equation. We start with $\mathfrak{M} = \mathfrak{X}_1 = L^2(\mathbb{R}^n)$ and the Lagrangian

$$L(\phi, \dot{\phi}) = \tfrac{1}{2}\langle\dot{\phi}, \dot{\phi}\rangle - \tfrac{1}{2}\langle\nabla\phi, \nabla\phi\rangle$$

defined on the space $\mathcal{P} = H^1 \times L^2$. Regard $\mathcal{P}$ as the restriction of $L^2 \times L^2 = TL^2$ to $\mathcal{Q} = H^1 \subset L^2$. The fiber derivative of L is the map from $H^1 \times L^2$ to $L^2 \times (L^2)^* = T^*(L^2)$ given by the formula $FL(\phi, \dot{\phi}) = (\phi, \langle\dot{\phi}, \cdot\rangle)$. Hence we have

$$\omega_L((\phi_1, \dot{\phi}_1), (\phi_2, \dot{\phi}_2)) = \langle\dot{\phi}_2, \phi_1\rangle - \langle\dot{\phi}_1, \phi_2\rangle,$$

This is the same symplectic form we used in the previous section for the wave equation. We know that ω_L is weakly non-degenerate on $\mathcal{P}$. By a straightforward computation we find the energy to be

$$E(\phi, \dot{\phi}) = \tfrac{1}{2}\langle\dot{\phi}, \dot{\phi}\rangle + \tfrac{1}{2}\langle\nabla\phi, \nabla\phi\rangle,$$

whence we have

$$dE(\phi, \dot{\phi})\cdot(\psi, \dot{\psi}) = \langle\dot{\phi}, \dot{\psi}\rangle + \langle\nabla\phi, \nabla\psi\rangle.$$

Next, let us use Theorem 3.13 to determine the domain $\mathfrak{D}$ of X_E. We know that $\mathfrak{D}$ is contained in $\mathcal{Q} = H^1 \times H^1$. We must consider $\alpha(\psi) = -\langle\nabla\phi, \nabla\psi\rangle$ as a function of ψ, then does α lie in the range of $D_2D_2L(\phi, \dot{\phi}) = \langle\dot{\phi}, .\rangle$? This is so if and only if $\phi \in H^2$, in which case $\alpha(\psi) = \langle \nabla\phi, \psi \rangle$, and so

$$X_E(\phi, \dot{\phi}) = (\dot{\phi}, \Delta\phi)$$

on the domain $\mathfrak{D} = H^2 \times H^1$. This is then in accord with our treatment of the wave equation in the previous section.

We now generalize this example in a way that is relevant for elastodynamics.

3.15 Definitions Suppose that $\mathfrak{M}$ is a (weak) Riemannian manifold with metric $\langle \cdot, \cdot \rangle$. Let $V: \mathfrak{Q} \subset \mathfrak{M} \to \mathbb{R}$ be a smooth function on a manifold domain $\mathfrak{Q}$. Define a Lagrangian by

$$L(v_q) = \tfrac{1}{2}\langle v_q, v_q \rangle - V(q), \quad \text{where } v_q \in T_q M \text{ and } q \in \mathfrak{Q}.$$

Note that L is a smooth function on $\mathcal{P} = T\mathfrak{M} \,|\, \mathfrak{Q}$. Moreover, the fiber derivative of L does not involve V; in fact, it is just the map from $T\mathfrak{M}$ to $T^*\mathfrak{M}$ determined by the metric $\langle \cdot, \cdot \rangle$. Hence ω_L is the weak symplectic form induced by the metric; that is, (6) reduces to (5) in this case.

We say that the weak Riemannian metric has a *smooth spray* provided there is a *smooth* Hamiltonian vector field $X_K: T\mathfrak{M} \to T(T\mathfrak{M})$ associated to the kinetic energy function $K(v) = \tfrac{1}{2}\langle v, v \rangle$ on $T\mathfrak{M}$.

If $q \in \mathfrak{Q}$, we say that grad $V(q)$ *exists* provided there is a vector $u \in T_q\mathfrak{M}$ such that, for all $v \in T_q\mathfrak{Q}$,

$$dV(q) \cdot v = \langle u, v \rangle.$$

We write $u = \operatorname{grad} V(q)$ in this case, and let $\mathfrak{D}_0 = \{q \in \mathfrak{Q} \,|\, \operatorname{grad} V(q) \text{ exists}\}$, the *domain* of grad V.

Problem 3.5 (a) In local coordinates, show that the condition for existence of a spray is that there be a smooth map $G_2(x, v)$ satisfying the relation

$$\langle G_2(x, v), v_1 \rangle_x = \tfrac{1}{2} D_x \langle v, v \rangle_x \cdot v_1 - D_x \langle v, v_1 \rangle_x \cdot v,$$

and that the spray X_K is then given by the local formula $X_K(x, v) \times (x, G_2(x, v))$. Show that $G_2(x, v)$ depends quadratically on v; this is a characteristic property of sprays.

(b) Let $\mathfrak{C}$ be the space of embeddings $\phi: \mathcal{B} \to \mathcal{S}$. Let

$$\langle V_1, V_2 \rangle = \int_{\mathcal{B}} \rho V_1 \cdot V_2 \, dV$$

Check that this metric has the smooth spray given by

$$\frac{\partial \phi}{\partial t} = V, \quad \rho \frac{DV}{Dt} = 0$$

Rewrite in spatial form.

Even though a metric $\langle\ ,\ \rangle$ is weak, it may nevertheless possess a smooth spray. This occurs in elastodynamics and fluid mechanics as we shall see in Chapter 6.

We now generalize a familiar theorem of Lagrangian mechanics to the infinite-dimensional case by a straightforward calculation.

3.16 Theorem *Let $\langle \cdot, \cdot \rangle$ be a weak Riemannian metric on $\mathfrak{M}$ that has a smooth spray. Let $V: \mathfrak{Q} \subset \mathfrak{M} \longrightarrow \mathbb{R}$ be a smooth function with manifold domain $\mathfrak{Q}$, and define L on $\mathcal{P} = T\mathfrak{M} | \mathfrak{Q}$ as above: $L(v_q) = K(v_q) - V(q)$. Assume that the spray X_K exists and maps $\mathcal{P}$ into $T\mathcal{P}$.*

The domain $\mathfrak{D}$ of X_E, the associated Lagrangian vector field, is $\mathfrak{D} = T\mathfrak{Q} | \mathfrak{D}_0$, where $\mathfrak{D}_0$ is the domain of grad V. Furthermore, we have the formula

$$X_E(v_q) = X_K(v_q) - [\operatorname{grad} V(q)]^l_{v_q}, \qquad v_q \in \mathfrak{D},$$

where $[\ \ ]^l_{v_q}$ denotes the canonical injection ("vertical lift") of $T_q\mathfrak{M}$ into $T_{v_q}(T\mathfrak{M})$ defined by

$$[u_q]^l_{v_q} = \frac{d}{dt}(v_q + tu_q)\bigg|_{t=0}.$$

Problem 3.6 Prove Theorem 3.16.

To treat incompressible elasticity, we shall need some general facts about constrained systems in the context of the present discussion. A technically useful fact is that in many cases the property of having a smooth spray is not destroyed by the presence of constraints. For the case of an ideal fluid with the constraint of incompressibility this was shown by Ebin and Marsden [1970]. It has technical utility for analytical purposes as well as being of interest in its own right. In Chapter 6 we shall explain how this same result can be used in elasticity.

3.17 Theorem *Let $\mathfrak{M}$ be a weak Riemannian manifold possessing a smooth spray $S: T\mathfrak{M} \longrightarrow T^2\mathfrak{M}$. Let $\mathfrak{N}$ be a submanifold of $\mathfrak{M}$. Suppose that, for each $x \in \mathfrak{N}$, there is an orthogonal decomposition*[5] *$T_x\mathfrak{M} = T_x\mathfrak{N} \oplus \mathfrak{C}_x$. Using this decomposition, define a projection $\mathbb{P}: T\mathfrak{M} | \mathfrak{N} \longrightarrow T\mathfrak{N}$. Assume $\mathbb{P}$ is smooth. Then the restriction to $\mathfrak{N}$ of the Riemannian metric has a smooth spray given by*

$$S_{\mathfrak{N}} = T\mathbb{P} \circ S \text{ at points of } T\mathfrak{N}.$$

Problem 3.7 Prove Theorem 3.17. (See Abraham and Marsden [1978], p. 229 if you get stuck).

Problem 3.8 A vector $w \in T^2\mathfrak{M}$ is called *vertical* if $T\pi_M \cdot w = 0$, where $\pi_{\mathfrak{M}}: T\mathfrak{M} \longrightarrow \mathfrak{M}$ is the projection. (a) Show that vertical vectors can be identified with "ordinary" vectors—that is, tangent vectors to $\mathfrak{M}$ by way

[5]The existence of such decompositions is automatic for strong metrics. For weak metrics it usually relies on the Fredholm alternative and elliptic theory; see Chapter 6.

of vertical lift. (b) Show that $S_{\mathfrak{N}} - S$ is vertical at points of $\mathfrak{N}$. (c) Because of (b), $S_{\mathfrak{N}} - S$ may be identified with a vector field in $\mathfrak{M}$, with domain $\mathfrak{N}$. Show that $S_{\mathfrak{N}} - S$ is orthogonal to $\mathfrak{N}$. (d) Intuitively, $S_{\mathfrak{N}} - S$ gives the "forces of constraint" which ensure that particle trajectories remain in $\mathfrak{N}$. (For incompressible fluid dynamics or elasticity, this force of constraint is the pressure gradient, as we shall see later). Verify this explicitly in terms of centripetal force for a particle in $\mathbb{R}^2$ constrained to move on a circle of radius R.

Box 3.1 ***Summary of Important Formulas for Section 5.3***
[*Finite-dimensional analogues are given in brackets.*]

Phase Space

$(\mathcal{P}, \omega)$; $d\omega = 0$ and $\omega_x(v_x, w_x) = 0$ for all w_x implies $v_x = 0$.

Locally, $\mathcal{P}$ is a Banach space and ω is a skew-symmetric nondegenerate bilinear form.
$[\omega = dq^i \wedge dp_i]$

Hamiltonian Vector Field

$X_H: \mathfrak{D} \subset \mathcal{P} \to T\mathcal{P}$, $i_{X_H}\omega = dH$ $\qquad$ $\omega_x(X_H(x), u) = dH(x)\cdot u$

Flow of X_H

$F_t: \mathfrak{D} \to \mathfrak{D}$ satisfies

$$\frac{d}{dt}F_t(x) = X_H(F_t(x))$$

$$\left[X_H(q^i, p_j) = \left(\frac{\partial H}{\partial p_i}, \frac{-\partial H}{\partial q^i}\right)\right]$$

Poisson Bracket

$\{H, K\} = \omega(X_H, X_K)$ and

$$\frac{d}{dt}K(F_t(x)) = \{K, H\}(F_t(x))$$

where F_t is the flow of X_H.

$$\left[\{H, K\} = \frac{\partial H}{\partial q^i}\frac{\partial K}{\partial p_i} - \frac{\partial H}{\partial p_i}\frac{\partial K}{\partial q^i}\right]$$

Canonical Forms on $T^*\mathcal{Q}$

$\theta(\alpha)\cdot w = \alpha \cdot T\pi^*(w)$, $\qquad$ $\theta(x, \alpha)(u, \beta) = -\alpha(u)$ $[\theta = p_i\, dq^i]$

$\omega = -d\theta$ $\qquad$ $\omega(x, \alpha)((u, \alpha_1), (u_2, \alpha_2)) = \alpha_2(u_1) - \alpha_1(u_2)$

$[\omega = dq^i \wedge dp_i]$

Symplectic Form on $T\mathcal{Q}$ with Respect to a Metric $\langle\ ,\ \rangle$ on $\mathcal{Q}$

$\Omega = l^*\omega$, where $l: T\mathcal{Q} \to T^*\mathcal{Q}$ is defined by $l(v)(w) = \langle v, w\rangle$.

$$\Omega(x, u)((v_1, w_1), (v_2, w_2)) = D_x\langle u, v_1\rangle_x \cdot v_2 - D_x\langle u, v_2\rangle_x \cdot v_1 + \langle w_2, v_1\rangle_x - \langle w_1, v_2\rangle_x$$

$$\left[\Omega = g_{ij}\, dq^i \wedge d\dot{q}^j + \frac{\partial g_{ij}}{\partial q^k}\dot{q}^i\, dq^j \wedge dq^k\right]$$

Lagrangian

$L: T\mathcal{Q} \to \mathbb{R}$ $\qquad$ $L = L(x, u)$ $[L = L(q^i, \dot{q}^i)]$

Legendre Transform

$FL\colon T\mathcal{Q} \longrightarrow T^*\mathcal{Q}$ $\qquad FL(x, u)(v, w) = D_2L(x, u)\cdot w$

$FL(v)\cdot w = \dfrac{d}{dt}L(v + tw)\Big|_{t=0}$ $\qquad \left[p_i = \dfrac{\partial L}{\partial \dot{q}^i}\right]$

Lagrange Two-Form

$\omega_L = (FL)^*\omega \qquad \omega_L(x, u)\cdot((v_1, w_1), (v_2, w_2))$

$$= D_1[D_2L(x, u)\cdot v_1]\cdot v_2 - D_1[D_2L(x_1 u)\cdot v_2]\cdot v_1 + D_2D_2L(x, u)\cdot w_2\cdot v_1 - D_2D_2L(x, u)\cdot w_1\cdot v_2$$

$$\left[\omega_L = \frac{\partial^2 L}{\partial q^j\,\partial \dot{q}^i} dq^i \wedge dq^j + \frac{\partial^2 L}{\partial \dot{q}^i\,\partial \dot{q}^j} dq^i \wedge dq^j\right]$$

Action and Energy

$A(v) = FL(v)\cdot v,$ $\qquad A(x, u) = D_2L(x, u)\cdot u\left[A = \dfrac{\partial L}{\partial \dot{q}_i}\dot{q}^i\right]$

$E = A - L$

Lagrange Vector Field

$i_{X_E}\omega_L = dE \qquad X_E(x, v) = (v, [D_2D_2L(x, v)]^{-1}(D_1L(x, v) - D_2(D_1L(x, v)\cdot v))$

$$\left[X_E(q^i, \dot{q}^i) = \left(\dot{q}^i, \left[\frac{\partial^2 L}{\partial \dot{q}^i\,\partial \dot{q}^j}\right]^{-1}\left(\frac{\partial L}{\partial q^i} - \frac{\partial^2 L}{\partial \dot{q}^i\,\partial q^k}\dot{q}^k\right)\right)\right]$$

Lagrange's Equations (Integral Curves of X_E)

$\dfrac{d}{dt}D_2L(x, \dot{x}) = D_1L(x, \dot{x})$ $\qquad \left[\dfrac{d}{dt}\left(\dfrac{\partial L}{\partial \dot{q}^i}\right) = \dfrac{\partial L}{\partial q^i}\right]$

Spray of a Metric $\langle\, , \rangle$

$S = X_K$, where $K(v) = \frac{1}{2}\langle v, v\rangle$ $\qquad [X_K(q^i, \dot{q}^i) = (\dot{q}^i, \Gamma^i_{jk}\dot{q}^j\dot{q}^k)]$

Motion in a Potential V and Metric $\langle\ \ \rangle$

$X_E(v_q) = X_K(v_q) - [\text{grad } V(q)]^l_{v_q}$ $\qquad \left[X_E(q^i, \dot{q}^i) = \left(\dot{q}^i, \Gamma^i_{jk}\dot{q}^j\dot{q}^k - g^{ij}\dfrac{\partial V}{\partial q^j}\right)\right]$

5.4 LAGRANGIAN FIELD THEORY AND NONLINEAR ELASTICITY

This section begins by studying Lagrangian systems whose Lagrangian is defined in terms of a density. This is the subject of *classical field theory*. This theory can be extensively developed, but we shall not do so here (see, for example, Truesdell and Toupin [1960] and Kijowski and Tulczyjew [1979]). We then show how elasticity fits into this context.

The "fields" of elasticity are the configurations $\phi\colon \mathcal{B} \longrightarrow \mathcal{S}$. To include elasticity, tensor field theories (such as electromagnetism) and other field theories (such as gauge theories) in one context, we shall regard the fields as sections of a bundle $\pi\colon \mathcal{E} \longrightarrow \mathcal{B}$; in our case $\mathcal{E} = \mathcal{B} \times \mathcal{S}$ and $\pi\colon \mathcal{E} \longrightarrow \mathcal{B}$ is the projection to the first factor. We can regard a map $\phi\colon \mathcal{B} \longrightarrow \mathcal{S}$ as a section $\tilde{\phi}\colon \mathcal{B} \longrightarrow \mathcal{E}$ of $\mathcal{E}$ by

$\tilde{\phi}(X) = (X, \phi(X))$ so that $\pi \circ \tilde{\phi} = $ Identity. In general, $\pi: \mathcal{E} \longrightarrow \mathcal{B}$ or just $\mathcal{E}$ will denote a fiber bundle. For us, this means that $\mathcal{E}$ and $\mathcal{B}$ are manifolds, $\pi: \mathcal{E} \longrightarrow \mathcal{B}$ is a smooth map, $\pi \circ \pi = \pi$ and the derivative of π is surjective at each point. The *fiber over* $X \in \mathcal{B}$ is defined by $\mathcal{E}_X = \pi^{-1}(X)$. One can think of $\mathcal{E}$ as the collection of fibers $\mathcal{E}_X$, one at each point of $\mathcal{B}$. In fact, fiber bundles are often defined by specifying the fiber over a typical point in the base. Vector bundles (such as a tensor bundle) are a special case in which each $\mathcal{E}_X$ is a vector space. If this idea seems foreign, a review of Section 1.4 may be necessary.

Following elasticity notation, coordinates on $\mathcal{B}$ are denoted X^A and those on the fibers are denoted x^a. Thus, local coordinates for the whole bundle are (X^A, x^a). The reader should fill in the coordinate formulas while reading this section.

We shall assume for simplicity that the fiber bundle $\mathcal{E}$ carries a *splitting* (connection). This means that at each point $p \in \mathcal{E}_X$, there is a projection map $\mathbb{P}_p: T_p\mathcal{E} \longrightarrow T_p(\mathcal{E}_X)$. [It follows that $T_p\mathcal{E} = T_p(\mathcal{E}_X) \oplus \ker \mathbb{P}_p$ and that $T\pi(p)$ gives an isomorphism of $\ker \mathbb{P}_p$ with $T_X\mathcal{B}$, so we can regard $T_p\mathcal{E} = T_p(\mathcal{E}_X) \oplus T_X\mathcal{B}$.] For the case $\mathcal{E} = \mathcal{B} \times \mathcal{S}$, $\mathbb{P}_p$ is just the tangent of the projection onto the second factor.

If $\phi: \mathcal{B} \longrightarrow \mathcal{E}$ is a section of $\mathcal{E}$—that is, $\pi \circ \phi = $ Identity—we define the derivative of ϕ at X to be the linear map $D\phi(X): T_X\mathcal{B} \longrightarrow T_{\phi(X)}\mathcal{E}_X$ given by $D\phi(X) = \mathbb{P}_{\phi(X)} \circ T\phi(X)$. If $\tilde{\phi}$ is the section of $\mathcal{B} \times \mathcal{S}$ corresponding to a map $\phi: \mathcal{B} \longrightarrow \mathcal{S}$, then $D\tilde{\phi}(X) = T\phi(X): T_X\mathcal{B} \longrightarrow T_x\mathcal{S}$ is just the deformation gradient of the configuration ϕ. By analogy with this special case, we shall sometimes write $\boldsymbol{F}(X) = D\phi(X)$ for the derivative of ϕ at X.

4.1 Definition Let $\pi: \mathcal{E} \longrightarrow \mathcal{B}$ be a fiber boundle with a splitting. Its *first jet bundle* is the fiber bundle $J^1(\mathcal{E})$ over $\mathcal{B}$ whose fiber at $X \in \mathcal{B}$ is

$$J^1(\mathcal{E})_X = \{(\phi_X, \boldsymbol{F}_X) \mid \phi_X \in \mathcal{E}_X = \pi^{-1}(X)$$
$$\text{and } \boldsymbol{F}_X \text{ is a linear map of } T_X\mathcal{B} \text{ to } T_{\phi_X}\mathcal{E}_X\}.$$

If ϕ is a section of $\mathcal{E}$, then its *first jet* is the section $j(\phi)$ of $J^1(\mathcal{E})$ given by

$$j(\phi)(X) = (\phi(X), \boldsymbol{F}(X))$$

where $\boldsymbol{F} = D\phi(X)$ is the derivative of ϕ at X.

In the literature (for example, see Palais [1965]) the reader will find the appropriate definitions without the splitting assumption. However, for elasticity a natural splitting is available, and it simplifies matters somewhat to use it.

Next, we make up a bundle that incorporates the velocity variables as well (in a space–time context, these would already be included as part of $D\phi$).

4.2 Definitions Let $J^1(\mathcal{E})$ be as in 4.1. Define the bundle $\dot{\mathcal{E}}$ over $\mathcal{B}$ to be the bundle whose fiber $\dot{\mathcal{E}}_X$ at $X \in \mathcal{B}$ consists of pairs $(\phi_X, \dot{\phi}_X)$, where $\phi_X \in \mathcal{E}_X$ and $\dot{\phi}_X \in T_{\phi(X)}\,\mathcal{E}_X$. Let the bundle Ξ over $\mathcal{B}$ have fiber at $X \in \mathcal{B}$ consisting of triples $(\phi_X, \dot{\phi}_X, \boldsymbol{F}_X)$, where $(\phi_X, \dot{\phi}_X) \in \dot{\mathcal{E}}_X$ and $\boldsymbol{F}_X$ is a linear map of $T_X\mathcal{B}$ to $T_{\phi(X)}\mathcal{E}$.

Let $\mathcal{C}$ denote a space of sections $\phi: \mathcal{B} \longrightarrow \mathcal{E}$ so that $T_\phi\mathcal{C}$ may be identified with sections of the bundle whose fiber at $X \in \mathcal{B}$ is $T_{\phi(X)}\mathcal{E}_X$. Thus, $T\mathcal{C}$ itself can be identified with sections of $\dot{\mathcal{E}}$.

A *Lagrangian density* is a smooth map $\mathcal{L}: \Xi \longrightarrow \mathbb{R}$ and we shall, by abuse of notation, write $\mathcal{L} = \mathcal{L}(X, \phi, \dot{\phi}, \boldsymbol{F})$.

Define the Lagrangian $L: T\mathcal{C} \longrightarrow \mathbb{R}$ associated to $\mathcal{L}$ and a volume element $dV(X)$ on $\mathcal{B}$ by

$$L(\phi, \dot{\phi}) = \int_{\mathcal{B}} \mathcal{L}(X, \phi(X), \dot{\phi}(X), \boldsymbol{F}(X))\, dV(X).$$

The notation $\mathcal{C}$ for the space of sections is consistent with our earlier usage in case $\mathcal{E} = \mathcal{B} \times \mathcal{S}$. (It may be necessary to restrict the sections, for example, by imposing regularity conditions and boundary conditions of place.)

The configuration manifold is $\mathcal{Q} = \mathcal{C}$, with appropriate differentiability restrictions on elements of $\mathcal{Q}$. Using Theorem 1.13 of Chapter 3, one can show that if members of $\mathcal{C}$ are at least C^1, then $L(\phi, \dot{\phi})$ is a smooth function. Moreover, we have the following formula (suppressing the variable X):

$$DL(\phi, \dot{\phi})\cdot(h, \dot{h}) = D_{\dot{\phi}}L(\phi, \dot{\phi})\cdot \dot{h} + D_\phi L(\phi, \dot{\phi})\cdot h$$

$$= \int_{\mathcal{B}} \partial_{\dot{\phi}}\mathcal{L}(\phi, \dot{\phi}, D\phi)\dot{h}\, dV + \left(\int_{\mathcal{B}} \partial_\phi\mathcal{L}(\phi, \dot{\phi}, D\phi)\cdot h\, dV + \int_{\mathcal{B}} \partial_{D\phi}\mathcal{L}(\phi, \dot{\phi}, D\phi)\cdot Dh\, dV\right). \tag{1}$$

Consider now Lagrange's equations for L from the previous section:

$$\frac{d}{dt} D_{\dot{\phi}}L(\phi, \dot{\phi}) = D_\phi L(\phi, \dot{\phi}). \tag{2}$$

Using (1), (2) means that for any section $h \in T_\phi\mathcal{C}$, the relation

$$\frac{d}{dt}\int_{\mathcal{B}} \partial_{\dot{\phi}}\mathcal{L}(\phi, \dot{\phi}, D\phi)\cdot h\, dV = \int_{\mathcal{B}} \partial_\phi\mathcal{L}(\phi, \dot{\phi}, D\phi)\cdot h\, dV + \int_{\mathcal{B}} \partial_{D\phi}\mathcal{L}(\phi, \dot{\phi}, D\phi)\cdot Dh\, dV \tag{3}$$

holds. By definition, (3) is the *weak form* of the field equations. To get the strong form, assume h has compact support or vanishes on $\partial\mathcal{B}$ and that we have enough differentiability for the second integral on the right-hand side to be integrated by parts. We then get

$$\int_{\mathcal{B}} \left(\frac{\partial}{\partial t}\partial_{\dot{\phi}}\mathcal{L}\right)\cdot h\, dV = \int_{\mathcal{B}} \{\partial_\phi\mathcal{L} - \text{DIV}\, \partial_{D\phi}\mathcal{L}\}\cdot h\, dV.$$

Since h is arbitrary, we must have the *Lagrangian density equation*

$$\frac{\partial}{\partial t}(\partial_{\dot{\phi}}\mathcal{L}) = \partial_\phi\mathcal{L} - \text{DIV}(\partial_{D\phi}\mathcal{L}). \tag{4}$$

The expression on the right-hand side is often called the *functional derivative* of $\mathcal{L}$ and is denoted

$$\frac{\delta \mathcal{L}}{\delta \phi} = \partial_\phi \mathcal{L} - \text{DIV}\, \partial_{D\phi} \mathcal{L}.$$

The situation is summarized as follows:

4.3 Proposition *Let $\mathcal{L}$ and L be related as in 4.2. Then Lagrange's equations (2) for L are equivalent to the weak form of the field equations (3). If elements of $\mathcal{C}$ are prescribed on $\partial \mathcal{B}$, so elements h of $T_\phi \mathcal{C}$ (variations) vanish on $\partial \mathcal{B}$, and if $\partial_{D\phi} \mathcal{L}$ is C^1, then the weak form (3) is equivalent to the strong form (4).*

The above formalism is suitable for field theories in all of space or with Dirichlet boundary conditions—that is, in elasticity, the displacement problem. For the traction problem, however, integration by parts leaves a boundary term, so we shall have to modify things, as follows. Using elasticity terminology for general field theory, we make the following:

4.4 Definition Let $\boldsymbol{P} = -\partial \mathcal{L}/\partial D\phi$ be the *first Piola–Kirchhoff stress tensor.*

Suppose boundary conditions of place are prescribed on ∂_d and the traction $\boldsymbol{P}\cdot \boldsymbol{N} = \boldsymbol{\tau}$ is prescribed on ∂_τ, where $\boldsymbol{N}$ is the unit outward normal to $\partial \mathcal{B}$. Let $\mathcal{V}_\tau$ be a function of ϕ such that $-\nabla_\phi \mathcal{V}_\tau = \boldsymbol{\tau}$ (in Euclidean space, with "dead" loads, choose $\mathcal{V}_\tau = -\boldsymbol{\tau}\cdot\boldsymbol{\phi}$). Consider the Lagrangian

$$L(\phi, \dot{\phi}) = \int_{\mathcal{B}} \mathcal{L}(\phi, \dot{\phi}, D\phi)\, dV - \int_{\partial\tau} \mathcal{V}_\tau(\phi)\, dA. \tag{5}$$

The same procedure as above shows that (if we have enough differentiability to pass through the weak from of the equations) *Lagrange's equations*

$$\frac{d}{dt} D_{\dot{\phi}} L(\phi, \dot{\phi}) = D_\phi L(\phi, \dot{\phi})$$

are equivalent to the field equations

$$\frac{\partial}{\partial t} \partial_{\dot{\phi}} \mathcal{L} = \partial_\phi \mathcal{L} - \text{DIV}\, \partial_{D\phi} \mathcal{L} \text{ in } \mathcal{B}$$

and the boundary conditions

$$\boldsymbol{P}\cdot\boldsymbol{N} = \boldsymbol{\tau} \quad \text{on} \quad \partial_\tau.$$

Thus we conclude that with this modified Lagrangian, the boundary conditions of traction emerge as part of Lagrange's equations, in accord with our work in the first section of this chapter.

Problem 4.1 What is the analogue of the Cauchy stress, $\boldsymbol{\sigma}$?

4.5 Example Let us specialize to the case of elasticity. Let $\mathcal{C}$ be the space of all regular configurations $\phi\colon \mathcal{B} \to \mathcal{S}$ of a specified differentiability class, with displacement boundary conditions, if any, imposed. As we saw in Chapter 4, the tangent space $T_\phi\mathcal{C}$ consists of vector fields $\boldsymbol{u}: \mathcal{B} \longrightarrow T\mathcal{S}$ covering ϕ; that is, $\boldsymbol{u}(X) \in T_{\phi(X)}\mathcal{S}$. (Compare with the definition of the bundle $\dot{\mathcal{E}}$ in 4.2.)

Consider the basic equations of motion:

$$\rho_{\text{Ref}}\boldsymbol{A} = \text{DIV}\,\boldsymbol{P} + \rho_{\text{Ref}}\boldsymbol{B}. \tag{6}$$

We assume that $\boldsymbol{P} = \rho_{\text{Ref}}(\partial W/\partial \boldsymbol{F})$ for a stored energy function W, where $\boldsymbol{F} = T\boldsymbol{\phi}$ is the usual displacement gradient. Define the potential energy $V: \mathcal{C} \longrightarrow \mathbb{R}$ by

$$V(\phi) = \int_{\mathcal{B}} \rho_{\text{Ref}} W(\boldsymbol{F})\, dV + \int_{\mathcal{B}} \rho_{\text{Ref}} \mathcal{V}_{\boldsymbol{B}}(\phi)\, dV$$

$$\left(+\int_{\partial\tau} \mathcal{V}_{\tau}(\phi)\, dA \text{ for the traction problem}\right), \tag{7}$$

where $\nabla \mathcal{V}_{\boldsymbol{B}} = -\boldsymbol{B}$ (in Euclidean space with dead loads, we can choose $\mathcal{V}_{\boldsymbol{B}}(\phi) = -\boldsymbol{B}\cdot\boldsymbol{\phi}$). Define the kinetic energy $K: T\mathcal{C} \longrightarrow \mathbb{R}$ by

$$K(u) = \tfrac{1}{2}\int_{\mathcal{B}} \rho_{\text{Ref}}\, \|u\|^2\, dV,$$

and let

$$H = K + V, \qquad L = K - V,$$

and

$$\mathcal{L}(\phi, \dot\phi, D\phi) = \tfrac{1}{2}\rho_{\text{Ref}}\|\dot\phi\|^2 - \rho_{\text{Ref}} W(D\phi) - \rho_{\text{Ref}}\mathcal{V}_{\boldsymbol{B}}. \tag{8}$$

The Hamiltonian system corresponding to this energy function is defined by Lagrange's equations:

$$\frac{d}{dt} D_{\dot\phi} L(\phi, \dot\phi) = D_{\phi} L(\phi, \dot\phi).$$

As was explained in Proposition 4.3, these equations are precisely the weak form of the equations of motion (and boundary conditions in the traction case), and if $\boldsymbol{P}$ is C^1, they give the strong form (6) and the traction boundary conditions.

> ***Problem 4.2*** Make sense out of, and derive, under suitable covariance assumptions, the Doyle–Ericksen formula for the Cauchy stress $\boldsymbol{\sigma} = -\partial\mathcal{L}/\partial \boldsymbol{g}$ for a general Lagrangian field theory.

4.6 Example (Incompressible Elasticity). Here we impose the constraint $J = 1$; that is, $\text{div}\, \boldsymbol{v} = 0$. The equations of motion are modified by replacing the Cauchy stress $\boldsymbol{\sigma}$ by $\boldsymbol{\sigma} + p\boldsymbol{I}$, where $J\boldsymbol{\sigma}\boldsymbol{F}^{-T} = \rho_{\text{Ref}}(\partial W/\partial \boldsymbol{F})$, that part of the stress derived from a stored energy function and where p is to be determined by the incompressibility condition. Such models are commonly used for materials like rubber. Interestingly, the geometric ideas developed in the last section can be of technical benefit for the incompressible case.

As we saw in the previous example, the equations of elastodynamics may be regarded as a Hamiltonian system with configuration space $\mathcal{C}$. For incompressible elasticity we work with

$$\mathcal{C}_{\text{vol}} = \{\phi \in \mathcal{C} \mid J(\phi) = 1\}.$$

Recall that in the displacement problem ϕ is fixed on ∂_d, but no boundary conditions are imposed in $\mathcal{C}$ for the traction problem.

One can show that in suitable function spaces (Sobolev spaces; see Chapter 6), $\mathcal{C}_{\text{vol}}$ is a smooth submanifold of $\mathcal{C}$. Its tangent space at $\phi \in \mathcal{C}_{\text{vol}}$ is

$$T_\phi \mathcal{C}_{\text{vol}} = \{V \in T_\phi \mathcal{C} \,|\, \text{div}(V \circ \phi^{-1}) = 0\}$$

(The proof is given in Ebin–Marsden [1970] and relies on facts about elliptic operators that are given in Chapter 6.)

The main point is that the equations (and boundary conditions) of incompressible elasticity are equivalent to Lagrange's equations for the usual Lagrangian given by (7) and (8) on $T\mathcal{C}_{\text{vol}}$. The extra term $p\mathbf{I}$ in the stress may be regarded as a Lagrange multiplier giving the force of constraint.

Problem 4.3 Establish the last two statements above by using 3.17 and Problem 3.8.

For another approach to the Hamiltonian structure of nonlinear elasticity based on "Lie–Poisson structures," see Seliger and Whitham [1968], Holm and Kuperschmidt [1982] and Marsden, Ratiu and Weinstein [1982].

Box 4.1 ***Summary of Important Formulas for Section 5.4***

Lagrangian Density

$$\mathfrak{L}(X, \phi, \dot{\phi}, D\phi) \qquad\qquad \mathfrak{L}(X^A, \phi^a, \dot{\phi}^a, F^a{}_A)$$

Lagrangian

$$L(\phi, \dot{\phi}) = \int_{\mathfrak{B}} \mathfrak{L}(X, \phi(X), \dot{\phi}(X), F(X))\, dV(X) \left(-\int_{\partial\mathfrak{B}} \mathfrak{V}_\tau(\phi)\, dA(X)\right.$$

$$\left.\text{for traction boundary conditions, where} \quad \nabla \mathfrak{V}_\tau = -\boldsymbol{\tau}\right)$$

Piola–Kirchhoff Stress

$$\mathbf{P} = -\frac{\partial \mathfrak{L}}{\partial \mathbf{F}} \qquad\qquad P_a{}^A = -\frac{\partial \mathfrak{L}}{\partial F^a{}_A}$$

Weak Form of the Lagrange Density Equations for $\mathfrak{L}$ ($\Longleftrightarrow$ Lagrange's Equations for L): for all variations h,

$$\frac{d}{dt}\int_{\mathfrak{B}} \partial_{\dot{\phi}}\mathfrak{L}(\phi, \dot{\phi}, D\phi)\cdot h\, dV = \int_{\mathfrak{B}} \partial_\phi \mathfrak{L}(\phi, \dot{\phi}, D\phi)\cdot h\, dV$$

$$+ \int_{\mathfrak{B}} \partial_{D\phi}\mathfrak{L}(\phi, \dot{\phi}, D\phi) Dh\, dV + \int_{\partial_\tau} \boldsymbol{\tau}\cdot \boldsymbol{h}\, dV$$

Strong Form of the Lagrange Density Equations (equivalent to the weak form if the stress is C^1)

$$\frac{\partial}{\partial t}(\partial_{\dot{\phi}}\mathfrak{L}) = \partial_\phi \mathfrak{L} - \text{DIV}\, \partial_{D\phi}\mathfrak{L}; \qquad \mathbf{P}\cdot \mathbf{N} = \boldsymbol{\tau} \quad \text{on} \quad \partial_\tau$$

Elasticity

$$\mathcal{L}(\phi, \dot{\phi}, F) = \tfrac{1}{2}\rho_{\text{Ref}}\|\dot{\phi}\|^2 - \rho_{\text{Ref}}W(F) - \rho_{\text{Ref}}\mathcal{V}_B(\phi)$$

Incompressible Elasticity

Impose the constraint $J(\phi) = 1$ and replace the Cauchy stress $\boldsymbol{\sigma}$ by $\boldsymbol{\sigma} + pI$, where p is the pressure.

5.5 CONSERVATION LAWS

This section derives special conservation laws for Lagrangian systems. This is done first for finite-dimensional systems and then for field theory. These results, commonly known as *Noether's theorem*, play an important role in Hamiltonian systems. (See Abraham and Marsden [1978], Chapter 4 for this theory in a more general context.) These conservation laws are then applied to elasticity, reproducing as a special case, results obtained by Knowles and Sternberg [1972] and Fletcher [1976], who obtained them by other methods.

For orientation, we give an example from one-dimensional elasticity: consider the equations

$$\phi_{tt} = P(\phi_X)_X \qquad X \in \mathbb{R},$$

where $P(\phi_X) = W'(\phi_X)$ for the stored energy function W and subscripts denote differentiation. We assume W is homogeneous—that is, independent of $X \in \mathbb{R}$. This homogeneity is associated with the identity

$$\frac{d}{dt}(\phi_t\phi_X) = \frac{d}{dx}(P(\phi_X)\phi_X + \mathcal{L}),$$

where $\mathcal{L} = \frac{1}{2}(\phi_t)^2 - W(\phi_X)$. This conservation law may be directly checked; for an equilibrium solution, note the special case: $(d/dx)(P(\phi_X)\phi_X - W(\phi_X)) = 0$. In general, spatial and material symmetries lead to such conservation laws or identities.

We begin with the classical Noether theorem for finite-dimensional Lagrangian systems. Let $L(q^1, \ldots, q^n, \dot{q}^1, \ldots, \dot{q}^n)$ be a Lagrangian and suppose $(q^i(t), \dot{q}^i(t))$ satisfies Lagranges equations

$$\left.\begin{aligned} \frac{d}{dt}q^i &= \dot{q}^i, \\ \frac{d}{dt}\frac{\partial L}{\partial \dot{q}^i} &= \frac{\partial L}{\partial q^i} \quad (i = 1, \ldots, n). \end{aligned}\right\} \tag{1}$$

Suppose Y is a vector field in Q space; $Y = (Y^1(q), \ldots, Y^n(q))$ such that if ψ_s is the flow of Y—that is, $(\partial/\partial s)\psi_s^i(q) = Y^i(\psi_s(q))$—then ψ_s leaves L invariant;

that is, we have the identity

$$L(q^i, \dot{q}^i) = L\left(\psi_s^i(q), \frac{\partial \psi_s^i}{\partial q^j}\dot{q}^j\right). \tag{2}$$

In other words, Y is a symmetry of L in the sense that the transformation of phase space induced by the flow of Y leaves L invariant.

5.1 Proposition *If Equations (1) and (2) hold, then* $\mathfrak{P} = (\partial L/\partial \dot{q}^i) Y^i$ *is a constant of the motion; that is,* $d\mathfrak{P}/dt = 0$.

Proof Differentiating (2) with respect to s at $s = 0$ gives the identity

$$\frac{\partial L}{\partial q^i} Y^i + \frac{\partial L}{\partial \dot{q}^i}\frac{\partial Y^i}{\partial q^j}\dot{q}^j = 0. \tag{3}$$

By (1) we have

$$\frac{d}{dt}\mathfrak{P} = \frac{d}{dt}\left(\frac{\partial L}{\partial \dot{q}^i} Y^i\right) = \frac{\partial L}{\partial q^i} Y^i + \frac{\partial L}{\partial \dot{q}^i}\frac{\partial Y^i}{\partial q^j}\dot{q}^j,$$

which vanishes by (3). ∎

5.2 Examples

(a) $\mathfrak{Q} = \mathbb{R}^3$ and $Y(q^1, q^2, q^3) = e_1 = (1, 0, 0)$, the first basis vector. The flow of Y is translation in the q^1 direction; the induced transformation of phase space is

$$\begin{aligned} q^1 &\mapsto q^1 + s, & \dot{q}^1 &\mapsto \dot{q}^1, \\ q^2 &\mapsto q^2, & \dot{q}^2 &\mapsto \dot{q}^2, \\ q^3 &\mapsto q^3, & \dot{q}^3 &\mapsto \dot{q}^3. \end{aligned}$$

Thus the associated conserved quantity for Lagrangians independent of q^1 is $\mathfrak{P} = \partial L/\partial \dot{q}^1$, the momentum in the q^1-direction.

(b) Let $\mathfrak{Q} = \mathbb{R}^3$ and $Y(q^1, q^2, q^3) = (q^2, -q^1, 0)$, the vector field whose flow consists of rotations about the q^3-axis. If L is invariant under such rotations, then the angular momentum about the q^3-axis, $\mathfrak{P} = p_1 q^2 - q^1 p_2$ is conserved, where $p_i = \partial L/\partial \dot{q}^i$.

Proposition 5.1 may be generalized to infinite dimensions using the same proof as follows:

5.3 Proposition *In the context of 3.13, let* ψ_s *be a flow consisting of* C^1 *maps of* $\mathfrak{Q}$ *to* $\mathfrak{Q}$. *Let*

$$Y(x) = \frac{d}{ds}\psi_s(x)\bigg|_{s=0} \quad \text{so} \quad DY(x)\cdot v = \frac{d}{ds}D\psi_s(x)\cdot v\bigg|_{s=0}$$

exist for $(x, v) \in \mathfrak{D}$. *Suppose* $D\psi_s$ *leaves* L *invariant; that is,* $D\psi_s$ *leaves* $\mathfrak{D}$ *invariant and* $L(x, v) = L(\psi_s(s), D\psi_s(x)\cdot v)$, *for* $(x, v) \in \mathfrak{D}$. *Then* $\mathfrak{P}(x, v) = D_2L(x, v)\cdot Y(x)$ *is a constant of the motion; that is, if* $(x(t), v(t))$ *satisfies Lagranges equations*

$$\frac{d}{dt}x(t) = v(t),$$

$$\frac{d}{dt}D_2L(x(t), v(t)) = D_1L(x(t), v(t)),$$

then $(d/dt)\mathfrak{P} = 0$.

Next we turn to conservation laws for a Lagrangian field theory using the context and notation of Section 5.4. [*Note*: All of the results that follow are given in differential form and assume that the solutions are at least C^1. As usual, when shocks or other discontinuities are present, the integrated form is preferable: this is, what is obtained if 5.3 is used.] We begin by proving a conservation law for the energy density.

Let $\mathfrak{L}$ be a smooth Lagrangian density on a bundle $\pi: \mathcal{E} \longrightarrow \mathcal{B}$ and let $\phi(t)$ be a differentiable curve of fields (sections of $\mathcal{E}$) such that the Lagrange density equation of motion holds:

$$\frac{\partial}{\partial t}(\partial_{\dot\phi}\mathfrak{L}) = \partial_\phi\mathfrak{L} - \text{DIV}\,\partial_{D\phi}\mathfrak{L}.$$

5.4 Proposition *Define the energy density by* $\mathfrak{E} = \dot\phi\cdot\partial_{\dot\phi}\mathfrak{L} - \mathfrak{L}$. *Then* $\mathfrak{E}$ *obeys the following conservation equation ("continutiy equation"):*

$$\frac{\partial\mathfrak{E}}{\partial t} + \text{DIV}(\dot\phi\cdot\partial_{D\phi}\mathfrak{L}) = 0, \quad \text{that is,} \quad \frac{\partial\mathfrak{E}}{\partial t} - \text{DIV}(\dot\phi\cdot P) = 0.$$

Proof Indeed, using the chain rule together with the equation of motion, we find

$$\begin{aligned}\frac{\partial\mathfrak{E}}{\partial t} &= \frac{\partial}{\partial t}(\dot\phi\partial_{\dot\phi}\mathfrak{L}) - \frac{\partial\mathfrak{L}}{\partial t} = \ddot\phi\partial_{\dot\phi}\mathfrak{L} + \dot\phi\frac{\partial}{\partial t}(\partial_{\dot\phi}\mathfrak{L}) - \ddot\phi\partial_{\dot\phi}\mathfrak{L} - \dot\phi\partial_\phi\mathfrak{L} - D\dot\phi\cdot\partial_{D\phi}\mathfrak{L}\\ &= \dot\phi\frac{\partial}{\partial t}(\partial_{\dot\phi}\mathfrak{L}) - \dot\phi\partial_\phi\mathfrak{L} - D\dot\phi\cdot\partial_{D\phi}\mathfrak{L}\\ &= \dot\phi\{\partial_\phi\mathfrak{L} - \text{DIV}\,\partial_{D\phi}\mathfrak{L}\} - \dot\phi\partial_\phi\mathfrak{L} - D\dot\phi\cdot\partial_{D\phi}\mathfrak{L}\\ &= -\dot\phi\,\text{DIV}\,\partial_{D\phi}\mathfrak{L} - D\dot\phi\cdot\partial_{D\phi}\mathfrak{L} = -\text{DIV}(\dot\phi\partial_{D\phi}\mathfrak{L}). \quad \blacksquare\end{aligned}$$

One can similarly localize the conservation laws associated with general symmetries. This proceeds as follows. Let ψ_s be a flow on $\mathcal{B}$ and let $\tilde\psi_s$ be a flow on $\mathcal{E}$, preserving fibers and covering ψ_s. This extends to a flow on $J^1(\mathcal{E})$, called say $\tilde{\tilde\psi}_s$ determined by $\tilde{\tilde\psi}_s\circ j(\phi)\circ\psi_s^{-1} = j(\tilde\psi_s\circ\phi\circ\psi_s^{-1})$ for ϕ a smooth section of $\mathcal{E}$. In coordinates X^A on $\mathcal{B}$ and x^a on the fibers of $\mathcal{E}$,

$$\tilde{\tilde\psi}_s(X^A, x^a, F^a{}_A) = \left(\psi_s^A(X), \tilde\psi_s^a(X, x), \frac{\partial\tilde\psi_s^a}{\partial x^b}\left[\left(\frac{\partial\psi_s}{\partial X}\right)^{-1}\right]^B{}_A F^b{}_B\right).$$

Let $\xi_\mathcal{B}$ and $\xi_\mathcal{E}$ be the corresponding infinitesimal generators on $\mathcal{B}$ and $\mathcal{E}$, respectively. Assume that $\mathfrak{L}$ is invariant in the sense that $\mathfrak{L}$ is unchanged under transformation by $\tilde{\psi}_s \times \tilde{\tilde{\psi}}_s$ of $\dot{\phi}$ and $(\phi, D\phi)$, that is,

$$\mathcal{L}\circ(\tilde{\psi}_s \times \tilde{\tilde{\psi}}_s) = \mathcal{L},$$

and that ψ_s preserves the volume element dV on $\mathcal{B}$. If ϕ is a solution of the Lagrange density equations, set

$$T = \mathcal{L}\xi_{\mathcal{B}} + \partial_{D\phi}\mathcal{L}\cdot(\xi_{\mathcal{S}}\circ\phi - D\phi\cdot\xi_{\mathcal{B}}) \quad \text{(a vector field on } \mathcal{B})$$

$$\mathcal{J} = \partial_{\dot{\phi}}\mathcal{L}\cdot(\xi_{\mathcal{S}}\circ\phi - D\phi\cdot\xi_{\mathcal{B}}) \quad \text{(a scalar field on } \mathcal{B}),$$

where $\mathcal{L}$ stands for $\mathcal{L}(X, \phi, \dot{\phi}, D\phi)$, and so on.

5.5 Proposition (*Noether's Theorem*) *The following identity* (*conservation law*) *holds:*

$$\boxed{\frac{\partial \mathcal{J}}{\partial t} + \mathrm{DIV}\, T = 0.}$$

This is proved along the lines already indicated in the proof of the conservation law for the energy density.

Problem 5.1 Give the details of this proof, both invariantly and in coordinates (see the summary in Box 5.1 for the coordinate expressions).

Problem 5.2

(a) Show that 5.5 implies that the flux of the four-vector $(T, \mathcal{J})$ through any smooth bounded region in spacetime is zero.

(b) What is the rate of change of the integral $\int_U \mathcal{J}\, dV$ over a smooth bounded region U in $\mathcal{B}$?

(c) Show that under some hypotheses, $\mathfrak{P}$ in 5.3 is given for field theories by $\mathfrak{P} = \int_{\mathcal{B}} \mathcal{J}\, dV$.

We shall now use Noether's theorem to derive conservation laws for elasticity. Here we use the set-up and notation from the second half of Section 5.4. In carrying this out, it is important to keep straight spatial and material invariances —that is, invariances under transformation of $\mathcal{S}$ (the space) and of $\mathcal{B}$ (the body), respectively. Such ideas are implicit in the work of Arnold [1966], for instance, and it is in this respect that our treatment differs from that of Knowles and Sternberg [1972] and Fletcher [1976]. (These authors prove more in the cases that they consider. They show that the only transformations that produce the desired infinitesimal invariance of $\mathcal{L}$ are those with which they started.[6])

[6]These conservation laws can also be carried out for plates and shells in an analogous manner (see Naghdi [1972]). A convenient context is nonlinear *Kirchhoff shell theory* in which the stored energy function depends on $\boldsymbol{C}$ and the second fundamental form, $\boldsymbol{k}$. See also Green, Naghdi and Wainwright [1965] and Golubitsky, Marsden and Schaeffer [1983].

5.6 Example Let us begin with spatial invariance. Let ψ_s be a flow on $\mathcal{S}$ generated by a vector field $\boldsymbol{w}$. This gives a flow on the bundle $\mathcal{B} \times \mathcal{S} \to \mathcal{B}$ by holding $\mathcal{B}$ pointwise fixed and moving points in $\mathcal{S}$ by the mapping ψ_s. Invariance of $\mathfrak{L}$ in this case reads

$$\mathfrak{L}(X, \psi_s(\phi), D\psi_s(\phi)\cdot\dot{\phi}, D\psi_s\cdot \boldsymbol{F}) = \mathfrak{L}(X, \phi, \dot{\phi}, \boldsymbol{F}),$$

as an identity on $\mathfrak{L}$ in its arguments $X, \phi, \dot{\phi}, \boldsymbol{F}$. Noether's theorem now states that if ϕ satisfies the equations of motion, then we have the identity

$$\frac{\partial}{\partial t}(\partial_{\dot{\phi}}\mathfrak{L}\cdot \boldsymbol{w}) + \mathrm{DIV}(\partial_{\boldsymbol{F}}\mathfrak{L}\cdot \boldsymbol{w}) = 0.$$

That is,
$$\frac{\partial}{\partial t}\left(\frac{\partial \mathfrak{L}}{\partial \dot{\phi}^a} w^a\right) + \left(\frac{\partial \mathfrak{L}}{\partial(\partial \phi^a/\partial X^A)} w^a\right)_{|A} = 0.$$

(In the notation of Noether's theorem, $\xi_{\mathcal{B}} = 0$, and $\xi_{\mathcal{S}} = (0, \boldsymbol{w})$.) Specializing to $\mathcal{S} = \mathbb{R}^3$ and choosing:

(i) ψ_s an arbitrary translational flow—$\psi_s(\boldsymbol{x}) = \boldsymbol{x} + s\boldsymbol{w}$, $\boldsymbol{w}$ a constant vector—we recover the equations of motion

$$\frac{\partial}{\partial t}\frac{\partial \mathfrak{L}}{\partial \dot{\phi}} + \mathrm{DIV}\, \partial_{\boldsymbol{F}}\mathfrak{L} = 0,$$

that is, balance of momentum. (The assumed invariance of $\mathfrak{L}$ holds if $\mathfrak{L}$ does not depend on the point values of ϕ: cf. Chapter 3.)

(ii) ψ_s an arbitrary rotational flow. Here $\boldsymbol{w}(\boldsymbol{x}) = \boldsymbol{B}\boldsymbol{x}$, where $\boldsymbol{B}$ is an arbitrary skew symmetric matrix. Noether's theorem (together with the equations of motion) now states:

$$(\partial_{\dot{\phi}}\mathfrak{L}) \otimes \dot{\phi} + (\partial_{\boldsymbol{F}}\mathfrak{L})\cdot \boldsymbol{F} \quad \text{is symmetric.}$$

For elasticity,
$$(\partial_{\dot{\phi}}\mathfrak{L}) \otimes \dot{\phi} = \frac{\partial \mathfrak{L}}{\partial \dot{\phi}^a}\dot{\phi}^b = \dot{\phi}^a\dot{\phi}^b$$

is symmetric, so this reduces to the assertion that $\boldsymbol{\sigma} = J^{-1}\boldsymbol{P}\boldsymbol{F}^T$ is symmetric—that is, balance of moment of momentum. Again, this invariance assumption will hold if $\mathfrak{L}$ depends only on $\boldsymbol{F}$ through $\boldsymbol{C}$.

5.7 Remark Noether's theorem provides a natural link between balance laws and material frame indifference. The assumption of material frame indifference plus the above Hamiltonian structure *implies* the usual balance laws. Thus, from an abstract point of view, the foundations of elasticity theory written in terms of a Lagrangian (or Hamiltonian) field theory seem somewhat more satisfactory—certainly more covariant—than the usual balance laws. (See Section 3.4.)

5.8 Example Next we examine material invariance. Let Λ_s be a volume-preserving flow on $\mathcal{B}$ generated by a vector field $\boldsymbol{W}$ on $\mathcal{B}$. This induces a flow on the bundle $\mathcal{B} \times \mathcal{S}$ by holding $\mathcal{S}$ pointwise fixed. An important remark is that

Noether's theorem is purely local. Thus we may consider rotations about each point of $\mathcal{B}$ but restrict attention to a ball centered at each such point. The result of Noether's theorem is still valid since the proof is purely local. This is necessary since we wish to speak of isotropic materials without assuming $\mathcal{B}$ itself is invariant under rotations.

Invariance of $\mathcal{L}$ means that

$$\mathcal{L}(\Lambda_s(X), \phi, \dot{\phi}, D\phi \cdot D\Lambda_s) = \mathcal{L}(X, \phi, \dot{\phi}, D\phi)$$

as an identity on $\mathcal{L}$ in its arguments.

Noether's theorem in this case states that

$$\frac{\partial}{\partial t}(\partial_{\dot{\phi}}\mathcal{L} \cdot D\phi \cdot W) + \text{DIV}(\partial_F \mathcal{L} \cdot D\phi \cdot W - \mathcal{L} W) = 0,$$

that is, $$\frac{\partial}{\partial t}\left(\frac{\partial \mathcal{L}}{\partial \dot{\phi}^a} F^a{}_A W^A\right) + \left(\frac{\partial \mathcal{L}}{\partial F^a{}_A} F^a{}_B W^B - \mathcal{L} W^A\right)_{|A} = 0.$$

(In the terminology of Proposition 5.5, $\xi_{\mathcal{B}} = W$ and $\xi_{\mathcal{E}} = (W, 0)$. Note that the "field values" of $\xi_{\mathcal{E}}$ are zero.) Again this can be rewritten using the equations of motion, if desired.

Now specialize to the case in which $\mathcal{B}$ is open in $\mathbb{R}^3$ and make the following two choices:

(i) Λ_s is an arbitrary translation $\Lambda_s(X) = X + sW$, W a constant vector. Then $\mathcal{L}$ will be invariant if it is *homogeneous*—that is, independent of X. In this case, Noether's theorem yields the identity

$$\frac{\partial}{\partial t}\left(\frac{\partial \mathcal{L}}{\partial \dot{\phi}^a} F^a{}_A\right) + \left(\frac{\partial \mathcal{L}}{\partial F^a{}_B} F^a{}_A\right)_{|B} - \mathcal{L}_{|A} = 0,$$

That is, for any subbody $\mathcal{U} \subset \mathcal{B}$ with unit outward normal N_A,

$$\frac{\partial}{\partial t}\int_{\mathcal{U}} \frac{\partial \mathcal{L}}{\partial \dot{\phi}^a} F^a{}_A \, dV = \int_{\partial \mathcal{U}} \left(\mathcal{L} N_A - \frac{\partial \mathcal{L}}{\partial F^a{}_B} F^a{}_A N_B\right) dA.$$

The identity expresses *conservation of material momentum*; indeed, for elasticity,

$$\frac{\partial \mathcal{L}}{\partial \dot{\phi}^a} F^a{}_A = \rho_{\text{Ref}} \dot{\phi}_a F^a{}_A$$

is just the momentum density expressed in material coordinates. Thus,

$$\mathcal{L} N_A - \frac{\partial \mathcal{L}}{\partial F^a{}_B} F^a{}_A N_B = \mathcal{L} N_A + P_a{}^B F^a{}_A N_B$$

(where P is the first Piola–Kirchhoff stress tensor) may be interpreted as a momentum flux. (If $\mathcal{L}$ is independent of X, the momentum identity can be verified directly using the equations of motion and the chain rule on $\mathcal{L}_{|A}$.)

Problem 5.3 Show that the identity derived in the introduction to this section is a special case of this result.

(ii) If Λ_s is a rotation about the point $\boldsymbol{X}_0$, then $\boldsymbol{W} = \boldsymbol{B}(\boldsymbol{X} - \boldsymbol{X}_0)$, where $\boldsymbol{B}$ is skew-symmetric matrix. In vector notation,

$$\boldsymbol{W} = \boldsymbol{V} \times (\boldsymbol{X} - \boldsymbol{X}_0),$$

where $\boldsymbol{V}$ is a constant vector; in Euclidean coordinates,

$$W_A = \epsilon_{ABC} V^B (X^C - X_0^C),$$

where ϵ^{ABC} is the alternator. Noether's theorem becomes (in Euclidean coordinates)

$$\frac{\partial}{\partial t}\left(\frac{\partial \mathfrak{L}}{\partial \dot\phi^a} F^a{}_A \epsilon^{ABC} X_C\right) + \left(\frac{\partial \mathfrak{L}}{\partial F^a{}_D} F^a{}_A \epsilon^{ABC} X_C - \mathfrak{L}\epsilon^{DBC} X_C\right)_{|D} = 0.$$

This expresses a conservation law for the *material angular momentum of the body*. For it to hold, $\mathfrak{L}$ must be isotropic in the sense discussed in Chapter 3, Section 3.5.

If $\mathfrak{L}$ is *also* homogeneous, then, using the identity in (i), (ii) reduces to

$$\frac{\partial \mathfrak{L}}{\partial F^a{}_D} F^a{}_A \epsilon^{ABD} = 0, \quad \text{that is,} \quad P_{aD} F^a{}_A \epsilon^{ABD} = 0 \quad (B = 1, 2, 3).$$

Problem 5.4 Use the standard isotropic representation for $\boldsymbol{P}$ to show directly that this identity holds.

Remarks. All of this can equally well be done from a space–time point of view. Other symmetry groups (e.g., dilatations, etc.,) can be dealt with in the same way. See Olver [1982] for more information.

Box 5.1 ***Summary of Important Formulas for Section 5.5***

Noether's Theorem (Finite Dimensional)

If $L(\boldsymbol{q}, \dot{\boldsymbol{q}})$ is invariant under the transformations induced by the flow of a vector field $\boldsymbol{Y}(\boldsymbol{q})$, then $\mathfrak{P}$ is conserved along any solution of Lagrange's equations, where

$$\mathfrak{P} = FL \cdot Y, \qquad \mathfrak{P} = \frac{\partial L}{\partial \dot q^i} Y^i$$

Continuity Equation for Energy

$$\mathfrak{E} = \dot\phi \partial_{\dot\phi} \mathfrak{L} - \mathfrak{L} \qquad \mathfrak{E} = \dot\phi^a \frac{\partial \mathfrak{L}}{\partial \dot\phi^a} - \mathfrak{L}$$

$$\frac{\partial \mathfrak{E}}{\partial t} + \mathrm{DIV}(\dot\phi\, \partial_{D\phi}\mathfrak{L}) = 0 \qquad \frac{\partial \mathfrak{E}}{\partial t} + \left(\dot\phi^a \frac{\partial \mathfrak{L}}{\partial F^a{}_A}\right)_{|A} = 0$$

Noether's Theorem for Classical Field Theory

If $\mathcal{L}(\dot{\phi}, \phi, D\phi)$ is invariant under transformations induced by vector fields $\xi_{\mathcal{B}}$ (with components ξ^A) on the base $\mathcal{B}$ and $\xi_{\mathcal{E}}$ (with components ξ^a) on the fibers of $\mathcal{E}$, then

$$\frac{\partial \mathcal{J}}{\partial t} + \text{DIV}\, \boldsymbol{T} = 0, \qquad \frac{\partial \mathcal{J}}{\partial t} + T^A{}_{|A} = 0,$$

where

$$\mathcal{J} = \partial_{\dot{\phi}}\mathcal{L}\cdot(\xi_{\mathcal{E}}\circ\phi - D\phi\cdot\xi_{\mathcal{B}}), \qquad \mathcal{J} = \frac{\partial \mathcal{L}}{\partial \dot{\phi}^a}(\xi^a - F^a{}_A\xi^A),$$

and

$$\boldsymbol{T} = \mathcal{L}\xi_{\mathcal{B}} + \partial_{D\phi}\mathcal{L}\cdot(\xi_{\mathcal{E}}\circ\phi - D\phi\cdot\xi_{\mathcal{B}}), \qquad T^a = \mathcal{L}\xi^A + \frac{\partial \mathcal{L}}{\partial F^a{}_A}(\xi^a - F^a{}_B\xi^B).$$

Noether's Theorem Applied to Elasticity

Spatial invariance:

(i) Under translations gives balance of momentum.

(ii) Under rotations gives balance of moment of momentum.

Material invariance:

(i) Under translations (homogeneous material) gives the following identity (conservation of linear material momentum):

$$\frac{\partial}{\partial t}\left(\frac{\partial \mathcal{L}}{\partial \dot{\phi}^a}F^a{}_A\right) + \left(\frac{\partial \mathcal{L}}{\partial F^a{}_B}F^a{}_A\right)_{|B} - \mathcal{L}_{|A} = 0,$$

where $\mathcal{L} = \frac{1}{2}\dot{\phi}^a\dot{\phi}^b g_{ab} - W(F)$.

(ii) Under rotations (isotropic material) gives the following identity (conservation of angular material momentum):

$$\frac{\partial}{\partial t}\left(\frac{\partial \mathcal{L}}{\partial \dot{\phi}^a}F^a{}_A\epsilon^{ABC}X_C\right) + \left(\frac{\partial \mathcal{L}}{\partial F^a{}_D}F^a{}_A\epsilon^{ABC}X_C - \mathcal{L}\epsilon^{DBC}X_C\right)_{|D} = 0.$$

5.6 RECIPROCITY

We begin this section with a statement and proof of the reciprocal theorem of Betti and Rayleigh. This theorem states that "for a hyperelastic body subject to two infinitesimal systems of body and surface forces, the work done by the first system in the displacement caused by the second equals the work done by the second in the displacement caused by the first."

The following special case (due to Maxwell) will emphasize the interest of this statement. Consider a beam (not necessarily unstressed) and choose two points on the beam P and Q. Put a concentrated load $\boldsymbol{F}_P$ at the point P; this causes a proportional displacement denoted $\alpha_{QP} F_P$ at the point Q; α_{PQ} is called the *influence coefficient*. Likewise, a load F_Q at the point Q produces a displacement $\alpha_{PQ}F_Q$ at P. Reciprocity implies the equality of the two workings: $F_Q \cdot (\alpha_{QP}F_P) = F_P \cdot (\alpha_{PQ}F_Q)$; that is, that

$$\alpha_{PQ} = \alpha_{QP} \quad \text{(Maxwell relations).}$$

Following our discussion of the reciprocal theorem in elasticity, we give a brief discussion of reciprocity in terms of Lagrangian submanifolds. The point is that whenever there is a potential (variational principle) for a given problem, there is a corresponding reciprocity principle and vice versa. For example, in thermodynamics the reciprocity principle is called the Onsager relation.

We begin by deriving the classical reciprocal theorem. We recall from Proposition 2.8, Section 4.2 that the linearized equations about a given *solution* $\mathring{\phi}_t$ corresponding to incremental loads B^*, τ^* and boundary conditions of place (if any), are given in material form by

$$\left.\begin{aligned} &\rho_{\text{Ref}}(B^* - \ddot{U}) + \text{DIV}(\mathbf{A}\cdot\nabla U) = 0, \\ &\langle \mathbf{A}\cdot\nabla U, N\rangle = \tau^* \quad \text{on} \quad \partial_\tau, \qquad U = \mathbf{0} \quad \text{on} \quad \partial_d, \end{aligned}\right\} \tag{1}$$

where $\mathbf{A}$ is the elasticity tensor evaluated in the configuration $\mathring{\phi}$. We consider another such system for incremental loads $(\bar{B}^*, \bar{\tau}^*)$ satisfying

$$\left.\begin{aligned} &\rho_{\text{Ref}}(\bar{B}^* - \ddot{\bar{U}}) + \text{DIV}(\mathbf{A}\cdot\nabla \bar{U}) = 0, \\ &\langle \mathbf{A}\cdot\nabla \bar{U}, N\rangle = \bar{\tau}^* \quad \text{on} \quad \partial_\tau, \qquad \bar{U} = \mathbf{0} \quad \text{on} \quad \partial_d. \end{aligned}\right\} \tag{2}$$

Now multiply (1) by $\bar{U}$ and integrate over $\mathcal{B}$ to get

$$\begin{aligned} \int_{\mathcal{B}} \rho_{\text{Ref}}\langle B^* - \ddot{U}, \bar{U}\rangle\, dV &= -\int_{\mathcal{B}} \langle \text{DIV}(\mathbf{A}\cdot\nabla U), \bar{U}\rangle\, dV \\ &= \int_{\mathcal{B}} (\mathbf{A}\cdot\nabla U)\cdot\nabla\bar{U}\, dV - \int_{\partial\tau} \langle \tau^*, \bar{U}\rangle\, dA. \end{aligned} \tag{3}$$

The quantity $(\mathbf{A}\cdot\nabla U)\cdot\nabla\bar{U}$ is given in coordinates by $\mathsf{A}_a{}^A{}_b{}^B U^b{}_{|B} \bar{U}^a{}_{|A}$, which is symmetric in ∇U and $\nabla\bar{U}$ for hyperelastic materials. Thus, (3) yields the following:

6.1 Theorem (*Betti Reciprocity*) *Let U and $\bar{U}$ be two solutions of the linearized equations of hyperelastodynamics corresponding to incremental loads* (B^*, τ^*) *and* $(\bar{B}^*, \bar{\tau}^*)$, *respectively—that is, satisfy* (1) *and* (2). *Then*

$$\begin{aligned} &\int_{\mathcal{B}} \rho_{\text{Ref}}\langle B^* - \ddot{U}, \bar{U}\rangle\, dV + \int_{\partial\tau} \langle \tau^* \bar{U}\rangle\, dA \\ &= \int_{\mathcal{B}} \rho_{\text{Ref}}\langle \bar{B}^* - \ddot{\bar{U}}, U\rangle\, dV + \int_{\partial\tau} \langle \bar{\tau}^*, U\rangle\, dA. \end{aligned} \tag{4}$$

Notice that for the special case of elastostatics, Betti reciprocity states that

$$\int_{\mathcal{B}} \rho_{\text{Ref}}\langle B^*, \bar{U}\rangle\, dV + \int_{\partial\tau} \langle \tau^*, \bar{U}\rangle\, dA = \int_{\mathcal{B}} \rho_{\text{Ref}}\langle \bar{B}^*, U\rangle\, dV + \int_{\partial\tau} \langle \bar{\tau}^*, U\rangle\, dA. \tag{5}$$

Problem 6.1 Derive the Maxwell relations from Betti reciprocity.

Problem 6.2 Prove that Betti reciprocity implies hyperelasticity.

Problem 6.3 Show that if ϕ = identity is stress free and the material is isotropic and homogeneous, each side of (4) (or (5) for elastostatics) becomes

$$\int [2\mu u_c{}^{|d} \bar{u}^c{}_{|d} + \lambda(\operatorname{div} \boldsymbol{u})(\operatorname{div} \bar{\boldsymbol{u}})]\, dv$$

(identifying material and spatial quantities).

Betti's reciprocal theorem is important in several applications. One of these is in the theory of bifurcation of three-dimensional elastic bodies, where it limits the number of possible solutions. (This theory is described in Section 7.3.) For other applications and a historical discussion, see Truesdell and Noll [1965] and Sokolnikoff [1956].

Now we turn our attention to a reformulation of Betti's theorem in terms of Lagrangian submanifolds. (For additional information, see Abraham and Marsden [1978], Section 5.3, Guillemin and Sternberg [1977], Weinstein [1977], Oster and Perelson [1973], and Tulczyjew [1974]).*

We shall begin by discussing the linear case, and then we shall globalize to manifolds. Following this we shall give some simple examples and explain the connections with reciprocity.

6.2 Definitions Let $(\mathcal{E}, \boldsymbol{\omega})$ be a symplectic vector space (see Section 5.2) and $\mathfrak{F} \subset \mathcal{E}$ a subspace. The *ω-orthogonal complement* of $\mathfrak{F}$ is the subspace defined by

$$\mathfrak{F}^{\perp} = \{e \in \mathcal{E} \mid \boldsymbol{\omega}(e, e') = 0 \text{ for all } e' \in \mathfrak{F}\}.$$

We say:

(i) $\mathfrak{F}$ is *isotropic* if $\mathfrak{F} \subset \mathfrak{F}^{\perp}$, that is, $\boldsymbol{\omega}(e, e') = 0$ for all $e, e' \in \mathfrak{F}$;

(ii) $\mathfrak{F}$ is *co-isotropic* if $\mathfrak{F} \supset \mathfrak{F}^{\perp}$, that is, $\boldsymbol{\omega}(e, e') = 0$ for all $e' \in \mathfrak{F}$ implies $e \in \mathfrak{F}$;

(iii) $\mathfrak{F}$ is *Lagrangian* if $\mathfrak{F}$ is isotropic and has an isotropic complement, that is, $\mathcal{E} = \mathfrak{F} \oplus \mathfrak{F}'$, where $\mathfrak{F}'$ is isotropic.

(iv) $\mathfrak{F}$ is *symplectic* if $\boldsymbol{\omega}$ restricted to $\mathfrak{F} \times \mathfrak{F}$ is nondegenerate.

The terminology "Lagrangian subspace" was appatently first used by Maslov [1965], although the ideas were in isolated use before that date.

To get a feel for some of the ideas we shall assume temporarily that our vector spaces and manifolds are *finite dimensional*. The following collects some properties related to Definition 6.2.

---

*Marsden/Abraham, *Foundation of Mechanics*, (2d ed.) © 1978. Addison-Wesley, Reading, MA. Sec. 5.3. Reprinted with permission.

6.3 Proposition

(i) $\mathfrak{F} \subset \mathfrak{G}$ *implies* $\mathfrak{G}^{\perp} \subset \mathfrak{F}^{\perp}$.
(ii) $\mathfrak{F}^{\perp} \cap \mathfrak{G}^{\perp} = (\mathfrak{F} + \mathfrak{G})^{\perp}$.
(iii) $\dim \mathcal{E} = \dim \mathfrak{F} + \dim \mathfrak{F}^{\perp}$.
(iv) $\mathfrak{F} = \mathfrak{F}^{\perp\perp}$.
(v) $(\mathfrak{F} \cap \mathfrak{G})^{\perp} = \mathfrak{F}^{\perp} + \mathfrak{G}^{\perp}$.

Proof The assertions (i) and (ii) are simple verifications. To prove (iii), consider the linear map $\omega^{\flat}: \mathcal{E} \longrightarrow \mathcal{E}^*$. Now for $e \in \mathfrak{F}$, $\omega^{\flat}(e)$ annihilates $\mathfrak{F}^{\perp}$, so we get an induced linear map $\omega^{\flat}_{\mathfrak{F}}: \mathfrak{F} \longrightarrow (\mathcal{E}/\mathfrak{F}^{\perp})^*$. Since ω is nondegenerate, this map in injective. Thus, by linear algebra

$$\dim \mathfrak{F} \leq \dim (\mathcal{E}/\mathfrak{F}^{\perp})^* = \dim \mathcal{E} - \dim \mathfrak{F}^{\perp}.$$

Next consider $\omega^{\flat}: \mathcal{E} \longrightarrow \mathcal{E}^* \longrightarrow \mathfrak{F}^*$. As a linear map $\omega^{\flat}_{\mathfrak{F}}$ of $\mathcal{E}$ to $\mathfrak{F}^*$, this has kernel exactly $\mathfrak{F}^{\perp}$. Thus, by linear algebra again

$$\dim \mathfrak{F} \geq \dim \text{Range}\, \omega^{\flat}_{\mathfrak{F}} = \dim \mathcal{E} - \dim \mathfrak{F}^{\perp}.$$

These two inequalities give (iii).

For (iv), notice that $\mathfrak{F} \subset \mathfrak{F}^{\perp\perp}$ is clear. From (iii) applied to $\mathfrak{F}$ and to $\mathfrak{F}^{\perp}$ we get $\dim \mathfrak{F} = \dim \mathfrak{F}^{\perp\perp}$, so $\mathfrak{F} = \mathfrak{F}^{\perp\perp}$.

Finally, for (v), notice that, using (ii) and (iv).

$$(\mathfrak{F} \cap \mathfrak{G})^{\perp} = (\mathfrak{F}^{\perp\perp} \cap \mathfrak{G}^{\perp\perp})^{\perp} = \mathfrak{F}^{\perp} + \mathfrak{G}^{\perp})^{\perp\perp} = \mathfrak{F}^{\perp} + \mathfrak{G}^{\perp}. \quad \blacksquare$$

The next result is often used to define Lagrangian subspaces.

6.4 Proposition *Let* $(\mathcal{E}, \omega)$ *be a symplectic vector space and* $\mathfrak{F} \subset \mathcal{E}$ *a subspace. Then the following assertions are equivalent:*

(i) $\mathfrak{F}$ *is Lagrangian.*
(ii) $\mathfrak{F} = \mathfrak{F}^{\perp}$.
(iii) $\mathfrak{F}$ *is isotropic and* $\dim \mathfrak{F} = \frac{1}{2} \dim \mathcal{E}$.

Proof First we prove that (i) implies (ii). We have $\mathfrak{F} \subset \mathfrak{F}^{\perp}$ by definition. Conversely, let $e \in \mathfrak{F}^{\perp}$ and write $e = e_0 + e_1$, where $e_0 \in \mathfrak{F}$ and $e_1 \in \mathfrak{F}'$, where $\mathfrak{F}'$ is given by Definition 6.2 (iii). We shall show that $e_1 = \mathbf{0}$. Indeed, $e_1 \in \mathfrak{F}'^{\perp}$ by isotropy of $\mathfrak{F}'$, and similarly $e_1 = e - e_0 \in \mathfrak{F}^{\perp}$. Thus $e_1 \in \mathfrak{F}'^{\perp} \cap \mathfrak{F}^{\perp} = (\mathfrak{F}' + \mathfrak{F})^{\perp} = \mathcal{E}^{\perp} = \{\mathbf{0}\}$ by nondegeneracy of ω. Thus $e_1 = \mathbf{0}$, so $\mathfrak{F}^{\perp} \subset \mathfrak{F}$ and (ii) holds.

Secondly, (ii) implies (iii) follows at once from 6.3 (iii).

Finally, we prove that (iii) implies (i). First, observe that (iii) implies that $\dim \mathfrak{F} = \dim \mathfrak{F}^{\perp}$ by 6.3 (iii). Since $\mathfrak{F} \subset \mathfrak{F}^{\perp}$, we have $\mathfrak{F} = \mathfrak{F}^{\perp}$. Now we construct $\mathfrak{F}'$ as follows. Choose arbitrarily $v_1 \notin \mathfrak{F}$ and let $\mathcal{V}_1 = \text{span}(v_1)$; since $\mathfrak{F} \cap \mathcal{V}_1 = \{\mathbf{0}\}$, $\mathfrak{F} + \mathcal{V}_1^{\perp} = \mathcal{E}$ by 6.3(v). Now pick $v_2 \in \mathcal{V}_1^{\perp}$, $v_2 \notin \mathfrak{F} + \mathcal{V}_1$, let $\mathcal{V}_2 = \mathcal{V}_1 + \text{span}(v_2)$, and continue inductively until $\mathfrak{F} + \mathcal{V}_k = \mathcal{E}$. By construction, $\mathfrak{F} \cap \mathcal{V}_k$

$= \{\mathbf{0}\}$, so $\mathcal{E} = \mathcal{F} \oplus \mathcal{V}_k$. Also by construction,

$$\mathcal{V}_2^\perp = \mathcal{V}_1 + \operatorname{span}(v_2))^\perp = \mathcal{V}_1^\perp \cap \operatorname{span}(v_2)^\perp \supset \operatorname{span}(v_1, v_2) = \mathcal{V}_2$$

since $v_2 \in \mathcal{V}_1^\perp$. Inductively, $\mathcal{V}_k$ is isotropic as well. Thus we can choose $\mathcal{F}' = \mathcal{V}_k$. ∎

We can rephrase 6.4 by saying that Lagrangian subspaces are *maximal isotropic subspaces.*

6.5 Examples

(i) Any one-dimensional subspace of $\mathcal{E}$ is isotropic, so if $\mathcal{E}$ is two dimensional, any one-dimensional subspace is Lagrangian.

(ii) Let $\mathcal{E} = \mathbb{R}^2 \times \mathbb{R}^2$ with elements denoted $v = (v_1, v_2)$ and with the usual symplectic structure

$$\omega((v_1, v_2), (w_1, w_2)) = \langle v_1, w_2 \rangle - \langle v_2, w_1 \rangle,$$

where $\langle , \rangle$ denotes the Euclidean inner product. Then the subspace spanned by linearly independent vectors v and w is Lagrangian if and only if

$$\langle v_1, w_2 \rangle = \langle v_2, w_1 \rangle.$$

For instance $\mathbb{R}^2 \times \{\mathbf{0}\}$ and $\{\mathbf{0}\} \times \mathbb{R}^2$ are Lagrangian subspaces, as is span((1, 1, 1, 1), (0, 1, 0, 1)), and so forth.

(iii) Let $\mathcal{E} = \mathcal{V} \oplus \mathcal{V}^*$ with the canonical symplectic form

$$\omega_{\mathcal{V}}((v_1, \alpha_1), (v_2, \alpha_2)) = \alpha_2(v_1) - \alpha_1(v_2).$$

Then $\mathcal{V} \oplus \{\mathbf{0}\} \subset \mathcal{E}$ and $\{\mathbf{0}\} \oplus \mathcal{V}^*$ are Lagrangian, since $\omega_{\mathcal{V}}$ vanishes on them and they have half the dimension of $\mathcal{E}$.

(iv) Let $\mathcal{H}$ be a complex inner product space (regarded as a real vector space) with the symplectic form

$$\omega(z, z') = -\operatorname{Im}\langle z, z' \rangle.$$

(See Example 2.4(c), Section 5.2.) Thus a subspace $\mathcal{V} \subset \mathcal{H}$ is isotropic if and only if all inner products of pairs of elements of $\mathcal{V}$ are *real.* Let $\mathbb{J}: \mathcal{H} \longrightarrow \mathcal{H}$ be multiplication by $i = \sqrt{-1}$. Then if $\mathcal{V}$ is isotropic, so is $\mathcal{V}' = \mathbb{J} \cdot \mathcal{V}$. Also, $\mathcal{V} \cap \mathbb{J} \cdot \mathcal{V} = \{\mathbf{0}\}$, as is easily seen. Thus $\mathcal{V}$ is Lagrangian if and only if all pairs of inner products of elements of $\mathcal{V}$ are real and $\mathcal{V} + \mathbb{J}\mathcal{V} = \mathcal{H}$. This last decomposition of $\mathcal{H}$ identifies $\mathcal{H}$ with the complexification of a real inner product space $\mathcal{V}$ and within $\mathcal{H}$, the "purely real" and "purely imaginary" subspaces are Lagrangian. Thus, this example merely rephrases Example (iii).

The next proposition shows that Example 6.5(iii) is, in a sense, the most general example.

6.6 Proposition *Let* $(\mathcal{E}, \omega)$ *be a symplectic vector space and* $\mathcal{V} \subset \mathcal{E}$ *a Lagrangian subspace. Then there is a symplectic isomorphism* $A: (\mathcal{E}, \omega) \longrightarrow (\mathcal{V} \oplus \mathcal{V}^*, \omega_{\mathcal{V}})$ *taking* $\mathcal{V}$ *to* $\mathcal{V} \oplus \{\mathbf{0}\}$.

Proof Let $\mathcal{E} = \mathcal{V} \oplus \mathcal{V}'$, where $\mathcal{V}'$ is isotropic, and consider the map

$$T: \mathcal{V}' \longrightarrow \mathcal{V}^*, \qquad T(e_1)\cdot e = \omega(e, e_1)$$

We claim T is an isomorphism. Indeed, suppose that $T(e_1) = 0$; then $\omega(e_1, e) = 0$ for all $e \in \mathcal{V}$ and hence—as $\mathcal{V}'$ is isotropic and $\mathcal{E} = \mathcal{V} \oplus \mathcal{V}'$—for all $e \in \mathcal{E}$. Since ω is nondegenerate, $e_1 = 0$. Hence T is one-to-one, and since $\dim \mathcal{V} = \dim \mathcal{V}'$, it is an isomorphism.

Now let $A = \text{Identity} \oplus T$. It is now easy to verify that $A^*\omega_{\mathcal{V}} = \omega$; indeed,

$$\begin{aligned}(A^*\omega_{\mathcal{V}})((e, e_1), (e', e_1')) &= \omega_{\mathcal{V}}((e, Te_1), (e', Te_1')) \\ &= (Te_1')(e) - (Te_1)(e') \\ &= \omega(e, e_1') - \omega(e', e_1) \\ &= \omega(e + e_1, e' + e_1')\end{aligned}$$

since each of $\mathcal{V}$ and $\mathcal{V}'$ is isotropic. ∎

Now we move from the context of linear spaces to manifolds. We begin with the following.

6.7 Definitions Let $(\mathcal{P}, \omega)$ be a symplectic manifold and $i: \mathcal{L} \longrightarrow \mathcal{P}$ an immersion. We say $\mathcal{L}$ is an *isotropic* (*co-isotropic, symplectic*) *immersed submanifold* of $(\mathcal{P}, \omega)$ if $(T_x i)(T_x\mathcal{L}) \subset T_{i(x)}\mathcal{P}$ is an isotropic (co-isotropic, symplectic) subspace for each $x \in \mathcal{L}$. The same terminology is used for submanifolds of $\mathcal{P}$ and for subbundles of $T\mathcal{P}$ over submanifolds of $\mathcal{P}$.

A submanifold $\mathcal{L} \subset \mathcal{P}$ is called *Lagrangian* if it is isotropic and there is an isotropic subbundle $\mathcal{E} \subset T\mathcal{P}\,|\,\mathcal{L}$ such that $T\mathcal{P}\,|\,\mathcal{L} = T\mathcal{L} \oplus \mathcal{E}$.

Notice that $i: \mathcal{L} \longrightarrow \mathcal{P}$ is isotropic if and only if $i^*\omega = 0$. Also note, from the linear theory, that if $\mathcal{P}$ is finite dimensional and $\mathcal{L} \subset \mathcal{P}$ is Lagrangian, then $\dim \mathcal{L} = \frac{1}{2} \dim \mathcal{P}$ and $(T_x\mathcal{L})^{\perp} = T_x\mathcal{L}$.

6.8 Proposition *Let $(\mathcal{P}, \omega)$ be a finite-dimensional symplectic manifold and $\mathcal{L} \subset \mathcal{P}$ a submanifold. Then $\mathcal{L}$ is Lagrangian if and only if $\mathcal{L}$ is isotropic and* $\dim \mathcal{L} = \frac{1}{2} \dim \mathcal{P}$.

Proof The preceding remark proves the "only if" part. For the "if" part, we know $T_x\mathcal{L}$ has an isotropic complement $\mathcal{E}_x$ at each $x \in \mathcal{L}$. One can readily check that they can be chosen in a smooth manner. ∎

An important example of a Lagrangian submanifold is given in the next proposition.

6.9 Proposition *Let α be a one-form on a finite-dimensional manifold $\mathcal{Q}$ and $\mathcal{L} \subset T^*\mathcal{Q}$ be its graph. Then $\mathcal{L}$ is a Lagrangian submanifold if and only if α is closed.*

Proof Clearly $\mathcal{L}$ is a submanifold with dimension $\frac{1}{2}$ dim $T^*\mathbb{Q}$. However, from Definition 3.8, we have, $\boldsymbol{\alpha}^*\boldsymbol{\theta} = \boldsymbol{\alpha}$, so

$$d\boldsymbol{\alpha} = \boldsymbol{d}^*\boldsymbol{d\theta} = -\boldsymbol{\alpha}^*\boldsymbol{\omega}.$$

Thus $\boldsymbol{\alpha}$ is closed if and only if $\boldsymbol{\alpha}^*\boldsymbol{\omega} = 0$, what is $\mathcal{L}$ is isotropic. ∎

In particular, note that $\mathbb{Q}$ itself, being the zero section, is Lagrangian. The argument also shows that the Lagrangian submanifolds of $T^*\mathbb{Q}$ which project diffeomorphically onto $\mathbb{Q}$ are in one-to-one correspondence with the closed one-forms on $\mathbb{Q}$.

Since $\boldsymbol{\alpha}$ is closed, locally $\boldsymbol{\alpha} = \boldsymbol{d}S$ for a function S by the Poincaré lemma. This remark leads into the next definition.

6.10 Definition Let $(\mathcal{P}, \boldsymbol{\omega})$ be a symplectic manifold, $\mathcal{L}$ a Lagrangian submanifold, and $i: \mathcal{L} \to \mathcal{P}$ the inclusion. If, locally, $\boldsymbol{\omega} = -\boldsymbol{d\theta}$, then $i^*\boldsymbol{\omega} = -\boldsymbol{d}i^*\boldsymbol{\theta} = \mathbf{0}$, so $i^*\boldsymbol{\theta} = \boldsymbol{d}S$ for a function $S: \mathcal{L} \to \mathbb{R}$ (locally defined). We call S a *generating function* or a *potential function* for $\mathcal{L}$.

If $\mathcal{L} \subset T^*Q$ is the graph of $\mathbf{d}S$, where $S: Q \to \mathbb{R}$, then $\mathcal{L}$ is Lagrangian and we can identify the generating function of $\mathcal{L}$ with S.

The idea of generating functions really goes back to Hamilton and Jacobi. However, the definition in the above general context is due to Sniatycki and Tulczyjew [1972].

Most of the concepts of analytical mechanics such as canonical transformations (and their generating functions) can be expressed entirely in terms of Lagrangian submanifolds. See Abraham and Marsden [1978], Section 5.3 for details.

We now turn to the ideas involved in reciprocity. First we consider an example,[7] namely, a 3- (or generally n-) port nonlinear DC electric network, schematically shown in Figure 5.6.1. Let q^i denote voltages applied to each terminal and let p_i denote the currents flowing into the terminal (in specified directions, as in the figure). The applied voltages determine the currents, so we have

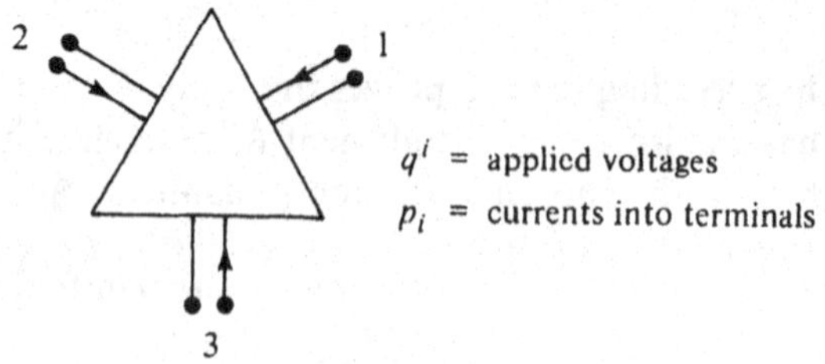

Figure 5.6.1

[7]Many of the ideas here are due to W. Tulczyjew and G. Oster.

relations

$$p_i = f_i(q^1, \ldots, q^n) \quad (i = 1, \ldots, n).$$

Thus small changes in the q's, Δq^i produce small changes Δp_i in the p's. Reciprocity means that

$$\frac{\Delta p_2}{\Delta q_1} = \frac{\Delta p_1}{\Delta q_2},$$

that is, the proportional change of current/voltage induced in port 2 by activating port 1 is the same if instead we activate port 1 and look at the current changed in port 2. Precisely, this means

$$\frac{\partial f_i}{\partial q^j} = \frac{\partial f_j}{\partial q^i} \quad (i, j = 1, \ldots, n).$$

Another way of saying this equality is that the manifold which is the graph of $f = (f_1, \ldots, f_n)$ in $\mathbb{R}^{2n}$ with its usual symplectic structure $\boldsymbol{\omega} = \sum dq^i \wedge dp_i$ is isotropic and hence Lagrangian (since its dimension is n). This fits in with 6.9 for the above relations say exactly that f regarded as a one-form, $\boldsymbol{f} = f_i\, dq^i$, is closed.

For a 2-port network, reciprocity (plus a nondegeneracy condition) is equivalent to having a function $F(q^1, q^2)$ such that $p_1 = -\partial F/\partial q^1$ and $p_2 = \partial F/\partial q^2$; that is, F is the generating function of a canonical transformation of (q^1, p_1) to (q^2, p_2). For three or higher ports, we still have generating functions (6.10).

Other physical examples are reciprocal relationships between generalized forces and displacements in an elastic system (see below), thermodynamics and thermostatics. The generating functions for these examples are the "internal energy" and the "free energy." In the first case, $\boldsymbol{\theta} = p_i\, dq^i$ is usually called the "virtual work." In thermostatics, the reciprocity relations are called the *Maxwell relations*, while in thermodynamics they are called the *Onsager relations*. Thus, *pairs of variables in a definite relationship are called reciprocal when they define a Lagrangian submanifold of the corresponding phase space.* The fact that these submanifolds have generating functions means that there will be a potential associated to any reciprocity relation, although its physical meaning will depend on the particular system. One can carry these ideas further, as Kijowski and Tulczyjew [1979] have done, and regard Lagrangian submanifolds as basic entities describing systems.

Let us now rephrase Betti's reciprocal theorem for hyperelastostatics in terms of Lagrangian submanifolds. Let $\mathcal{C}$ denote the space of all configurations $\phi: \mathcal{B} \to \mathcal{S}$ satisfying boundary conditions of place (if any). We identify $T^*\mathcal{C}$ with loads (generalized forces) as follows: an element of $T^*\mathcal{C}$ is a linear map l: $T_\phi\mathcal{C} \to \mathbb{R}$. We can identify a pair of (dead) loads $(\boldsymbol{B}, \boldsymbol{\tau})$ with such an l by mapping the infinitesimal displacement $\boldsymbol{U} \in T_\phi\mathcal{C}$ to the corresponding virtual work:

$$\int_{\mathcal{B}} \boldsymbol{B} \cdot \boldsymbol{U}\, dV + \int_{\partial\tau} \boldsymbol{\tau} \cdot \boldsymbol{U}\, dA.$$

(One sees again that forces should "really" be one-forms and not vectors.)

Given constitutive functions for the stress, let $\mathcal{L} \subset T^*\mathcal{C}$ be defined to be the set of all $(\phi, (\boldsymbol{B}, \boldsymbol{\tau})) \in T^*\mathcal{C}$ such that the equations of elastostatics are satisfied.

6.11 Theorem *$\mathcal{L}$ is a Lagrangian submanifold of $T^*\mathcal{C}$ if and only if the stress is derived from an internal energy function $\hat{W}$. The function*

$$\mathfrak{W}(\phi) = \int_{\mathcal{B}} \hat{W}(F)\, dX$$

is the generating function for $\mathcal{L}$ in the sense of 6.10.

Problem 6.4 Verify this theorem and that it yields the reciprocal relations (5).

Problem 6.5 Derive a dynamic version of 6.11. (Consult Abraham and Marsden [1978]).

Problem 6.6 (Tulczyjew) *An Elastic Beam.* Consider an elastic beam in Euclidean space. The equilibrium configuration of the beam with no external forces is a straight line l. Small deflections induced by external forces and torques can be represented by points of a plane $\mathfrak{M}$ perpendicular to the line l. The distance measured along l from an arbitrary reference point is denoted by s. We select a section of the beam corresponding to an interval $[s_1, s_2]$ and assume that external forces and bending torques are applied only to the ends of the section. The configuration manifold $\mathcal{Q}$ of the section of the beam is the product $T\mathfrak{M} \times T\mathfrak{M}$ with coordinates $(q_2^i, q'^j_2, q_1^k, q'^l_1)$ $(i, j, k, \mathring{l} = 1, 2)$. The force bundle $\mathfrak{F} = T^*T\mathfrak{M} \times T^*T\mathfrak{M}$ has coordinates $(q_2^i, q'^j_2, f^2{}_k, t^2{}_l, q_1{}^m, q'^n_1, f^1{}_p, t^1{}_r)$. The coordinates $f^2{}_k$ and $t^2{}_l$ are components of the reaction force and the reaction moment respectively at q_2^i; $f^1{}_p$ and $t^1{}_r$ are components of the force and the torque applied to the end of the beam section at $q_1{}^m$. If $(\delta q_2{}^i, \delta q'^j_2, \delta f^2{}_k, \delta t^2{}_l, \delta q_1{}^m, \delta q'^n_1, \delta f^1{}_p, \delta t^1{}_r)$ are components of an infinitesimal "displacement" u in $\mathfrak{F}$ at $(q_2{}^i, q'^j_2, f^2{}_k, t^2{}_l, q_1{}^m, q'^n_1, f^1{}_p, t^1{}_r)$ then the virtual work is

$$w = f^1{}_i\, \delta q_1{}^i + t^1{}_i\, \delta q'^j_1 - f^2{}_i\, \delta q_2{}^i - t^2{}_i\, \delta q'^j_2 = -\langle u, \vartheta \rangle,$$

where

$$\vartheta = f^2{}_i\, dq_2{}^i + t^2{}_i\, dq'^i_2 - f^1{}_i\, dq_1{}^i - t^1{}_i\, dq'^i_1$$

In the limit $s_2 \longrightarrow s_1$ the configuration manifold $\mathcal{Q}'$ is the bundle $TT\mathfrak{M}$ with coordinates $(q^i, q'^j, \dot{q}^k, \dot{q}'^l)$ and the force bundle $\mathfrak{F}'$ is the bundle $TT^*T\mathfrak{M}$ with coordinates $(q^i, q'^j, f_k, t_l, \dot{q}^m, \dot{q}'^n, \dot{f}_p, \dot{t}_r)$. The form ϑ becomes

$$\dot{\vartheta} = \dot{f}_i\, dq^i + f_i\, d\dot{q}^i + \dot{t}_i\, dq'^i + t_i\, d\dot{q}'^i.$$

Equilibrium conditions are

$$\dot{f}_i = 0, \qquad f_i + \dot{t}_i = 0, \qquad t_i = k_{ij}\dot{q}'^j,$$

where k_{ij} is a tensor characterizing the elastic properties of the beam. These conditions express the vanishing of the total force and the total moment, and also Hooke's law. In addition to these conditions there is a constraint condition $\dot{q}^i = q'^i$. This condition defines a constraint submanifold

$$\mathcal{C} = \{w \in TT\mathfrak{M} \mid \tau_{T\mathfrak{M}}(w) = T\tau_m(w)\}.$$

We use on $\mathcal{C}$ coordinates $(q^i, \dot{q}^j, \ddot{q}^k)$ related to coordinates $(q^i, q'^j, \dot{q}^k, \dot{q}'^l)$ by $\dot{q}'^l = \ddot{q}^l$. Show that these conditions define a Lagrangian submanifold $\mathcal{S}' \subset \mathfrak{F}'$ generated by $-L$, where

$$L(q^i, \dot{q}^j, \ddot{q}^k) = \tfrac{1}{2} k_{ij}\ddot{q}^i\ddot{q}^j$$

is the potential energy per unit length of the beam and is defined on $\mathcal{C}$. Finally, show that the equations

$$\dot{q}^i = q'^i, \qquad \dot{q}'^i = k^{ij}t_j, \qquad (k^{ij}k_{jl} = \delta^i_l), \qquad \dot{f}_i = 0, \qquad t_i = -f_i$$

define a Hamiltonian vector field X on $\mathfrak{F}'$ and that the Lagrangian submanifold $\mathcal{S}'$ is the image of the field X. (See Tulczyjew [1976] for more information).

Problem 6.7 (*Harmonic Maps*) A map $\phi: \mathcal{B} \to \mathcal{S}$ between Riemannian manifolds is called *harmonic* if it minimizes the energy function $E(\phi) = \frac{1}{2}\int_{\mathcal{B}} |d\phi(X)|^2\, dV(X)$. (a) Compute the first Piola–Kirchhoff stress and write down the Euler–Lagrange equations. (b) Find a Lagrangian submanifold that contains the harmonic maps. (c) Read the introduction to Eells and Lemaire [1978] and relate the notions to those of this book. Identify their "tension field" as one of the stress tensors. (d) Transcribe the results of Tanyi [1978] into the notations of this book.

Box 6.1 ***Summary of Important Formulas for Section 5.6***

Betti Reciprocity Principle

If $\boldsymbol{U}$ and $\bar{\boldsymbol{U}}$ solve the linearized equations of hyperelastodynamics with incremental loads $(\boldsymbol{B}^*, \boldsymbol{\tau}^*)$ and $(\bar{\boldsymbol{B}}, \bar{\boldsymbol{\tau}}^*)$, respectively, then

$$\int_{\mathcal{B}} \rho_{\text{Ref}}\langle(\boldsymbol{B}^* - \ddot{\boldsymbol{U}}), \bar{\boldsymbol{U}}\rangle\, dV + \int_{\partial\tau} \langle\boldsymbol{\tau}^*, \bar{\boldsymbol{U}}\rangle\, dA$$

is unchanged if the barred and unbarred quantities are swapped.

Lagrangian Manifolds

A submanifold $\mathcal{L}$ of a symplectic manifold $(\mathcal{P}, \boldsymbol{\omega})$ is called Lagrangian when the form $\boldsymbol{\omega}$ vanishes when pulled back to $\mathcal{L}$ and $\mathcal{L}$ is maximal with respect to this property (if $\mathcal{P}$ is finite dimensional this means $\dim \mathcal{L} = \frac{1}{2}\dim \mathcal{P}$).

Reciprocity and Lagrangian Submanifolds

If $\mathcal{L} \subset T^*\mathcal{C}$ denotes the set of all solutions $(\phi, (\boldsymbol{B}, \boldsymbol{\tau}))$ of the equations of hyperelastics, then $\mathcal{L}$ is Lagrangian and has the internal energy as its generating function.

5.7 RELATIVISTIC ELASTICITY[8]

This section outlines some of the basic ingredients that go into elasticity in the context of general relativity. This theory is believed to be important in a number of astrophysical situations, such as neutron stars. The theory is also of academic interest since related continuum theories such as fluids, magnetohydrodynamics, and plasmas are of great astrophysical importance.

We shall begin by discussing the kinematics of relativistic continua and then discuss very briefly the dynamics. We cannot be exhaustive here, as this subject should properly take an entire textbook. The reader is assumed to have a nodding acquaintance with the basic tools of relativity; our main source is Misner, Thorne, and Wheeler [1973] (and its companion problem book by Lightman, Press, Price, and Teukolsky).

There are many references relevant to relativistic elasticity, although the literature is discouragingly diffuse. For us, the most useful sources have been Carter and Quintana [1972] and Carter [1973]. Other standard references are Bennoun [1965], Bressan [1978], Lianis and Rivlin [1972], and Maugin [1975]. Our aim is to bring out the unity of the relativistic and nonrelativistic principles. Because of the development of covariant methods in this book, our burden is somewhat lightened.

Let us begin now with *kinematics*. The reader should consult Section 2.4 for a comparison with the nonrelativistic case. In that section $\mathcal{V}$ denoted a classical spacetime. Now it denotes a spacetime in the sense of general relativity; that is, $\mathcal{V}$ is a four-manifold with a pseudo-Riemannian metric ${}^{(4)}g$ of Lorentz signature $(+, +, +, -)$.

7.1 Definitions

(1) A *world-tube* is a one-to-one map $\Phi: \mathcal{B} \times \mathbb{R} \to \mathcal{V}$, where $\mathcal{B}$ is a three-dimensional reference body (i.e., a *particle-label space*).

(2) A *slicing* of $\mathcal{V}$ is a diffeomorphism $i: \mathcal{S} \times \mathbb{R} \to \mathcal{V}$, where $\mathcal{S}$ is a three-dimensional *reference manifold* or *physical space*, such that the hypersurface $\mathcal{S}_t \equiv i(\mathcal{S} \times \{t\}) \equiv i_t(\mathcal{S})$ (called a *constant-t-slice*) is spacelike.

(3) A *motion* of $\mathcal{B}$ in $\mathcal{S}$ is, as usual, a curve of embeddings from $\mathcal{B}$ into $\mathcal{S}$, namely, $\phi: \mathcal{B} \times \mathbb{R} \to \mathcal{S}$.

(4) Given a world tube Φ and a slicing i, we define a *corresponding motion* ϕ^i of $\mathcal{B}$ in $\mathcal{S}$ exactly as in Section 2.4 by

$$\Phi(X, t) = i(\phi^i(X, t), t).$$

---

[8]This section was written in collaboration with David Bao.

[9]For a Hamiltonian treatment of relativistic fluids, see Taub [1949] and Hawking and Ellis [1973]; for magnetohydrodynamics, see Morrison and Green [1980] and Marsden, Ratiu and Weinstein [1982]; for electrofluid dynamics, see Spencer [1982]; for plasmas, see Morrison [1980], Marsden and Weinstein [1982], and Bialynicki-Birula and Hubbard [1982]. For a recent Hamiltonian treatment of general relativity, see Fischer and Marsden [1979a], and references therein.

(See Section 2.4 for figures going with these definitions.) As in Section 2.4, note that the superscript i refers to a slicing i and **not** to tensor components.

This definition of ϕ^i makes sense if the world tube and the slicing are *synchronized*; that is, $\Phi(X, t) \in \mathcal{S}_t$ for all $X \in \mathcal{B}$ and $t \in \mathbb{R}$. We shall assume this is the case without loss of generality, since the world-tube can always be reparametrized if necessary.

7.2 Definitions We define $V^\Phi = \partial\Phi/\partial t\colon \mathcal{B} \times \mathbb{R} \longrightarrow T\mathcal{V}$ to be the *material* 4-*velocity of* Φ and let $v^\Phi = V^\Phi \circ \Phi^{-1}$ be the corresponding *spatial velocity*. Likewise, $V^i = \partial i/\partial t\colon \mathcal{S} \times \mathbb{R} \longrightarrow T\mathcal{V}$ is the *material frame velocity* and $v^i = V^i \circ i^{-1}$ is the *spatial frame velocity*. Finally, $V^{\Phi,i} = \partial\phi^i/\partial t\colon \mathcal{B} \times \mathbb{R} \longrightarrow T\mathcal{S}$ is the *apparent material velocity* and $v_t^{\Phi,i} = V_t^{\Phi,i} \circ (\phi_t^i)^{-1}$ is the *apparent spatial velocity*.

By differentiating the defining relation for ϕ^i with respect to t, we find, exactly as in Section 2.4, that:

7.3 Proposition *At* $p = i(x, t)$, $v^\Phi - v^i$ *is, tangent to* $\mathcal{S}_t$ *and* $i_{t*}v_t^{\Phi,i} = v^\Phi - v^i$, *where* $i_{t*} = Ti_t$ *is the tangent of* i_t.

Now since $\mathcal{V}$ has a connection, namely, the Levi–Civita connection ${}^{(4)}\nabla$ of ${}^{(4)}g$, we can define accelerations as well. We let ∇^t be the Levi–Civita connection of g_t on $\mathcal{S}_t$, where g_t is the metric ${}^{(4)}g$ restricted (pulled back) to $\mathcal{S}_t$. We know that

$$\nabla^t = {}^{(4)}\nabla - n_t \otimes k_t = \text{the piece of } {}^{(4)}\nabla \text{ tangent to } \mathcal{S}_t,$$

where n_t is the unit normal vector field on $\mathcal{S}_t$ and k_t is the second fundamental form of $\mathcal{S}_t$ (see Section 1.5 for a discussion of the second fundamental form). Finally, let ∇_{app}^t denote the connection on $\mathcal{S}_t$, which is the Levi–Civita connection of $g_t^i = i_t^* g_t$. Corresponding to the pull-back of the metric we get a pull-back relation for the connections, $\nabla_{\text{app}}^t = i_t^* \nabla^t$, which is defined as $(i_t^*\nabla^t)_v w \equiv (i_{t*})^{-1} \nabla^t_{i_{t*}v}\, i_{t*}w$. ("app" stands for "apparent").

In classical spacetimes, the condition ${}^{(4)}\nabla\, d\tau = 0$ ensures that ${}^{(4)}\nabla_u v$ is tangent to $\mathcal{S}_t$ whenever u and v are; so even though the notion of a unit normal does not make sense there, the second fundamental form—which measures the amount of non-tangency of ${}^{(4)}\nabla_u v$—is effectively zero.

All the above connections are, by construction, torsion-free and satisfy $\nabla_u v - \nabla_v u = [u, v]$.

Our *index conventions* are as follows: Latin indices (say, a, b, c) range from 1 to 3, while Greek indices (say, α, β, γ) range from 0 to 3.

7.4 Definitions Let i be a slicing and let Φ be a world-tube that is synchronized with i. The *world-tube material 4-acceleration* is defined by

$$A^\Phi = \frac{{}^{(4)}D}{dt} V^\Phi\colon \mathcal{B} \times \mathbb{R} \longrightarrow T\mathcal{V},$$

In a coordinate chart,

$$(A^\Phi(X, t))^\alpha = \frac{d}{dt}(V^\Phi(X, t))^\alpha + {}^{(4)}\gamma^\alpha_{\beta\gamma}|_{\Phi(X,t)}(V^\Phi(X, t))^\beta(V^\Phi(X, t))^\gamma.$$

The *world-tube spatial 4-acceleration* is defined by

$$\boldsymbol{a}^\Phi = A^\Phi \circ \Phi^{-1} = {}^{(4)}\nabla_{\boldsymbol{v}^\Phi}\boldsymbol{v}^\Phi.$$

The *frame material 4-acceleration* A^i and the *frame spatial 4-acceleration* $\boldsymbol{a}^i$ are similarly defined as above. The *material apparent 3-acceleration* of the motion ϕ^i is denoted by $A_t^{\Phi,i}\colon \mathcal{B} \to T\mathcal{S}$, where, in a coordinate chart,

$$(A_t^{\Phi,i}(X))^a = \frac{d}{dt}(V_t^{\Phi,i}(X))^a + \gamma^a_{bc}(x, t)(V_t^{\Phi,i}(X))^b(V_t^{\Phi,i}(X))^c.$$

Finally, the *spatial apparent 3-acceleration* of the motion ϕ^i is $\boldsymbol{a}_t^{\Phi,i} = A_t^{\Phi,i}\circ(\phi_t^i)^{-1}\colon \mathcal{S} \to T\mathcal{S}$, which is a time-dependent vector field on $\mathcal{S}$.

If we write $(V_t^{\Phi,i}(X))^a$ as $(\boldsymbol{v}^{\Phi,i}(\phi^i(X, t), t))^a$, use the chain rule to compute $(d/dt)(V_t^{\Phi,i}(X))^a$, and the definitions of spatial objects, we get

$$\boldsymbol{a}_t^{\Phi,i} = \frac{\partial}{\partial t}\boldsymbol{v}_t^{\Phi,i} + \nabla^t_{\text{app } \boldsymbol{v}_t^{\Phi,i}}\boldsymbol{v}_t^{\Phi,i}$$

To compute $\boldsymbol{a}^{\Phi,i}$ in terms of $\boldsymbol{a}^\Phi$ and $\boldsymbol{a}^i$ we need to generalize what we did in Section 2.4. As we did there, we use:

7.5 Lemma *On the slice* $\mathcal{S}_t$,

$$i_{t*}\left(\frac{\partial}{\partial t}\boldsymbol{v}_t^{\Phi,i}\right) = [\boldsymbol{v}^i, \boldsymbol{v}^\Phi - \boldsymbol{v}^i].$$

Proof Let $x = \phi^i(X, t)$. Then

$$\left[i_{t*}\left(\frac{\partial}{\partial t}\boldsymbol{v}_t^{\Phi,i}\right)\right]\bigg|_{i_t(x)} = i_{t*}\left[\left(\frac{\partial}{\partial t}\boldsymbol{v}_t^{\Phi,i}\right)\bigg|_{(x)}\right]$$

$$= i_{t*}\lim_{h\to 0}\frac{1}{h}(\boldsymbol{v}_{t+h}^{\Phi,i}(x) - \boldsymbol{v}_t^{\Phi,i}(x)) = \lim_{h\to 0}\frac{1}{h}(i_{t*}\boldsymbol{v}_{t+h}^{\Phi,i}(x) - i_{t*}\boldsymbol{v}_t^{\Phi,i}(x))$$

$$= \lim_{h\to 0}\frac{1}{h}(i_{t*}(i_{t+h})_*^{-1}(i_{t+h})_*\boldsymbol{v}_{t+h}^{\Phi,i}(x) - i_{t*}\boldsymbol{v}_t^{\Phi,i}(x))$$

$$= \lim_{h\to 0}\frac{1}{h}[(i_{t+h}\circ i_t^{-1})_*^{-1}(\boldsymbol{v}^\Phi - \boldsymbol{v}^i)|_{i_{t+h}(x)} - (\boldsymbol{v}^\Phi - \boldsymbol{v}^i)|_{i_t(x)}],$$

Now if F_h is the flow on $\mathcal{V}$ generated by the time-independent vector field $\boldsymbol{v}^i$, then definition-chasing shows that, on $\mathcal{S}_t$, F_h agrees with $i_{t+h}\circ i_t^{-1}$, hence F_{h*}^{-1} and $(i_{t+h}\circ i_t^{-1})_*^{-1}$ have the same effect on vectors tangent to $\mathcal{S}_{t+h}$, and therefore the above limit equals

$$\lim_{h\to 0}\frac{1}{h}[F_h^*(\boldsymbol{v}^\Phi - \boldsymbol{v}^i) - (\boldsymbol{v}^\Phi - \boldsymbol{v}^i)]|_{i_t(x)}$$

$$= \left[\frac{d}{ds}F_s^*(\boldsymbol{v}^\Phi - \boldsymbol{v}^i)\right]\bigg|_{s=0}(i_t(x)) = [\mathcal{L}_{\boldsymbol{v}^i}(\boldsymbol{v}^\Phi - \boldsymbol{v}^i)](i_t(x))$$

$$= [\boldsymbol{v}^i, \boldsymbol{v}^\Phi - \boldsymbol{v}^i](i_t(x)). \quad \blacksquare$$

7.6 Proposition *On the slice* $\mathcal{S}_t$, *we have*

$$\boxed{i_{t*}a_t^{\Phi,i} = a^\Phi - a^i - 2\,{}^{(4)}\nabla_{(v^\Phi - v^i)}v^i - k_t(v^\Phi - v^i, v^\Phi - v^i)n_t.}$$

Proof

$$\begin{aligned}
i_{t*}a_t^{\Phi,i} &= i_{t*}\left(\frac{\partial}{\partial t}v_t^{\Phi,i} + \nabla^t_{\mathrm{app}\, v_t^{\Phi,i}}\, v_t^{\Phi,i}\right)\\
&= [v^i, v^\Phi - v^i] + \nabla^t_{i_{t*}v_t^{\Phi,i}}\, i_{t*}v_t^{\Phi,i}\\
&= [v^i, v^\Phi] + ({}^{(4)}\nabla - n_t k_t)_{(v^\Phi - v^i)}(v^\Phi - v^i)\\
&= [v^i, v^\Phi] + (a^\Phi + a^i - 2{}^{(4)}\nabla_{v^\Phi}v^i - [v^i, v^\Phi]) - k_t(v^\Phi - v^i, v^\Phi - v^i)n_t\\
&= [v^i, v^\Phi] + (a^\Phi - a^i - 2{}^{(4)}\nabla_{(v^\Phi - v^i)}v^i - [v^i, v^\Phi]) - k_t(v^\Phi - v^i, v^\Phi - v^i)n_t\\
&= a^\Phi - a^i - 2{}^{(4)}\nabla_{(v^\Phi - v^i)}v^i - k_t(v^\Phi - v^i, v^\Phi - v^i)n_t \quad \blacksquare
\end{aligned}$$

In the computation above, there is an intermediate formula worth noting:

$$i_{t*}a_t^{\Phi,i} = [v^i, v^\Phi] + \nabla^t_{(v^\Phi - v^i)}(v^\Phi - v^i).$$

Next we discuss the transformation laws for velocity and acceleration under a change of framing.

Let i, j be slicings that generate the same constant t-slices $\mathcal{S}_t$; that is, i and j are synchronized. We have a motion $\xi: \mathcal{S} \times \mathbb{R} \to \mathcal{S}$ defined by $\xi_t = j_t^{-1} \circ i_t$. On the other hand, by viewing *i as a world-tube and j as a slicing*, we can construct the motion $i^j: \mathcal{S} \times \mathbb{R} \to \mathcal{S}$.

The definitions show that $(i^j)_t = \xi_t$; some straightforward computations then give

$$V_t^{i,j} = V_t^\xi \quad \text{and} \quad v_t^{i,j} = v_t^\xi;$$

Building on these formulas and with a little care, one gets

$$A_t^{i,j} = A_t^\xi \quad \text{and} \quad a_t^{i,j} = a_t^\xi.$$

7.7 Proposition *Let* Φ *be a world-tube and i, j,* ξ *be as above. Then*

$$\begin{aligned}
\phi_t^j &= \xi_t \circ \phi_t^i,\\
v^{\Phi,j} &= v_t^\xi + \xi_{t*}v_t^{\Phi,i},\\
a_t^{\Phi,j} &= a_t^\xi + \xi_{t*}a_t^{\Phi,i} + 2\nabla^t_{\mathrm{app}\,\xi_{t*}v_t^{\Phi,i}}\, v_t^\xi.
\end{aligned}$$

Problem 7.1 Prove these three formulas.

Box 7.1 ***The Correspondence with Newtonian Mechanics***

In 7.3 and 7.6, we saw, respectively, that $i_{t*}v_t^{\Phi,i} = v^\Phi - v^i$ and

$$i_{t*}a_t^{\Phi,i} = a^\Phi - a^i - 2\,{}^{(4)}\nabla_{(v^\Phi - v^i)}v^i - k_t(v^\Phi - v^i, v^\Phi - v^i)n_t.$$

For any four-vector $\boldsymbol{u}$ on $\mathcal{V}$, let us denote by $\boldsymbol{u}_\parallel$ the component of $\boldsymbol{u}$

tangent to $\mathcal{S}_t$; the projection $\boldsymbol{u} \longrightarrow \boldsymbol{u}_\parallel$ is linear. The preceding formulas, after a projection and a rearrangement, read

$$(\boldsymbol{v}^\Phi)_\parallel = i_{t*}\boldsymbol{v}_t^{\Phi,t} + (\boldsymbol{v}^t)_\parallel,$$

$$(\boldsymbol{a}^\Phi)_\parallel = i_{t*}\boldsymbol{a}_t^{\Phi,t} + 2({}^{(4)}\nabla_{(\boldsymbol{v}^\Phi - \boldsymbol{v}^t)}\boldsymbol{v}^t)_\parallel + (\boldsymbol{a}^t)_\parallel.$$

Note that these have a poetic resemblance to the classical statements we saw in Section 2.4, namely,

$$\frac{d\boldsymbol{x}}{dt} = \frac{d^*\boldsymbol{x}}{dt} + \boldsymbol{\omega} \times \boldsymbol{x},$$

$$\frac{d^2\boldsymbol{x}}{dt^2} = \frac{d^{*2}\boldsymbol{x}}{dt^2} + 2\left(\boldsymbol{\omega} \times \frac{d^*\boldsymbol{x}}{dt}\right) + \left[\boldsymbol{\omega} \times (\boldsymbol{\omega} \times \boldsymbol{x}) + \frac{d^*\boldsymbol{\omega}}{dt} \times \boldsymbol{x}\right],$$

where one has, at some fixed origin Θ, a frame $\{\boldsymbol{f}_1, \boldsymbol{f}_2, \boldsymbol{f}_3\}$ rotating with a time-dependent angular velocity $\boldsymbol{\omega}$ relative to a frozen background, say the distant stars.

Any vectorial quantity $\boldsymbol{r}$ can be expressed as $r^a\boldsymbol{f}_a$. We introduce the following notation:

$$\frac{d^*\boldsymbol{r}}{dt} \text{ means } \left(\frac{dr^a}{dt}\right)\boldsymbol{f}_a \quad \text{and} \quad \frac{d\boldsymbol{r}}{dt} \text{ means } \left(\frac{dr^a}{dt}\right)\boldsymbol{f}_a + r^a\frac{d}{dt}\boldsymbol{f}_a,$$

which is simply $d^*\boldsymbol{r}/dt + \boldsymbol{\omega} \times \boldsymbol{r}$ because $(d/dt)\boldsymbol{f}_a = \boldsymbol{\omega} \times \boldsymbol{f}_a$.

We want to verify that, in some slicing i to be explicitly constructed, first-order approximations of some sort yield the correspondence table 5.7.1.

Table 5.7.1

	Relativistic Object	Classical Object
V1	$(\boldsymbol{v}^\Phi)_\parallel$	$\frac{d\boldsymbol{x}}{dt}$
V2	$i_{t*}\boldsymbol{v}_t^{\Phi,t}$	$\frac{d^*\boldsymbol{x}}{dt}$
V3	$(\boldsymbol{v}^t)_\parallel$	$\boldsymbol{\omega} \times \boldsymbol{x}$
A1	$(\boldsymbol{a}^\Phi)_\parallel$	$\frac{d^2\boldsymbol{x}}{dt^2}$
A2	$i_{t*}\boldsymbol{a}_t^{\Phi,t}$	$\frac{d^{*2}\boldsymbol{x}}{dt^2}$
A3	$2({}^{(4)}\nabla_{(\boldsymbol{v}^\Phi - \boldsymbol{v}^t)}\boldsymbol{v}^t)_\parallel$	$2\left(\boldsymbol{\omega} \times \frac{d^*\boldsymbol{x}}{dt}\right)$
A4	$(\boldsymbol{a}^t)_\parallel$	$\boldsymbol{\omega} \times (\boldsymbol{\omega} \times \boldsymbol{x}) + \frac{d^*\boldsymbol{\omega}}{dt} \times \boldsymbol{x}$

To find a suitable slicing i, we will find an appropriate coordinate system i^{-1} (for $\mathcal{V}$) which is, in some sense, rotating because the classical formulas we are trying to make contact with were derived by expressing things in terms of rotating coordinates.

We shall construct this "rotating" coordinate system i^{-1} by translating the classical set-up into terms that make relativistic sense. We do that, step-by-step, as follows:

(1) The orgin Θ is fixed, say attached to a distant star; being under no acceleration, its history must be a timelike geodesic $t \mapsto \Lambda(t)$ in $\mathcal{V}$. For convenience, let us require Λ to have unit speed; that is, $\langle d\Lambda/dt, d\Lambda/dt\rangle = -1$ for all t. Also, denote $d\Lambda/dt$ by $\boldsymbol{e}_0$.

(2) At $t = 0$, one selects a preferred frame $\{f_1, f_2, f_3\}$ attached to Θ; the orientations of this frame at other times t are governed by the differential equations

$$\frac{d}{dt}\begin{bmatrix} f_1 \\ f_2 \\ f_3 \end{bmatrix} = \begin{bmatrix} \boldsymbol{\omega}(t) \times f_1 \\ \boldsymbol{\omega}(t) \times f_2 \\ \boldsymbol{\omega}(t) \times f_3 \end{bmatrix} = \begin{bmatrix} 0 & \omega^3 & -\omega^2 \\ -\omega^3 & 0 & \omega^1 \\ \omega^2 & -\omega^1 & 0 \end{bmatrix}\begin{bmatrix} f_1 \\ f_2 \\ f_3 \end{bmatrix}.$$

Relativistically, one goes to the rest space of $\Lambda(0)$, namely the three-dimensional subspace orthogonal to $\boldsymbol{e}_0$, pick out an orthonormal basis $\{\boldsymbol{e}_1, \boldsymbol{e}_2, \boldsymbol{e}_3\}$ (this corresponds to $\{f_1, f_2, f_3\}$ at $t = 0$), and evolve it along Λ with the help of the differential equations

$$\frac{^{(4)}D}{dt}\begin{bmatrix} \boldsymbol{e}_1 \\ \boldsymbol{e}_2 \\ \boldsymbol{e}_3 \end{bmatrix} = \begin{bmatrix} 0 & \omega^3 & -\omega^2 \\ -\omega^3 & 0 & \omega^1 \\ \omega^2 & -\omega^1 & 0 \end{bmatrix}\begin{bmatrix} \boldsymbol{e}_1 \\ \boldsymbol{e}_2 \\ \boldsymbol{e}_3 \end{bmatrix}.$$

We may suggestively abbreviate the above system as

$$\frac{^{(4)}D}{dt}\boldsymbol{e}_a = \boldsymbol{\omega}(t) \times \boldsymbol{e}_a$$

indicating "spatial" rotation. Incidentally, note that $(^{(4)}D/dt)\boldsymbol{e}_0 = 0$ because Λ is a geodesic. One must check that the above evolution equations preserve the orthonormality of $\{\boldsymbol{e}_0, \boldsymbol{e}_1, \boldsymbol{e}_2, \boldsymbol{e}_3\}$ at all times. This is not difficult; in fact, using the Leibniz rule

$$\frac{d}{dt}\langle \boldsymbol{u}(t), \boldsymbol{v}(t)\rangle = \left\langle \frac{^{(4)}D}{dt}\boldsymbol{u}, \boldsymbol{v}\right\rangle + \left\langle \boldsymbol{u}, \frac{^{(4)}D}{dt}\boldsymbol{v}\right\rangle$$

and letting

$$P = \begin{bmatrix} \langle e_0, e_1 \rangle \\ \langle e_0, e_2 \rangle \\ \langle e_0, e_3 \rangle \end{bmatrix}, \qquad Q = \begin{bmatrix} \langle e_1, e_1 \rangle \\ \langle e_2, e_2 \rangle \\ \langle e_3, e_3 \rangle \\ \langle e_1, e_2 \rangle \\ \langle e_1, e_3 \rangle \\ \langle e_2, e_3 \rangle \end{bmatrix},$$

one obtains

$$\frac{d}{dt}P = \begin{bmatrix} 0 & \omega^3 & -\omega^2 \\ -\omega^3 & 0 & \omega^1 \\ \omega^2 & -\omega^1 & 0 \end{bmatrix} P, \qquad P(0) = \begin{bmatrix} 0 \\ 0 \\ 0 \end{bmatrix},$$

and

$$\frac{d}{dt}Q = \left[\begin{array}{ccc|ccc} 0 & 0 & 0 & 2\omega^3 & -2\omega^2 & 0 \\ 0 & 0 & 0 & -2\omega^3 & 0 & 2\omega^1 \\ 0 & 0 & 0 & 0 & 2\omega^2 & -2\omega^1 \\ \hline -\omega^3 & \omega^3 & 0 & 0 & \omega^1 & -\omega^2 \\ \omega^2 & 0 & -\omega^2 & -\omega^1 & 0 & \omega^3 \\ 0 & -\omega^1 & \omega^1 & \omega^2 & -\omega^3 & 0 \end{array}\right] Q,$$

$$Q(0) = \begin{bmatrix} 1 \\ 1 \\ 1 \\ 0 \\ 0 \\ 0 \end{bmatrix},$$

Now $P(t) \equiv P(0)$ and $Q(t) \equiv Q(0)$ are constant solutions satisfying the initial data, so by uniqueness they must be the only solutions. Similar considerations lead to the fact that $\langle e_0, e_0 \rangle \equiv -1$ along Λ, which is consistent with our initial specification that Λ has unit speed. See Figure 5.7.1.

(3) Now we can define i^{-1} for points close to Λ. If a point p of $\mathcal{V}$ can be reached by traveling for one unit of parameter time along a geodesic emanating from $\Lambda(t)$ with initial tangent $r^a e_a$, we define $i^{-1}(p)$ to be (r^1, r^2, r^3, t). i^{-1} is unambiguously definable for all points close to Λ. This completes our construction of the slicing i.

Readers seasoned in relativity may have already recognized the previous construction as an important special case of the Generalized

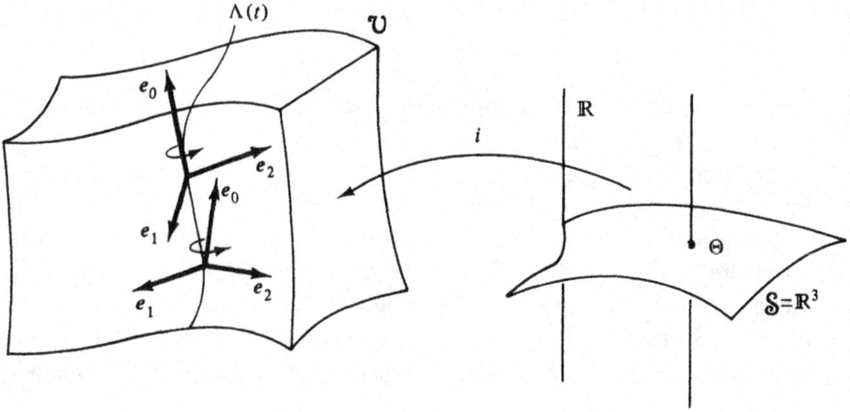

Figure 5.7.1

Fermi–Walker Transport; see Misner, Thorne, and Wheeler [1973], p. 329.

Next we need to familiarize ourselves with *the connection* ${}^{(4)}\boldsymbol{\gamma}$ *along* Λ *and the metric* ${}^{(4)}\boldsymbol{g}$ *near* Λ: Geodesics emanating from $\Lambda(t)$ with tangents orthogonal to $\boldsymbol{e}_0$ must, in the coordinate system i^{-1}, have the form $s \mapsto (r^1 s, r^2 s, r^3 s, t)$; the geodesic equation

$$\frac{d^2x^\alpha}{ds^2} + {}^{(4)}\gamma^\alpha_{\beta\gamma}\frac{dx^\beta}{ds}\frac{dx^\gamma}{ds} = 0$$

for these curves reduces to ${}^{(4)}\gamma^\alpha_{ab} r^a r^b = 0$, which implies that ${}^{(4)}\gamma^\alpha_{ab} = 0$ *along* Λ (in fact, along those curves). Also, the geodesic equation for Λ tells us that ${}^{(4)}\gamma^\alpha_{00} = 0$ *along* Λ. Similarly, using our prescriptions for $({}^{(4)}D/dt)\boldsymbol{e}_a$, we obtain

$${}^{(4)}\gamma^0_{b0} = 0 \quad \text{and} \quad ({}^{(4)}\gamma^a_{b0}) = \begin{bmatrix} 0 & -\omega^3 & \omega^2 \\ \omega^3 & 0 & -\omega^1 \\ -\omega^2 & \omega^1 & 0 \end{bmatrix} \quad \text{along} \quad \Lambda.$$

This information, together with the identity

$${}^{(4)}\gamma^\sigma_{\beta\gamma} = \frac{{}^{(4)}g^{\sigma\alpha}}{2}({}^{(4)}g_{\alpha\beta,\gamma} - {}^{(4)}g_{\beta\gamma,\alpha} + {}^{(4)}g_{\gamma\alpha,\beta})$$

and the fact that ${}^{(4)}g_{\alpha\beta} = \eta_{\alpha\beta}$ along Λ, yields

$${}^{(4)}g_{00,c} = 0, \qquad {}^{(4)}g_{ab,c} = 0, \quad \text{and}$$

$$({}^{(4)}g_{0b,c}) = \begin{bmatrix} 0 & -\omega^3 & \omega^2 \\ \omega^3 & 0 & -\omega^1 \\ -\omega^2 & \omega^1 & 0 \end{bmatrix}, \text{ along } \Lambda.$$

The last equality may be compactly rewritten as ${}^{(4)}g_{0b,c} = -\omega^a \epsilon_{0abc}$, where ϵ_{0abc} is totally antisymmetric and $\epsilon_{0123} = 1$.

The above information on ${}^{(4)}g_{\alpha\beta,c}$ along Λ will allow us to approximate ${}^{(4)}\boldsymbol{g}$ near Λ by a first-order Taylor expansion.

From now on, let us focus on a fixed particle world-line, which will be of the form $t \mapsto (x^1(t), x^2(t), x^3(t), t)$ in the coordinate system provided by i^{-1} because our world-tube Φ is synchronized with respect to the slices $\mathcal{S}_t$ generated by the slicing i. Also, let $\boldsymbol{x} = (x^1(t), x^2(t), x^3(t))$, $\boldsymbol{\omega} = (\omega^1(t), \omega^2(t), \omega^3(t))$. A Taylor expansion in $\boldsymbol{x}$, to first order, shows that

$$\begin{aligned} {}^{(4)}\boldsymbol{g} \cong &-dx^0 \otimes dx^0 + (\boldsymbol{\omega} \times \boldsymbol{x})_a\, dx^a \otimes dx^0 \\ &+ (\boldsymbol{\omega} \times \boldsymbol{x})_a\, dx^0 \otimes dx^a + \delta_{ab}\, dx^a \otimes dx^b, \end{aligned}$$

where $(\boldsymbol{\omega} \times \boldsymbol{x})_a = (\boldsymbol{\omega} \times \boldsymbol{x})^a =$ the ath component of the usual cross product in $\mathbb{R}^3$, the lowered index a allows us to use the automatic summation convention. This determines ${}^{(4)}\boldsymbol{g}$ near Λ.

Before we apply these preliminaries to derive the correspondence table, let us make an observation that will prove to be repeatedly useful later, and that is: the above approximation of ${}^{(4)}\boldsymbol{g}$ says that, to first order in $\boldsymbol{x}$, the vectors $\{\partial/\partial x^1, \partial/\partial x^2, \partial/\partial x^3\}$ remain orthonormal; hence standard linear algebra implies that

$$\left(\frac{\partial}{\partial x^0}\right)_\parallel \cong -\frac{{}^{(4)}g_{0a}}{{}^{(4)}g_{00}} \frac{\partial}{\partial x^a} \cong (\boldsymbol{\omega} \times \boldsymbol{x})^a \frac{\partial}{\partial x^a} \quad (\cong \text{means "to first order in } \boldsymbol{x}\text{"}).$$

Now we are prepared to establish *the correspondence of velocities.* Recall that our particle's world-line has coordinates $(x^1(t), x^2(t), x^3(t), t)$; hence

$$(V^\Phi)_\parallel = \left(\frac{\partial}{\partial x^0} + \frac{dx^a}{dt}\frac{\partial}{\partial x^a}\right)_\parallel \cong (\boldsymbol{\omega} \times \boldsymbol{x})^a \frac{\partial}{\partial x^a} + \frac{dx^a}{dt}\frac{\partial}{\partial x^a},$$

$$(v^i)_\parallel = \left(\frac{\partial}{\partial x^0}\right)_\parallel \cong (\boldsymbol{\omega} \times \boldsymbol{x})^a \frac{\partial}{\partial x^a}.$$

Also, examination of the definitions yields

$$i_{t*}v_t^{\Phi,i} = \frac{dx^a}{dt}\frac{\partial}{\partial x^a}.$$

This verifies items V1 to V3 of the correspondence table.

The next and final task is to establish *the correspondence of accelerations*. Let

$$\boldsymbol{x}_{,0} = \left(\frac{dx^1}{dt}, \frac{dx^2}{dt}, \frac{dx^3}{dt}\right) \quad \text{and} \quad \boldsymbol{\omega}_{,0} = \left(\frac{d\omega^1}{dt}, \frac{d\omega^2}{dt}, \frac{d\omega^3}{dt}\right).$$

All the following approximations will be done, without explicit mention, to first order in $\boldsymbol{x}$ and $\boldsymbol{x}_{,0}$. We will give the details for approximating $(\boldsymbol{a}^\Phi)_\text{I}$ and will simply state the answers for the other terms, leaving their verifications as instructive exercises. We compute at the point $(x^1(t), x^2(t), x^3(t), t)$ on the particle's world-line. First we have

$$(\boldsymbol{a}^\Phi)_\text{I} = (a^\Phi)^0\left(\frac{\partial}{\partial x^0}\right)_\text{I} + (a^\Phi)^a \frac{\partial}{\partial x^a}.$$

But

$$\begin{aligned}(a^\Phi)^0\left(\frac{\partial}{\partial x^0}\right)_\text{I} &= \left(\frac{d^2x^0}{dt^2} + {}^{(4)}\gamma^0_{\beta\gamma}\frac{dx^\beta}{dt}\frac{dx^\gamma}{dt}\right)\left(\frac{\partial}{\partial x^0}\right)_\text{I} \\ &\cong \left({}^{(4)}\gamma^0_{00} + 2\,{}^{(4)}\gamma^0_{b0}\frac{dx^b}{dt}\right)(\boldsymbol{\omega}\times\boldsymbol{x})^a\frac{\partial}{\partial x^a} \\ &\cong {}^{(4)}\gamma^0_{00|\Lambda(t)}(\boldsymbol{\omega}\times\boldsymbol{x})^a\frac{\partial}{\partial x^a} = 0,\end{aligned}$$

so

$$\begin{aligned}(\boldsymbol{a}^\Phi)_\text{I} \cong (a^\Phi)^a\frac{\partial}{\partial x^a} &= \left(\frac{d^2x^a}{dt^2} + {}^{(4)}\gamma^a_{\beta\gamma}\frac{dx^\beta}{dt}\frac{dx^\gamma}{dt}\right)\frac{\partial}{\partial x^a} \\ &\cong \left(\frac{d^2x^a}{dt^2} + {}^{(4)}\gamma^a_{00} + 2\,{}^{(4)}\gamma^a_{b0}\frac{dx^b}{dt}\right)\frac{\partial}{\partial x^a} \\ &\cong \left(\frac{d^2x^a}{dt^2} + {}^{(4)}\gamma^a_{00|\Lambda(t)} + {}^{(4)}\gamma^a_{00,c|\Lambda(t)}x^c + 2\,{}^{(4)}\gamma^a_{b0|\Lambda(t)}\frac{dx^b}{dt}\right)\frac{\partial}{\partial x^a} \\ &= \left(\frac{d^2x^a}{dt^2} + {}^{(4)}\gamma^a_{00,c|\Lambda(t)}x^c + 2(\boldsymbol{\omega}\times\boldsymbol{x}_{,0})^a\right)\frac{\partial}{\partial x^a}.\end{aligned}$$

Next note that

$${}^{(4)}R^a_{0c0} = {}^{(4)}\gamma^a_{00,c} - {}^{(4)}\gamma^a_{0c,0} + {}^{(4)}\gamma^a_{\alpha c}\,{}^{(4)}\gamma^\alpha_{00} - {}^{(4)}\gamma^a_{\alpha 0}\,{}^{(4)}\gamma^\alpha_{0c}.$$

Evaluating at $\Lambda(t)$ and contracting with x^c, we get

$${}^{(4)}\gamma^a_{00,c|\Lambda(t)}x^c = {}^{(4)}R^a_{0c0|\Lambda(t)}x^c + (\boldsymbol{\omega}_{,0}\times\boldsymbol{x})^a + (\boldsymbol{\omega}\times(\boldsymbol{\omega}\times\boldsymbol{x}))^a.$$

Hence

$$\begin{aligned}(\boldsymbol{a}^\Phi)_\text{I} \cong \Bigg(&\frac{d^2x^a}{dt^2} + 2(\boldsymbol{\omega}\times\boldsymbol{x}_{,0})^a + {}^{(4)}R^a_{0c0|\Lambda(t)}x^c \\ &+ (\boldsymbol{\omega}_{,0}\times\boldsymbol{x})^a + (\boldsymbol{\omega}\times(\boldsymbol{\omega}\times\boldsymbol{x}))^a\Bigg)\frac{\partial}{\partial x^a}.\end{aligned}$$

Similarly, one finds that

$$(\boldsymbol{a}^{\prime})_{\parallel} \cong ({}^{(4)}R^{a}_{0c0|\Lambda(t)}x^{c} + (\boldsymbol{\omega}_{,0} \times \boldsymbol{x})^{a} + (\boldsymbol{\omega} \times (\boldsymbol{\omega} \times \boldsymbol{x}))^{a})\frac{\partial}{\partial x^{a}}$$

and

$$2({}^{(4)}\nabla_{(v^{\Phi}-v^{\prime})}v^{\prime})_{\parallel} \cong 2(\boldsymbol{\omega} \times \boldsymbol{x}_{,0})^{a}\frac{\partial}{\partial x^{a}}$$

while definition-chasing establishes that

$$i_{t*}a^{\Phi,t}_{t} = \frac{d^{2}x^{a}}{dt^{2}}\frac{\partial}{\partial x^{a}}.$$

This verifies items A1 to A4 of the correspondence table.

We hasten to point out that the term ${}^{(4)}R^{a}_{0c0|\Lambda(t)}\, x^{c}(\partial/\partial x^{a})$ in $(\boldsymbol{a}^{\Phi})_{\parallel}$ and $(\boldsymbol{a}^{\prime})_{\parallel}$ has no classical analogue and may be interpreted as the first order "truly gravitational" acceleration; see Misner, Thorne and Wheeler [1973], p. 1008.

The *dynamics* of relativistic elasticity concerns two basic unknowns: the world-tube Φ and the metric ${}^{(4)}\boldsymbol{g}$. Our objective is to introduce the most fundamental tensors and equations studied by Carter and Quintana [1972], and to point out their nonrelativistic analogues within the context of Chapters 1, 2, and 3 of this book; then, following Carter [1973], we write relativistic perfect elasticity as a classical Lagrangian field theory. To begin the dynamics, we present:

7.8 The Basic Conservation Postulate $\mathcal{B}$ is endowned with a function—the local number density of idealized particles in the reference body. Out of this is constructed a function n on $\mathcal{V}$, which at each event along a particle's world-line represents the local number density as seen in the particle's rest space (the three-dimensional subspace orthogonal to the particle's velocity four-vector).

By *construction*, n satisfies the continuity equation $(nu^{\alpha})_{;\alpha} = 0$, where $\boldsymbol{u}$ is the unit 4-velocity field of the world-tube and ";" denotes covariant differentiation with ${}^{(4)}\nabla$. At a typical event $(x^{a}(t), t)$ along a particle world-line, $\boldsymbol{u}$ has components $u^{a} = dx^{a}/d\tau$, $u^{0} = dt/d\tau$, where τ denotes proper time.

Such a contrived continuity equation is analogous to our conservation of mass postulate. This is easily seen by taking its Newtonian limit in Minkowski spacetime. There the equation reads $(nu^{\alpha})_{,\alpha} = 0$ in Euclidean coordinates, and becomes $\partial n/\partial t + \operatorname{div}(n\boldsymbol{v}) = 0$ upon using the Newtonian approximations

$$u^{0} = \frac{dt}{d\tau} = \frac{1}{\sqrt{1 - v^{2}}} = 1 + \tfrac{1}{2}v^{2} + \tfrac{3}{8}v^{4} + \cdots \cong 1,$$

$$u^a = \frac{dx^a}{d\tau} = \frac{v^a}{\sqrt{1 - v^2}} \cong v^a,$$

where

$$v^a = \frac{dx^a}{dt} \quad \text{and} \quad v(\text{the Newtonian speed}) = \sqrt{\sum_{a=1}^{3} (v^a)^2}.$$

Such approximations are valid only when $v \ll 1$ (1 is the speed of light in geometric units).

To complete the analogy, one only needs to multiply the above equation by μ_0—the typical rest mass per particle—and then identify the quantity $n\mu_0$—the rest mass density as observed in rest spaces—with our time-dependent mass density $\rho(x, t)$.

Next we shall discuss some *fundamental objects of elasticity:* In Table 5.7.2 below, we compare the material and spatial versions of some of the most popular functions and tensors with their relativistic analogues, as given by Carter and Quintana. An intuitive appreciation of the table hinges on one's awareness that our spatial objects correspond, sometimes with the help of an approximation scheme, to the so-called "orthogonal" tensors (i.e., any slot, when contracted with $\boldsymbol{u}$, gives 0). These tensors live on spacetime but only carry the information about the physics that is present in the local rest spaces of the particles. We have already seen this philosophy at work in the case of our ρ versus Carter and Quintana's n (or $n\mu_0$). A slightly more sophisticated example is the following: the spatial version of our right Cauchy–Green tensor, namely, the metric $\boldsymbol{g}$ in our physical space, is the object that determines the distances between neighboring particles; the spacetime object that performs the same duty in the local rest spaces is therefore the standard projection tensor $\boldsymbol{\gamma}$ (with both indices down) where invariantly $\boldsymbol{\gamma}(\boldsymbol{v}, \boldsymbol{w}) = \langle \boldsymbol{v}^\perp, \boldsymbol{w}^\perp \rangle$ ($\boldsymbol{v}$ and $\boldsymbol{w}$ are four-vectors and $\boldsymbol{v}^\perp$, $\boldsymbol{w}^\perp$ denote their components orthogonal to the particle's velocity four-vector; e.g., $\boldsymbol{v}^\perp = \boldsymbol{v} + \langle \boldsymbol{v}, \boldsymbol{u} \rangle \boldsymbol{u}$, where the plus sign is used because $\langle \boldsymbol{u}, \boldsymbol{u} \rangle = -1$) and, in coordinates, $\gamma_{\alpha\beta} = {}^{(4)}g_{\alpha\beta} + u_\alpha u_\beta$. Clearly, $\boldsymbol{\gamma}$ is an orthogonal tensor.

After this preliminary discussion, the table on the next page will hopefully seem believable.

To reinforce our understanding of Table 5.7.2, let us now briefly study *the equations of motion.* Relativistic perfect elasticity is described by the symmetric stress-energy tensor $t^{\alpha\beta} = (n\xi)\, u^\alpha u^\beta + p^{\alpha\beta}$; note that $n\xi$ gives the total energy density and that this $\boldsymbol{t}$ generalizes the case of a relativistic perfect fluid, hence the forementioned name.

The tensor $\boldsymbol{t}$ is to obey the standard conservation law $t^{\alpha\beta}{}_{;\beta} = 0$ provided that external forces (e.g., electromagnetism) are absent. This embodies our balance laws for energy and momentum, and is consistent with taking the Newtonian limit in Minkowski spacetime, as we shall outline below.

As a preliminary remark, note that the orthogonality of $\boldsymbol{p}$ (i.e., $p^{\alpha\beta}u_\beta = 0$) together with the approximations $u_0 \cong -1$, $u_a \cong v_a$ (provided that $v \ll 1$),

Table 5.7.2

	Our Objects: Material	Our Objects: Spatial	Carter–Quintana's Orthogonal Generalizations
Rest-mass density	ρ_{Ref}	ρ	$n\mu_0$ $\quad \mu_0 =$ rest-mass per particle
Deformation, separation between particles	C right Cauchy–Green tensor	g	$\boldsymbol{\gamma} = {}^{(4)}\boldsymbol{g} + \boldsymbol{u} \otimes \boldsymbol{u}$
Equation of state	$E = E(C, G)$ internal energy per unit rest-mass	$e = e(g, c)$	$\xi = \xi(\boldsymbol{\gamma})$ relativistic total energy per particle
Non-relativistic total energy per unit rest-mass	$E + \frac{1}{2}\langle V, V\rangle$	$e + \frac{1}{2}\langle v, v\rangle$	$\dfrac{\xi - \mu_0}{\mu_0}$
Pressure and stress tensors	$S = 2\rho_{\text{Ref}} \dfrac{\partial E}{\partial C}$ second Piola–Kirchhoff stress tensor	$\boldsymbol{\sigma} = 2\rho \dfrac{\partial e}{\partial g}$ Cauchy stress tensor	symmetric pressure tensor $(-\boldsymbol{p}) = 2n \dfrac{\partial \xi}{\partial \boldsymbol{\gamma}} = 2(n\mu_0) \dfrac{\partial}{\partial \boldsymbol{\gamma}}\left(\dfrac{\xi - \mu_0}{\mu_0}\right)$ and is identifiable with $2\rho \dfrac{\partial}{\partial g}(e + \frac{1}{2}\langle v, v\rangle) = \boldsymbol{\sigma} + \rho v \otimes v$
The elasticity tensor and its symmetries	$\mathsf{C} = 4\rho_{\text{Ref}} \dfrac{\partial^2 E}{\partial C\, \partial C}$ C^{ABCD} is symmetric under: $A \leftrightarrow B$, $C \leftrightarrow D$, $AB \leftrightarrow CD$ (hyperelasticity)	$\mathsf{c} = 4\rho \dfrac{\partial^2 e}{\partial g\, \partial g}$ c^{abcd} is symmetric under: $a \leftrightarrow b$, $c \leftrightarrow d$, $ab \leftrightarrow cd$ (hyperelasticity)	$\mathbf{e} = 4n \dfrac{\partial^2 \xi}{\partial \boldsymbol{\gamma}\, \partial \boldsymbol{\gamma}} = 4(n\mu_0) \dfrac{\partial^2}{\partial \boldsymbol{\gamma}\, \partial \boldsymbol{\gamma}}\left(\dfrac{\xi - \mu_0}{\mu_0}\right)$ and is then identifiable with $4\rho \dfrac{\partial^2}{\partial g\, \partial g}(e + \frac{1}{2}\langle v, v\rangle) = 4\rho \dfrac{\partial}{\partial g}\left(\dfrac{\partial e}{\partial g} + \frac{1}{2} v \otimes v\right) = 4\rho \dfrac{\partial^2 e}{\partial g\, \partial g} = \mathsf{c}$ $\mathbf{e}^{\alpha\beta\gamma\delta}$ is symmetric under: $\alpha \leftrightarrow \beta$, $\gamma \leftrightarrow \delta$, $\alpha\beta \leftrightarrow \gamma\delta$ (perfect elasticity)

imply that $p^{a0} \cong p^{ab}v_b$ and $p^{00} \cong p^{ab}v_a v_b$, hence

$$t^{00} \cong n\xi + p^{ab}v_a v_b,$$
$$t^{0a} = t^{a0} \cong n\xi v^a + p^{ab}v_b, \text{ and}$$
$$t^{ab} \cong n\xi v^a v^b + p^{ab}.$$

In flat space using Euclidean coordinates, the four-vector conservation equation reads $t^{\alpha\beta}{}_{,\beta} = 0$. Geometrically, one expects its component along $\boldsymbol{u}$ to yield energy balance and that orthogonal to $\boldsymbol{u}$ to give balance of momentum; in other words, one needs to take suitable linear combinations of the coordinate equations $t^{0\beta}{}_{,\beta} = 0$ and $t^{\alpha\beta}{}_{,\beta} = 0$ in order to give the whole procedure geometric meaning. This can be unnecessarily cumbersome! Fortunately, it turns out that in the $v \ll 1$ limit, $t^{0\beta}{}_{,\beta} = 0$ alone already gives balance of energy, and likewise, $t^{a\beta}{}_{,\beta} = 0$ suffices for balance of momentum. So this is what we will present a brief outline of. An exceedingly crude justification (which borders on being incorrect) for this simplification is that at low speeds (i.e., $v \ll 1$), $\boldsymbol{u}$ is practically parallel to $\partial/\partial t$ and therefore the rest space $\boldsymbol{u}^\perp$ is essentially spanned by $\{\partial/\partial x^1, \partial/\partial x^2, \partial/\partial x^3\}$.

We substituted our approximations of $t^{\alpha\beta}$ into the equations $t^{0\beta}{}_{,\beta} = 0$ and $t^{a\beta}{}_{,\beta} = 0$; then, with the help of particle number conservation (i.e., $\partial n/\partial t + \operatorname{div}(n\boldsymbol{v}) = 0$), the symmetry of $\boldsymbol{p}$, and the material derivative notation $\dot{} = \partial/\partial t + \boldsymbol{v}\cdot\nabla$, the above equations respectively become

$$n\dot{\xi} = -p^{ba}v_{b,a} - p^{ba}{}_{,a}v_b - \frac{\partial}{\partial t}(p^{ab}v_a v_b)$$

and

$$\xi n\dot{v}^a = -p^{ab}{}_{,b} - n\dot{\xi}v^a - \frac{\partial}{\partial t}(p^{ab}v_b).$$

The correct way to reduce these equations to our balance laws is to reexpress all the terms in terms of our classical objects and then do a case study of them in order to decide which terms are negligible compared to the others. This is too ambitious for our purposes. Instead, we shall take advantage of the following controversial but consistent short-cut:

(1) Recover suppressed factors of c (the speed of light) from each term by simply replacing $\partial/\partial t$ by $(1/c)(\partial/\partial t)$, $\boldsymbol{v}$ by $(1/c)\boldsymbol{v}$, and hence the operator $\dot{}$ by $(1/c)\dot{}$; for example, $\dot{v}^a$ becomes $(1/c^2)\dot{v}^a$.

(2) Using our correspondence table, reexpress the equations in terms of our classical objects. Straightforward examples are $(-\boldsymbol{p}) \longrightarrow \boldsymbol{\sigma}$ and

$$n\dot{\xi} \longrightarrow \frac{1}{c}n\dot{\xi} = \frac{1}{c}(n\mu_0)\left(\frac{\xi - \mu_0}{\mu_0}\right)^{\cdot}$$
$$\longrightarrow \frac{1}{c}\rho(e + \tfrac{1}{2}\langle\!\langle v, v\rangle\!\rangle)^{\cdot} = \frac{1}{c}\rho\dot{e} + \frac{1}{c}\rho\dot{v}^b v_b.$$

A not so straightforward example would be

$$\xi n\dot{v}^a \longrightarrow \frac{1}{c^2}\xi n\dot{v}^a = \left(\frac{\text{energy density}}{c^2}\right)\dot{v}^a = \left(\frac{\text{rest-mass density}}{\sqrt{1-(v/c)^2}}\right)\dot{v}^a$$

$$\cong (\text{rest-mass density})\dot{v}^a = \rho\dot{v}^a,$$

where the superficial factor $1/c^2$ was canceled by using the famous "$E = mc^2$" equation.

(3) In each equation, keep only terms of lowest order in $1/c$. Equivalently, cancel off common factors of $1/c$ and then let c tend to infinity, assuming that all classical objects and their derivatives are bounded.

Note that steps (1) and (2) modify our equations to read

$$\frac{1}{c}\rho\dot{e} + \frac{1}{c}\rho\dot{v}^b v_b = \frac{1}{c}\sigma^{ba}v_{b,a} + \frac{1}{c}\sigma^{ba}{}_{,a}v_b + \frac{1}{c^3}\frac{\partial}{\partial t}(\sigma^{ab}v_a v_b)$$

$$\rho\dot{v}^a = \sigma^{ab}{}_{,b} - \left(\frac{1}{c}\rho\dot{e} + \frac{1}{c}\rho\dot{v}^b v_b\right)\frac{v^a}{c} + \frac{1}{c^2}\frac{\partial}{\partial t}(\sigma^{ab}v_b).$$

Now step (3) simplifies these to

$$\rho\dot{e} + \rho\dot{v}^b v_b = \sigma^{ba}v_{b,a} + \sigma^{ba}{}_{,a}v_b$$

$$\rho\dot{v}^a = \sigma^{ab}{}_{,b}.$$

The second equation is already balance of momentum; substituting it into the first equation, we obtain $\rho\dot{e} = \sigma^{ba}v_{b,a}$, which is balance of energy.

Finally, we will outline how one could view *relativistic perfect elasticity as a Lagrangian field theory*. The basic "fields" are Φ—the world-tube—and ${}^{(4)}\boldsymbol{g}$. An important unknown $\boldsymbol{\gamma} = {}^{(4)}\boldsymbol{g} + \boldsymbol{u} \otimes \boldsymbol{u}$ is formed from them; this is the generalization of the nonrelativistic right Cauchy–Green tensor.

The particle number density is specified materially and its spatial analogue $n = n(\boldsymbol{\gamma})$ is to obey the conservation law $(nu^\alpha)_{;\alpha} = 0$.

As a constitutive assumption, one proposes an equation of state $\xi = \xi(\boldsymbol{\gamma})$ for the energy ξ, computes from it the pressure tensor $-\boldsymbol{p} = 2n(\partial\xi/\partial\boldsymbol{\gamma})$, and specifies the stress-energy tensor as

$$\boldsymbol{t} = (n\xi)\boldsymbol{u} \otimes \boldsymbol{u} + \boldsymbol{p}.$$

Varying the action $\int_{\mathcal{V}} ({}^{(4)}R - 16\pi n\xi)\sqrt{-{}^{(4)}g}\, d^4x$ with respect to the basic fields Φ and ${}^{(4)}\boldsymbol{g}$, one obtains the Euler–Lagrange equations

$$\boxed{\operatorname{Ein}({}^{(4)}\boldsymbol{g}) = 8\pi\boldsymbol{t} \quad \text{and} \quad t^{\alpha\beta}{}_{;\beta} = 0.}$$

Here ${}^{(4)}R$ is the scalar curvature of the four metric ${}^{(4)}\boldsymbol{g}$ and $\operatorname{Ein}({}^{(4)}\boldsymbol{g})_{\mu\nu} = R_{\mu\nu} - \frac{1}{2}Rg_{\mu\nu} = G_{\mu\nu}$ is the Einstein tensor of ${}^{(4)}\boldsymbol{g}$. These are the basic field equations of the coupled system—general relativity and perfect elasticity. The second set of equations $t^{\alpha\beta}{}_{;\beta} = 0$ follows from the first using the Bianchi identities. The Hamiltonian structure of this coupled system can be obtained by combining the methods of this chapter with those of general relativity in, for instance, Fischer and Marsden [1979a].

In relativity the stress-energy-momentum tensor is usually taken to be the derivative of the matter Lagrangian with respect to the spacetime metric $^{(4)}\boldsymbol{g}$, as in Misner-Thorne and Wheeler [1973], p. 491–504 and Hawking and Ellis [1973], p. 66. Let us check that this is the case here. Namely, we claim that with $\boldsymbol{t}$ as defined above,

$$\boxed{\boldsymbol{t} = \frac{2}{\sqrt{-{}^{(4)}g}} \frac{\partial(-n\xi\sqrt{-{}^{(4)}g})}{\partial {}^{(4)}\boldsymbol{g}}}$$

whereas the pressure tensor can be written

$$\boldsymbol{p} = 2n\frac{\partial \xi}{\partial {}^{(4)}\boldsymbol{g}}$$

It is the latter which is analogous to $\boldsymbol{\sigma} = 2\rho(\partial e/\partial \boldsymbol{g})$.

To prove our claim, note that

$$\frac{2}{\sqrt{-{}^{(4)}g}} \frac{\partial(-n\xi\sqrt{-{}^{(4)}g})}{\partial {}^{(4)}g_{\alpha\beta}}$$

$$= \frac{-2}{\sqrt{-{}^{(4)}g}}\left[\frac{\partial n}{\partial {}^{(4)}g_{\alpha\beta}}\xi\sqrt{-{}^{(4)}g} + n\frac{\partial \xi}{\partial {}^{(4)}g_{\alpha\beta}}\sqrt{-{}^{(4)}g} + n\xi\frac{\partial\sqrt{-{}^{(4)}g}}{\partial {}^{(4)}g_{\alpha\beta}}\right]$$

now both n and ξ depend on $^{(4)}\boldsymbol{g}$ through $\boldsymbol{\gamma} = {}^{(4)}\boldsymbol{g} + \boldsymbol{u} \otimes \boldsymbol{u}$, so

$$\frac{\partial n}{\partial {}^{(4)}\boldsymbol{g}} = \frac{\partial n}{\partial \boldsymbol{\gamma}}\frac{\partial \boldsymbol{\gamma}}{\partial {}^{(4)}\boldsymbol{g}} = \frac{\partial n}{\partial \boldsymbol{\gamma}}\cdot \boldsymbol{I} = \frac{\partial n}{\partial \boldsymbol{\gamma}}$$

and likewise, $\partial\xi/\partial {}^{(4)}\boldsymbol{g} = \partial \xi/\partial \boldsymbol{\gamma}$; also,

$$\frac{\partial\sqrt{-{}^{(4)}g}}{\partial {}^{(4)}g_{\alpha\beta}} = \tfrac{1}{2}\sqrt{-{}^{(4)}g}\,{}^{(4)}g^{\alpha\beta} = \tfrac{1}{2}\sqrt{-{}^{(4)}g}(\gamma^{\alpha\beta} - u^\alpha u^\beta);$$

using these, our expression becomes

$$-2\left[\frac{\partial n}{\partial \gamma_{\alpha\beta}}\xi + n\frac{\partial \xi}{\partial \gamma_{\alpha\beta}} + \frac{n\xi}{2}(\gamma^{\alpha\beta} - u^\alpha u^\beta)\right].$$

Finally, one learns from Carter [1973] that $\partial n/\partial \gamma_{\alpha\beta} = -\frac{1}{2}n\gamma^{\alpha\beta}$, hence the above expression has an internal cancellation, reducing it to $-2[n(\partial\xi/\partial\gamma_{\alpha\beta}) - (1/2)\, n\xi u^\alpha u^\beta]$ which is $(n\xi)\,\boldsymbol{u}\otimes\boldsymbol{u} - 2n(\partial\xi/\partial\boldsymbol{\gamma}) = \boldsymbol{t}$ indeed.

Box 7.2 ***Summary of Important Formulas for Section 5.7***

World Tube

$\Phi\colon \mathcal{B} \times \mathbb{R} \to \mathcal{V}$ ($\mathcal{V}$ = spacetime)

Slicing

$i\colon \mathcal{S} \times \mathbb{R} \to \mathcal{V}$

Motion

$$\Phi(X, t) = i(\phi^t(X, t), t)$$

Velocity

$$V^\Phi = \frac{\partial \Phi}{\partial t}, \quad (V^\Phi)^\alpha = \frac{\partial \Phi^\alpha}{\partial t}$$

Material Frame Velocity

$$V^i = \frac{\partial i}{\partial t}, \quad (V^i)^\alpha = \frac{\partial i^\alpha}{\partial t}$$

Material Apparent Velocity

$$V_t^{\Phi,i} = \frac{\partial \phi^i}{\partial t}$$

Material Acceleration

$$A^\Phi = \frac{{}^{(4)}D}{dt} V^\Phi, \quad (A^\Phi)^\alpha = \frac{\partial (V^\Phi)^\alpha}{\partial t} + \gamma^\alpha_{\beta\gamma}(V^\Phi)^\beta (V^\Phi)^\gamma$$

Material Frame Acceleration

$$A^i = \frac{{}^{(4)}Dv^i}{dt}$$

Material Apparent Acceleration

$$A_t^{\Phi,i} = \frac{{}^{(3)}DV^{\Phi,i}}{dt}$$

Spatial Quantities

$\boldsymbol{v}, \boldsymbol{a}$; compose material ones with Φ^{-1} or $(\phi^i)^{-1}$.

Absolute vs. Apparent Acceleration

$$i_{t*}\boldsymbol{a}_t^{\Phi,i} = \boldsymbol{a}^\Phi - \boldsymbol{a}^i - 2\,{}^{(4)}\nabla_{(v^\Phi - v^i)}\boldsymbol{v}^i - \boldsymbol{k}_t(\boldsymbol{v}^\Phi - \boldsymbol{v}^i, \boldsymbol{v}^\Phi - \boldsymbol{v}^i)\boldsymbol{n}_t$$

Change of Framing

$$\phi_t^j \circ (\phi_t^i)^{-1} = \xi_t$$

$$\boldsymbol{a}_t^{\Phi,j} = \boldsymbol{a}_t^\xi + \xi_{t*}\boldsymbol{a}_t^{\Phi,i} + 2\nabla^t_{\mathrm{app}\,\xi_{t*}v_t^{\Phi,i}}\,\boldsymbol{v}_t^\xi$$

Correspondence with Newtonian Mechanics

See Table 5.7.1

Relativistic vs. Non-Relativistic

Correspondence between stress, energy, and so on; see Table 5.7.2.

Stress-energy tensor

$$\boldsymbol{t} = n\xi \boldsymbol{u} \otimes \boldsymbol{u} + \boldsymbol{p} = \frac{2}{\sqrt{-{}^{(4)}g}} \frac{\partial}{\partial\, {}^{(4)}g}[-n\xi\sqrt{-{}^{(4)}g}]$$

Action for Relativistic Elasticity

$$\int_{\mathcal{V}} ({}^{(4)}R - 16\pi n\xi)\sqrt{-{}^{(4)}g}\, d^4x$$

Field Equations

$$\mathrm{Ein}({}^{(4)}g) = 8\pi \boldsymbol{t}, \quad G_{\alpha\beta} = 8\pi t_{\alpha\beta}$$

$$\mathrm{div}\, \boldsymbol{t} = 0, \quad t^{\alpha\beta}{}_{;\beta} = 0$$

CHAPTER 6

METHODS OF FUNCTIONAL ANALYSIS IN ELASTICITY

The purpose of this chapter is to present some basic theorems in elasticity concerning the boundary value problems of elastostatics and elastodynamics. The techniques are primarily those of linear and nonlinear functional analysis. The first five sections are motivated by questions of existence and uniqueness, but the results turn out to bear on basic questions such as: what constitutive inequalities should one impose? Section 6.6 discusses what is currently known (to the authors) concerning a longstanding problem: are minima of the energy stable? This turns out to be a very subtle yet significant point. The final section gives an application of nonlinear analysis to a control problem for a beam as a sample of how the general machinery can be used in a problem arising from nonlinear elasticity.

This chapter is not intended to be comprehensive. Some of the important topics omitted are the methods of variational inequalities (see Duvaut and Lions [1972]) and, except for a few illustrative examples, the existence theory for rods, plates, and shells. The topics omitted include both the approximate models such as the von Karmen equations and the full nonlinear models (see, for example, Ciarlet [1983], Berger [1977], Antman [1978a], [1979b], [1980c], and references therein).

6.1 ELLIPTIC OPERATORS AND LINEAR ELASTOSTATICS

This section discusses existence and uniqueness for linear elastostatics. The methods used are based on elliptic theory. Basic results in this subject are stated without proof; for these proofs, the reader should consult, for example, Agmon

[1965], Friedman [1969], Morrey [1966], or Wells [1980]. These results are then applied to linear elastostatics. The methods and emphasis here differ slightly from those in the important reference work of Fichera [1972a].

To simplify notation and the spaces involved, we shall explicitly assume the body $\mathcal{B}$ is a bounded open set Ω in $\mathbb{R}^n$ with piecewise smooth boundary. The reader is forewarned that standard usage and a number of notational conflicts have resulted in us adopting various notations for the volume element throughout this chapter.

The linearized equations under consideration are as follows (see Chapter 4):

$$\operatorname{div}(\mathbf{a}\cdot\nabla\boldsymbol{u})(x) = \boldsymbol{f}(x), \qquad x \in \Omega, \tag{1}$$

where $\boldsymbol{f} = -\rho\boldsymbol{b} - \operatorname{div}\boldsymbol{\sigma}$, $\boldsymbol{b}$ is the body force, $\boldsymbol{\sigma}$ is the Cauchy stress in the configuration we are linearizing about, and $\mathbf{a} = \boldsymbol{\sigma} \otimes \boldsymbol{\delta} + \mathbf{c}$ is the corresponding elasticity tensor; we assume $\mathbf{a}$ is C^∞. We assume boundary conditions of displacement or traction are used and shall assume them to be homogeneous. [This is no real loss of generality, for if the boundary conditions are not homogeneous, say $\boldsymbol{u}$ equals some displacement $\tilde{\boldsymbol{u}}$ on $\partial\Omega$, replace $\boldsymbol{f}$ by $\boldsymbol{f} + \operatorname{div}(\mathbf{a}\cdot\nabla\tilde{\boldsymbol{u}})$ and $\boldsymbol{u}$ by $\boldsymbol{u} - \tilde{\boldsymbol{u}}$.]

1.1 Definitions Let L^2 denote the Hilbert space of all displacements $\boldsymbol{u}$: $\Omega \longrightarrow \mathbb{R}^n$ that are square integrable:

$$\|\boldsymbol{u}\|_{L^2}^2 = \int_\Omega \|\boldsymbol{u}\|^2\, dv < \infty.$$

Let H^1 denote the Hilbert space of all displacements $\boldsymbol{u}\colon \Omega \longrightarrow \mathbb{R}^n$ that belong to L^2 and such that the gradient $\nabla\boldsymbol{u}$ (initially only a distribution) is also an L^2 tensor. The H^1-norm is defined by

$$\|\boldsymbol{u}\|_{H^1} = \left(\int_\Omega \|\boldsymbol{u}\|^2\, dv + \int_\Omega \|\nabla\boldsymbol{u}\|^2\, dv\right)^{1/2}.$$

Similarly, define H^s for positive integers s, with the convention $H^0 = L^2$. (One can show that H^1 is a Hilbert space: basic facts about these *Sobolev spaces* will be given shortly without proof; we shall not require these detailed proofs, which are contained in the references cited earlier and Adams [1975]. See Box 1.1 for further information and some sample proofs.)

If p is a real number $(1 \leq p \leq \infty)$, L^p denotes the space of displacements $\boldsymbol{u}$ such that

$$\|\boldsymbol{u}\|_{L^p} = \left(\int_\Omega \|\boldsymbol{u}\|^p\, dv\right)^{1/p} < \infty$$

(with $\|\boldsymbol{u}\|_{L^\infty} = \operatorname{ess\,sup}_{x\in\Omega}\|\boldsymbol{u}(x)\|$). Define $W^{1,p}$ to be the space of $\boldsymbol{u} \in L^p$ such that $\nabla\boldsymbol{u}$ is L^p as well, with norm

$$\|\boldsymbol{u}\|_{W^{1,p}} = (\|\boldsymbol{u}\|_{L^p}^p + \|\nabla\boldsymbol{u}\|_{L^p}^p)^{1/p}.$$

Similarly, define the Sobolev spaces $W^{s,p}$ for s a positive integer.[1] In this

---

[1]For s not an integer, these may still be defined by means of the Fourier transform; see Box 1.1.

notation, we have the following coincidences:

$$H^0 = L^2, \qquad W^{s,2} = H^s.$$

Let H^2_∂ denote the set of $\boldsymbol{u} \in H^2$ with homogeneous boundary conditions of displacement or traction (or both) imposed; that is,

$$\boldsymbol{u} = \mathbf{0} \quad \text{on} \quad \partial_d \quad \text{and} \quad (\mathbf{a} \cdot \nabla \boldsymbol{u}) \cdot \boldsymbol{n} = 0 \quad \text{on} \quad \partial_\tau$$

(where one of ∂_d or ∂_τ may be empty). (One has to show that H^2_∂ is a well-defined closed subspace of H^2; this is true and relies on the so-called "trace theorems"; cf. Adams [1975], Section 5.22 and p. 330 below.)

Define the linear operator

$$A\colon H^2_\partial \longrightarrow L^2 \quad \text{by} \quad A(\boldsymbol{u}) = \operatorname{div}(\mathbf{a} \cdot \nabla \boldsymbol{u}). \tag{2}$$

Problem (1) can be phrased as follows: *given $f \in L^2$, can we solve $A\boldsymbol{u} = f$?*

Throughout this section, we shall make the following:

1.2 Assumptions (i) The elasticity tensor $\mathbf{a}$ is *hyperelastic;* that is,

$$\mathrm{a}^{abcd} = \mathrm{a}^{cdab}.$$

(ii) The elasticity tensor $\mathbf{a}$ is *strongly elliptic;* that is, there is an $\epsilon > 0$ such that

$$\mathrm{a}^{abcd}(x)\xi_a\xi_c\eta_b\eta_d \geq \epsilon \|\boldsymbol{\xi}\|^2 \|\boldsymbol{\eta}\|^2$$

for all vectors $\boldsymbol{\xi}, \boldsymbol{\eta} \in \mathbb{R}^n$ and all $x \in \Omega$.

1.3 Proposition *Assumption* (i) *is equivalent to symmetry of A; that is,*

$$\langle A\boldsymbol{u}, \boldsymbol{v} \rangle = \langle \boldsymbol{u}, A\boldsymbol{v} \rangle$$

for $\boldsymbol{u}, \boldsymbol{v} \in H^2_\partial$, where $\langle\ ,\ \rangle$ is the L^2 inner product.

Proof Assume (i). Then

$$\langle A\boldsymbol{u}, \boldsymbol{v} \rangle = \int_\Omega (\mathrm{a}^{abcd} u_{c|d})_{|b} v_a \, dv.$$

We can integrate by parts noting that the boundary terms are zero, giving the following:

$$\langle A\boldsymbol{u}, \boldsymbol{v} \rangle = -\int_\Omega \mathrm{a}^{abcd} u_{c|d} v_{a|b} \, dv.$$

(This requires some further justification since $\boldsymbol{u}$ and $\boldsymbol{v}$ are not smooth; this is done by an approximation argument which is omitted.) This expression is symmetric in $\boldsymbol{u}$ and $\boldsymbol{v}$. The converse is left as an exercise. ∎

1.4 Definition The mapping B from $H^1 \times H^1$ to $\mathbb{R}$ defined by

$$B(\boldsymbol{u}, \boldsymbol{v}) = \int_\Omega \mathrm{a}^{abcd} u_{c|d} v_{a|b} \, dv \tag{3}$$

(which is a continuous symmetric bilinear form) is called the *Dirichlet form.*

The next proposition, discussed in Box 1.1, relates the Dirichlet form and strong ellipticity.

1.5 Proposition *Let A be defined as above, and let* (*i*) *of 1.2 hold. Then strong ellipticity is equivalent to* Gårding's inequality:

$$\begin{cases} \text{There are constants } c > 0 \text{ and } d > 0 \text{ such that} \\ \qquad B(\boldsymbol{u}, \boldsymbol{u}) \geq c\,\|\boldsymbol{u}\|^2_{H^1} - d\,\|\boldsymbol{u}\|^2_{L^2} \\ \text{for all } \boldsymbol{u} \in H^1. \end{cases}$$

Problem 1.1 Use Gårding's inequality and the inequality $2ab \leq \epsilon a^2 + (1/\epsilon)b^2$ to show that $-\langle A\boldsymbol{u}, \boldsymbol{u}\rangle \geq c_1\|\boldsymbol{u}\|^2_{H^1} - d_1\|\boldsymbol{u}\|^2_{L^2}$ for all $\boldsymbol{u} \in H^2_\partial$, for suitable positive constants c_1 and d_1.

There are some technical difficulties with the mixed problem that will force us to make hypotheses on ∂_d and ∂_τ that are stronger than one would ideally like. Specifically, for the validity of certain theorems, we shall need to assume that the *closures* of ∂_d and ∂_τ do not intersect. This means, in effect, that ∂_d and ∂_τ do not touch. An example of something that *is* allowed by this is when $\mathcal{B}$ is an annulus, with ∂_d the outer boundary and ∂_τ the inner boundary. For the pure displacement or traction problem, we likewise require $\partial\mathcal{B}$ to be smooth, say C^1. Again this is an unpleasant assumption, for it eliminates bodies with corners.

The assumptions in the previous paragraph may well be just technical and not necessary, but to eliminate them would require nontrivial modifications of what follows. The difficulty lies with the *regularity* of solutions; for example, if one is solving $\Delta\phi = f$ with Dirichlet boundary conditions and $\partial\mathcal{B}$ has corners, the issue of whether or not ϕ is in H^2 when f is in L^2 is delicate. If $\mathcal{B}$ is a square or a cube, this is true and can be seen by using Fourier series. If the angles are not 90°, however, this assertion need not be true in Sobolev spaces, but analogous results may be true in other spaces. As far as we know, this theory has not yet been developed, except for isolated cases. (See also the remarks on p. 371.)

For both the linear and nonlinear theories, these problems do not really affect the existence question if one seeks a generalized solution with less regularity. However, if regularity is desired, or if the inverse function theorem is to be used, the above assumptions must be made, or else the spaces must be modified. The situation calls for further research to see if these difficulties are just technical or are of physical interest. Presumably the modifications in the spaces needed correspond to known asymptotic solutions near corners, cracks, etc.

For the rest of this section we shall make the assumptions of 1.2 and those on $\partial\mathcal{B}$ above without explicit mention.

1.6 Basic Facts about Elliptic Operators

(i) *Elliptic estimates.* If $\boldsymbol{u} \in H^2_\partial$, then for each $s \geq 2$ there is a constant K such that

$$\|\boldsymbol{u}\|_{H^s} \leq K(\|A\boldsymbol{u}\|_{H^{s-2}} + \|\boldsymbol{u}\|_{L^2}),$$

More generally, if $1 < p < \infty$,

$$\|u\|_{W^{s,p}} \leq K(\|Au\|_{W^{s-2,p}} + \|u\|_{L^p}).$$

The proof is omitted. See Box 1.1 for the proof in a simple case.

(ii) *The kernel of A is finite dimensional.* [*Proof:* The elliptic estimates show that $\|u\|_{H^2} \leq C\|u\|_{L^2}$ on Ker A. Rellich's theorem, which states that the inclusion $H^s \longrightarrow H^r$ $(s > r)$ is compact, then implies that the unit ball in Ker A is compact. Hence it is finite dimensional.[2]] We also get $\|u\|_{H^s} \leq C\|u\|_{L^2}$ so $H^\infty = \bigcap_{s \geq 0} H^s \subset$ Ker A. Since $H^s \subset C^k$ if $s > n/2 + k$ (see Box 1.1), $H^\infty = C^\infty$, so *elements of* Ker A *are smooth* in this case. Similarly, *the spectrum of A is discrete* and *each eigenvalue has finite multiplicity.*

(iii) *The range of A is closed in* L^2. The outline of the proof of this for those knowing some functional analysis requires the following facts:

(1) Let T be a closed linear operator in a Banach space $\mathfrak{X}$ with domain $\mathfrak{D}(T)$ (T closed means its graph is closed). Then if T is continuous, $\mathfrak{D}(T)$ is closed.

This follows from the definition.

(2) Let $\mathfrak{X}$ and $\mathfrak{Y}$ be Banach spaces, $C: \mathfrak{X} \longrightarrow \mathfrak{Y}$ a $1-1$ continuous linear map with closed range, and $B: \mathfrak{X} \longrightarrow \mathfrak{Y}$ a compact linear operator. Suppose $C + B$ is $1 - 1$. Then $C + B$ has closed range.

Proof The operator $(C + B)^{-1}$ defined on Range$(C + B)$ is closed since the inverse of a $1 - 1$ closed operator is closed (consider the graphs). Thus, by (1), it suffices to show that $(C + B)^{-1}$ is continuous. Suppose that $(C + B)^{-1}y_n = x_n$ and $y_n \longrightarrow 0$. Suppose $x_n \not\rightarrow 0$. By passing to a subsequence, we can suppose $\|x_n\| \geq \epsilon > 0$. Let $\bar{x}_n = x_n/\|x_n\|$. Then

$$\|(C + B)(\bar{x}_n)\| = \frac{1}{\|x_n\|}\|(C + B)x_n\| = \frac{\|y_n\|}{\|x_n\|} \leq \frac{1}{\epsilon}\|y_n\|,$$

so $(C + B)(\bar{x}_n) \longrightarrow 0$. Since B is compact and $\|\bar{x}_n\| = 1$, we can suppose $B(\bar{x}_n)$ converges. Thus $C(\bar{x}_n)$ converges too. Since C has closed range, $C(\bar{x}_n) \longrightarrow C\bar{x}$ for some $\bar{x} \in \mathfrak{X}$. By the closed graph theorem, C has a bounded inverse on its range. Thus $\bar{x}_n \longrightarrow \bar{x}$, so $\|\bar{x}\| = 1$. But $(C + B)(\bar{x}_n) \longrightarrow 0$ and so $(C + B)(\bar{x}) = 0$ and thus $\bar{x} = 0$, a contradiction.

(3) Now, let $\tilde{\mathfrak{X}} = H^2_\partial$, $\mathfrak{Y} = L^2 \times L^2$, $Cu = (Au, u)$, and $Bu = (0, -u)$. Let $\mathfrak{X}$ be the Hilbert space orthogonal complement of ker $(C + B)$ in $\tilde{\mathfrak{X}}$, and restrict C and B to $\mathfrak{X}$. By construction, $C + B$ is $1 - 1$. Clearly C is $1 - 1$; moreover, using the sum norm on $\mathfrak{Y}$,

$$\|Cu\|_{\mathfrak{Y}} = \|Au\|_{L^2} + \|u\|_{L^2} > \alpha\|u\|_{\mathfrak{X}}$$

by the elliptic estimate (i) with $\alpha = 1/K$. This estimate shows directly that the

[2]A Banach space is finite dimensional if and only if its unit ball is compact (another theorem of Rellich).

range of C is closed. Finally, B is compact by Rellich's theorem. Thus, by (2), $(C+B)(u) = (Au, 0)$ has closed range. Hence A has closed range. ∎

(iv) *Weak solutions are strong solutions.* Suppose $\boldsymbol{u} \in L^2, \boldsymbol{f} \in L^2$, and $\langle \boldsymbol{u}, A\boldsymbol{v}\rangle = \langle \boldsymbol{f}, \boldsymbol{v}\rangle$ for all $\boldsymbol{v} \in H^2_\partial$. Then $\boldsymbol{u} \in H^2_\partial$ and $A\boldsymbol{u} = \boldsymbol{f}$.

Proof We sketch the proof of (iv) as follows: This is a *regularity* result. The statement $\langle \boldsymbol{u}, A\boldsymbol{v}\rangle = \langle \boldsymbol{f}, \boldsymbol{v}\rangle$ implies that $A\boldsymbol{u} = \boldsymbol{f}$ in the sense of distributions, so $A\boldsymbol{u} \in L^2$. Thus, as in the basic elliptic estimate (i), $\boldsymbol{u} \in H^2$. In fact, the most delicate part here is that $\boldsymbol{u}$ is H^2 near the boundary, a relatively deep fact. That $\boldsymbol{u}$ must satisfy the boundary conditions comes about *formally* as follows. Integrating by parts,

$$\begin{aligned}\langle \boldsymbol{u}, A\boldsymbol{v}\rangle &= \int_\Omega \boldsymbol{u}\cdot \operatorname{div}(\mathbf{a}\cdot\nabla \boldsymbol{v})\, dv \\ &= -\int_\Omega \nabla \boldsymbol{u}\cdot\mathbf{a}\cdot\nabla\boldsymbol{v}\, dv + \int_{\partial\Omega} \boldsymbol{u}\cdot[\mathbf{a}\cdot\nabla\boldsymbol{v}\cdot\boldsymbol{n}]\, da \\ &= \int_\Omega \operatorname{div}(\mathbf{a}\cdot\nabla\boldsymbol{u})\boldsymbol{v}\, dv + \int_{\partial\Omega} \{\boldsymbol{u}\cdot[\mathbf{a}\cdot\nabla\boldsymbol{v}\cdot\boldsymbol{n}] - \boldsymbol{v}\cdot[\mathbf{a}\cdot\nabla\boldsymbol{u}\cdot\boldsymbol{n}]\}\, da.\end{aligned}$$

Suppose the boundary conditions are $\boldsymbol{v} = \mathbf{0}$ on $\partial\Omega$. If this is to equal $\langle \boldsymbol{f}, \boldsymbol{v}\rangle$ for all $\boldsymbol{v} \in H^2_\partial$, then first take $\boldsymbol{v}$ of compact support in Ω to get $\operatorname{div}(\mathbf{a}\cdot\nabla\boldsymbol{u}) = \boldsymbol{f}$. Thus, we are left with the identity

$$\int_{\partial\Omega} \boldsymbol{u}\cdot[\mathbf{a}\cdot\nabla\boldsymbol{v}\cdot\boldsymbol{n}]\, da = 0.$$

Since no condition is imposed on $\mathbf{a}\cdot[\nabla\boldsymbol{v}\cdot\boldsymbol{n}]$ on $\partial\Omega$, $\boldsymbol{u}$ must vanish on $\partial\Omega$. A similar argument holds for the traction or mixed case. ∎

From these facts we can deduce the following crucial result for symmetric elliptic operators:

1.7 Fredholm Alternative Theorem $L^2 = \text{Range } A \oplus \text{Ker } A$, *an* L^2 *orthogonal sum.*

Proof By 1.6(iii), Range A is closed. Let $\mathfrak{N}$ be its L^2 orthogonal complement. We claim that $\mathfrak{N} = \text{Ker } A$. This will prove the result. First of all, if $\boldsymbol{u} \in \text{Ker } A$ and $\boldsymbol{f} = A\boldsymbol{v} \in \text{Range } A$, then as A is symmetric, $\langle \boldsymbol{u}, \boldsymbol{f}\rangle = \langle \boldsymbol{u}, A\boldsymbol{v}\rangle = \langle A\boldsymbol{u}, \boldsymbol{v}\rangle = 0$. Conversely, if $\boldsymbol{u} \in \mathfrak{N}$, then $\langle \boldsymbol{u}, A\boldsymbol{v}\rangle = 0$ for all $\boldsymbol{v} \in H^2_\partial$. Thus by 1.6(iv) $\boldsymbol{u} \in H^2_\partial$ and $A\boldsymbol{u} = \mathbf{0}$; that is, $\boldsymbol{u} \in \text{Ker } A$. ∎

Remarks From the elliptic estimates, the decomposition in 1.7 has a regularity property:

$$W^{s,p} = \text{Range}(A\,|\,H^2_\partial \cap W^{s+2,p}) \oplus \text{Ker } A.$$

For non-symmetric elliptic operators, the Fredholm alternative reads

$$L^2 = \text{Range } A \oplus \text{Ker } A^*,$$

where A^* is the adjoint of A.

Problem 1.2 Use the Fredholm alternative applied to $\lambda I - A$ and Gårding's inequality to show that $\lambda I - A\colon H^2_\partial \to L^2$ is an isomorphism, where $\lambda > d_1$, with d_1 given in Problem 1.1.

From 1.7 we obtain the following main result for our boundary value problem.

1.8 Theorem *Let $f \in L^2$. Then there exists a $\boldsymbol{u} \in H^2_\partial$ such that $A\boldsymbol{u} = f$ if and only if*

$$\langle f, \boldsymbol{h} \rangle = 0 \qquad \textit{for all } \boldsymbol{h} \in \operatorname{Ker} A.$$

In this case, (i) $\boldsymbol{u}$ is unique up to addition of elements in Ker A, *and (ii) if f is of class $W^{s,p}$, $\boldsymbol{u}$ is of class $W^{s+2,p}$ (up to and including the boundary).*

Theorem 1.8 gives complete information on when the boundary value problem

$$A\boldsymbol{u} = f, \qquad \boldsymbol{u} \in H^2_\partial$$

is solvable for $\boldsymbol{u}$. To be useful, one must be able to compute Ker A in a specific instance. We show how this may be done for the important case of stable classical elasticity (see Section 4.3). Thus we now deal with the case $A\boldsymbol{u} = \operatorname{div}(\mathbf{c}\cdot\nabla\boldsymbol{u})$ on $\mathcal{B} = \Omega \subset \mathbb{R}^3$.

1.9 Definition Let $\mathbf{c}^{abcd}$ be a classical elasticity tensor on Ω. We say $\mathbf{c}$ is *uniformly pointwise stable* if there is an $\eta > 0$ such that

$$\epsilon = \tfrac{1}{2}\boldsymbol{e}\cdot\mathbf{c}\cdot\boldsymbol{e} \geq \eta \|\boldsymbol{e}\|^2$$

for all symmetric $\boldsymbol{e}$ (see 3.5 and 3.6 in Section 4.3).

The argument of 3.9, Section 4.3 shows that *uniform pointwise stability implies strong ellipticity, but not conversely.*

1.10 Lemma *If $\mathbf{c}$ is uniformly pointwise stable and if $\boldsymbol{u} \in$* Ker A, *then $\boldsymbol{u}$ is an infinitesimal Euclidean motion; i.e. a rotation or translation.*

Proof If $\boldsymbol{u} \in$ Ker A, then $\operatorname{div}(\mathbf{c}\cdot\nabla\boldsymbol{u}) = \mathbf{0}$. Thus, multiplying by $\boldsymbol{u}$ and integrating,

$$0 = \int_\Omega \boldsymbol{u}\cdot\operatorname{div}(\mathbf{c}\cdot\nabla\boldsymbol{u})\,dv = -\int_\Omega \nabla\boldsymbol{u}\cdot\mathbf{c}\cdot\nabla\boldsymbol{u}\,dv.$$

Here we integrated by parts using the boundary conditions. By $\mathbf{c}$'s symmetries, $\nabla\boldsymbol{u}\cdot\mathbf{c}\cdot\nabla\boldsymbol{u} = \boldsymbol{e}\cdot\mathbf{c}\cdot\boldsymbol{e}$, where $\boldsymbol{e} = \frac{1}{2}\mathfrak{L}_{\boldsymbol{u}}\boldsymbol{g}$; that is, $e_{ab} = \frac{1}{2}(u_{a|b} + u_{b|a})$. By 1.9 we conclude that $\boldsymbol{e} = \mathbf{0}$, so $\boldsymbol{u}$ is an infinitesimal rotation or translation. ∎

For the displacement or mixed problem, any infinitesimal rotation must vanish since it vanishes on a portion of $\partial\Omega$ containing three linearly independent points. For the traction problem any infinitesimal rotation or translation lies in Ker A. Thus, we have proved the following:

1.11 Theorem *For classical elasticity, assume* **c** *is uniformly pointwise stable, and the hypotheses on the boundary conditions preceding 1.6 hold. Then*

(i) *For displacement or mixed boundary conditions and any $f \in L^2$, the problem* $\mathrm{div}(\mathbf{c}\cdot\nabla u) = f$, *has a unique solution $u \in H^2_\partial$. If $f \in W^{s,p}$, then $u \in W^{s+2,p}$ for $s \geq 0$, $1 < p < \infty$.*

(ii) *(Traction Problem) The equation*

$$\mathrm{div}(\mathbf{c}\cdot\nabla u) = f, \qquad f \in L^2$$

with $\mathbf{c}\cdot\nabla u = \mathbf{0}$ *on $\partial\Omega$ has a solution $u \in H^2_\partial$ if and only if*

$$\int_\Omega f(x)\cdot(a + bx)\, dv = 0,$$

where a is any constant vector and b is any skew-symmetric matrix. If this holds, u is unique up to the addition of a term of the form $a + bx$, b a skew matrix. If $f \in W^{s,p}$ ($s \geq 0$, $1 < p < \infty$), then $u \in W^{s+2,p}$.

Sometimes Korn's inequalities are used in studying results like those in Theorem 1.11; however, our presentation did not require or use them. They are very relevant for questions of stability in linear elastodynamics, as we shall see in Section 6.3. We shall state these without detailed proof (see Fichera [1972a], Friedrichs [1947], Payne and Weinberger [1961], the remark below and Box 1.1 for the proofs).

1.12 Korn's Inequalities

(i) *First Inequality.* For $u \in H^2_\partial$ satisfying displacement boundary conditions on $\partial_d \subset \partial\Omega$, we have

$$\int_\Omega \|e\|^2\, dv \geq c\,\|u\|^2_{H^1}$$

for a susitable constant $c > 0$ independent of u.

(ii) *Second Inequality.* There is a constant $\bar{c} > 0$ such that

$$\int_\Omega \|e\|^2\, dv + \int_\Omega \|u\|^2\, dv \geq \bar{c}\,\|u\|^2_{H^1}$$

for all $u \in H^1$.

The first inequality is fairly straightforward (see Box 1.1) while the second is more subtle. For the displacement or mixed problem, uniform pointwise stability implies that we have

$$-\langle Au, u\rangle = \int_\Omega e\cdot\mathbf{c}\cdot e\, dv \geq 2\eta \int \|e\|^2\, dv \geq 2\eta c\,\|u\|^2_{H^1}.$$

So it follows that Ker $A = \{0\}$ in this case, reproducing what we found above. As indicated in 2.8, Section 5.2, this inequality will guarantee dynamic stability.

Korn's inequalities are actually special cases of Gårdings inequality for (not necessarily square) elliptic systems. The Lie derivative operator

$$u \mapsto e$$

is in fact elliptic in the sense of *systems* of partial differential equations (see, for example, Berger and Ebin [1969]). Gårdings inequality applied to L is exactly Korn's second inequality. The first inequality comes from the general fact that for elliptic operators with constant coefficients satisfying zero boundary conditions, Gårdings inequality reads $B(\boldsymbol{u}, \boldsymbol{u}) \geq c\|\boldsymbol{u}\|_{H^1}^2$—that is, stability. (This is seen by examining the proof; see Box 1.1.)

Box 1.1 ***Some Useful Inequalities***

This box discusses four topics: (1) Gårdings inequality (see 1.5); (2) Korn's first inequality (see 1.12); (3) a sample elliptic estimate (see 1.6(i)); and (4) some key Sobolev inequalities.

(1) *Garding's inequality*

(a) Let us prove the inequality

$$B(\boldsymbol{u}, \boldsymbol{u}) \geq c\|\boldsymbol{u}\|_{H^1}^2 - d\|\boldsymbol{u}\|_{L^2}^2$$

in case $\mathbf{a}$ is strongly elliptic and is constant (independent of x). We shall also assume $\boldsymbol{u}$ is C^∞ with compact support in $\mathbb{R}^n$.

Let $\hat{\boldsymbol{u}}$ be the Fourier transform of $\boldsymbol{u}$:

$$\hat{\boldsymbol{u}}(\boldsymbol{\xi}) = \frac{1}{(2\pi)^{n/2}} \int_{\mathbb{R}^n} e^{-i\boldsymbol{\xi}\cdot x}\boldsymbol{u}(x)\,dx \quad (i = \sqrt{-1}).$$

The H^1-norm of $\boldsymbol{u}$ is given by

$$\|\boldsymbol{u}\|_{H^1}^2 = \sum_{i,j} \int_{\mathbb{R}^n} \left(\frac{\partial u^i}{\partial x^j}\right)^2 dx + \sum_i \int_{\mathbb{R}^n} (u^i)^2\,dx.$$

Since the Fourier transform preserves the L^2-norm (Plancherel's theorem), and $(\partial u^i/\partial x^j)\hat{} = \xi_j \hat{u}^i/i$, we have

$$\|\boldsymbol{u}\|_{H^1}^2 = \int_{\mathbb{R}^n} |\boldsymbol{\xi} \otimes \hat{\boldsymbol{u}}(\boldsymbol{\xi})|^2\,d\xi + \int_{\mathbb{R}^n} |\hat{\boldsymbol{u}}(\boldsymbol{\xi})|^2\,d\boldsymbol{\xi}.$$

By strong ellipticity,

$$\begin{aligned} B(\boldsymbol{u}, \boldsymbol{u}) &= \int_{\mathbb{R}^n} a_i{}^j{}_k{}^l \frac{\partial u^i}{\partial x^j}\frac{\partial u^k}{\partial x^l}\,dx \\ &= \int_{\mathbb{R}^n} \xi_j \xi_l a_i{}^j{}_k{}^l \hat{u}^i \hat{u}^k\,d\boldsymbol{\xi} \geq \int_{\mathbb{R}^n} \epsilon|\boldsymbol{\xi}|^2|\hat{\boldsymbol{u}}|^2\,d\boldsymbol{\xi} \\ &= \epsilon \int_{\mathbb{R}^n} |\boldsymbol{\xi} \otimes \hat{\boldsymbol{u}}|^2\,d\boldsymbol{\xi}. \end{aligned}$$

Thus, we can take $c = \epsilon$ and $d = \epsilon$. The case of variable coefficients and a general domain requires a modification of this basic idea (cf. Yosida [1971] or Morrey [1966]).

(b) Relevant to Gårding's inequality is the *Poincaré inequality*, one version of which states that if $\boldsymbol{u} = \mathbf{0}$ on $\partial\Omega$, then

$$\|\boldsymbol{u}\|_{L^2} \leq C\| D\boldsymbol{u}\|_{L^2}.$$

In this case, H_∂^1 can be normed by $\| D\boldsymbol{u}\|_{L^2}$, and so if $\mathbf{a}$ has constant coefficients, we can choose $d = 0$ in Gårding's inequality.

(c) Next we discuss *Hadamard's theorem* [1902], which states that *Gårding's inequality implies strong ellipticity.* Let us again just prove a simple case. Suppose $\mathbf{a}$ is constant and assume $\mathbf{0} \in \Omega$. Choose $\boldsymbol{u}(x) = \boldsymbol{\xi}\phi(\boldsymbol{\lambda}\cdot x)$, where $\boldsymbol{\xi}$ and $\boldsymbol{\lambda}$ are constant vectors in $\mathbb{R}^n$ and $\phi\colon \mathbb{R} \to \mathbb{R}$ is smooth with compact support. Then

$$B(\boldsymbol{u}, \boldsymbol{u}) = \int_\Omega a^{ijkl}\xi_i\lambda_j\xi_k\lambda_l\,|\phi'(\boldsymbol{\lambda}\cdot x)|^2\, dx$$
$$= \left(\int_\Omega |\phi'(\boldsymbol{\lambda}\cdot x)|^2\, dx\right)(a^{ijkl}\xi_i\xi_k\lambda_j\lambda_l).$$

By assumption, we get

$$\left(\int_\Omega |\phi'(\boldsymbol{\lambda}\cdot x)|^2\, dx\right)(a^{ijkl}\xi_i\xi_k\lambda_j\lambda_l)$$
$$\geq c\left(\int_\Omega |\boldsymbol{\xi}|^2|\boldsymbol{\lambda}|^2\phi'(\boldsymbol{\lambda}\cdot x)\, dx\right)$$
$$+ (c - d)\left(\int_\Omega |\boldsymbol{\xi}|^2|\phi(\boldsymbol{\lambda}\cdot x)|^2\, dx\right).$$

It is easy to see that we can choose a sequence of ϕ_n's such that

$$\left(\int_\Omega |\phi_n(\boldsymbol{\lambda}\cdot x)|^2\, dx\right)\Big/\int_\Omega |\phi_n'(\boldsymbol{\lambda}\cdot x)|^2\, dx \to 0 \quad \text{as} \quad n \to \infty.$$

Dividing the preceding inequality by $\int_\Omega |\phi_n'(\boldsymbol{\lambda}\cdot x)|^2\, dx$ and letting $n \to \infty$, we see that strong ellipticity holds with $\epsilon = c$.

(2) *Korn's first inequality* Let us prove that

$$\int |\boldsymbol{e}|^2\, dx \geq c\|\boldsymbol{u}\|_{H^1}^2$$

for $\boldsymbol{u}$ a C^∞ displacement in $\mathbb{R}^n$ with compact support. As in Gårding's inequality, the general case can be reduced to this one. Let $\hat{\boldsymbol{u}}(\boldsymbol{\xi})$ be the Fourier transform of $\boldsymbol{u}$ as defined above. Thus

$$\hat{\boldsymbol{e}}(\boldsymbol{\xi}) = \tfrac{1}{2}(\boldsymbol{\xi} \otimes \hat{\boldsymbol{u}} + \hat{\boldsymbol{u}} \otimes \boldsymbol{\xi}).$$

Therefore, by Plancherel's theorem for Fourier transforms of tensor fields,

$$\int_{\mathbb{R}^n} |e|^2\, dx = \int_{\mathbb{R}^n} |\hat{e}(\xi)|^2\, d\xi$$
$$= \tfrac{1}{4} \int \sum_{i,j} (\xi_i \hat{u}_j + \xi_j \hat{u}_i)^2\, d\xi.$$

However,

$$\sum_{i,j} (\xi_i \hat{u}_j + \xi_j \hat{u}_i)^2 = \sum_{i,j} (\xi_i^2 \hat{u}_j^2 + \xi_j^2 \hat{u}_i^2 + 2\xi_i \hat{u}_i \xi_j \hat{u}_j)$$
$$= 4 \sum_i (\xi_i^2 \hat{u}_i^2) + \sum_{i \neq j} (\xi_i^2 \hat{u}_j^2 + \xi_j^2 \hat{u}_i^2)$$
$$+ 2(\sum_{i \neq j} \xi_i \hat{u}_i \xi_j \hat{u}_j).$$

Using the inequality $2\xi_i \hat{u}_i \xi_j \hat{u}_j \geq -(\xi_i^2 \hat{u}_i^2 + \xi_j^2 \hat{u}_j^2)$, we get

$$\sum_{i,j} (\xi_i \hat{u}_j + \xi_j \hat{u}_i)^2 \geq 2 \sum_i \xi_i^2 \hat{u}_i^2 + \sum_{i \neq j} (\xi_i^2 \hat{u}_j^2 + \xi_j^2 \hat{u}_i^2)$$
$$= 2 \sum_{i,j} \xi_i^2 \hat{u}_j^2.$$

Thus,

$$\int_{\mathbb{R}^n} |e|^2\, dx \geq 2 \int_{\mathbb{R}^n} |Du|^2\, dx$$

which gives Korn's first inequality (cf. the Poincaré inequality above).

As pointed out in the remark preceding this box, this inequality is, secretly, Gårding's inequality for the elliptic operator $\boldsymbol{u} \mapsto \boldsymbol{e}$.

(3) *A sample elliptic estimate* Suppose **a** is constant and strongly elliptic and $\boldsymbol{u}$ is smooth with compact support on $\mathbb{R}^n$. We will prove the inequality

$$\|\boldsymbol{u}\|_{H^2}^2 \leq C(\|A\boldsymbol{u}\|_{L^2}^2 + \|\boldsymbol{u}\|_{L^2}^2)$$

for a constant C. We have

$$\|\boldsymbol{u}\|_{H^2}^2 = \sum_{i,j,k} \int_{\mathbb{R}^n} \xi_i^2 \xi_j^2 \hat{u}_k^2\, d\xi + \sum_{i,j} \int_{\mathbb{R}^n} \xi_i^2 \hat{u}_j^2\, d\xi + \sum_i \int_{\mathbb{R}^n} \hat{u}_i^2\, d\xi.$$

It suffices to deal with the top order terms by Gårding's inequality. Now by strong ellipticity,

$$\|A\boldsymbol{u}\|_{L^2}^2 = \sum_{i,j,k,l} \int_{\mathbb{R}^n} |a^{ijkl} \xi_j \xi_l \hat{u}_k|^2\, d\xi$$
$$\geq \int_{\mathbb{R}^n} \epsilon (\sum_i \xi_i^2)^2 (\sum_k \hat{u}_k^2)\, d\xi \geq \epsilon \int_{\mathbb{R}^n} \sum_{i,k} \xi_i^4 \hat{u}_k^2\, d\xi.$$

Using $\xi_i^2 \xi_j^2 \leq \frac{1}{2}(\xi_i^4 + \xi_j^4)$, we get

$$\sum_{i,j,k} \int_{\mathbb{R}^n} \xi_i^2 \xi_j^2 \hat{u}_k^2\, d\xi \leq \frac{1}{2} \sum_{i,j,k} \int_{\mathbb{R}^n} (\xi_i^4 + \xi_j^4) \hat{u}_k^2\, d\xi$$
$$\leq n \sum_{i,k} \int_{\mathbb{R}^n} \xi_i^4 \hat{u}_k^2\, d\xi \leq \frac{n}{\epsilon} \|A\boldsymbol{u}\|_{L^2}^2.$$

This gives the desired inequality.

(4) *Some key Sobolev inequalities* Let us begin by being a little more precise in our definitions of Sobolev spaces. Let Ω be an open set in $\mathbb{R}^n$ with piecewise smooth boundary. Let $C^\infty(\Omega, \mathbb{R}^l)$ denote the maps $f: \Omega \to \mathbb{R}^l$ that have C^∞ extensions to maps $\bar{f}$, which are C^∞ on all of $\mathbb{R}^n$. If Ω is bounded, we set

$$W^{k,p}(\Omega, \mathbb{R}^l) = \begin{cases} \text{completion of } C^\infty(\Omega, \mathbb{R}^l) \text{ in the norm,} \\ \|f\|_{k,p} = \sum_{0 \leq l \leq k} \|D^l f\|_{L^p}, \end{cases}$$

where $\|\phi\|_{L^p} = \left(\int_\Omega \|\phi(x)\|^p \, dx\right)^{1/p}$ is the L^p-norm on Ω, $D^i f$ is the ith derivative of f, and we take its norm in the usual way. For $p = 2$ we set $H^s(\Omega, \mathbb{R}^l) = W^{s,2}(\Omega, \mathbb{R}^l)$. Thus H^s is a Hilbert space. (One can show that H^s consists of those L^2 functions whose first s derivatives, in the sense of distribution theory, lie in L^2. This is called the *Meyer–Serrin theorem.* A convenient reference for the proof is Friedman [1969].)

For general Ω, set $C_0^\infty(\Omega, \mathbb{R}^l)$ = the C^∞ functions from Ω to $\mathbb{R}^n$ that have compact support in Ω. The completion of this space in the $\|\cdot\|_{k,p}$ norm is denoted $W_0^{k,p}$, and the corresponding H^s space is denoted H_0^s. For $\Omega = \mathbb{R}^n$ we just write $H^s = H_0^s$. Again $H^s(\mathbb{R}^n, \mathbb{R}^l)$ consists of those L^2 functions whose first s derivatives are in L^2.

In order to obtain useful information concerning the Sobolev spaces $W^{k,p}$, we need to establish certain fundamental relationships between these spaces. To do this, one uses the following fundamental inequality of Sobolev, as generalized by Nirenberg and Gagliardo. We give a special case (the more general case deals with Hölder norms as well as $W^{k,p}$ norms).

1.13 Theorem *Let $1 \leq q \leq \infty$, $0 \leq r \leq \infty$, $0 \leq j < m$, $j/m \leq a \leq 1$, $0 < p < \infty$, with j, m integers ≥ 0; assume that*

$$\frac{1}{p} = \frac{j}{n} + a\left(\frac{1}{r} - \frac{m}{n}\right) + (1 - a)\frac{1}{q} \tag{1}$$

(if $1 < r < \infty$ and $m - j - n/r$ is an integer ≥ 0, assume $j/m \leq a < 1$). Then there is a constant C such that for any smooth $\mathbf{u}: \mathbb{R}^n \to \mathbb{R}^l$, we have

$$\|D^j \mathbf{u}\|_{L^p} \leq C \|D^m \mathbf{u}\|_{L^r}^a \|\mathbf{u}\|_{L^q}^{1-a}. \tag{2}$$

(If $j = 0$, $rm < n$, and $q = \infty$, assume $\mathbf{u} \to 0$ at ∞ or $\mathbf{u}$ lies in $L^{\tilde{q}}$ for some finite $\tilde{q} > 0$.)

Below we shall prove some special cases of this result. (The arguments given by Nirenberg [1959] are geometric in flavor in contrast to the usual Fourier transform proofs and therefore are more suitable for generalization to manifolds; cf. Cantor [1975a] and Aubin [1976].)

The above theorem remains valid for $\boldsymbol{u}$ defined on a region with piecewise smooth boundary, or more generally if the boundary satisfies a certain "cone condition."

Note: If one knows an inequality of the form (2) exists, one can infer that (1) must hold as follows: replace $\boldsymbol{u}(x)$ by $\boldsymbol{u}(tx)$ for a real $t > 0$. Then writing $\boldsymbol{u}_t(x) = \boldsymbol{u}(tx)$, one has

$$\| D^j \boldsymbol{u}_t \|_{L^p} = \| D^j \boldsymbol{u} \|_{L^p} \cdot t^{j-n/p}, \qquad \| D^m \boldsymbol{u}_t \|_{L^r}^a = \| D^m \boldsymbol{u} \|_{L^r}^a \cdot t^{a(m-n/r)}, \quad \text{and}$$
$$\| \boldsymbol{u}_t \|_{L^q}^{1-a} = \| \boldsymbol{u} \|_{L^q}^{1-a} \cdot t^{-n(1-a)/q}.$$

Thus if (2) is to hold for $\boldsymbol{u}_t$ (with the constant independent of t), we must have $j - n/p = a(m - n/r) - n(1 - a)/q$, which is exactly the relation (1).

The following corollary is useful in a number of applications:

1.14 Corollary *With the same relations as in Theorem 1.13, for any* $\epsilon > 0$ *there is a constant* K_ϵ *such that*

$$\| D^j \boldsymbol{u} \|_{L^p} \leq \epsilon \| D^m \boldsymbol{u} \|_{L^p} + K_\epsilon \| \boldsymbol{u} \|_{L^q}$$

for all (smooth) functions $\boldsymbol{u}$.

Proof This follows from 1.13 and Young's inequality: $x^a y^{1-a} \leq ax + (1 - a)y$, which implies that $x^a y^{1-a} = (\epsilon x)^a (K_\epsilon y)^{1-a} \leq a\epsilon x + (1 - a)K_\epsilon y$ where, $K_\epsilon = 1/\epsilon^{a/(1-a)}$. ∎

Let us illustrate how Fourier transform techniques can be used to directly prove the special case of 1.14 in which $n = 3$, $l = 1$, $j = 0$, $p = \infty$, $m = 2$, $r = 2$, and $q = 2$.

1.15 Proposition *There is a constant* $c > 0$ *such that for any* $\epsilon > 0$ *and function* $f: \mathbb{R}^3 \longrightarrow \mathbb{R}$ *smooth with compact support, we have*

$$\| f \|_\infty \leq c(\epsilon^{3/2} \| f \|_{L^2} + \epsilon^{-1/2} \| \Delta f \|_{L^2}).$$

(It follows that if $f \in H^2(\mathbb{R}^3)$*, then* f *is uniformly continuous and the above inequality holds.)*

Proof Let

$$\hat{f}(\boldsymbol{k}) = \frac{1}{(2\pi)^{3/2}} \int_{\mathbb{R}^3} e^{-i\boldsymbol{k}\cdot\boldsymbol{x}} f(\boldsymbol{x})\, d\boldsymbol{x}$$

denote the Fourier transform. Recall that $(\widehat{\Delta f})(\boldsymbol{k}) = -\| \boldsymbol{k} \|^2 \hat{f}(\boldsymbol{k})$. From

Schwarz' inequality, we have

$$\left(\int |\hat{f}(k)|\,dk\right)^2 \leq \left(\int \frac{dk}{(\epsilon^2 + \|k\|^2)^2}\right)\left(\int (\epsilon^2 + \|k\|^2)^2 |\hat{f}(k)|^2\,dk\right)$$

$$= \frac{c_1}{\epsilon}\|(\epsilon^2 - \Delta)f\|^2_{L^2},$$

where
$$c_1 = \int_{\mathbb{R}^3} \frac{d\xi}{(1 + \|\xi\|^2)^2} < \infty.$$

Here we have used the fact that $h \mapsto \hat{h}$ is an isometry in the L^2-norm (Plancherel's theorem). Thus, from $f(x) = \frac{1}{(2\pi)^{3/2}}\int_{\mathbb{R}^3} e^{ik\cdot x}\hat{f}(k)\,dk$,

$$(2\pi)^{3/2}\|f\|_\infty \leq \|\hat{f}\|_{L^1} \leq \frac{c_2}{\sqrt{\epsilon}}\|(\epsilon^2 - \Delta)f\|_{L^2}$$

$$\leq c_2(\epsilon^{3/2}\|f\|_{L^2} + \epsilon^{-1/2}\|\Delta f\|_{L^2}). \quad ∎$$

Thus we have shown that $H^2(\mathbb{R}^3) \subset C^0(\mathbb{R}^3)$ and that the inclusion is continuous. More generally one can show by similar arguments that $H^s(\Omega) \subset C^k(\Omega)$ provided $s > n/2 + k$ and $W^{s,p}(\Omega) \subset C^k(\Omega)$ if $s > n/p + k$. This is one of the celebrated *Sobolev embedding theorems.*

Problem 1.3 Prove that the last assertion is a special case of the result in Theorem 1.13.

For Ω bounded, the inclusion $W^{s,p}(\Omega) \to C^k(\Omega)$, $s > n/p + k$ is compact; that is, the unit ball in $W^{s,p}(\Omega)$ is compact in $C^k(\Omega)$. This is proved in a manner similar to the classical Arzela–Ascoli theorem, one version of which states that the inclusion $C^1(\Omega) \subset C^0(\Omega)$ is compact (see Marsden [1974a], for instance). Also, $W^{s,p}(\Omega) \subset W^{s',p'}(\Omega)$ is compact if $s > s'$ and $p = p'$ or if $s = s'$ and $p > p'$. (See Friedman [1969] for the proofs.)

We already saw one application of Rellich's theorem in our proof of the Fredholm alternative. It is often used this way in existence theorems, using compactness to extract convergent sequences.

As we shall see later, compactness comes into existence theory in another crucial way when one seeks weak solutions. This is through the fact that the unit ball in a Banach space is weakly compact—that is, compact in the weak topology. See, for example, Yosida [1971] for the proof (and for refinements, involving weak *sequential* compactness).

We shall give another illustration of Theorem 1.13 through a special case that is useful in the study of nonlinear wave equations. This is the following important inequality in $\mathbb{R}^3$:

$$\|u\|_{L^6} \leq C\|Du\|_{L^2}.$$

1.16 Proposition *Let $u: \mathbb{R}^3 \to \mathbb{R}$ be smooth and have compact support. Then*

$$\int_{\mathbb{R}^3} u^6\, dx \leq 48\left(\int_{\mathbb{R}^3} \|\operatorname{grad} u\|^2\, dx\right)^3.$$

so $C = \sqrt[6]{48}$.

Proof[3] From

$$u^3(x, y, z) = 3\int_{-\infty}^{x} u^2 \frac{\partial u}{\partial x}\, dx$$

one gets

$$\sup_x |u^3(x, y, z)| \leq 3\int_{-\infty}^{\infty} \left|u^2 \frac{\partial u}{\partial x}\right| dx.$$

Set $I = \int_{\mathbb{R}^3} u^6\, dx$ and write

$$\begin{aligned} I &= \int_{-\infty}^{\infty} \left(\iint |u^3|\,|u^3|\, dy\, dz\right) dx \\ &\leq \int_{-\infty}^{\infty} \left[\left(\sup_y \int_{-\infty}^{\infty} |u^3|\, dz\right)\left(\int_{-\infty}^{\infty} \sup_z |u^3|\, dy\right)\right] dx \\ &\leq 9\int_{-\infty}^{\infty} \left[\left(\iint \left|u^2 \frac{\partial u}{\partial y}\right| dy\, dz\right)\left(\iint \left|u^2 \frac{\partial u}{\partial z}\right| dy\, dz\right)\right] dx. \end{aligned}$$

Using Schwarz' inequality on this gives

$$I \leq 9\int_{-\infty}^{\infty} \left[\left(\iint u^4\, dy\, dz\right)\left(\iint_{-\infty}^{\infty} \left(\frac{\partial u}{\partial y}\right)^2 dy\, dz\right)^{1/2}\left(\iint_{-\infty}^{\infty} \left(\frac{\partial u}{\partial z}\right)^2 dy\, dz\right)^{1/2}\right] dx$$

$$\leq 9 \max_x \left(\iint_{-\infty}^{\infty} u^4\, dy\, dz\right)\left(\iiint_{-\infty}^{\infty} \left(\frac{\partial u}{\partial y}\right)^2 dx\, dy\, dz\right)^{1/2}\left(\iiint_{-\infty}^{\infty} \left(\frac{\partial u}{\partial z}\right)^2 dx\, dy\, dz\right)^{1/2}$$

$$\leq 36\int_{\mathbb{R}^3} \left|u^3 \frac{\partial u}{\partial x}\right| dx\, dy\, dz\left(\int_{\mathbb{R}^3} \left(\frac{\partial u}{\partial y}\right)^2\right)^{1/2}\left(\int_{\mathbb{R}^3} \left(\frac{\partial u}{\partial z}\right)^2\right)^{1/2}$$

$$\leq 36\sqrt{I}\left(\int_{\mathbb{R}^3} \left(\frac{\partial u}{\partial x}\right)^2\right)^{1/2}\left(\int_{\mathbb{R}^3} \left(\frac{\partial u}{\partial y}\right)^2\right)^{1/2}\left(\int_{\mathbb{R}^3} \left(\frac{\partial u}{\partial z}\right)^2\right)^{1/2}.$$

Now using the arithmetic geometric mean inequality $\sqrt[3]{a}\sqrt[3]{b}\sqrt[3]{c} < (a + b + c)/3$ gives

$$I \leq 36\sqrt{I}\left(\int_{\mathbb{R}^3} \|\operatorname{grad} u\|^2\right)^{3/2} \Big/ 3^{3/2}.$$

[3]Following Ladyzhenskaya [1969].

That is,

$$I \leq \frac{(36)^2}{3^3} \cdot \left(\int_{\mathbb{R}^3} \| \operatorname{grad} u \|^2 \right)^3 . \quad \blacksquare$$

One can use the same technique to prove similar inequalities.

There is another important corollary of Theorem 1.13 that we shall prove. The techniques can be used to determine to which $W^{s,p}$ space a product belongs.

1.17 Corollary *For $s > n/2$, $H^s(\mathbb{R}^n)$ is a Banach algebra (under pointwise multiplication). That is, there is a constant $K > 0$ such that for $u, v \in H^s(\mathbb{R}^n)$,*

$$\| u \cdot v \|_{H^s} \leq K \| u \|_{H^s} \| v \|_{H^s}.$$

This is an important property of H^s not satisfied for low s. It certainly is not true *that L^2 forms an algebra under multiplication.*

Proof Choose $a = j/s$, $r = 2$, $q = \infty$, $p = 2s/j$, $m = s$ $(0 \leq j \leq s)$ to obtain

$$\| D^j u \|_{L^{2s/j}} \leq \text{const.} \| D^s u \|_{L^2}^{j/s} \| u \|_{\infty}^{1-j/s} \leq \text{const.} \| u \|_{H^s}.$$

(See 1.15.) Let $j + k = s$. From Holder's inequality we have

$$\| D^j u \cdot D^k v \|_{L^2}^2 \leq \| D^j u \|_{L^{2s/j}}^2 \| D^k v \|_{L^{2s/k}}^2 \leq \text{const.} \| u \|_{H^s}^2 \| v \|_{H^s}^2.$$

Now $D^s(uv)$ consists of terms like $D^j u \cdot D^k v$, so we obtain

$$\| D^s uv \|_{L^2} \leq \text{const.} \| u \|_{H^s} \| v \|_{H^s}.$$

Similarly for the lower-order terms. Summing gives the result. $\blacksquare$

The trace theorems have already been mentioned in 1.1. Generally, they state that the restriction map from Ω to a submanifold $\mathfrak{M} \subset \Omega$ of codimension m induces a bounded operator from $W^{s,p}(\Omega)$ to $W^{s-(1/mp),p}(\mathfrak{M})$. Adams [1975] and Morrey [1966] are good references; the latter contains some useful refinements of this.

There are also some basic extension theorems that are right inverses of restriction maps. Thus, for example, the *Calderon extension theorem* asserts that there is an extension map $T: W^{s,p}(\Omega) \to W^{s,p}(\mathbb{R}^n)$ that is a bounded operator and "restriction to Ω" $\circ T =$ Identity. This is related to a classical C^k theorem due to Whitney. See, for example, Abraham and Robbin [1967], Stein [1970], and Marsden [1973a].

Finally, we mention that all these $\mathbb{R}^n$ results carry over to manifolds in a straightforward way. See, for example, Palais [1965] and Cantor [1979].

Problem 1.4 Prove a $W^{s,p}$ version of the ω-lemma given in Box 1.1, Chapter 4, by using the results of this box.

Box 1.2 ***Summary of Important Formulas for Section 6.1***

Sobolev Spaces on $\Omega \subset \mathbb{R}^n$

$$L^2 = \left\{ u : \Omega \to \mathbb{R}^n \,\middle|\, \int_\Omega \| u^2 \| \, dv < \infty \right\}$$

$$\langle u, v \rangle = \int_\Omega u \cdot v \, dv = \int_\Omega u^a(x) v_a(x) \, dx$$

$H^s = \{u : \Omega \to \mathbb{R}^n \mid u, Du, \ldots, D^s u \text{ are in } L^2\}$

norm: $\| u \|_{H^s}^2 = \| u \|_{L^2}^2 + \cdots + \| D^s u \|_{L^2}^2$.

$H_\partial^2 = \{u \in H^2 \mid u$ satisfies the boundary conditions (displacement, traction or mixed)$\}$

$W^{s,p} = \{u : \Omega \to \mathbb{R}^n \mid u, Du, \ldots, D^s u \in L^p\}$

Symmetric Elliptic Operator

Form: $Au = \operatorname{div}(\mathbf{a} \cdot \nabla u)$ $\quad (Au)^a = (\mathrm{a}^{abcd} u_{c|d})_{|b}$

Symmetry: $\mathrm{a}^{abcd} = \mathrm{a}^{cdab}$, $\langle Au, v \rangle = \langle u, Av \rangle$ for $u, v \in H_\partial^2$

Strong ellipticity: $\mathrm{a}^{abcd} \xi_a \xi_c \eta_b \eta_d \geq \epsilon \| \boldsymbol{\xi} \|^2 \| \boldsymbol{\eta} \|^2$ for some $\epsilon > 0$ and all $\boldsymbol{\xi}, \boldsymbol{\eta} \in \mathbb{R}^n$

Gårding's Inequality ($\Leftrightarrow$ Strong Ellipticity)

$B(u, u) \geq c \| u \|_{H^1}^2 - d \| u \|_{L^2}^2$ for all $u \in H^1$, where

$$B(u, v) = \int_\Omega \nabla u \cdot \mathbf{a} \cdot \nabla v \, dv = \int_\Omega u_{a|b} \mathrm{a}^{abcd} v_{c|d} \, dv$$

$(= -\langle Au, v \rangle$ for $u, v \in H_\partial^2)$.

Fredholm Alternative

(1) $L^2 = \text{Range } A \oplus \text{Ker } A$, an L^2 orthogonal sum.

(2) $Au = f$ is solvable for $u \in H_\partial^2$ if and only if $f \perp \text{Ker } A$.

Classical Elasticity

Uniform pointwise stability: $\frac{1}{2} e \cdot \mathbf{c} \cdot e \geq \eta \| e \|^2$, $\frac{1}{2} e_{ab} \mathrm{c}^{abcd} e_{cd} \geq \eta e_{ab} e^{ab}$, implies strong ellipticity. In $\Omega \subset \mathbb{R}^3$,

$\operatorname{div}(\mathbf{c} \cdot \nabla u) = f$ is solvable for $u \in H_\partial^2$ using displacement or mixed boundary conditions for any f and for traction boundary conditions if $\int_\Omega f(x)(a + bx) \, dx = 0$ for a any constant vector and b any 3×3 skew matrix.

Korn's Inequalities

(1) Displacement (or mixed) boundary conditions:

$$\int_\Omega \| e \|^2 \, dv \geq c \| u \|_{H^1}^2, \quad e_{ab} = \tfrac{1}{2}(u_{a|b} + u_{b|a})$$

(2) General:

$$\int_\Omega \| e \|^2 \, dv + \int_\Omega \| u \|^2 \, dv \geq c \| u \|_{H^1}^2$$

6.2 ABSTRACT SEMIGROUP THEORY

This section gives an account of those parts of semigroup theory that are needed in the following section for applications to elastodynamics. Although the account is self contained and gives fairly complete proofs of most of the theorems, it is not exhaustive. For example, we have omitted details about the theory of analytic semigroups, since it will be treated only incidentally in subsequent sections. The standard references for semigroup theory are Hille and Phillips [1957], Yosida [1971], Kato [1966], and Pazy [1974]. This theory also occurs in many books on functional analysis, such as Balakrishnan [1976].

We shall begin with the definition of a semigroup. The purpose is to capture, under the mildest possible assumptions, what we mean by solvability of a linear evolution equation

$$\frac{du}{dt} = Au \quad (t \geq 0), \qquad u(0) = u_0. \tag{1}$$

Here A is a linear operator in a Banach space $\mathfrak{X}$. We are interested in when (1) has unique solutions and when these solutions vary continuously in $\mathfrak{X}$ as the initial data varies in the $\mathfrak{X}$ topology. When this holds, one says that Equation (1) is *well-posed.* If A is a bounded operator in $\mathfrak{X}$, solutions are given by

$$u(t) = e^{tA}u_0 = \sum_{k=0}^{\infty} \frac{(tA)^k}{k!} u_0.$$

For partial differential equations, however, A will usually be unbounded, so the problem is to make sense out of e^{tA}. Instead of power series, the operator analogue of the calculus formula $e^x = \lim_{n\to\infty} (1 - x/n)^{-n}$ will turn out to be appropriate.

2.1 Definitions A (C^0) *semigroup* on a Banach space $\mathfrak{X}$ is a family $\{U(t) \mid t \geq 0\}$ of bounded linear operators of $\mathfrak{X}$ to $\mathfrak{X}$ such that the following conditions hold:

(i) $U(t+s) = U(t) \circ U(s)$ $(t, s \geq 0)$ (semigroup property);
(ii) $U(0) =$ Identity; and
(iii) $U(t)x$ is t-continuous at $t = 0$ for each $x \in \mathfrak{X}$; that is, $\lim_{t\downarrow 0} U(t)x = x$. (This pointwise convergence is also expressed by saying strong $\lim_{t\downarrow 0} U(t) = I$.)

The *infinitesimal generator* A of $U(t)$ is the (in general unbounded) linear operator given by

$$Ax = \lim_{t\downarrow 0} \frac{U(t)x - x}{t} \tag{2}$$

on the domain $\mathfrak{D}(A)$ defined to be the set of those $x \in \mathfrak{X}$ such that the limit (2) exists in $\mathfrak{X}$.

We now derive a number of properties of semigroups. (Eventually we will prove an existence and uniqueness theorem for semigroups *given* a generator A.)

For Propositions 2.2–2.12, assume that $U(t)$ is a given C^0 semigroup with infinitesimal generator A.

2.2 Proposition *There are constants $M > 0$, $\beta \geq 0$ such that $\|U(t)\| \leq Me^{t\beta}$ for all $t \geq 0$. In this case we write $A \in \mathcal{G}(\mathfrak{X}, M, \beta)$ and say A is the generator of a semigroup of type (M, β).*

Proof We first show that $\|U(t)\|$ is bounded on some neighborhood of zero. If not, there would be a sequence $t_n \downarrow 0$ such that $\|U(t_n)\| \geq n$. But $U(t_n)x \longrightarrow x$ as $n \longrightarrow \infty$, so $U(t_n)$ is pointwise bounded as $n \longrightarrow \infty$, and therefore by the Uniform Boundedness Theorem[4] $\|U(t_n)\|$ is bounded, which is a contradiction.

Thus for some $\delta > 0$ there is a constant M such that $\|U(t)\| \leq M$ for $0 \leq t \leq \delta$. For $t \geq 0$ arbitrary, let n be the largest integer in t/δ so $t = n\delta + \tau$, where $0 \leq \tau < \delta$. Then by the semigroup property,

$$\|U(t)\| = \|U(n\delta)U(\tau)\| \leq \|U(\tau)\| \|U(\delta)\|^n \leq M \cdot M^n \leq M \cdot M^{t/\delta} \leq Me^{t\beta}.$$

where $\beta = (1/\delta) \log M$. ∎

2.3 Proposition *$U(t)$ is strongly continuous[5] in t; that is, for each fixed $x \in \mathfrak{X}$, $U(t)x$ is continuous in $\mathfrak{X}$ as a function of $t \in [0, \infty)$.*

Proof Let $s > 0$. Since $U(\tau + s)x = U(s)U(\tau)x$, 2.1(iii) gives

$$\begin{aligned} \lim_{t \downarrow s} U(t)x &= \lim_{\tau \downarrow 0} U(\tau + s)x \\ &= U(s) \lim_{\tau \downarrow 0} U(\tau)x = U(s)x, \end{aligned}$$

so we have right continuity in t at $t = s$. For left continuity let $0 \leq \tau \leq s$, and write

$$\|U(s - \tau)x - U(s)x\| = \|U(s - \tau)(x - U(\tau)x)\| \leq Me^{\beta(s-\tau)} \|x - U(\tau)x\|,$$

which tends to zero as $\tau \downarrow 0$. ∎

2.4 Proposition

(i) *$U(t)\mathfrak{D}(A) \subset \mathfrak{D}(A)$;*
(ii) *$U(t)Ax = AU(t)x$ for $x \in \mathfrak{D}(A)$; and*
(iii) *$(d/dt)U(t)x_0 = A(U(t)x_0)$ for all $x_0 \in \mathfrak{D}(A)$ and $t \geq 0$. In other words,*

$$x(t) = U(t)x_0 \quad \textit{satisfies} \quad \frac{dx}{dt} = Ax \textit{ and } x(0) = x_0.$$

---

[4]This theorem states that if $\{T_\alpha\}$ is a family of bounded linear operators on $\mathfrak{X}$ and if $\{T_\alpha x\}$ is bounded for each $x \in \mathfrak{X}$, then the norms $\|T_\alpha\|$ are bounded. See, for example, Yosida [1971], p. 69.

[5]One can show that strong continuity at $t = 0$ can be replaced by weak continuity at $t = 0$ and strong continuity in $t \in [0, \infty)$ can be replaced by strong measurability in t. See Hille and Phillips [1957] for details.

Proof From $[U(h)U(t)x - U(t)x]/h = U)t)[(U(h)x - x)/h]$ we get (i) and (ii). We get (iii) by using the fact that if $x(t) \in \mathfrak{X}$ has a continuous right derivative, then $x(t)$ is differentiable—from the right at $t = 0$ and two sided if $t > 0$.[6] ∎

From (i) and (ii) we see that if $x \in \mathfrak{D}(A^n)$, then $U(t)x \in \mathfrak{D}(A^n)$. This is often used to derive *regularity* results, because if A is associated with an elliptic operator, $\mathfrak{D}(A^n)$ may consist of smoother functions for larger n. Notice that we have now shown that the concept of semigroup given here and that given in 2.5, Chapter 5, agree.

2.5 Proposition *$\mathfrak{D}(A)$ is dense in $\mathfrak{X}$.*

Proof Let $\phi(t)$ be a C^∞ function with compact support in $[0, \infty)$, let $x \in \mathfrak{X}$ and set

$$x_\phi = \int_0^\infty \phi(t)U(t)x\,dt.$$

Noting that

$$U(s)x_\phi = \int_0^\infty \phi(t)U(t + s)x\,dt = \int_0^\infty \phi(\tau - s)U(\tau)x\,d\tau$$

is differentiable in s, we find that $x_\phi \in \mathfrak{D}(A)$. On the other hand, given any $\epsilon > 0$ we claim that there is a ϕ (close to the "δ function") such that $\|x_\phi - x\| < \epsilon$. Indeed, by continuity, choose $\delta > 0$ such that $\|U(t)x - x\| < \epsilon$ if $0 \leq t \leq \delta$. Let ϕ be C^∞ with compact support in $(0, \delta)$, $\phi \geq 0$ and $\int_0^\infty \phi(t)\,dt = 1$. Then

$$\|x_\phi - x\| = \left\| \int_0^\infty \phi(t)(U(t)x - x)\,dt \right\| \leq \int_0^\delta \phi(t)\|U(t)x - x\|\,dt$$
$$< \epsilon \int_0^\delta \phi(t)\,dt = \epsilon. \quad ∎$$

The same argument in fact shows that $\bigcap_{n=1}^\infty \mathfrak{D}(A^n)$ *is dense in* $\mathfrak{X}$.

2.6 Proposition *A is a closed operator; that is, its graph in $\mathfrak{X} \times \mathfrak{X}$ is closed.*[7]

Proof Let $x_n \in \mathfrak{D}(A)$ and assume that $x_n \to x_0$ and $Ax_n \to y$. We must show that $x \in \mathfrak{D}(A)$ and $y = Ax$. By 2.4,

$$U(t)x_n = x_n + \int_0^t U(s)Ax_n\,ds.$$

[6]This follows from the corresponding real variables fact by considering $l(u(t))$ for $l \in \mathfrak{X}^*$. See Yosida [1971], p. 235.

[7]We shall prove more than this in Proposition 2.12 below, but the techniques given here are more direct and also apply to certain nonlinear semigroups as well. See Chernoff and Marsden [1974].

Since $U(s)Ax_n \longrightarrow U(s)y$ uniformly for $s \in [0, t]$, we have

$$U(t)x = x + \int_0^t U(s)y\, ds.$$

It follows that $(d/dt+)\, U(t)x\,|_{t=0}$ exists and equals y. ▮

Next we show that integral curves are unique. (Compare 2.15, Chapter 5.)

2.7 Proposition *Suppose $c(t)$ is a differentiable curve in $\mathfrak{X}$ such that $c(t) \in \mathfrak{D}(A)$ and $c'(t) = A(c(t))$ $(t \geq 0)$. Then $c(t) = U(t)c(0)$.*

Proof Fix $t_0 > 0$ and define $h(t) = U(t_0 - t)c(t)$ for $0 \leq t \leq t_0$. Then for τ small,

$$\begin{aligned}\|h(t+\tau) - h(t)\| &= \|U(t_0 - t - \tau)c(t+\tau) - U(t_0 - t - \tau)U(\tau)c(t)\| \\ &\leq Me^{\beta(t_0 - t - \tau)}\|c(t+\tau) - U(\tau)c(t)\|.\end{aligned}$$

However,

$$\frac{1}{\tau}[c(t+\tau) - U(\tau)c(t)] = \frac{1}{\tau}[c(t+\tau) - c(t)] - \frac{1}{\tau}[U(\tau)c(t) - c(t)],$$

which converges to $Ac(t) - Ac(t) = 0$, as $\tau \longrightarrow 0$. Thus, $h(t)$ is differentiable for $0 < t < t_0$ with derivative zero. By continuity, $h(t_0) = \lim_{t \downarrow t_0} h(t) = c(t_0) = \lim_{t \downarrow t_0} h(t) = U(t_0)c(0)$. (The last limit is justified by the fact that $\|U(t)\| \leq Me^{t\beta}$.) This is the result with t replaced by t_0. ▮

One also has uniqueness in the class of weak solutions as is explained in the optional Box 2.1.

Box 2.1 ***Adjoints and Weak Solutions (Balakrishnan*** [*1976*] ***and Ball*** [*1977c*])

Let the adjoint $A^*: \mathfrak{D}(A^*) \subset \mathfrak{X}^* \longrightarrow \mathfrak{X}^*$ be defined by $\mathfrak{D}(A^*) = \{v \in \mathfrak{X}^* \mid \text{there exists a } w \in \mathfrak{X}^* \text{ such that } \langle w, x\rangle = \langle v, Ax\rangle \text{ for all } x \in \mathfrak{D}(A)\}$, where $\langle\ ,\ \rangle$ denotes the pairing between $\mathfrak{X}$ and $\mathfrak{X}^*$. Set $A^*v = w$. If $c(t)$ is a continuous curve in $\mathfrak{X}$ and if, for every $v \in \mathfrak{D}(A^*)$, $\langle c(t), v\rangle$ is absolutely continuous and

$$\frac{d}{dt}\langle c(t), v\rangle = \langle c(t), A^*v\rangle \quad \text{almost everywhere}$$

that is, $\langle c(t), v\rangle = \langle c(0), v\rangle + \int_0^t \langle c(s), A^*v\rangle\, ds$, then $c(t)$ is called a *weak solution* of $dx/dt = Ax$.

2.8 Proposition *Let $\{U(t)\}$ be a C^0 semigroup on $\mathfrak{X}$. If $c(t)$ is a weak solution, then $c(t) = U(t)c(0)$. Conversely, for $x_0 \in \mathfrak{X}$ (not necessarily in the domain of A), then $c(t) = U(t)x_0$ is a weak solution.*

Proof If $x_0 \in \mathfrak{D}(A)$, then $U(t)x_0$ is a solution in $\mathfrak{D}(A)$ and hence a weak solution. Since $U(t)$ is continuous and $\mathfrak{D}(A)$ is dense, the same is true for $x_0 \in \mathfrak{X}$; that is, we can pass to the limit in

$$\langle U(t)x_n, v\rangle = \langle x_n, v\rangle + \int_0^t \langle U(\tau)x_n, v\rangle\, d\tau$$

for $x_n \in \mathfrak{D}(A)$, $x_n \to x_0 \in \mathfrak{X}$.

Now suppose $c(t)$ is a weak solution. Let $w(t) = c(t) - U(t)c(0)$. Then $w(0) = 0$ and for $v \in \mathfrak{D}(A^*)$,

$$\langle w(t), v\rangle = \int_0^t \langle w(\tau), A^*v\rangle\, d\tau = \left\langle \int_0^t w(\tau)\, d\tau, A^*v\right\rangle.$$

Thus, $\int_0^t w(\tau)\, d\tau \in \mathfrak{D}(A)$ since A is closed (see 2.6). Here we have used the fact that if A is closed, then $A^{**} \subset A$, where we identify $\mathfrak{X}$ with a subspace of $\mathfrak{X}^{**}$. (If $\mathfrak{X}$ is reflexive, $A^{**} = A$; cf. Kato [1966], p. 168.) It follows that $z(t) = \int_0^t w(\tau)\, d\tau$ satisfies $\dot{z} = Az$, and since $z(0) = 0$, z is identically zero by 2.7. ∎

Ball [1977c] also shows that if the equation $\dot{x} = Ax$ admits unique weak solutions and A is densely defined and closed, then A is a generator.

We continue now to develop properties of a given semigroup of type (M, β). If $\beta = 0$, we say the semigroup is *bounded*, and if $M = 1$, we say it is *quasi-contractive*. If $M = 1$ and $\beta = 0$, it is called *contractive*.

2.9 Proposition *If $U(t)$ is (a C^0 semigroup) of type (M, β) on $\mathfrak{X}$, then: (i) $T(t) = e^{-t\beta}U(t)$ is a bounded semigroup with generator $A - \beta I$; (ii) there is an equivalent norm $|||\cdot|||$ on $\mathfrak{X}$ relative to which $U(t)$ is quasi-contractive.*

We shall leave the proof as an exercise. For (ii), use the norm $|||x||| = \sup_{t\geq 0} \|e^{-t\beta}U(t)x\|$.

2.10 Example Let $\mathfrak{X} = L^2(\mathbb{R})$ with the norm

$$\|f\|^2 = \int_{-1}^1 |f(x)|^2\, dx + \tfrac{1}{2}\int_{\mathbb{R}\setminus[-1,1]} |f(x)|^2\, dx$$

and let $(U(t)f)(x) = f(t + x)$. Then $U(t)$ is a C^0 semigroup and $Af = df/dx$

with domain $H^1(\mathbb{R})$ (absolutely continuous functions with derivatives in L^2). Here, $\| U(t) \| \leq 2$. If we form the norm $|||\ |||$, we get the usual L^2-norm and a contraction semigroup.

2.11 Proposition *$U(t)$ is norm continuous at $t = 0$ if and only if A is bounded.*

Proof Choose $\epsilon > 0$ so that $\|U(t) - I\| < \frac{1}{2}$ if $0 \leq t \leq \epsilon$ and pick ϕ to be a C^∞ function with compact support in $[0, \epsilon)$ such that $\phi \geq 0$ and $\int_0^\epsilon \phi(t)\, dt = 1$.

Let $J_\phi(x) = \int_0^\infty \phi(\tau)U(\tau)x\, d\tau$ and note that

$$J_\phi(U(t)x) = \int_0^\infty \phi(\tau)U(\tau + t)x\, d\tau = \int_t^\infty \phi(\tau - t)U(\tau)x\, d\tau.$$

However,

$$\|(J_\phi - I)(x)\| = \left\| \int_0^\infty \phi(\tau)(U(\tau)x - x)\, d\tau \right\| \leq \tfrac{1}{2} \int_0^\infty \phi(\tau) \| x \|\, d\tau = \tfrac{1}{2} \| x \|,$$

so $\|J_\phi - I\| \leq \frac{1}{2}$ and hence J_ϕ is invertible. By construction

$$U(t)x = J_\phi^{-1}\left(\int_t^\infty \phi(\tau - t)U(\tau)x\, d\tau\right),$$

which is therefore differentiable in t for all x and also shows $A \in \mathcal{B}(\mathfrak{X})$ (the set of all bounded linear operators on $\mathfrak{X}$). The converse is done by noting that $e^{tA} = \sum_{n=0}^{\infty}(tA)^n/(n!)$ is norm continuous in t. ▮

Next we give a proposition that will turn out to be a complete characterization of generators.

2.12 Proposition *Let $A \in \mathcal{G}(\mathfrak{X}, M, \beta)$. Then:*

(i) *$\mathfrak{D}(A)$ is dense in $\mathfrak{X}$;*
(ii) *$(\lambda - A)$ is one-to-one and onto $\mathfrak{X}$ for each $\lambda > \beta$ and the* resolvent *$R_\lambda = (\lambda - A)^{-1}$ is a bounded operator; and*
(iii) *$\|(\lambda - A)^{-n}\| \leq M/(\lambda - \beta)^n$ for $\lambda > \beta$ and $n = 1, 2, \ldots$.*

Note. Here and in what follows, $\lambda - A$ stands for $\lambda I - A$, where I is the identity operator.

Proof Given $x \in \mathfrak{X}$, let

$$y_\lambda = \int_0^\infty e^{-\lambda t}U(t)x\, dt, \qquad \lambda > \beta.$$

Then

$$\begin{aligned}(U(s) - I)y_\lambda &= \int_0^\infty e^{-\lambda t}U(t + s)x\, dt - y_\lambda \\ &= e^{\lambda s}\int_s^\infty e^{-\lambda \tau}U(\tau)x\, d\tau - y_\lambda \\ &= (e^{\lambda s} - 1)y_\lambda - e^{\lambda s}\int_0^s e^{-\lambda t}U(t)x\, dt.\end{aligned}$$

Hence $y_\lambda \in \mathfrak{D}(A)$ and $Ay_\lambda = \lambda y_\lambda - x$. Thus $(\lambda - A)$ is surjective. (Taking $\lambda \longrightarrow \infty$ shows that $\lambda y_\lambda \longrightarrow x$, which also shows $\mathfrak{D}(A)$ is dense, reproducing 2.5.) The formula

$$u = \int_0^\infty e^{-\lambda t}U(t)(\lambda - A)u\,dt, \qquad u \in \mathfrak{D}(A),$$

which follows from $-(d/dt)e^{-\lambda t}U(t)u = e^{-\lambda t}U(t)(\lambda - A)u$, shows that $(\lambda - A)$ is one-to-one.

Thus we have proved the Laplace transform relation

$$R_\lambda x = (\lambda - A)^{-1}x = \int_0^\infty e^{-\lambda t}U(t)x\,dt, \quad (\lambda > \beta),$$

from which it follows that

$$\|(\lambda - A)^{-1}\| \leq \int_0^\infty e^{-\lambda t}Me^{\beta t}\,dt = \frac{M}{\lambda - \beta}.$$

The estimate (iii) follows from the formulas

$$(n-1)!\,(\lambda - A)^{-n}x = \int_0^\infty e^{-\lambda t}t^{n-1}U(t)x\,dt, \tag{3}$$

$$\int_0^\infty e^{-\mu t}t^{n-1}\,dt = \frac{(n-1)!}{\mu^n}. \tag{4}$$

Equation (4) is proved by integration by parts and (3) follows from the relation

$$\left(\frac{d}{d\lambda}\right)^{n-1}(\lambda - A)^{-1} = (-1)^{n-1}(n-1)!\,(\lambda - A)^{-n}. \quad \blacksquare$$

Problem 2.1 Show that the *resolvent identity* $R_\lambda - R_\mu = (\mu - \lambda)R_\lambda R_\mu$ holds and that $\mu R_\lambda \longrightarrow$ Identity strongly as $\lambda \longrightarrow \infty$.

The following *Hille–Yosida theorem* asserts the converse of 2.12. It is, in effect, an existence and uniqueness theorem. Uniqueness was already proved in 2.7.

2.13 Theorem *Let A be a linear operator in $\mathfrak{X}$ with domain $\mathfrak{D}(A)$. Assume there are positive constants M and β such that:*

(i) *$\mathfrak{D}(A)$ is dense;*

(ii) *$(\lambda - A)$ is one-to-one and onto $\mathfrak{X}$ for $\lambda > \beta$ and $(\lambda - A)^{-1} \in \mathfrak{B}(\mathfrak{X})$; and*

(iii) $\|(\lambda - A)\|^{-n} \leq M/(\lambda - \beta)^n$ $(\lambda > \beta, n = 1, 2, \ldots)$.

Then $A \in \mathcal{G}(\mathfrak{X}, M, \beta)$; that is, there exists a C^0 semigroup of type (M, β) whose generator is A. We shall often write e^{tA} for the semigroup generated by A.

Proof If $(A - \beta I)$ generates the semigroup U, then A generates the semigroup $e^{t\beta}U$ (see 2.9). Thus it suffices to prove the theorem for $\beta = 0$.

Rewrite (iii) as $\|(1 - \alpha A)^{-n}\| \leq M$ $(\alpha > 0, n = 1, \ldots)$ by taking $\alpha = 1/\lambda$. Thus, if $x \in \mathfrak{D}(A)$, then $(1 - \alpha A)^{-1}x - x = \alpha(1 - \alpha A)^{-1}Ax$, so $(1 - \alpha A)^{-1} - I$

$\to 0$ strongly on $\mathfrak{D}(A)$ as $\alpha \downarrow 0$. Since $(1 - \alpha A)^{-1} - I \in \mathfrak{B}(\mathfrak{X})$, convergence also holds on $\mathfrak{X}$.

Let $U_n(t) = (1 - (t/n)A)^{-n}$, a uniformly bounded sequence of operators. We shall show it converges on a dense set. Write

$$U_n(t)x - U_m(t)x = U_m(t-s)U_n(s)x \Big|_{s=0}^{s=t} = \operatorname{s-lim}_{\epsilon \downarrow 0} \int_\epsilon^t \frac{d}{ds} U_m(t-s)U_n(s)x\, ds$$

$$= \operatorname{s-lim}_{\epsilon \downarrow 0} \int_\epsilon^t \left(\frac{s}{n} - \frac{t-s}{m}\right) A^2 \left(1 - \frac{t-s}{m}A\right)^{-m-1} \left(1 - \frac{s}{n}A\right)^{-n-1} x\, ds.$$

Thus, if $x \in \mathfrak{D}(A^2)$, we get,

$$\| U_n(t)x - U_m(t)x \| \leq M^2 \| A^2 x \| \tfrac{1}{2}\left(\frac{1}{n} + \frac{1}{m}\right) t^2.$$

Thus $U_n(t)x$ converges for $x \in \mathfrak{D}(A^2)$. But

$$\mathfrak{D}(A^2) = \mathfrak{D}((1 - A)^2) = \operatorname{Range}(1 - A)^{-2} = (1 - A)^{-1}\mathfrak{D}(A).$$

Now $(1 - A)^{-1} : \mathfrak{X} \to \mathfrak{D}(A)$ is bounded and surjective. Since $\mathfrak{D}(A) \subset \mathfrak{X}$ is dense, $(1 - A)^{-1}(\mathfrak{D}(A)) \subset \mathfrak{D}(A)$ is dense; that is, $\mathfrak{D}(A^2) \subset \mathfrak{X}$ is dense.

Let $U(t)x = \operatorname{s-lim}_{n \to \infty} U_n(t)x$. Clearly, $\| U(t) \| \leq M$, $U(0)x = x$, and $U(t + s) = U(t) \circ U(s)$. Since $U_n(t)x \to U(t)x$ uniformly on compact t-intervals for $x \in \mathfrak{D}(A^2)$ and this is dense, $U(t)x$ is t-continuous. Thus we have a C^0 semigroup.

Let A' be the generator of $U(t)$. We need to show that $A' = A$. For $x \in \mathfrak{D}(A)$,

$$\frac{d}{dt} U_n(t)x = A\left(1 - \frac{t}{n}A\right)^{-1} U_n(t)x.$$

Thus

$$U_n(t)x = x + \int_0^t \left(1 - \frac{s}{n}A\right)^{-1} U_n(s)Ax\, ds,$$

and so

$$U(t)x = x + \int_0^t U(s)Ax\, ds.$$

Therefore $x \in \mathfrak{D}(A')$ and $A' \supset A$. But $(1 - A')^{-1} \in \mathfrak{B}(\mathfrak{X})$ by the previous proposition and $(1 - A)^{-1} \in \mathfrak{B}(\mathfrak{X})$ by assumption, so they must agree. ▮

We shall give some concrete examples of how to check these hypotheses below and in the next section. In this regard, note that if $M = 1$, we have a quasi-contractive semigroup and verification of (iii) for $n = 1$ is sufficient. Also, as the proof shows, it suffices to verify (ii) and (iii) for *some* sufficiently large λ. Finally, we note that if (ii) and (iii) hold for $|\lambda| > \beta$, then $U(t)$ is a *group*—that is, is defined for all $t \in \mathbb{R}$, not just $t \geq 0$.

For applications, there are two special verisions of the Hille–Yosida theorem that are frequently used. These involve the notion of the closure of an operator. Namely, A is called *closable* if the closure of the graph of A in $\mathfrak{X} \times \mathfrak{X}$ is the graph of an operator; this operator is called the *closure* of A and is denoted $\bar{A}$. In practice, this often enlarges the domain of A. (For example, the Laplacian

∇^2 in $\mathfrak{X} = L^2$ may originally be defined on smooth functions satisfying desired boundary conditions; its closure will extend the domain to H^2_∂.)

2.14 First Corollary *A linear operator A has a closure $\bar{A}$ that is the generator of a quasi-contractive semigroup on $\mathfrak{X}$ if and only if (i) $\mathfrak{D}(A)$ is dense and (ii) for λ sufficiently large, $(\lambda - A)$ has dense range and $\|(\lambda - A)x\| \geq (\lambda - \beta)\|x\|$.*

Proof Necessity follows Proposition 2.12. For sufficiency, we use the following:

2.15 Lemma (*a*) *Let B be a closable linear operator with a densely defined bounded inverse B^{-1}. Then $(\overline{B^{-1}})$ is injective, and $(\bar{B})^{-1} = (\overline{B^{-1}})$.*

(*b*) *Suppose that A is a densely defined linear operator such that $(\lambda - A)^{-1}$ exists, is densely defined, with $\|(\lambda - A)^{-1}\| \leq K/\lambda$ for large λ. Then A is closable. (Hence, by part (a), $(\lambda - \bar{A})$ is invertible, with $(\lambda - \bar{A})^{-1} = (\overline{\lambda - A})^{-1}$.)*

Proof (a) Since B^{-1} is bounded, $\overline{B^{-1}}$ is a bounded, everywhere-defined operator. Suppose that $\overline{B^{-1}}y = 0$. We will show that $y = 0$. Let $y_n \in$ Range of B and $y_n \to y$. Then $y_n = Bx_n$ for $x_n \in \mathfrak{D}(B)$, and $\|x_n\| \leq \|B^{-1}\|\|y_n\| \to 0$. Since B is closable, we must have $y = 0$. Thus $\overline{B^{-1}}$ is injective and (a) follows.

(b) We shall first show that $\lambda R_\lambda \to I$ as $\lambda \to \infty$, where $R_\lambda = (\overline{\lambda - A})^{-1}$ by definition. By assumption, $\|R_\lambda\| \leq K/\lambda$. Now pick any $x \in \mathfrak{D}(A)$. Then $x = R_\lambda(\lambda - A)x$, so $x = \lambda R_\lambda x - R_\lambda Ax$, and $\|R_\lambda Ax\| \leq (K/\lambda)\|Ax\| \to 0$ as $\lambda \to \infty$. Thus $\lambda R_\lambda \to I$ strongly on $\mathfrak{D}(A)$. But $\mathfrak{D}(A)$ is dense and $\|\lambda R_\lambda\| \leq K$ for all large λ, so $\lambda R_\lambda \to I$ on the whole of $\mathfrak{X}$.

To prove that A is closable, we suppose $x_n \in \mathfrak{D}(A)$, $x_n \to 0$, and $Ax_n \to y$. We claim that $y = 0$. Indeed, choose a sequence $\lambda_n \to \infty$ with $\lambda_n x_n \to 0$. Then $(\lambda_n - A)x_n + y \to 0$. Since $\|\lambda_n R_{\lambda_n}\| \leq K$, we have $\lambda_n R_{\lambda_n}[(\lambda_n - A)x_n + y] \to 0$. Thus, $\lambda_n x_n + \lambda_n R_{\lambda_n} y \to 0$. But $\lambda_n x_n \to 0$ and $\lambda_n R_{\lambda_n} y \to y$, so $y = 0$. ∎

The rest of 2.14 now follows. Indeed, A satisfies the conditions of part (b) of the lemma, and hence $\bar{A}$ satisfies the hypothesis of the Hille–Yosida theorem with $M = 1$). ∎

Now we give a result in Hilbert space. We will sometimes refer to this result as the *Lumer–Phillips Theorem*.[8] For applications we shall give in the next section, it will be one of the most useful results of this section.

2.16 Second Corollary *Let A be a linear operator in a Hilbert space $\mathfrak{H}$. Then A has a closure $\bar{A}$ that is the generator of a quasi-contractive semigroup on X (that is, $\bar{A} \in \mathcal{G}(\mathfrak{H}, 1, \beta)$) if and only if:*

(i) *$\mathfrak{D}(A)$ is dense in $\mathfrak{H}$;*

[8]See Lumer and Phillips [1961] for the case of Banach spaces. It proceeds in a similar way, using a duality map in place of the inner product.

(ii) $\langle Ax, x\rangle \leq \beta\langle x, x\rangle$ *for all* $x \in \mathfrak{D}(A)$ *(If* β *is zero, we call* A *dissipative); and*

(iii) $(\lambda - A)$ *has dense range for sufficiently large* λ.

If in (iii) $(\lambda - A)$ *is onto, then* A *is closed and* $A \in \mathcal{G}(\mathcal{H}, 1, \beta)$.

Proof First suppose (i), (ii), and (iii) hold. Then $\langle(\lambda - A)x, x\rangle \geq (\lambda - \beta)\cdot \|x\|^2$, and so by Schwarz's inequality, $\|(\lambda - A)x\| \geq (\lambda - \beta)\|x\|$. Thus $\bar{A} \in \mathcal{G}(\mathcal{H}, 1, \beta)$ by 2.15.

Conversely, assume $\bar{A} \in \mathcal{G}(\mathcal{H}, 1, \beta)$. We need only show that

$$\langle \bar{A}x, x\rangle \leq \beta\langle x, x\rangle \quad \text{for all } x \in \mathfrak{D}(\bar{A}).$$

By 2.9 we can assume that $\beta = 0$ and $U(t)$ is contractive. Now $\langle x, U(t)x\rangle \leq \|x\|\|U(t)x\| \leq \|x\|^2$ and therefore $\langle x, U(t)x - x\rangle \leq 0$. Dividing by t and letting $t \downarrow 0$ gives $\langle x, \bar{A}x\rangle \leq 0$ as desired. ▮

2.17 Further Results Some additional useful results that we just quote are as follows:

1. *Bounded Perturbations* If $A \in \mathcal{G}(\mathfrak{X}, M, \beta)$ and $B \in \mathcal{B}(\mathfrak{X})$, then $A + B \in \mathcal{G}(\mathfrak{X}, M, \beta + \|B\|M)$ (Kato [1966], p. 495).

2. *Trotter–Kato Theorem* If $A_n \in \mathcal{G}(\mathfrak{X}, M, \beta)$ $(n = 1, 2, 3, \ldots)$ $A \in \mathcal{G}(\mathfrak{X}, M, \beta)$ and for λ sufficiently large, $(\lambda - A_n)^{-1} \longrightarrow (\lambda - A)^{-1}$ strongly, then $e^{tA_n} \longrightarrow e^{tA}$ strongly, uniform on bounded t-intervals (Kato [1966], p. 502).

> ***Problem 2.2*** Show that if $\mathfrak{D}(A_n)$ and $\mathfrak{D}(A)$ all have a common *core* $\mathcal{Y} \subset \mathfrak{X}$—that is, A_n, A are the closures of their restrictions to $\mathcal{Y}$, and $A_n \longrightarrow A$ strongly on $\mathcal{Y}$—then $(\lambda - A_n)^{-1} \longrightarrow (\lambda - A)^{-1}$ strongly, from the resolvent identity (see Problem 2.1).

3. *Lax Equivalence Theorem* If $A \in \mathcal{G}(\mathfrak{X}, M, \beta)$, and $K_\epsilon \in \mathcal{B}(\mathfrak{X})$ is a family of bounded operators defined for $\epsilon \geq 0$, with $K_0 = I$, we say $\{K_\epsilon\}$ is:

(i) *stable* if $\|K^n_{t/n}\|$ is bounded on bounded t-intervals $(n = 1, 2, \ldots)$;

(ii) *resolvent consistent* if for λ sufficiently large

$$(\lambda - A)^{-1} = \operatorname*{s-lim}_{\epsilon \downarrow 0} \left(\lambda - \frac{1}{\epsilon}(K_\epsilon - I)\right)^{-1} \quad \text{(strong limit)};$$

(iii) *consistent* if $(d/d\epsilon+)K_\epsilon(x)|_{\epsilon=0} = Ax$, $x \in$ a core of A.

The Lax equivalence theorem states that $e^{tA} = \text{s-lim}_{n\to\infty} K^n_{t/n}$ uniformly on bounded t-intervals if and only if $\{K_\epsilon\}$ is stable and resolvent consistent (see Chorin, Hughes, McCracken, and Marsden [1978] for a proof and applications). Assuming stability, consistency implies resolvent consistency.

4. *Trotter Product Formula* If A, B are generators of quasi-contractive semigroups and $C = \overline{A + B}$ is a generator, then

$$e^{tC} = \operatorname*{s-lim}_{n\to\infty} (e^{tA/n}e^{tB/n})^n.$$

(This is a special case of 3.)

5. *Inhomogeneous Equations* Let $A \in \mathcal{G}(\mathfrak{X}, M, \beta)$ and consider the following initial value problem: Let $f(t)$ $(0 \leq t \leq T)$ be a continuous $\mathfrak{X}$-valued function. Find $x(t)$ $(0 \leq t \leq T)$ with $x(0)$ a given member of $\mathfrak{D}(A)$, such that

$$x'(t) = Ax(t) + f(t). \tag{I}$$

If we solve (I) formally by the variation of constants formula, we get

$$x(t) = e^{tA}x(0) + \int_0^t e^{(t-\tau)A}f(\tau)\,d\tau \quad (0 \leq t \leq T).$$

However, $x(t)$ need not lie in $\mathfrak{D}(A)$; but it will if f is a C^1 function from $[0, T]$ to $\mathfrak{X}$. Then (I) is satisfied in the classical sense (Kato [1966], p. 486). For uniqueness, suppose $y(t)$ is another solution of (I), with $y(0) = x(0)$. Let $z(t) = x(t) - y(t)$. Then $z'(t) = Az(t)$ and $z(0) = 0$, so $z(t) \equiv 0$ by uniqueness. Thus $x(t) = y(t)$.

6. *Trend to Equilibrium* Let $A \in \mathcal{G}(\mathfrak{X}, 1, \beta)$ and suppose there is a $\delta > 0$ such that the spectrum of e^A lies inside the unit disk a positive distance δ from the unit circle. Then for any $x \in \mathfrak{X}$,

$$e^{tA}x \longrightarrow 0 \quad \text{as} \quad t \longrightarrow +\infty.$$

Indeed, if $0 < \delta' < \delta$, we can, by way of the spectral theorem, find a new norm in which

$$A \in \mathcal{G}(\mathfrak{X}, 1, -\delta'),$$

from which the result follows. (See Marsden and McCracken, [1976], §2A, Slemrod [1976], and Dafermos [1968].) Abstract conditions under which spectrum $e^A = e^{\text{spectrum}\,A}$ are unfortunately rather complex (see Carr [1981] and Roh [1982]).

7. *Analytic Semigroups.* If $\|(\zeta - A)^{-1}\| < M/|\zeta|$ for ζ complex and belonging to a sector $|\arg \zeta| \leq \pi/2 + \omega$ $(\omega > 0)$, then A generates a bounded semigroup $V(t)$ that can be extended to complex t's as an analytic function of t $(t \neq 0)$. For real $t > 0$, $x \in \mathfrak{X}$ one has $V(t)x \in \mathfrak{D}(A)$ and

$$\left\| \frac{d^n V(t)x}{dt^n} \right\| \leq (\text{const.}) \cdot \|x\| \cdot t^{-n}.$$

Consult one of the standard references for details.

2.18 Comments on Operators in Hilbert Space and Semigroups The results here are classical ones due to Stone and von Neumann, which may be found in several of the aforementioned references. A densely defined operator in Hilbert space is called *symmetric* if $A \subset A^*$; that is, $\langle Ax, y \rangle = \langle x, Ay \rangle$ for all $x, y \in \mathfrak{D}(A)$ (see Box 2.1 for the definition of A^*; replace $\mathfrak{X}^*$ by $\mathfrak{X}$ in Hilbert space), *self-adjoint* if $A = A^*$ (i.e., A is symmetric and A and A^* have the the same domain), and *essentially self-adjoint* if $\bar{A}$ is self-adjoint.

For the first two results following, $\mathfrak{X}$ is assumed to be a *complex* Hilbert space.

1. *Let A be closed and symmetric. Then A is self-adjoint if and only if $A + \lambda I$ is surjective when* $\operatorname{Im} \lambda \neq 0$.

> ***Problem 2.3*** (a) Show that a symmetric operator is closable. (b) By consulting a book on functional analysis, prove 1.

2. (*Stone's Theorem*) *A is self-adjoint if and only if iA generates a one-parameter unitary group.*

Proof From the symmetry of A and $\|(A+\lambda)x\|^2 \geq 0$, we get $\|(A+\lambda)x\| \geq |\text{Im}\,\lambda|\,\|x\|$, and so for λ real, $\|(\lambda - iA)x\| \geq |\lambda|\,\|x\|$. Thus 2 results from 1 and the Hille–Yosida theorem. ∎

3. (*Real version of Stone's Theorem*) *Let A be a skew-adjoint operator on a real Hilbert space (i.e., $A = -A^*$). Then A generates a one-parameter group of isometries and conversely.* (This follows by an argument similar to 2.)

4. *Let A be closed, symmetric, and $A \leq 0$; that is, $\langle Ax, x\rangle \leq 0$ for all $x \in \mathfrak{D}(A)$. Then A is self-adjoint if and only if $(\lambda - A)$ is onto $(\lambda \geq 0)$.*

> ***Problem 2.4*** Use this and the Fredholm alternative to show that the symmetric elliptic operator A in Section 6.1 is self-adjoint. See Problem 1.2.

5. *If A dissipative $(\langle Ax, x\rangle \leq 0)$ and self-adjoint, then A generates a contraction semigroup.* (This follows from 4 and the Lumer–Phillips theorem.)

2.19 Example (*Heat Equation*) Let $\Omega \subset \mathbb{R}^n$ be an open region with smooth boundary, $\mathfrak{X} = L^2(\Omega)$, $Au = \Delta u$, and $\mathfrak{D}(A) = C_0^\infty(\Omega)$, where $C_0^\infty(\Omega)$ are the C^∞ functions with compact support in Ω. Then $\bar{A}$ generates a contraction semigroup in $\mathfrak{X}$. ($\bar{A}$ will turn out to be the Laplacian on $\mathfrak{D}(\bar{A}) = H_0^2(\Omega)$, the H^2 functions with 0 boundary conditions.)

Proof Obviously A is symmetric. It follows that it is closable (Problem 2.3). Moreover, for $u \in \mathfrak{D}(A)$,

$$\langle Au, u\rangle = \int_\Omega \Delta u \cdot u\, dx = -\int_\Omega \nabla u \cdot \nabla u\, dx \leq 0,$$

so A is dissipative. By the second corollary of the Hille–Yosida theorem, it suffices to show that for $\lambda > 0$, $(\lambda - A)$ has dense range; that is, A is self-adjoint. Suppose $v \in L_2(\Omega)$ is such that

$$\langle(\lambda - A)u, v\rangle = 0 \quad \text{for all} \quad u \in \mathfrak{D}(A).$$

Then

$$\langle(\lambda - \Delta)u, v\rangle = 0 \quad \text{for all} \quad u \in C_0^\infty(\Omega).$$

By the regularity of weak solutions of elliptic equations, v is in fact C^∞ and $v = 0$ on $\partial\Omega$. (See 1.6(iv).) Thus, setting $u = v$ and integrating by parts,

$$\lambda \int_\Omega |v|^2\, dx + \int_\Omega |\nabla v|^2\, dx = 0,$$

and so $v = 0$. ∎

The semigroup produced here is actually analytic.

Problem 2.5 Show directly, using the elliptic theory in Section 6.1 (see, especially, Problem 1.2) that Δ on H_∂^2 generates a semigroup. Conclude that this is $\bar{A}$.

Problem 2.6 Generalize 2.19 to the equation $du/dt = Au$, where A is the operator discussed in Section 6.1.

The above example concerns the *parabolic* equation $\partial u/\partial t = Au$. The *hyperbolic* equation $\partial^2 u/\partial t^2 = Au$ is directly relevant to linear elastodynamics and will be considered in the next section.

Box 2.2 ***Summary of Important Formulas for Section 6.2***

Semigroup in a Banach Space

$U(t)\colon \mathfrak{X} \longrightarrow \mathfrak{X}$ is defined for $t \geq 0$, $U(t+s) = U(t) \circ U(s)$, $U(0) = I$, and $U(t)$ is strongly continuous in t at $t = 0+$ (and hence for all $t \geq 0$).

Infinitesimal Generator

$$Ax = \lim_{t \to 0+} \frac{U(t)x - x}{t}$$

on the domain $\mathfrak{D}(A)$ for which the limit exists ($\mathfrak{D}(A)$ is always dense). We write $U(t) = e^{tA}$.

Class (M, β)

$A \in \mathcal{G}(\mathfrak{X}, M, \beta)$ means A is a generator of a $U(t)$ satisfying $\| U(t) \| \leq Me^{t\beta}$. (Bounded if $\beta = 0$; quasi-contractive if $M = 1$, and contractive if both.)

Evolution Equation

If $x_0 \in \mathfrak{D}(A)$, then $x(t) = U(t)x_0 \in \mathfrak{D}(A)$, $x(0) = x_0$ and $dx/dt = Ax$. Solutions are unique if a semigroup exists.

Hille–Yosida Theorem Necessary and sufficient conditions on an operator A to satisfy $A \in \mathcal{G}(\mathfrak{X}, M, \beta)$ are:

(i) $\mathfrak{D}(A)$ is dense;
(ii) $(\lambda - A)^{-1} \in \mathfrak{B}(\mathfrak{X})$ exists for $\lambda > \beta$; and
(iii) $\|(\lambda - A)^{-n}\| \leq M/(\lambda - \beta)^n$ $(n = 1, 2, \ldots)$.

Useful Special Case (Lumer–Phillips Theorem) On Hilbert space $\mathcal{H}$, $\bar{A} \in \mathcal{G}(\mathcal{H}, 1, \beta)$ if and only if $\mathfrak{D}(A)$ is dense, $\langle Ax, x\rangle \leq \beta\langle x, x\rangle$ for all $x \in \mathfrak{D}(A)$, and $(\lambda - A)$ has dense range. (If $(\lambda - A)$ is onto, there is no need to take the closure.)

6.3 LINEAR ELASTODYNAMICS

In this section we discuss various aspects of the initial value problem in linear elastodynamics using the theory of semigroups developed in the previous section. We begin with the main result for hyperelasticity: *strong ellipticity is necessary and sufficient for the equations to generate a quasi-contractive semigroup* in $\mathfrak{X} = H^1 \times L^2$.[9] Following this, we discuss stability and show that definiteness of the energy implies dynamical stability (the linear energy criterion). It is also shown, using a result of Weiss [1967], that if Cauchy elasticity generates stable dynamics, then it is necessarily hyperelastic. Box 3.1 describes some general abstract results for Hamiltonian systems, linear hyperelasticity being a special case. Box 3.2 shows how semigroup techniques can be used in a problem of panel flutter, and in Box 3.3 various linear dissipative mechanisms are considered, again using semigroup techniques. Finally, Box 3.4 considers symmetric hyperbolic systems and how they may be used in linear elasticity.

Table 6.3.1 shows the interrelationships between some of the topics considered in this section.

Table 6.3.1

	A. *Linear Hyperelasticity*	
Strong ellipticity $\Uparrow$ (Gårding)	$\Longleftrightarrow$	The equations of motion generate a quasi-contractive semigroup in $H^1 \times L^2$ (relative to some Hilbert space structure).
		$\Uparrow$
Stability (classical elasticity) $\Uparrow$ (Korn) Elasticity tensor positive-definite	$\Longleftrightarrow$	Generates a *contractive* semigroup on $H^1 \times L^2$ (relative to some Hilbert space norm).
	B. *Cauchy Elasticity*	
Generation of a contractive semigroup on a Hilbert space of the form $\mathcal{Y} \times L^2$, $\mathcal{Y} \subset L^2$	$\Longrightarrow$	Hyperelasticity
	C. *Logarithmic Convexity*	
Strong ellipticity "strictly" fails	$\Longleftrightarrow$	No semigroup on any space of the form $\mathcal{Y} \times L^2$, $\mathcal{Y} \subset L^2$.

Consider then the equations of linearized elasticity on a region Ω that is (for simplicity) a bounded open set in $\mathbb{R}^n$ with piecewise smooth boundary or else

[9]The fact that the equations generate a semigroup in $\mathfrak{X}$ embodies the idea that we have a continuous linear dynamical system in $\mathfrak{X}$. In particular, the solutions depend continuously in $\mathfrak{X}$ as the initial conditions are varied in the *same* space $\mathfrak{X}$. This is to be compared with other types of continuous data dependence where the solution and initial data vary in *different* spaces. For the latter, strong ellipticity is not required. See Knops and Payne [1971] for an extensive discussion of this point.

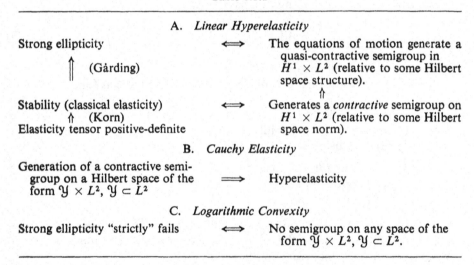

$\Omega = \mathbb{R}^n$. These equations are

$$\left.\begin{aligned} \rho \ddot{u} &= \operatorname{div}(\mathsf{a}\cdot\nabla u) \quad \text{on} \quad \Omega, \\ \text{that is,} \qquad \rho \ddot{u}^a &= (\mathsf{a}^{abcd} u_{c|d})_{|b} \quad \text{on} \quad \Omega. \end{aligned}\right\} \tag{1}$$

We have dropped the inhomogeneous terms $f = -\rho \mathring{a} + \rho b + \operatorname{div}(\mathring{\sigma})$, for they cause no added difficulty in questions of existence and uniqueness; see item 5 in 2.17.

The boundary conditions are assumed to be either

$$\left.\begin{aligned} \text{displacement:} \quad u &= 0 \text{ on } \partial\Omega, \\ \text{traction:} \quad \mathsf{a}\cdot\nabla u\cdot n &= 0 \text{ on } \partial\Omega, \\ \text{or mixed:} \quad u = 0 \text{ on } \partial_d \text{ and } \mathsf{a}\cdot\nabla u &= 0 \text{ on } \partial_\tau \end{aligned}\right\} \tag{2}$$

again taken to be homogeneous without loss of generality. We assume the elasticity tensor $\mathsf{a}^{abcd}(x)$ is C^1 in x and $\rho(x) \geq \delta > 0$ is C^0.

We recall that the material in question is *hyperelastic* when $\mathsf{a}^{abcd} = \mathsf{a}^{cdab}$. This is equivalent to *symmetry* of the operator $Au = (1/\rho)\operatorname{div}(\mathsf{a}\cdot\nabla u)$ in L^2 as in 1.3, where we put on L^2 a modified inner product corresponding to the $1/\rho$ factor in A, namely, $\langle u, v\rangle = \int_\Omega \rho(x) u^a(x) v_a(x)\, dv(x)$.

Rewrite the equations of motion as $\ddot{u} = Au$ or

$$\frac{d}{dt}\begin{pmatrix} u \\ \dot{u} \end{pmatrix} = A'\begin{pmatrix} u \\ \dot{u} \end{pmatrix}, \qquad \text{where} \quad A' = \begin{pmatrix} 0 & I \\ A & 0 \end{pmatrix}. \tag{3}$$

The domain of A is taken to be $H^2_\partial(\Omega)$, defined in 1.1.

The main existence theorem for linear elastodynamics is as follows:

3.1 Theorem *Let* a *be hyperelastic and let the symmetric operator* $A\colon \mathfrak{D}(A) = H^2_\partial \to L^2$ *be defined by* $Au = [\operatorname{div}(\mathsf{a}\cdot\nabla u)]/\rho$. *Let* $\mathfrak{X} = H^1 \times L^2$ *and let* A' *be defined by* (3) *with* $\mathfrak{D}(A') = \mathfrak{D}(A) \times H^1$. *The following assertions are equivalent:*

(i) *There is a Hilbert space norm on* H^1 *such that* A' *generates a quasi-contractive group on* $\mathfrak{X}$.

(ii) a *is strongly elliptic.*

Proof First, assume a is strongly elliptic. For $v \in \mathfrak{D}(A)$ and $u \in H^1(\Omega)$, the Dirichlet from is given by $B(u, v) = -\langle u, Av\rangle$, where the L^2 inner product is weighted with ρ, as above. By Gårding's inequality,

$$B(u, u) + d\,\|u\|^2_{L^2} = |||u|||^2_{H^1}$$

is equivalent to the H^1-norm. We use norm $(|||u|||^2_{H^1} + \|\dot{u}\|^2_{L^2})^{1/2}$ on $\mathfrak{X} = H^1 \times L^2$. If $\langle\!\langle \cdot, \cdot \rangle\!\rangle_{H^1}$ and $\langle\!\langle \cdot, \cdot \rangle\!\rangle$ denote the corresponding inner products on H^1 and $\mathfrak{X}$, respectively, then

$$\begin{aligned} \langle\!\langle A'(u, \dot{u}), (u, \dot{u}) \rangle\!\rangle &= \langle\!\langle u, \dot{u} \rangle\!\rangle_{H^1} + \langle Au, \dot{u}\rangle_{L^2} \\ &= B(u, \dot{u}) + d\langle u, \dot{u}\rangle_{L^2} + \langle Au, \dot{u}\rangle_{L^2} = d\langle u, \dot{u}\rangle_{L^2} \\ &\leq \tfrac{1}{2} d(\|u\|^2_{L^2} + \|\dot{u}\|^2_{L^2}) \leq \tfrac{1}{2} d(|||u|||^2_{H^1} + \|\dot{u}\|^2_{L^2}). \end{aligned}$$

The same estimate holds for $-A'$ since $\dot{u}$ can be replaced by $-\dot{u}$. By Problem 1.2 of Section 6.1, $\lambda u - Au = f$ is solvable for u if $\lambda > d_1$. The solution of

$$(\lambda - A')(u, \dot{u}) = (f, \dot{f})$$

is readily checked to be

$$u = (\lambda^2 - A)^{-1}\dot{f} + \lambda f, \qquad \dot{u} = -f + \lambda u,$$

so $(\lambda - A')$ is onto for $|\lambda|$ sufficiently large. Thus by the second corollary of the Hille–Yosida theorem (2.16), A' is the generator of a quasi-contractive semigroup.

Next we prove the converse. If A' generates a quasi-contractive semigroup, then 2.16 gives the estimate

$$\langle\!\langle u, \dot{u}\rangle\!\rangle_{H^1} + \langle Au, \dot{u}\rangle_{L^2} \leq \beta\{\langle\!\langle u, u\rangle\!\rangle_{H^1} + \langle \dot{u}, \dot{u}\rangle_{L^2}\}$$

for all $u \in \mathfrak{D}(A)$ and $\dot{u} \in H^1$. Letting $\dot{u} = \alpha u$ ($\alpha > 0$) and using the equivalence of the $\|\cdot\|_{H^1}$ and $|||\cdot|||_{H^1}$ norms,

$$\begin{aligned} -\langle Au, u\rangle_{L^2} &\geq \frac{1}{\alpha}\{\alpha\langle\!\langle u, u\rangle\!\rangle_{H^1} - \beta\langle\!\langle u, u\rangle\!\rangle_{H^1} - \beta\alpha^2\langle u, u\rangle_{L^2}\} \\ &\geq \frac{1}{\alpha}\{\alpha\|u\|^2_{H^1} - \gamma\|u\|^2_{H^1} - \beta\alpha^2\|u\|^2_{L^2}\} \end{aligned}$$

for a constant $\gamma > 0$. Choosing $\alpha > \gamma^2$, we get

$$-\langle Au, u\rangle_{L^2} \geq c\|u\|^2_{H^1} - d\|u\|^2_{L^2},$$

where $c = (\alpha - \gamma^2)/\alpha\gamma$, $d = \beta\alpha$. Thus Garding's inequality holds and so by 1.5, we have strong ellipticity. ∎

Remarks

(1) This argument also gives a sharp regularity result. Recall from 1.6(i) that if $u \in H^2(\Omega)$, u satisfies non-mixed boundary conditions, and Ω has a smooth boundary, then for $s \geq 2$,

$$\|u\|_{H^s} < C(\|Au\|_{H^{s-2}} + \|u\|_{L^2}).$$

This shows that $\mathfrak{D}(A^m) \subset H^{2m}$ (which, by the Sobolev embedding theorem, lies in C^k if $k < 2m - n/2$). From the abstract theory of semigroups in Section 6.2, we know that if $(u(0), \dot{u}(0)) \in \mathfrak{D}((A')^m)$, then $(u(t), \dot{u}(t)) \in \mathfrak{D}((A')^m)$ as well. For instance $(u(0), \dot{u}(0)) \in \mathfrak{D}((A')^3)$ means that

$$u(0) \in \mathfrak{D}(A^2) \quad \text{and} \quad \dot{u}(0) \in \mathfrak{D}(A).$$

Note that this automatically means $u(0)$ and $\dot{u}(0)$ must satisfy extra boundary conditions; in general, these extra conditions for $(u, \dot{u})$ to belong to $\mathfrak{D}((A')^m)$ are called the *compatibility conditions.*

In particular, if $u(0)$ and $\dot{u}(0)$ are C^∞ in x and belong to the domain of every power of A, then the solutions are C^∞ in (x, t) in the classical sense.

(2) If we have

$$a^{abcd}(x)\xi_a\xi_c\eta_b\eta_d \geq \epsilon(x)|\xi|^2|\eta|^2,$$

and $\epsilon(x)$ vanishes at some points, then one can still, under technical conditions sufficient to guarantee A is self-adjoint (see, e.g., Reed and Simon [1975], and references therein), get a quasi-contractive semigroup on $\mathcal{Y} \times L^2$, where $\mathcal{Y}$ is the completion of $H^1(\Omega)$ in the energy norm. One can show along the lines of 3.1, that if A' generates a quasi-contractive semigroup on $\mathcal{Y} \times L^2$, then we must have

$$\mathsf{a}^{abcd}(x)\xi_a\xi_c\eta_b\eta_d \geq 0.$$

If $\mathsf{a}^{abcd}\xi_a\xi_c\eta_b\eta_d < 0$ somewhere in Ω, we say strong ellipticity *strictly fails*. In 3.7 it is shown that in this case no semigroup is possible on any space $\mathcal{Y} \times L^2$.

Next we mention the abstract version of 3.1 (see also Box 3.1).

3.2 Theorem (*Weiss* [1967] *and Goldstein* [1969]) *Let $\mathcal{H}$ be a real Hilbert space and A a self-adjoint operator on $\mathcal{H}$ satisfying $\langle Ax, x\rangle \geq c\langle x, x\rangle$ for a constant $c > 0$. Let $A^{1/2}$ be the positive square root of A and let $\mathcal{H}_1$ be the domain of $A^{1/2}$ with the graph norm. Then the operator*

$$A' = \begin{pmatrix} 0 & 1 \\ -A & 0 \end{pmatrix}$$

generates a one-parameter group on $\mathcal{H}_1 \times \mathcal{H}$ with domain $\mathfrak{D}(A) \times \mathcal{H}_1$. The semigroup $e^{tA'}$ solves the abstract wave equation $(\partial^2 x)/(\partial t^2) = -Ax$.

Proof Our condition on A means that the graph norm of $A^{1/2}$ is equivalent to the norm $|||x||| = \langle A^{1/2}x, A^{1/2}x\rangle$. Thus on $\mathcal{H}_1 \times \mathcal{H}$ we can take the Hilbert space norm

$$\|(x, y)\|^2 = \langle A^{1/2}x, A^{1/2}x\rangle + \langle y, y\rangle.$$

We will show that A' is skew-adjoint on $\mathcal{H}_1 \times \mathcal{H}$, so the result follows from the real form of Stone's theorem (2.18(3)). Let us first check skew symmetry. Let $x_1, x_2 \in \mathfrak{D}(A)$ and $y_1, y_2 \in \mathcal{H}_1$. Then

$$\begin{aligned}\langle A'(x_1, x_2), (y_1, y_2)\rangle &= \langle (x_2, -Ax_1), (y_1, y_2)\rangle \\ &= \langle A^{1/2}x_2, A^{1/2}y_1\rangle + \langle -Ax_1, y_2\rangle \\ &= \langle Ax_2, y_1\rangle - \langle Ax_1, y_2\rangle\end{aligned}$$

since $x_2 \in \mathfrak{D}(A)$. Similarly, for $x_1, x_2 \in \mathcal{H}$, and $y_1, y_2 \in \mathfrak{D}(A)$,

$$\langle (x_1, x_2), A'(y_1, y_2)\rangle = \langle x_1, Ay_2\rangle - \langle x_2, Ay_1\rangle,$$

so A' is skew-symmetric.

To show A' is skew-adjoint, let $(y_1, y_2) \in \mathfrak{D}(A'^{+})$, where A'^{+} denotes the skew-adjoint of A'. This means there is $(z_1, z_2) \in \mathcal{H}_1 \times \mathcal{H}$ such that

$$\langle A'(x_1, x_2), (y_1, y_2)\rangle = -\langle (x_1, x_2), (z_1, z_2)\rangle$$

for all $(x_1, x_2) \in \mathfrak{D}(A) \times \mathcal{H}_1$. This assertion is equivalent to

$$\langle A^{1/2}x_2, A^{1/2}y_1\rangle = -\langle x_2, z_2\rangle \quad \text{for all} \quad x_2 \in \mathfrak{D}(A^{1/2})$$

and

$$\langle Ax_1, y_2\rangle = \langle A^{1/2}x_1, A^{1/2}z_1\rangle \quad \text{for all} \quad x_1 \in \mathfrak{D}(A).$$

The first statement implies $A^{1/2}y_1 \in \mathfrak{D}(A^{1/2})$ or $y_1 \in \mathfrak{D}(A)$ and the second implies $y_2 \in \mathfrak{D}(A^{1/2})$. Hence $\mathfrak{D}(A'^{+}) = \mathfrak{D}(A')$ so A' is skew-adjoint. ∎

Remarks

(1) The group generated by A' can be written explicitly in terms of that generated by $C = A^{1/2}$ as

$$e^{tA'} = \cosh(tC)(\text{Identity}) + \frac{\sinh(tC)}{C}A',$$

where for example $\cosh tC = (e^{tC} + e^{-tC})/2$. Division by C is in terms of the operational calculus.

(2) The condition $c > 0$ in the hypothesis can be relaxed to $c = 0$ if the spaces are modified as follows. Let A be self-adjoint and non-negative with trivial kernel and let $\mathcal{H}_A$ be the completion of $\mathcal{H}$ with respect to the norm $\|x\|_A^2 = \langle Ax, x\rangle$. Let $\mathfrak{X} = \mathcal{H}_A \times \mathcal{H}$ and let $A'(x, y) = (y, -Ax)$. Then the closure of A' is a generator in $\mathfrak{X}$. The argument follows the lines above (see Weiss [1967]).

(3) The wave equation $u_{tt} = \nabla^2 u$ does *not* generate a semigroup in $W^{1,p} \times L^p$ if $p \neq 2$ and $n > 1$. See Littman [1973].

Problem 3.1 Show that A' cannot be a generator on $\mathcal{H} \times \mathcal{H}$ unless A is a bounded operator. (*Hint:* $\begin{pmatrix} 0 & I \\ 0 & 0 \end{pmatrix}$ is bounded on $\mathcal{H} \times \mathcal{H}$; so if A' is a generator, so is $A_0 = \begin{pmatrix} 0 & 0 \\ A & 0 \end{pmatrix}$. Now compute e^{tA_0}.)

The next abstract theorem will show that stable dynamics implies hyperelasticity. Here, we say A' is *dynamically stable* if it generates a contractive semigroup on $\mathfrak{X}$ (relative to some norm on H^1).

3.3 Theorem (*Weiss* [1967]) *Let A be a linear operator in a Hilbert space $\mathcal{H}$ with domain $\mathfrak{D}(A)$. Let $\mathcal{Y}$ be a Hilbert space, with $\mathfrak{D}(A) \subset \mathcal{Y} \subset \mathcal{H}$. Assume that*

$$A' = \begin{pmatrix} 0 & I \\ A & 0 \end{pmatrix}, \qquad \mathfrak{D}(A') = \mathfrak{D}(A) \times \mathcal{Y}$$

generates a contractive semigroup on $\mathfrak{X} = \mathcal{Y} \times \mathcal{H}$. Then A is a self-adjoint operator, and in particular is symmetric.

Proof By the Lumer–Phillips theorem 2.16, with $\beta = 0$, we have

$$\langle (u, \dot{u}), A'(u, \dot{u})\rangle_{\mathfrak{X}} \leq 0, \qquad \text{that is,} \quad \langle u, \dot{u}\rangle_{\mathcal{Y}} + \langle \dot{u}, Au\rangle_{\mathcal{H}} \leq 0.$$

Since this holds for all $u \in \mathfrak{D}(A)$, $\dot{u} \in \mathcal{Y}$, we can replace $\dot{u}$ by $-\dot{u}$. The left-hand side changes sign, so we must have $\langle u, \dot{u}\rangle_{\mathcal{Y}} + \langle \dot{u}, Au\rangle_{\mathcal{H}} = 0$. Thus, $\langle \dot{u}, Au\rangle_{\mathcal{H}} = -\langle \dot{u}, u\rangle_{\mathcal{Y}}$, and so A is symmetric and non-positive. It is also self-adjoint since $(\lambda - A)$ is surjective for $\lambda > 0$ (see 2.18(4)). ∎

This result shows that *linear Cauchy elasticity cannot give a stable dynamical system in* $\mathcal{Y} \times L^2$ *unless it is hyperelastic.* This is, presumably, an undesirable situation for Cauchy elasticity. It is a semigroup analogue of the work theorems that are used to cast doubt on Cauchy elasticity (cf. Gurtin [1972b], p. 82).

The proof of the above theorem has another interesting corollary. It shows that, under the hypothesis given, the $\mathcal{Y}$-norm is necessarily the energy norm, $\|u\|_{\mathcal{Y}}^2 = -\langle u, Au\rangle_{\mathcal{X}}$, and that our semigroups are forced to be *groups of isometries.*

Remarks

(1) Weiss [1967] also shows that one is forced into working on Hilbert space as opposed to general Banach spaces (cf. Remark 3 following 3.2).

(2) Related to the contractive assumption is an abstract result of Nagy (Riesz and Nagy [1955], p. 396), namely, that a bounded one-parameter *group* on Hilbert space is actually unitary in an equivalent Hilbert norm.

Next we discuss the *energy criterion* for linear elastodynamics.

3.4 Definition We say that $\mathbf{a}$ is *stable* provided that there is a $c > 0$ such that

$$B(\boldsymbol{u}, \boldsymbol{u}) \geq c\|\boldsymbol{u}\|_{H^1}^2 \quad \text{for all} \quad \boldsymbol{u} \in H^1(\Omega).$$

That is, the elastic potential energy is positive-definite relative to the H^1-norm. We recall from 1.5 that stability implies strong ellipticity.

3.5 Theorem *A' is dynamically stable (in the sense that A' generates a contractive group on $H^1 \times L^2$ relative to some inner product on H^1) if and only if $\mathbf{a}$ is stable.*[10]

Proof In the proof of sufficiency in 3.1 we can take $\beta = 0$ by stability, so we get a contractive group.

Conversely, if we get a contractive group relative to some equivalent inner product $\langle\!\langle\ ,\ \rangle\!\rangle_{H^1}$ on H^1, we saw in the proof of Weiss' theorem 3.3 that we must have $B(\boldsymbol{u}, \boldsymbol{u}) = \langle\!\langle \boldsymbol{u}, \boldsymbol{u}\rangle\!\rangle_{H^1}$, which implies stability. ▮

3.6 Corollary *For the displacement problem in classical elasticity, uniform pointwise stability of the elasticity tensor, that is, there is a $\delta > 0$ such that*

$$\mathbf{e} \cdot \mathbf{c} \cdot \mathbf{e} \geq \delta \|\mathbf{e}\|^2$$

for all symmetric e_{ab}, implies stability.

---

[10] In 2.7, Chapter 5 we defined a semigroup to be stable when it is bounded. Note that there, stable means contractive relative to *some* Hilbert space structure. There is a slight technical difference.

Proof This follows by virtue of Korn's first inequality

$$\int_\Omega \|\boldsymbol{e}\|^2 \, dx \geq c \|\boldsymbol{u}\|^2_{H^1} \quad (c > 0),$$

where $e_{ab} = \frac{1}{2}(u_{a,b} + u_{b,a})$ and $\boldsymbol{u} = \boldsymbol{0}$ on $\partial\Omega$. (See 1.12.) ▮

For the traction problem, Korn's first inequality cannot hold since $\boldsymbol{e}$ is invariant under the Euclidean group. Instead we have *Korn's second inequality* (see 1.12),

$$\int_\Omega \|\boldsymbol{e}\|^2 \, dx + \int_\Omega \|\boldsymbol{u}\|^2 \, dx \geq c \|\boldsymbol{u}\|^2_{H^1}.$$

As it stands, this shows that positive-definiteness of the elasticity tensor implies Gårding's inequality. However, we already know Gårding's inequality is true from strong ellipticity alone. Thus Korn's second inequality is not required for existence. However, there is a deeper reason for Korn's second inequality. Namely, if we view the traction problem as a Hamiltonian system (as in Chapter 5) and move into center of mass and constant angular momentum "coordinates,"[11] then in the appropriate quotient space of $H^1 \times L^2$, we get a new Hamiltonian system and *in this quotient space*, Korn's second inequality can be interpreted as saying that *positive-definiteness of the elasticity tensor implies stability* and hence dynamical stability.

Problem 3.2 (On the level of a masters thesis.) Carry out the details of the remarks just given.

Finally, we sketch an argument due to Wilkes [1980] based on logarithmic convexity (see Knops and Payne [1971]) to show that dynamics is not possible when strong ellipticity fails, even when H^1 is replaced by some other space $\mathcal{Y}$ in 3.1.

3.7 Theorem *If the strong ellipticity condition strictly fails then A' cannot generate a semigroup on $\mathcal{Y} \times L^2$, where $\mathfrak{D}(A) \subset \mathcal{Y} \subset L^2$ and $\mathcal{Y}$ is a Banach space.*

Proof Suppose A' generates a semigroup $U(t)$ of type (M, β). Since strong ellipticity strictly fails, the argument used to prove Hadamard's theorem (Box 2.2) shows that

$$\inf_{\|\boldsymbol{u}\|_{L^2}=1} -\langle \boldsymbol{u}, A\boldsymbol{u} \rangle = -\infty.$$

(Roughly speaking, one can rescale $\boldsymbol{u}(x)$ keeping its L^2-norm constant, but blowing up its H^1-norm.) We can thus choose $\boldsymbol{u}(0)$, $\dot{\boldsymbol{u}}(0)$ such that

$$2\langle \dot{\boldsymbol{u}}(0), \boldsymbol{u}(0) \rangle > \beta, \qquad \|\boldsymbol{u}(0)\|^2_{L^2} = 1$$

and

$$\tfrac{1}{2}\langle \dot{\boldsymbol{u}}(0), \dot{\boldsymbol{u}}(0) \rangle - \tfrac{1}{2}\langle \boldsymbol{u}(0), A\boldsymbol{u}(0) \rangle = c < 0.$$

[11] In Hamiltonian systems language this is a special case of the general procedure of "reduction." See Abraham and Marsden [1978] and Marsden and Weinstein [1974].

Here $\langle\, , \rangle$ is the L^2 inner product. (For example, we can let $\dot{\boldsymbol{u}}(0) = \mu \boldsymbol{u}(0)$ where $\mu > \beta/2$ and $\boldsymbol{u}(0)$ has unit L^2-norm.) Let $(\boldsymbol{u}(t), \dot{\boldsymbol{u}}(t)) = U(t)(\boldsymbol{u}(0), \dot{\boldsymbol{u}}(0))$ and $F(t) = \frac{1}{2}\langle \boldsymbol{u}(t), \boldsymbol{u}(t)\rangle$. Then clearly

$$\dot{F} = \langle \dot{\boldsymbol{u}}, \boldsymbol{u}\rangle \quad \textit{and} \quad \ddot{F} = \langle \boldsymbol{u}, A\boldsymbol{u}\rangle + \langle \dot{\boldsymbol{u}}, \dot{\boldsymbol{u}}\rangle.$$

Note that c is the initial energy, and the energy is constant in time. By Schwarz's inequality,

$$\frac{\dot{F}^2}{F} = \frac{2\langle \dot{\boldsymbol{u}}, \boldsymbol{u}\rangle^2}{\langle \boldsymbol{u}, \boldsymbol{u}\rangle} \leq 2\|\dot{\boldsymbol{u}}\|_{L^2}^2 = 2c + \ddot{F}.$$

Thus, $F\ddot{F} - \dot{F}^2 \geq -2cF \geq 0$. Hence, $(d^2/dt^2)(\log F) \geq 0$, and so

$$F(t) \geq F(0) \exp\left(\frac{\dot{F}(0)}{F(0)}t\right), \qquad \text{that is,} \quad \|\boldsymbol{u}\|_{L^2}^2 \geq \|\boldsymbol{u}(0)\|_{L^2}^2\, e^{\gamma t},$$

where $\gamma = 2\langle \dot{\boldsymbol{u}}(0), \boldsymbol{u}(0)\rangle / \langle \boldsymbol{u}(0), \boldsymbol{u}(0)\rangle$. Because $U(t)$ is a semigroup of type (M, β), the $\mathcal{Y}$ topology is stronger than the L^2 topology and $\gamma > \beta$, such an inequality is impossible. ∎

Box 3.1 ***Hamiltonian One-Parameter Groups***

In Section 5.2 we studied some properties of linear Hamiltonian systems. Now we re-examine a few of these topics in the light of semigroup theory. The main result is an abstract existence theorem that applies to hyperelasticity under the assumption of stability.

Let $\mathfrak{X}$ be a Banach space and ω a (weak) symplectic form (see Definition 2.1 in Chapter 5). We recall that a linear operator A on $\mathfrak{X}$ is called *Hamiltonian* when it is ω-skew; that is, $\omega(Ax, y) = -\omega(x, Ay)$ for all $x, y \in \mathfrak{D}(A)$. Let A^+ denote the ω-adjoint of A; that is,

$$\mathfrak{D}(A^+) = \{w \in \mathfrak{X} \mid \text{there is a } z \in \mathfrak{X} \text{ such that } \omega(z, x) = \omega(w, Ax) \text{ for all } x \in \mathfrak{D}(A)\}$$

and $A^+w = z$. We call A *ω-skew-adjoint* when $A = -A^+$. This is a stronger condition than ω-skew symmetry.

In 2.6, Chapter 5, we saw that if A is Hamiltonian and generates a semigroup $U(t)$, then each $U(t)$ is a canonical transformation. The next result shows that A is necessarily ω-skew-*adjoint*. The result, whose proof is based on an idea of E. Nelson, is due to Chernoff and Marsden [1974] (see also Marsden [1968b]).

3.8 Proposition *Let $U(t)$ be a one-parameter group of canonical transformations on $\mathfrak{X}$ with generator A. Then A is ω-skew-adjoint.*

Proof Let A^+ be the ω-adjoint of A. Since $\omega(U(t)x, U(t)y) = \omega(x, y)$, we have

$$\omega(Ax, y) + \omega(x, Ay) = 0 \quad \text{for} \quad x, y \in \mathfrak{D}(A),$$

so $A^+ \supset -A$. Now let $f \in \mathfrak{D}(A^+)$ and $A^+f = g$. For $x \in \mathfrak{D}(A)$, write

$$U(t)x = x + \int_0^t AU(s)x\,ds,$$

so

$$\omega(U(t)x, f) = \omega(x, f) + \int_0^t \omega(AU(s)x, f)\,ds.$$

Thus

$$\omega(x, U(-t)f) = \omega(x, f) + \int_0^t \omega(x, U(-s)A^+f)\,ds.$$

Since $\mathfrak{D}(A)$ is dense,

$$U(-t)f = f + \int_0^t U(-s)A^+ f\,ds.$$

It follows that $f \in \mathfrak{D}(A)$ and $-Af = A^+f$. ∎

Problem 3.3 Deduce that the generator of a one-parameter unitary group on complex Hilbert space is i times a self-adjoint operator (one-half of Stone's theorem).

In 3.8, ω-skew-adjointness is not sufficient for A to be a generator. (This is seen from the ill-posed problem $\ddot{\phi} = -\nabla^2\phi$, for example.) However, it does become sufficient if we add a positivity requirement.

3.9 Theorem (*Chernoff and Marsden* [1974]) *Let $\mathfrak{X}$ be a Banach space and ω a weak symplectic form on $\mathfrak{X}$.*

Let A be an ω-skew-adjoint operator in $\mathfrak{X}$ and set

$$[x, y] = \omega(Ax, y),$$

the energy inner product. Assume the following stability condition:

$$[x, x] \geq c\,\|x\|_{\mathfrak{X}}^2$$

for a constant $c > 0$.

Let $\mathcal{H}$ be the completion of $\mathfrak{D}(A)$ with respect to [,], *let*

$$\mathfrak{D}(\tilde{A}) = \{x \in \mathfrak{D}(A) \mid Ax \in \mathcal{H}\} \text{ and set } \tilde{A}x = Ax \text{ for } x \in \mathfrak{D}(\tilde{A}).$$

Then $\tilde{A}$ generates a one-parameter group of canonical transformations in $\mathcal{H}$ (relative to $\tilde{\omega}$, the restriction of $\tilde{\omega}$ to $\mathcal{H}$) and these are, moreover, isometries relative to the energy inner product on $\mathcal{H}$.

Proof Because the energy inner product satisfies $[x, x] \geq c\,\|x\|_{\mathfrak{X}}^2$, we can identify $\mathcal{H}$ with a subspace of $\mathfrak{X}$. Relative to [,] we note that $\tilde{A}$ is skew-symmetric: for $x, y \in \mathfrak{D}(A)$,

$$[x, \tilde{A}y] = \omega(Ax, Ay) = -\omega(Ay, Ax) = -[y, \tilde{A}x] = -[Ax, y].$$

We next shall show that $\tilde{A}$ is skew-adjoint. To do this, it is enough to show $\tilde{A}: \mathfrak{D}(\tilde{A}) \longrightarrow \mathfrak{H}$ is onto. This will follow if we can show that A: $\mathfrak{D}(A) \longrightarrow \mathfrak{X}$ is onto; see 2.18.

Let $w \in \mathfrak{X}$. By the Riesz representation theorem, there is an $x \in \mathfrak{H}$ such that

$$\omega(w, y) = [x, y] \quad \text{for all} \quad y \in \mathfrak{H}.$$

In particular,

$$\omega(w, y) = \omega(Ay, x) = -\omega(x, Ay) \quad \text{for all} \quad y \in \mathfrak{D}(A).$$

Therefore, $x \in \mathfrak{D}(A^+) = \mathfrak{D}(A)$ and $Ax = w$. Thus A is onto.

It remains to show that $\tilde{\omega}$ is invariant under $U(t) = e^{t\tilde{A}}$. By 2.6, Chapter 5, we need only verify that $\tilde{A}$ is $\tilde{\omega}$-skew. Indeed for $x, y \in \mathfrak{D}(\tilde{A})$,

$$\tilde{\omega}(\tilde{A}x, y) = \omega(Ax, y) = -\omega(x, Ay) = -\tilde{\omega}(x, Ay). \quad \blacksquare$$

3.10 Example (*Abstract Wave Equation*) (See Examples 2.9, Chapter 5 and Theorem 3.2 above.) Let $\mathfrak{H}$ be a real Hilbert space and B a self-adjoint operator satisfying $B \geq c > 0$. Then

$$A = \begin{pmatrix} 0 & I \\ -B & 0 \end{pmatrix}$$

is Hamiltonian on $\mathfrak{X} = \mathfrak{D}(B^{1/2}) \times \mathfrak{H}$ with $\mathfrak{D}(A) = \mathfrak{D}(B) \times \mathfrak{D}(B^{1/2})$, $\omega((x_1, y_1), (x_2, y_2)) = \langle y_2, x_1 \rangle - \langle x_2, y_1 \rangle$, and energy $H(x, y) = \frac{1}{2}\|y\|^2 + \frac{1}{2}\langle Bx, x \rangle$. Theorem 3.9 then reproduces Theorem 3.2, which we proved above concerning the abstract wave equation $\ddot{x} = -Bx$. It follows from 2.6, Chapter 5 that the corresponding one-parameter group consists of canonical transformations that preserve energy. The dynamics is thus stable.

Problem 3.4 Show that if **a** is stable in the sense of 3.4, then Theorem 3.9 applies to A' and reproduces one direction of the result obtained in 3.5.

Problem 3.5 (Converse of 3.9.) Suppose A is Hamiltonian on $\mathfrak{X}$ and generates a one-parameter group $U(t)$. Suppose, relative to some complex Hilbert space structure $\langle\ ,\ \rangle$ on $\mathfrak{X}$, that $U(t)$ is contractive and $\omega(x, y) = -\text{Im}\langle x, y \rangle$. Show that $\omega(Ax, y) = \text{Re}\langle x, y \rangle$ and hence that A is stable (cf. 3.3).

Box 3.2 ***A Semigroup Arising in Panel Flutter***

Consider small vibrations of a panel, as shown in Figure 6.3.1. Neglecting nonlinear and two-dimensional effects, the equations for the panel deflection $v(x, t)$ are

$$\ddot{v} + \alpha\dot{v}'''' + v'''' - \Gamma v'' + \rho v' + \sqrt{\rho}\,\delta\dot{v} = 0, \qquad \text{(PF)}$$

where $\dot{} = \partial/\partial t$ and $' = \partial/\partial x$. Here α is a viscoelastic structural damping constant, ρ is an aerodynamic pressure, Γ is an in-plane tensile load, and $\sqrt{\rho}\,\delta$ is aerodynamic damping. We assume $\alpha > 0$, $\delta > 0$, and $\rho > 0$. If the edges of the plate are simply supported, we impose the boundary conditions $v = 0$, $v'' + \alpha\dot{v}'' = 0$ at $x = 0, 1$.

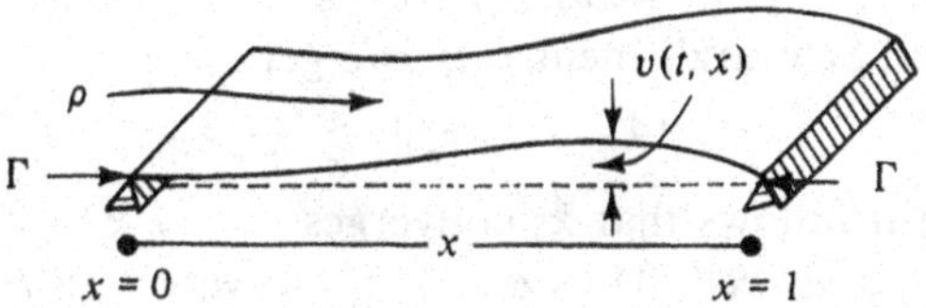

Figure 6.3.1

The equations (PF) are derived in Dowell [1975]. Nonlinear versions of (PF) are discussed in Chapter 7. Let $H^2_\partial = \{u \in H^2([0, 1]) \mid u = 0$ at $x = 0, 1\}$ and $\mathfrak{X} = H^2_\partial \times L^2$. Define the operator A on $\mathfrak{X}$ by

$$A\begin{pmatrix} v \\ \dot{v} \end{pmatrix} = \begin{pmatrix} \dot{v} \\ -\alpha\dot{v}'''' - v'''' \end{pmatrix}$$

with $\mathfrak{D}(A) = \{(v, \dot{v}) \in H^2_\partial \times L^2 \mid v + \alpha\dot{v} \in H^4, v'' + \alpha\dot{v}'' = 0$ at $x = 0, 1$, and $\dot{v} \in H^2_\partial\}$. On $\mathfrak{X}$ define the inner product

$$\langle (v, \dot{v}), (w, \dot{w}) \rangle = \langle v'', w'' \rangle_{L^2} + \langle \dot{v}, \dot{w} \rangle_{L^2},$$

where $\langle\ ,\ \rangle_{L^2}$ denotes the L^2 inner product.

Let $B\colon \mathfrak{X} \to \mathfrak{X}$ be defined by

$$B\begin{pmatrix} v \\ \dot{v} \end{pmatrix} = \begin{pmatrix} 0 \\ \Gamma v'' - \rho v' - \sqrt{\rho}\,\delta\dot{v} \end{pmatrix}$$

and observe that equations (PF) may be written

$$\frac{d}{dt}\begin{pmatrix} v \\ \dot{v} \end{pmatrix} = A\begin{pmatrix} v \\ \dot{v} \end{pmatrix} + B\begin{pmatrix} v \\ \dot{v} \end{pmatrix}. \qquad \text{(PF)}'$$

3.11 Proposition *The equations (PF)′ generate a contraction semigroup on $\mathfrak{X}$.*

Proof Since B is a bounded operator, it suffices to show that A is a generator. (See 2.17(1).) To do this, observe that

$$\left\langle A\begin{pmatrix} v \\ \dot{v} \end{pmatrix}, \begin{pmatrix} v \\ \dot{v} \end{pmatrix} \right\rangle = \left\langle \begin{pmatrix} \dot{v} \\ -\alpha \dot{v}'''' - v'''' \end{pmatrix}, \begin{pmatrix} v \\ \dot{v} \end{pmatrix} \right\rangle = \langle \dot{v}'', v'' \rangle_{L^2} - \langle \alpha \dot{v}'''' + v'''', \dot{v} \rangle_{L^2}$$
$$= \langle \dot{v}'', v'' \rangle_{L^2} - \langle \alpha \dot{v}'' + v'', \dot{v}'' \rangle_{L^2}$$
$$= -\alpha \langle \dot{v}'', \dot{v}'' \rangle_{L^2} \leq 0.$$

Thus A is dissipative. Next we show that $(\lambda - A)$ is onto for $\lambda > 0$. We do this in two steps.

First of all, the range of $\lambda - A$ is closed; let $x_n = (v_n, \dot{v}_n) \in \mathfrak{D}(A)$ and suppose $y_n = (\lambda - A)(v_n, \dot{v}_n) \longrightarrow y \in \mathfrak{X}$. From the above dissipative estimate, and Schwarz's inequality, we get

$$\| \lambda x_n - A x_n \|_x \geq \lambda \| x_n \|,$$

from which it follows that x_n converges to say x in $\mathfrak{X}$. Since $y_n \longrightarrow y$, Ax_n converges as well. Thus $\alpha \dot{v}_n + v_n$ converges in H^4, so $x \in \mathfrak{D}(A)$ and $Ax = y$.

Secondly, the range of $\lambda - A$ is dense. Suppose that there is a $y \in \mathfrak{X}$ such that $\langle (\lambda - A)x, y \rangle = 0$ for all $x \in \mathfrak{D}(A)$. Thus if $x = (v, \dot{v})$, $y = (w, \dot{w})$, then

$$\langle \lambda v'' - \dot{v}'', w'' \rangle_{L^2} = 0 \quad \text{and} \quad \langle \lambda \dot{v} + \alpha \dot{v}'''' + v'''', \dot{w} \rangle_{L^2} = 0.$$

Set $\dot{v} = 0$; the first equation gives $w'' = 0$, so $w = 0$. The second equation with $\dot{v} = 0$ shows $\dot{w}$ satisfies $\dot{w}'''' = 0$ in the weak sense, so is smooth. Setting $v = 0$ and $\dot{v} = \dot{w}$ then shows $\dot{w} = 0$ since $\lambda > 0$ and $\alpha > 0$.

The result now follows from 2.16. ∎

Problem 3.6

(a) Show that the operator

$$f_0\begin{pmatrix} v \\ \dot{v} \end{pmatrix} = \begin{pmatrix} \dot{v} \\ -v'''' + \Gamma v'' \end{pmatrix}$$

is Hamiltonian on $\mathfrak{X}$ and generates a group. Show that the energy is

$$H(v, \dot{v}) = \tfrac{1}{2} \| \dot{v} \|_{L^2}^2 - \frac{\Gamma}{2} \| v' \|_{L^2}^2 + \tfrac{1}{2} \| v'' \|_{L^2}^2.$$

(b) Prove the inequality $\| v' \|_{L^2}^2 \leq \| v'' \|_{L^2}^2 / \pi^2$ using Fourier series. (A Wirtinger inequality.)

(c) Show that f_0 is stable if $0 \leq \Gamma \leq \pi^2$. (Buckling occurs for $\Gamma > \pi^2$.)

We have shown that e^{tA} is a contractive semigroup. Actually, the origin is globally attracting in the sense that in a suitable equivalent norm

$$\| e^{tA} \|_{\mathfrak{X}} \leq e^{\epsilon t} \qquad (\epsilon > 0).$$

This is, roughly speaking, because the spectrum of A (computed by separation of variables) is discrete and consists of eigenvalues

$$\lambda_j = -\frac{j^4\pi^4\alpha}{2}\left(1 \pm \sqrt{1 - \frac{4}{\alpha^2\pi^4 j^4}}\right) \quad (j = 1, 2, \ldots)$$

with Re $\lambda_j \leq \max\{-\alpha\pi^4/2, -1/\alpha\}$. (See 2.17(vi).)

For additional information, see Chapter 7, Parks [1966], Holmes and Marsden [1978a], and Walker [1980].

Box 3.3 ***Linear Elastodynamics with Dissipative Mechanisms***

We give three examples of how semigroup theorems can be applied to modifications of the equations of linear elastodynamics. These are: (1) *viscoelasticity with dissipation of rate type;* (2) *thermal dissipation;* and (3) *viscoelasticity of memory* (*or Boltzmann*) *type.* References for these topics, where related topics may be found, are Weiss [1967], Dafermos [1976], and Navarro [1978].

1. *Dissipation of Rate Type* The form of the abstract equations is $\ddot{u} = Au + B\dot{u}$, and the relevant abstract theorem is as follows.

3.12 Proposition *Suppose A and B generate (quasi)-contractive semigroups on Hilbert space $\mathfrak{H}$ and $\mathfrak{D}(B) \subset \mathfrak{D}(A)$. Suppose $A' = \begin{pmatrix} 0 & I \\ A & 0 \end{pmatrix}$ generates a (quasi)-contractive semigroup on $\mathfrak{Y} \times \mathfrak{H}$ with domain $\mathfrak{D}(A) \times \mathfrak{Y}$, where $\mathfrak{D}(A) \subset \mathfrak{Y}$. Let $C = \begin{pmatrix} 0 & I \\ A & B \end{pmatrix}$ with $\mathfrak{D}(C) = \mathfrak{D}(A) \times \mathfrak{D}(B)$. Then $\bar{C}$, the closure of C, generates a (quasi)-contractive semigroup on $\mathfrak{Y} \times \mathfrak{H}$.*

The proof will use the following result.

3.13 Lemma *Let A and B generate (quasi)-contractive semigroups on a Banach space $\mathfrak{X}$ and let $\mathfrak{D}(B) \subset \mathfrak{D}(A)$. Then there is a $\delta > 0$ such that $cA + B$ generates a quasi-contractive semigroup if $0 \leq c \leq \delta$.*

This result is due to Trotter [1959]. The interested reader should look up the original article for the proof, or else deduce it from 2.17(4).

Proof of 3.12 From 2.16, there is a $\beta > 0$ such that

$$\langle (u, \dot{u}), A'(u, \dot{u}) \rangle_{\mathcal{Y} \times \mathcal{X}} \leq \beta \| (u, \dot{u}) \|^2_{\mathcal{Y} \times \mathcal{X}}$$

and a $\gamma > 0$ be such that $\langle B\dot{u}, \dot{u} \rangle_{\mathcal{X}} \leq \gamma \| \dot{u} \|^2_{\mathcal{X}}$. Setting $B' = \begin{pmatrix} 0 & 0 \\ 0 & B \end{pmatrix}$, we get

$$\begin{aligned} &\langle (u, \dot{u}), (A' + B')(u, \dot{u}) \rangle_{\mathcal{Y} \times \mathcal{X}} \\ &\qquad \leq \beta(\| u \|^2_{\mathcal{Y}} + \| \dot{u} \|^2_{\mathcal{X}}) + \gamma \| \dot{u} \|^2_{\mathcal{X}} \leq \rho \| (u, \dot{u}) \|^2_{\mathcal{Y} \times \mathcal{X}}, \end{aligned}$$

where $\rho = \beta + \gamma$. By 2.16 it is sufficient to show that $\lambda - C = \lambda - A' - B'$ has dense range for λ sufficiently large. Suppose $(v, \dot{v})$ is orthogonal to the range of $\lambda - C$. Then

$$\langle \lambda u - \dot{u}, v \rangle_{\mathcal{Y}} + \langle \lambda \dot{u} - Au - B\dot{u}, \dot{v} \rangle_{\mathcal{X}} = 0$$

$$\text{for} \quad u \in \mathfrak{D}(A) \quad \text{and} \quad \dot{u} \in \mathfrak{D}(A).$$

Setting $\dot{u} = \lambda u$, we get

$$\langle \lambda^2 u - Au - \lambda B u, \dot{v} \rangle = 0, \qquad \text{that is,} \quad \left\langle \lambda u - \frac{1}{\lambda} Au - Bu, \dot{v} \right\rangle = 0.$$

If $\lambda > \delta^{-1}$, where δ is given in the lemma, we conclude that $\lambda - A/\lambda - B$ is onto, so $\dot{v} = 0$. From the original orthogonality condition, we get $v = 0$. ∎

If A is symmetric, then $\ddot{u} = Au$ is Hamiltonian. Furthermore, if $B \leq 0$, then the energy is decreasing:

$$\tfrac{1}{2} \frac{d}{dt} (\langle \dot{u}, \dot{u} \rangle - \langle u, Au \rangle) = \langle \dot{u}, B\dot{u} \rangle \leq 0.$$

This is the usual situation for rate-type dissipation.

3.14 Example If a^{abcd} is strongly elliptic, then

$$\rho \ddot{u}^a = (\mathsf{a}^{abcd} u_{c|d})_{|b} + \dot{u}^a{}_{|b|b},$$

that is,

$$\rho \ddot{u} = \operatorname{div}(\mathbf{a} \cdot \nabla u) + \nabla^2 \dot{u}$$

with, say, displacement boundary conditions, generates a quasi-contractice semigroup on $\mathcal{H} = H^1 \times L^2$. If the elastic energy is positive-definite—that is, stability holds—then the semigroup is contractive. One can establish trend to equilibrium results either by spectral methods (see 2.17(vi)) or by Liapunov techniques (see Dafermos [1976] and Potier–Ferry [1978a], Ch. 11).

Problem 3.6 Show that 3.12 applies to the panel flutter equations (PF) in the preceding box, with A' Hamiltonian. For what parameter values is the energy decreasing?

2. *Dissipation of Thermal Type* Now the equations take the form

$$\ddot{u} = Au + B\theta, \qquad \dot{\theta} = C\theta + D\dot{u}. \tag{TE}$$

We make these assumptions:

(i) $A' = \begin{pmatrix} 0 & I \\ A & 0 \end{pmatrix}$ generates a quasi-contractive semigroup on $\mathcal{Y} \times \mathcal{H}$.

(ii) C is a non-positive self-adjoint operator on a Hilbert space $\mathcal{H}_\theta$.

(iii) B is an operator from $\mathcal{H}_\theta$ to $\mathcal{H}$, is densely defined; $D = -B^*$, and is densely defined.

(iv) $\mathfrak{D}(A) \subset \mathfrak{D}(D) \subset \mathcal{Y}$.

(v) $\mathfrak{D}(C) \subset \mathfrak{D}(B)$.

(vi) $B(1 - C)^{-1}D$, a non-positive symmetric operator, has self-adjoint closure (i.e., is essentially self-adjoint). (In Example 3.16 below $B(1 - C)^{-1}D$ is bounded.)

Let $G = \begin{pmatrix} 0 & I & 0 \\ A & 0 & B \\ 0 & D & C \end{pmatrix}$ with domain $\mathfrak{D}(A) \times \mathfrak{D}(D) \times D(C) \subset \mathcal{Y} \times \mathcal{H} \times \mathcal{H}_\theta = \mathfrak{X}$, so that $\dfrac{d}{dt}\begin{pmatrix} u \\ \dot{u} \\ \theta \end{pmatrix} = G\begin{pmatrix} u \\ \dot{u} \\ \theta \end{pmatrix}$ represents the thermoelastic system (TE).

3.15 Proposition *Under assumptions* (i)–(vi), $\bar{G}$ *generates a quasi-contractive semigroup on* $\mathfrak{X}$. *If* A' *generates a contractive semigroup, so does* $\bar{G}$.

Proof In the inner product on $\mathcal{Y} \times \mathcal{H} \times \mathcal{H}_\theta$, we have

$$\begin{aligned}\langle (u, \dot{u}, \theta), G(u, \dot{u}, \theta)\rangle &= \langle (u, \dot{u}, \theta), (\dot{u} + Au + B\theta, C\theta + D\dot{u})\rangle \\ &= \langle u, \dot{u}\rangle_{\mathcal{Y}} + \langle \dot{u}, Au + B\theta\rangle_{\mathcal{H}} + \langle \theta, C\theta + D\dot{u}\rangle_{\mathcal{H}_\theta}\end{aligned}$$

$$\leq \beta \|(u, \dot{u})\|^2_{\mathcal{Y}\times\mathcal{H}} + \langle \dot{u}, B\theta \rangle + \langle \theta, D\dot{u} \rangle + \langle \theta, C\theta \rangle$$
$$= \beta \|(u, \dot{u})\|^2_{\mathcal{Y}\times\mathcal{H}} + \langle \theta, C\theta \rangle$$
$$\leq \beta \|(u, \dot{u})\|^2_{\mathcal{Y}\times\mathcal{H}} \leq \beta \|(u, \dot{u}, \theta)\|^2_{\mathfrak{X}},$$

so $(G - \beta)$ is dissipative. By 2.16, it remains to show that for λ sufficiently large, $(\lambda - G)$ has dense range. Let $(v, \dot{v}, g)$ be orthogonal to the range:

$$\langle \lambda u - \dot{u}, v \rangle + \langle \lambda \dot{u} - Au - B\theta, \dot{v} \rangle + \langle \lambda\theta - C\theta - D\dot{u}, g \rangle = 0.$$

For $u \in \mathfrak{D}(A)$, let $\dot{u} = \lambda u$ and $\theta = \lambda(\lambda - C)^{-1} Du$. Then

$$\langle \lambda^2 u - Au - \lambda B(\lambda - C)^{-1} Du, \dot{v} \rangle = 0.$$

Using 3.13 and the same argument as in the preceding proposition, we see that if λ is sufficiently large, $\lambda^2 - A - \lambda B(\lambda - C)^{-1} D$ has dense range. Thus $\dot{v} = 0$. Taking $\dot{u}$ and $\theta = 0$, one sees that $v = 0$ and taking $u = 0 = \dot{u}$, one gets $\theta = 0$. ∎

3.16 Example If $\mathbf{a}$ is strongly elliptic, then the system

$$\begin{cases} \rho\ddot{u} = \operatorname{div}(\mathbf{a}\cdot\nabla u) + m\nabla\theta, \\ c\dot{\theta} = k\nabla^2\theta + \dfrac{m}{\rho}\nabla\cdot\dot{u} \quad (u, \theta = 0 \quad \text{on} \quad \partial\Omega), \end{cases}$$

generates a quasi-contractive semigroup on $H^1 \times L^2 \times L^2$ (with the L^2 spaces appropriately weighted), where $c, k > 0$, $m > 0$.[12]

3. *Viscoelasticity of Memory (or Boltzmann) Type* The equations now have the form

$$\ddot{u} = Au + Bw, \qquad \dot{w} = Cw. \tag{BE}$$

We make these assumptions:

(i) $A' = \begin{pmatrix} 0 & I \\ A & 0 \end{pmatrix}$ generates a quasi-contractive semigroup on $\mathcal{Y} \times \mathcal{H}$.
(ii) C generates a contractive semigroup on $\tilde{\mathcal{H}}$.
(iii) There is an injection $i\colon \mathcal{H} \to \tilde{\mathcal{H}}$ (corresponding to constant histories) such that $C \circ i = 0$.
(iv) B is a densely defined operator of $\tilde{\mathcal{H}}$ to $\mathcal{H}$ such that $i \circ B$ is symemtric and non-negative, $\mathfrak{D}(B) \subset \mathfrak{D}(C)$, $\mathfrak{D}(B)$ is a core for C, and B is one-to-one.

---

[12] In Theorem 3.1 we saw that "well-posedness" of the elastic part implies $\mathbf{a}$ is strongly elliptic. The Clausius–Duhem inequality implies $k \geq 0$. Well-posedness of the heat part then implies $c > 0$, and then of the whole implies $m \geq 0$.

(v) Let $\mathfrak{A}(w_1, w_2) = \langle i \cdot Bw_1, w_2 \rangle_{\tilde{\mathfrak{K}}}$ for $w_1, w_2 \in \mathfrak{D}(B)$. Suppose that on $\mathfrak{X} = \mathcal{Y} \times \mathcal{K} \times \tilde{\mathcal{K}}$,

$$\|(u, \dot{u}, w)\|_{\mathfrak{X}}^2 = \|(u, \dot{u})\|_{\mathcal{Y}\times\mathcal{K}}^2 + \mathfrak{A}(iu - w, iu - w) - \mathfrak{A}(iu, iu)$$

is an inner product equivalent to the original one. Let

$$G = \begin{pmatrix} 0 & I & 0 \\ A & 0 & B \\ 0 & 0 & C \end{pmatrix},$$

which is the operator on $\mathfrak{X}$ corresponding to the equations (BE), with domain $\mathfrak{D}(A) \times \mathcal{Y} \times \mathfrak{D}(B)$.

(vi) $\langle iBw, Cw \rangle \geq 0$ for all $w \in \mathfrak{D}(B)$.

3.17 Proposition *Under assumptions* (i)–(vi), $\bar{G}$ *generates a quasi-contractive semigroup on* $\mathfrak{X}$. *Moreover,* $\exp(t\bar{G})$ *is contractive if* A' *generates a contractive semigroup.*

Proof Using the $\mathfrak{X}$ inner product of (*v*),

$$\begin{aligned}
&\langle (u, \dot{u}, w), (\dot{u}, Au + Bw, Cw) \rangle \\
&\quad = \langle u, \dot{u} \rangle + \langle \dot{u}, Au \rangle + \langle \dot{u}, Bw \rangle + \langle iB(iu - w), i\dot{u} - Cw \rangle \\
&\qquad\qquad - \langle iBiu, iu \rangle \\
&\quad = \langle u, \dot{u} \rangle + \langle \dot{u}, Au \rangle + \langle iB(iu - w), - Cw \rangle \\
&\quad = \langle u, \dot{u} \rangle + \langle \dot{u}, Au \rangle - \langle iB(iu - w), C(iu - w) \rangle \\
&\quad \leq \langle u, \dot{u} \rangle + \langle \dot{u}, , Au \rangle \leq \beta \|u, \dot{u}\|_{\mathcal{Y}\times\mathcal{K}}^2 \leq \beta \|(u, \dot{u}, w)\|_{\mathfrak{X}}^2
\end{aligned}$$

for a constant β.

It remains to show that $(\lambda - G)$ has dense range for λ sufficiently large. If $(v, \dot{v}, h)$ is orthogonal to the range in the original inner product,

$$\langle \lambda u - \dot{u}, v \rangle_{\mathcal{Y}} + \langle \lambda u - Au - Bw, \dot{v} \rangle_{\mathcal{K}} + \langle \lambda w - Cw, h \rangle_{\tilde{\mathcal{K}}} = 0,$$

for all $u \in \mathfrak{D}(A)$, $\dot{u} \in \mathcal{Y}$ and $w \in \mathfrak{D}(B)$. Taking $\dot{u} = \lambda u$ and $w = 0$, and using the fact that $(\lambda - A)$ is surjective, we get $\dot{v} = 0$. Choosing $u, \dot{u} = 0$, and using the fact that $\mathfrak{D}(B)$ is a core for C, we find that $h = 0$, and finally choosing $\dot{u} = 0$ and $w = 0$ gives $v = 0$. ∎

3.18 Example (See Coleman and Mizel [1966] and Navarro [1978].) Suppose that the body occupies a bounded region $\Omega \subset \mathbb{R}^n$ with smooth boundary $\partial\Omega$, and that the reference configuration is a natural state in which stress is zero and base temperature θ_0 is a strictly positive constant. Let $x \in \Omega$ be the position of a material point, $\mathbf{u}(x, t)$ the displacement, and $\theta(x, t)$ the temperature difference from θ_0.

We assume that the Cauchy stress $\boldsymbol{\sigma}$ and specific entropy difference

η are given by functionals depending upon both displacement and temperature difference history in the following form:

$$\boldsymbol{\sigma}(x,t) = \mathbf{g}(x,0)\cdot\nabla\boldsymbol{u}(x,t) - \theta(x,t)\boldsymbol{l}(x,0) + \int_0^\infty \{\mathbf{g}'(x,s)\cdot\nabla\boldsymbol{u}(x,t-s) - \boldsymbol{l}'(x,s)\theta(x,t-s)\}\,ds$$

$$\rho(x)\eta(x,t) = \boldsymbol{l}(x,0)\cdot\nabla\boldsymbol{u}(x,t) + \rho(x)c(x,0)\theta(x,t)/\theta_0 + \int_0^\infty \{\boldsymbol{l}'(x,s)\cdot\nabla\boldsymbol{u}(x,t-s) + \rho(x)c'(x,s)\theta(x,t-s)/\theta_0\}\,ds,$$

where $\rho(x)$ is the mass density in the natural state and $\partial a/\partial s$ is denoted by a'. The material functions $\mathbf{g}(x,s)$, $\boldsymbol{l}(x,s)$, and $c(x,s)$ $(s \geq 0)$ are the relaxation tensors of fourth, second, and zero order, respectively ($\mathbf{g}$ is assumed to have the symmetry $\mathrm{g}^{abcd} = \mathrm{g}^{cdab}$). We call $\mathbf{g}(x,0)$, $\boldsymbol{l}(x,0)$, and $c(x,0)$ the *instantaneous elastic modulus*, *instantaneous stress-temperature tensor*, and *instantaneous specific heat*, respectively. We also assume Fourier's law for the heat flux vector $\boldsymbol{q}(x,t)$;

$$\boldsymbol{q}(x,t) = -\boldsymbol{\kappa}(x)\cdot\nabla\theta(x,t), \qquad \text{that is,} \quad q^a = -k^{ab}\theta_{/b},$$

where $\boldsymbol{\kappa}(x)$ is the *thermal conductivity* in the reference configuration. The local equation for balance of momentum is then

$$\rho(x)\dot{\boldsymbol{v}}(x,t) = \operatorname{div}\boldsymbol{\sigma}(x,t), \quad \text{where} \quad \boldsymbol{v} = \dot{\boldsymbol{u}},$$

while the linearized energy balance equation becomes

$$\theta_0\rho(x)\dot{\eta}(x,t) + \operatorname{div}\boldsymbol{q}(x,t) = 0.$$

Substitution of σ, η, and $\boldsymbol{q}$ into these equations then yields the system of coupled equations for the linear theory of thermoeleastic materials with memory (with no body forces or heat supply):

$$\left.\begin{aligned}
\dot{\boldsymbol{v}}(x,t) &= \frac{1}{\rho(x)}\operatorname{div}(\mathbf{g}(x,0)\cdot\nabla\boldsymbol{u}(x,t)) - \theta(x,t)\boldsymbol{l}(x,0) \\
&\quad + \int_0^\infty \{\mathbf{g}'(x,s)\cdot\nabla\boldsymbol{u}(x,t-s) - \boldsymbol{l}'(x,s)\theta(x,t-s)\}\,ds) \\
\dot{\theta}(x,t) &= \theta_0(\operatorname{div}\{\boldsymbol{\kappa}(x)\cdot\nabla\theta(x,t)\}/\theta_0 - \boldsymbol{l}(x,0)\cdot\nabla\boldsymbol{v}(x,t) \\
&\quad + \int_0^\infty \{\boldsymbol{l}'(x,s)\cdot\nabla\dot{\boldsymbol{u}}(x,t-s) \\
&\quad + (\rho(x)\theta_0)c'(x,s)\dot{\theta}(x,t-s)\}\,ds)/(\rho(x)c(x,0))
\end{aligned}\right\}\ \text{(TEM)}$$

The boundary conditions are assumed to be

$$\boldsymbol{u}(x,t) = \mathbf{0}, \qquad \theta(x,t) = 0 \quad \text{on} \quad \partial\Omega \times [0,\infty),$$

while the prescribed initial histories for the displacement and temperature difference are given by

$$\boldsymbol{u}(x,-s) = \boldsymbol{w}^0(x,-s), \qquad \theta(x,-s) = \alpha^0(x,-s) \quad (0 \leq s < \infty,\ x \in \bar{\Omega}.)$$

We shall now state the main hypothesis on the material properties.

First, we assume that $\boldsymbol{\kappa}(x)$, and $\mathbf{g}(x, s)$, $\boldsymbol{l}(x, s)$, and $c(x, s)$ for fixed $s \geq 0$, are Lebesgue measurable and essentially bounded on Ω.

In addition, we also postulate the following conditions:

(i) $0 < \rho_0 \leq \operatorname*{ess.\,inf}_{x \in \Omega} \rho(x)$.

(ii) $0 < c_0 < \operatorname*{ess.\,inf}_{x \in \Omega} c(x, 0)$.

(iii) $\mathbf{g}$ is stable:

$$\int_\Omega \nabla y(x) \cdot \mathbf{g}(x, \infty) \cdot \nabla y(x)\, dV \geq g \int_\Omega |\nabla y|^2\, dV,$$

where g is a positive constant and $\mathbf{g}(x, \infty) = \lim_{s \to \infty} \mathbf{g}(x, s)$ is called the *equilibrium elastic modulus*.

(iv) $\boldsymbol{\kappa}$ is positive-definite:

$$\int_\Omega \nabla \beta(x) \cdot \boldsymbol{\kappa}(x) \cdot \nabla \beta(x)\, dV \geq k \int_\Omega |\nabla \beta|^2,$$

where k is a positive constant.

(v) $\mathbf{g}''$ is stable:

$$\int_\Omega \nabla y(x) \cdot \mathbf{g}''(x, s) \cdot \nabla y(x)\, dV \geq g_2(s) \int_\Omega |\nabla y|^2\, dV,$$

where $0 \leq s < \infty$ and $g_2(s) > 0$.

(vi) $-\int_\Omega c''(x, s) \beta^2(x)\, dV \geq c_2(s) \int_\Omega |\beta(x)|^2\, dV,$

where $0 \leq s < \infty$ and $c_2(s) > 0$.

(vii) For all $s \geq 0$,

$$\|\boldsymbol{l}'(s)\| = \operatorname*{ess.\,sup}_{x \in \Omega} \|\boldsymbol{l}'(x, s)\| \in L^1(0, \infty)$$

$$\|\boldsymbol{l}''(s)\| \leq \left(\frac{\rho_0}{\theta_0}\right)^{1/2} [c_2(s)]^{1/2} [g_2(s)]^{1/2}.$$

Let $\boldsymbol{w}(x, t, s)$ and $\alpha(x, t, s)$ denote the displacement and temperature difference histories; that is, $\boldsymbol{w}(x, t, s) = \boldsymbol{u}(x, t - s)$ and $\alpha(x, t, s) = \theta(x, t - s) (0 \leq s \leq \infty)$. Denote by $\mathfrak{X}$ the Hilbert space obtained as the completion of the space

$$\begin{aligned}(\boldsymbol{u}, \boldsymbol{v}, \theta, \boldsymbol{w}, \alpha) \in C_0^\infty(\Omega; \mathbb{R}^3) &\times C_0^\infty(\Omega; \mathbb{R}^3) \times C_0^\infty(\Omega) \\ &\times C^\infty([0, \infty); H_\partial^1(\Omega; \mathbb{R}^3)) \\ &\times C^\infty([0, \infty); H_\partial^1(\Omega))\end{aligned}$$

under the norm induced by the inner product

$$\begin{aligned}&\langle (\boldsymbol{u}, \boldsymbol{v}, \theta, \boldsymbol{w}, \alpha), (\bar{\boldsymbol{u}}, \bar{\boldsymbol{v}}, \bar{\theta}, \bar{\boldsymbol{w}}, \bar{\alpha}) \rangle \\ &= \int_\Omega \left\{ \nabla \boldsymbol{u} \cdot \mathbf{g}(\infty) \cdot \nabla \bar{\boldsymbol{u}} + \rho \boldsymbol{v} \cdot \bar{\boldsymbol{v}} + \frac{\rho c(0)}{\theta_0} \theta \bar{\theta} \right\} dV \\ &\quad - \iint_\Omega^\infty{}_{\!0} \Big\{ [\nabla \boldsymbol{u} - \nabla \boldsymbol{w}(s)] \cdot \mathbf{g}'(s) \cdot [\nabla \bar{\boldsymbol{u}} - \nabla \bar{\boldsymbol{w}}(s)] + \bar{\alpha}(s) \boldsymbol{l}'(s) \cdot [\nabla \boldsymbol{u} \\ &\quad - \nabla \boldsymbol{w}(s)] + \alpha(s) \boldsymbol{l}'(s) \cdot [\nabla \bar{\boldsymbol{u}} - \nabla \bar{\boldsymbol{w}}(s)] - \frac{\rho}{\theta_0} c'(s) \alpha(s) \bar{\alpha}(s) \Big\}\, ds\, dV.\end{aligned}$$

Define the operator

$$G\begin{pmatrix} u \\ v \\ \theta \\ w \\ \alpha \end{pmatrix} = \begin{pmatrix} v \\ \frac{1}{\rho}\operatorname{div}\left\{\mathbf{g}(0)\cdot\nabla u - \theta l(0) + \int_0^\infty [\mathbf{g}'(s)\cdot\nabla w(s) - l'(s)\alpha(s)]\,ds\right\} \\ \frac{\theta_0}{\rho c(0)}\left\{-l'(0)\cdot\nabla u - l(0)\cdot\nabla v + \frac{\operatorname{div}(\boldsymbol{\kappa}\cdot\nabla\theta)}{\theta_0} - \frac{\rho c'(0)}{\theta_0} - \int_0^\infty [l''(s)\cdot\nabla w(s) + (\rho/\theta_0)c''(s)\alpha(s)]\,ds\right\} \\ -w'(s) \\ -\alpha'(s) \end{pmatrix}$$

with domain $\mathfrak{D}(G)$ given by those $(u, v, \theta, w, \alpha)$ such that the right-hand side of the above equation lies in $\mathfrak{X}$. Thus, we obtain the abstract evolutionary equation

$$\frac{d}{dt}\begin{pmatrix} u(t) \\ v(t) \\ \theta(t) \\ w(t) \\ \alpha(t) \end{pmatrix} = G\begin{pmatrix} u(t) \\ v(t) \\ \theta(t) \\ w(t) \\ \alpha(t) \end{pmatrix}, \quad \begin{pmatrix} u(0) \\ v(0) \\ \theta(0) \\ w(0) \\ \alpha(0) \end{pmatrix} = \begin{pmatrix} w^0(0) \\ v^0 \\ \alpha^0(0) \\ w^0 \\ \alpha^0 \end{pmatrix}. \qquad \text{(TEM)}'$$

The existence and uniqueness result for equation (TEM) is as follows:

The operator G is the infinitesimal generator of a C^0 contractive semigroup on $\mathfrak{X}$.

In fact, 3.17 shows that $\bar{G}$ generates a contractive semigroup. A slightly more careful analysis shows $G = \bar{G}$; that is, G is already closed. (See Navarro [1978].)

Box 3.4 ***Symmetric Hyperbolic Systems***

Here we study the symmetric hyperbolic systems of Friedrichs [1954], [1958]. This type of system occurs in many problems of mathematical physics—for example, Maxwell's equations, as is shown in Courant and Hilbert [1962]. As we shall see below, this includes the equations of linear elasticity. As Friedrichs has shown, many nonlinear equations are also covered by systems of this type. For elasticity, see Section 6.5 below, John [1977], Hughes and Marsden [1978], and, for general relativity, see Fischer and Marsden [1972b]. For time-dependent and non-linear cases, see Section 6.5 below, Dunford and Schwartz [1963], and Kato [1975a]. We consider the linear equations in all of space ($\Omega = \mathbb{R}^n$) for simplicity. For general Ω, see Rauch and Massey [1974].

Let $U(x, t) \in \mathbb{R}^N$ for $x \in \mathbb{R}^m$, $t \geq 0$ and consider the following evolution equation

$$\boldsymbol{a}_0(x)\frac{\partial U}{\partial t} = \sum_{j=1}^{m} \boldsymbol{a}_j(x)\frac{\partial U}{\partial x^j} + \boldsymbol{b}(x)U + \boldsymbol{f}(x), \qquad (SH)$$

where $\boldsymbol{a}_0$, $\boldsymbol{a}_j$ and $\boldsymbol{b}$ are $N \times N$ matrix functions. We assume $\boldsymbol{a}_0$ and $\boldsymbol{a}_j$ are *symmetric* and $\boldsymbol{a}_0$ is uniformly positive-definite; that is, $\boldsymbol{a}_0(x) \geq \epsilon$ for some $\epsilon > 0$. (This is a matrix inequality; it means $\langle \boldsymbol{a}_0(x)\xi, \xi \rangle \geq \epsilon \|\xi\|^2$ for all $\xi \in \mathbb{R}^N$.) To simplify what follows we shall take $\boldsymbol{a}_0 =$ Identity. The general case is dealt with in the same way by weighting the L^2-norm by $\boldsymbol{a}_0$. We can also assume $\boldsymbol{f} = \boldsymbol{0}$ by the remarks 2.17(v) on inhomogeneous equations.

We make the following technical assumptions. The functions $\boldsymbol{a}_j$ and $\boldsymbol{b}$ are to be of class C^1, uniformly bounded and with uniformly bounded first derivatives.

3.19 Theorem *Let the assumptions just stated hold and let $A_{\min}$: $C_0^\infty \longrightarrow L^2(\mathbb{R}^m, \mathbb{R}^N)$ (C_0^∞ denotes the C^∞ functions U: $\mathbb{R}^m \longrightarrow \mathbb{R}^N$ with compact support) be defined by*

$$A_{\min} U = \sum_{j=1}^{m} \boldsymbol{a}_j(x)\frac{\partial U}{\partial x^j} + \boldsymbol{b}(x)U(x).$$

Let A be a closure of $A_{\min}$. Then A generates a quasi-contractive one-parameter group in $L^2(\mathbb{R}^m, \mathbb{R}^N)$.

Proof Define $B_{\min}$ on C_0^∞ by

$$B_{\min} U = -\sum \frac{\partial}{\partial x^j}(\boldsymbol{a}_j(x)U) + \boldsymbol{b}(x)U.$$

Integration by parts shows that $B_{\min}$ is a restriction of the adjoint of $A_{\min}$ on C_0^∞; $A_{\min}^* \supset B_{\min}$. Let $A_{\max} = B_{\min}^*$. (In distribution language, $A_{\max}$ is just $A_{\min}$ defined on all U for which $A_{\min}\, U$ lies in L^2 with derivatives in the sense of distributions.)

We shall need the following:

3.20 Lemma *$A_{\max}$ is the closure of $A_{\min}$.*

Proof We shall sketch out the main steps. The method is often called that of the "Friedrichs Mollifier."

Let $U \in \mathfrak{D}(A_{\max})$. We have to show there is $U_n \in C_0^\infty$ such that $U_n \longrightarrow U$ and $A_{\min} U_n \longrightarrow A_{\max} U \in L^2$. Let $\rho\colon \mathbb{R}^m \longrightarrow \mathbb{R}$ be C^∞ with support in the unit ball, $\rho \geq 0$ and $\int \rho \, dx = 1$. Set

$$\rho_\epsilon(x) = \frac{1}{\epsilon^n}\rho\left(\frac{x}{\epsilon}\right) \quad \text{for} \quad \epsilon > 0.$$

Let $U_\epsilon = \rho_\epsilon * U$ (componentwise convolution). We assert that $U_\epsilon \longrightarrow U$ as $\epsilon \longrightarrow 0$ (in L^2). Indeed, $\| U_\epsilon \| \leq \| U \|$, so it is not enough to check this for $U \in C_0^\infty$. Then it is a standard (and easy) argument; one actually obtains uniform convergence.

Now each U_ϵ is C^∞. Let L denote the differential operator

$$L = \sum a_j(x)\frac{\partial}{\partial x_j} + b(x).$$

Then one computes that

$$\begin{aligned} L(U_\epsilon) = &\int \left\{-\sum_j \frac{\partial}{\partial y^j}[a_j(y)\rho_\epsilon(x-y)] + b(y)\rho_\epsilon(x-y)\right\} U(y)\, dy \\ &+ \int \Big\{\sum \frac{\partial}{\partial y^j}([a_j(y) - a_j(x)]\rho_\epsilon(x-y)) \\ &- [b(y) - b(x)]\rho_\epsilon(x-y)\Big\} U(y)\, dy. \end{aligned}$$

The first term is just $\rho_\epsilon *(A_{\max} U)$ and thus we have proved that $L(\rho_\epsilon * U) - \rho_\epsilon * A_{\max} U \longrightarrow 0$ as $\epsilon \longrightarrow 0$. It follows that $C_0^\infty \cap L^2$ is a core of $A_{\max}$. That is, $A_{\max}$ restricted to $C_0^\infty \cap L^2 \cap \mathfrak{D}(A)_{\max}$ has closure $A_{\max}$.

Let $\omega \in C_0^\infty(\mathbb{R}^m)$, ω with support in a ball of radius 2, and $\omega \equiv 1$ on a ball of radius 1, and let $\omega_n(x) = \omega(x/n)$. Then $\omega_n U_\epsilon \in C_0^\infty$ and

$$L(\omega_n U_\epsilon) = \omega_n L U_\epsilon + \sum a_j(x)\frac{\partial \omega_n}{\partial x^j} U_\epsilon.$$

As $n \longrightarrow \infty$, this converges to $L(U_\epsilon)$, which proves the lemma. ▼

Now we shall complete the proof of 3.19. Let $A = A_{\max}$. For $U \in C_0^\infty$,

$$\begin{aligned} \langle AU, U\rangle &= \int_\Omega \sum_j \left\langle a_j \frac{\partial U}{\partial x^j}, U\right\rangle + \langle bU, U\rangle\, dx \\ &= \int_\Omega \left\{\sum \tfrac{1}{2}\frac{\partial}{\partial x_j}\langle a_j U, U\rangle - \tfrac{1}{2}\left\langle \frac{\partial a_j}{\partial x^j}U, U\right\rangle + \langle bU, U\rangle\right\} dx \\ &\qquad\qquad \text{(by symmetry of } a_j) \\ &= \int_\Omega -\tfrac{1}{2}\left\langle \frac{\partial a_j}{\partial x^j}U, U\right\rangle + \langle bU, U\rangle\, dx. \end{aligned}$$

Thus,

$$\langle AU, U\rangle \leq \beta_1 \int_\Omega \langle U, U\rangle\, dx, \quad \text{where} \quad \beta_1 = \sup\left(\tfrac{1}{2}\left|\frac{\partial a_j}{\partial x^j}\right| + |b|\right).$$

By the lemma, this same inequality holds for all $U \in \mathfrak{D}(A)$. Thus, $\langle(\lambda - A)U, U\rangle \geq (\lambda - \beta_1)\langle U, U\rangle$ from which it follows that $\|(\lambda - A)U\| \geq (\lambda - \beta_1)\|U\|$. Thus $(\lambda - A)$ has closed range if $\lambda \geq \beta_1$, and is one-to-one. To show the range is the whole space we must show that $(\lambda - A)^*\boldsymbol{\omega} = \mathbf{0}$ implies $\boldsymbol{\omega} = \mathbf{0}$. ($(\lambda - A)^*\boldsymbol{\omega} = \mathbf{0}$ means $\boldsymbol{\omega}$ is orthogonal to the range.) But $B =$ closure of $B_{\min}$ (defined in the proof of the lemma) equals A^*. Thus $(\lambda - B)\boldsymbol{\omega} = \mathbf{0}$. As above, we have $\|(\lambda - B)\boldsymbol{\omega}\| \geq (\lambda - \beta_2)\|\boldsymbol{\omega}\|$ so $(\lambda - B)\boldsymbol{\omega} = \mathbf{0}$ implies $\boldsymbol{\omega} = \mathbf{0}$ for $\lambda > \beta_2$. For $\beta = \sup(\beta_1, \beta_2)$ and $\lambda > \beta$, we have

$$\|(\lambda - A)^{-1}\| < 1/(\lambda - \beta).$$

Since conditions on A are unaffected by replacing A with $-A$, we see that

$$\|(A + \lambda)^{-1}\| \leq 1/(|\lambda| - \beta) \quad (|\lambda| > \beta).$$

Hence A generates a quasi-contractive group. ▮

Provided the coefficients are smooth enough, one can also show that A generates a semigroup on H^s as well as on $H^0 = L^2$. (This follows by using Gronwall's inequality to show that the H^s norm remains bounded under the flow on L^2.)

3.21 Example (The Wave Equation) Consider the system:

$$\left.\begin{aligned}
\frac{\partial U^0}{\partial t} &= U^{n+1}, \\
\frac{\partial U^1}{\partial t} &= \frac{\partial U^{n+1}}{\partial x^1}, \\
&\;\;\vdots \\
\frac{\partial U^n}{\partial t} &= \frac{\partial U^{n+1}}{\partial x^n}, \\
\frac{\partial U^{n+1}}{\partial t} &= \frac{\partial U^1}{\partial x^1} + \cdots + \frac{\partial U^n}{\partial x^n}.
\end{aligned}\right\} \tag{W}$$

Here

$$U = (U^0, U^1, \ldots, U^{n+1}), \qquad \boldsymbol{a}_1 = \begin{pmatrix} 0 & 0 & \cdots & 0 \\ 0 & 0 & \cdots & 1 \\ 0 & 0 & \cdots & 0 \\ \cdot & & & \\ \cdot & & & \\ \cdot & & & \\ 0 & 1 & \cdots & 0 \end{pmatrix}, \quad \text{and so on}$$

so our system is symmetric. By Theorem 3.19, it generates a group. Let $u, \dot{u} \in H^1 \times L^2$ and consider the initial data

$$U^0 = u, \qquad U^1 = \frac{\partial u}{\partial x^1}, \ldots, U^n = \frac{\partial u}{\partial x^n}, \qquad U^{n+1} = \dot{u}.$$

Then the equations for U reduce exactly to the wave equation for u, so $\partial^2 u/\partial t^2 = \Delta u$ generates a group on $H^1 \times L^2$, reproducing the result we found by second-order methods in 3.2.

3.22 Example (Linear Hyperelasticity) We will use symmetric hyperbolic methods to reproduce the implication (ii) $\Rightarrow$ (i) in Theorem 3.1 for $\Omega = \mathbb{R}^n$. Consider, then, the system $\rho \ddot{u} = \operatorname{div}(\mathbf{a} \cdot \nabla u)$, where $\mathbf{a}$ is symmetric (hyperelastic) and strongly elliptic.

This is easiest to carry out in the case of stable classical elasticity, so we consider it first. Thus, we begin by dealing with

$$\rho \frac{\partial^2 u^a}{\partial t^2} = (\mathsf{c}^{abcd} u_{c|d})_{|b} = \mathsf{c}^{abcd} u_{c|d|b} + \mathsf{c}^{abcd}{}_{|b} u_{c|d}.$$

Following John [1977], let U be defined by $U_{ab} = u_{a|b}$ and $U_{a0} = \dot{u}_a$. The system under consideration is thus (in Euclidean coordinates)

$$\begin{cases} \rho \dfrac{\partial}{\partial t} U_{i0} = \mathsf{c}_{ikjm} \dfrac{\partial}{\partial x_k} U_{jm} + \mathsf{c}_{ikjm,k} U_{jm}, \\ \mathsf{c}_{ikjm} \dfrac{\partial}{\partial t} U_{jm} = \mathsf{c}_{ikjm} \dfrac{\partial}{\partial x_m} U_{j0}. \end{cases}$$

This has the form (SH) and a_0 is positive-definite if $\mathbf{c}$ is uniformly pointwise stable. Thus Theorem 3.19 applies.

In the general strongly elliptic case, one can replace a_{ikjm} by

$$\mathfrak{A}_{ikjm} = \mathsf{a}_{ikjm} + \gamma(\delta_{ik}\delta_{jm} - \delta_{im}\delta_{jk})$$

for a suitable constant γ, so $\mathfrak{A}_{ikjm}$ becomes positive-definite: $\mathfrak{A}_{ikjm}\zeta^{ik}\zeta^{jm} \geq \epsilon |\zeta|^2$ for all 3×3 matrices ζ (not necessarily symmetric). This may be proved by the arguments in 3.10, Chapter 4.*

> ***Problem 3.7*** Carry out the details of this, and for isotropic classical elasticity, show that one can choose $\gamma = c_2^2$, where c_2 is the wave velocity defined in Problem 3.4, Chapter 4.

A calculation shows that the added term, miraculously, does not affect the equations of motion.[13] Therefore, the preceding reduction applies and we get a symmetric hyperbolic system.

---

[13] As was pointed out by J. Ball, this is because it is the elasticity tensor of a *null Lagrangian* —that is, a Lagrangian whose Euler–Lagrange operator vanishes identically. Apparently this trick of adding on γ was already known to Hadamard (cf. Hill [1957]). See Ball [1977a] for a general discussion of the role of null-Lagrangians in elasticity. They will appear again in our discussion of the energy criterion in Section 6.6.

*As pointed out by S. Spector, this trick does not universally work. See also *Arch. Rat. Mech. An.* 98 (1987):1-30.

Box 3.5 ***Summary of Important Formulas for Section 6.3***

Linear Elastodynamics

$$\text{(LE)}\qquad \begin{cases} \rho\ddot{u} = \operatorname{div}(\mathsf{a}\cdot\nabla u) & \rho\ddot{u}^a = (\mathsf{a}^{abcd}u_{c|d})_{|b} \\ u = 0 \text{ on } \partial_d, & u^a = 0 \text{ on } \partial_d, \\ \mathsf{a}\cdot\nabla u\cdot n = 0 \text{ on } \partial_\tau & \mathsf{a}^{abcd}u_{c|d}n_b = 0 \text{ on } \partial_\tau \end{cases}$$

Initial Value Problem (Hyperelasticity)

The equations (LE) are well posed (give a quasi-contractive semigroup in $H^1 \times L^2$) if and only if a is strongly elliptic.

Cauchy Elasticity

If (LE) generates a contraction semigroup in $H^1 \times L^2$, then a is hyperelastic.

Energy Criterion

If a is hyperelastic and stable:

$$\int_\Omega \nabla u\cdot\mathsf{a}\cdot\nabla u\, dv = \int_\Omega u_{a|b}\mathsf{a}^{abcd}u_{c|d}\, dv$$

$$\geq c\int_\Omega u_{a|b}u^{a|b}\, dv \quad (c = \text{const.} > 0),$$

then a is strongly elliptic and (LE) generates stable dynamics (a contractive semigroup in a suitable norm) in $H^1 \times L^2$.

Korn's Lemma (Classical Elasticity)

If c is uniformly pointwise stable, then c is stable.

Hamiltonian Existence Theorem

If $(\mathfrak{X}, \omega)$ is a Banach space with symplectic form ω and A is ω-skew-adjoint and if the energy $H(x) = \omega(Ax, x)$ satisfies the stability condition $H(x) \geq c\|x\|^2$, then in the completion of $\mathfrak{D}(A)$ in the energy norm, A generates a one-parameter group.

Panel Flutter Equations

The equations

$$\ddot{v} + \alpha\dot{v}'''' + v'''' + \Gamma'' + \rho v' + \sqrt{\rho}\,\delta\dot{v} = 0,$$

$$v = v'' + \alpha\dot{v}'' = 0 \quad \text{at} \quad x = 0, 1.$$

generate a semigroup in $\mathfrak{X} = H_0^2 \times L^2$.

Dissipative Mechanisms

(1) Rate type:

$$\ddot{u} = Au + B\dot{u} \quad \text{or} \quad \rho\ddot{u} = \operatorname{div}(\mathsf{a}\cdot\nabla u) + \nabla^2\dot{u}$$

(2) Thermal type:

$$\ddot{u} = Au + B\theta \quad \text{or} \quad \rho\ddot{u} = \operatorname{div}(\mathsf{a}\cdot\nabla u) + m\nabla\theta$$

$$\dot{\theta} = C\theta + D\dot{u}, \qquad c\dot{\theta} = k\nabla^2\theta + \frac{m}{\rho}\nabla\cdot\dot{u}$$

(3) Memory:

$$\ddot{u} = Au + Bw$$

(In models, $w(x, t, s) = u(x, t - s)$ is the retarded value of u.)

$$\dot{w} = Cw$$

Symmetric Hyperbolic Systems for $U(x, t) \in \mathbb{R}^N$, $x \in \mathbb{R}^m$

$$a_0 \frac{\partial U}{\partial t} = a_j \frac{\partial U}{\partial x^j} + bU + f, \tag{SH}$$

where a_0, a_j, b are $N \times N$ matrices with a_0, a_j symmetric and a_0 uniformly positive-definite. (SH) generates a quasi-contractive group in $L^2(\mathbb{R}^m, \mathbb{R}^N)$.

For linear elasticity use the vector U given by $U_{ij} = u_{i,j}$, $U_{i0} = \dot{u}_i$ to write (LE) in the form (SH).

6.4 NONLINEAR ELASTOSTATICS

This section begins by giving the perturbation theory for nonlinear elastostatics in cases where solutions of the linear problem correspond faithfully to those of the nonlinear problem—that is, when there is no bifurcation. Bifurcation problems are studied in Chapter 7. In particular, we show that if the linearized problem has unique solutions, then so does the nonlinear one, nearby. This is done using the linear theory of Section 6.1 and the implicit function theorem (see Section 4.1). These results are due essentially to Stoppelli [1954]. This procedure fails in the important case of the pure traction problem because of its rotational invariance; this case is treated in Section 7.3. We also give an example from Ball, Knops, and Marsden [1978] showing that care must be taken with the function spaces.

Following this, we briefly describe some aspects of the global problem for three-dimensional elasticity following Ball [1977a, b] and state some of the open problems.

For the perturbation theorem (4.2 below), we make the following assumptions;

4.1 Assumptions

(i) The material is hyperelastic; $P = \partial W/\partial F$ and P is a smooth function of x and F (C^1 will do).

(ii) The boundary of $\mathcal{B}$ is smooth (C^1 will do). Here we use $\mathcal{B} \subset \mathbb{R}^3$ to simplify the function spaces involved, but this is not an essential assumption.

(iii) ∂_d; the portion of $\partial\mathcal{B}$ on which the displacement is prescribed is a connected component of $\partial\mathcal{B}$ (or a union of components), and ∂_τ the portion of $\partial\mathcal{B}$ on which the traction is prescribed is a (union of) component(s) of $\partial\mathcal{B}$. Assume $\partial_d \neq \varnothing$.

(iv) $\phi_0\colon \mathcal{B} \longrightarrow \mathbb{R}^3$ is a given regular deformation (C^3 will do) and the elasticity tensor at ϕ_0 is strongly elliptic.

(v) The equations of the linearized theory have unique solutions; that is,

$$\mathrm{DIV}[\mathbf{A}\cdot\nabla U] = 0$$

$U = 0$ on ∂_d and $\langle \mathbf{A}\cdot\nabla U, N\rangle = 0$ on ∂_τ implies $U = 0$ ($\mathbf{A}$ is the elasticity tensor evaluated at ϕ_0).

There are several remarks to be made on these assumptions.

(a) Condition (ii) is unpleasant but is necessitated by the function spaces we use. Bodies with corners are excluded by these methods.[14] The example given below in 4.4 shows that in function spaces such as $W^{1,p}$ (suitable for regions with corners), the results predicted by a formal application of the inverse function theorem simply are not true. It is possible that the source of the difficulties is stress concentrations in corners and that this may be a real problem. (See also p. 318.)

(b) Similar remarks hold for (iii). We do *not* allow genuine mixed problems where ∂_d and ∂_τ have common boundaries.

(c) These assumptions imply existence and uniqueness for the linearized problem by the basic existence theorem for linear elasticity, 1.8. Indeed, assumption (v) states exactly that Ker $A = \{0\}$.

Next define the Banach space $\mathfrak{X}$ to be all $W^{s,p}$ maps $\phi\colon \mathcal{B} \longrightarrow \mathbb{R}^3$ and let $\mathcal{C} \subset \mathfrak{X}$ be the regular ϕ's. If $s > 3/p + 1$, then $\mathcal{C}$ is open in $\mathfrak{X}$ because $W^{s,p} \subset C^1$. Let $\mathcal{Y}$ consist of all triples $(\boldsymbol{B}, \psi, \boldsymbol{\tau})$, where $\boldsymbol{B}$ is a $W^{s-2,p}$ vector function $\boldsymbol{B}$: $\mathcal{B} \longrightarrow \mathbb{R}^3$, ψ is a $W^{s-1/p,p}$ map of ∂_d to $\mathbb{R}^3$, and $\boldsymbol{\tau}$ is a $W^{s-1-1/p,p}$ map of ∂_τ to $\mathbb{R}^3$. Define $F\colon \mathcal{C} \subset \mathfrak{X} \longrightarrow \mathcal{Y}$ by

$$F(\phi) = (-\mathrm{DIV}\, \boldsymbol{P}(F), \phi\,|\,\partial_d, \boldsymbol{P}\cdot N\,|\,\partial_\tau).$$

4.2 Theorem (*Stoppelli* [1954])[15] *Make the assumptions (i)–(v) above and that* $s > 3/p + 1$ $(1 < p < \infty)$ (*e.g., for* $p = 2$, $s > \frac{5}{2}$). *Then there are neighborhoods* $\mathcal{U}$ *of* ϕ_0 *in* $\mathfrak{X}$ *and* $\mathcal{V}$ *of* ($\rho_{\text{Ref}}\mathring{B} = -\mathrm{DIV}\,\mathrm{P}(\mathring{F}), \phi_0\,|\,\partial_d, \mathring{P}\cdot N\,|\,\partial_\tau$) *such that* $F\colon \mathcal{U} \longrightarrow \mathcal{V}$ *is one-to-one and onto.*

Proof By the ω-lemma for composition (see Box 1.1, Chapter 3 and Box 1.1, this chapter), F is a C^1 mapping. By Theorem 1.8, and elliptic regularity,

---

[14]The situations in Figure 4.2, page 13 are examples where the results do not apply unless the corners are "smoothed out" a bit or meet at 90°. There is a clear call for better techniques here. See Kellogg and Osborn [1976] and references therein for related work.

[15]Related references are Van Buren [1968], John [1972], and Wang and Truesdell [1973]. The work of John [1972] is especially interesting since he is able to prove uniqueness under a hypothesis of small *stress*.

the derivative of F at ϕ_0 is a linear isomorphism. Therefore the result follows by the implicit function theorem (Box I.1, Chapter 4). ▮

The assumption $s > 3/p + 1$ (or $n/p + 1$ in n dimensions) is crucial for F to be smooth. Indeed, Example 4.4 below shows that the conclusion is not true without this assumption.

Rephrased, 4.2 states that if the linearized problem is strongly elliptic, hyperelastic,[16] and has unique solutions, then for slight perturbations of the load or boundary conditions from their values at the given displacement, the nonlinear problem

$$\mathrm{DIV}\,\boldsymbol{P} + \rho_{\mathrm{Ref}}\boldsymbol{B} = \boldsymbol{0}, \qquad \boldsymbol{\phi} = \boldsymbol{\phi}_d \quad \text{on} \quad \partial_d, \qquad \boldsymbol{P}\cdot\boldsymbol{N} = \boldsymbol{\tau} \quad \text{on} \quad \partial_\tau$$

has a unique solution $\boldsymbol{\phi}$ near $\boldsymbol{\phi}_0$. Moreover, $\boldsymbol{\phi}$ is a smooth function of $\boldsymbol{B}$, $\boldsymbol{\phi}_d$ and $\boldsymbol{\tau}$ (in the topologies above).

4.3 Corollary *Assume (i), (ii), (iii) with $\partial_d \neq \varnothing$ in 4.1, take $\boldsymbol{\phi}_0$ = Identity and assume that $\boldsymbol{\phi}_0$ is a natural state; that is $\boldsymbol{P} = \boldsymbol{0}$. Assume that the elasticity tensor $\mathbf{C}$ is uniformly pointwise stable. Then conditions (iv) and (v) are satisfied and so Theorem 4.2 applies.*

The proof of 4.2 shows that the existence of a unique solution remains valid if the constitutive function $\boldsymbol{P}(\boldsymbol{F})$ is also perturbed. This is important philosophically since in any given situation $\boldsymbol{P}(\boldsymbol{F})$ is only known approximately and the "real" material will have various imperfections. This apparently obvious remark takes on deeper significance in the context of bifurcation theory in Chapter 7.

It is clear that in most situations, the neighborhoods $\mathcal{U}$ and $\mathcal{V}$ in Theorem 4.2 really are small. Indeed, a bar in a natural state has unique solutions if the boundary is given a small prescribed displacement. If, however, the prescribed displacements are large, buckling and non-uniqueness can occur. (The fact that Theorem 4.2 cannot handle the traction or mixed problems makes examples of this a bit artificial.) One example, due to Fritz John, is shown in Figure 6.4.1(a) and another for a nonhomogeneous material pointed out by John Ball is shown in Figure 6.4.1(b).

Problem 4.1 Let $\boldsymbol{\phi}_0$ = Identity be a stress-free (natural) configuration for a strongly elliptic hyperelastic material, satisfying material frame indifference. Consider the problem DIV $\boldsymbol{P} = \boldsymbol{0}$ with $\boldsymbol{P}\cdot\boldsymbol{N} = \boldsymbol{0}$ on $\partial\mathcal{B}$. We have the solution $\boldsymbol{\phi} = \boldsymbol{\phi}_0$. We wish to show that all solutions near $\boldsymbol{\phi}_0$ are obtained by translating and rotating $\boldsymbol{\phi}_0$. The case $\boldsymbol{B} \neq \boldsymbol{0}$ and $\boldsymbol{P}\cdot\boldsymbol{N} \neq \boldsymbol{0}$ is the most interesting and is the one discussed in Section 7.3.

(a) Let $\mathcal{C}_{\mathrm{tr}} = \{\boldsymbol{\phi} \in \mathcal{C} \,|\, \boldsymbol{P}\cdot\boldsymbol{N} = \boldsymbol{0}$ on $\partial\mathcal{B}\}$. Show that $\mathcal{C}_{\mathrm{tr}} \subset \mathcal{C}$ is a smooth submanifold of $W^{s,p}$, $s < n/p + 1$. (*Hint:* Show

[16]If the problem is not hyperelastic, one must also assume the adjoint A^* has trivial kernel.

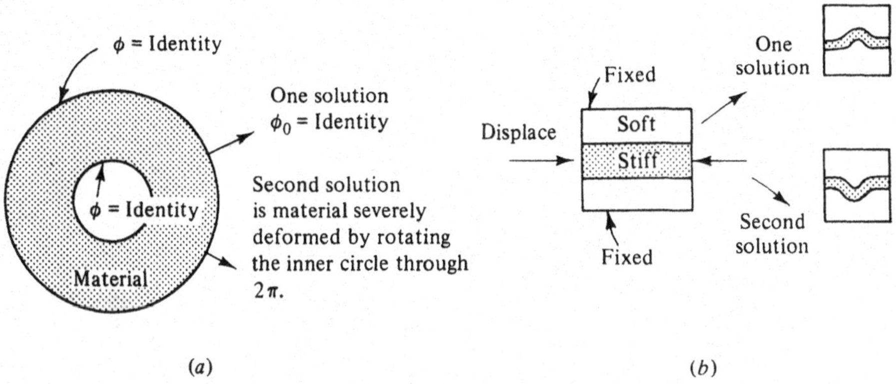

Figure 6.4.1 Global non-uniqueness of solutions for the displacement problem.

that $\boldsymbol{\phi} \mapsto \boldsymbol{P}\cdot\boldsymbol{N}|\partial\boldsymbol{B}$ has a surjective derivative by using strong ellipticity.)

(b) Let $\mathfrak{X}$ be the space of $W^{s-2,p}$ vector fields on $\mathcal{B}$ and let $\mathfrak{X}_0 \subset \mathfrak{X}$ be the six-dimensional subspace of vector fields of the form $\boldsymbol{a} + \boldsymbol{b}\boldsymbol{x}$, as in 1.11 (at $\boldsymbol{\phi} = \boldsymbol{\phi}_0$, identify material and spatial coordinates). Define $F: \mathcal{C}_{\mathrm{tr}} \to \mathfrak{X}$ by $F(\boldsymbol{\phi}) = \mathrm{DIV}\,\boldsymbol{P}$. Use 1.11(ii) to show that the range of $DF(\boldsymbol{\phi}_0)$ complements $\mathfrak{X}_0$. (We say F is *transversal* to $\mathfrak{X}_0$.)

(c) Deduce from (b) and the implicit function theorem that $F^{-1}(\mathfrak{X}_0)$ is, near $\boldsymbol{\phi}_0$, a smooth manifold of dimension six. Compute its tangent space at $\boldsymbol{\phi}_0$.

(d) Let $\mathcal{O}_{\phi_0}$ denote all rigid motions of $\boldsymbol{\phi}_0$. Show that $\mathcal{O}_{\phi_0} \subset \mathcal{C}_{\mathrm{tr}}$ is six dimensional and, by material frame indifference, show that $F(\mathcal{O}_{\phi_0}) = 0$. Deduce that every zero of F near $\boldsymbol{\phi}_0$ lies in $\mathcal{O}_{\phi_0}$.

The problems with mixed boundary conditions discussed above may be just technicalities that could be overcome by means of a more powerful differential calculus. For example, it seems intuitive that the inverse function theorem and bifurcation theory should work for situations like that in Figure 6.4.2. (Formally, the operator A remains strongly elliptic but picks up a kernel when buckling occurs; this kernel is associated with the axial symmetry of the problem; see Chapter 7.) One way to handle such situations to allow, for example, large stress concentrations in the corners, is to use function spaces like $W^{1,p}$ for $\boldsymbol{\phi}$. Then the stress can be (pointwise) unbounded. These function spaces are used in the global theory of Ball, described below. However, we shall now give an example which shows that in $W^{1,p}$, the conclusions of 4.2 are not true. In fact, we do it just in one dimension ($n = 1$) and show that while the formal linearization is an isomorphism and so by 4.2 the solution is isolated in $W^{2,p}$, it

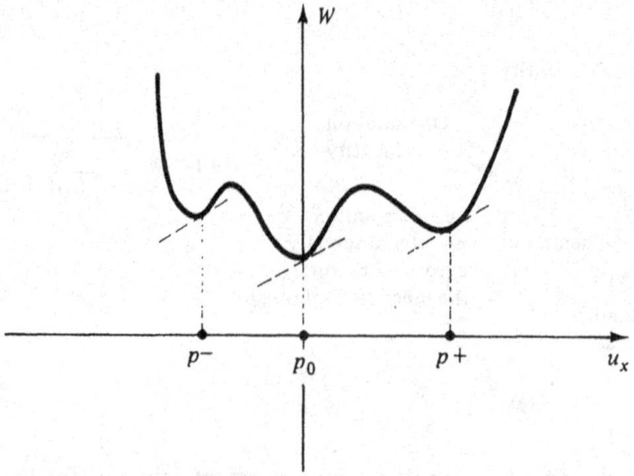

Figure 6.4.2

is *not isolated* in $W^{1,p}$. The difficulty is precisely that the nonlinear map is *not* C^1. (See Morrey [1966] and Martini [1979] for related results.)

4.4 Example (Ball, Knops, and Marsden [1978]). We consider the displacement problem in one dimension, writing u for the nonlinear displacement: $u = \phi - \text{Identity}$. On [0, 1] we consider a stored energy function $W(u_x)$, suppose there are no external forces, and assume the boundary conditions $u(0) = u(1) = 0$. Assume W is smooth and let $p_- < 0 < p_+$ be such that

$$W'(p_-) = W'(0) = W'(p_+) \quad \text{and} \quad W''(0) > 0.$$

(See Figure 6.4.2.)

In $W^{2,p}$ (with the boundary conditions $u(0) = 0$, $u(1) = 0$), the trivial solution $u_0 \equiv 0$ is isolated because the map $u \mapsto W(u_x)_x$ from $W^{2,p}$ to L^p is smooth and its derivative at u_0 is the linear isomorphism $v \mapsto W''(0)v_{xx}$. Therefore, by the inverse function theorem, zeros of $W(u_x)_x$ are isolated in $W^{2,p}$, as above. Note that the second variation of the energy $V(u) = \int_0^1 W(u_x)\,dx$ is positive-definite (relative to the $H^1 = W^{1,2}$ topology) at u_0 because if v is in $W^{1,2}$ and vanishes at $x = 0, 1$, then

$$\frac{d^2}{d\epsilon^2} V(u_0 + \epsilon v)_{|\epsilon=0} = W''(0) \int_0^1 v_x^2\,dx \geq c\,\|v\|^2_{W^{1,2}}.$$

Now we show that u_0 *is not isolated in* $W^{1,p}$. Given $\epsilon > 0$, let

$$u_\epsilon(x) = \begin{cases} p_+ x & \text{for } 0 \leq x \leq \epsilon, \\ p_+\epsilon + p_-(x - \epsilon) & \text{for } \epsilon \leq x \leq \dfrac{p_- - p_+}{p_-}\epsilon, \\ 0 & \text{for } \dfrac{p_- - p_+}{p_-}\epsilon \leq x \leq 1. \end{cases}$$

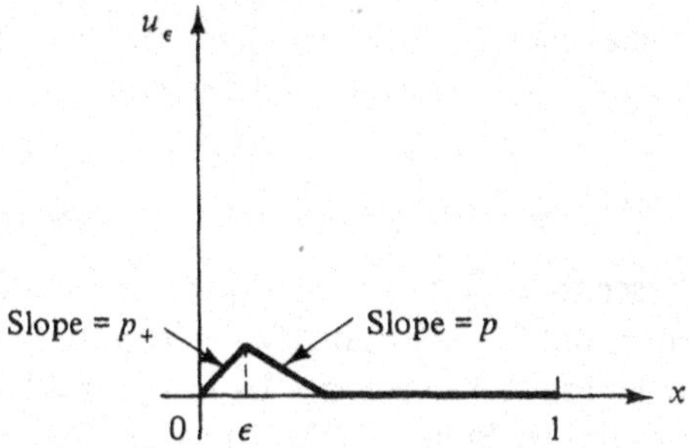

Figure 6.4.3

(See Figure 6.4.3.) Since $W'(u_{\epsilon x})$ is constant, each u_ϵ is an extremal. Also

$$\int_0^1 |u_{\epsilon x} - u_{0x}|^p \, dx = \epsilon |p_+|^p + \epsilon |p_-|^p |p_+|^p,$$

which tends to zero as $\epsilon \longrightarrow 0$. Thus u_0 is not isolated in $W^{1,p}$.

Remarks If $W(p_-) = W(p_+) = W(0)$ and if $W(p) \geq W(0)$ for all p, the same argument shows that there are absolute minima of V arbitrarily close to u_0 in $W^{1,p}$.

Phenomena like this seem to have first been noticed by Weierstrass. See Bolza [1904], footnote 1, p. 40. The "pathology" could possibly be eliminated by making good constitutive assumptions on W.

Now we turn to a discussion of global existence of solutions for the elastotatics problem. Thus, we wish to remove the restriction of locality in Theorem 4.2 and prove the existence of solutions for arbitrary loads. This is, then, a *global* nonlinear elliptic boundary value problem for a system. There have been at least four (overlapping) general methods used for dealing with such problems:

1. Convexity methods and monotone operators
2. Continuity methods
3. Topological methods and degree theory
4. Calculus of variations and minimizers

Method 1 considers an abstract operator $F\colon \mathcal{Y} \longrightarrow \mathcal{X}$, where $\mathcal{X}$ is a Hilbert space and $\mathcal{Y} \subset \mathcal{X}$ is a compactly embedded Banach space. We seek to solve $F(u) = f$ assuming:

(i) F is C^1; $F(0) = 0$.

(ii) $DF(u)$ is symmetric, so $F(u) = f$ is the Euler–Lagrange operator of $J\colon \mathcal{Y} \longrightarrow \mathbb{R}$,

$$J(u) = \int_0^1 \langle F(tu), u\rangle \, dt - \langle f, u\rangle \quad \text{(see Box 7.2, Chapter 1).}$$

(iii) $\langle DF(u)\cdot h, h\rangle > c\|h\|^2_{\mathcal{Y}}$ for all $h \in \mathcal{Y}$.

Condition (iii) implies that J is strictly convex: if $u \neq v$ and $0 < t < 1$, then $J(tu + (1 - t)v) < tJ(u) + (1 - t)J(v)$, that J is bounded below and that F is strictly monotone:

$$\langle F(u) - F(v), u - v \rangle > 0 \quad \text{if} \quad u \neq v.$$

From this it is easy to see that solutions of $F(u) = f$ are unique. Also, by selecting a minimizing sequence, one shows that it converges to a unique (generalized) solution of $F(u) = f$. This idea is developed by Langenbach [1961] and is applied to elasticity by, for example, Beju [1971]. However, the convexity assumption (iii) is unsatisfactory for three good reasons: (1) it implies uniqueness of solutions and so precludes buckling; (2) convexity is incompatible with material frame indifference (Coleman and Noll [1959]); and (3) even in ranges where solutions may be unique, convexity need not hold, as is seen from the stretching experiment discussed in Ball [1977a].

Thus, while method 1 is mathematically very powerful and is appropriate for a number of important situations, it is not appropriate in its present form for three-dimensional nonlinear elasticity.

Gurtin and Spector [1979] (see also Gurtin [1981a]) have looked for regions in the space of deformations where a convexity argument might be useful, and thereby have attempted to find sets of deformations in which uniqueness does hold.

Method 2 is based on the following "principle": to solve $F(u) = f$, we look at the range of F. Using the implicit function theorem as in 4.2, one can sometimes show that the range of F is open. Some additional estimates could be used to show the range is closed. By connectedness, this would imply that Range(F) $= \mathfrak{X}$, so $F(u) = f$ would be solvable. The following proposition illustrates the idea, although what is important is the method of proof. As far as we know, this technique has not yet been successfully applied in elasticity.[17]

4.5 Proposition *Suppose $\mathfrak{X}$, $\mathfrak{Y}$ are Banach spaces and $F\colon \mathfrak{Y} \to \mathfrak{X}$ is a C^1 map and that at each point $x \in \mathfrak{Y}$, $DF(x)$ is an isomorphism of $\mathfrak{Y}$ onto $\mathfrak{X}$. Assume that F is* proper; *that is, if $C \subset \mathfrak{X}$ is compact, then $F^{-1}(C) \subset \mathfrak{Y}$ is compact. Then $F\colon \mathfrak{Y} \to \mathfrak{X}$ is onto.*

Proof We can suppose without loss of generality that $F(0) = 0$. Let $x \in \mathfrak{X}$ and consider the curve $\sigma(t) = tx$. By the inverse function theorem, there is a unique C^1 curve $\rho(t)$ defined for $0 \leq t \leq \epsilon$ such that $F(\rho(t)) = \sigma(t)$. As in ordinary differential equations, extend ρ uniquely to its maximum domain of existence; say $0 \leq t < T \leq 1$. Let $t_n \to T$; then $F(\rho(t_n)) = \sigma(t_n) \to \sigma(T)$, so by the properness of F, $\rho(t_n)$ has a convergent subsequence, converging, say, to y_0. But as F is a local diffeomorphism in a neighborhood of y_0, the curve $\rho(t)$ can

[17]For another proof and references, see Wu and Desoer [1972].

be defined up to and including T and beyond T if $T < 1$. Thus ρ is defined for $0 \leq t \leq 1$ and $F(\rho(1)) = x$. ▮

Problem 4.2 (a) Show that F is proper if one has the estimate $\| DF(x)^{-1} \| \geq m > 0$ for a constant $m > 0$. (b) Show that F in 4.5 is, in fact, one-to-one by using the above proof and the fact that $\mathfrak{X}$ is simply connected.

This proposition may be regarded as a primitive version of method 3, as well as an illustration of method 2. However, in method 3 the most common device used is not so much the use of curves in the domains and ranges of F, but rather to join the map F to another one F_0 that can be understood; for example, F_0 can be a linear map and the curve could be a straight line: $F(t) = tF + (1 - t)F_0$. The idea is to now invoke topological tools of degree theory to show that questions of solvability of $F(u) = f$ can be continued back to those for solvability of $F_0(u) = f$. The details are not appropriate for us to go into at this point; however, the method is powerful in a number of contexts. It allows multiple solutions (indeed it is very useful in bifurcation theory) and does not require any convexity assumptions. For general background and some applications, see Choquet-Bruhat, DeWitte, and Morette [1977]. For applications to rod and shell theory in elasticity, see Antman [1976a, b]. A global uniqueness theorem in the same spirit is given by Meisters and Olech [1963].

Now we turn to method 4, which we shall discuss a bit more extensively, following parts of Ball [1977a, b]. Let $\mathcal{B} = \Omega$ be a region in $\mathbb{R}^3$ with piecewise C^1 boundary, and $\phi: \Omega \to \mathbb{R}^3$ a typical deformation. Let $W(\boldsymbol{F})$ be a given smooth stored energy function, $\mathcal{V}_B$ a potential for the body forces, and $\mathcal{V}_\tau$ a potential for the tractions. As usual, the displacement will be prescribed $\phi = \phi_d$ on $\partial_d \subset \partial\Omega$ and the traction will be prescribed on ∂_τ: $\mathbf{P} \cdot N = \boldsymbol{\tau}$. The energy functional whose critical points we seek is (see Section 5.1)

$$I(\phi) = \int_\Omega W(\boldsymbol{F})\, dV + \int_\Omega \mathcal{V}_B\, dV + \int_{\partial_\tau} \mathcal{V}_\tau\, dA.$$

In fact, we seek to minimize $I(\phi)$ over all ϕ satisfying $\phi = \phi_d$ on ∂_d. For dead loads we can take $\mathcal{V}_B = -\boldsymbol{B} \cdot \phi$ and $\mathcal{V}_\tau = -\boldsymbol{\tau} \cdot \phi$, which are linear functions of ϕ. The essential part of $I(\phi)$ is the stored energy function, so we shall assume that $I(\phi) = \int_\Omega W(\boldsymbol{F})\, dV$ for simplicity.

The method for minimizing $I(\phi)$ proceeds according to the following outline. On a suitable function space $\mathcal{C}$ of ϕ's:

(1) Show I is bounded below; then $m = \inf_{\psi \in \mathcal{C}} I(\psi)$ exists as a real number. Let ϕ_n be such that $I(\phi_n) \to m$ as $n \to \infty$; that is, select a minimizing sequence (this is possible as $m > -\infty$).

(2) Find a subsequence of ϕ_n that converges weakly; that is, $\phi_n \rightharpoonup \phi$ (this notion is explained in Box 4.1).

(3) Show that I is weakly sequentially lower semicontinuous; that is, $\phi_n \rightharpoonup \phi$ implies that $I(\phi) \leq \liminf_{n \to \infty} I(\phi_n)$.

If each of these steps can be effected, then ϕ is the minimizer of I. [*Proof:* Clearly $I(\phi) \geq m$ by definition of m. Also, if $I(\phi_n) \longrightarrow m$, then $\liminf_{n\to\infty} I(\phi_n) = m$, so by 3, $I(\phi) \leq m$. Thus $I(\phi) = m$.]

In one dimension, this method can only work if W is convex. Indeed for this case, a theorem of Tonelli states that if $I(\phi)$ is weakly sequentially semi-continuous, then W is convex, and conversely. This is actually not difficult and is proved in Box 4.1. The related "relaxation theorem" is discussed in this box as well.

In elasticity, even in one dimension, two of the properties of minimizers ϕ that have to be carefully considered are smoothness and invertibility of ϕ. These are not simple; in one dimension, however, the situation is fairly well understood (mostly due to Antman [1976a, b]) and is discussed in Box 4.1.

In three dimensions, convexity of W is *not* necessary for this method to work. Indeed, it can be made to work under hypotheses that are reasonable for elasticity. The analog of Tonelli's result is due to Morrey [1952]; assume throughout that $|W(\boldsymbol{F})| \leq C_1 + C_2|\boldsymbol{F}|^p$ for constants C_1 and C_2, so $I(\phi)$ is defined for $\phi \in W^{1,p}(\Omega)$.

4.6 Proposition *If* $I(\phi) = \int_\Omega W(\boldsymbol{F})\,dV$ *is weakly sequentially lower semi-continuous on* $W^{1,p}(\Omega)$, *then* W *is* quasi-convex; *that is, for all* (*constant*) 3×3 *matrices* $\boldsymbol{F}$ *with* $\det \boldsymbol{F} > 0$ *and all* $\psi\colon \Omega \longrightarrow \mathbb{R}^3$ *that are* C^∞ *with compact support in* Ω,

$$\int_\Omega W(\boldsymbol{F} + \nabla\psi(X))\,dV(X) \geq W(\boldsymbol{F}) \times volume(\Omega). \qquad \text{(QC)}$$

This also implies strong ellipticity.

The proof is actually similar to Tonelli's theorem proved in Box 4.1, and reduces to it in one dimension, so is omitted. The inequality (QC) says essentially that if ϕ_h is a homogeneous deformation, $I(\phi_h)$ is a minimum among deformations ϕ with the same boundary conditions. Morrey also shows that (QC) and growth conditions imply sequential weak lower semicontinuity. However, this does not apply to elasticity because of the condition $\det \boldsymbol{F} > 0$ that we must be aware of.

To understand which stored energy functions give sequentially weakly lower semicontinuous (s.w.l.s) I's, first consider the question of which ones give sequential weak continuity (s.w.c.). We state the following without proof:

4.6′ Proposition (*Ericksen, Edelen, Reshetnyak, Ball*) *Let* $W\colon M^{m\times n}$ (*the* $m \times n$ *matrices*) $\longrightarrow \mathbb{R}$ *be continuous,* $|W(\boldsymbol{F})| \leq C_1 + C_2|\boldsymbol{F}|^p$ $(1 < p < \infty)$, *and let* $L(\phi) = W(\nabla\phi)$; $L\colon W^{1,p} \longrightarrow L^1$. *The following are equivalent:*

(1) *L is s.w.c.*

(2) *L is a null Lagrangian; that is,* $L(\phi + \psi) = L(\phi)$ *for all* ψ *smooth with*

compact support in Ω *and all* C^1 *maps* $\phi\colon \Omega \to \mathbb{R}^3$. *(If W is smooth, this means the Euler–Lagrange equations for W are identically satisfied for any* ϕ.*)*

(3) $W(F) =$ *Constant + linear combination of* $r \times r$ *minors of* F *for* $1 \leq r \leq \min(m, n)$.

For example, in three dimensions the null Lagrangians are $W(\boldsymbol{F}) = C + \boldsymbol{A}\cdot\boldsymbol{F} + \boldsymbol{B}\cdot(\operatorname{adj}\boldsymbol{F}) + D(\det \boldsymbol{F})$, where $(\operatorname{adj}\mathbf{F})$ is the matrix of cofactors of $\boldsymbol{F}$ and C, $\boldsymbol{A}$, $\boldsymbol{B}$, and D are constant. Some feeling for what is involved in 4.6 can be gained by working the following problem. (See Ball [1977a] for the complete proof.)

Problem 4.3 (Ball) Prove that $\det \boldsymbol{F}$ is s.w.c. for 2×2 matrix functions by using the identity

$$\det \nabla\phi = \phi^1_{,1}\phi^2_{,2} - \phi^1_{,2}\phi^2_{,1} = (\phi^1\phi^2_{,2})_{,1} - (\phi^1\phi^2_{,1})_{,2}.$$

(This sort of trick is used in "compensated compactness"; cf. Tarter [1979].)

A good hypothesis on constitutive functions must be invariant under various transformations of coordinates. The following is one that is invariant under adding on null Lagrangians and the inversion transformation of fields $\phi \mapsto \phi^{-1}$.

4.7 Definition W is called (strictly) *polyconvex* if there is a (strictly) convex function $g\colon M^{3\times 3}_+ \times M^{3\times 3} \times (0, \infty) \to \mathbb{R}$ such that $W(\boldsymbol{F}) = g(\boldsymbol{F}, \operatorname{adj}\boldsymbol{F}, \det \boldsymbol{F})$ for all $\boldsymbol{F} \in M^{3\times 3}_+$.

One has the following chain of implications

Strict Convexity ⇒ Strict Polyconvexity ⇒ Quasi-Convexity
↘ (Strict Polyconvexity) s.w.l.s. ⇒ Strong Ellipticity ↙ (Quasi-Convexity)

As remarked before, convexity is not a useful assumption. However, polyconvexity is valid for many specific materials, such as the Mooney–Rivlin and Ogden materials.

The following is a sample of one of Ball's results.

4.8 Theorem (*Ball* [1977b]) *Suppose:*

(H1) *W is polyconvex.*

(H2) $g(\boldsymbol{F}, \boldsymbol{H}, \delta) \leq C + K(|\boldsymbol{F}|^p + |\boldsymbol{H}|^q + \delta^r)$ *for constants* $K > 0$, $p \geq 2$, $q \geq p/(p-1)$, *and* $r > 1$.

(H3) *if* $(\boldsymbol{F}_n, \boldsymbol{H}_n, \delta_n) \to (\boldsymbol{F}, \boldsymbol{H}, 0)$, *then* $g(\boldsymbol{F}_n, \boldsymbol{H}_n, \delta_n) \to \infty$.

Let $\mathfrak{a} = \{\phi \in W^{1,p}(\Omega) \mid \operatorname{adj}(\nabla\phi) \in L^q, \det \nabla\phi \in L^r, I(\phi) < \infty,$ *and* $\phi = \phi_d$ *on* $\partial_d\}$ *and suppose* $\mathfrak{a} \neq \varnothing$. *Then there exists a* $\phi \in \mathfrak{a}$ *that minimizes I in* $\mathfrak{a}$.

Sketch of Proof Take a minimizing sequence ϕ_n. By (H2), we see that $\nabla\phi_n$ is bounded in L^p, $\operatorname{adj}(\nabla\phi_n)$ is bounded in L^q, and $\det(\nabla\phi_n)$ is bounded in L^r. By the Poincaré inequality and weak compactness of the unit ball we get a subsequence ϕ_μ such that

$$\phi_\mu \rightharpoonup \phi_0 \quad \text{in} \quad W^{1,p}, \qquad \operatorname{adj} \nabla\phi_\mu \rightharpoonup H \quad \text{in} \quad L^q, \quad \text{and} \quad \det \nabla\phi_\mu \rightharpoonup \delta \quad \text{in} \quad L^r.$$

By 4.6′, adj and det are weakly continuous, so $H = \operatorname{adj}(\nabla\phi_0)$ and $\delta = \det(\nabla\phi_0)$. Since g is convex, $\phi \mapsto I(\phi) = \int_\Omega g(F, \operatorname{adj} F, \det F)\, dV$ is s.w.l.s. so ϕ_0 is the required minimizer. ∎

This sort of argument works well for the incompressible case, too, by weak continuity of det.

Open Problems

(a) Are minimizers weak solutions of the Euler–Lagrange equations?

(b) Are minimizers C^1? (Ball [1980] has shown that strong ellipticity is *necessary* for regularity; however it may be sufficient if $W(F)/|F|^3 \longrightarrow \infty$ as $|F| \longrightarrow \infty$. Without such a condition, Ball [1982] has shown by means of a very important example involving *cavitation* that minimizers need not be smooth.)

(c) Are minimizers 1–1 deformations? (Using a result of Meisters and Olech [1963], Ball [1981] shows that this is true in the incompressible case.)

Box 4.1 ***Some Facts About Weak Convergence***[18]

In this box we shall state a few basic properties and examples of weak convergence; prove Tonelli's theorem and discuss the related relaxation theorem of L.C. Young; and discuss the proof of existence and regularity for one-dimensional problems.

These results only hint at the extensive literature on uses of the weak topology. Besides the work of Ball already cited, the articles of Tartar [1979] and DiPerna [1982] are indicative of current research in this area.

If $\mathfrak{X}$ is a Banach space and x_n is a sequence in $\mathfrak{X}$, we write $x_n \rightharpoonup x$ and say x_n *converges weakly* to x if for all $l \in \mathfrak{X}^*$ (i.e., $l\colon \mathfrak{X} \longrightarrow \mathbb{R}$ is continuous and linear), $l(x_n) \longrightarrow l(x)$ in $\mathbb{R}$. If $\mathfrak{X} = L^p([0, 1])$, $(1 < p < \infty)$, then $u_n \rightharpoonup u$ means that

$$\int_0^1 u_n v\, dx \longrightarrow \int_0^1 uv\, dx$$

for all $v \in L^{p'}[0, 1]$, where $(1/p) + (1/p') = 1$. This is because $(L^p)^* =$

[18]We thank J. Ball for help with this box.

$L^{p'}$ (Riesz representation theorem). For $\mathfrak{X} = L^\infty$ and we choose $v \in L^1$, we speak of *weak* convergence*. Clearly ordinary convergence implies weak convergence.

Problem 4.4 Let $0 < \lambda < 1$, $a, b > 0$, and $u_n = a$ on $[0, \lambda/n]$, $u_n = b$ on $[\lambda/n, 1/n]$, and repeat on every subinterval $[i/n, (i+1)/n]$ $(i = 0, 1, \ldots, n-1)$. Prove that $u_n \rightharpoonup u = \lambda a + (1-\lambda)b$ weak* in L^∞ [and hence in $L^p[0, 1]$ $(1 < p < \infty)$], but $u_n \not\to u$.

In a reflexive separable Banach space $\mathfrak{X}$ (such as $W^{s,p}(\Omega)$, $1 < p < \infty$) the unit ball is weakly sequentially compact; that is, if $u_n \in \mathfrak{X}$ and $\|u_n\| = 1$, then there is a subsequence $u_{n_k} \rightharpoonup u \in \mathfrak{X}$. The unit ball (or in fact any closed convex set) is weakly closed, so $\|u\| \leq 1$. (This result may be found in Yosida [1971], p. 125.)

Problem 4.5 In $L^2[0, 2\pi]$ show that $(1/\sqrt{\pi}) \sin nx = u_n(x)$ satisfies $u_n \rightharpoonup 0$, yet $\|u_n\| = 1$. (*Hint:* Use the Riemann–Lebesgue lemma from Fourier series.)

Suppose $W\colon \mathbb{R}^+ \to \mathbb{R}$ is a given smooth function and for $\phi\colon [0, 1] \to \mathbb{R}$, $\phi' \geq 0$ almost everywhere, define

$$I(\phi) = \int_0^1 W(\phi'(X))\, dX.$$

Assume $|W(F)| \leq C_1 + C_2|F|^p$ so I maps $\{\phi \in W^{1,p} \mid \phi' \geq 0 \text{ a.e.}\}$ to $\mathbb{R}$.

4.9 Proposition (*Tonelli*) *I is weakly sequentially lower semicontinuous (w.s.l.s.) if and only if W is convex.*

Proof First of all, assume W is convex and let $\phi_n \rightharpoonup \phi$ in $W^{1,p}$. By Mazur's theorem (see Yosida [1971]) there is a sequence of finite convex combinations, say $\psi_n = \sum_j \lambda_n^j \phi_j$, $\sum_j \lambda_n^j = 1$ $(0 \leq \lambda_j \leq 1)$ such that $\psi_n \to \phi$ (strongly) in $W^{1,p}$. By going to a further subsequence we can suppose $\psi_n' \to \psi'$ a.e. By Fatou's lemma,

$$\int_0^1 W(\phi'(X))\, dX \leq \liminf_{n\to\infty} \int_0^1 W(\psi_n'(X))\, dX.$$

By convexity of W, the right-hand side does not exceed

$$\liminf_{n\to\infty} \sum_j \lambda_n^j \int_0^1 W(\phi_j'(X))\, dX \leq \liminf_{n\to\infty} \int_0^1 W(\phi_n'(X))\, dX.$$

Thus I is w.s.l.s.

Conversely, assume I is w.s.l.s. Let $\phi_n(0) = 0$ and let $\phi_n'(X) = u_n(X)$, given by Problem 4.4. By that problem, $\phi_n \rightharpoonup \phi(X) = (\lambda a + (1-\lambda)b)X$ and $W(\phi_n') \rightharpoonup \lambda W(a) + (1-\lambda) W(b)$. Thus by w.s.l.s. $I(\phi) \leq \liminf_{n\to\infty}$

$I(\phi_n)$ becomes $W(\lambda a + (1-\lambda)b) \leq \lambda W(a) + (1-\lambda)W(b)$; that is, W is convex. ∎

Problem 4.6 Use this argument to prove that I is w.s.c. if and only if $\phi(X) = aX + b$ for constants a and b. (This is the special case of Proposition 4.6 when $n = 1 = m$.)

In one dimension there is evidence that non-convex W's may be useful for describing phase transitions (Ericksen [1975]).[19] However, in static experiments the non-convexity of W may not be observable. Indeed, the relaxation theorem of L.C. Young states that the minimum of $\int_0^1 W(\phi'(X))\,dX$ is the same as that of $\int_0^1 W^*(\phi'(X)\,dX$, where W^* is the convex lower envelope of W (Figure 6.4.4). A convenient reference for this is Ekeland and Temam [1974]. In three dimensions the situation is far from settled.

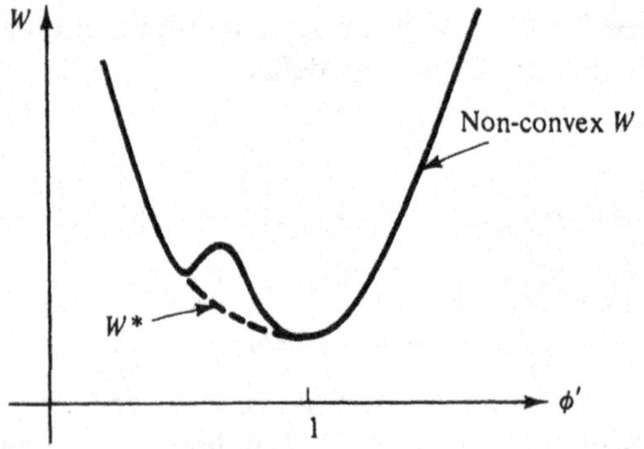

Figure 6.4.4

Remaining in one dimension, consider the following hypotheses on $W(X, F)$:

(H1) $W: [0, 1] \times (0, \infty) \to \mathbb{R}$ is C^1.
(H2) $W(X, F) \to +\infty$ as $F \to 0+$.
(H3) $W(X, F)/F \to \infty$ as $F \to +\infty$ uniformly in X.
(H4) $W(X, F)$ is convex in F.

---

[19]Phase transitions contemplated here may be seen if the polyethelene used in beer can packaging is stretched with your hands. It turns white, changing phase. Gentle heat will restore the original phase.

(H5) $V: [0, 1] \times [0, \infty) \longrightarrow \mathbb{R}$ is C^1. Let

$$I(\phi) = \int_0^1 W(X, \phi'(X))\, dX + \int_0^1 V(X, \phi(X))\, dX.$$

The following refines 4.8 slightly in one dimension.

4.10 Proposition *If (H1)–(H4) hold and α is given, then there is a member of*

$$\mathcal{A} = \{\phi \in W^{1,1} \mid I(\phi) < \infty, \phi(0) = 0, \phi(1) = \alpha\}$$

that minimizes I. (Note that $W^{1,1} \subset C^0$, so if $\phi \in \mathcal{A}$, then by (H2), $\phi' > 0$ a.e., so ϕ is one-to-one.)

Proof Note that $\mathcal{A} \neq \varnothing$ since $\phi(X) = \alpha X$ lies in $\mathcal{A}$. We first prove I is bounded below on $\mathcal{A}$.

By (H2) and (H3), W is bounded below, so $\int_0^1 W(X, \phi'(X)\, dX$ is bounded below. Since members of $\mathcal{A}$ are continuous and one-to-one, they are bounded between 0 and α. Thus $V(X, \phi(X))$ is uniformly bounded. Hence $I(\phi)$ is bounded below.

Let $\phi_n \in \mathcal{A}$ be such that $I(\phi_n) \searrow \inf\{I(\phi) \mid \phi \in \mathcal{A}\}$. (H3) implies ϕ_n are bounded in $W^{1,1}$. However, as $p = 1$, the space is not reflexive, so it is not obvious we can extract a weakly convergent subsequence. However, a result of de la Valleé Poussin shows that we can (see Morrey [1966]). The proof is then completed using w.s.l.s. of I from 4.9. ∎

The condition $W(X, F)/F \longrightarrow +\infty$ as $F \longrightarrow +\infty$ has direct physical meaning. Namely, consider a small piece of material that undergoes the homogeneous deformation $\phi(X) = FX$. This stretches a length l/F to the length l. The energy required to do this is $W(F)l/F$. Thus, (H3) means it takes more and more energy to stretch small lengths out to a prescribed length. The analogue in dimension three was mentioned in open problem (b) above.

Without the conditions of convexity or $W(X, F)/F \longrightarrow \infty$ as $F \longrightarrow \infty$, one runs into difficulties. The following examples (motivated by examples of L.C. Young) show this.

4.11 Examples (J. Ball) (a) Consider the (non-convex) problem of minimizing

$$\int_0^1 \left(\frac{(\phi' - 1)^2(\phi' - 2)^2}{\phi'} + (\phi - \tfrac{3}{2}X)^2\right) dX$$

for $x = \phi(X)$ with $\phi' \geq 0$, $\phi(0) = 0$, and $\phi(1) = \frac{3}{2}$. One sees by taking small broken segments as in Figure 6.4.5(a) that the minimum is zero.

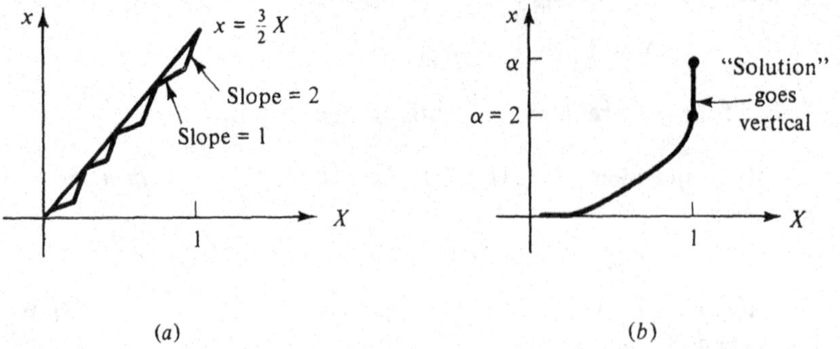

Figure 6.4.5

However, the minimum can never be attained. Roughly, it is a line with infinitely many zig-zags. If W does not depend explicitly on X, then examples like this are not possible (cf. Aubert and Tahraoui [1979]).

(b) An example violating $W(X, F)/F \longrightarrow \infty$ is

$$\int_0^1 \left(\frac{1}{\phi'} + \phi' + \phi\right) dX, \qquad \phi(0) = 0, \qquad \phi(1) = \alpha.$$

Direct calculation shows that the solution does not exist for $\alpha > 2$. See Figure 6.4.5(b). (This problem can be dealt with by making the transformation $\phi \mapsto \phi^{-1}$—that is, by interchanging the roles of x and X.)

In one dimension there is more known about regularity than in three dimensions. In fact, we have the following (Antman [1976b]):

4.12 Proposition *Suppose (H1), (H2), and (H5) hold and ϕ minimizes I in $\mathcal{A}$. Then the Euler–Lagrange equations*

$$\frac{d}{dX} W_F(X, \phi'(X)) = V_\phi(X, \phi(X))$$

hold a.e. on $[0, 1]$. *If, moreover, W is strictly convex in F and (H3) holds, then ϕ is C^1 and ϕ' is bounded away from zero.*

Proof The usual derivation of the Euler–Lagrange equations is not valid, so some care is needed; indeed, on $\mathcal{A}$, I is not differentiable. Let

$$\Omega_n = \{X \in [0, 1] \mid 1/n \leq \phi'(X) \leq n\}.$$

Thus $\Omega_n \subset \Omega_{n+1}$ and $\mu([0, 1] \backslash \cup \Omega_n) = 0$. Let $v \in L^\infty$, let χ_n be the characteristic function of Ω_n, and define $\psi(t, X)$ by

$$\psi'(t, X) = \phi'(X) + t\chi_n(X)v(X), \qquad \psi(0) = 0.$$

We will show that $(d/dt)I(\psi(t, \cdot))|_{t=0}$ exists. Since ϕ is a minimum, this is then zero. We have

$$\frac{1}{t}[I(\psi(t, \cdot)) - I(\psi(0, \cdot))] =$$

$$\frac{1}{t}\int_0^1 \{W(X, \phi'(X) + t\chi_n(X)v(X)) - W(X, \phi'(X))\}\, dX$$

$$+ \frac{1}{t}\int_0^1 \left\{V\left(X, \int_0^X [\phi'(\bar{X}) + t\chi_n(\bar{X})v(\bar{X})]\, d\bar{X}\right) - V(X, \phi(X)\right\} dX$$

From the mean value theorem and the definition of Ω_n we see that the integrands are bounded, so we can pass to the limit $t \longrightarrow 0$ by the dominated convergence theorem to get

$$0 = \frac{d}{dt} I(\psi(t, \cdot))\Big|_{t=0} = \int_{\Omega_n} W_F(X, \phi'(X))v(X)\, dX$$
$$+ \int_{\Omega_n} V_\phi(X, \phi(X))\left(\int_0^X \chi_n(\bar{X})v(\bar{X})\, d\bar{X}\right) dX.$$

Integration by parts gives

$$0 = \int_{\Omega_n} W_F(X, \phi'(X))v(X)\, dX - \int_{\Omega_n}\left(\int_0^X V_\phi(\bar{X}, \phi(\bar{X}))\, d\bar{X}\right) v(X)\, dX.$$

Since v and n are arbitrary, we get

$$W_F(X, \phi'(X)) = \int_0^X V_\phi(\bar{X}, \phi(\bar{X}))\, d\bar{X} \text{ a.e.}$$

But this implies, by continuity of the integrand,

$$\frac{d}{dX} W_F(X, \phi'(X)) = V_\phi(X, \phi(X)).$$

The second part follows by implicitly solving

$$W_F(X, \phi'(X)) = \int_0^X V_\phi(\bar{X}, \phi(\bar{X}))\, d\bar{X}$$

for ϕ'. ∎

Box 4.2 ***Summary of Important Formulas for Section 6.4***

Local Existence Theory for Elastostatics

If about a given configuration ϕ_0 the linearized problem has unique solutions (see Section 6.1, so the problem is not pure traction and no bifurcation occurs), and the boundaries of $\mathcal{B}$ are smooth and no mixed boundary conditions with contiguous parts occur, then a small change in any of the data (boundary conditions, forces, or constitutive functions) produces a corresponding unique configuration ϕ in $W^{s,p}$, $s > 3/p + 1$ depending smoothly on the data.

Non-Applicability of the Inverse Function Theorem in $W^{1,p}$

In $W^{1,p}$ solutions to the elastostatics equations need not be isolated, even though the formal linearization of the equations is an isomorphism.

Convexity

The assumptions of convexity and monotonicity are not appropriate for the operators in three-dimensional elasticity.

Topological Methods

Degree theory, Morse theory, and so on may be useful in three-dimensional elasticity, but so far have not been successfully applied because of technical difficulties.

Minimizers

Global solutions in $W^{1,p}$ for elastostatics can be found by using weak convergence and minimizers. Under the assumption of polyconvexity and growth conditions on the stored energy function, minimizers exist. Their regularity is not known, except in one dimension.

6.5 NONLINEAR ELASTODYNAMICS

This section surveys some results that are relevant to elastodynamics. The only part of this theory that is well understood is that dealing with *semilinear equations*—that is, equations that are linear plus lower-order nonlinear terms. This theory, due to Jorgens [1961] and Segal [1962], will be presented and applied to an example—the equations of a vibrating panel. The theory for *quasi-linear* equations—equations whose leading terms are nonlinear but depend linearly on the highest derivative—is appropriate for three-dimensional nonlinear elasticity. This will be briefly sketched, but much less is known. The primary difficulty is the problem of shock waves. The recent literature will be briefly discussed concerning this problem.

Elastostatics is imbedded in elastodynamics; each solution of the elastostatic equations is a fixed point for the equations of elastodynamics. Eventually, the dynamical context provides important additional information and intuition. For example, we may wish to know if the fixed points are stable, unstable, or are saddle points. We may also wish to find periodic orbits and examine their stability. For ordinary differential equations this leads to the large subject of dynamical systems (cf. Abraham and Marsden [1978] for more information and references). For partial differential equations a good deal is known for semilinear equations and we shall give some examples in Section 6.6 and in Chapter 7. However, for the quasi-linear equations of three-dimensional elasticity, much less is known about qualitative dynamics.

Let us begin by recalling some general terminology (see Definition 3.3, Chapter 5).

A *continuous local semiflow* on a Banach space $\mathcal{Y}$ is a continuous map $F: \mathfrak{D} \subset \mathcal{Y} \times \mathbb{R}^+ \longrightarrow \mathcal{Y}$, where $\mathfrak{D}$ is open, such that: (i) $\mathcal{Y} \times \{0\} \subset \mathfrak{D}$ and $F(x, 0) = x$; and (ii) if $(x, t) \in \mathfrak{D}$, then $(x, t + s) \in \mathfrak{D}$ if and only if $(F(x, t), s) \in \mathfrak{D}$ and in this case $F(x, t + s) = F(F(x, t), s)$.

Suppose $\mathcal{Y}$ is continuously included in another Banach space $\mathfrak{X}$ and G maps $\mathcal{Y}$ (or an open subset $\mathfrak{D}$ of $\mathcal{Y}$) to $\mathfrak{X}$. We say G *generates* the semiflow if for $t \geq 0$ and for each $x \in \mathcal{Y}$, $F(x, t)$ is t-differentiable and

$$\frac{d}{dt} F(x, t) = G(F(x, t)). \tag{1}$$

As usual, if $F(x, t)$ is defined for all $t \in \mathbb{R}$, we call F a *flow*.

There is a slight philosophical difference with the linear case. For the latter, we started with $F_t: \mathfrak{X} \longrightarrow \mathfrak{X}$ and constructed the generator by looking at where F_t is t-differentiable at $t = 0+$. The domain of the generator is a Banach space $\mathcal{Y}$ with the graph norm, and F_t maps it to itself. This is compatible with the above definition. In the nonlinear case, it is better to start right off with F_t defined on the smaller space $\mathcal{Y}$. Then F_t may or may not extend to all of $\mathfrak{X}$. As we shall see, it does in the semilinear case. In this case we can use the phrase "G is the generator of F_t."

Often we are given G and want to construct $F(x, t)$ such that (1) holds. If an F exists, satisfying 5.1, we say the equations $dx/dt = G(x)$ are *well-posed*. If solutions exist for all time $t \geq 0$ (or all t for flows), we say the equations generate *global solutions*.

A crucial part of well-posedness is the continuous dependence of the solution $F(x, t)$ on the initial data x—that is, continuity of the map $x \mapsto F(x, t)$ from (an open subset of) $\mathcal{Y}$ to $\mathcal{Y}$ for each $t \geq 0$. The satisfactory answer to such problems can depend on the choice of $\mathcal{Y}$ made.

Sometimes it is necessary to study the case in which G depends explicitly on time. Then the flow is replaced by evolution operators $F_{t,s}: \mathcal{Y} \longrightarrow \mathcal{Y}$ satisfying $F_{s,s} =$ Identity and $F_{t,s} \circ F_{s,r} = F_{t,r}$. This is just as in Definition 6.5, Chapter 1. We replace (1) by

$$\frac{d}{dt} F_{t,s}(x) = G(F_{t,s}(x), t) \tag{1'}$$

with initial condition $F_{s,s}(x) = x$ (so "s" is the *starting time*).

We begin now with a discussion of the semilinear case. The main method is based on the Duhamel, or variation of constants, formula; namely, the fact that the solution of

$$\frac{du}{dt} = Au + f, \qquad u(0) = u_0$$

is

$$u(t) = e^{tA}u_0 + \int_0^t e^{(t-s)A}f(s)\,ds.$$

(See 2.17(5) in Section 6.2.) If B depends on t and u, we conclude that the solution $u(t)$ of

$$\frac{du}{dt} = Au + B(t, u), \qquad u(0) = u_0 \tag{2}$$

satisfies the implicit equation

$$u(t) = e^{tA}u_0 + \int_0^t e^{(t-s)A}B(s, u(s))\,ds. \tag{3}$$

The point now is that if B is a Lipschitz operator on the Banach space $\mathfrak{X}$ (on which e^{tA} defines a semigroup), then the Picard iteration technique from ordinary differential equations applies to (3). If B is actually a C^1 map of $\mathfrak{X}$ to $\mathfrak{X}$, then we will show that solutions of (3) are in fact in $\mathfrak{D}(A)$ if $u_0 \in \mathfrak{D}(A)$ and satisfy (2) in the strict sense.

The part of the analysis of (3) that is the same as that in ordinary differential equations is outlined in the following problem.

Problem 5.1 (*Existence*) Let A generate a C^0 semigroup on $\mathfrak{X}$ of type (M, β) and let $\mathfrak{F} = \{u \in C([0, t_0], \mathfrak{X}) \mid \|u(t) - u_0\| \leq R\}$, where $t_0 > 0$ and $R > 0$ are constants and $C([0, t_0], \mathfrak{X})$ is the set of continuous maps of $[0, t_0]$ to $\mathfrak{X}$. Suppose $B(t, u)$ is continuous and $\|B(t, u)\| \leq Cp(t)$ if $\|u - u_0\| \leq R$ and $0 \leq t \leq t_0$. Define

$$T: \mathfrak{F} \to C([0, t_0], \mathfrak{X}); (Tu)(t) = e^{tA}u_0 + \int_0^t e^{(t-s)A}B(s, u(s))\,ds.$$

(i) Show that if

$$\|e^{tA}u_0 - u_0\| + Me^{\beta t_0}C\int_0^{t_0} |p(s)|\,ds \leq R \quad (0 \leq t \leq t_0),$$

then T maps $\mathfrak{F}$ to $\mathfrak{F}$.

(ii) If B satisfies $\|B(t, u_1) - B(t, u_2)\| \leq Kp(t)\|u_1 - u_2\|$ for u_1 and u_2 in the ball $\|u - u_0\| \leq R$ and $0 \leq t \leq t_0$, where K is a (Lipschitz) constant (depending on R), then T satisfies

$$\|Tu_1 - Tu_2\| \leq \alpha\|u_1 - u_2\|,$$

where the distance function on $\mathfrak{F} \subset C([0, t_0], \mathfrak{X})$ is

$$\|u_1 - u_2\| = \sup_{0 \leq t \leq t_0} \|u_1(t) - u_2(t)\|$$

and α is defined by

$$\alpha = KMe^{\beta t_0}\int_0^{t_0} |p(s)|\,ds.$$

(iii) Deduce that if $\alpha < 1$, then T has a unique fixed point by the contraction mapping principle (see Lemma 1.3 of Section 4.1).

(iv) Show that instead of continuity of B, it is sufficient for $B(t, u)$ to be L^1 in t and Lipschitz in u.

Problem 5.2 (Uniqueness and Continuous Dependence on Initial Data)

(i) Prove *Gronwall's inequality:* if p is integrable on $[a, b]$ and v is non-negative, bounded, and measurable on $[a, b]$ and there is a constant $C \geq 0$ such that for all $t \in [a, b]$,

$$v(t) \leq C + \int_a^t |p(s)| v(s)\, ds,$$

then
$$v(t) \leq C \exp\left(\int_a^t |p(s)|\right) ds.$$

(See any book on ordinary differential equations; the solution is on p. 124 of Carroll [1969].)

(ii) In the setting of Problem 5.1, show that solutions of (3) satisfy

$$\| u(t) \| \leq Me^{\beta t_0} \| u_0 \| \exp\left(CMe^{\beta t_0} \int_0^t |p(s)|\, ds\right)$$

and
$$\| u(t) - v(t) \| \leq Me^{\beta t_0} \| u_0 - v_0 \| \exp\left(KMe^{\beta t_0} \int_0^t |p(s)|\, ds\right).$$

(iii) Use (ii) to show that solutions of (3) are Lipschitz functions of the initial data.

(iv) Use (ii) to give another proof of local uniqueness of solutions.

(v) Prove that any two solutions of (3) are globally unique on their common domain of definition if B is locally Lipschitz. (*Hint:* Show that the set of t where the two solutions coincide is both open and closed.)

Now we shall state the main theorem for semilinear equations in a version due to Segal [1962]. (Further important information is contained in subsequent corollaries.) To simplify the exposition, we shall assume B is independent of s. (The reader should do the general case.)

5.1 Theorem *Let $\mathfrak{X}$ be a Banach space and U_t a linear semigroup on $\mathfrak{X}$ with generator A having domain $\mathfrak{D}(A)$. Suppose $B\colon \mathfrak{X} \to \mathfrak{X}$ is a C^k* map ($k \geq 1$). *Let $G(u) = Au + B(u)$ on $\mathfrak{D}(A)$. Then (3) defines a unique local semiflow $u(t) = F_t(u_0)$ on $\mathfrak{X}$ ($t \geq 0$); F_t is a local flow if A generates a group. If $u_0 \in \mathfrak{D}(A)$, then $F_t(u_0) \in \mathfrak{D}(A)$ and (1) holds. Moreover, for each fixed t, F_t is a C^k mapping of an open set in $\mathfrak{X}$ to $\mathfrak{X}$.*

Proof Problems 5.1 and 5.2 show that (3) defines a local flow and that F_t is locally Lipschitz. We next show that $F_t(x)$ is differentiable in x. To do so one can appeal to a general theorem on ordinary differential equations in Hale [1969] (Theorem 3.2, p. 7). We can also show that F_t is C^1 by a direct calculation as follows. For $x \in \mathfrak{X}$, let $\theta_t(x) \in \mathcal{B}(\mathfrak{X})$ (the bounded linear operators from $\mathfrak{X}$ to $\mathfrak{X}$) satisfy the linearized equations:

$$\theta_t(x) = U_t + \int_0^t U_{t-s} DB(F_s(x)) \cdot \theta_s(x)\, ds.$$

$\theta_t(x)$ is defined as long as $F_t(x)$ is defined. It is easy to check that $t \mapsto \theta_t(x)$ is continuous in the strong operator topology and that (for fixed t), $x \mapsto \theta_t(x)$ is norm continuous. We claim that $DF_t(x) = \theta_t(x)$, which will thus prove F_t is C^1. Let

$$\lambda_t(x, h) = \| F_t(x + h) - F_t(x) - \theta_t(x) \cdot h \|.$$

Then

$$\begin{aligned} \lambda_t(x,h) &= \left\| \int_0^t U_{t-s} \{B(F_s(x + h)) - B(F_s(x)) - DB(F_s(x)) \cdot \theta_s(x) \cdot h\}\, ds \right\| \\ &\leq M \exp(\beta |t|) \left\{ \int_0^t \| B(F_s(x+h)) - B(F_s(x)) - DB(F_s(x)) \cdot [F_s(x+h) - F_s(x)] \|\, ds \right. \\ &\quad \left. + \int_0^t \| DB(F_s(x)) \cdot [F_s(x + h) - F_s(x) - \theta_s(x) \cdot h] \|\, ds \right\}. \end{aligned}$$

Thus, given $\epsilon > 0$, there is a $\delta > 0$ such that $\|h\| < \delta$ implies

$$\lambda_t(x, h) \leq (\text{const.}) \cdot \left\{ \|h\| \epsilon + \int_0^t \lambda_s(x, h)\, ds \right\}.$$

Hence (by Gronwall's inequality), $\lambda_t(x, h) \leq C(t) \|h\| \epsilon$. Hence, by definition of the derivative, $DF_t(x) = \theta_t(x)$. Thus F_t is C^1. An induction argument can be used to show F_t is C^k.

Now we prove that F_t maps $\mathfrak{D}(A)$ to $\mathfrak{D}(A)$ and

$$\frac{d}{dt} F_t(u_0) = G(F_t(u_0))$$

is continuous in t. Let $u_0 \in \mathfrak{D}(A)$. Then, setting $u(t) = F_t(u_0)$, (3) gives

$$\begin{aligned} \frac{1}{h}[u(t + h) - u(t)] &= \frac{1}{h}(U_{t+h}u_0 - U_t u_0) + \frac{1}{h}\int_0^t (U_{t+h-s} - U_{t-s}) B(u(s))\, ds \\ &\quad + \frac{1}{h}\int_0^{t+h} U_{t+h-s} B(u(s))\, ds \\ &= \frac{1}{h}[U_h(u(t)) - u(t)] + \frac{1}{h}\int_t^{t+h} U_{t+h-s} B(u(s))\, ds. \qquad (4) \end{aligned}$$

The second term approaches $B(u(t))$ as $h \to 0$. Indeed,

$$\begin{aligned} \left\| \frac{1}{h}\int_t^{t+h} U_{t+h-s} B(u(s))\, ds - B(u(t)) \right\| &\leq \frac{1}{h}\int_t^{t+h} \| U_{t+h-s} B(u(s)) - B(u(t)) \|\, ds \\ &\leq \frac{1}{h}\int_t^{t+h} \| U_{t+h-s} B(u(s)) - U_{t+h-s} B(u(t)) \|\, ds \\ &\quad + \frac{1}{h}\int_t^{t+h} \| U_{t+h-s} B(u(t)) - B(u(t)) \|\, ds \end{aligned}$$

$$\leq \frac{1}{h} \cdot C(t) \cdot \int_t^{t+h} \| B(u(s)) - B(u(t)) \| \, ds$$

$$+ \frac{1}{h} \int_t^{t+h} \| U_{t+h-s} B(u(t)) - B(u(t)) \| \, ds$$

and each term $\longrightarrow 0$ as $h \longrightarrow 0$.

It follows that $F_t(u_0)$ is right differentiable at $t = 0$ and has derivative $G(u_0)$. To establish the formula at $t \neq 0$ we first prove that $u(t) \in \mathfrak{D}(A)$. But

$$\frac{1}{h}(F_{t+h}u_0 - F_t u_0) = \frac{1}{h}(F_t F_h u_0 - F_t u_0)$$

has a limit as $h \longrightarrow 0$ since F_t is of class C^1. Hence, from (4),

$$\frac{1}{h}[U_h(u(t)) - u(t)]$$

has a limit as $h \longrightarrow 0$. Thus, $u(t) \in \mathfrak{D}(A)$. It follows that

$$\frac{d}{dt} F_t(u_0) = G(F_t(u_0)) = DF_t(u_0) \cdot G(u_0).$$

Since the right derivative is continuous, the ordinary (two-sided) derivative exists as well. ▮

Next we give a criterion for global existence.

5.2 Proposition *Let the hypotheses of 5.1 hold. Suppose $u(t)$ is a solution of (3) defined for $0 \leq t < T$ and that $\| B(u(t)) \|$ is an integrable function of t on $[0, T]$. Then $u(t)$ can be continued to a solution for $0 \leq t \leq T + \epsilon$ for $\epsilon > 0$.*

Proof For $0 \leq t, s < T$, we have from (3),

$$\| u(t) - u(s) \| \leq \| e^{tA}u_0 - e^{sA}u_0 \| + \left\| \int_s^t e^{(t-\tau)A} B(u(\tau)) \right\| d\tau$$

$$\leq \| e^{tA}u_0 - e^{sA}u_0 \| + Me^{\beta T} \int_s^t \| B(u(\tau)) \| \, d\tau$$

Thus $u(t)$ is a Cauchy sequence as $t \longrightarrow T$, so converges, to, say u_T. The local existence theory applied in a neighborhood of u_T shows the time of existence is uniformly bounded away from zero as $t \longrightarrow T$; so the result follows. ▮

Here is an example of how this works.

5.3 Corollary *Suppose B satisfies $\| B(u) \| \leq C + K \| u \|$ for constants C and K. Then the semiflow F_t is global—that is, is defined for all $t \geq 0$.*

Proof Fix $T > 0$ and let $u(t)$ solve (3) for $0 \leq t < T$. Then

$$\| u(t) \| \leq Me^{\beta T} \| u_0 \| + \int_0^t Me^{\beta T}(C + K \| u(\tau) \|) \, d\tau.$$

From Grownwall's inequality (Problem 5.2),

$$\| u(t) \| \leq [M e^{\beta T} \| u_0 \| + CT] \exp(KMe^{\beta T})$$

so $\| u(t) \|$ and hence $\| B(u(t)) \|$ are bounded for $0 \leq t < T$. Thus u can be extended beyond T by 5.2. Hence u is defined for all $t \geq 0$. ∎

Problem 5.3 Prove a global existence theorem if

$$\| B(t, u) \| \leq p(t)[C + K \| u \|],$$

where $p(t)$ is L^1 on every finite interval.

Another way of obtaining global existence is through energy estimates. That is, if the equations are Hamiltonian (or are related to a Hamiltonian system), there may be a conserved (or decreasing) energy function that can be used to obtain the needed estimate in 5.2. We give an example in the following:

5.4 Corollary *Suppose the conditions of Theorem 5.1 hold. Suppose, moreover, that there is a C^1 function $H: \mathfrak{X} \longrightarrow \mathbb{R}$ such that:*

(i) *there is a monotone increasing function $\phi: [a, \infty) \longrightarrow [0, \infty)$, where $[a, \infty) \supset$ Range H, satisfying $\| x \| \leq \phi(H(x))$;*

(ii) *there is a constant $K \geqq 0$ such that if $u(t)$ satisfies (3), then*

$$\frac{d}{dt} H(u(t)) \leq KH(u(t)).$$

Then $F_t(u_0)$ is defined for all $t \geq 0$ and $u_0 \in \mathfrak{X}$; that is, the semiflow is global.

If, in addition, H is bounded on bounded sets and

(iii) $$\frac{d}{dt} H(u(t)) \leq 0 \quad \text{if} \quad \| u(t) \| \geq B,$$

then any solution of (3) remains uniformly bounded in $\mathfrak{X}$ for all time; that is, given $u_0 \in \mathfrak{D}(A)$, there is a constant $C = C(u_0)$ such that $\| u(t) \| \leqq C$ for all $t \geqq 0$.

Proof By (ii), $H(u(t)) \leq H(u_0) \exp(Kt)$ so by (i), $\| u(t) \| \leq \phi(H(u_0) \exp(Kt))$. Thus, global existence follows by 5.2. Let $H_B = \sup\{H(u) \,|\, \| u \| \leq B\}$, so, by (iii), $H(u(t)) \leq \max\{H(u_0), H_B\}$. Hence, by (i), we can take $C = \phi(\max\{H(u_0), H_B\})$. ∎

Next we have a criterion for asymptotic stability of a fixed point x_0; that is, $F_t(x_0) = x_0$ and $F_t(x) \longrightarrow x_0$ as $t \longrightarrow +\infty$ for all $x \in \mathfrak{X}$. (See Theorem 4.1 of Chapter 7 for the use of spectral methods to obtain a related result.)

5.5 Proposition *Let (i) and (ii) of 5.4 hold, and suppose $G(x_0) = 0$; that is, x_0 is a fixed point, and:*

(iii′) *There is a continuous monotone function $f: [0, \infty) \longrightarrow [0, \infty)$, locally Lipschitz on $(0, \infty)$ such that*

(a) $\frac{dH}{dt}(F_t(x)) \leq -f(H(F_t(x))$ *for* $x \in U$, *a neighborhood of* x_0,

and

(b) *solutions of* $\dot{r} = -f(r)$ *tend to zero as* $t \to +\infty$.

(iv) $H(x) \geq 0$ *and there is a strictly monotone continuous function* $\psi: [0, \infty) \to [0, \infty)$ *such that*

$$\|x - x_0\| \leq \psi(H(x)).$$

Then x_0 *is asymptotically stable.*

Proof Let $r(t)$ be the solution of $\dot{r} = -f(r)$ with $r(0) = H(x)$, $x \in U$, $x \neq x_0$. Then, by (iii′), $H(F_t(x)) \leq r(t)$. Hence $H(F_t(x)) \to 0$ as $t \to +\infty$. Thus by (iv), $F_t(x)$ converges to x_0 as $t \to +\infty$. ∎

Problem 5.4 Consider $\ddot{u} + \dot{u} + u^3 = 0$. Show that solutions decay to zero like $C/\sqrt{t}$ as $t \to +\infty$, by considering $H(u, \dot{u}) = (u + \dot{u})^2 + \dot{u}^2 + u^4$. (See Ball and Carr [1976] for more information.)

The following is a specific situation relevant for some semilinear wave equations. See 3.10 in Box 3.1, Section 6.3.

5.6 Proposition *Let* $\mathcal{H}$ *be a real Hilbert space and* B *a self-adjoint operator in* $\mathcal{H}$ *with* $B \geq c > 0$. *On* $\mathfrak{X} = \mathfrak{D}(B^{1/2}) \times \mathcal{H}$, *let* $A = \begin{bmatrix} 0 & I \\ -B & 0 \end{bmatrix}$ *with domain* $\mathfrak{D}(A) = \mathfrak{D}(B) \times \mathfrak{D}(B^{1/2})$. *Let* $V: \mathfrak{D}(B^{1/2}) \to \mathbb{R}$ *be a smooth function and suppose* V *has a smooth gradient; that is,* $\nabla V: \mathfrak{D}(B^{1/2}) \to \mathcal{H}$ *is smooth and satisfies*

$$\langle \nabla V(x), y \rangle = dV(x) \cdot y$$

for all $x, y \in \mathfrak{D}(B^{1/2})$. *Suppose* $V \geq 0$ *and* V *is bounded on bounded sets in* $\mathfrak{D}(B^{1/2})$. *Let*

$$G: \mathfrak{D}(A) \to \mathfrak{X}, \text{ defined by } G(u, \dot{u}) = A(u, \dot{u}) + (0, \nabla V(u)).$$

Then the flow of G *is globally defined and solutions are uniformly bounded for all time.*

Proof Clearly G satisfies the conditions of 5.1. Also, G is Hamiltonian with energy

$$H(u, u) = \tfrac{1}{2}\|u\|^2 + \tfrac{1}{2}\langle B^{1/2}u, B^{1/2}u \rangle + V(u)$$

and H is conserved along solutions. Since $V \geq 0$,

$$H(u, u) \geq \tfrac{1}{2}(\|u\|^2 + \|B^{1/2}u\|^2).$$

The result therefore follows by 5.4. ∎

Problem 5.5 Instead of $V \geq 0$, assume $V(0) = 0$, $DV(0) = 0$ and $D^2V(0) = 0$. Show that in a neighborhood of (0, 0), solutions are globally defined and remain in a ball about (0, 0). (*Hint*: Apply Taylor's theorem to

H to show that in a neighborhood of $(0, 0)$, $C_1\|(u, \dot{u})\|^2 \leq H(u, \dot{u}) \leq C_2\|(u, \dot{u})\|^2$ for constants C_1 and C_2.)

Problem 5.6 (*Regularity*) Let 5.1 hold and suppose that $B: \mathfrak{D}(A^{k-l}) \to \mathfrak{D}(A^{k-l})$ is C^{l-1}, $l = 1, \ldots, k-1$, where $\mathfrak{D}(A^{k-l})$ has the graph norm. Prove that F_t maps $\mathfrak{D}(A^{k-l})$ to itself and is C^{k-l}.

Problem 5.7 (*Singular Case*) Let A generate a C^0 semigroup on $\mathfrak{X}$ and let $\mathfrak{D}(A)$ have the graph norm. Suppose $B: \mathfrak{D}(A) \to \mathfrak{X}$ is C^1 and the tangent of B, $TB: (u, v) \mapsto (u, DB(u)\cdot v)$ extends to a C^1 map of $\mathfrak{D}(A) \times \mathfrak{X}$ to $\mathfrak{D}(A) \times \mathfrak{X}$. Then $G = A + B$ generates a unique local C^1 flow. (*Hint*: Apply 5.1 to the operator $TG(u, v) = (0, Av) + TB(u, v)$; the first component of this flow gives the flow of G; cf. Segal [1962].)

We will now give two applications of this theory. The first is to semilinear non-linear wave equations and the second is to a problem of panel flutter. The first is perhaps not directly relevant to elasticity, but it is a topic of current interest and illustrates the methods well. (See Reed [1976] for similar results from a different point of view. For a recent spectacular application of the semilinear theory, see Eardley and Moncrief [1982].)

5.7 Example (Semilinear Wave Equation) We consider the following equation for $\phi(x, t) \in \mathbb{R}$, where $x \in \mathbb{R}^n$ and $t \in \mathbb{R}$:

$$\left.\begin{aligned} \frac{\partial^2 \phi}{\partial t^2} &= \nabla^2\phi - m^2\phi - g\phi^p, \\ \phi(x, 0) &\text{ and } \dot{\phi}(x, 0) \text{ given} \end{aligned}\right\} \tag{5}$$

where $p \geq 2$ is an integer and $g \in \mathbb{R}$ is a constant. (One can also consider this problem on complete Riemannian manifolds, but we stick to $\mathbb{R}^n$ for simplicity.) To obtain the results, essential use is made of the Sobolev spaces $H^s(\mathbb{R}^n)$ and the Sobolev inequalities. To obtain global solutions, these inequalities must be applied with care. The relevant facts we need were given in Box 1.1, Section 6.1.

The results for Equation (5) are summarized as follows:

(i) *For $n = 1, 2$, $m \geq 0$, $g \geq 0$, p an odd integer, and $s \geq 1$, Equation (5) has unique global solutions in $H^{s+1}(\mathbb{R}^n, \mathbb{R}) \times H^s(\mathbb{R}^n, \mathbb{R})$ for any initial data in this space.*

(ii) *For $n = 3$, $m \geq 0$, $g \geq 0$, and $p = 3$, the conclusion (i) remains valid.*

(iii) *For $n = 1, 2$, $m \geq 0$, any g, and any p, there are unique global solutions as in (i) if the initial data is small enough in $H^1 \times L^2$ norm. The same holds for $n = 3$ if $p \leq 4$.*

(iv) *For any n, m, g, p, there are unique solutions local in time in $H^{s+1}(\mathbb{R}^n, \mathbb{R}) \times H^s(\mathbb{R}^n, \mathbb{R})$ if $s + 1 > n/2$.*

Note: From hyperbolicity arguments (Courant and Hilbert [1962]) it follows that if we start with C^∞ data with compact support, the solution will be C^∞ with compact support as well.

To establish (i) and (ii), we use Proposition 5.6 (with a slight modification if $m = 0$), and to prove (iii) we use Problem 5.5, and to prove (iv) we use Theorem 5.1. We shall give the details for case (ii).

First, suppose $m > 0$. Then as was discussed in Section 6.3 the Klein–Gordon operator

$$A = \begin{bmatrix} 0 & I \\ \Delta - m^2 & 0 \end{bmatrix}$$

is skew-adjoint on $H^1 \times L^2$ with the energy norm, which is equivalent to the usual Sobolev norm. We set

$$V: H^1 \longrightarrow \mathbb{R}, \qquad V(\phi) = \frac{g}{4}\int_{\mathbb{R}^3} \phi^4 \, dx.$$

Here is the key fact:

Lemma 1 *V is a well-defined smooth map with derivative*

$$DV(\phi)\cdot\psi = g\int_{\mathbb{R}^3} \phi^3 \psi \, dx.$$

Proof From the Sobolev–Nirenberg–Gagliardo inequality (see Box 1.1), there is a positive constant C such that for $\phi \in L^2(\mathbb{R}^3)$,

$$\|\phi\|_{L^p} \le C \|\nabla\phi\|_{L^2}^a \|\phi\|_{L^2}^{1-a}$$

for $2 \le p \le 6$ and $a = 3\cdot(\frac{1}{2} - 1/p)$. In particular, we have the estimate

$$\|\phi\|_{L^p} \le C \|\phi\|_{H^1}$$

for $2 \le p \le 6$. The main case, $p = 6$, was proved in Proposition 1.16. Taking $p = 4$ we see that V is well defined. Consider

$$\mathcal{V}: H^1 \times H^1 \times H^1 \times H^1 \longrightarrow \mathbb{R}, \qquad \mathcal{V}(\phi_1, \phi_2, \phi_3, \phi_4) = \int_{\mathbb{R}^3} \left(\prod_{i=1}^4 \phi_i\right) dx.$$

From the Schwartz inequality,

$$\|\mathcal{V}(\phi_1, \phi_2, \phi_3, \phi_4)\| \le \prod_{i=1}^4 \|\phi_i\|_{L^4} \le C \prod_{i=1}^4 \|\phi_i\|_{H^1}.$$

Hence $\mathcal{V}$ is a continuous multilinear map; so is C^∞. Hence V is as well, and so the lemma follows. ▮

Lemma 2 *Set* $Y: H^1 \longrightarrow L^2$; $Y(\phi) = (-g\phi^3)$. *Then* Y *is a* C^∞ *map and the conditions of Proposition 5.6 hold, with* A *as above,* $\mathcal{H} = L^2$, *and* $\nabla V = Y$.

Proof Since $L^6 \subset H^1$, Y is well defined. Consider $\mathfrak{Y}: H^1 \times H^1 \times H^1 \longrightarrow L^2$, $\mathfrak{Y}(\phi_1, \phi_2, \phi_3) = \prod_{i=1}^3 \phi_i$. Then by Hölder's inequality,

$$\|\mathfrak{Y}(\phi_1, \phi_2, \phi_3)\|_{L^2} \le \prod_{i=1}^3 \|\phi_i\|_{L^6} \le C \prod_{i=1}^3 \|\phi_i\|_{H^1}.$$

Thus $\mathfrak{Y}$ and hence Y is smooth.

A straightforward calculation shows that $\nabla V = Y$. ▮

That solutions which start in $H^{s+1} \times H^s$ for $s \geq 2$ stay in that space follows by regularity (Problem 5.6). We ask the reader to supply the details.

The case $m = 0$ may be dealt with as follows. Using the usual norm on $H^1 \times L^2$, conservation of energy and $V \geq 0$ implies that $\|\nabla\phi\|_{L^2} + \|\dot{\phi}\|_{L^2}$ is bounded, say, by M. But this implies that $\|\nabla\phi\|_{L^2} + \|\phi\|_{L^2} + \|\dot{\phi}\|_{L^2} \leq M + tM\|\phi_0\|_{L^2}$, so the $(H^1 \times L^2)$-norm is bounded on finite t-intervals. Thus we again get our result.

The result (iv) is perhaps the easiest of all. Indeed, for $s > n/2$, $H^s(\mathbb{R}^n, \mathbb{R})$ is a ring: $\|fg\|_{H^s} \leq C\|f\|_{H^s}\|g\|_{H^s}$, so multiplication is smooth. (See 1.17 in Box 6.1.1.) Therefore, we can apply Theorem 5.1 directly because any polynomial Y on H^s will be smooth.

5.8 Example (Panel Flutter) The linear problem was considered in Box 3.2. (See Figure 6.3.1.) Here we consider the nonlinear problem. The equations are

$$\alpha\dot{v}'''' + v'''' - \left\{\Gamma + \kappa\int_0^1 (v'(t,\xi))^2\,d\xi + \sigma\int_0^1 (v'(t,\xi)\dot{v}'(t,\xi))\,d\xi\right\}v''$$
$$+ \rho v' + \sqrt{\rho}\,\delta\dot{v} + \ddot{v} = 0. \tag{6}$$

(See Dowell [1975] and Holmes [1977a].) As in Box 3.1, $\dot{} = \partial/\partial t$, $' = \partial/\partial x$, and we have included viscoelastic structural damping terms α, σ as well as aerodynamic damping $\sqrt{\rho}\,\delta$; κ represents nonlinear (membrane) stiffness, ρ the dynamic pressure, and Γ an in-plane tensile load. We have boundary conditions at $x = 0, 1$ that might typically be simply supported ($v = (\dot{v} + \alpha v)'' = 0$) or clamped ($v = v' = 0$). To be specific, let us choose the simply supported condition.

To proceed with the methods above, we first write (6) in the form (2), choosing as our basic space $\mathfrak{X} = H^2_\partial([0, 1]) \times L^2([0, 1])$, where H^2_∂ denotes H^2 functions in $[0, 1]$ that vanish at 0, 1. Set $\|(v, \dot{v})\|_{\mathfrak{X}} = (\|\dot{v}\|^2 + \|v''\|^2)^{1/2}$, where $\|\cdot\|$ denotes the usual L^2-norm. This is equivalent to the usual norm because of the boundary conditions. In fact, the two Poincaré-type inequalities $\|v'\|^2 \geq \pi^2\|v\|^2$ and $\|v''\| \geq \pi^4\|v\|^2$ may be checked using Fourier series. Define

$$A = \begin{pmatrix} 0 & I \\ C & D \end{pmatrix}, \quad \text{where} \quad \begin{cases} Cv = -v'''' + \Gamma v'' - \rho v', \\ D\dot{v} = -\alpha\dot{v}'''' - \sqrt{\rho}\,\delta\dot{v}. \end{cases} \tag{7}$$

The domain $\mathfrak{D}(A)$ of A consists of all pairs $(v, \dot{v}) \in \mathfrak{X}$ such that $\dot{v} \in H^2_\partial$, $v + \alpha\dot{v} \in H^4$, and $v'' + \alpha\dot{v}'' = 0$ at $x = 0, 1$. Define the nonlinear operator $B(v, \dot{v}) = (0, [\kappa\|v'\|^2 + \sigma\langle v', \dot{v}'\rangle]v'')$, where $\langle\cdot\rangle$ denotes the L^2 inner product; so (6) can be rewritten as

$$\frac{dx}{dt} = Ax + B(x) \equiv G(x), \quad \text{where } x = (v, \dot{v}) \text{ and } x(t) \in \mathfrak{D}(A). \tag{8}$$

By Proposition 3.11 in Box 3.2, Section 6.2, A generates a semigroup on $\mathfrak{X}$. In one dimension H^1 forms a ring; so as in the previous example, $B\colon \mathfrak{X} \to \mathfrak{X}$ is a C^∞ mapping bounded on bounded sets. Thus, by Theorem 5.1, Equations (6) generate a local semiflow on $\mathfrak{X}$. We claim that this semiflow is global. To see this, we temporarily omit the dissipative terms in (6), namely, we consider

$$v'''' - \left\{\Gamma + \kappa \int_0^1 (v'(t, \xi))^2 \, d\xi\right\} v'' + \ddot{v} = 0.$$

As in Problem 3.6, Section 6.3, this is Hamiltonian with

$$H(v, \dot{v}) = \tfrac{1}{2}\|\dot{v}\|^2 + \tfrac{1}{2}\|v''\|^2 + \frac{\Gamma}{2}\|v'\|^2 + \frac{\kappa}{4}\|v'\|^4.$$

For the full equation (6) we find by a simple calculation that

$$\frac{dH}{dt} = -\rho\langle v', \dot{v}\rangle - \sqrt{\rho}\,\delta\|\dot{v}\|^2 - \alpha\|\dot{v}''\|^2 - \sigma\|(v, \dot{v})\|^2.$$

$$\leq -\rho\langle v', \dot{v}\rangle \leq \rho\|v'\|\|\dot{v}\| \leq \tfrac{1}{2}\rho(\|v'\|^2 + \|\dot{v}\|^2).$$

For $\|v'\|$ large enough,

$$H(v, \dot{v}) \geq C\|(v, \dot{v})\|_X^2.$$

[Note that Γ can be ≤ 0; write

$$\frac{\Gamma}{2}\|v'\|^2 + \frac{\kappa}{4}\|v'\|^4 = \frac{\kappa}{4}\|v'\|^2\left\{\|v'\|^2 + \frac{2\Gamma}{\kappa}\right\}$$

to see this.] Thus, for $\|v'\|$ large, hypotheses (i) and (ii) of 5.4 are satisfied, and so we have global solutions. We shall return to this example again in Chapter 7.

Problem 5.8 (Parks [1966], Holmes and Marsden [1978a]) Let $\tilde{H}(v, \dot{v}) = \frac{1}{2}\{\sqrt{\rho}\,\delta\|v\|^2 + \alpha\|v''\|^2 + 2\langle v, \dot{v}\rangle + (\sigma/2)\|v'\|^4\}$ and let $\mathcal{H}(v, \dot{v}) = H(v, \dot{v}) + \nu\tilde{H}(v, \dot{v})$, where $\nu = (\sqrt{\rho}\,\delta + \alpha\pi^4)/2$. If $\rho^2 < (\sqrt{\rho}\,\delta + \alpha\pi^4)^2 \times (\Gamma + \rho^2)$, show that $d\mathcal{H}/dt < 0$ along non-zero solutions. Use 5.5 to show that in these circumstances solutions tend to $(0, 0)$ as $t \longrightarrow +\infty$.

While the above theory does apply to a number of special situations in elasticity involving rod and plate approximations, it does not apply to the "full" theory of nonlinear elasticity, even in one dimension. Here the equations have the form

$$\dot{u} = A(t, u)u + f(t, u) \qquad (0 \leq t \leq T, \quad u(0) = u_0), \tag{9}$$

the point being that the linear operator A depends on u. For example, one-dimensional elasticity has this form:

$$\frac{\partial}{\partial t}\begin{pmatrix}\phi \\ \dot{\phi}\end{pmatrix} = \begin{pmatrix} 0 & I \\ \sigma'(\phi_x)\dfrac{\partial^2}{\partial x^2} & 0\end{pmatrix}\begin{pmatrix}\phi \\ \dot{\phi}\end{pmatrix}.$$

Such equations in which the highest derivatives occur linearly, but possibly multiplied by functions of lower derivatives, are called *quasi-linear*. The remainder of this section discusses the theory for these equations. This theory began with Schauder [1935] with contributions by Petrovskii [1937], Sobolev [1939], Choquet-Bruhat [1952], and many others. However, most of these treatments had a few loose ends, and none proved the continuous dependence on initial data (in the same space Y). We shall follow the formulations of Kato [1975b], [1977], and of Hughes, Kato, and Marsden [1977].

The abstract theory for (9) divides into the local theory and the global theory. We begin with a discussion of the local theory. Proofs will be omitted as they are rather technical.

We start from four (real) Banach spaces

$$\mathcal{Y} \subset \mathfrak{X} \subset \mathcal{Z}' \subset \mathcal{Z},$$

with all the spaces reflexive and separable and the inclusions continuous and dense. We assume that

(Z′) $\mathcal{Z}'$ is an interpolation space between $\mathcal{Y}$ and $\mathcal{Z}$; thus if $U \in \mathcal{B}(\mathcal{Y}) \cap \mathcal{B}(\mathcal{Z})$, then $U \in B(\mathcal{Z}')$ with $\|U\|_{\mathcal{Z}'} \leq c \max\{\|U\|_{\mathcal{Y}}, \|U\|_{\mathcal{Z}}\}$; $\mathcal{B}(\mathcal{Y})$ denotes the set of bounded operators on $\mathcal{Y}$.

Let $\mathfrak{N}(\mathcal{Z})$ be the set of all norms in $\mathcal{Z}$ equivalent to the given one $\|\ \|_{\mathcal{Z}}$. Then $\mathfrak{N}(\mathcal{Z})$ is a metric space with the distance function

$$d(\|\ \|_\mu, \|\ \|_\nu) = \log \max\{\sup_{0 \neq z \in \mathcal{Z}} \|z\|_\mu / \|z\|_\nu, \sup_{0 \neq z \in \mathcal{Z}} \|z\|_\nu / \|z\|_\mu\}.$$

We now introduce four functions, A, N, S, and f on $[0, T] \times \mathcal{W}$, where $T > 0$ and $\mathcal{W}$ is an open set in $\mathcal{Y}$, with the following properties:

For all $t, t', \ldots, \in [0, T]$ and all $w, w', \ldots \in \mathcal{W}$, there is a real number β and there are positive numbers $\lambda_N, \mu_N, \ldots$ such that the following conditions hold:

(N) $N(t, w) \in \mathfrak{N}(\mathcal{Z})$, with

$$d(N(t, w), \|\ \|_{\mathcal{Z}}) \leq \lambda_N,$$
$$d(N(t', w'), N(t, w)) \leq \mu_N(|t' - t| + \|w' - w\|_{\mathfrak{X}}).$$

(S) $S(t, w)$ is an isomorphism of $\mathcal{Y}$ onto $\mathcal{Z}$, with

$$\|S(t, w)\|_{\mathcal{Y},\mathcal{Z}} \leq \lambda_S, \quad \|S(t, w)^{-1}\|_{\mathcal{Z},\mathcal{Y}} \leq \lambda'_s,$$
$$\|S(t', w') - S(t, w)\|_{\mathcal{Y},\mathcal{Z}} \leq \mu_S(|t' - t| + \|w' - w\|_{\mathfrak{X}}).$$

(A1) $A(t, w) \in G(\mathcal{Z}_{N(t,w)}\ 1, \beta)$, where $\mathcal{Z}_{N(t,w)}$ denotes the Banach space $\mathcal{Z}$ with norm $N(t, w)$. This means that $A(t, w)$ is a C_0-generator in $\mathcal{Z}$ such that $\|e^{\tau A(t,w)} z\| \leq e^{\beta\tau}\|z\|$ for all $\tau \geq 0$ and $z \in \mathcal{Z}$.

(A2) $S(t, w)A(t, w)S(t, w)^{-1} = A(t, w) + B(t, w)$, where $B(t, w) \in \mathcal{B}(\mathcal{Z})$, $\|B(t, w)\|_{\mathcal{Z}} \leq \lambda_B$.

(A3) $A(t, w) \in \mathcal{B}(\mathcal{Y}, \mathfrak{X})$, with $\|A(t, w)\|_{\mathcal{Y},\mathfrak{X}} \leq \lambda_A$ and $\|A(t, w') - A(t, w)\|_{\mathcal{Y},\mathcal{Z}'} \leq \mu_A\|w' - w\|_{\mathcal{Z}'}$ and with $t \mapsto A(t, w) \in \mathcal{B}(\mathcal{Y}, \mathcal{Z})$ continuous in norm.

(f1) $f(t, w) \in \mathcal{Y}$, $\|f(t, w)\|_{\mathcal{Y}} \leq \lambda_f$, $\|f(t, w') - f(t, w)\|_{\mathcal{Z}'} \leq \mu_f\|w' - w\|_{\mathcal{Z}}$, and $t \mapsto f(t, w) \in \mathcal{Z}$ is continuous.

Remarks (i) If $N(t, w) = \text{const.} = \| \quad \|_{\mathbf{Z}}$, condition (N) is redundant. If $S(t, w) = \text{const.} = S$, condition (S) is trivial. If both are assumed, and $\mathfrak{X} = \mathbf{Z}' = \mathbf{Z}$, we have the case of Kato [1975b].

(ii) In most applications we can choose $\mathbf{Z}' = \mathbf{Z}$ and/or $\mathbf{Z}' = \mathfrak{X}$.

(iii) The paper of Hughes, Kato, and Marsden [1977] had an additional condition (A4) that was then shown to be redundant in Kato [1977].

(iv) The possibility of dropping (S) and related refinements are discussed in Graff [1981] and Altman [1982].

5.9 Theorem *Let conditions* (Z′), (N), (S), (A1) *to* (A3), *and* (f1) *be satified. Then there are positive constants* ρ' *and* $T' \leq T$ *such that if* $u_0 \in \mathcal{Y}$ *with* $\| u_0 - y_0 \|_{\mathcal{Y}} \leq \rho'$, *then* (9) *has a unique solution u on* $[0, T']$ *with*

$$u \in C^0([0, T']; \mathcal{W}) \cap C^1([0, T']; \mathfrak{X}).$$

Here ρ' depends only on λ_N, λ_S, λ'_S, and $R = \text{dist}\,(y_0, \mathcal{Y}\backslash\mathcal{W})$, while T' may depend on all the constants β, λ_N, $\mu_N, \ldots$ and R. When u_0 varies in $\mathcal{Y}$ subject to $\| u_0 - y_0 \|_{\mathcal{Y}} \leq \rho'$, the map $u_0 \mapsto u(t)$ is Lipschitz continuous in the $\mathbf{Z}'$-norm, uniformly in $t \in [0, T']$.

To establish well-posedness, we have to strengthen some of the assumptions. We assume the following conditions:

(B) $\| B(t, w') - B(t, w) \|_{\mathbf{Z}} \leq \mu_B \| w' - w \|_{\mathcal{Y}}$.
(f2) $\| f(t, w') - f(t, w) \| \mathcal{Y} \leq \mu'_f \| w' - w \|_{\mathcal{Y}}$.

5.10 Theorem *Let conditions* (Z′), (N), (S), (A1) *to* (A3), (B), (f1), *and* (f2) *be satisfied, where* $S(t, w)$ *is assumed to be independent of w. Then there is a positive constant* $T'' \leq T'$ *such that when* u_0 *varies in* $\mathcal{Y}$ *subject to* $\| u_0 - y_0 \|_{\mathcal{Y}} \leq \rho'$, *the map* $u_0 \mapsto u(t)$ *given by Theorem* 5.9 *is continuous in the* $\mathcal{Y}$*-norm, uniformly in* $t \in [0, T'']$.

As in Kato [1975b], one can prove a similar continuity theorem when not only the initial value u_0 but also the functions N, A, and f are varied; that is, the solution is "stable" when the equations themselves are varied. Variation of S is discussed in Graff [1981]. In summary, Theorem 5.10 guarantees the existence of (locally defined) evolution operators $F_{t,s}: \mathcal{Y} \longrightarrow \mathcal{Y}$ that are continuous in all variables.

The idea behind the proofs of the Theorems 5.9 and 5.10 is to fix a curve $v(t)$, in $\mathcal{Y}$ satisfing $v(0) = u_0$ and let $u(t)$ be the solution of the "frozen coefficient problem"

$$\dot{u} = A(t, v)u + f(t, v), \qquad u(0) = u_0,$$

which is guaranteed by linear theory (Kato [1970], [1973]). This defines a map $\Phi: v \mapsto u$ and we look for a fixed point of Φ. In a suitable function space and for T' sufficiently small, Φ is in fact a contraction, so has a unique fixed point.

However, it is not so simple to prove that u depends continuously on u_0 and detailed estimates from the linear theory are needed. The proof more or less has to be delicate since the dependence on u_0 is not locally Lipschitz in general. For details of these proofs, we refer to Kato [1975b], [1977] and Hughes, Kato, and Marsden [1977]. The continuous dependence of the solution on u_0 leads us to investigate if it is smooth in any sense. This is important in Hamiltonian systems, as we saw in Proposition 3.4, Chapter 5. This is explored in Box 5.1 below.

Next we give some specific systems to which 5.10 applies. These will be nonlinear versions of first-order symmetric hyperbolic systems and second-order hyperbolic systems. Each of these applies to nonlinear elasticity; to verify the hypotheses (details of which are omitted) one uses the linear theory presented in Theorems 3.1 (for the second-order version) and 3.19 of Section 6.3. We shall treat these equations in all of $\mathbb{R}^m$ as the machinery needed for boundary value problems is yet more complex (see Kato [1977] and Box 5.2 below).

5.11 Example (Quasi-linear Symmetric Hyperbolic Systems) Consider the equation

$$\boldsymbol{a}_0(t, x, \boldsymbol{u})\frac{\partial \boldsymbol{u}}{\partial t} = \sum_{j=1}^{m} \boldsymbol{a}_j(t, x, \boldsymbol{u})\frac{\partial \boldsymbol{u}}{\partial x^j} + \boldsymbol{a}(t, x, \boldsymbol{u}). \tag{10}$$

for $x \in \mathbb{R}^m$, $t \in \mathbb{R}$, $\boldsymbol{u}(t, x) \in \mathbb{R}^N$, and $\boldsymbol{a}_j$, $\boldsymbol{a}$ real $N \times N$ matrices. We assume:

(i) $s > \frac{1}{2}m + 1$ and $\boldsymbol{a}_0$, $\boldsymbol{a}$ are of class C^{s+1} in the variables $t, x, \boldsymbol{u}$ (possibly locally defined in $\boldsymbol{u}$);

(ii) there are constant matrices $\boldsymbol{a}_j^\infty$, $\boldsymbol{a}^\infty$ such that

$$\boldsymbol{a}_j - \boldsymbol{a}_j^\infty, \boldsymbol{a} - \boldsymbol{a}^\infty \in C([0, T], H^0(\mathbb{R}^m) \cap L^\infty([0, T], H^s(\mathbb{R}^m)) \qquad (j = 0, 1, \ldots, m),$$

$\boldsymbol{a}_0 - \boldsymbol{a}_0^\infty \in \operatorname{Lip}([0, T], H^{s-1}(\mathbb{R}^m))$,
locally uniformly in $\boldsymbol{u}$;

(iii) $\boldsymbol{a}_j$, $(j = 0, \ldots, m)$ are symmetric;

(iv) $\boldsymbol{a}_0(t, x, \boldsymbol{u}) \geq c\boldsymbol{I}$ for some $c > 0$ for all x and locally in t, $\boldsymbol{u}$.

Under these conditions, the hypotheses of Theorem 5.10 hold with

$$\mathfrak{X} = H^{s-1}(\mathbb{R}^m), \qquad \mathcal{Y} = H^s(\mathbb{R}^m), \qquad \mathcal{Z} = \mathcal{Z}' = L^2(\mathbb{R}^m),$$

$$S = (1 - \Delta)^{s/2}, \qquad A = \boldsymbol{a}_0^{-1}\left(\sum_{j=1}^{m} \boldsymbol{a}_j \frac{\partial}{\partial x^j} + \boldsymbol{a}\right).$$

Thus (10) generates a unique continuous evolution system $F_{t,s}$ on $\mathcal{Y}$.

5.12 Example (Second-order Quasi-linear Hyperbolic Systems) Consider the equation

$$\boldsymbol{a}_{00}(t, s, \boldsymbol{u}, \nabla \boldsymbol{u})\frac{\partial^2 \boldsymbol{u}}{\partial t^2} = \sum_{i,j=1}^{m} \boldsymbol{a}_{ij}(t, x, \boldsymbol{u}, \nabla \boldsymbol{u})\frac{\partial^2 \boldsymbol{u}}{\partial x^i\, \partial x^j} + 2\sum_{i=1}^{m} \boldsymbol{a}_{0i}(t, x, \boldsymbol{u}, \nabla \boldsymbol{u})\frac{\partial^2 \boldsymbol{u}}{\partial t\, \partial x^i} + \boldsymbol{a}(t, x, \boldsymbol{u}, \nabla \boldsymbol{u}). \tag{11}$$

Here $$\nabla u = \left(\frac{\partial u}{\partial x^1}, \ldots, \frac{\partial u}{\partial x^m}, \frac{\partial u}{\partial t}\right).$$

We assume:

(i) $s > (m/2) + 1$ and $a_{\alpha\beta}$, a are of class C^{s+1} in all variables (possibly locally defined in u);

(ii) there are constant matrices $a^\infty_{\alpha\beta}$, a^∞ such that

$$a_{\alpha\beta} - a^\infty_{\alpha\beta},\ a - a^\infty \in \mathrm{Lip}([0, T];\ H^{s-1}(\mathbb{R}^m)) \subset L^\infty([0, T];\ H^s(\mathbb{R}^m)),$$

locally uniformly in u;

(iii) $a_{\alpha\beta}$ is symmetric;

(iv) $a_{00}(t, x, u) \geq cI$ for some $c > 0$;

(v) strong ellipticity. There is an $\epsilon > 0$ such that

$$\sum_{i,j=1}^m a_{ij}(t, x, u)\xi^i\xi^j \geq \epsilon\left(\sum_{j=1}^m [\xi^j]^2\right)$$

(a matrix inequality) for all $\xi = (\xi^1, \ldots, \xi^m) \in \mathbb{R}^m$, $x \in \mathbb{R}^m$, and locally in t, u.

Under these conditions, Theorem 5.10 holds with

$$\mathfrak{X} = H^s(\mathbb{R}^m) \times H^{s-1}(\mathbb{R}^m),$$
$$\mathcal{Z} = Z' = H^1(\mathbb{R}^m) \times H^0(\mathbb{R}^m),$$
$$\mathcal{Y} = H^{s+1}(\mathbb{R}^m) \times H^s(\mathbb{R}^m),$$
$$S = (1 - \Delta)^{s/2} \times (1 - \Delta)^{s/2},$$

$$A(t) = \left(\begin{array}{c|c} 0 & I \\ \hline a_{00}^{-1}\left[\sum a_{ij}\dfrac{\partial^2}{\partial x^i\,\partial x^j}\right] & a_{00}^{-1}\left[2\sum a_{0j}\dfrac{\partial}{\partial x^j}\right] \end{array}\right).$$

Thus (11) generates a unique continuous evolution system on $\mathcal{Y}$.

From either 5.11 or 5.12 we conclude that the equations of nonlinear elastodynamics generate a unique continuous local evolution system on the space of ϕ, $\dot{\phi}$, which are sufficiently smooth; ϕ is at least C^2 and $\dot{\phi}$ is C^1.

There are several difficulties with theorems of this type: (A). The existence is only local in time. (B). The function spaces are too restrictive to allow shocks and other discontinuities.

With regard to (A), some global results for (11) have been proved by Klainermann [1978] (and simplified by Shatah [1982]) in four or higher dimensions for small initial data. However, these global Properties are not true in three dimensions (John [1979]). Much work has been done on (B), but the success is very limited. In fact, the problem in one dimension is not settled. Simple dissipative mechanisms sometimes relieve the situation, as is discussed below.

A few selected topics of current interest related to difficulties (A) and (B) are as follows:

1. Lax [1964] proved the non-existence of global smooth solutions to

$$u_{tt} = \sigma(u_x)_x \tag{12}$$

assuming that σ'' never vanishes. Equation (12) is studied by writing it as a system of conservation laws:

$$\begin{cases} w_t = v_x \\ v_t = \sigma(w)_x \end{cases}$$

where $w = u_x$ and $v = u_t$.

2. The assumption $\sigma'' \neq 0$ is unrealistic for nonlinear elasticity. Indeed it is reasonable that a stress-strain function $\sigma(u_x)$ should satisfy $\sigma(p) \longrightarrow +\infty$ as $p \longrightarrow +\infty$ and $\sigma(p) \longrightarrow -\infty$ as $p \longrightarrow 0+$. Strong ellipticity is the assertion that $\sigma'(p) \geq \epsilon > 0$. This does not imply that σ'' never vanishes. This hypothesis was overcome by MacCamy and Mizel [1967] under realistic conditions on σ for the displacement problem on [0, 1] (representing longitudinal displacements of a bar). These results have been improved by Klainermann and Majda [1980] who show that singularities develop in finite time for arbitrarily small initial data under rather general hypotheses. The results are still, however, one dimensional.

3. A general existence theorem for weak solutions of (12) was proved by Glimm [1965] and Dafermos [1973]. The crucial difficulty is in selecting out the "correct" solution by imposing an entropy condition. This problem has been recast into the framework of nonlinear contractive semigroups by several authors, such as Quinn [1971] and Crandall [1972].

Glimm constructed solutions in the class of functions of bounded variation by exploiting the classical solution to the Riemann problem for shocks. The proof that the scheme converges involves probabilistic considerations. DiPerna [1982], using ideas of Tartar [1979], obtains solutions in L^∞ as limits of solutions of a viscoelastic problem as the viscosity tends to zero. DiPerna's results involve acceptable hypotheses on the stress σ and impose correct entropy conditions on the solutions. (The solutions are not known to be of bounded variation).

4. Much has been done on equations of the form (12) with a dissipative mechanism added on. For example, viscoelastic-type dissipation is considered in MacCamy [1977], Matsumura and Nishida [1980] and Potier–Ferry [1980] and thermoeleastic dissipation is considered in Coleman and Gurtin [1965] and Slemrod [1981]. It is proved that smooth initial data with small norm has a global solution. Dafermos [1982], using the ideas in Andrews [1980], shows smooth, global existence for arbitrarily large initial data for thermovisco-elasticity.

5. The viscoelastic equations $u_{tt} = \sigma(u_x)_x + u_{xxt}$ *without* the assumption of strong ellipticity are shown to have unique global weak solutions in Andrews [1980]. This is of interest since, as we mentioned in the previous section, a σ without the restriction $\sigma' > 0$ may be relevant to phase transitions.

Box 5.1 ***Differentiability of Evolution Operators***

This box gives an abstract theorem which shows that in some sense the evolution operators $F_{t,s}$ for a system of the type (1)′ are differentiable. This is of interest for quasi-linear systems and applies, in particular, to the situation of 5.10. For semilinear systems, $F_{t,s}$ is smooth in the usual sense, as we already proved in Theorem 5.1. However, the example $u_t + uu_x = 0$ shows that $F_t\colon H^s \longrightarrow H^s$ is not even locally Lipschitz, although $F_t\colon H^s \longrightarrow H^{s-1}$ is differentiable (see Kato [1975a]), so the situation is more subtle for quasi-linear *hyperbolic* systems. Despite this, there is a notion of differentiability that is adequate for Proposition 3.4, Chapter 5, for example. For quasi-linear *parabolic* systems, the evolution operator will be smooth in the usual sense. For more details, see Marsden and McCracken [1976], Dorroh and Graff [1979], and Graff [1981].

First, we give the notion of differentiability appropriate for the generator. Let $\mathfrak{X}$ and $\mathcal{Y}$ be Banach spaces with $\mathcal{Y} \subset \mathfrak{X}$ continuously and densely included. Let $\mathfrak{U} \subset \mathcal{Y}$ be open and $f\colon \mathfrak{U} \longrightarrow \mathfrak{X}$ be a given mapping. We say f is *generator-differentiable* if for each $x \in \mathfrak{U}$ there is a bounded linear operator $Df(x)\colon \mathcal{Y} \longrightarrow \mathfrak{X}$ such that

$$\frac{\|f(x+h) - f(x) - Df(x)\cdot h\|_{\mathfrak{X}}}{\|h\|_{\mathfrak{X}}} \longrightarrow 0$$

as $\|h\|_{\mathcal{Y}} \longrightarrow 0$. If f is generator-differentiable and $x \mapsto Df(x) \in \mathfrak{B}(\mathcal{Y}, \mathfrak{X})$ is norm continuous, we call f C^1 *generator-differentiable.* Notice that this is *stronger* than C^1 in the Fréchet sense. If f is generator-differentiable and

$$\|f(x+h) - f(x) - Df(x)\cdot h\|_{\mathfrak{X}}/\|h\|_{\mathfrak{X}}$$

is uniformly bounded for x and $x + h$ in some $\mathcal{Y}$ neighborhood of each point, we say that f is *locally uniformly generator-differentiable.*

Most concrete examples can be checked using the following proposition:

5.13 Proposition *Suppose $f\colon \mathfrak{U} \subset \mathcal{Y} \longrightarrow \mathfrak{X}$ is of class C^2, and locally in the $\mathcal{Y}$ topology*

$$x \mapsto \frac{\|D^2 f(x)(h, h)\|_{\mathfrak{X}}}{\|h\|_{\mathcal{Y}}\|h\|_{\mathfrak{X}}}$$

is bounded. Then f is locally uniformly C^1 generator-differentiable.

This follows easily from the identity

$$f(x+h) - f(x) - Df(x)\cdot h = \int_0^1 \int_0^1 D^2 f(x + sth)(h, h)\, ds\, dt.$$

Next, we turn to the appropriate notion for the evolution operators.

A map $g: \mathfrak{U} \subset \mathcal{Y} \to \mathfrak{X}$ is called *flow-differentiable* if it is generator-differentiable and $Dg(x)$, for each $x \in \mathfrak{U}$, extends to a bounded operator $\mathfrak{X}$ to $\mathfrak{X}$. Flow-differentiable maps obey a chain rule. For example, if $g_1: \mathcal{Y} \to \mathcal{Y}$, $g_2: \mathcal{Y} \to \mathcal{Y}$ and each is flow-differentiable (as maps of $\mathcal{Y}$ to $\mathfrak{X}$) and are continuous from $\mathcal{Y}$ to $\mathcal{Y}$, then $g_2 \circ g_1$ is flow-differentiable with, of course,

$$D(g_2 \circ g_1)(x) = Dg_2(g_1(x)) \cdot Dg_1(x).$$

The proof of this fact is routine. In particular, one can apply the chain rule to $F_{t,s} \circ F_{s,r} = F_{t,s}$ if each $F_{t,s}$ is flow-differentiable. Differentiating this in s at $s = r$ gives the backwards equation:

$$\frac{\partial}{\partial s} F_{t,s}(x) = -DF_{t,s}(x) \cdot G(x).$$

Differentiation in r at $r = s$ gives

$$DF_{t,s}(x) \cdot G(x) = G(F_{t,s}(x)),$$

the flow invariance of the generator.

> ***Problem 5.9*** Use the method of Proposition 2.7, Section 6.2 to show that integral curves are unique if (1)′ has an evolution operator that is flow-differentiable.

For the following theorem we assume these hypotheses: $\mathcal{Y} \subset \mathfrak{X}$ is continuously and densely included and $F_{t,s}$ is a continuous evolution system on an open subset $\mathfrak{D} \subset \mathcal{Y}$ and is generated by a map $G(t): \mathfrak{D} \to \mathfrak{X}$. Also, we assume:

(H_1) $G(t): \mathfrak{D} \subset \mathcal{Y} \to \mathfrak{X}$ is locally uniformly C^1 generator-differentiable. Its derivative is denoted $D_xG(t, x)$ and is assumed strongly continuous in t.

(H_2) For $x \in \mathfrak{D}$, $s \geq 0$, let $T_{x,s}$ be the lifetime of x beyond s: that is, $T_{x,s} = \sup\{t \geq s \mid F_{t,s}(x) \text{ is defined}\}$. Assume there is a strongly continuous linear evolution system $\{U^{x,s}(\tau, \sigma) \mid 0 \leq \sigma \leq \tau \leq T_{x,s}\}$ in $\mathfrak{X}$ whose $\mathfrak{X}$-infinitesimal generator is an extension of $\{D_xG(t, F_{t,s}x) \in B(\mathcal{Y}, \mathfrak{X}) \mid 0 \leq t \leq T_{x,s}\}$; that is, if $y \in \mathcal{Y}$, then

$$\left.\frac{\partial}{\partial \tau_+} U^{x,s}(\tau, \sigma) \cdot y\right|_{\sigma=\tau} = D_xG(\tau, F_{\tau,s}(x)) \cdot y.$$

(Sufficient conditions for (H2) are given in Kato [1973].)

5.14 Theorem *Under the hypotheses above, $F_{t,s}$ is flow-differentiable at x and, in fact,*

$$DF_{t,s}(x) = U^{x,s}(t, s).$$

Proof Define $\varphi_t(x, y) = \varphi(t, x, y)$ by

$$G(t, x) - G(t, y) = D_x G(t, y)\cdot(x - y) + \|x - y\|_{\mathfrak{X}}\varphi_t(x, y)$$

(or zero if $x = y$) and notice that by local uniformity, $\|\varphi(t, x, y)\|_{\mathfrak{X}}$ is uniformly bounded if x and y are $\mathcal{Y}$-close. By joint continutity of $F_{t,s}(x)$, for $0 < t < T_{x,s}$, $\|\varphi(t, F_{t,s}(y), F_{t,s}(x))\|_{\mathfrak{X}}$ is bounded for $0 \le s \le T$ if $\|x - y\|_{\mathcal{Y}}$ is sufficiently small.

By construction, we have the equation

$$\frac{d}{dt}F_{t,s}(x) = G[F_{t,s}(x)] \quad (0 \le s \le t \le T_{x,s},\ x \in \mathfrak{D}).$$

Let
$$w(t, s) = F_{t,s}(y) - F_{t,s}(x)$$

so that
$$\frac{\partial w(t, s)}{\partial t} = G(t, F_{t,s}(y)) - G(t, F_{t,s}(x))$$
$$= D_x G(t, F_{t,s}(x))w(t, s) + \|w(t, s)\|_{\mathfrak{X}}\varphi(t, F_{t,s}y, F_{t,s}x).$$

Since $D_x G(t, F_{t,s}x)\cdot w(t, s)$ is continuous in t, s with values in $\mathfrak{X}$, and writing $U = U^{x,s}$, we have the backwards differential equation:

$$\frac{\partial}{\partial\sigma}U(t, \sigma)w(\sigma, s) = U(t, \sigma)\frac{\partial w(\sigma, s)}{\partial\sigma} - U(t, \sigma)D_x G(\sigma, F_{\sigma,s}(x))\cdot w(\sigma, s)$$
$$= U(t, \sigma)\cdot\|w(\sigma, s)\|_{\mathfrak{X}}\varphi(\sigma, F_{\sigma,s}(y), F_{\sigma,s}(x)).$$

Hence, integrating from $\sigma = s$ to $\sigma = t$,

$$w(t, s) = U(t, s)(y - x) + \int_s^t U(t, \sigma)\|w(\sigma, s)\|_{\mathfrak{X}}\varphi(\sigma, F_{\sigma,s}(y), F_{\sigma,s}(x))\, d\sigma.$$

Let $\|U(\tau, \sigma)\|_{\mathfrak{X},\mathfrak{X}} \le M_1$ and $\|\varphi(\sigma, F_{\sigma,s}(y), F_{\sigma,s}(x))\|_{\mathfrak{X}} \le M_2$, $(0 \le s \le \sigma \le \tau \le T)$. Thus, by Gronwall's inequality,

$$\|w(t, s)\|_{\mathfrak{X}} \le M_1 e^{M_1 M_2 T}\|y - x\|_{\mathfrak{X}} = M_3\|y - x\|_{\mathfrak{X}}.$$

In other words,

$$\frac{\|F_{t,s}(y) - F_{t,s}(x) - U(t, s)(y - x)\|_{\mathfrak{X}}}{\|y - x\|_{\mathfrak{X}}}$$
$$\le M_1 M_3 \int_s^t \|\varphi(\sigma, F_{\sigma,s}(y), F_{\sigma,s}(x))\|_{\mathfrak{X}}\, d\sigma.$$

From the Lebesgue bounded convergence theorem, we conclude that $F_{t,s}$ is flow-differentiable at x and $DF_{t,s}(x) = U(t, s)$; ($\varphi(t, F_{t,s}(y), F_{t,s}(x))$ is strongly measurable in s since $\varphi(x, y)$ is continuous for $x \ne y$.) ∎

Box 5.2 ***Remarks on Continuity in the Initial Data for the Initial Boundary Value Problem for Quasi-linear Systems***

As stated, Example 5.12 applies only to an elastic body filling all of space. For the initial boundary value problem, Kato [1977] has shown local existence and uniqueness, but the continuous dependence on the initial data is not yet known. The method is to replace $\mathcal{Y}$ by the domain of a power of A, such as A^3. The major complication that needs to be dealt with is the compatability conditions and the possibility that the domains of powers of A will be time dependent, even though that of A is not. All of this is caused by the degree of smoothness required in the methods and the dependence of A on u.

We shall now sketch a method that may be used in proving the continuous dependence in certain cases.[20] Write the equations this way:

$$\dot{u} + \mathfrak{A}(u) = 0, \qquad u_0 = \phi,$$

where $\mathfrak{A}(u) = A(u)\cdot u$ and we have dropped the term $f(u)$ for simplicity. Suppose that the boundary conditions are written

$$B(u) = 0, \qquad u \in \mathfrak{D}(\mathfrak{A}).$$

If we seek solutions in the domain of $[A(u)]^3$, then the compatability conditions for the initial data are obtained by differentiating $B(u) = 0$ twice:

(i) $B(\phi) = 0$;
(ii) $B'(\phi)\cdot\mathfrak{A}(\phi) = 0$;
(iii) $B''(\phi)\cdot(\mathfrak{A}(\phi), \mathfrak{A}(\phi)) + B'(\phi)\cdot\mathfrak{A}'(\phi)\cdot\mathfrak{A}(\phi) = 0$.

The difficulty is that even if $B(\phi) = 0$ are linear boundary conditions, this is a nonlinear space of functions in which we seek the solution. Let C denote the space of functions satisfying the compatability conditions. It seems natural to try to show C is a smooth manifold. We assume B itself is *linear* for simplicity and assume our function spaces are Hilbert spaces, so closed subspaces will split.

5.15 Proposition *Let $\phi \in C$ and assume:*
(a) *B is surjective;*
(b) *the linear boundary value problem*

$$\mathfrak{A}'(\phi)\cdot\psi = \rho, \qquad B\cdot\psi = 0,$$

has a solution ψ for any ρ; and

[20] M. Ortiz has pointed out to us that this result may also be provable using product formula methods. C. Dafermos has noted that for the second order hyperbolic case, which includes elasticity, classical energy and elliptic estimates may give the result.

(c) *the linear boundary value problem*

$$\mathfrak{A}''(\phi)(\mathfrak{A}(\phi), \psi) + (\mathfrak{A}'(\phi))^2\psi = \rho \qquad B\cdot\psi = 0,$$
$$B\cdot\mathfrak{A}'(\phi)\cdot\psi = 0$$

has a solution ψ *for any* ρ.

Then C is a smooth manifold in the neighborhood of ϕ.

Proof Let $C_1 = \operatorname{Ker} B$, the "first" boundary conditions. Map

$$\phi \in C_1 \mapsto B(\phi)\cdot\mathfrak{A}(\phi) = B\cdot\mathfrak{A}(\phi).$$

This has derivative at ϕ given by $\psi \mapsto B\cdot\mathfrak{A}'(\phi)\cdot\psi$. Since B is surjective and $\mathfrak{A}'(\phi)$: Ker $B \dashrightarrow$ (range space) is surjective by (b), this map has a surjective derivative, so $C_2 = \{\phi \in C_1 \mid B\cdot\mathfrak{A}(\phi) = 0\}$ is a submanifold of C_1 with tangent space $T_\phi C_2 = \{\psi\colon B\psi = 0 \text{ and } B\cdot\mathfrak{A}'(\phi)\cdot\psi = 0\}$, by the implicit function theorem.

Finally, map

$$C_2 \longrightarrow \text{(range space)}, \qquad \phi \mapsto B\cdot\mathfrak{A}'(\phi)\cdot\mathfrak{A}(\phi),$$

which has the derivative $\psi \mapsto B\cdot[\mathfrak{A}''(\phi)(\mathfrak{A}(\phi), \psi) + (\mathfrak{A}'(\phi))^2\cdot\psi]$. Thus, by assumption (c), this is surjective on $T_\phi C_2$. Thus,

$$C_3 = \{\phi \in C_2 \mid B\cdot\mathfrak{A}'(\phi)\cdot\mathfrak{A}(\phi) = 0\} = C$$

is a submanifold by the implicit function theorem. ∎

We want to solve

$$\frac{du}{dt} + \mathfrak{A}(u) = 0$$

for $u(t) \in C$ with a given initial condition $\phi \in C$. To do so, we can use the local diffeomorphism

$$\Phi\colon C \longrightarrow \text{(linear space)} = \gamma,$$

with $\Phi(\phi) = \phi$ mapping a neighborhood of ϕ in C to a ball about ϕ in a linear space obtained from the proof above. (So Φ is only defined implicitly.) Let $v = \Phi(u)$. Thus

$$\frac{dv}{dt} = \Phi'(u)\frac{du}{dt} = \Phi'(\Phi^{-1}(v))\cdot(-\mathfrak{A}(\Phi^{-1}(v))),$$

so v satisfies the modified equation

$$\frac{dv}{dt} + \tilde{\mathfrak{A}}(v) = 0,$$

where $\tilde{\mathfrak{A}}(v) = \Phi'(\Phi^{-1}(v))\cdot\mathfrak{A}(\Phi^{-1}(v))$. (In geometry notation, $\tilde{\mathfrak{A}} = \Phi_*\mathfrak{A}$.) If the modified problem is well-posed, then clearly the original one is as well. We can choose $Y = \{\psi \mid B\psi = 0,\ B\cdot\mathfrak{A}'(\phi)\cdot\psi = 0$ and

$B \cdot [\mathfrak{A}''(\phi)(\mathfrak{A}(\phi), \psi) + (\mathfrak{A}'(\phi))^2\psi] = 0\}$ and let Φ be the projection of C onto Y; see Figure 6.5.1.

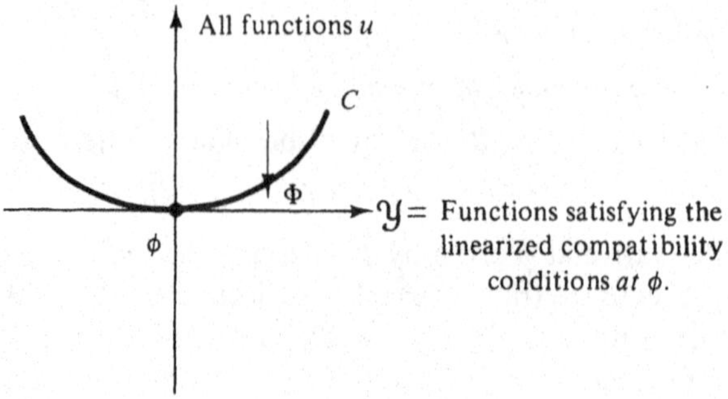

Figure 6.5.1

Now let $\mathfrak{A}$ have the form $\mathfrak{A}(u) = A(u) \cdot u$. Since Φ is a *linear* projection, the equation

$$\tilde{\mathfrak{A}}(v) = \Phi \cdot A(\Phi^{-1}(v)) \cdot \Phi^{-1}(v)$$

is still quasi-linear. It seems reasonable to ask that the abstract quasi-linear Theorem 5.11 applies to the new system $\mathfrak{A}$. If it does, then continuous dependence for the initial boundary value problem follows.

Box 5.3 ***Remarks on Incompressible Elasticity***

In Example 4.6 Section 5.4 we remarked that the configuration space for incompressible elasticity is $\mathfrak{C}_{\text{vol}}$, the deformations ϕ with $J = 1$. (This requires careful interpretation if ϕ is not C^1.) The equations of motion are modified by replacing the first Piola–Kirchhoff stress $P_a{}^A = \rho_{\text{Ref}}(\partial W/\partial F^a{}_A)$ by $\rho_{\text{Ref}}(\partial W/\partial F^a{}_A) + Jp(F^{-1})_a{}^A$, where p is a scalar function to be determined by the condition of incompressibility. A fact discovered by Ebin and Marsden [1970] is that this extra term $P_{\text{presure}}(\phi, \dot{\phi}) = Jp(F^{-1})$ in material coordinates is a C^∞ function of $(\phi, \dot{\phi})$ in the $H^s \times H^s$ (and hence $H^{s+1} \times H^s$) topology. (In fact, there is one order of smoothing in the $H^{s+1} \times H^s$ topology; see also Cantor [1975b].) This remark enables one to deduce that in all of space the equations of incompressible elasticity are well-posed as a consequence of that for the compressible equations. (One can use, for example, Marsden

[1981] for this, or the abstract Theorem 5.10.) Again the boundary value problem requires further technical requirements.

The smoothness (as a function of ϕ) of the pressure term $\boldsymbol{P}_{\text{pressure}}$ in the stress seems not to have been exploited fully. For example, it may help with studies of weak (Hopf) solutions in fluid mechanics. In particular, it may help to prove additional regularity or uniqueness of these solutions. In general (as has been emphasized by J. Ball) it is a good idea in fluid mechanics not to forget that material coordinates could be very useful and natural in studies of existence and uniqueness.

We conclude with a few remarks on the incompressible limit, based on Rubin and Ungar [1957], Ebin [1977], and Klainerman and Majda [1918]. If the "incompressible pressure" p is replaced by a constitutive law $p_k(\rho)$, where $(dp_k/d\rho) = k$, so $1/k$ is the *compressibility*, then a potential V_k is added to our Hamiltonian which has the property:

$$V_k(\phi) = 0 \quad \text{if} \quad \phi \in \mathcal{C}_{\text{vol}} \quad \text{and} \quad V_k(\phi) \longrightarrow \infty \text{ as } k \longrightarrow +\infty$$
$$\text{if } \phi \notin \mathcal{C}_{\text{vol}}.$$

In such a case, it is intuitively clear from conservation of energy that this ought to force compressible solutions with initial data in $\mathcal{C}_{\text{vol}}$ to converge to the incompressible solutions as $k \longrightarrow \infty$.

Such convergence in the *linear* case can be proved by the Trotter–Kato theorem (see Section 6.2). Kato [1977] has proved analogues of this for non-linear equations that, following Theorem 5.10, are applicable to nonlinear elastodynamics. These approximation theorems may now be used in the proofs given by Ebin [1977]. Although all the details have not been checked, it seems fairly clear that this method will enable one to prove the convergence of solutions in the incompressible limit, at least for a short time. (See Klainerman and Kohn [1982].)

We also mention that the smoothness of the operator $\boldsymbol{P}_{\text{pressure}}$ and the convergence of the constraining forces as $k \longrightarrow \infty$ should enable one to give a simple proof of convergence of solutions of the stationary problem merely by employing the implicit function theorem. Rostamian [1978] gives some related results.

Problem 5.10 Formulate the notion of the "rigid limit" by considering $\text{SO}(3) \subset \mathcal{C}$ and letting the rigidity $\longrightarrow \infty$.

Problem 5.11 A compressible fluid may be regarded as a special case of elasticity, with stored energy function $W(\boldsymbol{F}) = h(\det \boldsymbol{F})$, where h is a strictly convex function.

(a) Show that the first Piola–Kirchhoff stress is $\boldsymbol{P} = h'(\det \boldsymbol{F}) \operatorname{adj} \boldsymbol{F}$ (or $\boldsymbol{\sigma} = h'(\det \boldsymbol{F})\boldsymbol{I}$) and the elasticity tensor is

$$\mathbf{A} = h''(\det F) \operatorname{adj} F \otimes \operatorname{adj} F + h'(\det F) \cdot \frac{\partial(\operatorname{adj} F)}{\partial F}.$$

(b) Show explicitly in two dimensions using suitable coordinates, that **A** is *not* strongly elliptic.

(c) Despite (b), show that the equations are locally well-posed by using Ebin [1977] or a direct argument to check the abstract hypotheses of 5.9 and 5.10.

Box 5.4 ***Summary of Important Formulas for Section 6.5***

Semiflow on $\mathcal{Y}$

Autonomous: $F_t\colon \mathcal{Y} \to \mathcal{Y}$; $F_0 =$ Identity, $F_{t+s} = F_t \circ F_s$

Time-dependent: $F_{t,s}\colon \mathcal{Y} \dashrightarrow \mathcal{Y}$; $F_{s,s} =$ Identity, $F_{t,s} = F_{t,r} \circ F_{r,s}$

Generator

$$\frac{d}{dt} F_t(x) = G(F_t(x)) \quad \text{(autonomous)}$$

$$\frac{d}{dt} F_{t,s}(x) = G(F_{t,s}(x), t) \quad \text{(time-dependent)}$$

Variation of Constants Formula

If $du/dt = Au + f(t)$, then $u(t) = e^{tA}u_0 + \int_0^t e^{(t-s)A} f(s)\, ds$.

Semilinear Equations

If A is a linear generator on $\mathfrak{X}$ and $B\colon \mathfrak{X} \to \mathfrak{X}$ is C^∞, then $G(x) = Ax + B(x)$ generates a smooth local semiflow on $\mathfrak{X}$. If an a priori bound for $\| B(x) \|$ can be found, the semiflow is global.

Panel Flutter

The equations

$$\alpha \dot{v}'''' + v'''' - \{\Gamma + \kappa \| v' \|^2_{L^2} + \sigma \langle v', \dot{v}' \rangle_{L^2}\} v'' + \rho v' + \sqrt{\rho}\, \delta \dot{v} + \ddot{v} = 0 \quad \text{on } [0, 1]$$

with boundary conditions

$$v = (\dot{v} + \alpha v)'' = 0 \quad \text{at} \quad x = 0, 1$$

generate a unique smooth global semiflow on $H^2_\partial \times L^2$.

Quasilinear Equations

Hyperbolic (symmetric first-order or strongly elliptic second-order) systems define local flows on spaces of sufficiently smooth functions.

The evolution operators are differentiable in the technical sense of "flow differentiability."

Little is known about global solutions for quasi-linear equations, although some results that necessarily involve entropy conditions are known in one dimension.

6.6 THE ENERGY CRITERION

The energy criterion states that *minima of the potential energy are dynamically stable*. A major problem in elasticity (that is not yet settled) is to find conditions appropriate to nonlinear elasticity under which this criterion can be proved as a theorem. The purpose of this section is to discuss some of what is known about the problem for nonlinear elasticity. The linearized case has been discussed in Theorem 3.5. There are both positive and negative results concerning the criterion for nonlinear elasticity; examples discussed below show that the criterion is probably false as a sweeping general criterion; but there are theorems to indicate that these counterexamples can be eliminated with reasonable hypotheses.

The energy criterion has been extensively discussed in the literature. It seems to have first been explicitly recognized as a genuine difficulty by Koiter [1945], [1965b], [1976a], although it has been very successfully used in engineering for much longer. Relevant mathematical theory is closely related to bifurcation theory and goes back at least to Poincaré [1885]. A detailed account of the history and basic results available up until 1973 are contained in Knops and Wilkes [1973], which should be consulted by any readers who have a serious interest in pursuing this subject. *Some* of the other more recent references that are important are (chronologically): Ericksen [1966a, b], Coleman [1973], Naghdi and Trapp [1973], Knops, Levine, and Payne [1974], Gurtin [1975], Payne and Sattinger [1975], Mejlbro [1976], Ball, Knops, and Marsden [1978], Knops and Payne [1978], Ball and Marsden [1980], and Ball [1982]. There are many references that deal with stability in the context of bifurcation theory such as Ziegler [1968] and Thompson and Hunt [1973]; these aspects will be discussed in Chapter 7.

The difficulties with the energy criterion for conservative infinite-dimensional quasi-linear systems may be genuine ones. Some possible alternatives are:

(i) Use some kind of averaging technique to mask the higher frequency motions, thereby making semilinear techniques applicable (see 6.10 below). As far as we know this possibility has not been explored.

(ii) Employ a dissipative mechanism in addition to a conservative minimum of the potential. It is possible that a suitable mechanism will move the spectrum into the left half-plane so that Liapunov's theorem (4.1 of Chapter 7) becomes applicable. In Potier–Ferry [1980] this is established for viscoelasticity of Kelvin-Voigt type, as in Proposition 3.12, Section 6.3, generalizing the work of Dafermos [1969], [1976]. For a dissipative mechanism of thermal type, see Slemrod [1981]. For both thermal and viscous dissipation, see Dafermos [1982].

The contents of this section are as follows.

1. First we give (in 6.3) sufficient conditions for validity of the energy criterion. This involves the notion of a *potential well.*
2. The applicability of this theorem to elasticity is then criticized and examples are presented to show that it is at best difficult to satisfy the hypotheses.
3. A positive result is proved showing that potential wells can be obtained in $W^{1,p}$ (spaces in which the existence of dynamics is questionable).
4. Semilinear equations are discussed and the validity of the energy criterion is proved under acceptable hypotheses in this case.
5. Some discussion of the role of dissipative mechanisms is given.

Let us begin with sufficient conditions for the validity of the energy criterion. This version goes back at least to Lagrange and Dirichlet and is satisfactory for finite degree of freedom systems and for some infinite degree of freedom systems (the semilinear ones), as we shall see.

The general context is that of Hamiltonian systems with energy of the form kinetic-plus-potential. To be a bit more concrete than the context of Section 5.3, let us adopt the following set-up:

(a) $\mathfrak{X}$ and $\mathfrak{Y}$ are Banach spaces.
(b) $K: \mathfrak{Y} \to \mathbb{R}$, called a kinetic energy function, is a given continuous function that satisfies $K(y) \geq 0$, with $K(y) = 0$ only if $y = 0$.
(c) $V: \mathfrak{X} \to \mathbb{R}$ is a given continuous potential energy function. Let $H: \mathfrak{X} \times \mathfrak{Y} \to \mathbb{R}$ be defined by $H(x, y) = K(y) + V(x)$.

We are contemplating the dynamical system associated with Hamilton's equations, so let us write $\dot{x}$ for y, although it need not be a time derivative. We consider the following two assumptions.

6.1 Assumptions Let $x_0 \in \mathfrak{X}$ be fixed. For each $\eta > 0$ and $\delta > 0$, let the *K-neighborhood* about $(x_0, 0)$ of radius (η, δ) be defined by

$$\mathcal{S}_{\eta,\delta} = \{(x, \dot{x}) \in \mathfrak{X} \times \mathfrak{Y} \mid \|x - x_0\| \leq \eta \quad \text{and} \quad K(\dot{x}) \leq \delta\}.$$

Assume:

(S1) There is an $\eta_0 > 0$, $\delta_0 > 0$, and $\tau > 0$ and a continuous local semiflow F_t defined on $\mathcal{S}_{\eta_0,\delta_0}$ for $0 \leq t \leq \tau$ such that (i) $(x_0, 0)$ is a fixed point for F_t—that is, $F_t(x_0, 0) = (x_0, 0)$—and (ii) $H(x(t), \dot{x}(t)) \leq H(x(0)\ \dot{x}(0))$ for all $(x(0), \dot{x}(0)) \in \mathcal{S}_{\eta_0,\delta_0}$, where $(x(t), \dot{x}(t)) =$

$F_t(x(0), \dot{x}(0))$. [This implies that there is a uniform time for which $(x(t), \dot{x}(t))$ is defined if $(x(0), \dot{x}(0)) \in \mathcal{S}_{\eta_0, \delta_0}$ and that H decreases along orbits for as long as they are defined.]

(S2) x_0 lies in a *potential well*; that is, there is an $\epsilon_0 > 0$ such that (a) $V(x_0) < V(x)$ if $0 < \| x - x_0 \| < \epsilon_0$, and (b) for each $0 < \epsilon \leq \epsilon_0$, $\rho(\epsilon) > 0$, where $\rho(\epsilon) = \inf_{\| x - x_0 \| = \epsilon} V(x) - V(x_0)$.

(If $\mathfrak{X}$ is finite dimensional, (b) follows from (a), but this need not be true in infinite dimensions.)

6.2 Definition We say $(x_0, 0)$ is (Liapunov) *K-stable* if for any $\eta, \delta > 0$ there is an (η_1, δ_1) such that if $(x(0), \dot{x}(0)) \in \mathcal{S}_{\eta_1, \delta_1}$, then $(x(t), \dot{x}(t))$ is defined for all $t \geq 0$ and lies in $\mathcal{S}_{\eta, \delta}$. If $K(\dot{x}) = \| \dot{x} \|^2/2$, we will just say "stable" for "*K*-stable."

6.3 Theorem *If* (S1) *and* (S2) *hold, then* $(x_0, 0)$ *is K-stable.*

Proof Let $\eta > 0$ and $\delta > 0$ be given. Choose $\bar{\delta} = \min(\delta, \epsilon_0)$ and $\bar{\eta} = \min(\eta, \rho(\bar{\delta}))$. By (S2), $\bar{\eta} > 0$. Next, choose $\delta_1 < \bar{\delta}$ such that $V(x(0)) - V(x_0) < \bar{\eta}/2$ if $\| x(0) - x_0 \| < \delta_1$ and choose $\eta_1 = \bar{\eta}/2$.

We will show that if $(x(0), \dot{x}(0)), \in \mathcal{S}_{(\eta_1, \delta_1)}$, then $(x(t), \dot{x}(t))$ is defined for all $t \geq 0$ and lies in $\mathcal{S}_{\bar{\eta}, (\bar{\delta})}$. Suppose $(x(t), \dot{x}(t))$ lies in $\mathcal{S}_{\bar{\eta}, \bar{\delta}}$ for a maximal interval $[0, T)$. By (S1), $(x(t), \dot{x}(t))$ is defined on $[0, T + \tau)$, and so $(x(T), \dot{x}(T))$ is defined and lies on the boundary of $\mathcal{S}_{\bar{\eta}, \bar{\delta}}$. Thus, $\| x(T) - x_0 \| = \bar{\delta}$ or $K(\dot{x}(T)) = \bar{\eta}$. We will derive a contradiction that will prove the proposition. We have

$$K(\dot{x}(T)) + V(x(T)) \leq K(\dot{x}(0)) + V(x(0)),$$

so

$$K(\dot{x}(T)) + V(x(T)) - V(x_0) \leq K(\dot{x}(0)) + V(x(0)) - V(x_0) < \bar{\eta}/2 + \bar{\eta}/2 = \bar{\eta}.$$

But the left-hand side is at least $\bar{\eta}$. ∎

Now we need to discuss whether or not this "potential well approach" is applicable to nonlinear elasticity. (One should bear in mind that a totally different approach based on more than simple energy estimates may be ultimately required.)

In discussing this, the choice of topologies is crucial. Let us first suppose that fairly strong topologies are chosen so that results from the previous section guarantee that a local semiflow F_t exists. We shall show in the example below (6.4) by means of a one-dimensional example that the choice of a strong topology implies that while (S2)(a) is satisfied, (S2)(b) *cannot* be satisfied. (See Knops and Payne [1978] for some related three-dimensional examples.)

Conclusion: Theorem 6.3 is not applicable in function spaces for which the elastodynamic equations are known to be (locally) well-posed. (In such spaces, the deformations are at least C^1.)

6.4 Example (Ball, Knops, and Marsden [1978]) Let $\mathcal{B} = [0, 1]$ and consider displacements $\phi: [0, 1] \to \mathbb{R}$ with $\phi(0) = 0$ and $\phi(1) = \lambda > 0$ prescribed. The potential is $V(\phi) = \int_0^1 W(\phi_X)\, dX$. Suppose that W is C^2, $W'(\lambda) = 0$, and $W''(\lambda) > 0$. Then $\phi_0(X) = \lambda X$ is an extremal for V. Let $\mathfrak{X}$ be a Banach space continuously included in $W^{1,\infty}$.

1. *Under these conditions,* (S2)(a) *holds; that is, there is an* $\epsilon > 0$ *such that* if $0 < \|\phi - \phi_0\|_{\mathfrak{X}} < \epsilon$, then $V(\phi) > V(\phi_0)$. That is, ϕ_0 *is a strict local minimum for* V.

Proof This follows from the fact that λ is a local minimum of W and that the topology on $\mathfrak{X}$ is as strong as that of $W^{1,\infty}$. ∎

2. *We necessarily have failure of* (S2)(b) *in* $\mathfrak{X}$; that is,

$$\inf_{\|\phi-\phi_0\|_{\mathfrak{X}}=\epsilon} V(\phi) = V(\phi_0).$$

Proof By Taylor's theorem,

$$\begin{aligned} V(\phi) - V(\phi_0) &= \int_0^1 (W(\phi_X) - W(\lambda))\, dX \\ &= \int_0^1 \int_0^1 (1 - s)W''(s\phi_X + (1 - s)\lambda)\ \ ds\, dX. \end{aligned}$$

Thus, as $s\phi_X$ is uniformly bounded ($\mathfrak{X} \subset W^{1,\infty}$) and W'' is continuous, there is $C > 0$ such that

$$V(\phi) - V(\phi_0) \leq C \int_0^1 (\phi_X - \lambda)^2\, dX.$$

However, the topology on $\mathfrak{X}$ is *strictly* stronger than the $W^{1,2}$ topology, and so

$$\inf_{\|\phi-\phi_0\|_{\mathfrak{X}}=\epsilon} \int_0^1 (\phi_X - \lambda)^2\, dX = 0,$$

which proves our claims. ∎

3. *In* $W^{1,p}$ *one cannot necessarily conclude that* ϕ_0 *is a local minimum even though the second variation of* V *at* ϕ_0 *is positive-definite.*

Proof The example $W(\phi_X) = \frac{1}{2}(\phi_X^2 - \phi_X^4)$ shows that in any $W^{1,p}$ neighborhood of ϕ_0, $V(\phi)$ can be unbounded below. ∎

There is a Morse lemma available which enables one to verify S2(a) and to bring V into a normal form. See Tromba [1976], Golubitsky and Marsden [1983] and Buchner, Marsden and Schecter [1983]. The hypotheses of this theorem are verified in item 1 above but fail in item 3.

A more important example of the failure of the energy criterion has been constructed by Ball [1982]. He shows that a sphere undergoing a radial tension will eventually rupture due to cavitation. This is done within the framework of minimizers in $W^{1,p}$ discussed in Section 6.4. The "rupture" solution corresponds to a change in topology from $\mathcal{B}$, a sphere, to $\phi(\mathcal{B})$, a hollow sphere; clearly ϕ cannot be continuous at the origin, but one can find such a ϕ in $W^{1,p}$ for a suitable $p < 3$. Formally, the energy criterion would say that the trivial unruptured radial solution is stable, but in fact it is unstable to rupturing, a phenomenon not "detected" by the criterion.

Since the hypothesis (S2)(b) fails if the topology is too strong, it is natural to ask if it is true in weaker topologies, especially $W^{1,p}$. In fact, we shall prove that *with reasonable hypotheses in* $W^{1,p}$, (S2)(b) *follows from* (S2)(a). *However, as there is no existence theorem for elastodynamics in* $W^{1,p}$, (S1) *is unknown* (and presumably is a difficult issue).

The heart of the argument already occurs in one dimension, so we consider it first.

Let $W: \mathbb{R}^+ \longrightarrow \mathbb{R}$ be a *strictly convex* C^1 function satisfying the growth condition

$$c_0 + \alpha_0 |s|^p \leq W(s)$$

for constants $c_0 \geq 0, \alpha_0 > 0, p > 1$. Fix $\lambda > 0$ and let $W_\lambda^{1,p}$ denote those $\phi \in W^{1,p}([0, 1])$ such that $\phi(0) = 0$, and $\phi(1) = \lambda$. Define $V: W_\lambda^{1,p} \longrightarrow [0, \infty]$ by

$$V(\phi) = \int_0^1 W(\phi_X)\, dX$$

and let

$$\|\phi\|_{1,p} = \left(\int |\phi_X|^p\, dX\right)^{1/p}$$

Let $\phi_0(X) = \lambda X$, so $\phi_0 \in W_\lambda^{1,p}$ and $V(\phi_0) = W(\lambda)$.

6.5 Proposition (*Ball and Marsden* [1980]) *V has a potential well at ϕ_0; that is, (a) $V(\phi) > V(\phi_0)$ for $\phi \neq \phi_0, \phi \in W_\lambda^{1,p}$, and (b) for $\epsilon > 0$,*

$$\inf_{\substack{\|\phi - \phi_0\|_{1,p} = \epsilon \\ \phi \in W_\lambda^{1,p}}} V(\phi) > V(\phi_0).$$

Proof

(a) Integration by parts and the boundary conditions give

$$0 = \int_0^1 W'(\lambda)(\phi_X - (\phi_0)_X)\, dX$$

and so

$$\int_0^1 [W(\phi_X) - W(\lambda)]\, dX$$
$$= \int_0^1 \{W(\phi_X) - W(\lambda) - W'(\lambda)(\phi_X - \lambda)\}\, dX > 0$$

since W lies strictly above its tangent line. This proves (a).

(b) We prove (b) by contradiction. By (a),

$$\inf_{\|\phi - \phi_0\|_{1,p} = \epsilon} V(\phi) \geq V(\phi_0).$$

Suppose equality held. Then there would be a squence $\phi_n \in W^{1,p}_\lambda$ satisfying

$$\|\phi_n - \phi_0\|_{1,p} = \epsilon \tag{1}$$

and

$$\int_0^1 W((\phi_n)_X)\, dX \longrightarrow W(\lambda). \tag{2}$$

From (1) we can extract a weakly convergent subsequence of the ϕ_n (See Box 4.1, Section 6.4). Thus we may suppose that $\phi_n \rightharpoonup \bar{\phi}$ in $W^{1,p}$. ∎

6.6 Lemma $\bar{\phi} = \phi_0$.

Proof By weak lower sequential continuity of V (Tonelli's theorem; see Box 4.1),

$$\int_0^1 W(\bar{\phi}_X)\, dX \leq \underline{\lim} \int_0^1 W(\phi_{nX})\, dX.$$

By (2) we get

$$\int_0^1 W(\bar{\phi}_X)\, dX \leq W(\lambda),$$

so by part (a), $\bar{\phi} = \phi_0$. ∎

6.7 Lemma *If $v_n \rightarrow v$ in L^p,*

$$\int_0^1 W(v)\, dX < \infty \quad \text{and} \quad \int_0^1 W(v_n)\, dX \longrightarrow \int_0^1 W(v)\, dX,$$

then there is a subsequence $v_{n_k} \rightarrow v$ a.e.

Proof Fix some $\theta \in (0, 1)$ and let

$$f_n = \theta W(v_n) + (1 - \theta) W(v) - W(\theta v_n + (1 - \theta) v).$$

Then $f_n(X) \geq 0$ by convexity of W. Notice that

$$\int_0^1 W(\theta v_n + (1 - \theta) v)\, dX < \infty$$

from $f_n(X) \geq 0$ and the finiteness of $\int_0^1 W(v_n)\, dX$ and $\int_0^1 W(v)\, dX$. Now

$$\begin{aligned}
0 &\leq \overline{\lim} \int_0^1 f_n(X)\, dX \\
&= \theta \int_0^1 W(v)\, dX + (1 - \theta) \int_0^1 W(v)\, dX - \underline{\lim} \int_0^1 W(v_n + (1 - \theta) v)\, dX \\
&= \int_0^1 W(v)\, dX - \underline{\lim} \int_0^1 W(\theta v_n + (1 - \theta) v)\, dX.
\end{aligned}$$

By weak lower sequential continuity of $\int_0^1 W(v)\, dX$, $v_n \rightharpoonup v$ implies

$$\begin{aligned}
\underline{\lim} \int_0^1 W(\theta v_n + (1 - \theta) v)\, dX &\geq \int_0^1 W(\theta v + (1 - \theta) v)\, dX \\
&= \int_0^1 W(v)\, dX.
\end{aligned}$$

Thus $0 \leq \underline{\lim} \int_0^1 f_n(X)\,dX \leq 0$ and so $\lim \int_0^1 f_n(X)\,dX = 0$. There is a subsequence such that $f_{n_k}(X) \longrightarrow 0$ a.e. Since W is strictly convex, this implies $v_{n_k} \longrightarrow v$ a.e. ∎

From Lemmas 6.6 and 6.7 we can pass to a subsequence and obtain $\phi_{nX} \longrightarrow \phi_{0X} = \lambda$ a.e. By the growth condition, $W(s) - c_0 - \alpha|s|^p \geq 0$, and so by Fatou's lemma

$$\int_0^1 \underline{\lim}\,[W(\phi_{nX}) - c_0 - \alpha|\phi_{nX}|^p]\,dX \leq \underline{\lim} \int_0^1 [W(\phi_{nX}) - c_0 - \alpha|\phi_{nX}|^p]\,dX,$$

that is,

$$\int_0^1 [W(\lambda) - c_0 - \alpha|\lambda|^p]\,dX \leq W(\lambda) - c_0 + \underline{\lim} \int_0^1 \alpha|\phi_{nX}|^p\,dX$$

by $\phi_{nX} \longrightarrow \lambda$ a.e. and (2). Thus

$$-\alpha|\lambda|^p \leq \underline{\lim} \int_0^1 -\alpha_0|\phi_{nX}|^p\,dX = -\alpha\,\overline{\lim} \int_0^1 |\phi_{nX}|^p\,dX,$$

and so

$$\overline{\lim} \int_0^1 |\phi_{nX}|^p\,dX \leq \int_0^1 |\phi_{0X}|^p\,dX$$
$$\leq \underline{\lim} \int_0^1 |\phi_{nX}|^p\,dX$$

(again using Fatou's lemma).

Thus $\|\phi_{nX}\|_p \longrightarrow \|\phi_{0X}\|_p$. But in L^p, convergence of the norms and weak convergence implies strong convergence (see Riesz and Nagy [1955], p. 78), so $\phi_n \longrightarrow \phi_0$ strongly in $W_\lambda^{1,p}$. This contradicts our assumption (1) that $\|\phi_n - \phi_0\|_{1,p} = \epsilon > 0$. (We do not know if the infimum in (b) of the theorem is actually attained.) ∎

Now we discuss the three-dimensional case. Following the results of Section 6.4 we assume that W is strictly polyconvex; that is,

$$W(F) = g(F, \operatorname{adj} F, \det F), \tag{3}$$

where

$$g\colon M_+^{3\times 3} \times M^{3\times 3} \times (0, \infty) \longrightarrow \mathbb{R}$$

is strictly convex; $M^{3\times 3}$ denotes the space of 3×3 matrices and F denotes the deformation gradient, $F = D\phi$. Assume g satisfies the growth conditions

$$g(F, H, \delta) \geq c_0 + k(|F|^p + |H|^q + \delta^r), \tag{4}$$

where $k > 0$ and, say, $c_0 > 0$. Assume that p, q, r satisfy

$$p \geq 2, \qquad q \geq \frac{p}{p-1}, \qquad r > 1. \tag{5}$$

Let $\Omega \subset \mathbb{R}^3$ be a bounded open domain with, say, piecewise C^1 boundary, and let displacement (and/or traction boundary conditions) be fixed on $\partial\Omega$. Denote by $\mathfrak{X}$ the space of $W^{1,p}$ maps $\phi\colon \Omega \longrightarrow \mathbb{R}^3$ subject to the given boundary conditions and satisfying

$$F \in L^p,\ \operatorname{adj} F \in L^q \quad \text{and} \quad \det F \in L^r$$

with the metric induced from $L^p \times L^q \times L^r$. (Note that $\mathfrak{X} \cong W^{1,p}_\partial$ if $q \leq p/2$, $1/p + 1/q < 1/r$). Let $d_{\mathfrak{X}}(\phi, \psi)$ denote the distance between ϕ and ψ in $\mathfrak{X}$.

6.8 Theorem (*Ball and Marsden* [1980]) *Suppose conditions* (3), (4), *and* (5) *above hold and that* $\phi_0 \in \mathfrak{X}$ *is a strict local minimizer; that is, for some* $\epsilon > 0$,

$$\int_\Omega W(D\phi)\, dX > \int_\Omega W(D\phi_0)\, dX.$$

if $0 < d_{\mathfrak{X}}(\phi, \phi_0) \leq \epsilon$. *Then there is a potential well at* ϕ_0; *that is,*

$$\inf_{\substack{d_{\mathfrak{X}}(\phi,\phi_0)=\epsilon \\ \phi\in\mathfrak{X}}} \int_\Omega W(D\phi)\, dX > \int_\Omega W(D\phi_0)\, dX.$$

The one-dimensional proof readily generalizes to this case, so we can omit it.

Remarks

1. There is a similar two-dimensional theorem for $W(F) = g(F, \det F)$.
2. Examples of stored energy functions W appropriate for natural rubber satisfying (3)–(5) and having a unique natural state $F = I$ (up to rotations) are the Ogden materials:

$$W(F) = \lambda_1^\alpha + \lambda_2^\alpha + \lambda_3^\alpha + (\lambda_2\lambda_3)^\beta + (\lambda_3\lambda_1)^\beta + (\lambda_1\lambda_2)^\beta + h(\lambda_1\lambda_2\lambda_3),$$

where $\alpha \geq 3$, $\beta \geq \frac{3}{2}$, $h'' > 0$, $\alpha + 2\beta + h'(1) = 0$, and where $\lambda_1, \lambda_2, \lambda_3$ are the principal stretches (eigenvalues of $(F^TF)^{1/2}$).
3. Homogeneous deformations provide basic examples of strict local minimizers.
4. The method shows that minimizing sequences actually converge strongly.
5. An obvious question concerns when (S2)(a) holds in three dimensions—that is, with $V(\phi) = \int_\Omega W(D\phi)\, dX$, if $V'(\phi_0) = 0$ and $V''(\phi_0) > 0$ under conditions of polyconvexity—is ϕ_0 a strict local minimum of V in $W^{1,p}$? (Example 6.4 shows that some condition like polyconvexity is needed.) Also, Ball's example on cavitation shows that the answer is generally "no" without additional growth conditions on W. However, Weierstrass' classical work in the calculus of variations (see Bolza [1904]) indicates that this problem is tractable. (Weierstrass made the big leap from C^1 to C^0 for the validity of (S2)(a) in one-dimensional variational problems.)

In practice the energy criterion has great success, according to Koiter [1976a]. However, this is consistent with the possibility that the energy criterion may fail for nonlinear elastodynamics. Indeed, "in practice" one usually does not observe the very high frequency motions. Masking these high frequencies may amount to an averaging process in which the quasi-linear equations of elastodynamics are replaced by finite-dimensional or semilinear approximations. For the latter, the energy criterion is valid under reasonable conditions. Indeed, for semilinear equations there usually are function spaces in which (S1) holds by using the semilinear existence theorem 5.1 and in which (S2) can be checked by just using differential calculus. The method was already applied to nonlinear wave equations in the previous section.[21] We isolate the calculus lemma as follows.

6.9 Proposition *Let $\mathfrak{X}$ be a Banach space and $V: \mathfrak{X} \to \mathbb{R}$ be C^2 in a neighborhood of $x_0 \in \mathfrak{X}$. Suppose that (i) $DV(x_0) = 0$, and (ii) there is a $c > 0$ such that $D^2V(x_0)\cdot(v, v) \geq c\|v\|^2$. Then x_0 lies in a potential well for V.*

Proof By Taylor's theorem and (*i*),

$$\begin{aligned} V(x) - V(x_0) &= \int_0^1 (1-s)D^2V(sx + (1-s)x_0)(x - x_0, x - x_0)\, ds \\ &= \tfrac{1}{2}D^2V(x_0)(x - x_0, x - x_0) \\ &\quad + \int_0^1 (1-s)[D^2V(sx + (1-s)x_0) - D^2V(x_0)](x - x_0, x - x_0)\, ds \\ &= \tfrac{1}{2}D^2V(x_0)(x - x_0, x - x_0) + R(x, x_0). \end{aligned}$$

By continuity of D^2V, there is a $\delta > 0$ such that $\|x - x_0\| < \delta$ implies $|R(x, x_0)| \leq (c/4)\|x - x_0\|^2$. Since the first term is $\geq c\|x - x_0\|^2/2$, we get

$$V(x) - V(x_0) \geq \frac{c}{4}\|x - x_0\|^2 \quad \text{if} \quad \|x - x_0\| < \delta.$$

This inequality gives (S2). ∎

Note that condition (ii) states that $\langle v, w\rangle = D^2V(x_0)\cdot(v, w)$ is an inner product on $\mathfrak{X}$ whose topology is the same as the $\mathfrak{X}$-topology.

The following theorem shows how the semilinear theory works. We take $x_0 = 0$ for simplicity.

6.10 Theorem *Let $\mathcal{H}$ be a Hilbert space and let A be a positive-definite ($A \geq c > 0$) self-adjoint operator and let $\mathcal{H}_1$ be the domain of $\sqrt{A}$ with the graph norm. Let $V: \mathcal{H}_1 \to \mathbb{R}$ be given by $V(x) = \frac{1}{2}\langle\sqrt{A}x, \sqrt{A}x\rangle + \tilde{V}(x)$, where $\tilde{V}$ is C^2 and $\tilde{V}(0) = 0$, $D\tilde{V}(0) = 0$, and $D^2\tilde{V}(0) \geq 0$; that is, $D^2\tilde{V}(0)(v, v) \geq 0$. Suppose that $\nabla\tilde{V}(x)$ exists and is a C^1 map of $\mathcal{H}_1$ to $\mathcal{H}$* (cf. 5.6). *Then the hypotheses* (S1) *and* (S2) *hold for the Hamiltonian equations*

$$\frac{\partial}{\partial t}\begin{pmatrix} x \\ \dot{x} \end{pmatrix} = \begin{pmatrix} 0 & I \\ -A & 0 \end{pmatrix}\begin{pmatrix} x \\ \dot{x} \end{pmatrix} + \begin{pmatrix} 0 \\ -\nabla\tilde{V}(x) \end{pmatrix} \tag{6}$$

with $H(x, \dot{x}) = \frac{1}{2}\|\dot{x}\|^2 + \frac{1}{2}\|\sqrt{A}x\|^2 + \tilde{V}(x)$ and so $(0, 0)$ is dynamically stable.

Proof This follows directly from 3.2, 5.1, 5.6, and 6.9, with $V(x) = \langle\sqrt{A}x, \sqrt{A}x\rangle + \tilde{V}(x)$. ∎

Problem 6.1 Use this theorem to re-derive the results of Example 5.7(iii).

[21] These methods appear in a number of references, such as Marsden [1973b] and Payne and Sattinger [1975].

6.11 Example (R. Knops) Consider the following equations:

$$u_{tt} = (P(u_x))_x - u_{xxxx}, \tag{7}$$

where $P(u_x) = W'(u_x)$, $W(0) = W'(0) = 0$, $W''(0) \geq 0$, and W is smooth. We work in one dimension on an interval, say [0, 1], with boundary conditions, say $u = 0$ and $u_{xx} = 0$, at the ends. We claim that Theorem 6.10 applies to this example with $\mathcal{H} = L^2[0, 1]$,

$$A(u) = -u_{xxxx}, \quad \sqrt{A}u = u_{xx}$$

$$\mathcal{H}_1 = \mathfrak{D}(\sqrt{A}) = H^2_\partial = \{u \in H^2 \mid u = 0 \text{ at } x = 0, 1\}$$

and

$$\mathfrak{D}(A) = H^4_\partial = \{u \in H^4 \mid u = u_{xx} = 0 \text{ at } x = 0, 1\}.$$

We have $\hat{V}(u) = \int W(u_x)\,dx$. Then since $H^2 \subset C^1$ in one dimension, $\hat{V}$ is smooth, with

$$D\hat{V}(u)\cdot v = \int W'(u_x)\cdot v_x\,dx = -\int (W'(u_x))_x\cdot v\,dx \tag{8a}$$

and

$$D^2\hat{V}(u)\cdot(u, w) = \int W''(u_x)\cdot u_x\cdot w_x\,dx. \tag{8b}$$

From (8a) $-\nabla\hat{V}(u) = (W'(u_x))_x$, so (6) and (7) coincide and $-\nabla\hat{V}$ is a smooth function from H^2_∂ to L^2. Finally, it is clear from (8a) that $D\hat{V}(0) = 0$ and as $W''(0) > 0$, (8b) gives $D^2\hat{V}(0)(v, v) \geq 0$. Thus, the trivial solution (0, 0) is (dynamically) stable in $H^2_\partial \times L^2$.

Problem 6.2 Modify (7) to $u_{tt} = P(u_x)_x + u_{xxxx}$ where boundary conditions $u(0) = 0$, $u(1) = \lambda$, $u_{xx}(0) = 0$, $u_{xx}(1) = 0$ are imposed and we insist $u'(x) > 0$.

Box 6.1 ***Summary of Important Formulas for Section 6.6***

Energy Criterion

"Minima of the potential energy are stable." This assertion "works," but a satisfactory theorem justifying it for nonlinear elasticity is not known.

Potential Well Conditions

(a) $V(x) > V(x_0)$; $x \neq x_0$, x near x_0.

(b) $\left(\inf_{\|x - x_0\| = \epsilon} V(x)\right) > V(x_0)$.

Stability Theorem

Well-posed dynamics and a potential well implies stability.

Applicability

1. The potential well condition (b) cannot hold for nonlinear elas-

ticity in topologies as strong as C^1, but in $W^{1,p}$ (b) follows from (a) and the assumption of polyconvexity. Dynamics however is not known to exist in $W^{1,p}$.

2. The stability theorem does apply to situations in which the basic equations are semilinear (rather than the quasi-linear equations of non-linear elasticity).

6.7 A CONTROL PROBLEM FOR A BEAM EQUATION[22]

This section discusses the abstract problem of controlling a semilinear evolution equation and applies the formalism to the case of a vibrating beam. The beam equation in question is

$$w_{tt} + w_{xxxx} + p(t)w_{xx} = 0 \quad (0 \leq x \leq 1)$$

with boundary conditions $w = w_{xx} = 0$ at $x = 0, 1$; see Figure 6.7.1. Here w represents the transverse deflection of a beam with hinged ends, and $p(t)$ is an axial force. The control question is this: given initial conditions $w(x, 0)$, $\dot{w}(x, 0)$, can we find $p(t)$ that controls the solution to a prescribed w, $\dot{w}$ after a prescribed time T?

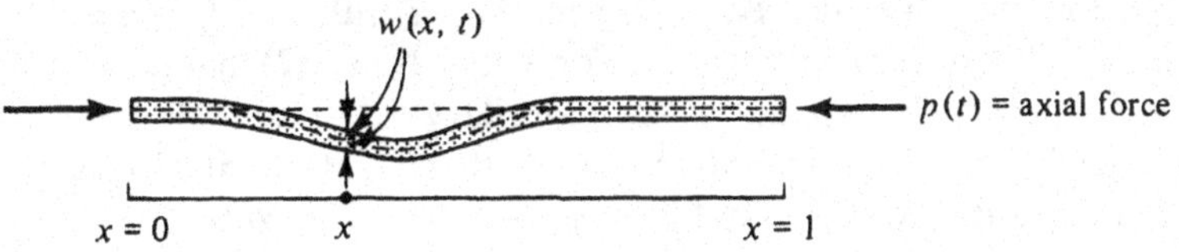

Figure 6.7.1

The beam equation is just one illustration of many in control theory, but it has a number of peculiarities that point out the caution needed when setting up a general theory for the control of partial, rather than ordinary, differential equations. In particular, we shall see that it is easy to prove controllability of any finite number of modes at once, but it is very delicate to prove controllability of *all* the modes simultaneously. The full theory needed to prove the latter is omitted here, but the points where the "naive" method breaks down will be discussed. For the more sophisticated theory needed to deal with all the modes

[22]This section was done in collaboration with J. M. Ball and M. Slemrod and is based on Ball, Marsden, and Slemrod [1982].

of the beam equation, see Ball, Marsden, and Slemrod [1982]. The paper of Ball and Slemrod [1979] considers some related stabilization problems. For background in control theory, see, for example, Brockett [1982] and Russell [1979]. References related to the work in this section are Sussman [1977], Jurdjevic and Quinn [1978] and Hermes [1979]. For control in the Hamiltonian context, see for example, Van der Schaft [1981].

We begin by treating the situation for abstract semilinear equations. Thus, we consider an evolution equation of the form

$$\dot{u}(t) = \mathfrak{A}u(t) + p(t)\mathfrak{B}(u(t)), \tag{1}$$

where $\mathfrak{A}$ generates a C^0 semigroup on a Banach space $\mathfrak{X}$, $p(t)$ is a real valued function of t that is locally L^1 and $\mathfrak{B}: \mathfrak{X} \longrightarrow \mathfrak{X}$ is C^k $(k \geq 1)$. Let u_0 be given initial data for u and let $T > 0$ be given. For $p = 0$, the solution of (1) after time T is just $e^{T\mathfrak{A}}u_0$, which we call the *free solution.*

7.1 Definition If there exists a neighborhood $\mathcal{V}$ of $e^{\mathfrak{A}T}u_0$ in $\mathfrak{X}$ with the property that for any $v \in V$, there exists a p such that the solution of (1) with initial data u_0 reaches v after time T, we say (1) is *locally controllable* about the free solution $e^{\mathfrak{A}T}u_0$.

One way to tackle this problem of local controllability is to use the implicit function theorem. Write (1) in integrated form:

$$u(t) = e^{\mathfrak{A}t}u_0 + \int_0^t e^{(t-s)\mathfrak{A}}p(s)\mathfrak{B}(u(s))\,ds. \tag{2}$$

Let p belong to a specified Banach space $\mathbb{Z} \subset L^1([0, T], \mathbb{R})$. The techniques used to prove Theorem 5.1 show that for short time, (2) has a unique solution $u(t, p, u_0)$ that is C^k in p and u_0. By Corollary 5.3, if $\|\mathfrak{B}(x)\| \leq C + K\|x\|$ (for example, $\mathfrak{B}$ linear will be of interest to us), then solutions are globally defined, so we do not need to worry about taking short time intervals.

Let $L: \mathbb{Z} \longrightarrow \mathfrak{X}$ denote the derivative of $u(T, p, u_0)$ with respect to p at $p = 0$. This may be found by implicitly differentiating (2). One gets

$$Lp = \int_0^T e^{(T-s)\mathfrak{A}}p(s)\mathfrak{B}(e^{s\mathfrak{A}}u_0)\,ds. \tag{3}$$

The implicit function theorem then gives:

7.2 Theorem *If $L: \mathbb{Z} \longrightarrow \mathfrak{X}$ is a surjective linear map, then* (1) *is locally controllable around the free solution.*

If $\mathfrak{A}$ generates a group and $\mathfrak{B}$ is linear, then surjectivity of L is clearly equivalent to surjectivity of $\hat{L}p = \int_0^T p(s)\, e^{-s\mathfrak{A}}\mathfrak{B}e^{s\mathfrak{A}}u_0\,ds.$

7.3 Example If $\mathfrak{X} = \mathbb{R}^n$ and $\mathfrak{B}$ is linear, we can expand

$$e^{-s\mathfrak{A}}\mathfrak{B}e^{s\mathfrak{A}} = \mathfrak{B} + s[\mathfrak{A}, \mathfrak{B}] + \frac{s^2}{2}[\mathfrak{A}, [\mathfrak{A}, \mathfrak{B}]] + \cdots$$

to recover a standard controllability criterion: if

$$\dim \operatorname{span}\{\mathfrak{B}u_0, [\mathfrak{A}, \mathfrak{B}]u_0, [\mathfrak{A}, [\mathfrak{A}, \mathfrak{B}]]u_0, \ldots\} = n,$$

then (1) is locally controllable. These ideas lead naturally to differential geometric aspects of control theory using Lie brackets of vector fields and the Frobenius theorem. See, for example, Sussman [1977] and Brockett [1982].

Next, suppose we wish only to control a finite-dimensional piece of u or a finite number of modes of u. To do so, we should "observe" only a finite-dimensional projection of u. This idea leads to the following:

7.4 Definition We say that (1) is *locally controllable about the free solution* $e^{T\mathfrak{A}}u_0$ *for finite-dimensional observations* if for any surjective continuous linear map $G\colon \mathfrak{X} \longrightarrow \mathbb{R}^n$ there exists a neighborhood $\mathfrak{U}$ of $G(e^{\mathfrak{A}T}u_0)$ in $\mathbb{R}^n$ with the property that for any $v \in \mathfrak{U}$ there exists a $p \in \mathfrak{Z}$ such that

$$G(u(T, p, u_0)) = v.$$

As above, we have:

7.5 Theorem *Suppose L, defined by* (3), *has dense range in $\mathfrak{X}$. Then* (1) *is locally controllable about the free solution for finite-dimensional observations.*

Proof The map $p \mapsto G(u(T, p, u_0))$ has derivative $G \circ L\colon \mathfrak{Z} \longrightarrow \mathbb{R}^n$ at $p = 0$. Since L has dense range, $G \circ L$ is surjective, so the implicit function theorem applies. ∎

As above, if $\mathfrak{A}$ generates a group, it is enough to show that $\hat{L}$ has dense range to get the conclusion of 7.5. *For the beam equation we shall see that $\hat{L}$ does have dense range but in the space $\mathfrak{X}$ corresponding to the energy norm and with $\mathfrak{Z} = L^2$, is not surjective.* To see this, we shall need some more detailed computations concerning hyperbolic systems in general.

Let A be a positive self-adjoint operator on a real Hilbert space $\mathfrak{H}$ with inner product $\langle\ ,\ \rangle_{\mathfrak{H}}$. Let A have spectrum consisting of eigenvalues λ_n^2 $(0 < \lambda_1 \leq \lambda_2 \leq \lambda_3 \leq \cdots)$ with corresponding orthonormalized eigenfunctions ϕ_n. Let $B\colon \mathfrak{D}(A^{1/2}) \longrightarrow \mathfrak{H}$ be bounded. We consider the equation

$$\ddot{w} + Aw + pBw = 0. \tag{4}$$

This in the form (1) with

$$u = \begin{pmatrix} w \\ \dot{w} \end{pmatrix}$$

and

$$\mathfrak{A} = \begin{pmatrix} 0 & I \\ -A & 0 \end{pmatrix}, \quad \mathfrak{B} = \begin{pmatrix} 0 & 0 \\ -B & 0 \end{pmatrix}.$$

Here $\mathfrak{X} = \mathfrak{D}(A^{1/2}) \times \mathfrak{H}$ and, by 3.2, $\mathfrak{A}$ generates a C^0 group of isometries on $\mathfrak{X}$. The inner product on $\mathfrak{X}$ is given by the "energy inner product":

$$\langle (y_1, z_1), (y_2, z_2) \rangle_{\mathfrak{X}} = \langle A^{1/2}y_1, A^{1/2}y_2 \rangle_{\mathfrak{H}} + \langle z_1, z_2 \rangle_{\mathfrak{H}}.$$

Write
$$u_0 = \begin{pmatrix} \sum_{m=1}^{\infty} b_m\phi_m \\ \sum_{m=1}^{\infty} -\lambda_m c_m \phi_m \end{pmatrix} \in \mathfrak{X},$$
where
$$\sum_{m=1}^{\infty} \lambda_m^2(b_m^2 + c_m^2) < \infty.$$
If we set $a_m = \frac{1}{2}(b_m + ic_m)$, we have
$$e^{\mathfrak{A}s}u_0 = \begin{pmatrix} \sum_{m=1}^{\infty} [a_m \exp(i\lambda_m s) + \bar{a}_m \exp(-i\lambda_m s)]\phi_m \\ \sum_{m=1}^{\infty} i\lambda_m[a_m \exp(i\lambda_m s) - \bar{a}_m \exp(-i\lambda_m s)]\phi_m \end{pmatrix}.$$
In particular, this gives an explicit formula for the free solution. Applying $\mathfrak{B}$, we get
$$\mathfrak{B}e^{\mathfrak{A}s}u_0 = \begin{pmatrix} 0 \\ \sum_{m=1}^{\infty} [a_m \exp(i\lambda_m s) + \bar{a}_m \exp(-i\lambda_m s)]B\phi_m \end{pmatrix}.$$

To simplify matters, let us *assume that* B is diagonal; that is, $\langle B\phi_m, \phi_n\rangle_{\mathfrak{X}} = d_m\delta_{mn}$. Then a little computation gives
$$e^{-s\mathfrak{A}}\mathfrak{B}e^{s\mathfrak{A}}u_0 = \begin{bmatrix} \sum_{n=1}^{\infty} \frac{-id_n}{2\lambda_n}\{a_n \exp(2i\lambda_n s) - \bar{a}_n \exp(-2i\lambda_n s) - (a_n - \bar{a}_n)\}\phi_n \\ \sum_{n=1}^{\infty} -\frac{d_n}{2}\{a_n \exp(2i\lambda_n s) + \bar{a}_n \exp(-2i\lambda_n s) + (a_n + \bar{a}_n)\}\phi_n \end{bmatrix} \tag{5}$$
This formula can now be inserted into (3) to give Lp in terms of the basis ϕ_n. Since $\mathfrak{A}$ generates a group, surjectivity of L comes down to solvability of
$$\hat{L}p = \int_0^T p(s)e^{-s\mathfrak{A}}\mathfrak{B}(e^{s\mathfrak{A}}u_0)\,ds = h \tag{6}$$
for $p(s)$ given $h \in \mathfrak{X}$. Also, (4) is locally controllable for finite-dimensional observers if $\hat{L}$ has dense range.

Let us now turn to our vibrating beam with hinged ends and an axial load $p(t)$ as a control:
$$\begin{aligned} w_{tt} + w_{xxxx} + p(t)w_{xx} &= 0 \quad (0 \le x \le 1), \\ w = w_{xx} &= 0 \quad \text{at} \quad x = 0, 1. \end{aligned} \tag{7}$$
Here, $Aw = w_{xxxx}$, $Bw = w_{xx}$, $\lambda_n = n^2\pi^2$, $\phi_n = (1/\sqrt{2})\sin(n\pi x)$, $d_n = -n^2\pi^2$, and $\mathfrak{X} = H_\partial^2 \times L^2$ (see Box 3.2).

7.6 Theorem *If $T \ge 1/\pi$, $\mathfrak{X} = H_\partial^2 \times L^2$, $\mathfrak{Z}$ is L^2 or more generally is dense in L^1 and the Fourier coefficients of the initial data u_0 satisfy $a_n \neq 0$ ($n = 1, 2, 3, \ldots$) (i.e., u_0 is active in all modes), then (7) is locally controllable about the free solution $e^{T\mathfrak{A}}u_0$ for finite-dimensional observations.*

Proof We have to show that $\hat{L}$ defined by (5) and (6) has dense range. To do this we show that any $h \in \mathfrak{X}$ orthogonal to the range of $\hat{L}$ must vanish.

If $h \in \mathfrak{X}$ is orthogonal to the range of $\hat{L}$, then

$$\int_0^T p(s)\langle e^{-s\mathfrak{A}}\mathfrak{B}e^{s\mathfrak{A}}u_0, h\rangle\, ds = 0$$

for all $p \in L^2$. This implies that

$$\langle e^{-s\mathfrak{A}}\mathfrak{B}e^{s\mathfrak{A}}u_0, h\rangle = 0, \qquad s \in [0, T]. \tag{8}$$

Write

$$h = \begin{pmatrix} \sum \alpha_m \phi_m \\ \sum -\lambda_m \beta_m \phi_m \end{pmatrix} \tag{9}$$

so $h \in \mathfrak{X}$ means $\sum n^4\pi^4(\alpha_m^2 + \beta_m^2) < \infty$. Using (5) and (9), (7) simplifies to

$$\begin{aligned} \sum_{n=1}^{\infty} i\frac{n^4\pi^4}{2}\{[a_n e^{2in^2\pi^2 s} - \bar{a}_n e^{-2in^2\pi^2 s} - (a_n - \bar{a}_n)]\,\alpha_n \\ + i[a_n e^{2in^2\pi^2 s} + \bar{a}_n e^{-2in^2\pi^2 s} + (a_n + \bar{a}_n)]\beta_n\} = 0. \end{aligned} \tag{10}$$

However, this represents some of the terms of a convergent Fourier series on $[0, 1/\pi]$, so all the coefficients must vanish. Since $a_n \neq 0$, this implies α_n and $\beta_n = 0$. ∎

Problem 7.1 Redo the above computations for general hyperbolic equations using complex notation by writing $z = \sqrt{A}w + i\dot{w}$. Show that (4) becomes

$$i\dot{z} = \sqrt{A}z + pBA^{-1/2}\,\mathrm{Re}\, z \quad \text{so} \quad \mathfrak{A} = -i\sqrt{A}, \quad \mathfrak{B} = -iB\circ A^{-1/2}\circ\mathrm{Re}$$

and (5) becomes

$$e^{-s\mathfrak{A}}\mathfrak{B}e^{s\mathfrak{A}}z = \sum_{n=1}^{\infty} \frac{-b_n}{2\lambda_n}(\bar{z}_n e^{2i\lambda_n s} + z_n)\,\phi_n.$$

Use this to give a cleaner looking expression for (10).

One can now ask if $\hat{L}$ is surjective. For $\mathbf{Z} = L^2$, we can see that it is *not*, as follows. To solve $\hat{L}p = h$ we loose no generality by seeking p's on $[0, 1/\pi]$ in the form

$$p(s) = p_0 + \sum\{p_{n^2}\exp(2in^2\pi^2 s) + \bar{p}_{n^2}\exp(-2in^2\pi^2 s)\} \tag{11}$$

and suppressing the remaining coefficients. Inserting (5) and (11) into (6), we can determine the Fourier coefficients of h. As in (9), the condition for h to be in $\mathfrak{X}$ is $\sum \lambda_m^2(\alpha_m^2 + \beta_m^2) < \infty$. But the condition for h to be in the range of $\hat{L}$ with an L^2 function p is that $\{a_n d_n p_{n^2}\} \in \ell_2$. This is, however, a stronger condition than $h \in \mathfrak{X}$. Thus, we conclude that $\hat{L}$ and hence L has range that is not equal to $\mathfrak{X}$.

In fact, one can show that not only is L not surjective, but that (7) is *not* locally controllable in the energy norm. To overcome this difficulty one can contemplate more sophisticated inverse function theorems and indeed these may be necessary in general. However, for a certain class of equations another trick works. Namely, instead of the $\mathfrak{X}$-norm, make up a new space related to the range of $\hat{L}$ and the graph norm. Miraculously, the solution $u(t, p, u_0)$ stays in this space and is still a smooth function of p in the new topology. Consequently, in this stronger norm the implicit function theorem can, in effect still be used. For the beam example, this trick (combined with a certain averaging method) actually works. In the end one finds a new space that differs from the space of the finite energy solution in that a more severe condition on the asymptotic decay rate of the modal amplitudes is required. In this space, control *is* possible. See Ball, Marsden, and Slemrod [1982] for details.

Problem 7.2 Show that the problem

$$u_{tt} - u_{xx} + pu = 0 \quad (0 < x < 1),$$
$$u = 0 \quad \text{at} \quad x = 0, 1,$$

is controllable for finite-dimensional observers provided the initial data is active in all modes.

Box 7.1 ***Summary of Important Formulas for Section 6-7***

Local controllability of $\dot{u} = \mathfrak{A}u + p\mathfrak{B}u$ about the free solution $e^{T\mathfrak{A}}u_0$ means that we can hit a whole neighborhood of $e^{T\mathfrak{A}}u_0$ by starting at u_0 and varying the control p in a neighborhood of 0.

If we ask only for Gu to hit a neighborhood of $G(e^{T\mathfrak{A}}u_0)$, where G is any surjective linear map to $\mathbb{R}^n$, we says the equation is locally controllable for *finite-dimensional observations.*

Criteria

1. If $Lp = \int_0^T e^{(T-s)\mathfrak{A}}p(s)\mathfrak{B}e^{s\mathfrak{A}}u_0\,ds$ is surjective, then the equation is locally controllable.

2. If L has dense range, then the equation is controllable for finite-dimensional observations.

3. If the equation is on $\mathbb{R}^n$, $\mathfrak{B}$ is linear, and

$$\dim \operatorname{span}\{\mathfrak{B}u_0, [\mathfrak{A}, \mathfrak{B}]u_0, [\mathfrak{A}, [\mathfrak{A}, \mathfrak{B}]]u_0, \ldots\} = n,$$

then the equation is locally controllable.

The Beam Equation

Equation (7) is controllable for finite-dimensional observations provided the initial data are active in all modes. It is never controllable to a full neighborhood of the free solution in the energy norm; however, it is controllable to a dense subset.

CHAPTER 7

SELECTED TOPICS IN BIFURCATION THEORY

This chapter introduces some basic ideas in bifurcation theory and gives a few examples of current interest. This subject is very large and the chapter cannot pretend to cover anything but a sampling. For this reason two sections—7.2 and 7.5—are written in the nature of surveys; we hope these will alert the reader to some of the important current literature. Sections 7.1 and 7.4 introduce basic ideas in the static and dynamic theory, respectively, and Sections 7.3 and 7.6 select a nontrivial example from each theory to work in detail. These examples are biased towards the authors' interests, so readers may wish to substitute one of their own choosing from the literature.

7.1 BASIC IDEAS OF STATIC BIFURCATION THEORY

This section presents a few sample results in static bifurcation theory. The reader should understand that there are many different points of view in this subject and that the results can be refined in several directions. Some of the books that the serious reader should consult after reading this section are Keller and Antman [1969], Sattinger [1973], Nirenberg [1974], Berger [1977], Iooss and Joseph [1980], and Antman [1983].

We begin with a few introductory remarks. Consider a beam free to move in a plane, distorted from its natural state by the application of a load λ, as shown in Figure 7.1.1. For small λ, the beam slightly compresses, but after a critical load λ_c is reached, it buckles into one of two possible states. The compressed state is still there, but it has become unstable; the stability has been transferred from the original trivial solution to the stable buckled solutions. The situation

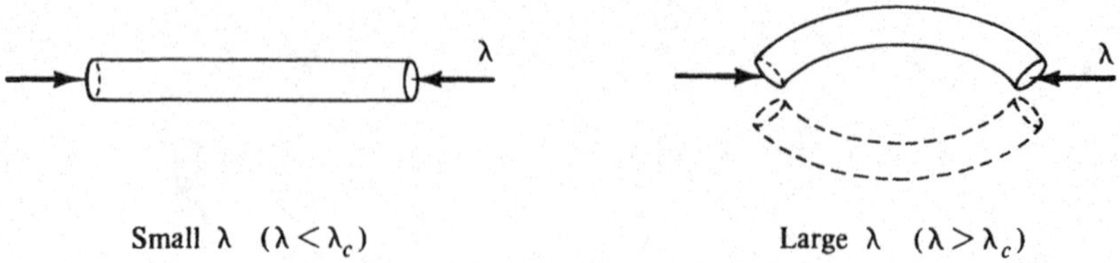

Figure 7.1.1

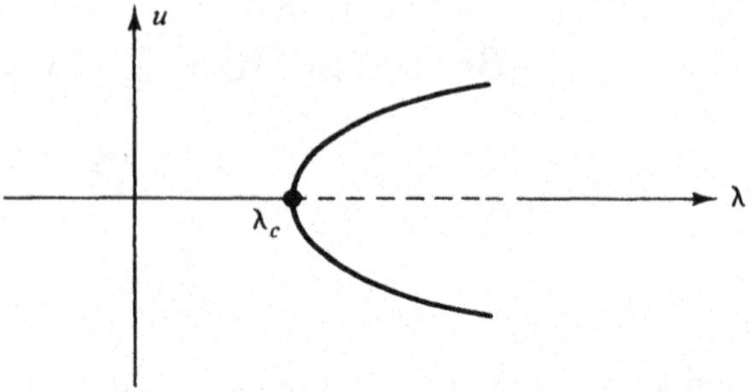

Figure 7.1.2

can be summarized by drawing the pitchfork bifurcation diagram, as in Figure 7.1.2. The vertical axis u represents the displacement of the center line of the beam. In this diagram, stable solutions are drawn with a solid line and unstable ones with a dashed line. One usually concentrates on stable solutions since these are the only ones one will "see."

There are many ways one can model the beam depicted in Figure 7.1.1. First, one could use a full model of three-dimensional elasticity. Second, one could use a rod-model and take into account shearing and extensibility. Perhaps the simplest, however, is the original model adopted by Euler in 1744, which effectively started the subject of bifurcation theory. He assumed the beam is an "elastica"; inextensible and unshearable. If s, ranging from 0 to 1, represents arc length along the beam (so s is a material variable) and $\theta(s)$ is the angle of deflection of the tangent at s, Euler derived the equation

$$EI\theta'' + \lambda \sin \theta = 0, \qquad \theta(0) = \theta(1) = 0,$$

where EI is a constant. There is extensive literature on this equation; we recommend the introductory article in Keller and Antman [1969] for an account. The problem was largely solved by Euler and one gets a bifurcation diagram as shown in Figure 7.1.3. As we shall see shortly, the points of bifurcation $\lambda/EI = k^2\pi^2$ $(k = 1, 2, 3, \ldots)$ on the λ-axis can be readily computed; they are the eigenvalues of the linearized problem about the trivial solution $\theta = 0$; that is,

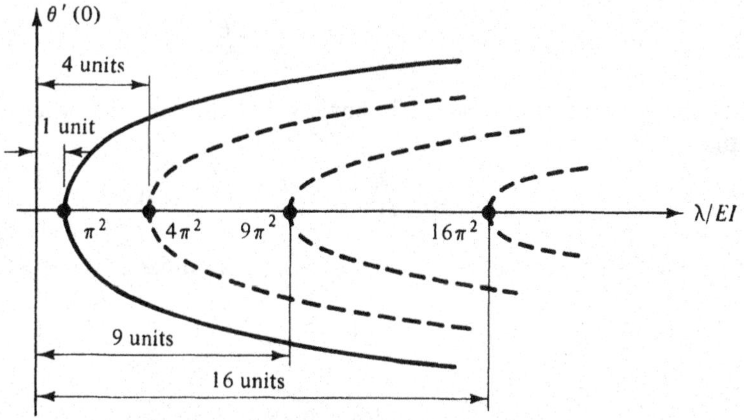

Figure 7.1.3

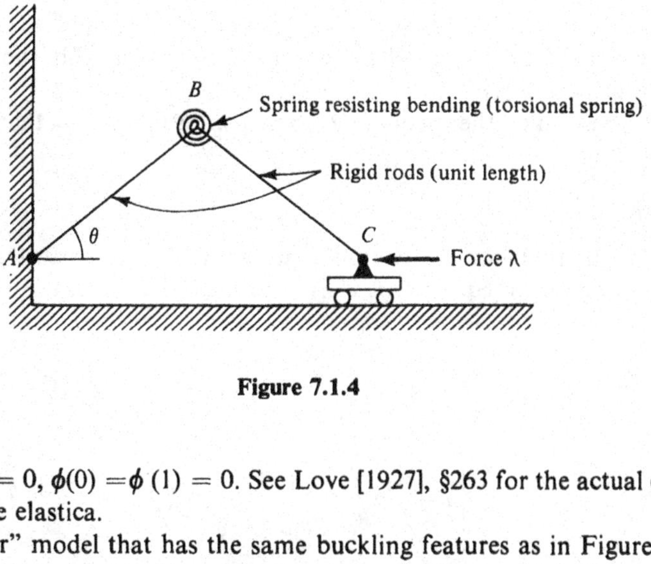

Figure 7.1.4

$EI\phi'' + \lambda\phi = 0$, $\phi(0) = \phi(1) = 0$. See Love [1927], §263 for the actual configurations of the elastica.

A "simpler" model that has the same buckling features as in Figure 7.1.2 is obtained by restricting to a "one mode" model, as in Figure 7.1.4. If the torsional spring has a spring constant κ, the potential energy is $V = \kappa\theta^2/2 + 2\lambda(\cos\theta - 1)$. The equilibria are the critical points of V:

$$\kappa\theta - 2\lambda \sin\theta = 0.$$

Near $\lambda/\kappa = \frac{1}{2}$, the trivial solution $\theta = 0$ bifurcates into two solutions, as in Figure 7.1.2. Note that this can be predicted by the inverse function theorem; if $F(\theta, \lambda) = \kappa\theta - 2\lambda \sin\theta$, then $(\partial F/\partial\theta)(0, \lambda) = \kappa - 2\lambda \cos 0 = \kappa - 2\lambda$, which vanishes when $\kappa - 2\lambda = 0$; that is, $\lambda/\kappa = \frac{1}{2}$. Thus it is near this point that the implicit function theorem fails to show that $\theta = 0$ is the unique solution. In terms of a series,

$$F(\theta, \lambda) = \kappa\theta - 2\lambda\left(\theta - \frac{\theta^3}{3!} + \frac{\theta^5}{5!} - \cdots\right)$$

and the first term vanishes if $\kappa = 2\lambda$. For $\lambda/\kappa > \frac{1}{2}$ there are two solutions, $\theta = 0$ and the solutions of

$$(\kappa - 2\lambda) + \frac{2\lambda\theta^2}{3!} - \frac{2\lambda\theta^4}{4!} + \cdots = 0.$$

Clearly, if θ is a solution, so is $-\theta$, and so θ is approximately the solution of

$$(\kappa - 2\lambda) + \frac{2\lambda}{3!}\theta^2 = 0,$$

that is,

$$\theta = \pm[3(2\lambda - \kappa)]^{1/2} + \text{higher-order terms.}$$

Graphically, we see the two solutions in Figure 7.1.5. From this graph we can also obtain the *global* bifurcation picture, as in Figure 7.1.6. Note the differences with Figure 7.1.3. The stability of solutions can be examined by looking at whether or not the solution is a maximum or minimum of V. For example, as λ/κ is increased beyond $\frac{1}{2}$, what happens to V in a neighborhood of $\theta = 0$ is depicted in Figure 7.1.7.

Before beginning the theory, we summarize a few things that examples like this teach us.

1.1 Remarks

(a) The problem of static bifurcation may be stated abstractly as that of solving an equation $f(x, \lambda) = 0$, where λ denotes one or more param-

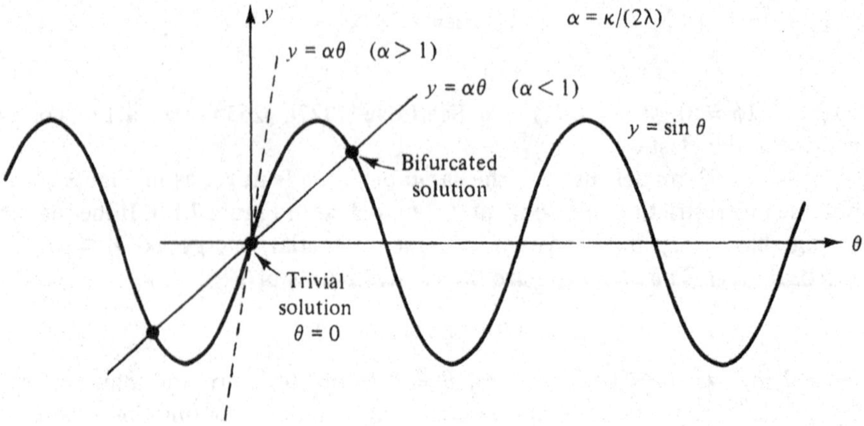

Figure 7.1.5

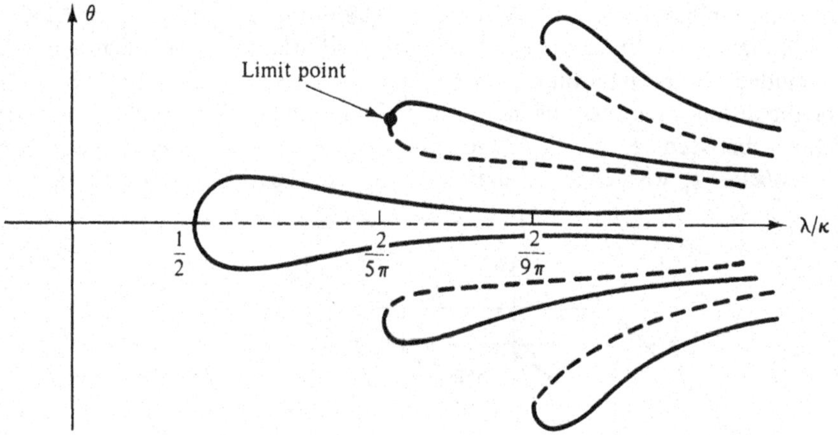

Figure 7.1.6

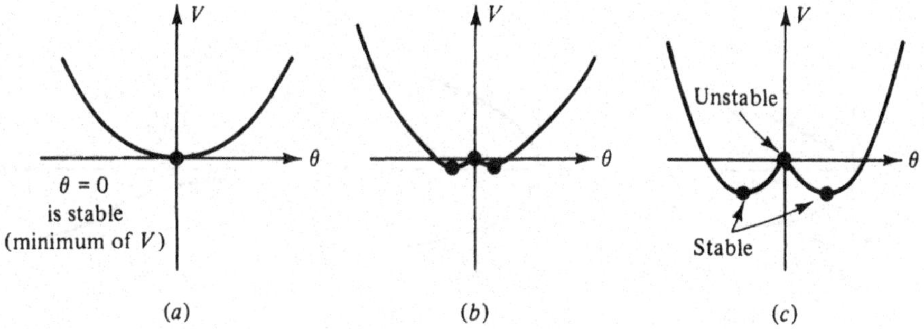

Figure 7.1.7 (a) $\lambda/\kappa < \frac{1}{2}$. (b) $\lambda/\kappa = \frac{1}{2}+$. (c) $\lambda/\kappa > \frac{1}{2}$.

eters to be varied and x is a variable representing the state of the system.

(b) When solutions are located, it is important to decide which are stable and which are unstable; this may be done by determining the spectrum of the linearization or by testing for maxima or minima of a potential.

(c) Is the bifurcation diagram sensitive to small perturbations of the equations or the addition of further parameters? A bifurcation diagram that is insensitive to such changes is called *structurally stable*.

(d) Before any declaration is made that "the complete global bifurcation diagram is obtained," the following criteria should be fulfilled: (i) Are you sure you have *all* the essential parameters (see (c))? (ii) Does the model you have chosen remain a good one for large values of the parameter and the variable?

Let us comment briefly on (c) and (d). The bifurcation diagram in 7.1.6 near $\theta = 0$, $\lambda/\kappa = \frac{1}{2}$ is *structurally unstable*. If an additional imperfection parameter is included, the bifurcation diagram changes. For example, in Figure 7.1.4, let ϵ be the distance between the direction of λ and the point A—that is, the vertical distance between A and C. If the solutions are plotted in ϵ, θ space, where $\mu = (\lambda/\kappa) - \frac{1}{2}$, we get the situation shown in Figure 7.1.8. This is generally

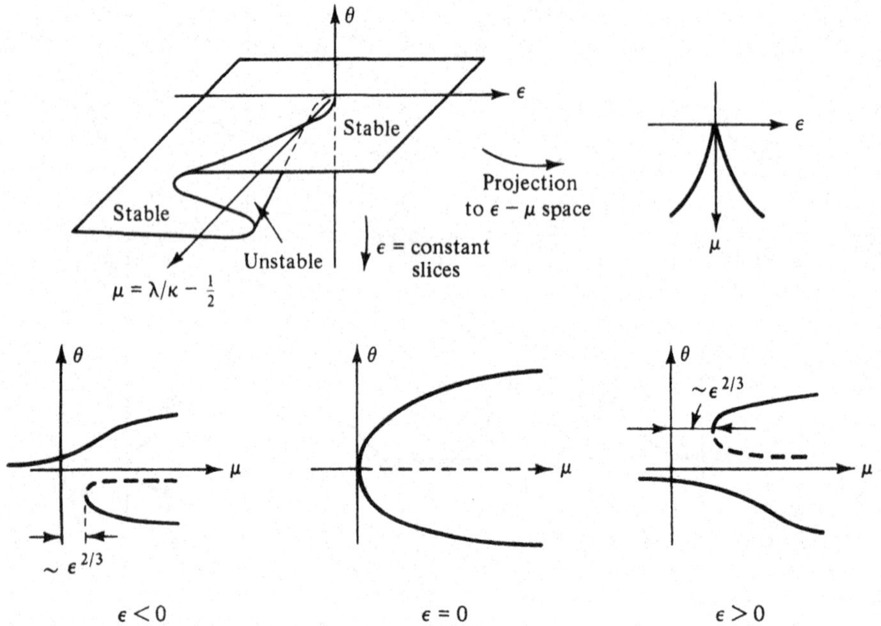

Figure 7.1.8

called an *imperfection-sensitivity diagram*. (The $\frac{2}{3}$ power law of Koiter [1945] is noted.) We discuss these points in greater depth in Box 1.1, and especially the important point: is one extra parameter like ϵ sufficient to completely capture all possible perturbations? (It is not, even for this basic example.)

Comment d(ii) is also relevant; suppose one goes to the trouble to produce the global bifurcation diagram in Figure 7.1.6. Are these extra branches meaningful? They correspond to θ beyond the range $[0, 2\pi]$, where the torsional spring has been "wound up" a number of extra times. For very large windings the linear spring law presumably breaks down, or, due to other constraints, large windings may be prohibited (the mechanism may not allow it). It then requires some work to decide which portions of Figure 7.1.6 are actually relevant to the problem at hand.

Now we begin the mathematical development of static bifurcation theory. Let us start with the simplest situation in which we have a trivial solution available and have one parameter. Thus, let $\mathfrak{X}$ and $\mathcal{Y}$ be Banach spaces and let $f: \mathcal{Y} \times \mathbb{R} \to \mathfrak{X}$ be a given C^∞ mapping; assume that $f(0, \lambda) = 0$ for all λ.

1.2 Definition We say that $(0, \lambda_0)$ is a *bifurcation point* of the equation $f(x, \lambda) = 0$ if every neighborhood of $(0, \lambda_0)$ contains a solution (x, λ) with $x \neq 0$.

The following gives a necessary condition for bifurcation.

1.3 Proposition *Suppose that* $A_\lambda = D_x f(0, \lambda)$ *(the derivative with respect to* x*) is an isomorphism from* $\mathcal{Y}$ *to* $\mathfrak{X}$*. Then* $(0, \lambda)$ *is not a bifurcation point.*

Proof By the implicit function theorem (see Section 4.1) $f(x, \lambda) = 0$ is uniquely solvable for $x(\lambda)$ near $(0, \lambda)$; since $x = 0$ is a solution, no others are possible in a neighborhood of $(0, \lambda)$. ∎

1.4 Example Suppose $f(x, \lambda) = Lx - \lambda x + g(x, \lambda)$, where $g(0, \lambda) = 0$ and $D_x g(0, \lambda) = 0$. For this to make sense, we assume L is a linear operator in a Banach space $\mathfrak{X}$ and let $\mathcal{Y}$ be its domain. Here $A_\lambda = D_x f(0, \lambda) = L - \lambda I$, so this is an isomorphism precisely when λ is not in the spectrum of L. (This is the definition of the spectrum.) Thus, loosely speaking (and this is correct if A_λ has discrete spectrum), *bifurcation can occur only at eigenvalues of* L.

> ***Problem 1.1*** Verify that this criterion correctly predicts the bifurcation points in Figure 7.1.3.

It is desirable to have a more general definition of bifurcation point than 1.2, for bifurcations do not always occur off a known solution. The *limit point* in Figure 7.1.6 is an example; limit points also occur in Figure 7.1.8. Limit points are sometimes called *fold points*, *turning points*, or *saddle-node bifurcations* in the literature.

A general definition of bifurcation point suitable for our purposes is this: we call (x_0, λ_0) a *bifurcation point* of f if for every neighborhood $\mathfrak{U}$ of λ_0, and $\mathcal{V}$ of (x_0, λ_0), there are points λ_1 and λ_2 in $\mathfrak{U}$ such that the sets $\Sigma_{\lambda_1} \cap \mathcal{V}$ and $\Sigma_{\lambda_2} \cap \mathcal{V}$, where $\Sigma_\lambda = \{x \in \mathcal{Y} \mid f(x, \lambda) = 0\}$, are not homeomorphic (e.g., contain different numbers of points). However, there is a sense in which even this is not general enough; for example, consider $f(x, \lambda) = x^3 + \lambda^2 x = 0$. According to the above definitions this does not have a bifurcation point at $(0, 0)$. However, bifurcations do occur in slight perturbations of f (such as imperfections). For these reasons, some authors may wish to call any point where $D_x f$ is not an isomorphism a bifurcation point. It may be useful, however, to call it a *latent bifurcation point*.

We will now give a basic bifurcation theorem for $f: \mathbb{R} \times \mathbb{R} \longrightarrow \mathbb{R}$. Below we shall reduce a more general situation to this one. This theorem concerns the simplest case in which $(0, \lambda_0)$ could be a bifurcation point [so $(\partial f/\partial x)(0, \lambda_0)$ must vanish], $x = 0$ is a trival solution [$f(0, \lambda) = 0$ for all λ, so $(\partial f/\partial \lambda)(0, \lambda_0) = 0$], and in which f has some symmetry such as $f(x, \lambda) = -f(-x, \lambda)$, which forces $f_{xx}(0, \lambda) = 0$.

There are many proofs of this result available and the theorem has a long history going back to at least Poincaré. See Nirenberg [1974] for an alternative proof (using the Morse lemma) and Crandall and Rabinowitz [1971] and Iooss and Joseph [1980] for a "bare hands" proof. The proof we have selected is based on the method of Lie transforms—that is, finding a suitable coordinate change by integrating a differential equation. These ideas were discussed in Section 1.7 (see the proof of the Poincaré lemma in Box 7.2, Chapter 1). This method turns out to be one that generalizes most easily to complex situations.[1]

1.5 Theorem *Let $f: \mathbb{R} \times \mathbb{R} \longrightarrow \mathbb{R}$ be a smooth mapping and satisfy the following conditions:*

(i) $f(x_0, \lambda_0) = 0$, $f_x(x_0, \lambda_0) = 0$, $f_\lambda(x_0, \lambda_0) = 0$, *and* $f_{xx}(x_0, \lambda_0) = 0$; *and*
(ii) $f_{xxx}(x_0, \lambda_0) \neq 0$ *and* $f_{x\lambda}(x_0, \lambda_0) \neq 0$.

Then (x_0, λ_0) is a bifurcation point. In fact, there is a smooth change of coordinates in a neighborhood of (x_0, λ_0) of the form

$$x = \phi(\bar{x}, \lambda) \quad \textit{with} \quad \phi(0, \lambda_0) = x_0$$

and a smooth nowhere zero function $T(\bar{x}, \lambda)$ with $T(0, \lambda_0) = +1$ such that[2]

$$T(\bar{x}, \lambda) f(\phi(\bar{x}, \lambda), \lambda) = \bar{x}^3 \pm \lambda \bar{x}$$

with $\pm$ depending on the sign of $[f_{x\lambda}(x_0, \lambda_0) \cdot f_{xxx}(x_0, \lambda_0)]$. See Figure 7.1.9.

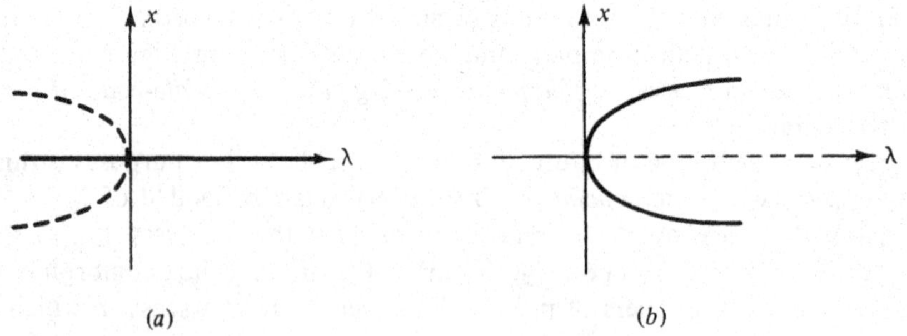

Figure 7.1.9 (a) The "+" case: $x^3 + \lambda x = 0$. (b) The "−" case: $x^3 - \lambda x = 0$.

[1]We thank M. Golubitsky for suggesting this proof.

[2]This kind of coordinate change (called *contact equivalence*), suggested by singularity theory, is the most general coordinate change preserving the structure of the zero set of f. See Box 1.1 for the general definitions.

Proof We can assume that $(x_0, \lambda_0) = (0, 0)$. By an initial rescaling and multiplication by -1 if necessary, we can assume that $f_{xxx}(0, 0) = 6$ and $f_{x\lambda}(0, 0) = \pm 1$, say $+1$. We seek a time-dependent family of coordinate transformations $\phi(\bar{x}, \lambda, t)$ and $T(\bar{x}, \lambda, t)$ $(0 \le t \le 1)$ such that

$$T(\bar{x}, \lambda, t)h(\phi(\bar{x}, \lambda, t), \lambda, t) = \bar{x}^3 + \lambda\bar{x} \equiv g(\bar{x}, \lambda), \tag{1}$$

where $h(x, \lambda, t) = (1 - t)g(x, \lambda) + tf(x, \lambda)$. If (1) can be satisfied, then at $t = 1$ it is the conclusion of the theorem. To solve (1), differentiate it in t:

$$\dot{T}h + T\dot{h} + Th_x\dot{\phi} = 0,$$

that is,

$$\frac{\dot{T}}{T}h + [f - g] + \dot{\phi}h_x = 0. \tag{2}$$

Now we need the following:

1.6 Lemma *Let $k(x, \lambda)$ be a smooth function of x and λ satisfying $k(0, 0) = 0$, $k_x(0, 0) = 0$, and $k_{xx}(0, 0) = 0$ and $k_\lambda(0, 0) = 0$. Then there are smooth functions $A(x, \lambda)$ and $B(x, \lambda)$ with B vanishing at $(0, 0)$ satisfying*

$$k(x, \lambda) = A(x, \lambda)(x^3 + \lambda x) + B(x, \lambda)(3x^2 + \lambda).$$

Moreover, if $k_{xxx}(0, 0) = 0$ and $k_{x\lambda}(0, 0) = 0$, then $A(0, 0) = 0$ and $B_x(0, 0) = 0$.

Proof By Taylor's theorem we can write

$$k(x, \lambda) = \lambda^2 a_1(x, \lambda) + x^3 b_1(x, \lambda) + x\lambda c_1(x, \lambda).$$

But
$$x^3 = \tfrac{1}{2}[x(3x^2 + \lambda) - (x^3 + \lambda x)],$$
$$\lambda x = \tfrac{1}{2}[3(x^3 + \lambda x) - x(3x^2 + \lambda)],$$
and
$$\lambda^2 = \lambda(3x^2 + \lambda) - 3x(\lambda x).$$

Substituting these expressions into the Taylor expansion for k gives the desired form for k. We have $B(0, 0) = k_\lambda(0, 0) = 0$ by assumption. If, in addition, $k_{xxx}(0, 0) = 0$ and if $k_{x\lambda}(0, 0) = 0$, then we can use Taylor's theorem to write

$$k(x, \lambda) = \lambda^2 a_2(x, \lambda) + x^4 b_2(x, \lambda) + x^2\lambda c_2(x, \lambda).$$

Substituting the above expressions for x^3 and $x\lambda$ into $x^4 = x \cdot x^3$ and $x^2\lambda = x \cdot x\lambda$ we get the desired form of k with $A(0, 0) = 0$ and $B(0, 0) = 0$. We then compute that $k_{xxx}(0, 0) = 6B_x(0, 0)$ and so $B_x(0, 0) = 0$ as well. ∎

1.7 Lemma (*Special Case of Nakayama's Lemma*) *Let $g(x, \lambda) = x^3 + \lambda x$ and $h(x, \lambda, t) = g(x, \lambda) + tp(x, \lambda)$, where $p(x, \lambda) = f(x, \lambda) - g(x, \lambda)$. Then for $0 \le t \le 1$ and (x, λ) in a neighborhood of $(0, 0)$, we can write $p(x, \lambda) = a(x, \lambda, t)h(x, \lambda, t) + b(x, \lambda, t)\, h_x(x, \lambda, t)$, where $a(0, 0, t) = 0$, $b(0, 0, t) = 0$* and $b_x(0, 0, t) = 0$.

Proof By 1.6 we can write $p(x, \lambda) = A(x, \lambda)g(x, \lambda) + B(x, \lambda)g_x(x, \lambda)$. Thus $h = (1 + tA)g + tBg_x$ and hence $h_x = tA_x g + (1 + tA + tB_x)g_x + tBg_{xx}$. Since

$B = B_x = 0$ at $(0, 0)$, 1.6 can be used to write $6xB = Bg_{xx} = Eg + Fg_x$. Thus, h_x has the form $h_x = tCg + (1 + tD)g_x$, where $D(0, 0) = 0$. Hence

$$\begin{pmatrix} h \\ h_x \end{pmatrix} = \begin{pmatrix} 1 + tA & tB \\ tC & 1 + tD \end{pmatrix} \begin{pmatrix} g \\ g_x \end{pmatrix}$$

At $(x, \lambda) = (0, 0)$ this matrix has the form $\begin{pmatrix} 1 & 0 \\ tC(0, 0) & 1 \end{pmatrix}$, so it is invertible in a neighborhood of $(0, 0)$. Hence g and g_x can be written as a linear combination of h and h_x. Substitution gives the result claimed. ∎

Let us now use 1.7 to solve (1) and (2). First find ϕ by solving the ordinary differential equation

$$\dot{\phi}(\bar{x}, \lambda, t) = -b(\phi(\bar{x}, \lambda, t), \lambda, t), \qquad \phi(x, \lambda, 0) = x.$$

This can be integrated for the whole interval $0 \leq t \leq 1$ in a neighborhood of $(0, 0)$ because b vanishes at $(0, 0)$. Next solve

$$\dot{T}(\bar{x}, \lambda, t) = -a(\phi(\bar{x}, \lambda, t), \lambda, t)T(\bar{x}, \lambda, t), \qquad T(x, \lambda, 0) = 1.$$

This is linear, so can be integrated to $t = 1$. This produces ϕ, T satisfying (2) and so, by integration, (1). Moreover, $a(0, 0, t) = 0$, so $T(0, 0, t) = 1$ and $b(0, 0, t) = 0$, $b_x(0, 0, t) = 0$ so $\phi(0, 0, t) = 0$, $\phi_x(0, 0, t) = 1$. Thus, the transformation is of the form $T(x, \lambda) = 1 +$ higher order terms and $\phi(x, \lambda) = x +$ higher order terms. ∎

One calls the function $g(x, \lambda) = x^3 \pm \lambda x$ into which f has been transformed, a *normal form*. The transformation of coordinates allowed preserves all the qualitative features we wish of bifurcation diagrams (note that the λ-variable was unaltered). Furthermore, once a function has been brought into normal form, the stability of the branches can be read off by a direct computation (stability in the context of the dynamical theory is discussed in Section 7.3 below). In Figure 7.1.9 note that the *subcritical* branch in (a) is unstable, while the *supercritical* branch in (b) is stable.

Problem 1.2 Let $f(0, 0) = 0$, $f_x(0, 0) = 0$ and $f_{xx}(0, 0) \neq 0$, $f_\lambda(0, 0) \neq 0$. Show that f has the normal form $x^2 \pm \lambda$ (limit point).

These techniques lead to the results shown in Table 7.1.1 classifying some of the simple cases in one variable. (The "index" equals the number of negative eigenvalues.) Methods of singularity theory, a special case of which was given in 1.5, allow one to do the same analysis for more complex bifurcation problems.

In Box 1.1 we describe the imperfection-sensitivity analysis of the pitchfork. Next, however, we shall describe how many bifurcation problems can be reduced to one of the above cases by means of the *Liapunov–Schmidt procedure*.

Suppose $f\colon \mathcal{Y} \times \Lambda \rightarrow \mathcal{X}$ is a smooth (or C^1) map of Banach spaces. Let $f(x_0, \lambda_0) = 0$ and suppose that (x_0, λ_0) is a candidate bifurcation point; thus the linear operator $A = A_{x_0} = D_x f(x_0, \lambda_0)\colon \mathcal{Y} \rightarrow \mathcal{X}$ will in general have a

Table 7.1.1

Defining Conditions at (0, 0)	Nondegeneracy Conditions at (0, 0)	Normal Form	Picture (+ case)	
(1) $f = f_x = 0$	$f_{xx} \neq 0, f_\lambda \neq 0$	$x^2 \pm \lambda$	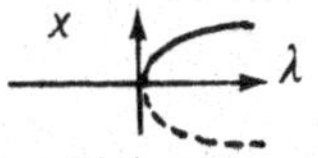	(limit point)
(2) $f = f_x = f_\lambda = 0$	$f_{xx} \neq 0$, $D^2 f$ has index 0 or 2	$x^2 + \lambda^2$	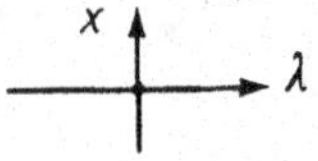	(isola)
(3) $f = f_x = f_\lambda = 0$	$f_{xx} \neq 0$, $D^2 f$ has index 1	$x^2 - \lambda^2$	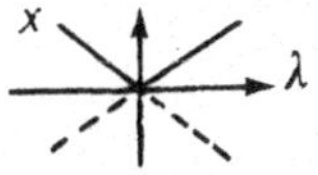	(transcritical bifurcation)
(4) $f = f_x = f_{xx} = 0$	$f_{xxx} \neq 0, f_\lambda \neq 0$	$x^3 \pm \lambda$	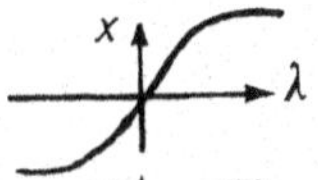	(hysteresis)
(5) $f = f_x = f_\lambda = f_{xx} = 0$	$f_{xxx} \neq 0, f_{x\lambda} \neq 0$	$x^3 \pm \lambda x$	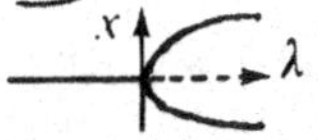	(pitchfork)

kernel Ker $A \neq \{0\}$ and a range Range $A \neq \mathfrak{X}$. Assume these spaces have closed complements. Keeping in mind the Fredholm alternative discussed in Section 6.1, let us write the complements in terms of adjoints even though they could be arbitrary at this point:

$$\mathcal{Y} = \text{Ker } A \oplus \text{Range } A^*,$$
$$\mathfrak{X} = \text{Range } A \oplus \text{Ker } A^*.$$

Recall that A is Fredholm when Ker A and Ker A^* are finite dimensional, for example, this is the case for the operator A of linear elastostatics; then A is actually self-adjoint: $A = A^*$.

Now let $\mathbb{P}: \mathfrak{X} \longrightarrow$ Range A denote the orthogonal projection to Range A and split up the equation $f(x, \lambda) = 0$ into two equations:

$$\mathbb{P}f(x, \lambda) = 0 \quad \text{and} \quad (I - \mathbb{P})f(x, \lambda) = 0.$$

The map $\mathbb{P}f(x, \lambda)$ takes $\mathcal{Y} \times \Lambda$ to Range A and has a surjective derivative at (x_0, λ_0). Therefore, by the implicit function theorem the set of solutions of $\mathbb{P}f(x, \lambda) = 0$ form the graph of a smooth mapping ψ: (a neighborhood of 0 in Ker A translated to x_0) $\times$ (a neighborhood of λ_0 in Λ) $\longrightarrow$ Range A^* (translated to x_0). See Figure 7.1.10. By construction, $\psi(x_0, \lambda_0) = (x_0, \lambda_0)$ and $D_u\psi(x_0, \lambda_0) = 0$ (u is the variable name in Ker A). This information can be substituted into the equation $(I - \mathbb{P})f(x, \lambda) = 0$ to produce the following theorem.

1.8 Theorem *The set of solutions of $f(x, \lambda) = 0$ equals, near (x_0, λ_0), the set of solutions of the bifurcation equation:*

$$(I - \mathbb{P})f((u, \psi(u, \lambda)), \lambda) = 0,$$

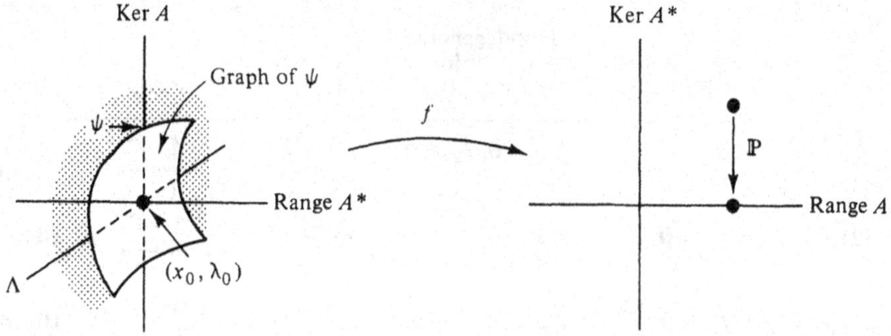

Figure 7.1.10

where ψ is implicitly defined by $\mathbb{P}f((u, \psi(u, \lambda)), \lambda) = 0$ *and where* $(u, \psi(u, \lambda)) \in \mathcal{Y} = \text{Ker } A \oplus \text{Range } A^* + \{x_0\}$.

Sometimes it is convenient to think of the Liapunov–Schmidt procedure this way: the equation $\mathbb{P}f(x, \lambda) = 0$ defines a smooth submanifold $\Sigma_{\mathbb{P}}$ of $\mathcal{Y} \times \Lambda$ (with tangent space Ker $A \oplus$ Ker $\mathbb{P}D_\lambda f(x_0, \lambda_0)$ at (x_0, λ_0)); the bifurcation equation is just the equation $(I - \mathbb{P})f \mid \Sigma_{\mathbb{P}} = 0$. For computations it is usually most convenient to actually realize $\Sigma_{\mathbb{P}}$ as a graph, as in 1.7, but for some abstract considerations the manifold picture can be useful (such as the following: if the original equation has a compact symmetry group, so does the bifurcation equation). Sometimes f is to be thought of as a vector field, depending parametrically on λ. This suggests replacing $\Sigma_{\mathbb{P}}$ by a manifold C tangent to Ker A and such that f is everywhere tangent to C. The bifurcation equation now is just $f \mid C = 0$. This has the advantage that if f is a gradient, so is the bifurcation equation. The manifold C is called a *center manifold* and is discussed in Section 7-4. The relationship between the center manifold and Liapunov–Schmidt approaches is discussed there and in Chow and Hale [1982] and in Schaeffer and Golubitsky [1981]. In Rabinowitz [1977a] it is shown how to preserve the gradient character directly in the Liapunov–Schmidt procedure. Another closely related procedure is the "splitting lemma" of Gromoll and Meyer; cf. Golubitsky and Marsden [1983].

Let us now apply the Liapunov–Schmidt procedure to the pitchfork. This is called *bifurcation at a simple eigenvalue* for reasons that will be explained below. (See Golubitsky and Schaeffer [1984], Ch. 4 for a generalization.)

1.9 Proposition *Assume* $f: \mathcal{Y} \times \mathbb{R} \to \mathcal{X}$ *is smooth,* $f(x_0, \lambda_0) = 0$, *and:*

(i) dim Ker $A = 1$, dim Ker $A^* = 1$;
(ii) $D_\lambda f(x_0, \lambda_0) = 0$, *and* $D_x^2 f(x_0, \lambda_0) \cdot (u_0, u_0) = 0$, *where* u_0 *spans* Ker A;

(iii)[3] $\langle (D_x^3 f(x_0, \lambda_0)(u_0, u_0, u_0)), v_0 \rangle \neq 0$ *and* $\langle (D_x D_\lambda f(x_0, \lambda_0) \cdot u_0), v_0 \rangle \neq 0$, *where* v_0 *spans* Ker A.

Then near $(x_0. \lambda_0)$, *the set of solutions of* $f(x, \lambda) = 0$ *consists of a pitchfork lying in a two-dimensional submanifold of* $\mathcal{Y} \times \mathbb{R}$.

Proof We can suppose that $x_0 = 0$ and $\lambda_0 = 0$. Identify Ker A and Ker A^* with $\mathbb{R}$ by writing elements of Ker A as $u = zu_0$, $z \in \mathbb{R}$ and elements of Ker A^* as wv_0, $w \in \mathbb{R}$. Define $F: \mathbb{R} \times \mathbb{R} \to \mathbb{R}$ by $F(z, \lambda) = \langle f(zu_0 + \psi(zu_0, \lambda), \lambda), v_0 \rangle$. By the Liapunov-Schmidt procedure, it suffices to verify the hypotheses of 1.5 for F. Since $\psi(0, 0) = 0$ and $f(0, 0) = 0$, clearly $F(0, 0) = 0$. Also

$$F_z(z, \lambda) = \langle (D_x f(zu_0 + \psi(zu_0, \lambda), \lambda) \cdot (u_0 + D_u \psi(zu_0, \lambda) \cdot u_0), v_0 \rangle,$$

which vanishes at (0, 0) since v_0 is orthogonal to the range of $D_x f$. Similarly,

$$F_\lambda(z, \lambda) = \langle D_\lambda f(zu_0 + \psi(zu_0, \lambda), \lambda), v_0 \rangle + \langle D_x f(zu_0 + \psi(zu_0, \lambda), \lambda) \cdot D_x \psi(zu_0, \lambda), v_0 \rangle \qquad (3)$$

vanishes at (0, 0) since $\langle D_\lambda f(0, 0), v_0 \rangle = 0$ and v_0 is orthogonal to the range of $D_x f(0, 0)$. Next,

$$F_{zz}(z, \lambda) = \langle D_x^2 f(zu_0 + \psi(zu_0, \lambda), \lambda) \cdot [u_0 + D_u \psi(zu_0, \lambda) \cdot u_0]^2, v_0 \rangle + \langle D_x f(zu_0 + \psi(zu_0, \lambda), \lambda) \cdot D_u^2 \psi(zu_0, \lambda) \cdot u_0^2, v_0 \rangle \qquad (4)$$

At (0, 0) the first term vanishes since $D_u \psi(0, 0) = 0$ and by our hypothesis on $D_x^2 f$. The second term vanishes since v_0 is orthogonal to the range of $D_x f(0, 0)$. By implicit differentiation, note that

$$\mathbb{P} D_x f(u + \psi(u, \lambda), \lambda) \cdot (w + D_u \psi(u, \lambda) \cdot w) = 0$$

so differentiating in u again,

$$\mathbb{P} D_x^2 f(u + \psi(u, \lambda), \lambda) \cdot [w + D_u \psi(u, \lambda) \cdot w]^2 + \mathbb{P} D_x f(u + \psi(u, \lambda), \lambda) \cdot D_u^2 \psi(u, \lambda) \cdot [w]^2 = 0.$$

Thus, $D_u^2 \psi(0, 0) = 0$ since $D_x^2 f(0, 0) = 0$ by assumption. Similarly from $D_\lambda f(0, 0) = 0$ we get $D_\lambda \psi(0, 0) = 0$. Therefore, we compute from (3) and (4),

$$F_{z\lambda}(0, 0) = \langle D_{z\lambda} f(0, 0) \cdot u_0, v_0 \rangle \neq 0$$

and

$$F_{xxx}(0,) = \langle D_x^3 f(0, 0) \cdot [u_0]^3, v_0 \rangle \neq 0. \blacksquare$$

Condition (ii) is sometimes associated with the assumption of $\mathbb{Z}_2$-symmetry $f(-x, \lambda) = -f(x, \lambda)$.

1.10 Example Suppose $f(x, \lambda) = Lx - \lambda x + g(x, \lambda)$, as in 1.4. Suppose $L^* = L$ and λ_0 is a *simple* non-zero eigenvalue of L. Then Ker A = Ker A^* =

---

[3] In the abstract context, $\langle \alpha, v \rangle = (I - \mathbb{P})\alpha$, but in actual problems, where A is an elliptic operator, $\mathbb{P}$ is the L^2-orthogonal projection and $\langle , \rangle$ is the L^2-inner product.

span u_0, where u_0 satisfies $Lu_0 = \lambda u_0$. Then condition (i) of 1.9 holds. Condition (ii) holds if, for example, $g(-x, \lambda) = -g(x, \lambda)$, and condition (iii) holds if $\langle D_x^3 g(0, \lambda_0)\cdot[u_0^3], u_0\rangle \neq 0$ (i.e., the leading term in g is a nontrivial cubic term). Note that $\langle D_x D_\lambda f(0, \lambda_0)\cdot u_0, v_0\rangle = -\lambda_0 \|u_0\|^2 \neq 0$ automatically. For instance, these conditions apply to the problem

$$\Delta\phi - \lambda\phi + \lambda\phi^3 = 0$$

on a region $\Omega \subset \mathbb{R}^3$, where Dirichlet boundary conditions hold, choosing $\mathcal{X} = L^2(\Omega)$, $\mathcal{Y} = H_0^2(\Omega)$, provided λ_0 is a simple eigenvalue for Δ on Ω.[4]

Problem 1.3 For the Euler beam, prove that a pitchfork bifurcation occurs at $\lambda/EI = k^2\pi^2$ ($k = 1, 2, \ldots$); see Figure 7.1.3.

Problem 1.4 Derive criteria for a transcritical bifurcation in a Banach space under condition (i) of 1.9 by imposing conditions (3) of Table 7.1.1 on the bifurcation equation.

For bifurcation at multiple eigenvalues one can in principle go through similar procedures, although the algebra becomes more complex. There are a number of observations to be made concerning multiple eigenvalues.

1.11 Remarks

(a) Bifurcation at multiple eigenvalues is often associated with symmetries of f and of the bifurcation point, just as the pitchfork is associated with the reflection symmetry $f(-x, \lambda) = -f(x, \lambda)$. When studying the bifurcation problem or imperfection-sensitivity analysis of it, this symmetry group must be taken into account. (See Golubitsky and Schaeffer [1979b] and Box 1.1 for more information.)

(b) Many problems of *secondary bifurcation* (further branching, appearance of limit points, etc.) can be dealt with by perturbing a bifurcation problem with a multiple eigenvalue (see, e.g., Bauer, Keller and Reiss [1975], Chow, Hale, and Mallet-Paret [1975], and Golubitsky and Schaeffer [1979a, b]; some examples are sketched in Section 7.2).

(c) Simple rules for the pitchfork such as "supercritical branches are stable" do not necessarily apply at multiple eigenvalues (see, e.g., McLeod and Sattinger [1973]). However, this information can usually be filled in using symmetries, considering the perturbed situation and applying the simple rules (from Table 7.1.1) to its component parts; see Schaeffer and Golubitsky [1981] and Section 7.2 for examples.

(d) See Nirenberg [1974] for a simple example of a problem with a double eigenvalue for which no bifurcation occurs; it is, however, a "latent" bifurcation point. Using degree theory, Krasnoselskii [1964] has shown, under some hypotheses, that an eigenvalue of *odd* multiplicity

[4]The first eigenvalue of a self-adjoint elliptic operator on scalars is always simple.

of a problem of the form

$$f(x, \lambda) = x - \lambda Tx + g(x, \lambda),$$

where T and g are compact, is a bifurcation point. This applies to Example 1.9 if L has a compact inverse by rewriting $Lx - \lambda x + g(x, \lambda) = 0$ as $x - \lambda L^{-1}x + L^{-1}g(x, \lambda) = 0$. Notice, however, that the details of the bifurcation (how many branches, their stability and structural stability) require further analysis. Krasnoselskii's theorem is discussed in Box 1.2. Conditions for an eigenvalue of *even* multiplicity to be a bifurcation point are given in Buchner, Marsden, and Schechter [1982] and references therein.

For problems in elasticity, most applications have been made to rods, plates, and shells as we shall outline in Section 7.2. Three-dimensional elasticity problems for pure displacement can be dealt with by the techniques outlined above, although no seminal examples have been computed. Most interesting examples are pure traction or have mixed boundary conditions. The mixed case is complicated by technical problems (with function spaces). Some basic examples for the pure traction problem are, however, available. Rivlin's example of homogeneous deformations of an incompressible cube will be given in Section 7.2 and the Signorini–Stoppelli problem of a natural state subjected to small loads will occupy the whole of 7.3. These problems both require great care with the symmetry group.

Box 1.1 ***Imperfection-Sensitivity Analysis of the Pitchfork***

That small imperfections can perturb a bifurcation diagram and, for example, bring about the onset of buckling significantly earlier than that predicted by the ideal theory has a long history in the engineering literature, going back at least to Koiter [1945]. For accounts in the engineering literature, Ziegler [1968] and references therein may be consulted.

Recent history developed along three more or less separate lines; in pure mathematics, the subject of catastrophe theory and more generally singularity theory was developed, starting with R. Thom around 1955, and is now a large subject; see, for example, Golubitsky and Guillemin [1973]. Secondly, in applied mathematics, the subject of perturbed bifurcation theory was developed by many authors, such as Keener and Keller [1973] and Benjamin [1978]. Thirdly, in engineering the subject was developed by Roorda [1965], Sewell [1966b], and Thompson and Hunt [1973], [1975]. These three lines of development are now merging through the works of people in all these branches; the papers of Chow,

Hale, and Mallet-Paret [1975] and Golubitsky and Schaeffer [1979a] have been especially important in making the unification. We shall present a few ideas of Golubitsky and Schaeffer [1979a] to indicate how the modern theory goes. Their theory differs from previous works in that they distinguish between bifurcation parameters and imperfection parameters; for example, catastrophe theory does not make this distinction explicit. This important point was already hinted at in the work of Thompson and Hunt.

The mathematical theory centers around two notions, called *contact equivalence* and *universal unfolding.* Let $f: \mathbb{R}^n \times \mathbb{R} \to \mathbb{R}^m$ be given and suppose f is C^∞ and $f(0, 0) = 0$. For example, f may be the map obtained from the bifurcation equation in the Liapunov–Schmidt procedure. Everything will be restricted to a small neighborhood of $(0, 0)$ without explicit mention.

1.12 Definition We say f_1 and f_2 are *contact equivalent* at $(0, 0)$ if there is a (local) diffeomorphism of $\mathbb{R}^n \times \mathbb{R}$ to itself of the form $(x, \lambda) \mapsto (\phi(x, \lambda), \Lambda(\lambda))$ such that $\phi(0, 0) = 0$, $\Lambda(0) = 0$, and a (smooth, local) map $(x, \lambda) \mapsto T(x, \lambda)$ from $\mathbb{R}^n \times \mathbb{R}$ to the invertible $m \times m$ matrices[5] such that

$$f_1(x, \lambda) = T(x, \lambda)\cdot f_2(\phi(x, \lambda), \Lambda(\lambda)).$$

Notice that the change of coordinates on $\mathbb{R}^n \times \mathbb{R}$ maps sets on which $\lambda =$ constant, to themselves. In this sense, this notion of equivalence recognizes the special role played by the bifurcation parameter, λ. It should be clear that the zero sets of f_1 and f_2 can then be said to have the "same" bifurcation diagram. See Figure 7.1.11.

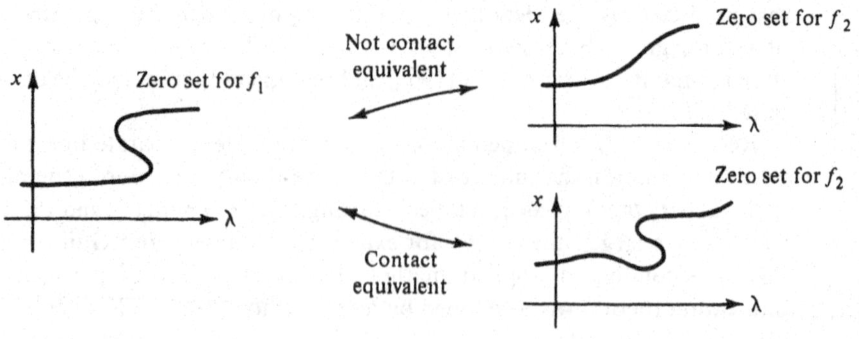

Figure 7.1.11

[5]Allowing *nonlinear* changes of coordinates on the range turns out not to increase the generality (cf. Golubitsky and Schaeffer [1979a], p. 23).

We can rephrase Theorem 1.5 by saying that if $n = m = 1$ and $f(0, 0) = 0$, $f_x(0, 0) = 0$, $f_\lambda(0, 0) = 0$, $f_{xx}(0, 0) = 0$, and $f_{xxx}(0, 0) \times f_{x\lambda}(0, 0) \neq 0$, then f is contact equivalent to $g(x, \lambda) = x^3 \pm \lambda x$.

Now we consider perturbations (or imperfections) of f.

1.13 Definition Let $f: \mathbb{R}^n \times \mathbb{R} \to \mathbb{R}^m$ be smooth and $f(0, 0) = 0$. An l-parameter *unfolding* of f is a smooth map $F: \mathbb{R}^n \times \mathbb{R} \times \mathbb{R}^l \to \mathbb{R}^m$ such that $F(x, \lambda, 0) = f(x, \lambda)$ for all x, λ (in a neighborhood of $(0, 0)$).

Let F_1 be an l_1-parameter unfolding of f and F_2 be an l_2-parameter unfolding. We say that F_1 *factors through* F_2 if there is a smooth map $\psi: \mathbb{R}^{l_1} \to \mathbb{R}^{l_2}$ such that for every $\beta \in \mathbb{R}^{l_1}$, $F_1(\cdot, \cdot, \beta)$ (i.e., β is held fixed) is contact equivalent to $F_2(\cdot, \cdot, \psi(\beta))$.

An l-parameter unfolding F of f is called a *universal unfolding of f* if every unfolding of f factors through F.

Roughly speaking, a universal unfolding F is a perturbation of f with l extra parameters that captures *all* possible perturbations of the bifurcation diagram of f (up to contact equivalence). Thus, if one can find F, one has solved the problem of imperfection-sensitivity of the bifurcation diagram for f. The number of *extra* parameters l required is unique and is called the *codimension* of f.

The complete theory for how to compute the universal unfolding would require too much space for us to go into here; see Golubitsky and Schaeffer [1979a]. However, we can indicate what is going on for the pitchfork. If we return to the proof of 1.5, we see that a general unfolding of $g(x, \lambda) = x^3 \pm \lambda x$ will have a Taylor expansion of the form

$$F(x, \lambda, \alpha_1, \ldots, \alpha_l) = x^3 \pm \lambda x + \alpha_1 + \alpha_2 x + \alpha_3 \lambda + \alpha_4 x^2 + \alpha_5 \lambda^2 + \alpha_6 x^2 \lambda + \alpha_7 x \lambda^2 + \alpha_8 \lambda^3 + \text{Remainder}.$$

A more difficult argument than the one given in the proof of 1.5 (though similar in spirit) shows that under contact equivalence, we can transform away all the terms except $\alpha_4 x^2$ and α_1 (these, roughly, correspond to the fact that before, we had $f(0, 0) = 0$ and $f_{xx}(0, 0) = 0$, so these terms were absent in its Taylor expansion).[6] This is the idea of the method behind the proof of the following.

1.14 Proposition

(a) *A universal unfolding of* $x^3 \pm \lambda x$ *is* $F(x, \lambda, \alpha, \beta) = x^3 + \beta x^2 \pm \lambda x + \alpha$.

[6]A subtlety is that after transformation the new α_4 will depend on the old α_4 and α_3. To properly deal with the Taylor expansion in this case requires the "Malgrange preparation theorem."

(b) *Let $f(x, \lambda)$ satisfy the hypotheses of* 1.5 *and let $F(x, \lambda, a, b)$ be a two-parameter unfolding of f. Then F is universal if*

$$\det\begin{bmatrix} 0 & 0 & f_{x\lambda} & f_{xxx} \\ 0 & f_{\lambda x} & f_{\lambda\lambda} & f_{\lambda xx} \\ F_a & F_{ax} & F_{a\lambda} & F_{axx} \\ F_b & F_{bx} & F_{b\lambda} & F_{bxx} \end{bmatrix} \neq 0,$$

Problem 1.5 Show that another universal unfolding of the pitchfork is $x^3 - \lambda x + \beta\lambda + \alpha$.

Part (b) of 1.14 is useful since one may wish to put on a variety of imperfections. For example, in the buckling of a beam one may wish to give it slight inhomogeneities, a slight transverse loading, and so on. The criterion above guarantees that one has enough extra parameters.

The perturbed bifurcation diagrams that go with the universal unfolding F in 1.14(a) are shown in Figure 7.1.12. Note that transcritical bifurcations and hysteresis are included, unlike Figure 7.1.8.

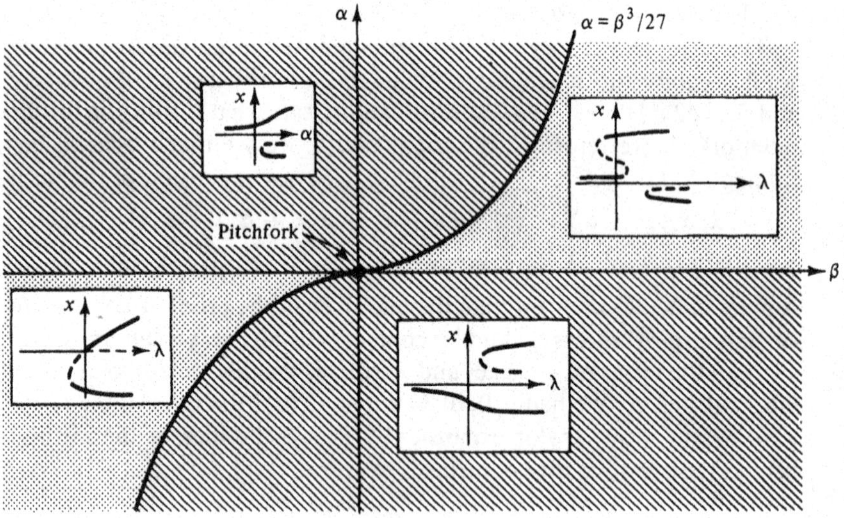

Figure 7.1.12

Problem 1.6 Show that the hysteresis in Figure 7.1.12 can be obtained by passing through the cusp (Fig. 7.1.8) along various lines for 1.14(a) and straight lines through the origin in Problem

1.5. (See Golubitsky and Schaeffer [1979a], p. 53 for the answer to the first part.)

Problem 1.7 Write an essay on imperfection-sensitivity analysis of the Euler beam using Zeeman [1976] and Golubitsky and Schaeffer [1979a]. Utilize the function spaces from Chapter 6.

Box 1.2 ***Remarks on Global Bifurcation***

There are some results on global bifurcation available that are useful in elasticity. The main result is a globalization of Krasnoselski's theorem mentioned above due to Rabinowitz [1971]. There have been important variants (useful for operators preserving positivity) due to Dancer [1973] and Turner [1971]. Most of the applications in elasticity under realistic global assumptions are due to Antman and are described in the next section. However, there are also a number of other intersting applications to, for instance, solitary water waves by Keady and Norbury [1978] and by Amick and Toland [1981].

We shall just state the results; the works of Nirenberg [1974] and Ize [1976] should be consulted for proofs. It is to be noted that *global* imperfection-sensitivity results are not available (to our knowledge).

One considers mappings of $\mathfrak{X} \times \mathbb{R}$ to $\mathfrak{X}$ of the form

$$f(x, \lambda) = x - \lambda Tx + g(x, \lambda),$$

where $T\colon \mathfrak{X} \longrightarrow \mathfrak{X}$ is compact, g is compact and $g(x, \lambda) = o(\|x\|)$, uniformly on compact λ-intervals. The proof of the following theorem is based on the notion of topological degree.

1.15 Theorem (*Krasnoselskii* [1964]) *If* $1/\lambda_0$ *is an eigenvalue of* T *of odd multiplicity, then* $(0, \lambda_0)$ *is a bifurcation point.*

Let $\mathcal{S} = \{(x, \lambda) \mid f(x, \lambda) = 0 \text{ and } x \neq 0\} \cup (0, 1/\lambda_0)$ (the nontrivial solutions) and let $\mathcal{C}$ be the maximal connected subset of $\mathcal{S}$ containing $(0, 1/\lambda_0)$. The theorem of Rabinowitz basically states that $\mathcal{C}$ cannot "end in mid-air."

1.16 Theorem (*Rabinowitz* [1971]) *Let* λ_0 *and* $\mathcal{C}$ *be as above. Then either* $\mathcal{C}$ *is unbounded or it intersects the* λ*-axis at a finite number of points* $0, 1/\lambda_j$*, where* λ_j *are eigenvalues of* T*; the number of* λ_j *with odd multiplicity is even.*

The two alternatives are shown schematically, in Figure 7.1.13.

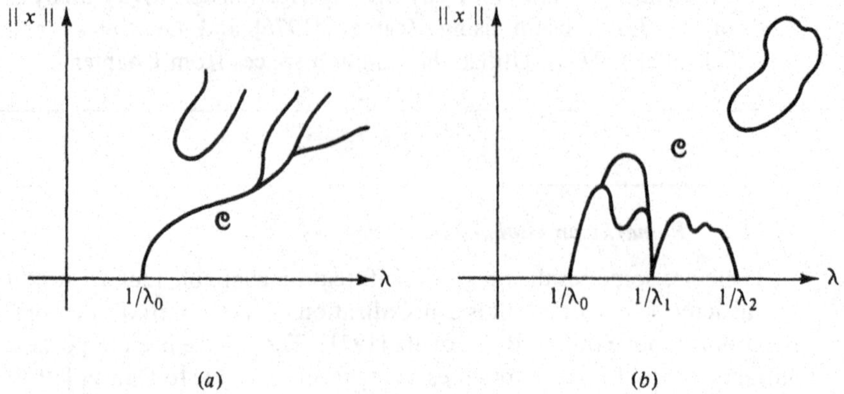

Figure 7.1.13 (a) $\mathcal{C}$ unbounded. (b) $\mathcal{C}$ returns to λ-axis.

Similar results for dynamic bifurcations (see Section 7.3) have been obtained by Alexander and Yorke [1978] and Chow and Mallet-Paret [1978].

Box 1.3 ***Summary of Important Formulas for Section 7.1***

Necessary Condition for Bifurcation

The necessary condition for bifurcation of $f(x, \lambda) = 0$ from a trivial solution $x = 0$ at λ_0 is that $D_x f(0, \lambda_0)$ not be an isomorphism.

Pitchfork Bifurcation

If $f(x, \lambda)$, $x \in \mathbb{R}$, $\lambda \in \mathbb{R}$ satisfies $f(x_0, \lambda_0) = 0$, $f_x(x_0, \lambda_0) = 0$, $f_\lambda(x_0, \lambda_0) = 0, f_{xx}(x_0, \lambda_0) = 0$, and $f_{xxx}(x_0, \lambda_0)\cdot f_{x\lambda}(x_0, \lambda_0) \neq 0$, then the zero set of f near (x_0, λ_0) is a pitchfork: f looks like $x^3 \pm \lambda x$ near $(0, 0)$.

Imperfection Sensitivity

The imperfection sensitivity analysis of $x^3 \pm \lambda x$ requires two extra imperfection parameters and is completely described by $F(x, \lambda, \alpha, \beta) = x^3 \pm \lambda x + \beta x^2 + \alpha$.

Liapunov Schmidt Procedure

If $\mathbb{P}$ is the projection onto Range $D_x f(x_0, \lambda_0)$, then solve $f(x, \lambda) = 0$ by solving $\mathbb{P} f(x, \lambda) = 0$ implicitly for $x = u + \phi(u, \lambda)$, $u \in$ Ker

$D_x f(x_0, \lambda_0)$, and substituting into $(I - \mathbb{P})f(x, \lambda) = 0$. The resulting equation, $(I - \mathbb{P})f(u + \phi(u, \lambda), \lambda)) = 0$ is the bifurcation equation. The pitchfork criterion may be applied to this if dim Ker $D_x f(x_0, \lambda_0) = 1$ and if $D_x f(x_0, \lambda_0)$ is self-adjoint.

7.2 A SURVEY OF SOME APPLICATIONS TO ELASTOSTATICS

This section is divided into three parts. First of all we present a basic example due to Rivlin. This concerns bifurcations that occur in an incompressible cube subject to a uniform tension on its faces. This is of interest because it is one of the few *three*-dimensional examples that can be computed explicitly. Furthermore, it is a seminal example for seeing how imperfection-sensitivity and symmetry can affect examples. We recommend reading Section 4 of the introductory chapter to review the context of the example. Secondly, we shall review *some* of the literature on the buckling of rods, plates, and shells. This literature is vast and our review is selective and biased towards the papers relevant to those current theoretical research directions that we know about and think are the most promising. Thirdly, we discuss (in Boxes 2.1 to 2.3) the following three points in conjunction with examples:

1. global versus local bifurcation analyses and exact verses approximate theories;
2. imperfection-sensitivity (Are there enough parameters?);
3. the role of symmetry.

In the next section we give a relatively detailed discussion of an important example: the traction problem near an unstressed state. This example was chosen for its interest to us and because it is in line with our emphasis in this book on *three*-dimensional problems. However, it might be of benefit to some readers to replace it by one of the examples mentioned in this section's survey, depending on interest.

We begin now with a discussion of Rivlin's [1948b] example of homogeneous deformations of a cube of incompressible neo-Hookean material. We thank John Ball and David Schaeffer for their help with this problem. The (dead load) traction problem is considered. The prescribed traction $\boldsymbol{\tau}$ is normal to each face of the cube with a magnitude τ, the same for each face, as in Figure 7.2.1.

We take a stored energy function for a homogeneous isotropic hyperelastic material; that is, of the form

$$W(\boldsymbol{F}) = \Phi(\lambda_1, \lambda_2, \lambda_3),$$

where $\lambda_1, \lambda_2, \lambda_3$ are the principal stretches and Φ is a symmetric function of

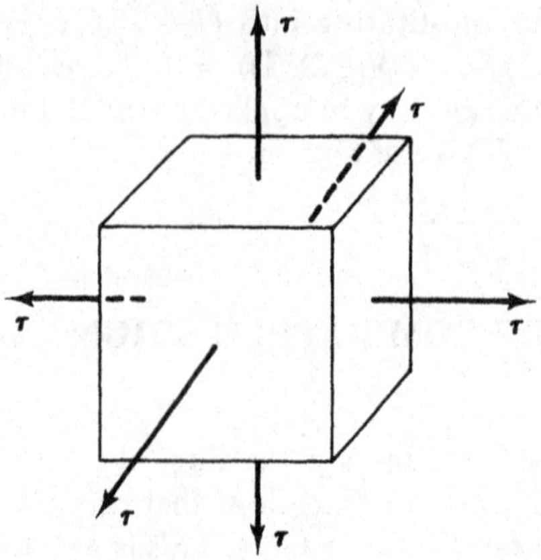

Figure 7.2.1

$\lambda_1, \lambda_2, \lambda_3$. Recall that the first Piola–Kirchhoff stress tensor P is given by

$$P_a{}^A = \rho_{\mathrm{Ref}} \frac{\partial W}{\partial F^a{}_A}.$$

We shall choose $\rho_{\mathrm{Ref}} = 1$.

Place the center of the cube at the origin and consider homogeneous deformations; that is, $x = F \cdot X$, where F is a constant 3×3 matrix. In particular, we seek solutions with $F = \mathrm{diag}(\lambda_1, \lambda_2, \lambda_3)$ relative to a rectangular coordinate system whose axes coincide with the axes of the block; the spatial and material coordinate systems are coincident. (This turns out to be the most general homogeneous solution; cf. Problem 2.2 below.)

> ***Problem 2.1*** Introduce an off diagonal entry δ into F and show that the new principal stretches $\bar{\lambda}_1, \bar{\lambda}_2, \bar{\lambda}_3$ satisfy $\bar{\lambda}_1^2 + \bar{\lambda}_2^2 + \bar{\lambda}_3^2 = \lambda_1^2 + \lambda_2^2 + \lambda_3^2 + \delta^2$. Show that $\partial \bar{\lambda}_i / \partial \delta = 0$ at $\delta = 0$ $(i = 1, 2, 3)$. Conclude that P is diagonal.

Since P is diagonal by this problem, we find that

$$P = \mathrm{diag}\left(\frac{\partial \Phi}{\partial \lambda_1}, \frac{\partial \Phi}{\partial \lambda_2}, \frac{\partial \Phi}{\partial \lambda_3}\right).$$

For a neo-Hookean material,

$$\Phi = \alpha(\lambda_1^2 + \lambda_2^2 + \lambda_3^2 - 3), \quad \alpha > 0 \text{ a constant.}$$

The equilbrium equations for an incompressible material are obtained from the usual ones by replacing P by $P - pF^{-1}$, where p is the pressure, to be determined from the incompressibility condition $J = 1$; that is, $\lambda_1\lambda_2\lambda_3 = 1$. Thus we must have

$$\mathrm{DIV}(P - pF^{-1}) = 0 \quad \text{in} \quad \mathcal{B}$$

and

$$(P - pF^{-1}) \cdot N = \tau \quad \text{on} \quad \partial\mathcal{B}.$$

We must solve these for the unknowns $\lambda_1, \lambda_2, \lambda_3$. For $W = \Phi$, these equations read

$$\frac{\partial}{\partial X^i}\left(\frac{\partial \Phi}{\partial \lambda_i} - \frac{p}{\lambda_i}\right) = 0 \quad \text{in} \quad \mathcal{B}$$

and

$$\frac{\partial \Phi}{\partial \lambda_i} - \frac{p}{\lambda_i} = \tau \text{ on the face along the } i\text{th axis} \quad (i = 1, 2, 3).$$

For a neo-Hookean material, $\partial\Phi/\partial\lambda_i = 2\alpha\lambda_i$, a constant, so the first equation is equivalent to the assertion that p is a constant in $\mathcal{B}$. The second equation becomes

$$2\alpha\lambda_1^2 - p = \tau\lambda_1,$$
$$2\alpha\lambda_2^2 - p = \tau\lambda_2,$$
$$2\alpha\lambda_3^2 - p = \tau\lambda_3.$$

Eliminating p gives

$$[2\alpha(\lambda_1 + \lambda_2) - \tau](\lambda_1 - \lambda_2) = 0, \tag{1}$$
$$[2\alpha(\lambda_2 + \lambda_3) - \tau](\lambda_2 - \lambda_3) = 0, \tag{2}$$
$$[2\alpha(\lambda_3 + \lambda_1) - \tau](\lambda_3 - \lambda_1) = 0. \tag{3}$$

Case 1 The λ_i's are distinct. Then (1), (2), and (3) yield $\tau = 2\alpha(\lambda_1 + \lambda_2) = 2\alpha(\lambda_2 + \lambda_3) = 2\alpha(\lambda_3 + \lambda_1)$, which implies $\lambda_1 = \lambda_2 = \lambda_3$, a contradiction. Thus, *there are no solutions with the λ_i's distinct.*

Case 2 $\lambda_1 = \lambda_2 = \lambda_3$. Since $\lambda_1\lambda_2\lambda_3 = 1$, we get $\lambda_i = 1$ $(i = 1, 2, 3)$ (and $p = 2\alpha - \tau$). This is a solution for all α, the trivial one.

Case 3 Two λ_i's equal. Suppose $\lambda_2 = \lambda_3 = \lambda$, so $\lambda_1 = \lambda^{-2}$. Then (1) and (3) coincide, giving

$$2\alpha(\lambda^{-2} + \lambda) - \tau = 0.$$

Thus, we need to find the positive roots of the cubic

$$f(\lambda) = \lambda^3 - \frac{\tau}{2\alpha}\lambda^2 + 1 = 0.$$

Since $f(0) = 1$ and $f'(\lambda) = 3\lambda(\lambda - \tau/3\alpha)$, a positive root requires $\tau > 0$. There will be none if $f(\tau/3\alpha) > 0$, one if $f(\tau/3\alpha) = 0$, and two if $f(\tau/3\alpha) < 0$; see Figure 7.2.2.

Since $f(\tau/3\alpha) = -\frac{1}{2}(\tau/3\alpha)^3 + 1$, there are no positive roots if $\tau < 3\sqrt[3]{2}\,\alpha$, one if $\tau = 3\sqrt[3]{2}\,\alpha$, and two if $\tau > 3\sqrt[3]{2}\,\alpha$. The larger of these two positive roots is always greater than unity; the smaller is greater than unity or less than unity according as $3\sqrt[3]{2}\,\alpha < \tau < 4\alpha$ or $4\alpha < \tau$, respectively. These solutions are graphed in Figure 7.2.3, along with the trivial solution $\lambda_i = 1$, τ artibrary. Thus taking permutations of $\lambda_1, \lambda_2, \lambda_3$ into account, we get:

(a) *One solution*, namely, $\lambda_1 = \lambda_2 = \lambda_3 = 1$ *if* $\tau < 3\sqrt[3]{2}\,\alpha$.

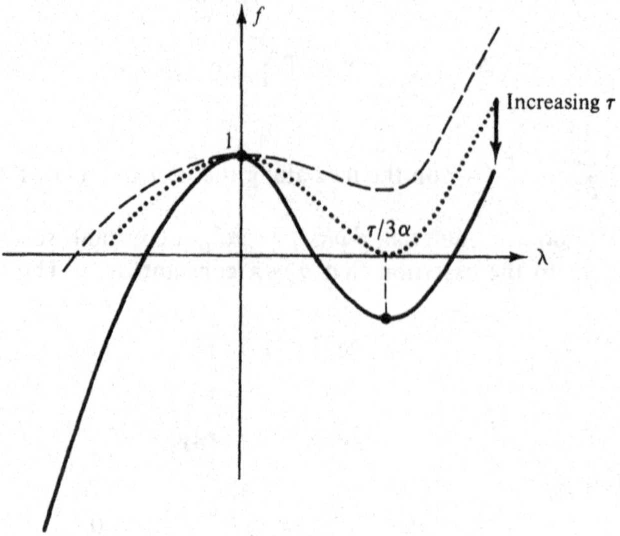

Figure 7.2.2

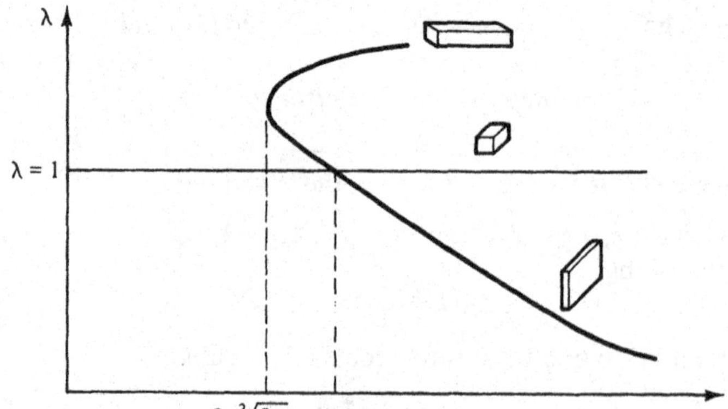

Figure 7.2.3

(b) *Four solutions* if $\tau = 3\sqrt[3]{2}\,\alpha$ or $\tau = 4\alpha$.
(c) *Seven solutions* if $\tau > 3\sqrt[3]{2}\,\alpha$, $\tau \neq 4\alpha$.

If we regard τ as a bifurcation parameter, we see that six new solutions are produced in "thin air" as τ crosses the critical value $\tau = 3\sqrt[3]{2}\,\alpha$. This is clearly a bifurcation phenomenon. Bifurcation of a more traditional sort occurs at $\tau = 4\alpha$. For unequal forces, see Sect. 7.3 and Sawyers [1976].

Rivlin [1948b], [1974b] shows that the trivial solution is stable for $0 \leq \tau/\alpha < 4$ and unstable for $\tau/\alpha > 4$; the trivial solution loses its stability when it is crossed by the nontrivial branch at $\tau = 4\alpha$. The three solutions corresponding to the larger root of f are always stable, and the three solutions corresponding to the smaller root are never stable. Beatty [1967b] established instability for $\tau < 0$.

Symmetry plays a crucial role in this problem. The two solutions found above led to six solutions when permutations of $\lambda_1, \lambda_2, \lambda_3$ were considered. This suggests that the basic symmetry group for the problem is S_3, and this is essentially correct—although the cube admits a much larger group of symmetries, most elements act trivially in the problem at hand, leaving only the group S_3. The same group and similar mathematics occurs in a convection problem studied by Golubitsky and Scheaffer [1981].

Because of the presence of this symmetry group, the transcritical bifurcation in Figure 7.2.3 at $\tau = 4\alpha$ is structurally stable. Without the symmetry the bifurcation would be imperfection sensitive; that is, a generic small perturbation would split the diagram into two distinct components. However, the bifurcation *cannot* be destroyed by a small perturbation that preserves the symmetry. Moreover, the usual rules about exchange of stability are completely modified by the symmetry. In particular, the nontrivial branch of solutions that crosses the trivial solution at $\tau = 4\alpha$ is *unstable both below and above* the bifurcation point.

Interesting new phenomena appear if a more general stored energy function is considered. Consider the Mooney–Rivlin material for which

$$\Phi(\lambda_1, \lambda_2, \lambda_3) = \alpha(\lambda_1^2 + \lambda_2^2 + \lambda_3^2 - 3) + \beta(\lambda_1^{-2} + \lambda_2^{-2} + \lambda_3^{-2} - 3),$$

where α and β are positive constants. (This reduces to the neo-Hookean case if $\beta = 0$.) If $0 < \beta/\alpha < \frac{1}{3}$, there are new fully asymmetric solutions of the equations (corresponding to Case 1 above) that bifurcate from the nontrivial symmetric solutions (Case 3). These new bifurcations cause some surprising changes of stability of the symmetric branches. As $\beta/\alpha \longrightarrow \frac{1}{3}$ from below, the fully asymmetric solutions collapse into the original bifurcation from the trivial solution, and for $\beta/\alpha > \frac{1}{3}$, they move off into the complex plane. This transition provides an example of a bifurcation problem that itself is structurally unstable but occurs stably in a one-parameter family of bifurcation problems; that is, it is of codimension one. See Ball and Schaeffer [1982] for details.

Problem 2.2 Consider the traction problem with $\boldsymbol{\tau} = \tau \boldsymbol{N}$ for a constant τ and an isotropic material. (a) Show that if ϕ_0 is a solution, then so is $\boldsymbol{Q}\circ\phi_0\circ\boldsymbol{Q}^{-1}$ for $\boldsymbol{Q} \in SO(3)$. (b) Conclude that nontrivial solutions can never be *strict* local minima of the energy (cf. Adeleke [1980]). {Thus, stability in this traction problem refers to neutral or conditional stability; see Ball and Schaeffer [1982] for more information.} (c) If $\phi_0(X) = \mathrm{diag}(\lambda^{-2}, \lambda, \lambda)(X)$ for $\lambda \neq 1$, show that (a) yields a set of solutions identifitable with $\mathbb{RP}^2$, real projective 2-space; i.e. the space of lines in $\mathbb{R}^3$.

Another example that can be worked out fairly explicitly is anti-plane shear, due to Knowles, and is described in Gurtin [1981a]. This example is important for light it may shed on the role of strong ellipticity and phase transitions. Phase transitions are of current interest in continuum mechanics (see Box 4.1, Chapter 6) but there seems to be a large gap to be bridged to the concepts one hears from physicists on the subject: symmetry breaking, renormalization group, and chaotic dynamics. The only thing one sees in common are energy functions with dimples, as in Figure 7.2.4. It is true that such pictures are common to the pitchfork bifurcation and its attendant symmetry breaking, to a loss of strong ellipticity, to chaotic dynamics, and to the Maxwell rule in thermodynamics. However, such observations are shallow. What are the deeper connections?

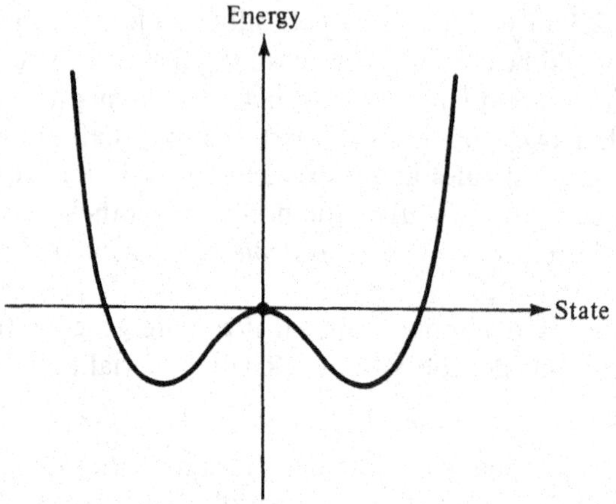

Figure 7.2.4

We turn next to surveying some of the literature on applications of bifurcation theory to elastostatics. Apart from Rivlin's example above, the Signorini–Stoppelli problem treated in the next section, and some semi-inverse methods (see below), virtually all the remaining examples are for rod, shell, and plate theories. The literature is massive and we can pretend to do no more than give a brief indication of some of the papers. We shall simply organize our selections in loosely grouped categories and make a few comments as we proceed. The following boxes discuss a few directions of current research and unifying treads. Of course it is impossible to be sure exactly in what directions current research will become most active.

Warning: Our list is not comprehensive, and many important papers are left out. We apologize for this, but a selection was necessary because of the massive undertaking involved.

Mathematical Theory. This in itself is a vast topic. The methods range from analytical to very geometric. Stackgold [1971], Sattinger [1973], Nirenberg

[1974], Berger [1977], Stuart [1979], and Iooss and Joseph [1980] are good places to begin. There the basic theorems on bifurcation at simple eigenvalues are proved by various methods, all different from ours in the previous section. Crandall and Rabinowitz [1971], [1973] is a standard reference, noted for clean and complete proofs. Nirenberg and Berger also discuss Kransoselski's [1964] basic results and methods based on topological degree. Nirenberg [1974] and Ize [1976] are good references for the global theory of Rabinowitz [1971].

The literature for bifurcation at multiple eigenvalues is more specialized, but still large. Some samples are McLeod and Sattinger [1973], Magnus [1976b], Shearer [1976, 7] and Marsden [1978], and Buchner, Marsden, and Schechter [1982].

Imperfection-sensitivity questions spawned a whole series of papers, especially recent ones using singularity theory. Three papers using classical methods that were among the first to exploit the fact that multiple eigenvalue problems can be very sensitive unless enough parameters are included and that including them can yield new solutions and explanations, are Keener and Keller [1973], Keener [1974], and Bauer, Keller, and Reiss [1975]. These new solutions are called "secondary bifurcations." Examples were given in the previous section. The first papers to attempt to systematically use singularity theory in bifurcation problems are those of Chow, Hale, and Mallet-Paret [1975], [1976]. See also Hale [1977]. This theory was advanced considerably by Golubitsky and Schaeffer [1979a, b], whose methods were outlined in the previous section.

Applications Using Analytic Methods. Again the literature here is massive. We shall only discuss some of the recent references. First, the collections Keller and Antman [1969] and Rabinowitz [1977b] contain many valuable articles. Also, there are numerous papers that were very important for the recent history such as Friedrichs [1941], Keller, Keller, and Reiss [1962] and Bauer and Reiss [1965].

Sturmian theory is used in Kolodner [1955] who analyzed the states of a rotating string. This was improved considerably by exploiting Rabinowitz' global theory as well as Sturmain theory by Antman [1980a].

For plate theory, Yanowitch [1956] is important for it is one of the first to discuss symmetry breaking. The von Karman and related equations are often used to model plates, and many papers are written on this, such as Knightly [1967], Wolkowski [1967] (circularly symmetric bifurcations), Berger and Fife [1967] (global analysis using calculus of variations and degree theory; see also Berger [1974]), Bauer, Keller, and Reiss [1970], Knightly and Sather [1970], [1974], [1975] (local analysis using the Liapunov–Schmidt procedure), and Matkowsky and Putnik [1974]. There are a number of similar papers for shell models, such as Knightly and Sather [1975], [1980], Sather [1977], and Shearer [1977].

In a series of important papers—Antman [1977], [1978a], [1979a], [1980c], Antman and Rosenfeld [1978], Antman and Nachman [1979], and Antman and Dunn [1980]—use Sturmian theory together with Rabinowitz' global theory to prove global bifurcation and preservation of nodal structure for geometrically exact models; see Box 2.1 below. This extends earlier work of, for example, Greenberg [1967]. Antman and Kenney [1981] use an extension of the Rabinowitz theory due to Alexander and Antman [1981] to study a two-parameter problem. Antman and Carbone [1977] show that shear- and necking-type bifurcations with hysteresis can occur within the context of hyperelasticity without plasticity-type assumptions. Maddocks [1982] considers non-planar configurations of the elastica.

A number of three-dimensional problems can be done by semi-inverse methods—that is, assuming symmetry and looking for solutions of a particular form. Perhaps the most famous is Antman [1978b], [1979b], where it is shown that thick spherical shells admit everted solutions (a tennis ball cut in half and then pushed inside out). Some barelling solutions in three dimensions for a traction displacement problem of a compressed cylinder were found by Simpson and Spector [1982].

Applications Using Imperfection-Sensitivity or Singularity Theory. Historically this really started with Koiter [1945], Roorda [1965], Sewell [1966b], and Thompson and Hunt [1973]. More systematic, but still "bare hands" methods were applied in Keener and Keller [1972] and Keller [1973].

Singularity theory was applied to the von Karman equations in Chow, Hale, and Mallet-Paret [1975]. Papers in a similar spirit are Mallet-Paret [1977] (inspired by Knightly and Sather [1975]), Magnus and Poston [1977], Vanderbauwhede [1978], Potier–Ferry [1979], Hunt [1982] and Golubitsky, Marsden and Schaeffer [1983].

A paper that explains mode jumping in the buckling of a rectangular plate is Schaeffer and Golubitsky [1980]. This is one of the most interesting uses of the imperfection-sensitivity approach in the previous section to a hard concrete example; see Box 2.2 below. For a survey of many other applications of a similar type, see Stewart [1981] and Thompson [1982].

Box 2.1 ***Global Bifurcation Analysis: Buckling of a Rod***

Usually a bifurcation analysis is called *global* when the structure of the solutions set is determined for all values of the parameter (or parameters) λ and the full range of the state variable x. If this determination is made only in a neighborhood of a given solution (x_0, λ_0), then the analysis is called *local.* In Section 7.1 we described some methods used in local bifurcation analysis and in Box 1.2 we mentioned some global techniques. Here we make a few additional remarks.

1. In the papers of Antman quoted above it is made clear that before one attempts a global bifurcation analysis, one should have a model

that is valid for large deformations. In particular, this is not the case for the von Karman equations and global results for them are of limited interest in the region of large deflections. In Antman's work geometrically exact models are used that are valid for large deformations.

2. For example, we briefly consider the deformations of nonlinearly elastic rods (Antman and Rosenfeld [1978], Antman [1980c]). The reference configuration is an interval $[s_1, s_2]$ on $\mathbb{R}$. A *configuration* is a map $\phi: [s_1, s_2] \rightarrow \mathbb{R}^3 \times S^2$ denoted $\phi(s) = (\boldsymbol{r}(s), \boldsymbol{d}_3(s))$. Here S^2 is the two-sphere and $\boldsymbol{d}_3 \in S^2$ represents the normal to a plane in $\mathbb{R}^3$ that describes shearing in the rod. See Figure 7.2.5, where we draw the rod with a thickness that has been suppressed in the mathematical model. Thus, we are considering rods that are capable of bending, elongating and shearing. One can also contemplate more complex situations allowing twisting

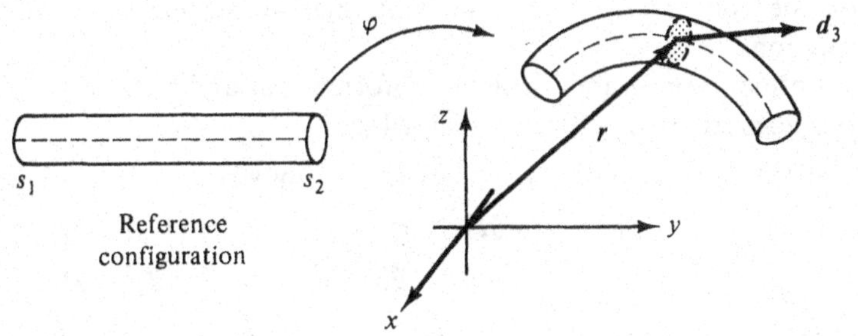

Figure 7.2.5

and necking. Note that bodies of this type fall into the general class of Cosserat continua, considered in Box 2.3, Chapter 2 with a reference $\boldsymbol{d}_3$ being $\boldsymbol{d}_3 = \boldsymbol{i}$, say. If we wished to take into account twisting, for example, it is not enough to specify $\boldsymbol{d}_3$, but we also need to specify twisting about $\boldsymbol{d}_3$ through some angle. For this situation one convenient way is to take a configuration to be a map $\phi: [s_1, s_2] \rightarrow \mathbb{R}^3 \times \mathbb{F}_3$, the oriented 3-frame bundle on $\mathbb{R}^3$; that is, $\phi(s)$ consists of a base point $\boldsymbol{r}(s)$ and an oriented orthonormal frame $(\boldsymbol{d}_1, \boldsymbol{d}_2, \boldsymbol{d}_3)$ at $\boldsymbol{r}(s)$. The plane of $\boldsymbol{d}_1$ and $\boldsymbol{d}_2$ (normal to $\boldsymbol{d}_3$) gives the shearing and the orientation of $\boldsymbol{d}_1$ and $\boldsymbol{d}_2$ within this plane gives the twisting. (Again the Cosserat theory requires a reference section of the frame bundle, which we can take to be the standard frame $(\boldsymbol{i}, \boldsymbol{j}, \boldsymbol{k})$.) In that notation, $\mathcal{B} = [s_1, s_2]$ and $\mathcal{S} = \mathbb{R}^3 \times \mathbb{F}_3$. The equation $\boldsymbol{d}_3 = \boldsymbol{d}_1 \times \boldsymbol{d}_2$ of course means we can write the equations just in terms of $\boldsymbol{r}$, $\boldsymbol{d}_1$, and $\boldsymbol{d}_2$. Analogous to the requirement $J > 0$ in three-dimensional elasticity, here we require that $\boldsymbol{r}$ be an embedding and that $\boldsymbol{r}' \cdot \boldsymbol{d}_3 > 0$; that is, the shearing is not infinitely severe.

The equilibrium equations for the rod are obtained by balancing forces and moments. One assumes there is a traction vector $\boldsymbol{n}(s)$ corresponding to contact forces in the rod. The balance equation for an external force $\boldsymbol{f}$ per unit length is then

$$\boldsymbol{n}' + \boldsymbol{f} = 0.$$

Likewise, one assumes a couple force field $\boldsymbol{m}(s)$ and an external couple $\boldsymbol{g}$ and derives the equation

$$\boldsymbol{m}' + \boldsymbol{r}' \times \boldsymbol{n} + \boldsymbol{g} = 0$$

by balancing torque.

Problem 2.3 Show that these balance equations are a special case of the Cosserat equations in Box 2.3, Chapter 2.

These equations together with boundary conditions and constitutive equations (i.e., $\boldsymbol{n}, \boldsymbol{m}$ as functions of $\boldsymbol{r}', \boldsymbol{d}_1, \boldsymbol{d}_2, \boldsymbol{d}_1', \boldsymbol{d}_2'$) are the equations for the rod. These are in general quasi-linear *ordinary* differential equations.

Antman's program for planar deformations and buckling of straight rods goes something like the following:

(a) introduce new variables ν, η, μ by writing

$$\boldsymbol{d}_3 = \cos\theta\boldsymbol{i} + \sin\theta\boldsymbol{j}, \qquad \boldsymbol{r}' = (1+\nu)\boldsymbol{d}_3 + \eta\boldsymbol{d}_1,$$
$$\boldsymbol{d}_1 = -\sin\theta\boldsymbol{i} + \cos\theta\boldsymbol{j}, \qquad \boldsymbol{n} = N\boldsymbol{d}_3 + H\boldsymbol{d}_1,$$
$$\boldsymbol{d}_2 = \boldsymbol{k}, \qquad \boldsymbol{m} = M\boldsymbol{k}.$$

Thus a configuration is specified by $\boldsymbol{r}(s)$ and $\theta(s)$. Let $\mu = \theta' - \theta_{\text{Ref}}$, so μ would be a curvature if s were arc length. The constitutive hypothesis is that N, H, M are functions ($\hat{N}, \hat{H}, \hat{M}$) of (ν, η, μ); the analogue of strong ellipticity is that the Jacobian matrix be positive-definite. Under suitable growth conditions one can globally invert this relationship to obtain

$$\nu = \hat{\nu}(N, H, \mu), \qquad \eta = \hat{\eta}(N, H, \mu).$$

With $f = 0$, $g = 0$, $s_1 = 0$, $s_2 = 1$, $\theta_{\text{Ref}} = 0$ and the boundary conditions $\theta(0) = 0$, $\theta(1) = 0$, $\boldsymbol{r}(0) = 0$, $\boldsymbol{n}(1) = -\lambda\boldsymbol{i}$, one gets $\hat{N} = -\lambda\cos\theta$ and $\hat{H} = \lambda\sin\theta$, so the problem reduces to the quasi-linear equation

$$[\hat{M}(\hat{\nu}, \hat{\eta}, \theta')]' + \lambda[(1+\hat{\nu})\sin\theta + \hat{\eta}\cos\theta] = 0, \qquad \text{(A)}$$

where $\hat{\nu}$ and $\hat{\eta}$ become functions of $-\lambda\cos\theta$, $\lambda\sin\theta$, and θ'.

For an inextensible ($\hat{\nu} = 0$) and unshearable ($\hat{\eta} = 0$) rod, with $\hat{M}(\mu) = EI\mu$, this equation becomes the Euler elastica equation:

$$EI\theta'' + \lambda\sin\theta = 0. \qquad \text{(E)}$$

(b) One analyzes (A) using Rabinowitz' global bifurcation theorem. One way to do this is to convert this quasi-linear equation to a semilinear one like (E). This can be done by regarding the basic variables as (λ, θ, M) and replacing (A) by the first-order semilinear system consisting of (A) and $\theta' = \mu$, a function of $(-\lambda \cos \theta, \lambda \sin \theta, M)$.

(c) Finally, one invokes elementary Sturmian theory to deduce that along the global solution branches found, the nodal properties do not change. Unlike the elastica, however, the bifurcated branch could rejoin the trivial solution at another eigenvalue. See Antman and Rosenfeld [1978] for details. For work related to Kirchhoff's problem on the loading and twisting of columns, where the geometrically exact theory produces quite different results from Kirchhoff's, see Antman [1974b] and Antman and Kenney [1981].

3. A major open problem connected with such global analyses is to see how they behave under an imperfection sensitivity analysis. As we indicated in the previous section, it is for such questions that the local theory is much more developed. In fact, often a complete local analysis can produce results that are in some sense global. For example, if a multiple eigenvalue bifurcation point is unfolded or perturbed, secondary bifurcations occur nearby and can be located quite precisely. Such things could be very difficult using currently known global techniques.

4. Even geometrically exact models can be criticized along the lines that approximate models such as the von Karman equations are criticized. Obviously for very severe deformations, the assumption that the rod can be realistically modeled in the manner indicated above is only an approximation, so is misleading unless it can be shown to be structurally stable. It also seems clear that the situation is much better for geometrically exact models than for geometrically approximate ones. Probably one should carefully investigate the range of validity for any model as part of the problem in any global bifurcation study. For example, the von Karman equations *do* successfully model many interesting bifurcation problems.

Box 2.2 ***Imperfection Sensitivity: Mode Jumping in the Buckling of a Plate***

In the previous section we indicated that singularity theory is a very powerful tool in an analysis of imperfection sensitivity. Such analyses, when fully done, produce bifurcation diagrams that are insensitive to further perturbations. It is therefore consistent to use any reasonable approximation to an exact model, valid near the bifurcation point of interest. Therefore, unlike the previous box, the use of approximate models such as the von Karman equations here is justified.

Carrying out a substantial singularity analysis can involve a variety of issues, some of which we wish to point out. We shall make some comments in the context of the beautiful paper of Schaeffer and Golubitsky [1980]. (Related work is found in Matkowsky and Putnik [1974], Chow, Hale and Mallet-Paret [1976], and Magnus and Poston [1977].)

The problem concerns the buckling of a rectangular plate (Figure 7.2.6). The aspect ratio l (i.e., length/width) used in experiments of

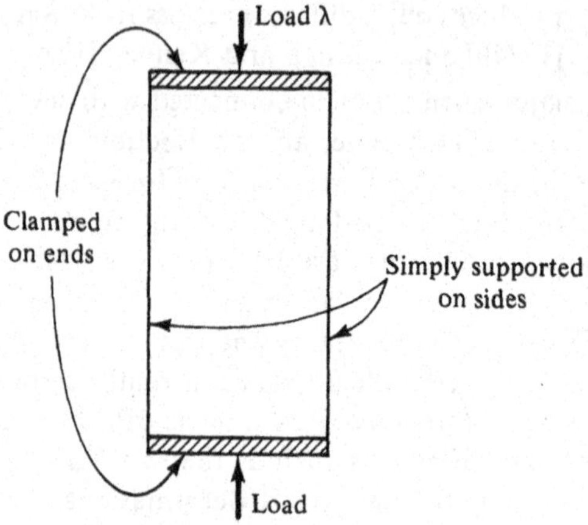

Figure 7.2.6

Stein [1959] was about 5.36. For the load λ exceeding a certain value λ_0, the plate buckles to a state with wave number 5. As λ increases further the plate undergoes a sudden and violent snap buckling to wave number 6. The phenomenon is called "mode jumping" and the problem is to explain it.

Parameterizing the plate by $\Omega = \{(z_1, z_2) | 0 \leq z_1 < l\pi \text{ and } 0 \leq z_2 \leq \pi\}$, the von Karman equations for w, the z_3 component of the deflection and ϕ, the Airy stress function, are

$$\Delta^2 w = [\phi, w] - \lambda w_{z_1 z_1},$$
$$\Delta^2 \phi = -\tfrac{1}{2}[w, w].$$

where Δ^2 is the biharmonic operator and [,] is the symmetric bilinear form defined by

$$[u, v] = u_{z_1z_1}v_{z_2z_2} + u_{z_2z_2}v_{z_1z_1} - 2u_{z_1z_2}v_{z_1z_2}.$$

The boundary conditions for w are $w = \partial w/\partial n = 0$ on the ends (clamped) and $w = \Delta w = 0$ on the sides (simply supported).

A few of the highlights of the procedures followed are given next:

(a) Bauer, Keller, and Riess [1975] used a spring model without boundary conditions. Matkowsky and Putnik [1974] and Matkowsky, Putnik and Reiss [1980] use simply supported boundary conditions. The type of boundary conditions used makes an important difference. Schaeffer and Stein noted that the clamped conditions for the ends makes more sense physically. In the present case, at $l_k = \sqrt{k(k+2)}$ there is a double eigenvalue λ_0. For $k = 5$, this is actually fairly close to the situation near wave numbers 5 and 6. Thus the strategy is to unfold the bifurcation near this double eigenvalue and see what secondary bifurcations arise.

(b) The Liapunov–Schmidt procedure is now done to produce a function $G: \mathbb{R}^2 \times \mathbb{R} \to \mathbb{R}^2$.

(c) There is a symmetry in the problem that is exploited. This symmetry on the $\mathbb{R}^2$ obtained in the Liapunov–Schmidt procedure is $\mathbb{Z}_2 \oplus \mathbb{Z}_2$, generated by $(x, y) \mapsto (-x, -y)$ and $(x, y) \mapsto (x, -y)$. These correspond to 2 of the 3 obvious symmetries of the original problem (the other gives no extra information). Also, G is the gradient (for each $\lambda \in \mathbb{R}$) of a function invariant under this action.

(d) The symmetry in (a) greatly simplifies the unfolding procedure, where now unfolding is done under the assumption of a symmetry group for the equations. The general theory for this is described in Golubitsky and Shaeffer [1979b].

(e) Mode jumping does not occur with all the boundary conditions simply supported. There one gets a bifurcation diagram like that in Figure 7.2.7(a); the wave number 5 solution never

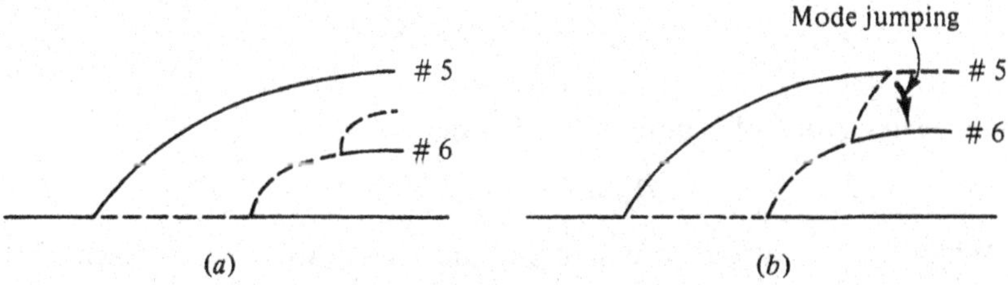

Figure 7.2.7

loses stability. With the boundary conditions above, the bifurcation diagram is like Figure 7.2.6(b); the wave number 5 solution loses stability and wave number 6 picks it up by way of a jump. These figures only show the orbits; to get all solutions one acts on the orbits by the symmetry group.

Examples like this show that the local analysis using singularity theory can produce rather sophisticated bifurcation diagrams. This kind of detailed explanation and computable complexity is beyond the reach of most global results known at present. The blending of techniques like this with those of the previous box represents a considerable challenge.

Box 2.3 ***The Role of Symmetry in Bifurcation Problems***

When studying bifurcation problems, questions of symmetry arise in many guises. This box discusses some of the ways symmetry can be exploited, and some of the tantalizing questions it raises.

If a bifurcation problem has a multiple eigenvalue, then the problem is usually non-generic. Sometimes this non-genericity is due to the invariance of the problem under a symmetry group. We indicated in the previous box that a bifurcation analysis including imperfection-sensitivity results can be obtained for such problems. If a (real) problem is anywhere near such a special point, it is often wise to regard it as an imperfection in a more ideal model. In fact, some otherwise simple eigenvalue problems may be better treated as belonging to a perturbation of a double eigenvalue problem. This whole philosophy of symmetrizing to bring eigenvalues together seems to be frutiful.

When we say a bifurcation problem has a certain symmetry group, we mean that it is covariant under the action of this group. For example, if $F\colon \mathfrak{X} \times \mathbb{R} \to \mathcal{Y}$ is a map whose zeros we wish to study and $\mathcal{G}$ is a group acting on $\mathfrak{X}$ and $\mathcal{Y}$, we say F is *covariant* when

$$F(gx, \lambda) = gF(x, \lambda)$$

for $g \in \mathcal{G}$, $x \in \mathfrak{X}$, $\lambda \in \mathbb{R}$, and where gx is the action of g on x. The *symmetry group* of a point $x \in \mathfrak{X}$ is defined by

$$\mathcal{G}_x = \{g \in \mathcal{G} \mid gx = x\}.$$

When a bifurcation occurs, often the trivial solution has symmetry group $\mathcal{G}$, but the bifurcating solution has a smaller symmetry group. We say that the bifurcation has *broken the symmetry*.

It is an important problem to study how symmetries are so broken and how they relate to pattern formation and related questions. If $\mathcal{G}_x$ gets smaller, the solution gets less symmetric, or more complex. What, if anything, does this have to do with entropy?

If $\mathcal{G}$ acts on $\mathfrak{X}$ linearly, methods of group representations can be used to analyze which "modes" go unstable and hence how the symmetry is broken. The idea is to break up $\mathfrak{X}$ into a direct sum (like a Fourier decomposition) on each piece of which $\mathcal{G}$ acts irreducibly and determine in which piece the eigenvalue crosses. Two basic references for this method are Ruelle [1973] and Sattinger [1979]. There are numerous related papers as well. (For example, Rodrigues and Vanderbauwhede [1978] give conditions under which the bifurcating solutions do not break symmetry.) This kind of phenomenon actually is abundant. It occurs in Taylor cells between rotating cylinders, in hexagonal cells in convection problems, and in many problems of chemical kinetics. For example, the breaking of S^1 symmetry to a discrete symmetry occurs in the Taylor problem in fluid muchanics (Rand [1982]) and the breaking of SO(3) to S^1 symmetry occurs in the blowing up of a balloon (Haughton and Ogden [1980]) and in convection in a spherical shell (Chossat [1979]). See, Sattinger [1980], Haken [1979], Buzano and Golubitsky [1982], and Golubitsky, Marsden and Schaeffer [1983] for more examples and references.

A much more serious kind of symmetry breaking is to allow imperfections that break the symmetry in various ways; that is, the equations themselves rather than the solutions break the symmetry. Here, not only is the mathematics difficult (it is virtually non-existent), but it is not as clear what one should allow physically.

In Arms, Marsden, and Moncrief [1981], a special class of bifurcation problems is studied where the structure of the bifurcation and its connection with symmetry can be nailed down. The problems studied are of the form $F(x, \lambda) = J(x) - \lambda = 0$, where J is the Noether conserved quantity for a symmetry group acting on phase space. It is shown that bifurcations occur precisely at points with symmetry; how the symmetry is broken is determined.

In the next section we shall see how symmetry in the form of SO(3) and material frame indifference comes into the analysis of the traction problem in an essential way. We shall see that bifurcation points are those with a certain symmetry, in accordance with the general philosophy exposed here. However the detailed way symmetry enters the problem is different from the examples mentioned so far since the trivial solutions are not fixed by the group and the group also acts on parameter space.

For more information on these points, see Golubitsky and Schaeffer [1982] and [1984].

7.3 THE TRACTION PROBLEM NEAR A NATURAL STATE (Signorini's Problem)[7]

In the 1930s Signorini discovered an amazing fact: the traction problem in nonlinear elasticity can have non-unique solutions even for small loads and near a natural state. Here non-unique means unequal up to a rigid body motion of the body and loads. What is even more amazing is that this non-uniqueness depends, in many cases, not on the whole stored energy function, but only on the elasticity tensor c_{abcd} for linearized elasticity, even though the traction problem for linearized elasticity has uniqueness up to rigid body motions, as we proved in Section 6.1. For example, the loads shown in Figure 7.3.1 can produce

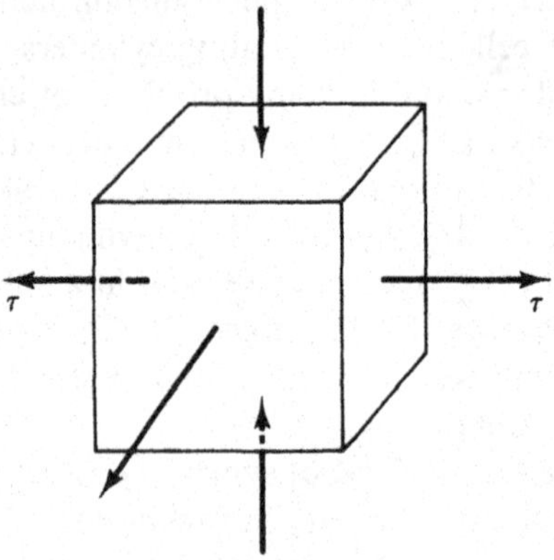

Figure 7.3.1

more than one solution, even for a (compressible) neo-Hookean material, and (arbitrarily small) loads near the one shown. The occurence of these extra solutions in the nonlinear theory and yet their absence in the linearized theory is not easy (for us) to understand intuitively, although it may be related to bulging or barelling solutions. Experiments for such situations are not easy to carry out; cf. Beatty and Hook [1968].

This state of affairs led to much work—much of it in the Italian school—and was the subject of some controversy concerning the validity of linearized elasticity. Some of the main contributions after Signorini were by Tolotti [1943], Stoppelli [1958], Grioli [1962], and Capriz and Podio-Guidugli [1974]. The problem is discussed at length and additional contributions given in Truesdell and Noll [1965]; see also Wang and Truesdell [1973] and Van Buren [1968].

---

[7]This section was done in collaboration with D. Chillingworth and Y. H. Wan and is based on Chillingworth, Marsden, and Wan [1982a, b].

Nowadays we do not see any contradictions, but rather we see a *bifurcation* in the space of solutions of the equations of elastostatics. Whenever there is a bifurcation, the correspondence with the linearized problem becomes singular; that is, the problem is *linearization unstable* in the sense of Section 4.4. In the framework of elastodynamics there is clearly no bifurcation or linearization instability, but this makes the bifurcation in the elastostatic problem no less interesting. This bifurcation in the space of solutions then takes its place alongside similar phenomena in other classical field theories such as general relativity and gauge theory (see Arms, Marsden, and Moncrief [1981], and references therein).

The most complete results in the literature before now are those of Stoppelli [1958]. His results are stated (without proof, but in English) in Grioli [1962]. However, this analysis is incomplete for three reasons. First, the load is varied only by a scalar factor. In a full neighborhood of loads with axes of equilibrium there are additional solutions missed by their analysis; thus, an imperfection-sensitivity-type analysis reveals more solutions. Second, their analysis is only local in the rotation group, so additional nearly stress-free solutions are missed by restricting to rotations near the identity. Third, some degenerate classes of loads were not considered. However, singularity theory can deal with these cases as well. The complexity of the problem is indicated by the fact that for certain types of loads one can find up to 40 geometrically distinct solutions that are nearly stress free, whereas Stoppelli's analysis produces at most 3.

These problems have recently been solved by Chillingworth, Marsden, and Wan [1982a]. This section gives a *brief introduction* to their methods. The paper should be consulted for the complete analysis. However, we do go far enough to include a complete and considerably simplified proof of the first basic theorem of Stoppelli. Apart from Van Buren [1968], whose proof is similar to Stoppelli's, a complete proof has not previously appeared in English.

3.1 Notation Let the reference configuration be a bounded region $\mathcal{B} = \Omega \subset \mathbb{R}^3$ with smooth boundary.[8] As we saw in Section 6.1, the linearized equations have a kernel consisting of infinitesimal rigid body motions. We can readily eliminate translations by assuming $0 \in \Omega$ and working with the configuration space $\mathcal{C}$ consisting of all deformations $\phi\colon \Omega \to \mathbb{R}^3$ that are of class $W^{s,p}$, $s > 3/p + 1$ and satisfy $\phi(0) = 0$. (Recall that such ϕ's are necessarily C^1.) The central difficulty of the problem is then the presence of the rotational covariance of the problem (material frame indifference).

Let $W(X, \boldsymbol{C})$ be a given smooth stored energy function, where $\boldsymbol{C}$ is, as usual,

---

[8] We believe that our results also hold when Ω has piecewise smooth boundary. This program depends on elliptic regularity for such regions. Except in special cases, this theory is nonexistent (as far as we know) and seems to depend on a modification of the usual Sobolev spaces near corners. However for simple shapes like cubes, where the linearized elastostatic equations can be solved explicitly, the necessary regularity can be checked by hand. See pp. 318 and 371.

the Cauchy–Green tensor. Let $\boldsymbol{P} = \partial W/\partial \boldsymbol{F}$ and $\boldsymbol{S} = 2\,\partial W/\partial \boldsymbol{C}$ be the first and second Piola–Kirchhoff stress tensors and $\mathbf{A} = \partial \boldsymbol{P}/\partial \boldsymbol{F}$ the elasticity tensor.

We make the following two assumptions.

3.2 Assumptions

(H1) When $\phi = I_\Omega$ (identity map on Ω), $\boldsymbol{P} = \mathbf{0}$; that is, the undeformed state is stress free, or natural.

(H2) Strong ellipticity holds at (and hence near) $\phi = I_\Omega$.

Since the undeformed state is stress free, the classical elasticity tensor for elasticity linearized about $\phi = I_\Omega$ is $\mathbf{c} = 2\,\partial^2 W/\partial \boldsymbol{C}\,\partial \boldsymbol{C}$ evaluated at $\phi = I_\Omega$.

Let $\boldsymbol{B}\colon \Omega \to \mathbb{R}^3$ denote a given body force (per unit volume) and $\boldsymbol{\tau}\colon \partial\Omega \to \mathbb{R}^3$ a given surface traction (per unit area). These are dead loads; in other words, the equilibrium equations for ϕ that we are studying are:

$$\left.\begin{aligned} &\mathrm{DIV}\,\boldsymbol{P}(X, \boldsymbol{F}(X)) + \boldsymbol{B}(X) = 0 \qquad &&\text{for } X \in \Omega, \\ &\boldsymbol{P}(X, \boldsymbol{F}(X))\cdot \boldsymbol{N}(X) = \boldsymbol{\tau}(X) &&\text{for } X \in \partial\Omega, \end{aligned}\right\} \tag{E}$$

where $\boldsymbol{N}(X)$ is the outward unit normal to $\partial\Omega$ at $X \in \partial\Omega$.

3.3 Definition Let $\mathcal{L}$ denote the space of all pairs $l = (\boldsymbol{B}, \boldsymbol{\tau})$ of loads (of class $W^{s-2,p}$ on Ω and $W^{s-1-1/p,p}$ on $\partial\Omega$) such that

$$\int_\Omega \boldsymbol{B}(X)\,dV(X) + \int_{\partial\Omega} \boldsymbol{\tau}(X)\,dA(X) = 0.$$

That is, the total force on Ω vanishes, where dV and dA are the respective volume and area elements on Ω and $\partial\Omega$.

Observe that if $(\boldsymbol{B}, \boldsymbol{\tau})$ are such that (E) holds for some $\phi \in \mathcal{C}$, then $(\boldsymbol{B}, \boldsymbol{\tau}) \in \mathcal{L}$.

The group $\mathrm{SO}(3) = \{\boldsymbol{Q} \in \boldsymbol{L}(\mathbb{R}^3, \mathbb{R}^3) \mid \boldsymbol{Q}^T\boldsymbol{Q} = I_{\mathbb{R}^3}$ and $\det \boldsymbol{Q} = +1\}$ of proper orthogonal transformations will play a key role. By (H1), $\phi = I_\Omega$ solves (E) with $\boldsymbol{B} = \boldsymbol{\tau} = \mathbf{0}$. By material frame indifference, $\phi = \boldsymbol{Q}|\Omega$ ($\boldsymbol{Q}$ restricted to Ω) is also a solution for any $\boldsymbol{Q} \in \mathrm{SO}(3)$. The map $\boldsymbol{Q} \mapsto \boldsymbol{Q}|\Omega$ embeds SO(3) into $\mathcal{C}$ and we shall identify its image with SO(3). Thus, the "trivial" solutions of (E) are elements of SO(3).

Our basic problem is as follows:

(*P1*) *Describe the set of all solutions of* (*E*) *near the trivial solutions* SO(3) *for various loads* $l \in \mathcal{L}$ *near zero.* Here, "describe" includes the following objectives:

(a) counting the solutions;
(b) determining the stability of the solutions;
(c) showing that the results are insensitive to small perturbations of the stored energy function and the loads; that is, the bifurcation diagram produced is structurally stable.

3.4 Notations and Facts about the Rotation Group SO(3) Let $\mathfrak{M}_3 = L(\mathbb{R}^3, \mathbb{R}^3)$ = linear transformations of $\mathbb{R}^3$ to $\mathbb{R}^3$; sym $= \{A \in \mathfrak{M}_3 \mid A^T = A\}$; skew $= \{A \in \mathfrak{M}^3 \mid A^T = -A\}$.

We identify skew with so(3), the Lie algebra of SO(3). $\mathbb{R}^3$ and skew are isomorphic by the mapping $\boldsymbol{v} \in \mathbb{R}^3 \mapsto \boldsymbol{W}_v \in$ skew, where $\boldsymbol{W}_v(\boldsymbol{w}) = \boldsymbol{w} \times \boldsymbol{v}$; relative to the standard basis, the matrix of $\boldsymbol{W}_v$ is

$$\boldsymbol{W}_v = \begin{bmatrix} 0 & r & -q \\ -r & 0 & p \\ q & -p & 0 \end{bmatrix}, \quad \text{where} \quad \boldsymbol{v} = (p, q, r).$$

The Lie bracket is $[\boldsymbol{W}_v, \boldsymbol{W}_w] = \boldsymbol{v} \otimes \boldsymbol{w} - \boldsymbol{w} \otimes \boldsymbol{v} = -\boldsymbol{W}_{v \times w}$, where $\boldsymbol{v} \otimes \boldsymbol{w} \in \mathfrak{M}_3$ is given by $(\boldsymbol{v} \otimes \boldsymbol{w})(\boldsymbol{u}) = \boldsymbol{v}\langle \boldsymbol{w}, \boldsymbol{u}\rangle$. The standard inner product on $\mathbb{R}^3$ is $\langle \boldsymbol{v}, \boldsymbol{w}\rangle = \frac{1}{2}\,\text{trace}(\boldsymbol{W}_v^T \boldsymbol{W}_w)$, called the *Killing form* on SO(3). Finally, $\exp(\boldsymbol{W}_v)$ is the rotation about the vector $\boldsymbol{v}$ in the positive sense, through the angle $\|\boldsymbol{v}\|$.

Now we turn to some preliminary facts about $\mathfrak{L}$ and $\mathfrak{C}$.

3.5 Definition Let $\boldsymbol{\phi} \in \mathfrak{C}$ and $l \in \mathfrak{L}$. We say that l is *equilibrated relative to* $\boldsymbol{\phi}$ if the total torque in the configuration $\boldsymbol{\phi}$ vanishes:

$$\int_\Omega \boldsymbol{\phi}(X) \times \boldsymbol{B}(X)\, dV(X) + \int_{\partial\Omega} \boldsymbol{\phi}(X) \times \boldsymbol{\tau}(X)\, dA(X) = 0,$$

where $l = (\boldsymbol{B}, \boldsymbol{\tau})$. Let $\mathfrak{L}_e$ denote the loads that are equilibrated relative to the identity.

Problem 3.1 Show that if $l = (\boldsymbol{B}, \boldsymbol{\tau})$ satisfies (E) for some $\boldsymbol{\phi} \in \mathfrak{C}$, then l is equilibrated relative to $\boldsymbol{\phi}$. (*Hint*: Use the Piola identity.)

3.6 Definition Define the *astatic load map* $\boldsymbol{k}\colon \mathfrak{L} \times \mathfrak{C} \to \mathfrak{M}_3$ by

$$\boldsymbol{k}(l, \boldsymbol{\phi}) = \int_\Omega \boldsymbol{B}(X) \otimes \boldsymbol{\phi}(X)\, dV(X) + \int_{\partial\Omega} \boldsymbol{\tau}(X) \otimes \boldsymbol{\phi}(X)\, dA$$

and write $\boldsymbol{k}(l) = \boldsymbol{k}(l, I_\Omega)$.

We have actions of SO(3) on $\mathfrak{L}$ and $\mathfrak{C}$ given by:

***Action of* SO(3) *on* $\mathfrak{L}$:** $\boldsymbol{Q}l(X) = (\boldsymbol{Q}\boldsymbol{B}(X), \boldsymbol{Q}\boldsymbol{\tau}(X))$.
***Action of* SO(3) *on* $\mathfrak{C}$:** $\boldsymbol{Q}\boldsymbol{\phi} = \boldsymbol{Q} \circ \boldsymbol{\phi}$.

Note that $\boldsymbol{Q}l$ means "the load arrows are rotated, keeping the body fixed." We shall write $\mathcal{O}_l$ and $\mathcal{O}_\phi$ for the SO(3) orbits of l and $\boldsymbol{\phi}$; that is,

$$\mathcal{O}_l = \{\boldsymbol{Q}l \mid \boldsymbol{Q} \in \text{SO(3)}\} \quad \text{and} \quad \mathcal{O}_\phi = \{\boldsymbol{Q}\boldsymbol{\phi} \mid \boldsymbol{Q} \in \text{SO(3)}\}.$$

Thus, $\mathcal{O}_{I_\Omega}$ consists of the trivial solutions corresponding to $l = 0$.

The following is a list of basic observations about the astatic load map, each of which may be readily verified.

3.7 Proposition

(A1) *l is equilibrated relative to ϕ if and only if $\mathbf{k}(l, \phi) \in$* sym. *In particular, $l \in \mathfrak{L}_e$ if and only if $\mathbf{k}(l) \in$* sym.

(A2) (*Equivariance*) *For $l \in \mathfrak{L}$, $\phi \in \mathfrak{C}$, and $\boldsymbol{Q}_1, \boldsymbol{Q}_2 \in$* SO(3),

$$\mathbf{k}(\boldsymbol{Q}_1 l, \boldsymbol{Q}_2 \phi) = \boldsymbol{Q}_1 \mathbf{k}(l, \phi) \boldsymbol{Q}_2^{-1}.$$

In particular, $\mathbf{k}(\boldsymbol{Q}l) = \boldsymbol{Q}\mathbf{k}(l)$.

(A3) (*Infinitesimal Equivariance*) *For $l \in \mathfrak{L}$, $\phi \in \mathfrak{C}$, $\boldsymbol{W}_1, \boldsymbol{W}_2 \in$* skew,

$$\mathbf{k}(\boldsymbol{W}_1 l, \phi) = \boldsymbol{W}_1 \mathbf{k}(l, \phi), \qquad \mathbf{k}(l, \boldsymbol{W}_2 \phi) = -\mathbf{k}(l, \phi)\boldsymbol{W}_2.$$

In particular, $\mathbf{k}(\boldsymbol{W}l) = \boldsymbol{W}\mathbf{k}(l)$.

Problem 3.2 Prove each of these assertions.

Later on, we shall be concerned with how the orbit of a given $l \in \mathfrak{L}$ meets $\mathfrak{L}_e$. The most basic result in this direction is the following.

3.8 DaSilva's Theorem Let $l \in \mathfrak{L}$. Then $\mathcal{O}_l \cap \mathfrak{L}_e \neq \varnothing$.

Proof By the polar decomposition, we can write $\mathbf{k}(l) = \boldsymbol{Q}^T \boldsymbol{A}$ for some $\boldsymbol{Q} \in$ SO(3) and $\boldsymbol{A} \in$ sym. By (A2), $\mathbf{k}(\boldsymbol{Q}l) = \boldsymbol{Q}\mathbf{k}(l) = \boldsymbol{A} \in$ sym, so by (A1), $\boldsymbol{Q}l \in \mathfrak{L}_e$. ∎

Similarly, any load can be equilibrated relative to any chosen configuration by a suitable rotation.

Solutions of (E) with an "axis of equilibrium" will turn out to coincide with the bifurcation points. The idea is to look for places where $\mathcal{O}_l$ meets $\mathfrak{L}_e$ in a degenerate way.

3.9 Definition Let $l \in \mathfrak{L}_e$ and $\boldsymbol{v} \in \mathbb{R}^3$, $\|\boldsymbol{v}\| = 1$. We say that $\boldsymbol{v}$ is an *axis of equilibrium* for l when $\exp(\theta \boldsymbol{W}_v) l \in \mathfrak{L}_e$ for all real θ—that is, when rotations of l through any angle θ about the axis $\boldsymbol{v}$ do not destroy equilibration relative to the identity.

There are a number of useful ways of reformulating the condition that $\boldsymbol{v}$ be an axis of equilibrium. These are listed as follows.

3.10 Proposition *Let $l \in \mathfrak{L}_e$ and $\boldsymbol{A} = \mathbf{k}(l) \in$* sym. *The following conditions are equivalent:*

(1) *l has an axis of equilibrium $\boldsymbol{v}$.*

(2) *There is a $\boldsymbol{v} \in \mathbb{R}^3$, $\|\boldsymbol{v}\| = 1$ such that $\boldsymbol{W}_v l \in \mathfrak{L}_e$.*

(3) *$\boldsymbol{W} \mapsto \boldsymbol{AW} + \boldsymbol{WA}$ fails to be an isomorphism of skew to itself.*

(4) Trace($\boldsymbol{A}$) *is an eigenvalue of $\boldsymbol{A}$.*

Proof

(1) ⇒ (2) Differentiate $\exp(\theta \boldsymbol{W}_v) l$ in θ at $\theta = 0$.

(2) ⇒ (1) Note that by (A2)

$$k(\exp(\theta W_v)l) = [I + W_{\theta v} + \tfrac{1}{2}(W_{\theta v})^2 + \cdots]\, k(l).$$

Since $k(W_v l) = W_v k(l)$ is symmetric, this is symmetric, term by term.

(2) $\Rightarrow$ (3) Since $k(W_v l) = W_v A$ is symmetric, $W_v A + A W_v = 0$, so $W \mapsto AW + WA$ is not an isomorphism.

(3) $\Rightarrow$ (2) There exists a $v \in \mathbb{R}^3$, $\|v\| = 1$ such that $W_v A + A W_v = \mathbf{0}$, so $k(W_v l) = W_v A$ is symmetric.

(3) $\Rightarrow$ (4) Define $L \in \mathfrak{M}_3$ by $L = (\text{trace } A)I - A$. Then one has the relationship

$$W_{Lv} = A W_v + W_v A,$$

as may be verified by considering a basis of eigenvectors for A. Therefore, $A W_v + W_v A = \mathbf{0}$ if and only if $Lv = \mathbf{0}$; that is, v is an eigenvector of A with eigenvalue trace(A). ∎

3.11 Corollary *Let $l \in \mathfrak{L}_e$ and $A = k(l) \in$ sym. Let the eigenvalues of A be denoted a, b, c. Then l has no axis of equilibrium if and only if*

$$(a + b)(a + c)(b + c) \neq 0.$$

Proof This condition is equivalent to saying that trace(A) is not an eigenvalue of A. ∎

3.12 Definition We shall say that $l \in \mathfrak{L}_e$ is a *type 0* load if l has no axis of equilibrium and if the eigenvalues of $A = k(l)$ are distinct.

The following shows how the orbits of type 0 loads meet $\mathfrak{L}_e$.

3.13 Proposition *Let $l \in \mathfrak{L}_e$ be a type 0 load. Then $\mathcal{O}_l \cap \mathfrak{L}_e$ consists of four type* 0 *loads.*

Proof We first prove that the orbit of A in $\mathfrak{M}_3$ under the action $(Q, A) \mapsto QA$ meets sym in four points. Relative to its basis of eigenvectors, we can write $A = \text{diag}(a, b, c)$. Then $\mathcal{O}_A \cap$ sym contains the four points

$$\begin{aligned} &\text{diag}(a, b, c) && (Q = I),\\ &\text{diag}(-a, -b, c) && (Q = \text{diag}(-1, -1, 1)),\\ &\text{diag}(-a, b, -c) && (Q = \text{diag}(-1, 1, -1)),\\ &\text{diag}(a, -b, -c) && (Q = \text{diag}(1, -1, -1)). \end{aligned}$$

These are distinct matrices since $(a + b)(a + c)(b + c) \neq 0$. Now suppose a, b, and c are distinct. Suppose $QA = S \in$ sym. Then $S^2 = A^2$. Let μ_i be an eigenvalue of S with eigenvector u_i. Then $S^2 u_i = \mu_i^2 u_i = A^2 u_i$, so μ_i^2 is an eigenvalue of A^2. Thus, as the eigenvectors of A^2 with a given eigenvalue are unique, u_i is an eigenvector of A and $\pm\mu_i$ is the corresponding eigenvalue. Since $\det Q = +1$, $\det S = \det A$, so we must have one of the four cases above.

By equivariance, $k(\mathcal{O}_l) \cap \text{sym} = \mathcal{O}_{k(l)} \cap \text{sym}$ consists of four points. Now $\mathcal{O}_l \cap \mathcal{L}_e = k^{-1}(\mathcal{O}_{k(l)} \cap \text{sym})$, so it suffices to show that k is one-to-one on $\mathcal{O}_l$. This is a consequence of the following and Property (A2) of 3.7.

3.14 Lemma *Suppose $A \in$ sym and dim Ker $A \leq 1$. Then A has no isotropy; that is, $QA = A$ implies $Q = I$.*

Proof Every $Q \neq I$ acts on $\mathbb{R}^3$ by rotation through an angle θ about a unique axis; say $l \subset \mathbb{R}^3$ (l is a line through the origin in $\mathbb{R}^3$). Now $QA = A$ means that Q is the identity on the range of A. Therefore, if $Q \neq I$ and $QA = A$, the range of A must be zero or one dimensional; that is, dim Ker $A > 2$. ∎

The next proposition considers the range and kernel of $k: \mathcal{L} \longrightarrow \mathfrak{M}_3$.

3.15 Proposition

1. Ker k *consists of those loads in $\mathcal{L}_e$ for which every axis is an axis of equilibrium.*
2. $k: \mathcal{L} \longrightarrow \mathfrak{M}_3$ *is surjective.*

Proof For 1, let $l \in$ Ker k. For $W \in$ skew, $k(Wl) = Wk(l) = 0$, so $Wl \in \mathcal{L}_e$; by 3.10 every axis is an axis of equilibrium. Conversely, if $Wl \in \mathcal{L}_e$ for all $W \in$ skew, then $k(Wl) = Wk(l)$ is symmetric for all W; that is, $k(l)W + Wk(l) = 0$ for all W. From $W_{Lv} = AW_v + W_vA$, where $A = k(l)$ and $L = (\text{trace } A)\, I - A$, we see that $L = 0$. This implies trace $A = 0$ and hence $A = 0$.

To prove 2, introduce the following SO(3)-invariant inner product on $\mathcal{L}$:

$$\langle l, \bar{l} \rangle = \int_\Omega \langle B(X), \bar{B}(X) \rangle \, dV(X) + \int_{\partial\Omega} \langle \tau(X), \bar{\tau}(X) \rangle \, dA(X).$$

Relative to this and the inner product $\langle A, B \rangle = \text{trace}(A^T B)$ on $\mathfrak{M}_3$, the adjoint $k^T: \mathfrak{M}_3 \longrightarrow \mathcal{L}$ of k is given by

$$k^T(A) = (B, \tau), \quad \text{where} \quad B(X) = AX - G, \tau(X) = AX,$$

and

$$G = \int_\Omega AX \, dV(X) + \int_{\partial\Omega} AX \, dA(X).$$

If $k^T(A) = (0, 0)$, then it is clear that $A = 0$. It follows that k is surjective. ∎

3.16 Corollary

1. Ker k *is the largest subspace of $\mathcal{L}_e$ that is* SO(3) *invariant.*
2. $k\,|\,(\text{Ker } k)^\perp: (\text{Ker } k)^\perp \longrightarrow \mathfrak{M}_3$ *is an isomorphism.*

Let $j = (k\,|\,(\text{Ker } k)^\perp)^{-1}$ and write

$$\text{Skew} = j \text{ skew}, \qquad \text{Sym} = j \text{ sym}.$$

These are linear subspaces of $\mathcal{L}$ of dimension three and six, respectively. Thus we have the decomposition:

$$\overbrace{\mathfrak{L} = \text{Skew} \oplus \text{Sym}}^{\text{SO(3)-invariant pieces}} \oplus \underbrace{\text{Ker}\, \boldsymbol{k}}_{\mathfrak{L}_e}$$

of $\mathfrak{L}$, corresponding to the decomposition $\mathfrak{M}_3 = \text{skew} \oplus \text{sym}$:

$$U = \tfrac{1}{2}(U - U^T) + \tfrac{1}{2}(U + U^T) \quad \text{of} \quad \mathfrak{M}_3.$$

Now we are ready to reformulate our problem in several ways that will be useful.

Define $\Phi\colon \mathcal{C} \longrightarrow \mathfrak{L}$ by $\Phi(\phi) = (-\text{DIV}\, \boldsymbol{P}, \boldsymbol{P}\cdot\boldsymbol{N})$; that is,

$$\Phi(\phi)(X) = (-\text{DIV}\, \boldsymbol{P}(X, \boldsymbol{F}(X)), \boldsymbol{P}(X, \boldsymbol{F}(X))\cdot\boldsymbol{N}(X))$$

so the equilibrium equations (E) become $\Phi(\phi) = l$. From material frame indifference, we have equivariance of Φ: $\Phi(Q\phi) = Q\Phi(\phi)$. The results of Boxes 1.1, Chapter 3 and 1.1, Chapter 6, show that Φ is a smooth mapping. The derivative of Φ is given by

$$D\Phi(\phi)\cdot\boldsymbol{u} = (-\text{DIV}(\mathbf{A}\cdot\nabla\boldsymbol{u}), (\mathbf{A}\cdot\nabla\boldsymbol{u})\cdot\boldsymbol{N})$$

and at $\phi = I_\Omega$ this becomes

$$D\Phi(I_\Omega)\cdot\boldsymbol{u} = 2(-\text{DIV}(\mathbf{c}\cdot\boldsymbol{e}), (\mathbf{c}\cdot\boldsymbol{e})\cdot\boldsymbol{N}),$$

where $\boldsymbol{e} = \frac{1}{2}[\nabla\boldsymbol{u} + (\nabla\boldsymbol{u})^T]$.

If $D\Phi(I_\Omega)\colon T_{I_\Omega}\mathcal{C} \longrightarrow \mathfrak{L}$ were an isomorphism, we could solve $\Phi(\phi) = l$ uniquely for ϕ near I_Ω and l small. The essence of our problem is that $D\Phi(I_\Omega)$ is *not* an isomorphism.

Define $\mathcal{C}_{\text{sym}} = \{\boldsymbol{u} \in T_{I_\Omega}\mathcal{C} \mid \boldsymbol{u}(0) = \mathbf{0} \text{ and } \nabla\boldsymbol{u}(0) \in \text{sym}\}$. From (H2) and Section 6.1, we have:

3.17 Lemma *$D\phi(I_\Omega)\,|\,\mathcal{C}_{\text{sym}}\colon \mathcal{C}_{\text{sym}} \longrightarrow \mathfrak{L}_e$ is an isomorphism.*

The connection between the astatic load map $\boldsymbol{k}\colon \mathfrak{L} \longrightarrow \mathfrak{M}_3$ and Φ is seen from the following computation of $\boldsymbol{k}\circ\Phi$.

3.18 Lemma *Let $\phi \in \mathcal{C}$ and $\boldsymbol{P}$ be the first Piola–Kirchhoff stress tensor evaluated at ϕ. Then*

$$\boldsymbol{k}(\Phi(\phi)) = \int_\Omega \boldsymbol{P}\, dV.$$

This follows by an application of Gauss' theorem to

$$\boldsymbol{k}(\Phi(\phi)) = \int_\Omega (-\text{DIV}\, \boldsymbol{P}) \otimes \boldsymbol{X}\, dV(X) + \int_{\partial\Omega} (\boldsymbol{P}\cdot\boldsymbol{N}) \otimes \boldsymbol{X}\, dA(X).$$

This should be compared with the astatic load relative to the configuration ϕ rather than I_Ω; one gets

$$\boldsymbol{k}(\Phi(\phi), \phi) = \int_{\phi(\Omega)} \boldsymbol{\sigma}\, dv.$$

which is symmetric, while $\boldsymbol{k}(\Phi(\phi)) = \boldsymbol{k}(\Phi(\phi), I_\Omega)$ need not be.

To study solutions of $\Phi(\phi) = l$ for ϕ near the trivial solutions and l near a given load l_0, it suffices to take $l_0 \in \mathfrak{L}_e$. This follows from DaSilva's theorem and equivariance of Φ.

Let $\mathfrak{C}_{sym}$ be regarded as an affine subspace of $\mathfrak{C}$ centered at I_Ω i.e. identify $\mathfrak{C}_{sym}$ and $\mathfrak{C}_{sym} + I_\Omega$. Let $\tilde{\Phi}$ be the restriction of Φ to $\mathfrak{C}_{sym}$. From the implicit function theorem we get:

3.19 Lemma *There is a ball centered at I_Ω in $\mathfrak{C}_{sym}$ whose image $\mathfrak{N}$ under $\tilde{\Phi}$ is a smooth submanifold of $\mathfrak{L}$ tangent to $\mathfrak{L}_e$ at 0 (see Figure 7.3.2). The manifold $\mathfrak{N}$ is the graph of a unique smooth mapping*

$$\boldsymbol{F}: \mathfrak{L}_e \longrightarrow \text{Skew}$$

such that $\boldsymbol{F}(0) = \mathbf{0}$ *and* $D\boldsymbol{F}(0) = \mathbf{0}$.

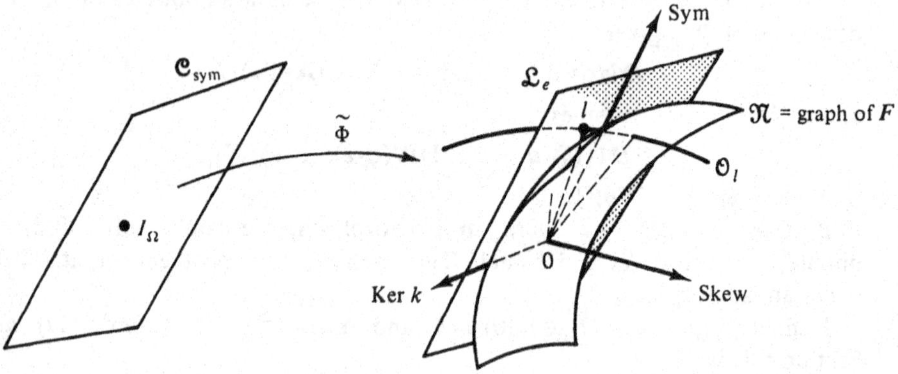

Figure 7.3.2

Later we shall show how to compute $D^2\boldsymbol{F}(0)$ in terms of $D\Phi(I_\Omega)^{-1}$ and $\mathbf{c}$ (see Proposition 3.34).

Now we are ready to reformulate Problem (***P1***).

> (***P2***) *For a given $l_0 \in \mathfrak{L}_e$ near zero, study how $\mathcal{O}_l$ meets the graph of $\boldsymbol{F}$ for various l near l_0.*

Problems (P1) and (P2) are related as follows. Let ϕ solve (E) with $l \in \mathfrak{L}$ and $\boldsymbol{Q}$ be such that $\bar{\phi} = \boldsymbol{Q}\phi \in \mathfrak{C}_{sym}$. Then $\Phi(\bar{\phi}) = \boldsymbol{Q}l$, so the orbit of l meets the graph of $\boldsymbol{F}$ at $\Phi(\bar{\phi})$. Conversely, if the orbit of l meets $\mathfrak{N}$ at $\Phi(\bar{\phi})$, then $\phi = \boldsymbol{Q}^{-1}\bar{\phi}$ solves (E).

We claim that near the trivial solutions, the numbers of solutions to each problem also correspond. This follows from the next lemma.

3.20 Lemma *There is a neighborhood $\mathfrak{U}$ of I_Ω in $\mathfrak{C}_{sym}$ such that $\phi \in \mathfrak{U}$ and $\boldsymbol{Q}\phi \in \mathfrak{U}$ implies $\boldsymbol{Q} = \boldsymbol{I}$.*

Proof Note that $\mathcal{C}_{\text{sym}}$ is transverse to $\mathcal{O}_{I_\Omega}$ at I_Ω and I_Ω has no isotropy. Thus, as SO(3) is compact, $\mathcal{O}_{I_\Omega}$ is closed, so there is a neighborhood $\mathcal{U}_0$ of I_Ω in $\mathcal{C}_{\text{sym}}$ such that $\boldsymbol{Q}|\Omega \in \mathcal{U}_0$ implies $\boldsymbol{Q} = \boldsymbol{I}$. The same thing is true of orbits passing through a small neighborhood of $\boldsymbol{I}$ by openness of transversality and compactness of SO(3). ∎

If $\mathcal{O}_l$ meets $\mathfrak{N}$ in k points $\boldsymbol{Q}_i l = \Phi(\bar{\phi}_i)$ $(i = 1, \ldots, k)$, then $\bar{\phi}_i$ are distinct as Φ is 1-1 on a neighborhood of I_Ω in $\mathcal{C}_{\text{sym}}$. If this neighborhood is also contained in $\mathcal{U}$ of 3.20, then the points $Q_i^{-1}\bar{\phi}_i = \phi_i$ are also distinct.

Hence problems (P1) and (P2) are equivalent.

In connection with the action $(\boldsymbol{Q}, \boldsymbol{A}) \mapsto \boldsymbol{QA}$ of SO(3) on $\mathfrak{M}_3$, we shall require some more notation. Let

$$\text{Skew}(\boldsymbol{QA}) = \tfrac{1}{2}(\boldsymbol{QA} - \boldsymbol{A}^T\boldsymbol{Q}^T) \in \text{skew}$$

and

$$\text{Sym}(\boldsymbol{QA}) = \tfrac{1}{2}(\boldsymbol{QA} + \boldsymbol{A}^T\boldsymbol{Q}^T) \in \text{sym}$$

be the skew-symmetric and symmetric parts of $\boldsymbol{QA}$, respectively.

We shall, by abuse of notation, suppress $\boldsymbol{j}$ and identify Sym with sym and Skew with skew. Thus we will write a load $l \in \mathcal{L}$ as $l = (\boldsymbol{A}, \boldsymbol{n})$, where $\boldsymbol{A} = \boldsymbol{k}(l) \in \mathfrak{M}_3$ and $n \in \ker \boldsymbol{k}$; hence $l \in \mathcal{L}_e$ precisely when $\boldsymbol{A} \in$ sym. The action of SO(3) on $\mathcal{L}$ is given by $\boldsymbol{Q}l = (\boldsymbol{QA}, \boldsymbol{Q}n)$.

Using this notation we can reformulate Problem (P2) as follows:

(P3) *For a given* $l_0 = (\boldsymbol{A}_0, \boldsymbol{n}_0) \in \mathcal{L}_e$ *near zero, and* $l = (\boldsymbol{A}, \boldsymbol{n})$ *near* l_0, *find* $\boldsymbol{Q} \in$ SO(3) *such that*

$$\text{Skew}(\boldsymbol{QA}) - \boldsymbol{F}(\text{Sym}(\boldsymbol{A}, \boldsymbol{Q}), \boldsymbol{Q}n) = \boldsymbol{0}.$$

Next define a rescaled map $\bar{\boldsymbol{F}}: \mathbb{R} \times \mathcal{L}_e \to$ Skew by

$$\bar{\boldsymbol{F}}(\lambda, l) = \frac{1}{\lambda^2}\boldsymbol{F}(\lambda l).$$

Since $\boldsymbol{F}(0) = \boldsymbol{0}$ and $D\boldsymbol{F}(0) = \boldsymbol{0}$, $\bar{\boldsymbol{F}}$ is smooth. Moreover, if $\boldsymbol{F}(l) = \frac{1}{2}\boldsymbol{G}(l) + \frac{1}{6}\boldsymbol{C}(l) + \cdots$ is the Taylor expansion of $\boldsymbol{F}$ about zero, then $\bar{\boldsymbol{F}}(\lambda, l) = \frac{1}{2}\boldsymbol{G}(l) + (\lambda/6)\boldsymbol{C}(l) + \cdots$.

In problem (E) let us measure the size of l by the parameter λ. Thus, replace $\Phi(\phi) = l$ for l near zero by $\Phi(\phi) = \lambda l$ for λ near zero. This scaling enables us to conveniently distinguish the size of l from its "orientation." In the literature l has always been fixed and λ taken small. Here we allow l to vary as well. These extra parameters are crucial for the complete bifurcation picture. Thus we arrive at the final formulation of the problem.

(P4) *For a given* $l_0 = (\boldsymbol{A}_0, \boldsymbol{n}_0) \in \mathcal{L}_e$, *for* l *near* l_0 *and* λ *small, find* $\boldsymbol{Q} \in$ SO(3) *such that*

$$\text{Skew}(\boldsymbol{QA}) - \lambda\bar{\boldsymbol{F}}(\lambda, \text{Sym}(\boldsymbol{QA}), \boldsymbol{Q}n) = \boldsymbol{0}.$$

The left-hand side of this equation will be denoted $H(\lambda, A, n, Q)$ or $H(\lambda, Q)$ if A, n are fixed.

With all these preliminaries at hand, we are ready to give a simple proof of one of the first of Stoppelli's basic theorems.[9]

3.21 Theorem *Suppose $l \in \mathfrak{L}_e$ has no axis of equilibrium. Then for λ sufficiently small, there is a unique $\bar{\phi} \in \mathfrak{C}_{sym}$ and a unique Q in a neighborhood of the identity in* SO(3) *such that $\phi = Q^{-1}\bar{\phi}$ solves the traction problem*

$$\Phi(\phi) = \lambda l.$$

Proof Define $H\colon \mathbb{R} \times \mathrm{SO}(3) \longrightarrow \mathrm{Skew}$ by

$$H(\lambda, Q) = \mathrm{Skew}(QA) - \lambda \bar{F}(\lambda, \mathrm{Sym}(QA), n),$$

where $l = (A, n) \in \mathfrak{L}_e = \mathrm{Sym} \oplus \mathrm{Ker}\, k$ is fixed. Note that $D_2H(0, I)\cdot W = \mathrm{Skew}(WA) = \frac{1}{2}(AW + WA)$. By Proposition 3.10, this is an isomorphism. Hence, by the implicit function theorem, $H(\lambda, Q) = \mathbf{0}$ can be uniquely solved for Q near $I \in \mathrm{SO}(3)$ as a function of λ near $0 \in \mathbb{R}$. ∎

The geometric reason "why" this proof works and the clue to treating other cases is the following:

3.22 Lemma *If $l \in \mathfrak{L}_e$ has no axis of equilibrium, then $\mathcal{O}_l$ intersects $\mathfrak{L}_e$ transversely at l* (i.e., $\mathfrak{L} = \mathfrak{L}_e \oplus T_l\mathcal{O}_l$), *and conversely.*

Proof The tangent space of $\mathcal{O}_l$ at $l \in \mathfrak{L}_e$ is $T_l\mathcal{O}_l = \{Wl \mid W \in \mathrm{Skew}\}$. Transversality means that the projection of $T_l\mathcal{O}_l$ to the complement Skew of $\mathfrak{L}_e$ is surjective. The projection is $Wl \mapsto \frac{1}{2}(WA + AW)$ where $A = k(l)$, so the result follows from 3.10, part 3. ∎

We have shown that there is only one solution to $\Phi(\phi) = \lambda l$ near the identity if λ is small and l has no axis of equilibrium. How many solutions are there near the set of all trivial solutions SO(3)? This problem has a complex answer that depends on the *type* of l. We analyze the simplest case now. Recall (see Definition 3.12) that a load $l \in \mathfrak{L}_e$ is said to be of type 0 if l has no axis of equilibrium and if $A = k(l)$ has distinct eigenvalues.

Loads with no axis of equilibrium occur for loads other than type 0 (see Box 3.1), and Stoppelli's theorem applies to them. However, the global structure of the corresponding set of solutions is quite different ("global" being relative to SO(3)).

3.23 Theorem *Let $l_0 \in \mathfrak{L}_e$ be of type* 0. *Then for λ sufficiently small the equation $\Phi(\phi) = \lambda l_0$ has exactly four solutions in a neighborhood of the trivial solutions* SO(3) $\subset \mathfrak{C}$ *(see Figure 7.3.3).*

---

[9]The only other *complete* proof in English we know of is given in Van Buren [1968], although sketches are available in Grioli [1962], Truesdell and Noll [1965], and Wang and Truesdell [1973]. Our proof is rather different; the use of the rescaled map $\bar{F}$ avoids a series of complicated estimates used by Stoppelli and Van Buren.

Proof By 3.13, $\mathcal{O}_{\lambda l_0}$ meets $\mathcal{L}_e$ in four points. By 3.21, in a neighborhood of 0 in $\mathcal{L}$, $\mathcal{O}_{\lambda l_0}$ meets $\mathfrak{N}$ in exactly four points, the images of $\bar{\phi}_1$, $\bar{\phi}_2$, $\bar{\phi}_3$, and $\bar{\phi}_4$, say. Thus Problem (P2) has four solutions. By the equivalence of (P1) and (P2), so does (P1). ∎

Let $A = k(l_0)$ and $\mathcal{S}_A = \{Q \mid QA \in \text{sym}\}$. From the proof of 3.13 we see that $\mathcal{S}_A$ is a four-element subgroup of SO(3), isomorphic to $\mathbb{Z}_2 \oplus \mathbb{Z}_2$. By our earlier discussions, ϕ_i are obtained from $\bar{\phi}_i$ by applying rotations close to elements of $\mathcal{S}_A$. In particular, as $\lambda \longrightarrow 0$, the solutions $\{\phi_i\}$ converge to the four element set $\mathcal{S}_A$ (regarded as a subset of $\mathcal{C}$).

For l sufficiently close to l_0 in 3.23, the problem $\Phi(\phi) = \lambda l$ will also have four solutions. Indeed by openness of transversality, $\mathcal{O}_{\lambda l}$ will also meet $\mathfrak{N}$ in four points. In other words, the picture for type 0 loads in Figure 7.3.3 is structurally stable under small perturbations of l_0.

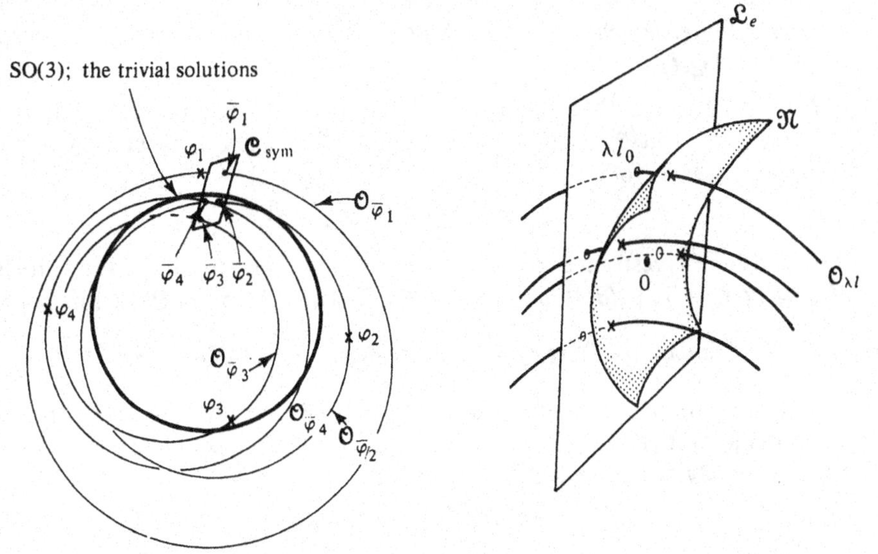

Figure 7.3.3

Next we study the dynamical stability of the four solutions found by Theorem 3.23. This is done under the hypothesis that the classical elasticity tensor is stable: we introduce the following condition.

(H3) *Assume there is an* $\eta > 0$ *such that for all* $e \in \text{Sym}(T_X\Omega, T_X\Omega)$,

$$\epsilon(e) = \tfrac{1}{2}\mathbf{c}(X)(e, e) \geq \eta \| e \|^2, \qquad \| \cdot \| = \text{pointwise norm.}$$

[$\epsilon(e)$ is the stored energy function for linearized elasticity; see Section 4.3.]

Because of the difficulties with potential wells and dynamical stability discussed in Section 6.6, we shall adopt the following "energy criteria" definition of stability.

3.24 Definition A solution ϕ of $\Phi(\phi) = l$ will be called *stable* if ϕ is a local minimum in $\mathcal{C}$ of the potential function

$$V_l(\phi) = \int_\Omega W(D\phi)\, dV - \langle l, \phi \rangle,$$

where
$$\langle l, \phi \rangle = \int_\Omega \boldsymbol{B}(X)\cdot\boldsymbol{\phi}(X)\, dV(X) + \int_{\partial\Omega} \boldsymbol{\tau}(X)\cdot\boldsymbol{\phi}(X)\, dA(X).$$

If ϕ is not stable, its *index* is the dimension of the largest subspace of vectors $\boldsymbol{u}$ tangent to $\mathcal{C}$ at ϕ with the property that V_l decreases along some curve tangent to $\boldsymbol{u}$ at ϕ. (Thus, index 0 corresponds to stability.)

3.25 Theorem *Assume (H1)–(H3) and let l_0 be as in 3.23. For λ sufficiently small, exactly one of the four solutions $\phi_1, \phi_2, \phi_3, \phi_4$ is stable; the others have indices 1, 2, and 3. More precisely, suppose ϕ_λ is a solution approaching $\boldsymbol{Q} \in \mathcal{S}_A$ as $\lambda \to 0$. Then for λ small, ϕ_l is stable if and only if $\boldsymbol{QA} - \operatorname{tr}(\boldsymbol{QA})\boldsymbol{I} \in$ sym is positive-definite. In general, the index of ϕ_λ is the number of negative eigenvalues of $\boldsymbol{QA} - \operatorname{tr}(\boldsymbol{QA})\boldsymbol{I}$.*

Proof Let $\phi_0 \in \mathcal{C}$ solve $\Phi(\phi) = \lambda l_0 = l$. Then ϕ_0 is a critical point of $V_{\lambda l_0}$. Consider the orbit $\mathcal{O}_{\phi_0} = \{\boldsymbol{Q}\phi_0 \mid \boldsymbol{Q} \in \mathrm{SO}(3)\}$ of ϕ_0. The tangent space to $\mathcal{C}$ at ϕ_0 decomposes as follows:

$$T_{\phi_0}\mathcal{C} = T_{\phi_0}\mathcal{O}_{\phi_0} \oplus (T_{\phi_0}\mathcal{O}_{\phi_0})^\perp.$$

First consider V_l restricted to $(T_{\phi_0}\mathcal{O}_{\phi_0})^\perp$. Its second derivative at ϕ in the direction of $\boldsymbol{u} \in (T_\phi \mathcal{O}_\phi)^\perp$ is $\int_\Omega (\partial^2 W/\partial \boldsymbol{F}\, \partial \boldsymbol{F})(\phi)\cdot(\nabla \boldsymbol{u}, \nabla \boldsymbol{u})\, dV$. At $\phi_0 = \boldsymbol{Q} \mid \Omega$, this becomes

$$\int_\Omega \mathsf{c}(X)\cdot(\boldsymbol{e}(X), \boldsymbol{e}(X))\, dV(X), \quad \text{where} \quad \boldsymbol{e} = \tfrac{1}{2}(\nabla \boldsymbol{u} + (\nabla \boldsymbol{u})^T).$$

This is larger than a positive constant times the L^2-norm of $\boldsymbol{e}$, by (H3). However, since $\boldsymbol{u}$ is in $(T_\phi\mathcal{O}_\phi)^\perp$, $\|\boldsymbol{e}\|_{L^2}^2 \geq (\text{const.})\, \|\boldsymbol{u}\|_{H^1}^2$ by Korn's inequality (see Box 1.1, Chapter 6). By continuity, if λ is small,

$$D^2 V_{\lambda l_0}(\phi_0)\cdot(\boldsymbol{u}, \boldsymbol{u}) \geq \delta \|\boldsymbol{u}\|_{H^1}^2$$

for all $\boldsymbol{u}$ orthogonal to $\mathcal{O}_{\phi_0}$ at ϕ_0. This implies ϕ_0 is a minimum for $V_{\lambda l_0}$ in directions transverse to $\mathcal{O}_{\phi_0}$ (cf. Section 6.6).

Next, consider $V_{\lambda l_0}$ restricted to $\mathcal{O}_{\phi_0}$. By material frame indifference, W is constant on $\mathcal{O}_{\phi_0}$ and as ϕ_0 must be a critical point for $V_{\lambda l_0}$ restricted to $\mathcal{O}_{\phi_0}$, it is also a critical point for $\lambda l_0 = l$ restricted to $\mathcal{O}_{\phi_0}$ (where $l: \mathcal{C} \to \mathbb{R}$ is defined by $l(\phi) = \langle l, \phi \rangle$). It suffices therefore to determine the index of $l \mid \mathcal{O}_{\phi_0}$ at ϕ_0. The result is a consequence of continuity and the limiting case $\lambda \to 0$ given in the following lemma about type 0 loads.

3.26 Lemma *Let l be type 0 and let $\boldsymbol{A} = \boldsymbol{k}(l)$. Then $\mathcal{S}_A$, regarded as a subset of $\mathcal{C}$ equals the set of critical points of $l \mid \mathcal{O}_{I_\Omega}$. These four critical points are non-degenerate with indices 0, 1, 2, and 3; the index of $\boldsymbol{Q}$ is the number of negative eigenvalues of $\boldsymbol{QA} - \operatorname{tr}(\boldsymbol{QA})\boldsymbol{I}$.*

Proof First note that $\mathcal{L}_e = (T_{I_\Omega}\,\mathrm{SO}(3))^\perp$ since $D\Phi(I_\Omega)$ has kernel $T_{I_\Omega}\,\mathrm{SO}(3)$ = skew, has range $\mathcal{L}_e$, and is self-adjoint. Thus $Ql \in \mathcal{L}_e$ if and only if $l \perp T_{Q^T}\,\mathrm{SO}(3)$. It follows that $Ql \in \mathcal{L}_e$ if and only if Q^T is a critical point of $l\,|\,\mathcal{O}_{I_\Omega}$. Recall that elements of $\mathcal{S}_A = \{Q \in \mathrm{SO}(3)\,|\,Ql \in \mathcal{L}_e\}$ are symmetric (see 3.13).

To compute the index of $l\,|\,\mathcal{O}_{I_\Omega}$ at $Q \in \mathcal{S}_A$, we compute the second derivative

$$\frac{d^2}{dt^2} l(\exp(tW)Q)\,|_{t=0} = l(W^2 Q).$$

Now
$$l(W^2Q) = \operatorname{trace} \mathbf{k}(l, W^2Q) = \operatorname{trace} W^2\mathbf{k}(l, Q)$$
$$= \operatorname{trace} W^2\mathbf{k}(Q^T l) = \operatorname{trace}[W^2 QA].$$

This quadratic form on skew is represented by the element $QA - \operatorname{tr}(QA)I$ of Sym as is seen by a simple computation. Using the representations for $\{QA\}$ given by Proposition 3.13, namely,

diag(a, b, c), diag(—a, —b, c), diag(—a, b, —c), and diag(a, —b, —c)

one checks that all four indices occur. ∎

In the following boxes we shall outline, omitting a number of proofs, the methods by which the analysis of the other types proceeds. Again, Chillingworth, Marsden, and Wan [1982a] should be consulted for details and the full results.

Box 3.1 ***Classification of Orbits in*** $\mathfrak{M}_3$

The classification of loads depends on a classification of the corresponding astatic loads. This will be done by classifying orbits in $\mathfrak{M}_3$ under the action $(Q, A) \mapsto QA$ of SO(3) on $\mathfrak{M}_3$ by the way the orbits meet sym. By the polar decomposition, we can assume $A \in$ sym. In 3.13 we proved:

3.27 Proposition (*Type 0*) *Suppose* $A \in$ sym *has no axis of equilibrium and has distinct eigenvalues. Then* $\mathcal{O}_A \cap$ sym *consists of four points at each of which the intersection is transversal.*

We shall let the eigenvalues of $A \in$ sym be denoted a, b, c; following 3.11 we shall say that A has no axis of equilibrium when $(a + b)(b + c)(a + c) \neq 0$; that is, $a + b + c \neq a, b,$ or c, and in this case $\mathcal{O}_A$ intersects sym transversly at A.

3.28 Definition We shall say A is of *type 1* if A has no axis of equilibrium and if exactly two of a, b, c are equal (say $a = b \neq c$).

3.29 Proposition *If A is type 1, then $\mathcal{O}_A \cap$ sym consists of two points (each with no axis of equilibrium) and an $\mathbb{RP}^1$ . . . or equivalently a circle (each point of which has one axis of equilibrium).*

$\mathbb{RP}^n$ denotes the set of lines through the origin in $\mathbb{R}^{n+1}$. The proof of 3.29 is essentially a straightforward exercise in linear algebra. Likewise, for type 2 we proceed as follows.

3.30 Definition We shall say A is of *type 2* if A has no axis of equilibrium and all three of a, b, c are equal (and so $\neq 0$).

3.31 Proposition *If A is type 2, then $\mathcal{O}_A \cap$ sym consists of one point (A itself) and an $\mathbb{RP}^2$ (each point of which has a whole circle of axes of equilibrium).*

Types 3 and 4 are treated next.

3.32 Definition We say A is *type 3* if dim Ker $A = 2$ and say A is *type* 4 if $A = \mathbf{0}$.

3.33 Proposition *If A is type 3, then $\mathcal{O}_A \cap$ sym consists of two points, A and $-A$. If A is type 4, $\mathcal{O}_A \cap$ sym $= \{\mathbf{0}\}$.*

Thus, orbits in $\mathfrak{M}_3$ fall into exactly five different types; 0, 1, 2, 3, and 4 with the properties above.

Stoppelli partially analyzed only types 0 and 1. We discussed type 0 in the text and shall briefly comment on type 1 in the next box. Types 2, 3, and 4 are also interesting; see Chillingworth, Marsden, and Wan [1982b] for details.

Problem 3.3 Use the results above and 3.21 to prove the existence of at least one solution of the traction problem for a load of types 1 or 2.

Problem 3.4 A load l is called *parallel* if there is a vector $\boldsymbol{a} \in \mathbb{R}^3$ and scalar functions $f\colon \Omega \longrightarrow \mathbb{R}$, $g\colon \partial\Omega \longrightarrow \mathbb{R}$ such that $l = (f\boldsymbol{a}, g\boldsymbol{a})$. Show that parallel loads are of type 3.

Problem 3.5 Suppose $\boldsymbol{B} = \mathbf{0}$ and $\boldsymbol{\tau} = \tau N$ for a constant τ. Show that this load is of type 2.

Figure 7.3.4 shows some examples of loads l where $\boldsymbol{k}(l)$ has types 1, 2, 3, and 4.

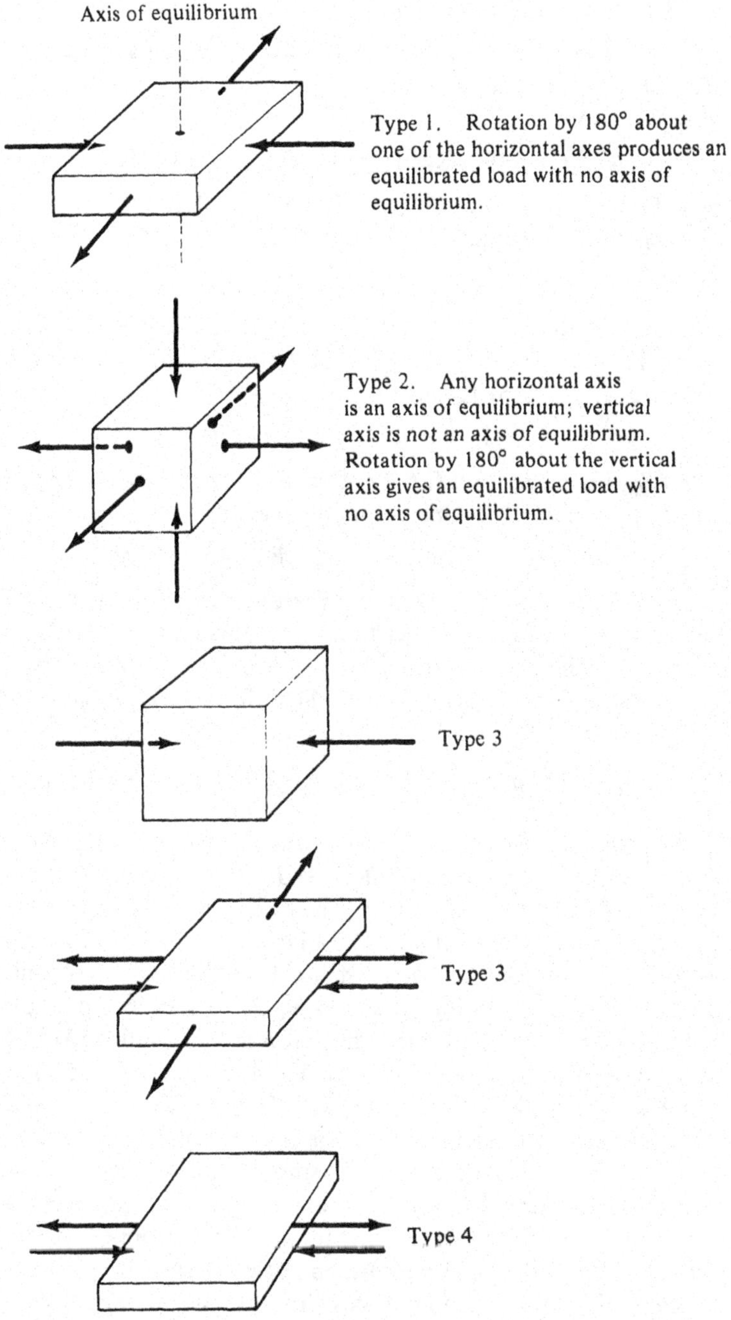

Axis of equilibrium
Type 1. Rotation by 180° about one of the horizontal axes produces an equilibrated load with no axis of equilibrium.
Type 2. Any horizontal axis is an axis of equilibrium; vertical axis is *not* an axis of equilibrium. Rotation by 180° about the vertical axis gives an equilibrated load with no axis of equilibrium.
Type 3
Type 3
Type 4

Figure 7.3.4

Box 3.2 ***The Bifurcation Equation for Type 1***

We shall now indicate briefly how the bifurcation analysis proceeds. According to the formulation (P4) of our problem, given $(A_0, n_0) \in \mathcal{L}_e$, we wish to solve

$$H(\lambda, A, n, Q) = \text{Skew}(QA) - \lambda F(\lambda, \text{Sym}(QA), n) = 0$$

for Q for various (A, n) near (A_0, n_0), and small λ.

Define the vector field X_{A_0} on SO(3) by

$$X_{A_0}(Q) = \text{skew}(QA)\cdot Q$$

(right translation of Skew(QA) from SO(3) to T_Q SO(3)). Likewise we regard H as a vector field $X(\lambda, A, n, Q)$ on SO(3) depending on the parameters λ, A, n by setting

$$X(\lambda, A, n, Q) = H(\lambda, A, n, Q)\cdot Q.$$

Let $\mathcal{S}_{A_0}$ be the zero set for X_{A_0}. For A_0 of type 1, $\mathcal{S}_{A_0}$ consists of two points and a circle $\mathcal{C}_{A_0}$. One computes that for $Q \in \mathcal{S}_{A_0}$,

$$T_Q\mathcal{S}_{A_0} = \{W_0 Q \mid W_0 \in \text{skew} \quad \text{and} \quad W_0(QA_0) + (QA_0)W_0 = 0\}.$$

From 3.10, $W_0 \mapsto W_0(QA_0)W_0$ corresponds to the linear transformation $QA_0 - \text{tr}(QA_0)I$ under the isomorphism of skew with $\mathbb{R}^3$. Symmetry of this linear transformation when QA_0 is symmetric is a reflection of the fact that X_{A_0} is a gradient field. In fact, X_{A_0} is the gradient of $l_0 | \mathcal{O}_{I\Omega}$—that is, of

$$l_0(Q) = \int_\Omega B_0(X)\cdot Q(X)\, dV(X) + \int_{\partial\Omega} \tau_0(X)\cdot Q(X)\, dA(X)$$

as one sees by a calculation. Thus, by the Fredholm alternative, at each point Q of $\mathcal{S}_{A_0}$, $DX_{A_0}(Q)\colon T_Q\,\text{SO}(3) \to T_Q\,\text{SO}(3)$ has range the orthogonal complement of $T_Q\mathcal{S}_{A_0}$. Thus, the range of DX_{A_0} over its zero set $\mathcal{S}_{A_0}$ is the normal bundle of $\mathcal{S}_{A_0}$. The Liapunov–Schmidt procedure now produces a unique section $\phi_{\lambda,A,n}$ of the normal bundle to $\mathcal{S}_{A_0}$ such that the orthogonal projection of $X(\lambda, A, n, Q)$ to the fiber of the normal bundle is zero. Let $\Gamma(\lambda, A, n)$ be the graph of $S_{\lambda,A,n}$ and let $\tilde{X}(\lambda, A, n, Q)$ be the projection of X to $T_Q\Gamma$. Thus, $\tilde{X}$ is a vector field on Γ. The equation $\tilde{X} = 0$ is our bifurcation equation. One can show that, essentially, $\tilde{X}$ is a gradient, so one is looking for critical points of a function on a circle. These can then be described using singularity theory. Cusp bifurcations (see Section 7.1) are, not surprisingly, present, so an isolated solution near the circle can bifurcate into three. Stoppelli [1958] found some of these extra solutions by analyzing a slice of the bifurcation diagram. For type 2 a double cusp occurs and a single solution can bifurcate into 9. These are also new solutions.

Box 3.3 ***Miscellany: Curvature and Linearization Stability***

The manifold $\mathfrak{N}$ was shown to be the graph of a map $\boldsymbol{F}$; see 3.19. Essentially, $\boldsymbol{F}$ is the skew-component of Φ. Now we will demonstrate an almost paradoxical fact: the *second* derivative of this map at I_Ω can be computed knowing only the classical elasticity tensor $\mathbf{c}$. Intuitively one would expect this second derivative to depend on higher nonlinearities. This second derivative tells us, essentially, the curvature of $\mathfrak{N}$.

3.34 Proposition *Let* $\mathfrak{F}\colon \mathcal{C} \dashrightarrow$ skew *be defined by* $\mathfrak{F}(\phi) =$ Skew$[\boldsymbol{k}(\Phi(\phi))]$. *Then* $\mathfrak{F}(I_\Omega) = 0$, $D\mathfrak{F}(I_\Omega) = 0$, *and*

$$D^2\mathfrak{F}(I_\Omega)(\boldsymbol{u}, \boldsymbol{u}) = 2\,\text{Skew}\left(\int_\Omega \nabla \boldsymbol{u}\, \mathbf{c}\cdot \nabla \boldsymbol{u}\, dV\right) = -2\,\text{Skew}\, \boldsymbol{k}(\boldsymbol{l}_u, \boldsymbol{u}),$$

where $\boldsymbol{l}_u = (\boldsymbol{b}_u, \boldsymbol{\tau}_u)$, $\boldsymbol{b}_u = -\text{DIV}(\mathbf{c}\cdot\boldsymbol{e})$, *and* $\boldsymbol{\tau}_u = (\mathbf{c}\cdot\boldsymbol{e})\cdot\boldsymbol{N}$. *Identifying skew with* $\mathbb{R}^3$, *this reads*

$$-D^2\mathfrak{F}(I_\Omega)(\boldsymbol{u}, \boldsymbol{u}) = \left(\int_\Omega \boldsymbol{b}_u \times \boldsymbol{u}\, dV + \int_{\partial\Omega} \boldsymbol{\tau}_u \times \boldsymbol{u}\, dA\right).$$

Proof By 3.17,

$$\mathfrak{F}(\phi) = \text{Skew}\left(\int_\Omega \boldsymbol{P}\, dV\right),$$

where $\boldsymbol{P}$ is the first Piola-Kirchhoff stress. We have $\boldsymbol{P}(I_\Omega) = \boldsymbol{0}$, so $\mathfrak{F}(I_\Omega) = 0$. Also,

$$D\mathfrak{F}(I_\Omega)\cdot\boldsymbol{u} = \text{Skew}\int_\Omega \frac{\partial \boldsymbol{P}}{\partial \boldsymbol{F}}\cdot\nabla\boldsymbol{u}\, dV = \text{Skew}\int_\Omega \mathbf{c}\cdot\nabla\boldsymbol{u}\, dV = 0,$$

as $\mathbf{c}\cdot\nabla\boldsymbol{u}$ is symmetric and $(\partial\boldsymbol{P}/\partial\boldsymbol{F})(I_\Omega) = \mathbf{c}$. Next, to compute $D^2\mathfrak{F}$, we need to use the fact that $\boldsymbol{S}$ is symmetric, so write $\boldsymbol{P} = \boldsymbol{F}\boldsymbol{S}$ and obtain $D_{\boldsymbol{F}}\boldsymbol{P}(\boldsymbol{F})\cdot\nabla\boldsymbol{u} = \nabla\boldsymbol{u}\boldsymbol{S}(\boldsymbol{F}) + \boldsymbol{F}D_{\boldsymbol{F}}\boldsymbol{S}(\boldsymbol{F})\cdot\nabla\boldsymbol{u}$. Thus,

$$\begin{aligned} D^2\boldsymbol{P}(I_\Omega)\cdot(\nabla\boldsymbol{u}, \nabla\boldsymbol{v}) &= \nabla\boldsymbol{u}\, D_{\boldsymbol{F}}\boldsymbol{S}(I_\Omega)\cdot\nabla\boldsymbol{v} + \nabla\boldsymbol{v} D_{\boldsymbol{F}}\boldsymbol{S}(I_\Omega)\cdot\nabla\boldsymbol{u} \\ &\quad + D_{\boldsymbol{F}}^2\boldsymbol{S}(I_\Omega)(\nabla\boldsymbol{u}, \nabla\boldsymbol{v}) \end{aligned}$$

Now $D_{\boldsymbol{F}}\boldsymbol{S}(I_\Omega)\cdot\nabla\boldsymbol{u} = D_{\boldsymbol{C}}\boldsymbol{S}(I_\Omega)\cdot(\nabla\boldsymbol{u} + \nabla\boldsymbol{u}^T) = \mathbf{c}\cdot\nabla\boldsymbol{u}$ and $D_{\boldsymbol{F}}^2\boldsymbol{S}(I_\Omega)$ is symmetric, so

$$\begin{aligned} D^2\mathfrak{F}(I_\Omega)\cdot(\boldsymbol{u}, \boldsymbol{v}) &= \text{skew}\int_\Omega D_{\boldsymbol{F}}^2\boldsymbol{P}(I_\Omega)(\nabla\boldsymbol{u}, \nabla\boldsymbol{v})\, dV) \\ &= \text{skew}\left(2\int_\Omega (\nabla\boldsymbol{u}\, \mathbf{c}\cdot\nabla\boldsymbol{v} + \nabla\boldsymbol{v}\, \mathbf{c}\cdot\nabla\boldsymbol{u})\, dV\right). \end{aligned}$$

Thus

$$\begin{aligned} D^2\mathfrak{F}(I_\Omega)(\boldsymbol{u}, \boldsymbol{u}) &= 2\,\text{skew}\left(\int_\Omega \nabla\boldsymbol{u}\, \mathbf{c}\cdot\nabla\boldsymbol{u}\, dV\right) \\ &= \text{skew}\int_\Omega \boldsymbol{b}\otimes\boldsymbol{u}\, dV + \int_{\partial\Omega} \boldsymbol{\tau}\otimes\boldsymbol{u}\, dA \end{aligned}$$

by the divergence theorem. ∎

3.35 Example For a homogeneous isotropic material, $\mathbf{c}\cdot e = \lambda(\text{trace } e)I + 2\mu e$ for constants λ and μ (the Lamé moduli). Thus

$$D^2\mathfrak{F}(I_\Omega)(u, u) = 2 \text{ skew}\left(\int_\Omega [2\mu \nabla u\, e + \lambda \text{ trace}(e)\, \nabla u]\, dV\right)$$

$$= 2 \text{ skew} \int_\Omega \{\mu\, \nabla u\, \nabla u + \lambda \text{ trace}(e)\, \nabla u\}\, dV$$

Finally we make a few remarks about linearization stability; see Section 4.4.

3.36 Definition Suppose a pair (u_1, l_1) satisfies the equations linearized about our stress free reference state I_Ω; that is,

$$D\Phi(I_\Omega)\cdot u_1 = l_1.$$

Let us call the pair (u_1, l_1) *linearization stable* (*or integrable*) if there exists a curve $(\phi(\lambda), l(\lambda)) \in \mathcal{C} \times \mathfrak{L}_e$ such that:

(i) $\phi(0) = I_\Omega$, $l(0) = 0$;
(ii) $\phi'(0) - u_1 \in \text{Ker } D\phi(I_\Omega)$, $l'(0) = l_1$; and
(iii) $\Phi(\phi(\lambda)) = l(\lambda)$.

Here $(\phi(\lambda), l(\lambda))$ should be defined in some interval; say $[0, \epsilon)$, $\epsilon > 0$. (We can, of course, do the same about any state, not just I_Ω.)

3.37 Proposition *Suppose $l_1 \in \mathfrak{L}_e$ has no axis of equilibrium and $D\Phi(I_\Phi)u_1 = l_1$. Then (u_1, l_1) is linearization stable.*

Proof Let $l(\lambda) = \lambda l_1$. Then there is a unique smooth curve $\phi(\lambda)$ through I_Ω such that $\Phi(\phi(\lambda)) = l(\lambda)$ by Theorem 3.21. Differentiating at $\lambda = 0$ gives $D\Phi(I_\Omega)\cdot\phi'(0) = l_1$, so $\phi'(0) - u_1 \in \text{Ker } D\Phi(I_\Omega)$. ∎

The following produces a potential obstruction to linearization stability. It is called the "Signorini compatability conditions." Let us use the notation

$$\int_\Omega u \times l \quad \text{for} \quad \int_\Omega u(X) \times B(X)\, dV(X) + \int_{\partial\Omega} u(X) \times \tau(X)\, dA(X).$$

Let us note that linearization stability really just involves l_1; let us say l_1 is *integrable* when there is a curve $(\phi(\lambda), l(\lambda)) \in \mathcal{C} \times \mathfrak{L}_e$ satisfying (i) and (iii) above with $l'(0) = l_1$. Then $D\Phi(I_\Omega)\cdot\phi'(0) = l_1$ is automatic.

3.38 Proposition *Suppose l_1 is integrable. Then there exists a u_1 such that*

$$D\Phi(I_\Omega)u_1 = l_1 \tag{L}$$

and

$$\int_\Omega u_1 \times l_1 = \mathbf{0}. \tag{C}$$

Proof Take $\boldsymbol{u}_1 = \phi'(0)$ and differentiate the identity $\int_\Omega \phi(\lambda) \times l(\lambda) = 0$ twice and set $\lambda = 0$; all that survives is $\int_\Omega \boldsymbol{u}_1 \times l_1 = \mathbf{0}$ since $l''(0) = l_2 \in \mathfrak{L}_e$ $\left(\text{so} \int_\Omega I_\Omega \times l_2 = 0\right)$ and $l(0) = 0$. ∎

Remarks

1. Note that $l_1 \in \mathfrak{L}_e$ is a necessary condition for integrability.

2. A basic question to be asked is when the compatability conditions (C) are sufficient for integrability and how much freedom there is in our choice of $l(\lambda)$. This is the spirit of the classical work, where $\phi(\lambda)$ and $l(\lambda)$ are expanded in power series. See Truesdell and Noll [1965] for extensive discussions.

Problem 3.6 Show that these compatability conditions coincide with those derived in Truesdell and Noll [1965].

The following major theorem of Marsden and Wan [1983], whose proof is omitted here, establishes a key link with and substantially improves upon the classical power series methods.

3.39 Theorem *Suppose* (u_1, l_1) *satisfy* (L) *and* (C). *Then* (u_1, l_1) *is linearization stable.*

In this result one cannot simply take $l(\lambda) = \lambda l_1$, the second term in the expansion $l(\lambda) = \lambda l_1 + \lambda^2 l_2 + \ldots$ plays a key role.

3. To give an example of a non-integrable $l_1 \in \mathfrak{L}_e$, one can find an l_1 such that for any $\boldsymbol{u}_1$ satisfying (L), condition (C) is violated. Such an example of Signorini is discussed in §9 of Capriz and Podio–Guidugli [1974].

4. One can carry out an analysis similar to this around a stressed state as well. The details of the computations and the possible bifurcation diagrams can be more complex, as one in effect has to deal with "genuine" three-dimensional buckling. This aspect is treated by Wan [1983]. The perturbation series approach of Signorini has been carried out in this case by Bharatha and Levinson [1978].

7.4 BASIC IDEAS OF DYNAMIC BIFURCATION THEORY

Dynamic bifurcation theory differs from the static theory in that we now concentrate on qualitative changes in phase portraits, such as the sudden appearance of periodic orbits. The static theory of Section 7.1 may be regarded as a subtheory, namely, the study of bifurcation of equilibrium points. One of the basic theorems in this subject is the "Hopf bifurcation theorem" on the appearance of closed orbits. We present a self-contained proof of this theorem in Box 4.1.

The dynamical framework in which we operate is described as follows. Let $\mathcal{Y} \subset \mathcal{X}$ be Banach spaces (or manifolds) and let

$$f\colon \mathcal{Y} \times \mathbb{R}^p \longrightarrow \mathcal{X}$$

be a given C^k mapping. Here $\mathbb{R}^p$ is the parameter space and f may be defined only on an open subset of $\mathcal{Y} \times \mathbb{R}^p$. The dynamics is determined by the evolution equation $dx/dt = f(x, \lambda)$, which will be assumed to define a local semiflow $F_t^\lambda\colon \mathcal{Y} \longrightarrow \mathcal{Y}$ by letting $F_t^\lambda(x_0)$ be the solution of $\dot{x} = f(x, \lambda)$ with initial condition $x(0) = x_0$. See Section 6.5 for instances when this is valid.

A *fixed point* is a point (x_0, λ) such that $f(x_0, \lambda) = 0$. Therefore, $F_t^\lambda(x_0) = x_0$; that is, x_0 is an equilibrium point of the dynamics.

A fixed point (x_0, λ) is called $\mathcal{X}$- (resp. $\mathcal{Y}$-) *stable* if there is an $\mathcal{X}$- (resp. $\mathcal{Y}$-) neighborhood $\mathcal{U}_0$ of x_0 such that for $x \in \mathcal{U}_0 \cap \mathcal{Y}$, $F_t^\lambda(x)$ is defined for all $t \geq 0$, and if for any neighborhood $\mathcal{U} \subset \mathcal{U}_0$, there is a neighborhood $\mathcal{V} \subset \mathcal{U}_0$ such that $F_t^\lambda(x) \in \mathcal{U}$ if $x \in \mathcal{V}$ and $t \geq 0$. The fixed point is called *asymptotically stable* if, in addition, $F_t^\lambda(x) \longrightarrow x_0$ in the $\mathcal{X}$-norm (resp. $\mathcal{Y}$-norm) as $t \longrightarrow +\infty$, for x in a neighborhood of x_0.

> ***Problem 4.1*** Discuss the relationship between this notion of stability and that in Section 6.6 (see Definition 6.2).

Many semilinear hyperbolic and most parabolic equations satisfy an additional smoothness condition; we say F_t^λ is a $\mathcal{Y}$–C^k *semiflow* if for each t and λ, $F_t^\lambda\colon \mathcal{Y} \longrightarrow \mathcal{Y}$ (where defined) is a C^k map and its derivatives are strongly continuous in t, λ. Similarly, we say F_t^λ is $\mathcal{X}$—C^k if it extends to a C^k map of $\mathcal{X}$ to $\mathcal{X}$. One especially simple case occurs when

$$f(x, \lambda) = A_\lambda x + B(x, \lambda),$$

where $A_\lambda\colon \mathcal{Y} \longrightarrow \mathcal{X}$ is a linear generator depending continuously on λ and $B\colon \mathcal{X} \times \mathbb{R}^p \longrightarrow \mathcal{X}$ is a C^k map. Then F_t^λ is C^k from $\mathcal{X}$ to $\mathcal{X}$ and if B is C^k from $\mathcal{Y}$ to $\mathcal{Y}$, so is F_t^λ. This result is readily proved by the variation of constants formula

$$x(t) = e^{tA_\lambda} x_0 + \int_0^t e^{(t-s)A_\lambda} B(x(s), \lambda)\, ds.$$

See Section 6.5 for details. For more general conditions under which a semiflow is smooth, see Marsden and McCracken [1976]; see also Box 5.1, Section 6.5. The stability of fixed points may often be determined by the following basic result. For example, it applies to the Navier–Stokes equations, reproducing Prodi [1962] as a special case.

4.1 Liapunov's Theorem *Suppose F_t is an $\mathcal{X}$–C^1 flow, x_0 is a fixed point and the spectrum of the linear semigroup*

$$\mathcal{U}_t = D_x F_t(x_0)\colon \mathcal{X} \longrightarrow \mathcal{X}$$

(the Fréchet derivative with respect to $x \in \mathfrak{X}$) is $e^{t\sigma}$, where σ lies in the left half-plane a distance $> \delta > 0$ from the imaginary axis. Then x_0 is asymptotically stable and for x sufficiently close to x_0 we have the estimate

$$\| F_t(x) - x_0 \| \leq Ce^{-t\delta}.$$

Proof We shall need to accept from spectral and semigroup theory that there is an $\epsilon > 0$ and an equivalent norm $|||\cdot|||$ on $\mathfrak{X}$ such that

$$||| DF_t(x_0) ||| < e^{-\epsilon t}.$$

(Indeed, if $\mathfrak{U}_t$ is a semigroup with spectral radius e^{rt}, set

$$||| x ||| = \sup_{t \geq 0} \| \mathfrak{U}_t x \| / e^{rt};$$

see Hille and Phillips [1957].) Thus, if $0 < \epsilon' < \epsilon$,

$$||| DF_t(x) ||| \leq \exp(-\epsilon' t) \quad \text{for} \quad 0 \leq t \leq 1$$

and x in a neighborhood of x_0, say $\mathfrak{U} = \{x \mid |||x - x_0||| < r\}$. This is because F_t is C^1 with derivative continuous in t.

We claim that if $x \in \mathfrak{U}$, and t is small, then $F_t x \in \mathfrak{U}$ and

$$||| F_t(x) - x_0 ||| \leq \exp(-\epsilon' t) |||x - x_0|||.$$

But it follows from this estimate:

$$\begin{aligned}
||| F_t(x) - x_0 ||| &= ||| F_t(x) - F_t(x_0) ||| \\
&= \left|\left|\left| \int_0^1 DF_t(sx + (1 - s)x_0)\cdot(x - x_0)\, ds \right|\right|\right| \\
&< \int_0^1 ||| DF_t(sx + (1 - s)x_0 ||| \, ||| (x - x_0) ||| \, ds \\
&< \exp(-\epsilon' t) |||x - x_0|||.
\end{aligned}$$

This result now holds for large t by using the facts that $F_t = F_{t/n}^n$ and $\exp(-\epsilon' t) = [\exp(-\epsilon' t/n)]^n$. Changing back to the original norm, the theorem is proved. ∎

Observe that the hypotheses do not explicitly involve the generator A, so the theorem can be used for $\mathfrak{Y}$-smooth flows as well. As we noted in Box 5.1, Section 6.5, the full equations of nonlinear elasticity cannot be expected to define smooth flows. However, 4.1 can be expected to apply when a semilinear model is used.

To locate fixed points in a bifurcation problem, we solve the equation $f(x, \lambda) = 0$. The stability of a fixed point x_0 is usually determined by the spectrum σ of the linearization at x_0:

$$A_\lambda = D_x f(x_0, \lambda).$$

(If the operator A_λ and its semigroup are non-pathological—for example, they have discrete spectrum—then $\sigma(e^{tA_\lambda}) = e^{t\sigma(A_\lambda)}$ or the closure of this set; see Carr [1981] and Roh [1982] for additional results.) Thus, if σ lies in the left half-plane, x_0 is stable. In critical cases where the spectrum lies on the imaginary axis, stability has to be determined by other means (see Problem 5.4, Section 6.5 for an example). It is at criticality where, for example, a curve of fixed points $x_0(\lambda)$ changes from being stable to unstable, a bifurcation can occur.

The second major point we wish to make is that within the context of smooth semiflows, the invariant manifold theorems from ordinary differential equations carry over.

In bifurcation theory it is often useful to apply the invariant manifold theorems to the *suspended flow*

$$F_t\colon \mathfrak{X} \times \mathbb{R}^p \longrightarrow \mathfrak{X} \times \mathbb{R}^p \quad \text{defined by} \quad (x, \lambda) \mapsto (F_t^\lambda(x), \lambda).$$

The invariant manifold theorem states that if the spectrum of the linearization A_λ at a fixed point (x_0, λ) splits into $\sigma_s \cup \sigma_c$, where σ_s lies in the left half-plane and σ_c is on the imaginary axis, then the flow F_t leaves invariant manifolds $\mathcal{W}^s$ and $\mathcal{W}^c$ tangent to the eigenspaces corresponding to σ_s and σ_c, respectively; $\mathcal{W}^s$ is the *stable* and $\mathcal{W}^c$ is the *center* manifold. (One can allow an *unstable* manifold too if that part of the spectrum is finite.) Orbits on $\mathcal{W}^s$ converge to (x_0, λ) exponentially. For suspended systems, note that we always have $1 \in \sigma_c$.

For bifurcation problems the center manifold theorem is the most relevant, so we summarize the situation. (See Marsden and McCracken [1976] and Hassard, Kazarinoff and Wan [1981] for details.)

4.2 Center Manifold Theorem for Flows *Let $\mathcal{Z}$ be a Banach space admitting a C^∞ norm away from 0 and let F_t be a C^0 semiflow defined on a neighborhood of 0 for $0 \leq t \leq \tau$. Assume $F_t(0) = 0$ and for each $t > 0$, $F_t\colon \mathcal{Z} \longrightarrow \mathcal{Z}$ is a C^{k-1} map whose derivatives are strongly continuous in t. Assume that the spectrum of the linear semigroup $DF_t(0)\colon \mathcal{Z} \longrightarrow \mathcal{Z}$ is of the form $e^{t(\sigma_s \cup \sigma_c)}$, where $e^{t\sigma_c}$ lies on the unit circle (i.e., σ_c lies on the imaginary axis) and $e^{t\sigma_s}$ lies inside the unit circle a nonzero distance from it, for $t > 0$; that is, σ_s is in the left half-plane. Let C be the (generalized) eigenspace corresponding to the part of the spectrum on the unit circle. Assume* $\dim C = d < \infty$.

Then there exists a neighborhood $\mathcal{V}$ of 0 in $\mathcal{Z}$ and a C^k submanifold $\mathcal{W}^c \subset \mathcal{V}$ of dimension d passing through 0 and tangent to C at 0 such that;

(a) *If $x \in \mathcal{W}^c$, $t > 0$ and $F_t(x) \in \mathcal{V}$, then $F_t(x) \in \mathcal{W}^c$.*
(b) *If $t > 0$ and $F_t^n(x)$ remains defined and in $\mathcal{V}$ for all $n = 0, 1, 2, \ldots$, then $F_t^n(x) \longrightarrow \mathcal{W}_c$ as $n \longrightarrow \infty$.*

See Figure 7.4.1 for a sketch of the situation.

For example, in the pitchfork bifurcation from Section 7.1, we have a curve of fixed points $x_0(\lambda)$ and $\lambda \in \mathbb{R}$, which become unstable as λ crosses λ_0 and two stable fixed points branch off. All three points lie on the center manifold for the suspended system. Taking $\lambda =$ constant slices yields an invariant manifold $\mathcal{W}^c$ for the parametrized system; see Figure 7.4.2.

Although the center manifold is only known implicity, it can greatly simplify the problem qualitatively by reducing an initially infinite-dimensional problem

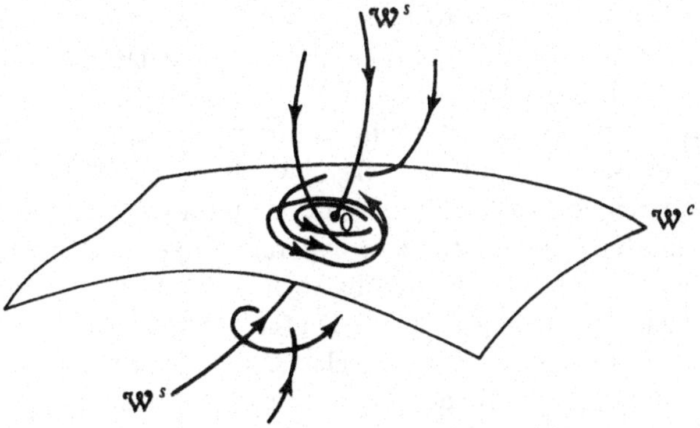

Figure 7.4.1

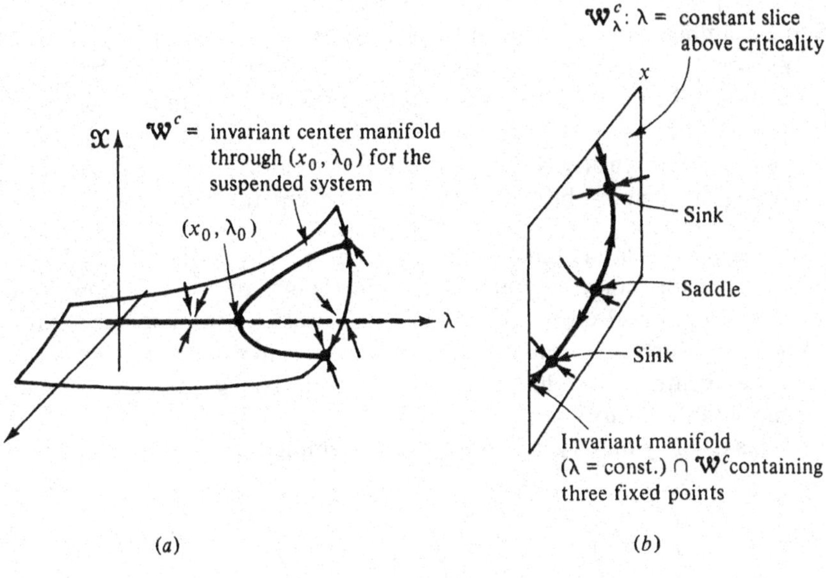

(*a*) (*b*)

Figure 7.4.2

to a finite-dimensional one. Likewise, questions of stability become questions on the center manifold itself. Thus, the center manifold theorem plays the same role in the dynamic theory that the Liapunov–Schmidt procedure plays in the static theory. However, as we shall see in the proof of Hopf's theorem in Box 4.1, sometimes the Liapunov–Schmidt procedure is applied directly in dynamic problems.

It turns out to be true rather generally that stability calculations done via the Liapunov-Schmidt procedure and via the center manifold approach are equivalent. This allows one to make dynamic deductions from the Liapunov-Schmidt procedure, which is convenient for calculations. See Schaeffer and Golubitsky [1981, §6] for details.

There are some important points to be made on the applicability of the preceeding theorems to nonlinear elasticity. First of all, dynamic elastic bifurcation phenomena often involve dissipation and forcing as well as the conservative elastic model. The equations of hyperelastodynamics (without dissipation) are such that the flow determined by them is probably not smooth. This has been indicated already in Section 6.5. On the other hand it is also not clear what dissipative mechanisms (such as viscoelasticity or thermo-elasticity) will produce smooth semiflows. As we already know, the situation is tractable for typical rod, beam, and plate models, for they give semilinear equations. Similar difficulties in delay equations can be overcome; cf. Hale [1981].

In short, for the full equations of three-dimensional nonlinear elasticity a dynamical bifurcation theory does not yet exist, for "technical reasons." For typical rod, beam, and plate models, however, the theory presented here does apply. (Some examples are discussed in the next section.)

We now turn our attention to a description of some of the basic dynamic bifurcations. Bifurcation theory for dynamical systems is more subtle than that for fixed points. Indeed the variety of bifurcations possible—their structure and an imperfection-sensitivity analysis—is much more complex. We begin by describing the simplest bifurcations for *one*-parameter system.

4.3 Saddle Node or Limit Point This is a bifurcation of fixed points; a saddle and a sink come together and annihilate one another, as shown in Figure 7.4.3. A simple real eigenvalue of the sink crosses the imaginary axis at the moment of bifurcation; one for the saddle crosses in the opposite direction. The suspended center manifold is two dimensional. The saddle-source bifurcation is similarly described.

If an axis of symmetry is present, then a symmetric pitchfork bifurcation can

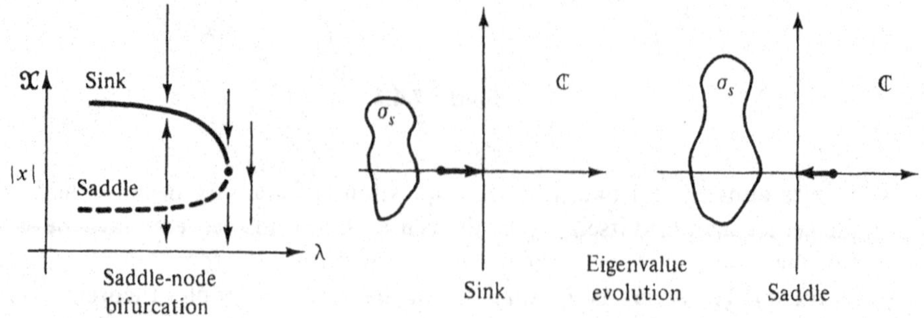

Figure 7.4.3

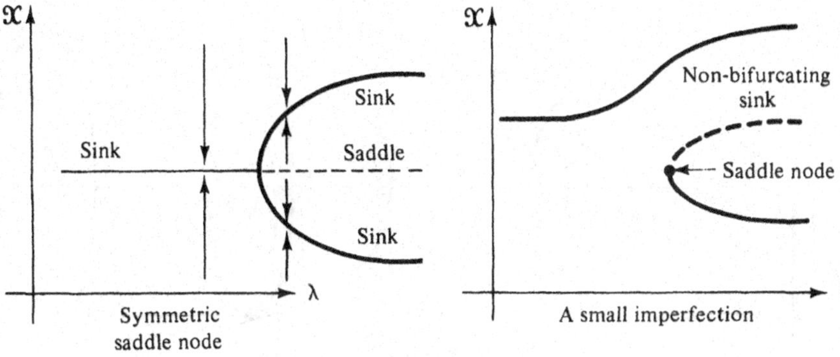

Figure 7.4.4

occur, as in Figure 7.4.4. As in our discussion of Euler buckling, in Section 7.1, small asymmetric perturbation or imperfection can "unfold" this in several ways, one of which is a simple non-bifurcating path and a saddle node.

4.4 Hopf Bifurcation This is a bifurcation of a fixed point to a periodic orbit; here a sink becomes a saddle by two complex conjugate non-real eigenvalues crossing the imaginary axis. As with the pitchfork, the bifurcation can be sub-critical (unstable closed orbits) or super-critical (stable closed orbits). Figure 7.4.5 depicts the supercritical attracting case in $\mathfrak{X} = \mathbb{R}^2$. Here the suspended center manifold is three dimensional.

The proof of the Hopf theorem will be sketched in Box 4.1. The use of center manifolds to prove it is due to Ruelle and Takens [1971]. For PDE's, many approaches are available; see the books of Marsden and McCracken [1976], Iooss and Joseph [1980], Henry [1981], and Hassard, Kazarinoff, and Wan [1981] for references and discussion.

These two bifurcations are local in the sense that they can be analyzed by linearization about a fixed point. There are, however, some *global* bifurcations that can be more difficult to detect. A *saddle connection* is shown in Figure 7.4.6. Here the stable and unstable separatrices of the saddle point pass through a state of tangency (when they are identical) and thus cause the annihilation of the attracting closed orbit.

These global bifurcations can occur as part of local bifurcations of systems with additional parameters. This approach has been developed by Takens [1974a, b], who has classified generic or "stable" bifurcations of two-parameter families of vector fields on the plane. This is an outgrowth of extensive work of the Russian school led by Andronov and Pontryagin [1937]. An example of one of Taken's bifurcations with a symmetry imposed is shown in Figure 7.4.7. (The labels will be used for reference in the next section.) In this bifurcation, rather

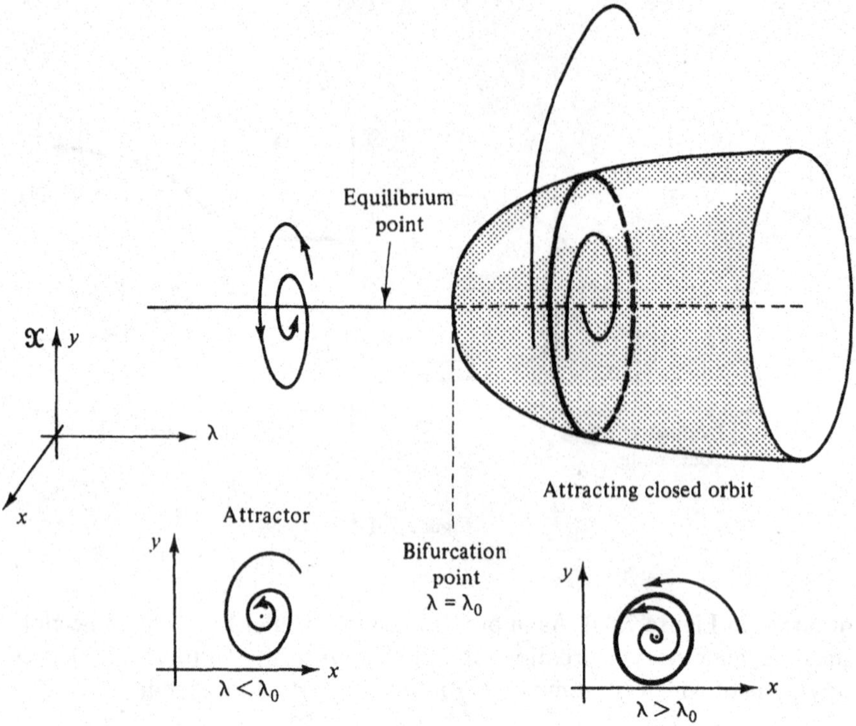

Figure 7.4.5

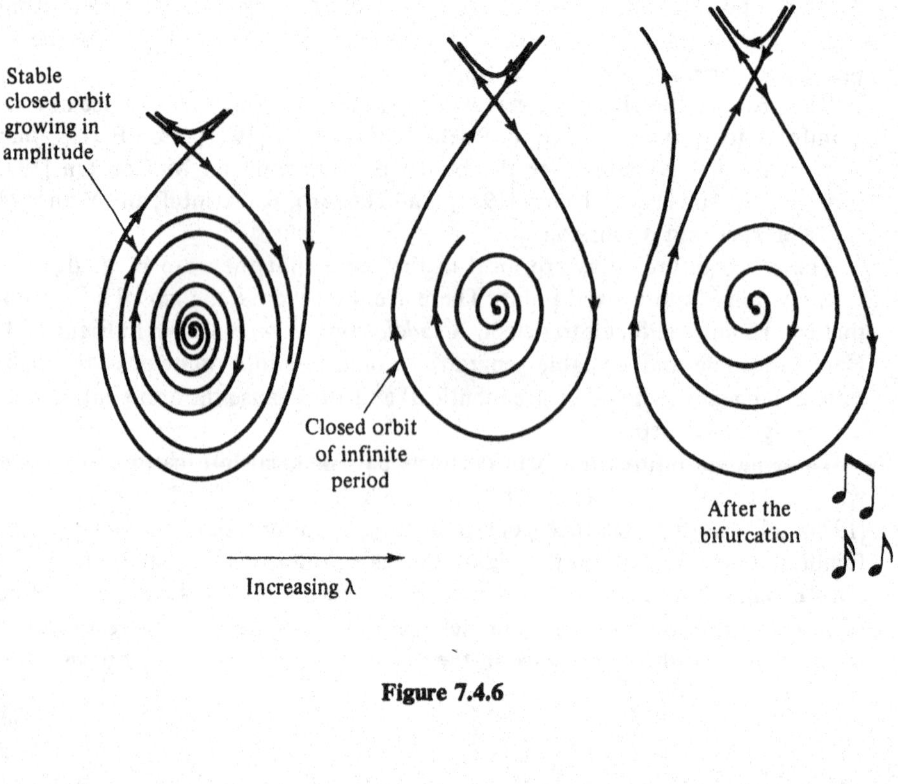

Figure 7.4.6

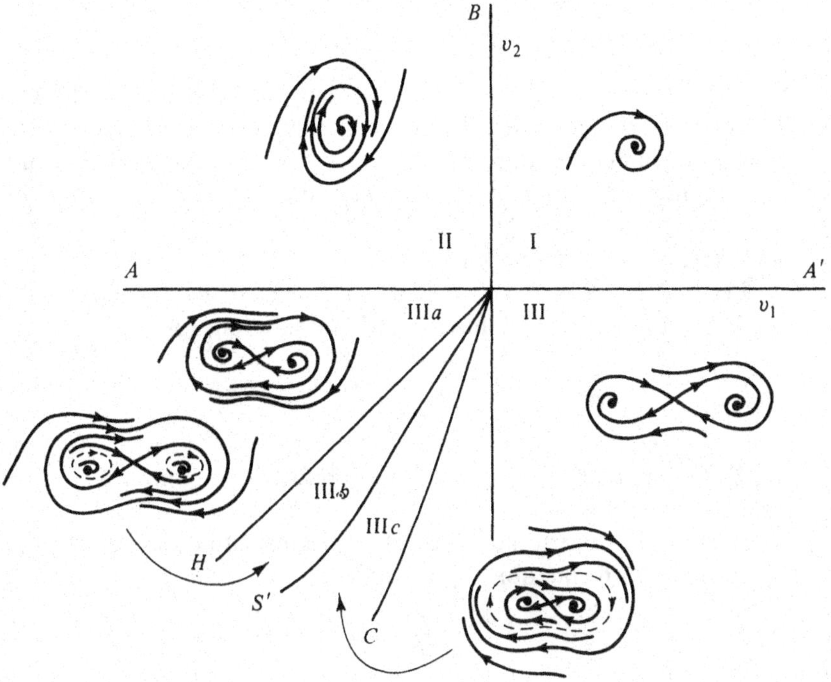

Figure 7.4.7 Takens' "(2, −) normal form" showing the local phase portrait in each region of parameter space.

than a single eigenvalue or a complex conjugate pair crossing the imaginary axis, a real double eigenvalue crosses at zero.

Many similar complex bifurcations are the subject of current research. For example, the eigenvalue configurations (a) one complex conjugate pair and one real zero and (b) two complex conjugate pairs, crossing the imaginary axis, are of interest in many problems. See, for example, Jost and Zehnder [1972], Cohen [1977], Takens [1973], Holmes [1980c], Guckenheimer [1980], and Langford and Iooss [1980]. A number of general features of dynamic bifurcation theory and additional examples are described in Abraham and Marsden [1978] and in Thompson [1982].

Some of the phenomena captured by the bifurcations outlined above have been known to engineers for many years. In particular, we might mention the jump phenomenon of Duffing's equation (see Timoshenko [1974], Holmes and Rand [1976]) and the more complex bifurcational behavior of the forced van der Pol oscillator (Hayashi [1964], Holmes and Rand [1978]; the latter contains a proof that the planar variational equation of the latter oscillator undergoes a saddle connection bifurcation as in Figure 7.4.6).

Box 4.1 ***The Hopf Bifurcation***

The references cited in the text contain many proofs of the Hopf bifurcation theorem. Here we give one that directly utilizes the Liapunov–Schmidt procedure rather than center manifolds. (It is similar to expositions of proofs known to Hale and Cesari, amongst others. The present version was told to us by G. Iooss, M. Golubitsky and W. Langford, whom we thank.)

Let $f\colon \mathbb{R}^n \times \mathbb{R} \to \mathbb{R}^n$ be a smooth mapping satisfying $f(0, \lambda) = 0$ for all λ. We are interested in finding periodic solutions for

$$\frac{dx}{dt} = f(x, \lambda). \tag{1}$$

Let $A_\lambda = D_x f(0, \lambda)$ be the linearization of f at the equilibrium point $(0, \lambda)$. For simplicity we can assume our bifurcation point will be $\lambda_0 = 0$ and we write $A = A_{\lambda_0}$.

Our search for periodic orbits for (1) begins with the assumption that the linearization equation

$$\frac{dv}{dt} = Av \tag{2}$$

has some. Normalizing the periods of (2) to be 2π, and eliminating resonance leads to the following condition:

(H1) *A has simple eigenvalues $\pm i$ and no eigenvalues equal to ki, where k is an integer other than ± 1.*

The period of a putative periodic orbit of (1) will drift from 2π to an unknown period when the nonlinear terms are turned on. Thus we can introduce a new variable s by rescaling time:

$$s = (1 + \tau)t \tag{3}$$

In terms of s, (1) becomes

$$(1 + \tau)\frac{dx}{ds} = f(x, \lambda). \tag{4}$$

We now seek a 2π-periodic function $x(s)$ and a number τ such that (4) holds. Thus, we let

$$\Lambda^0 = \text{all continuous } 2\pi\text{-periodic functions } x(s) \text{ in } \mathbb{R}^n$$

and Λ^1 be the corresponding C^1 functions. Now set

$$F\colon \Lambda^1 \times \mathbb{R} \times \mathbb{R} \to \Lambda^0 \qquad F(x, \tau, \lambda) = (1 + \tau)\frac{dx}{ds} - f(x, \lambda).$$

We seek zeros of F; these will be periodic orbits of period $(1 + \tau)2\pi$ (or in case $x = 0$, fixed points).

Now we apply the Liapunov–Schmidt procedure to F. The derivative of F with respect to its first argument at the trivial solution (0, 0, 0) is denoted L:

$$Lu = D_1F(0, 0, 0)\cdot u = \frac{du}{ds} - Au. \tag{5}$$

From (H1) we see that the kernel of L is spanned by two functions, say $\phi_1, \phi_2 \in \Lambda^1$. In fact, if $Aw = iw$, then we can choose $\phi_1(s) = \mathrm{Re}(e^{is}w)$ and $\phi_2(s) = \mathrm{Im}(e^{is}w)$. The space spanned by ϕ_1 and ϕ_2 can be identified with $\mathbb{R}^2$ by $(x, y) \leftrightarrow x\phi_1 + y\phi_2$. The kernel of the adjoint, L^*, which is orthogonal to the range of L (see Section 6.1) is likewise spanned by two functions, say $\phi_1^*, \phi_2^* \in \Lambda^1$; L^* is given by

$$L^*u = -\frac{du}{ds} + A^*. \tag{6}$$

The Liapunov–Schmidt procedure thus gives us an (implicitly defined) map

$$g\colon \mathbb{R}^2 \times \mathbb{R} \times \mathbb{R} \to \mathbb{R}^2$$

whose zeros we seek. The first $\mathbb{R}^2$ is the space spanned by ϕ_1 and ϕ_2 and the second is that spanned by ϕ_1^* and ϕ_2^*.

Now the circle S^1 acts on Λ^1 by $x(s) \mapsto \theta x(s) = x(s - \theta)$, where $\theta \in S^1$ (S^1 is regarded as real numbers modulo 2π). The function F is covariant with respect to this action, as is easily checked: $F(\theta x, \tau, \lambda) = \theta F(x, \tau, \lambda)$. Now in general when a function whose zeros we seek is covariant (or equivariant) with respect to a group action, preserving the norm, the function produced by the Liapunov–Schmidt procedure is also covariant.

Problem 4.2 Prove this assertion. (See Sattinger [1979] if you get stuck.)

From the form of ϕ_1 and ϕ_2, the action of S^1 on $\mathbb{R}^2$ is just given by rotations through an angle θ. Now a rotationally covariant function from $\mathbb{R}^2$ to $\mathbb{R}^2$ is determined by its restriction to a line through the origin in its domain. Thus, we can write g in the form

$$g(x, y, \tau, \lambda) = \begin{pmatrix} x & -y \\ y & x \end{pmatrix} \begin{pmatrix} \mu \\ \beta \end{pmatrix}, \tag{7}$$

where μ and β are smooth[10] functions of $u = \epsilon^2 = x^2 + y^2$, τ, and λ. We have $\mu(0, \tau, \lambda) = 0 = \beta(0, \tau, \lambda)$ corresponding to the trivial solu-

[10]It is clear that μ and β are smooth functions of $\epsilon = \sqrt{x^2 + y^2}$; one can show that their evenness on reflection through the origin implies they are smooth functions of ϵ^2, a classical result of Whitney; cf. Schwarz [1975] for a general study of such phenomena.

tions. If we find a zero of (μ, β) other than at $x = 0$, $y = 0$, we have a periodic orbit. Roughly speaking, (μ, β) defines the perturbations of the amplitude and period of the periodic orbits we seek. From the fact that the variable τ is directly proportional to the changes in period, we find that $(\partial\beta/\partial\tau)(0, 0, 0) = 1$.

Problem 4.3 Prove the preceeding assertion.

Thus, by the implicit function theorem we can solve $\beta = 0$ for $\tau(\epsilon^2, \lambda)$.

We still need to solve $\mu = 0$. By S^1 covariance it is enough to look at the function $\mu(u, \lambda) = \mu(u, \lambda, \tau(u, \lambda))$; that is, we can restrict to $y = 0$ and take $x \geq 0$; here $u = \epsilon^2$.

(H2) $\dfrac{\partial\mu}{\partial\lambda}(0, 0) \neq 0.$

This is often called the "Hopf condition." As stated, it is not very easy to check. However, it holds iff the eigenvalues of A_λ cross the imaginary axis with non-zero speed (with respect to the parameter λ).

Problem 4.4 Prove this assertion. Consult Marsden and McCracken [1976] or Iooss and Joseph [1980] if you get stuck.

The condition (H2) implies that $\mu = 0$ is solvable for $\lambda(u)$. Thus we have proved some key parts of the following important result of Hopf [1942]:

4.5 Hopf Theorem *If (H1) and (H2) hold, then there is a unique one-parameter family of periodic orbits of (1) in $\mathbb{R}^n \times \mathbb{R}$, that are tangent to $\mathbb{R}^n \times \{0\}$ at $\lambda = 0$. Moreover, if*

(H3) $\dfrac{\partial\mu}{\partial u}(0, 0) \neq 0,$

then $g(\epsilon, \lambda, \tau) = a(\epsilon^2, \lambda)\epsilon$ is contact equivalent (with a $\mathbb{Z}_2$-symmetry) to $(x^2 \pm \lambda)x$. [In the + case $(\partial\mu/\partial u > 0)$ the periodic orbits are supercritical and are stable and in the − case $(\partial\mu/\partial u < 0)$ they are subcritical and are unstable.] The Hopf and the saddle-node bifurcation are, in a sense, analogous to that explained in Box 1.1, the only one-parameter structurally stable dynamic bifurcations.

For the completion of the proof, methods for computing $\mu_2 = \partial\mu/\partial u$, and infinite-dimensional generalizations, we refer to one of the references already given. (See also Crandall and Rabinowitz [1978], Hassard, Kazarinoff and Wan [1981], and Gurel and Rossler [1979]. We also refer to Takens [1973] and Golubitsky and Langford [1981] for an imperfection-sensitivity analysis when (H2) or (H3) fail and, to Thompson and Lunn [1981a] for Hopf bifurcation with forcing, to Langford and Iooss [1980] for the interaction of the Hopf and pitchfork bifurcations and to Langford [1979] for the interaction of the Hopf and transcritical bifurcations. A "catalogue" of some of the important dynamic bifurcations is given in Abraham and Marsden [1978].

Box 4.2 ***Summary of Important Formulas for Section 7.4***

Dynamic Bifurcation

A bifurcation in a parameter-dependent dynamical system means a qualitative change in the phase portrait as the parameter(s) varies.

Liapunov's Theorem

A fixed point is stable if the eigenvalue of the linearized system lie in the left half-plane.

Bifurcation at a fixed point can occur only when eigenvalues cross the imaginary axis.

Center Manifold

An invariant manifold corresponding to the purely imaginary eigenvalues captures all the bifurcation behavior.

Limit Point

Bifurcation of fixed points occurring when a saddle and a sink self-destruct (or are spontaneously created).

Hopf Bifurcation

If conditions (H1), (H2), (H3) hold (see the previous box), then the fixed point bifurcates to a family of periodic orbits that are either supercritical (stable) or are subcritical (unstable); see Figure 7.4.5 for the stable case.

7.5 A SURVEY OF SOME APPLICATIONS TO ELASTODYNAMICS[11]

As with Section 7.2, we shall give a biased and incomplete survey. The number of papers dealing with dynamical bifurcation in systems related to elasticity is astronomical. Two examples are Hsu [1977] and Reiss and Matkowsky [1971].

[11]This section was written in collaboration with Philip Holmes.

We shall concentrate on the phenomena of *flutter* in various engineering systems. We begin by describing some general features of flutter.

A dynamical system is said to be *fluttering* if it has a stable closed orbit. Often flutter is suggested if a system linearized about a fixed point has two complex conjugate eigenvalues with positive real part. However, a general proclamation of this sort is certainly false, as shown in Figure 7.5.1. A theorem that can

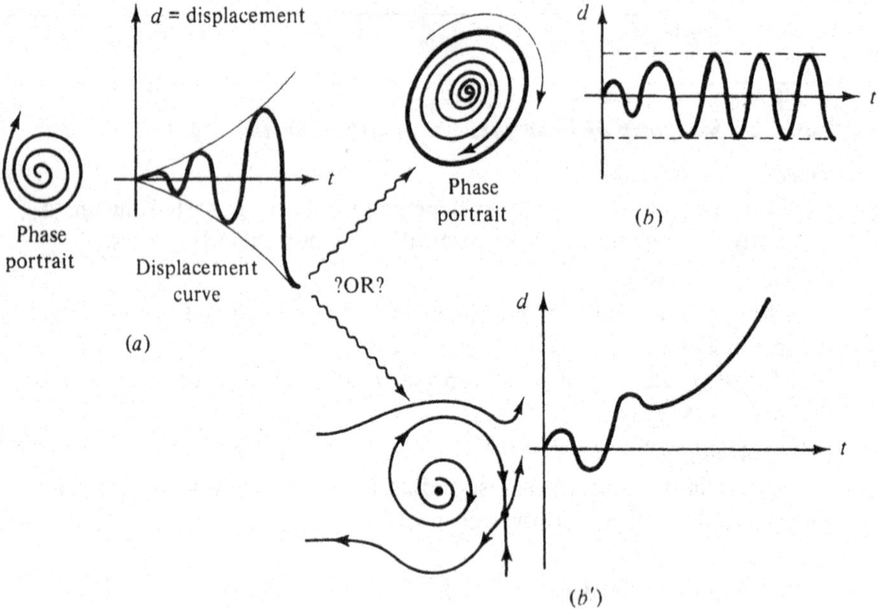

Figure 7.5.1 (a) Linear "flutter." (b) Nonlinear flutter (limit cycle). (b′) An example of linear, but not nonlinear, flutter (no limit cycle).

be used to substantiate such a claim is the Hopf bifurcation theorem, which was proved in the preceding section.

Similar remarks may be made about divergence (a saddle point or source) as shown in Figure 7.5.2.

There are, in broad terms, three kinds of flutter of interest to the engineer. Here we briefly discuss these types. Our bibliography is not intended to be exhaustive, but merely to provide a starting point for the interested reader.

5.1 Airfoil or Whole Wing Flutter on Aircraft Here linear stability methods *do* seem appropriate since virtually any oscillations are catastrophic. Control surface flutter probably comes under this heading also. See Bisplinghoff and Ashley [1962] and Fung [1955] for examples and discussion.

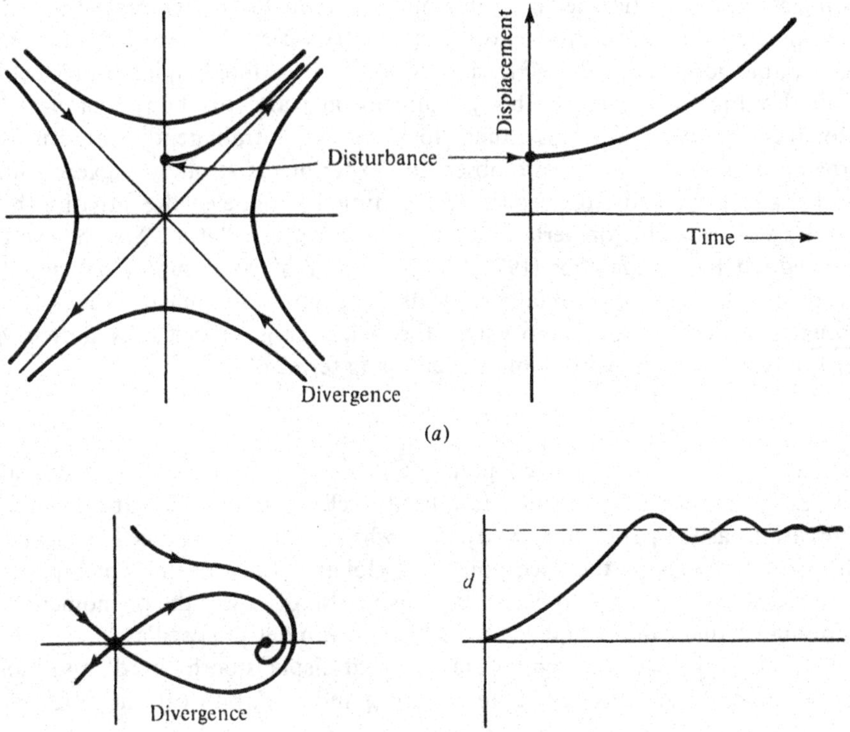

Figure 7.5.2 (a) Linear theory. (b) A nonlinear possibility.

5.2 Cross-Flow Oscillations The familiar flutter of sun-blinds in a breeze comes under this heading. The "galloping" of power transmission lines and of tall buildings and suspension bridges provide examples that are of more direct concern to engineers: the famous Tacoma Narrows bridge disaster was caused by cross-flow oscillations. In such cases (small) limit cycle oscillations are acceptable (indeed, they are inevitable), and so a nonlinear analysis is appropriate.

Cross-flow flutter is believed to be due to the oscillating force caused by "von Karman" vortex shedding behind the body; see Figure 7.5.3. The alter-

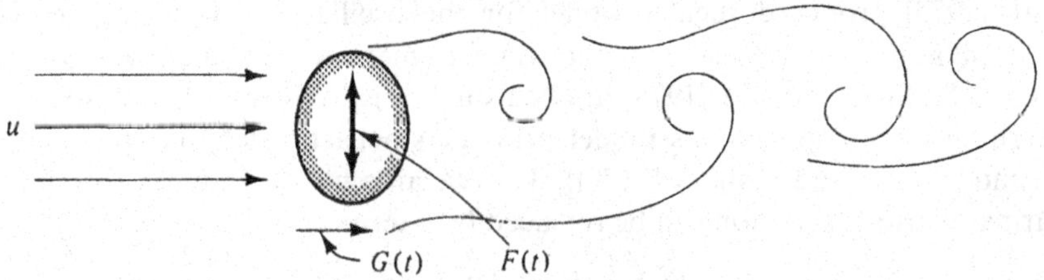

Figure 7.5.3

nating stream of vortices leads to an almost periodic force $F(t)$ transverse to the flow in addition to the in-line force $G(t)$; $G(t)$ varies less strongly than $F(t)$. The flexible body responds to $F(t)$ and, when the shedding frequency (a function of fluid velocity, u, and the body's dimensions) and the body's natural (or resonance) frequency are close, then "lock on" or "entrainment" can occur and large amplitude oscillations are observed. Experiments strongly suggest a limit cycle mechanism and engineers have traditionally modeled the situation by a van der Pol oscillator or perhaps a pair of coupled oscillators. See the symposium edited by Naudascher [1974] for a number of good survey articles; the review by Parkinson is especially relevant. In a typical treatment, Novak [1969] discusses a specific example in which the behavior is modeled by a free van der Pol type oscillator with nonlinear damping terms of the form

$$a_1\dot{x} + a_2\dot{x}^2 + a_3\dot{x}^3 + \cdots.$$

Such equations possess a fixed point at the origin $x = \dot{x} = 0$ and can also possess multiple stable and unstable limit cycles. These cycles are created in bifurcations as the parameters $a_1, a_2, \ldots$, which contain windspeed terms, vary. Bifurcations involving the fixed-point and global bifurcations in which pairs of limit cycles are created both occur. Parkinson also discusses the phenomenon of entrainment that can be modeled by the *forced* van der Pol oscillator.

Landl [1975] discusses such an example that displays both "hard" and "soft" excitation, or, in Arnold's [1972] term, *strong* and *weak* bifurcations. The model is

$$\ddot{x} + \delta\dot{x} + x = a\Omega^2 C_L,$$
$$\ddot{C}_L + (\alpha - \beta C_L^2 + \gamma C_L^4)\dot{C}_L + \Omega^2 C_L = b\dot{x}.$$

Here $\dot{x} \equiv d/dt$ and α, β, γ, δ, a, b are generally positive constants for a given problem (they depend upon structural dimensions, fluid properties, etc.) and Ω is the vortex shedding frequency. As Ω varies the system can develop limit cycles leading to a periodic variation in C_L, the *lift coefficient*. The term $a\Omega^2 C_L$ then acts as a periodic driving force for the first equation, which represents one mode of vibration of the structure. This model, and that of Novak, appear to display generalized Hopf bifurcations (see Takens [1973] and Golubitsky and Langford [1981]).

In related treatments, allowance has been made for the effects of (broad band) turbulence in the fluid stream by including stochastic excitations. Vacaitis et al. [1973] proposed such a model for the oscillations of a two degree of freedom structure and carried out some numerical and analogue computer studies. Holmes and Lin [1978] applied qualitative dynamical techniques to a deterministic version of this model prior to stochastic stability studies of the full model (Lin and Holmes [1978]). The Vacaitis model assumes that the von Karman vortex excitation can be replaced by a term

$$F(t) \equiv F\cos(\Omega t + \Psi(t)),$$

where Ω is the (approximate) vortex shedding frequency and $\Psi(t)$ is a random phase term. In common with all the treatments cited above, the actual mechanism of vortex generation is ignored and "dummy" drag and lift coefficients are introduced. These provide discrete analogues of the actual fluid forces on the body. Iwan and Blevins [1974] and St. Hilaire [1976] have gone a little further in attempting to relate such force coefficients to the fluid motion, but the problem appears so difficult that a rigorous treatment is still impossible. The major problem is, of course, our present inability to solve the Navier–Stokes equations for viscous flow around a body. Potential flow solutions are of no help here, but recent advances in numerical techniques may be useful. Ideally a rigorous analysis of the fluid motion should be coupled with a continuum mechanical analysis of the structure. For the latter, see the elegant Hamiltonian formulation of Marietta [1976], for example.

The common feature of all these treatments (with the exception of Marietta's) is the implicit reduction of an infinite-dimensional problem to one of finite dimensions, generally to a simple nonlinear oscillator. The use of center manifold theory and the concept structural stability suggests that in some cases this reduction might be rigorously justified. To illustrate this we turn to the third broad class of flutter, which we discuss in more detail.

5.3 Axial Flow-Induced Oscillations In this class of problems, oscillations are set up directly through the interaction between a fluid and a surface across which it is moving. Examples are oscillations in pipes and (supersonic) panel flutter; the latter is analyzed in 5.6 below. Experimental measurements (vibration records from nuclear reactor fuel pins, for example) indicate that axial flow-induced oscillations present a problem as severe as the more obvious one of cross-flow oscillations. See the monograph by Dowell [1975] for an account of panel flutter and for a wealth of further references. Oscillations of beams in axial flow and of pipes conveying fluid have been studied by Benjamin [1961], Paidoussis [1966], Paidoussis and Issid [1974] and Holmes [1980d]. Figure 7.5.4 shows the three situations. In addition to the effects of the fluid flow velocity ρ, the structural element might also be subject to mechanical tensile or compressive forces Γ, which can lead to buckling instabilities even in the absence of fluid forces.

The equations of motion of such systems, written in one-dimensional form and with all coefficients suitably nondimensionalized, can be shown to be of the type

$$\alpha\dot{v}'''' + v'''' - \left(K\int_0^1 (v'(\xi))^2\, d\xi + \sigma \int_0^1 (v'(\xi)\dot{v}'(\xi))\, d\xi\right)v'' + \ddot{v}$$
$$+ [\text{linear fluid and mechanical loading terms in } v'', \dot{v}', v', \dot{v}] = 0 \qquad (0)$$

Here $\alpha, \sigma > 0$ are structural viscoelastic damping coefficients and $K > 0$ is a (nonlinear) measure of membrane stiffness; $v = v(z, t)$ and $\cdot = \partial/\partial t$; $' = \partial/\partial z$. (Holmes [1977a], Benjamin [1961], Paidoussis [1966], and Dowell [1975], for

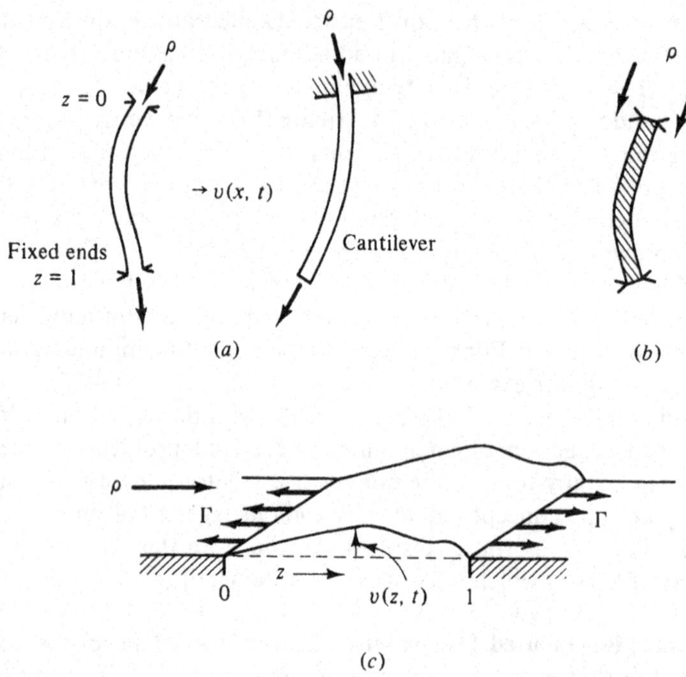

Figure 7.5.4 (a) Pipes conveying fluid. (b) Beam in axial flow. (c) Panel flutter.

example, provide derivations of specific equations of this type.) The fluid forces are again approximated, but in a more respectable manner.

In the case of panel flutter, if a static pressure differential exists across the panel, the right-hand side carries an additional parameter P. Similarly, if mechanical imperfections exist so that compressive loads are not symmetric, then the "cubic" symmetry of (0) is destroyed.

Problems such as those of Figure 7.5.4 have been widely studied both theoretically and experimentally, although, with the exception of Dowell and a number of other workers in the panel flutter area, engineers have for the most part concentrated on linear stability analyses. Such analyses can give misleading results. In many of these problems, engineers have also used low-dimensional models, even though the full problem has infinitely many degrees of freedom. Such a procedure can sometimes be justified if careful use is made of the center manifold theorem.

Often the location of fixed points and the evolution of spectra about them have to be computed by making a Galerkin or other approximation and then using numerical techniques. There are obvious convergence problems (see Holmes and Marsden [1978a]), but once this is done, the organizing centers and dimension of the center manifolds can be determined relatively simply.

5.4 Pipes Conveying Fluid and Supported at Both Ends Pipe flutter is an excellent illustration of the difference between the linear prediction of flutter and what actually happens in the nonlinear PDE model. The phase portrait on the center manifold in the nonlinear case is shown in Figure 7.5.5 at parameter values for which the linear theory predicts "coupled mode" flutter (cf. Paidoussis and Issid [1974] and Plaut and Huseyin [1975]). In fact, we see that the pipe merely settles to one of the stable buckled rest points with no nonlinear flutter. The presence of imperfections should not substantially change this situation.

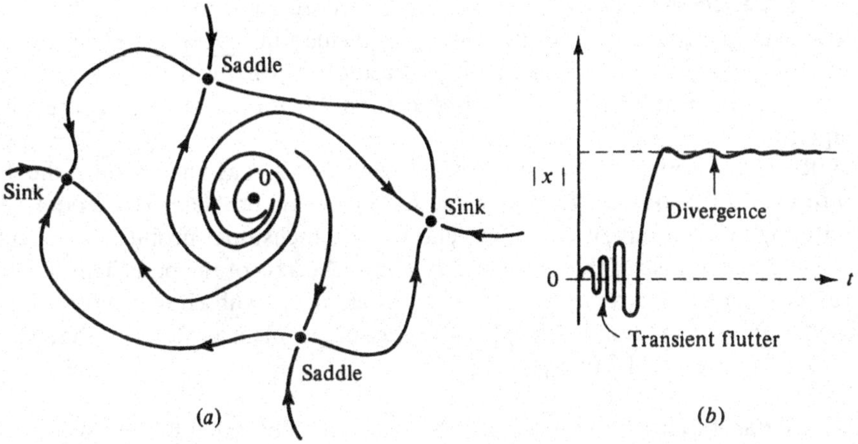

Figure 7.5.5 (a) Vector field. (b) Time evolution of a solution starting near {0}.

The absence of flutter in the nonlinear case can be seen by differentiating a suitable Liapunov function along solution curves of the PDE. In the pipe flutter case the PDE is

$$\alpha\dot{v}'''' + v'''' - \{\Gamma - \rho^2 + \gamma(1-z) + K\|v'\|^2 + \sigma\langle v', \dot{v}'\rangle\}v'' + 2\sqrt{\beta}\,\rho\dot{v}' + \gamma v' + \delta\dot{v} + \ddot{v} = 0.$$

Here $\|\cdot\|$ and $\langle\cdot,\cdot\rangle$ denote the usual L^2- norm and inner product and solutions $x = (v, \dot{v})$ lie in a Hilbert space $\mathfrak{X} = H_0^2([0, 1]) \times L^2([0, 1])$. (See Section 6.5 for the specific analytic framework for such a problem.) For our Liapunov function we choose the energy, in this case given by

$$H(x(t)) = \tfrac{1}{2}\|\dot{v}\|^2 + \tfrac{1}{2}\|v''\|^2 + \frac{\Gamma - \rho^2}{2}\|v'\|^2 + \frac{K}{4}\|v'\|^4 + \frac{\gamma}{2}\langle[1 - z]v', v'\rangle.$$

Differentiating $H(x(t))$ along solution curves yields

$$\frac{dH}{dt} = -\delta\|\dot{v}\|^2 - \alpha\|\dot{v}''\|^2 - \sigma\langle v', \dot{v}'\rangle^2 - 2\sqrt{\beta}\,\rho\langle\dot{v}', \dot{v}\rangle.$$

Since $\langle \dot{v}', \dot{v} \rangle = 0$ and $\delta, \alpha, \sigma > 0$, dH/dt is negative for all $v > 0$ and thus all solutions must approach rest points. In particular, for $\Gamma < \Gamma_0$, the first Euler buckling load, all solutions approach $x_0 = \{0\} \in \mathfrak{X}$ and the pipe remains straight. Thus a term of the type $p\dot{v}'$ cannot lead to nonlinear flutter. In the case of a beam in axial flow, terms of this type *and* of the type $p^2 v'$ both occur and nonlinear flutter evidently *can* take place (see Paidoussis [1966] for a linear analysis). Experimental observations actually indicate that fluttering motions more complex than limit cycles can occur.

5.5 Cantilevered Pipes Flexible pipes free at one end can flutter. Anyone who has played with a hose knows this. Benjamin [1961] has some excellent photographs of a two-link model. Here flutter is caused by the so-called follower force at the free end, which introduces an additional term into the energy equation.

Recently, Sethna [1980], and references therein, has shown how the Hopf bifurcation can be used to obtain the flutter in this problem. His model allows only planar motions of the pipe. The three-dimensional problem is especially interesting because of the S^1 symmetry about the axis of the pipe. This leads one to guess that the flutter will become modulated in a subsequent bifurcation, as in the analysis of Rand [1982] for the Taylor problem in fluid mechanics. See also Thompson and Lunn [1981b].

5.6 Panel Flutter Now we turn to an analysis of panel flutter. We consider the "one-dimensional" panel shown in Figure 7.5.4(c) and we shall be interested in bifurcations near the trivial zero solution. The equation of motion of such a thin panel, fixed at both ends and undergoing "cylindrical" bending (or spanwise bending) can be written as

$$\alpha \dot{v}'''' + v'''' - \left(\Gamma + K \int_0^1 (v'(\xi))^2 \, d\xi + \sigma \int_0^1 (v'(\xi)\dot{v}'(\xi)) \, d\xi\right) v'' + pv'' + \sqrt{p}\, \delta \dot{v} + \ddot{v} = 0. \tag{1}$$

See Dowell [1975] and Holmes [1977a]. Here $\cdot \equiv \partial/\partial t$, $' = \partial/\partial z$, and we have included viscoelastic structural damping terms α, σ as well as aerodynamic damping $\sqrt{p}\,\delta$. K represents nonlinear (membrane) stiffness, p the dynamic pressure, and Γ an in-plane tensile load. All quantities are nondimensionalized and associated with (1). We have boundary conditions at $z = 0, 1$, which might typically be simply supported ($v = v'' = 0$) or clamped ($v = v' = 0$). In the following we make the physically reasonable assumption that $\alpha, \sigma, \delta, K$ are fixed > 0 and let the control parameter $\mu = \{(p, \Gamma) \mid p \geq 0\}$ vary. In contrast to previous studies in which (1) and similar equations were analyzed for specific parameter values and initial conditions by numerical integration of a finite-dimensional Galerkin approximation, here we study the qualitative behavior of (1) under the action of μ.

As in Section 6.5, we first redefine (1) as an ODE on a Banach space, choosing as our basic space $\mathfrak{X} = H_\partial^2([0,1]) \times L^2([0,1])$, where H_∂^2 denotes H^2 functions in $[0, 1]$ that vanish at 0, 1. Set $\|\{v, \dot v\}\|_{\mathfrak{X}} = (\|\dot v\|^2 + \|v''\|^2)^{1/2}$, where $\|\cdot\|$ denotes the usual L^2-norm and define the linear operator

$$A_\mu = \begin{pmatrix} 0 & I \\ C_\mu & D_\mu \end{pmatrix} \qquad \begin{aligned} C_\mu v &= -v'''' + \Gamma v'' - \rho v', \\ D_\mu \dot v &= -\alpha \dot v'''' - \sqrt{\rho}\,\delta \dot v. \end{aligned} \tag{2}$$

The basic domain $\mathfrak{D}(A_\mu)$ of A_μ, consists of $(v, \dot v) \in \mathfrak{X}$ such that $\dot v \in H_\partial^2$ and $v + \alpha \dot v \in H^4$; particular boundary conditions necessitate further restrictions. After defining the nonlinear operator $B(v, \dot v) = (0, [K\|v'\|^2 + \sigma\langle v', \dot v'\rangle]v'')$, where $\langle\ ,\ \rangle$ denotes the L^2 inner product, (1) can be rewritten as

$$\frac{dx}{dt} = A_\mu x + B(x) \equiv G_\mu(x) \qquad x = (v, \dot v) \qquad x(t) \in \mathfrak{D}(A_\mu). \tag{3}$$

From Section 6.5 recall that we have an energy function $H\colon \mathfrak{X} \to \mathbb{R}$ defined by

$$H(v, \dot v) = \tfrac{1}{2}\|\dot v\|^2 + \tfrac{1}{2}\|v''\|^2 + \frac{\Gamma}{2}\|v'\|^2 + \frac{K}{4}\|v'\|^4 \tag{4}$$

and

$$\frac{dH}{dt} = -\rho\langle v', \dot v\rangle - \sqrt{\rho}\,\delta\|\dot v\|^2 - \alpha\|\dot v''\|^2 - \sigma\langle v', \dot v\rangle^2.$$

In Section 6.5 we showed that (3) and hence (1) defines a unique smooth global semi-flow F_t^μ on $\mathfrak{X}$.

By making two-mode and four-mode approximations, one finds that for $\sigma = 0.0005$, $\delta = 0.1$, the operator A_μ has a double zero eigenvalue at $\mu = (\rho, \Gamma) \approx (110, -22.6)$ (the point 0 in Figure 7.5.6), the remaining eigenvalues

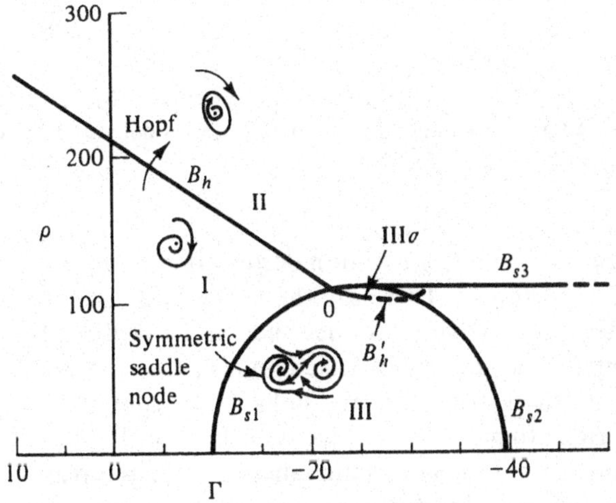

Figure 7.5.6 Partial bifurcation set for the two-mode panel ($\alpha = 0.005$, $\delta = 0.1$).

being in the left half-plane. (See Holmes [1977a] and Holmes and Marsden [1978a].) Thus around the zero solution we obtain a four-dimensional suspended center manifold. (Note that the control parameter μ is now two dimensional.) Referring to the eigenvalue evolution at the zero solution in Figure 7.5.7, which is obtained numerically, we are able to fill in the portions of the bifurcation diagram shown in Figure 7.5.6.

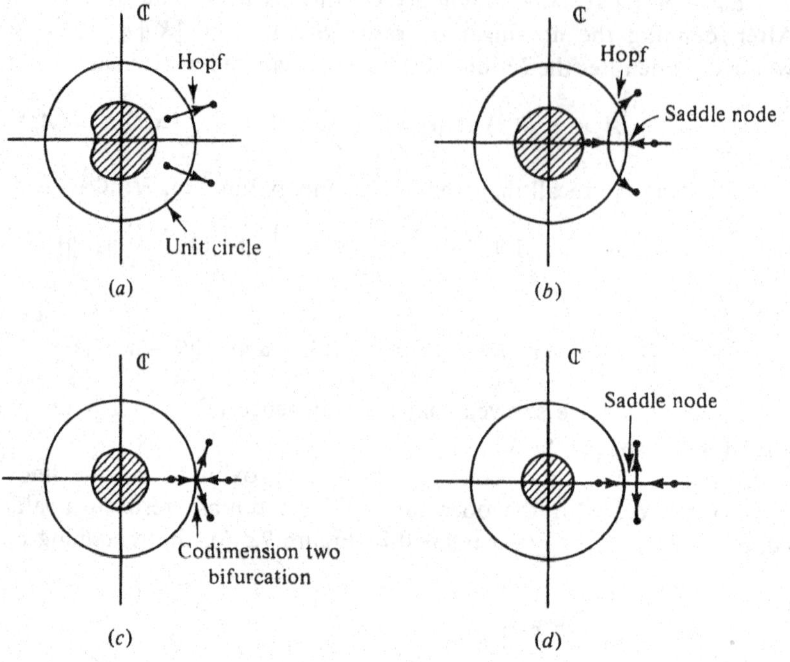

Figure 7.5.7 Eigenvalue evolutions for $DF_t^\mu(0): X \longrightarrow X$, Γ fixed, ρ increasing, estimated from two-mode model. (a) $\Gamma = 0$. (b) $\Gamma = -16$. (c) $\Gamma \simeq -21.91$. (d) $\Gamma = -24$.

A supercritical Hopf bifurcation occurs crossing B_h and a symmetrical saddle node on B_{s1}, as shown. These are the flutter and buckling or divergence instabilities detected in previous studies such as Dowell's. Moreover, finite-dimensional computations for the two fixed points $\{\pm x_0\}$ appearing on B_{s1} and existing in region III show that they are sinks ($|\text{spectrum}\,(DF_t^\mu(\pm x_0))| < 1$) below a curve B_h' originating at 0, which we also show on Figure 7.5.6. As μ crosses B_h' transversally, $\{\pm x_0\}$ undergo simultaneous Hopf bifurcations before coalescing with $\{0\}$ on B_{s1}. A fuller description of the bifurcations, including those occurring on B_{s2} and B_{s3}, is provided by Holmes [1977a]. First consider the case where μ crosses B_{s2} from region I to region III, not at 0. Here the eigenvalues indicate that a saddle-node bifurcation occurs. In Holmes [1977a] exact expressions are derived for the new fixed points $\{\pm x_0\}$ in the two-mode case. This then approximates the behavior of the full evolution equation and the associated semiflow $F_t^\mu: \mathfrak{X} \to \mathfrak{X}$ and we can thus assert that a symmetric saddle-node bifurcation occurs on a one-dimensional manifold as shown in Figure 7.5.6 and that the "new" fixed points are sinks in region III. Next consider μ crossing $B_h \backslash O$. Here the eigenvalue evolution shows that a Hopf bifurcation occurs on a two-manifold and use of the stability calculations from Hassard [1979] show that the family of closed orbits existing in region II are attracting.

Now let μ cross $B_{s2} \backslash O$ from region II to region IIIa. Here the closed orbits presumably persist, since they lie at a finite distance from the bifurcating fixed point $\{0\}$. In fact, the new points $\{\pm x_0\}$ appearing on B_{s2} are saddles in region IIIa, with two eigenvalues of spectrum $(DG_\mu(\pm x_0))$ in the right half-plane and all others in the left half-plane. As this bifurcation occurs one of the eigenvalues of spectrum $DF_t^\mu(0)$ passes into the unit circle so that throughout regions IIIa and III $\{0\}$ remains a saddle. Finally, consider what happens when μ crosses B_h' from region IIIa to III. Here $\{\pm x_0\}$ undergo simultaneous Hopf bifurcations and the stability calculations show that the resultant sinks in region III are surrounded by a family of repelling closed orbits. We do not yet know how the multiple closed orbits of region III interact or whether any other bifurcations occur, but we now have a *partial* picture of behavior near 0 derived from the two-mode approximation and from use of the stability criterion. The key to completing this analysis lies in the point 0, the "organizing center" of the bifurcation set at which B_{s2}, B_h, and B_h' meet.

According to our general scheme, we now postulate that our bifurcation diagram near 0 is stable to small perturbations in our (approximate) equations. Takens' bifurcation shown in Figure 7.4.7 is consistent with the information found in Figure 7.5.6. Thus we are led to the complete bifurcation diagram shown in Figure 7.5.8 with the oscillations in various regions as shown in Figure 7.4.7.

One can actually check this rigorously by proving that our vector field on

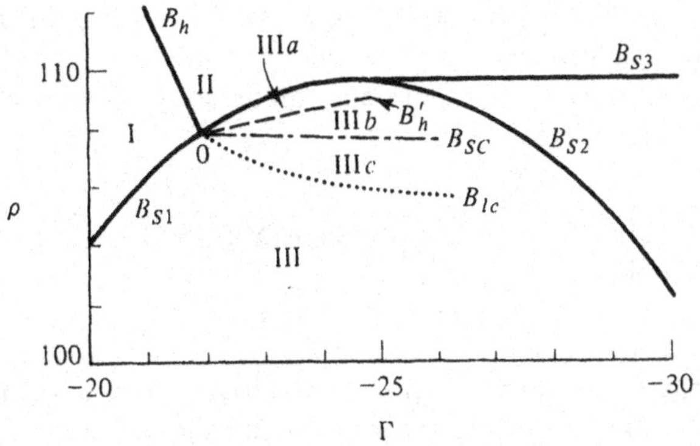

Figure 7.5.8 A local model for bifurcations of the panel near 0, $(\rho, \Gamma) \simeq (107.8, -21.91)$; $\alpha = 0.005$, $\delta = 0.1$.

the center manifold has the appropriate normal form. This calculation is rather long. See Holmes [1982].

Although the eigenvalue computations used in this analysis were derived from two and four mode models (in which A_μ of (2) is replaced by a 4×4 or 8×8 matrix and $\mathfrak{X}$ is replaced by a vector space isomorphic to $\mathbb{R}^4$ or $\mathbb{R}^8$), the convergence estimates of Holmes and Marsden [1978a] indicate that in the infinite-dimensional case the behavior remains qualitatively identical. In particular, for $\mu \in \mathfrak{U}$, a neighborhood of 0, all eigenvalues but two remain in the negative half-plane. Thus the dimension of the center manifold does not increase and our four-dimensional "essential model," a two-parameter vector field on a 2-manifold, provides a local model for the onset of flutter and divergence. We are therefore justified in locally replacing the infinite-dimensional semiflow F_t^μ: $\mathfrak{X} \to \mathfrak{X}$ by a finite-dimensional system. Moreover, the actual vector fields and bifurcation set shown in Figure 7.4.7. can be realized by the explicit nonlinear oscillator

$$\ddot{y} + \lambda_2\dot{y} + \lambda_1 y + \gamma y^2\dot{y} + \eta y^3 = 0 \qquad \gamma, \eta > 0$$

or

$$\dot{y}_1 = y_2 \qquad \dot{y}_2 = -\lambda_1 y_1 - \lambda_2 y_2 - \gamma y_1{}^2 y_2 - \eta y_1{}^3. \tag{5}$$

In engineering terms (5) might be thought of as a "nonlinear normal mode" of the system of Equation (1), with λ_1, λ_2 representing equivalent linear stiffness and damping. (See Rosenberg [1966].) Note, however, that the relationship between the coordinates y_1, y_2 and any conveniently chosen basis in the function space $\mathfrak{X}$ is likely to be nonlinear: in particular, a single "natural" normal mode model of the panel flutter problem *cannot* exhibit flutter, although it can diverge. (See Holmes [1977a]; flutter occurs through coupling between the natural (linear) normal modes.)

The bifurcation diagram in Figure 7.4.7 for panel flutter is derived under an assumption of symmetry. One would expect extremely complex dynamics to be possible if this symmetry is broken because the homoclinic orbits can be broken. The reason we say this is explained in the next section. In fact, Dowell [1980] has found numerically that in certain parameter regions, *chaotic* dynamics occurs in panel flutter. This indicates that breaking the symmetry in Figure 7.4.7—which amounts to an imperfection sensitivity analysis—gives a bifurcation to chaotic dynamics. A situation where one can actually prove such an assertion is described in the next section.

7.6 BIFURCATIONS IN THE FORCED OSCILLATIONS OF A BEAM[12]

In recent years many examples of dynamical systems have been found with the property that the equations of motion are relatively simple, yet the trajectories are very complex and depend very sensitively on the initial data. The literature

[12] This section was written jointly with Philip Holmes and is based on Holmes and Marsden [1981].

on this topic is vast, but some of the more accessible works are Temam [1976], Ratiu and Bernard [1977], Lorenz [1979], Gurel and Rossler [1979], Holmes [1980a], Collet and Eckman [1980], and Guckenheimer and Holmes [1983].

Our goal is to sketch a method that enables one to rigorously describe some of the complexity in the dynamics of a forced beam. Experimentally, aperiodic or apparently random motions have been observed by Tseng and Dugundjii [1971] and by Moon [1980a, b]. One sees in a power spectrum, periodicity (energy concentrated at certain frequencies) shift to aperiodicity (energy spread over a broad band of frequencies) as a parameter is increased. As we shall see, our analysis enables one to compute explicitly the bifurcation point where periodicity switches to aperiodicity for a special class of equations.

We shall consider a motivating example first and state the results for it. Following this we shall describe the methods by which they are obtained.

Consider a beam that is buckled by an external load Γ, so that there are two stable and one unstable equilibrium states (see Figure 7.6.1). The whole structure

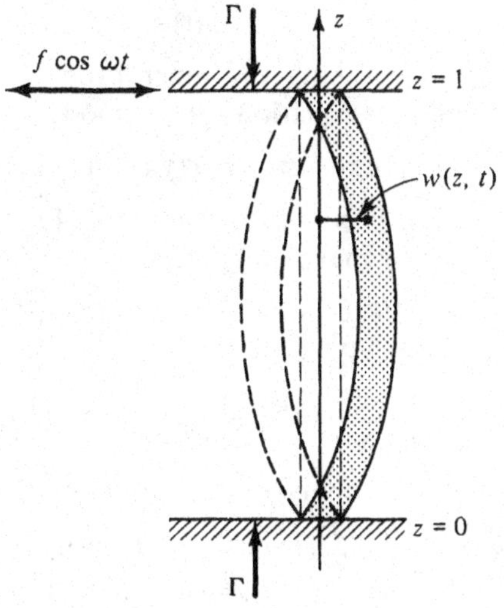

Figure 7.6.1 The forced, buckled beam.

is shaken with a transverse periodic displacement, $f \cos \omega t$ and the beam moves due to its inertia. One observes periodic motion about either of the two stable equilibria for small f, but as f is increased, the motion becomes aperiodic or chaotic.

A specific model for the transverse deflection $w(z, t)$ of the centerline of the beam is the following partial differential equation

$$\ddot{w} + w'''' + \Gamma w'' - \kappa\left(\int_0^1 [w']^2 \, d\zeta\right) w'' = \epsilon(f \cos \omega t - \delta \dot{w}), \qquad (1)$$

where $\cdot = \partial/\partial t$, $' = \partial/\partial z$, Γ = external load, κ = stiffness due to "membrane" effects, δ = damping, and ϵ is a parameter used to measure the size of f and δ. Among many possible boundary conditions we shall choose $w = w'' = 0$ at $z = 0, 1$—that is, simply supported, or hinged, ends. With these boundary conditions, the eigenvalues of the linearized, unforced equations—that is, complex numbers λ such that

$$\lambda^2 w + w'''' + \Gamma w'' = 0$$

for some non-zero w satisfying $w = w'' = 0$ at $z = 0, 1$—form a countable set

$$\lambda_j = \pm \pi j \sqrt{\Gamma - \pi^2 j^2} \quad (j = 1, 2, \ldots).$$

Assume that

$$\pi^2 < \Gamma < 4\pi^2,$$

in which case the solution $w = 0$ is unstable with one positive and one negative eigenvalue, and the nonlinear equation (1) with $\epsilon = 0$, $\kappa > 0$ has two nontrivial stable buckled equilibrium states.

A simplified model for the dynamics of (1) is obtained by seeking lowest mode solutions of the form

$$w(z, t) = x(t) \sin(\pi z).$$

Substitution into (1) and taking the inner product with the basis function $\sin(\pi z)$, gives a Duffing-type equation for the modal displacement $x(t)$:

$$\ddot{x} - \beta x + \alpha x^3 = \epsilon(\gamma \cos \omega t - \delta \dot{x}), \tag{2}$$

where $\beta = \pi^2(\Gamma - \pi^2) > 0$, $\alpha = \kappa\pi^4/2$, and $\gamma = 4f/\pi$.

Further assumptions we make on (1) follow:

1. (No resonance): $j^2\pi^2(j^2\pi^2 - \Gamma) \neq \omega^2$ $(j = 2, 3, 4, \ldots)$.
2. (Large forcing to damping ratio):

$$\frac{f}{\delta} > \frac{\pi}{3} \frac{\Gamma - \pi^2}{\omega\sqrt{\kappa}} \cosh\left(\frac{\omega}{2\sqrt{\Gamma - \pi^2}}\right) \quad \text{(bifurcation point)}.$$

3. (Small forcing and damping): ϵ is sufficiently small.

By the results of Section 6.5, (1) has well-defined smooth global dynamics on the Banach space $\mathfrak{X} = H^2_\partial \times L^2$ of pairs $(w, \dot{w})$. In particular, there is a time $2\pi/\omega$ map $P: \mathfrak{X} \to \mathfrak{X}$ that takes initial data and advances it in time by one period of the forcing function. The main result shows that the map P has complicated dynamics in a very precise sense.

6.1 Theorem *Under the above hypotheses, there is some power P^N of P that has a compact invariant set $\Lambda \subset \mathfrak{X}$ on which P^N is conjugate to a shift on two symbols. In particular, (1) has infinitely many periodic orbits with arbitrarily high period.*

This set Λ arises in a way similar to Smale's famous "horseshoe" that is described below. The statement that P^N is conjugate to a shift on two symbols means that there is a homeomorphism $h: \Lambda \to$ (space of bi-infinite sequences of

0's and 1's) $= \{(\ldots, a_{-2}, a_{-1}, a_0, a_1, a_2, \ldots) \mid a_j = 0 \text{ or } 1, j \in \mathbb{Z}\} = \{0, 1\}^{\mathbb{Z}}$ such that $h \circ P^N \circ h^{-1}$ is the shift map taking the sequence (a_j) to the sequence (b_j), where $b_j = a_{j-1}$. Any periodic sequence then gives a periodic point for P^N.

There are many results on periodic orbits for partial differential equations. Perhaps the best are due to Rabinowitz [1978]. However, the approach and results here are quite different.

Let us now explain briefly how the "horseshoe" comes about. We consider Equation (2) for simplicity although the basic idea is the same for (1). The key thing is that when $\epsilon = 0$ (no forcing or damping) the flow has homoclinic orbits—that is, an orbit connecting a saddle point to itself. See Figure 7.6.2. In fact, this equation is Hamiltonian on $\mathbb{R}^2$ with

$$H(x, \dot{x}) = \frac{(\dot{x})^2}{2} - \frac{\beta x^2}{2} + \frac{\alpha x^4}{4}.$$

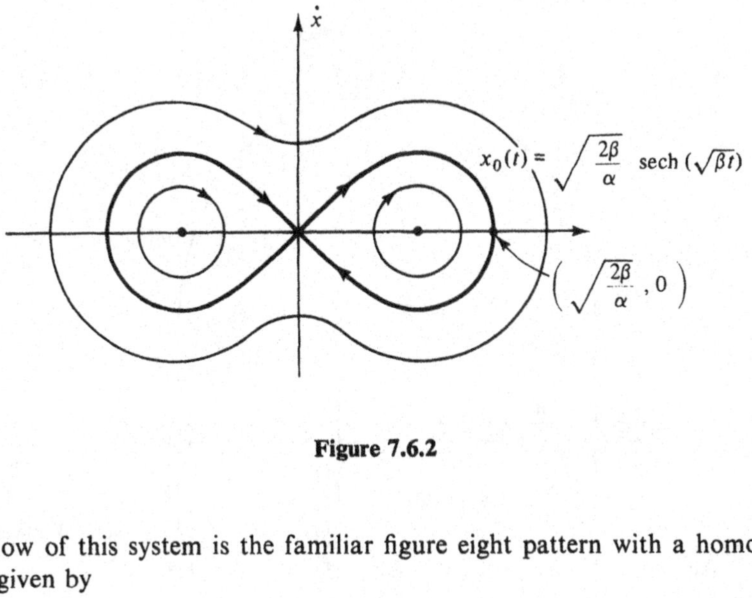

Figure 7.6.2

The flow of this system is the familiar figure eight pattern with a homoclinic orbit given by

$$x_0(t) = \sqrt{\frac{2\beta}{\alpha}}\,\mathrm{sech}(\sqrt{\beta}\,t).$$

When forcing and damping are turned on ($\epsilon > 0$) the idea is to use a technique of Melnikov [1963] (see also Arnold [1964]) to give a criterion for when the map P has stable and unstable manifolds that intersect transversally (see Figure 7.6.3). We shall go through this procedure shortly. Redrawing the situation, we have a map P of $\mathbb{R}^2$ to itself with stable and unstable manifolds as shown in Figure 7.6.4. It is plausible that the rectangle $\mathcal{R}$ is mapped as shown under a high power N of P. This is the reason for the name "horseshoe." Smale's basic work on this (Smale [1963], [1967]) was motivated by work of Cartwright,

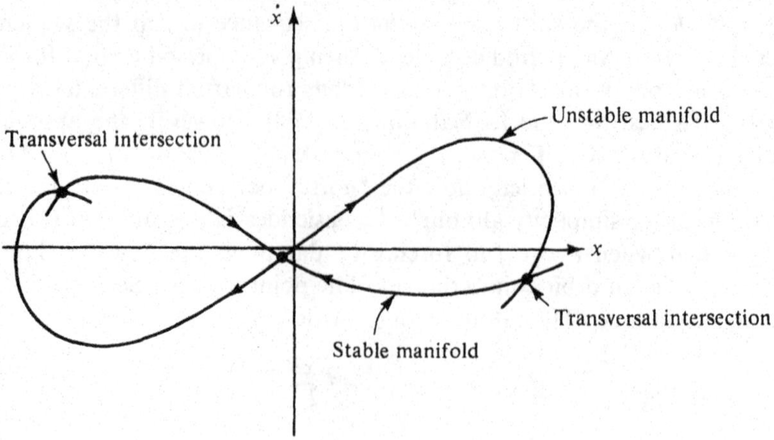

Figure 7.6.3

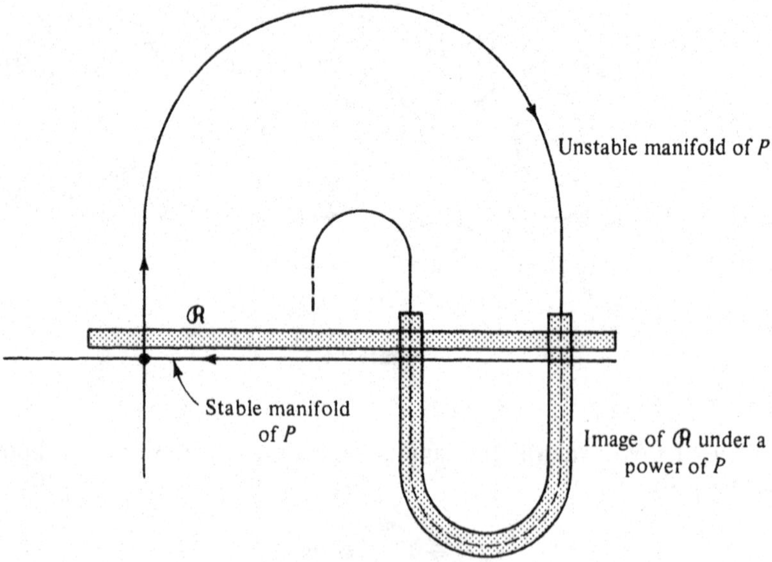

Figure 7.6.4

Littlewood, and Levinson on nonlinear oscillations. The invariant set Λ is obtained as $\Lambda = \bigcap_{n=-\infty}^{\infty}(P^N)^n(\mathcal{R})$.

For purposes of Equation (1) one requires an infinite-dimensional generalization of this situation. Using an elegant argument of Conley and Moser (see Moser [1973]) this was done by Holmes and Marsden [1981]. Some important refinements of this horseshoe picture for two-dimensional systems have been obtained by Chow, Hale and Mallet–Paret [1980] and Greenspan and Holmes [1982].

It is known that the time t-maps of the Euler and Navier–Stokes equations written in Lagrangian coordinates are smooth (Ebin and Marsden [1970]). Thus the methods described here can apply to these equations, in principle. On regions with no boundary, one can regard the Navier–Stokes equations with forcing as a perturbation of a Hamiltonian system (the Euler eauations). Thus, if one knew a homoclinic orbit for the Euler equations, then the methods of this section would produce infinitely many periodic orbits with arbitrarily high period, indicative of turbulence. No specific examples of this are known to us (one could begin by looking on the two-torus T^2 and studying Arnold [1966]).

Similar situations probably arise in traveling waves and the current-driven Josephson junction. For example, an unforced sine-Gordon equation with damping studied by M. Levi seems to possess a homoclinic orbit (cf. Levi, Hoppenstadt, and Miranker [1978]). Presumably the ideas will be useful for the *KdV* equation as well.

There has been considerable interest recently in chaotic dynamics and strange attractors (cf. Gurel and Rossler [1979] and Hellman [1980]). The methods described here do not prove they exist for (1) or (2), but they do provide evidence that they are there. For a discussion in connection with (2), see Holmes [1979b]. The difference between our set Λ and a true strange attractor $\mathcal{S}$ is that $\mathcal{S}$ is an attracting set and the flow near $\mathcal{S}$ has well-defined statistical properties. However, Λ is not an attractor; the flow near Λ is statistical for a long time, but eventually this leaks out. For a discussion of horseshoes in Hamiltonian systems, see Holmes and Marsden [1982a] and Kopell and Washburn [1981].

We shall now outline the abstract methods by which the result on transversal intersection is proved.

We consider an evolution equation in a Banach space $\mathfrak{X}$ of the form

$$\dot{x} = f_0(x) + \epsilon f_1(x, t), \tag{3}$$

where f_1 is periodic of period T in t. Our hypotheses on (3) are as follows:

(H1) (a) *Assume* $f_0(x) = Ax + B(x)$, *where* A *is an (unbounded) linear operator that generates a* C^0 *one-parameter group of transformations on* $\mathfrak{X}$ *and where* $B: \mathfrak{X} \longrightarrow \mathfrak{X}$ *is* C^∞. Assume that $B(0) = 0$ and $DB(0) = 0$.

(b) *Assume* $f_1: \mathfrak{X} \times S^1 \longrightarrow \mathfrak{X}$ *is* C^∞, *where* $S^1 = \mathbb{R}/(T)$, *the circle of length* T.

From the results of Section 6.5, Assumption (1) implies that the associated suspended autonomous system on $\mathfrak{X} \times S^1$,

$$\left.\begin{aligned} \dot{x} &= f_0(x) + \epsilon f_1(x, \theta), \\ \dot{\theta} &= 1, \end{aligned}\right\} \tag{4}$$

has a smooth local flow, F_t^ϵ. This means that $F_t^\epsilon: \mathfrak{X} \times S^1 \longrightarrow \mathfrak{X} \times S^1$ is a smooth map defined for small $|t|$, which is jointly continuous in all variables

ϵ, t, $x \in \mathfrak{X}$, $\theta \in S^1$, and for x_0 in the domain of A, $t \mapsto F_t^\epsilon(x_0, \theta_0)$ is the unique solution of (4) with initial condition x_0, θ_0.

The final part of Assumption (1) follows:

(c) *Assume that F_t^ϵ is defined for all $t \in \mathbb{R}$ for $\epsilon > 0$ sufficiently small.*

Our second assumption is that the unperturbed system is Hamiltonian. We recall from Chapter 5 that this means that $\mathfrak{X}$ carries a skew-symmetric continuous bilinear map $\Omega\colon \mathfrak{X} \times \mathfrak{X} \to \mathbb{R}$, which is weakly non-degenerate (i.e., $\Omega(u, v) = 0$ for all v implies $u = 0$), called the *symplectic form*, and there is a smooth function $H_0\colon \mathfrak{X} \to \mathbb{R}$ such that

$$\Omega(f_0(x), u) = dH_0(x)\cdot u$$

for all x in $\mathfrak{D}(A)$, the domain of A.

(H2) (a) *Assume that the unperturbed system $\dot{x} = f_0(x)$ is Hamiltonian with* energy $H_0\colon \mathfrak{X} \to \mathbb{R}$.

Problem 6.1 Verify that Equations (1) and (2) are Hamiltonian on $H_\partial^2 \times L^2$ and $\mathbb{R}^2$, respectively. The Hamiltonian for (1) is

$$H(w, \dot{w}) = \tfrac{1}{2}\|\dot{w}\|^2 - \frac{\Gamma}{2}\|w'\|^2 + \tfrac{1}{2}\|w''\|^2 + \frac{\kappa}{4}\|w'\|^2.$$

(b) *Assume there is a symplectic 2-manifold $\Sigma \subset \mathfrak{X}$ invariant under the flow F_t^0 and that on Σ the fixed point $p_0 = 0$ has a homoclinic orbit $x_0(t)$*; that is:

$$\dot{x}_0(t) = f_0(x_0(t))$$

and

$$\lim_{t\to+\infty} x_0(t) = \lim_{t\to-\infty} x_0(t) = 0.$$

Next we introduce a non-resonance hypothesis.

(H3) (a) *Assume that the forcing term $f_1(x, t)$ in* (3) *has the form*

$$f_1(x, t) = A_1 x + f(t) + g(x, t), \tag{5}$$

where $A_1\colon \mathfrak{X} \to \mathfrak{X}$ is a bounded linear operator, f is periodic with period T, $g(x, t)$ is t-periodic with period T and satisfies $g(0, t) = 0$, $D_x g(0, t) = 0$, so g admits the estimate

$$\|g(x, t)\| \le (\text{const.})\,\|x\|^2 \tag{6}$$

for x in a neighborhood of 0.

(b) *Suppose that the "linearized" system*

$$\dot{x}_L = Ax_L + \epsilon A_1 x_L + \epsilon f(t) \tag{7}$$

has a T-periodic solution $x_L(t, \epsilon)$ such that $x_L(t, \epsilon) = O(\epsilon)$.

For finite-dimensional systems, (H3) can be replaced by the assumption that 1 does not lie in the spectrum of e^{TA}; that is, none of the eigenvalues of A resonates with the forcing frequency.

Next, we need an assumption that A_1 contributes positive damping and that $p_0 = 0$ is a saddle.

(H4) (a) *For* $\epsilon = 0$, e^{TA} *has a spectrum consisting of two simple real eigenvalues* $e^{\pm\lambda T}$ $(\lambda \neq 0)$ *with the rest of the spectrum on the unit circle.*

(b) *For* $\epsilon > 0$, $e^{T(A+\epsilon A_1)}$ *has a spectrum consisting of two simple real eigenvalues* $e^{T\lambda_\epsilon^\pm}$ *(varying continuously in* ϵ *from perturbation theory of spectra) with the rest of the spectrum,* σ_R^ϵ, *inside the unit circle* $|z| = 1$ *and obeying the estimates*

$$C_2\epsilon \leq \text{distance}(\sigma_R^\epsilon, |z| = 1) \leq C_1\epsilon \tag{8}$$

for C_1, C_2 *positive constants.*

Finally, we need an extra hypothesis on the nonlinear term. We have already assumed B vanishes at least quadratically as does g. Now we assume B vanishes cubically.

(H5) $B(0) = 0$, $DB(0) = 0$, *and* $D^2B(0) = 0$.

This implies that in a neighborhood of 0, $\|B(x)\| \leq \text{const.}\ \|x\|^3$. (Actually $B(x) = o(\|x\|^2)$ would do.)

Consider the suspended system (4) with its flow $F_t^\epsilon\colon \mathfrak{X} \times S^1 \longrightarrow \mathfrak{X} \times S^1$. Let $P^\epsilon\colon \mathfrak{X} \longrightarrow \mathfrak{X}$ be defined by $P^\epsilon(x) = \pi_1 \cdot (F_T^\epsilon(x, 0))$, where $\pi_1\colon \mathfrak{X} \times S^1 \longrightarrow \mathfrak{X}$ is the projection onto the first factor. The map P^ϵ is just the Poincaré map for the flow F_t^ϵ. Note that $P^0(p_0) = p_0$, and that fixed points of P^ϵ correspond to periodic orbits of F_t^ϵ.

6.2 Lemma *For* $\epsilon > 0$ *small, there is a unique fixed point* p_ϵ *of* P^ϵ *near* $p_0 = 0$; *moreover,* $p_\epsilon - p_0 = O(\epsilon)$; *that is, there is a constant* K *such that* $\|p_\epsilon\| < K\epsilon$ *(for all small* ϵ*).*

For ordinary differential equations, Lemma 6.2 is a standard fact about persistence of fixed points, assuming 1 does not lie in the spectrum of e^{TA} (i.e., p_0 is hyperbolic). For general partial differential equations, the proof is similar in spirit, but is more delicate, requiring our assumptions. See Holmes and Marsden [1981] for details. An analysis of the spectrum yields the following.

6.3 Lemma *For* $\epsilon > 0$ *sufficiently small, the spectrum of* $DP^\epsilon(p_\epsilon)$ *lies strictly inside the unit circle with the exception of the single real eigenvalue* $e^{T\lambda_\epsilon^+} > 1$.

The next lemma deals with invariant manifolds.

6.4 Lemma *Corresponding to the eivengvalues* $e^{T\lambda_\epsilon^\pm}$, *there are unique invariant manifolds* $\mathcal{W}^{ss}(p_\epsilon)$ *(the strong stable manifold) and* $\mathcal{W}^u(p_\epsilon)$ *(the unstable manifold) of* p_ϵ *for the map* p_ϵ *such that:*

(i) $\mathcal{W}^{ss}(p_\epsilon)$ *and* $\mathcal{W}^u(p_\epsilon)$ *are tangent to the eigenspaces of* $e^{T\lambda_\epsilon^\pm}$, *respectively, at* p_ϵ.

(ii) *They are invariant under* P^ϵ.

(iii) *If* $x \in \mathcal{W}^{ss}(p_\epsilon)$, *then*

$$\lim_{n\to\infty} (P^\epsilon)^n(x) = p_\epsilon,$$

and if $x \in \mathcal{W}^u(p_\epsilon)$, *then*

$$\lim_{n\to-\infty} (P^\epsilon)^n(x) = p_\epsilon.$$

(iv) *For any finite* t^*, $\mathcal{W}^{ss}(p_\epsilon)$ *is* C^r *close as* $\epsilon \to 0$ *to the homoclinic orbit* $x_0(t)$ $(t^* \le t < \infty)$ *and for any finite* t_*, $\mathcal{W}^u(p_\epsilon)$ *is* C^r *close to* $x_0(t)$ $(-\infty < t \le t_*)$ *as* $\epsilon \to 0$. *Here, r is any fixed integer* $(0 \le r < \infty)$.

The Poincaré map P^ϵ was associated to the section $\mathfrak{X} \times \{0\}$ in $\mathfrak{X} \times S^1$. Equally well, we can take the section $\mathfrak{X} \times \{t_0\}$ to get Poincaré maps $P^\epsilon_{t_0}$. By definition, $P^\epsilon_{t_0}(x) = \pi_1(F^\epsilon_T(x, t_0))$. There is an analogue of Lemmas 6.2, 6.3, and 6.4 for $P^\epsilon_{t_0}$. Let $p_\epsilon(t_0)$ denote its unique fixed point and $\mathcal{W}^{ss}_\epsilon(p_\epsilon(t_0))$ and $\mathcal{W}^u_\epsilon(p_\epsilon(t_0))$ be its strong stable and unstable manifolds. Lemma 6.3 implies that the stable manifold $\mathcal{W}^s(p_\epsilon)$ of p_ϵ has codimension 1 in $\mathfrak{X}$. The same is then true of $\mathcal{W}^s(p_\epsilon(t_0))$ as well.

Let $\gamma_\epsilon(t)$ denote the periodic orbit of the (suspended) system with $\gamma_\epsilon(0) = (p_\epsilon, 0)$. We have

$$\gamma_\epsilon(t) = (p_\epsilon(t), t).$$

The invariant manifolds for the periodic orbit γ_ϵ are denoted $\mathcal{W}^{ss}(\gamma_\epsilon)$, $\mathcal{W}^s(\gamma_\epsilon)$, and $\mathcal{W}^u(\gamma_\epsilon)$. We have

$$\mathcal{W}^s(p_\epsilon(t_0)) = \mathcal{W}^s(\gamma_\epsilon) \cap (\mathfrak{X} \times \{t_0\}),$$

$$\mathcal{W}^{ss}(p_\epsilon(t_0)) = \mathcal{W}^{ss}(\gamma_\epsilon) \cap (\mathfrak{X} \times \{t_0\}),$$

$$\mathcal{W}^u(p_\epsilon(t_0) = \mathcal{W}^u(\gamma_\epsilon) \cap (\mathfrak{X} \times \{t_0\}).$$

See Figure 7.6.5.

We wish to study the structure of $\mathcal{W}^u_\epsilon(p_\epsilon(t_0))$ and $\mathcal{W}^s_\epsilon(p_\epsilon(t_0))$ and their intersections. To do this, we first study the perturbation of solution curves in $\mathcal{W}^{ss}_\epsilon(\gamma_\epsilon)$, $\mathcal{W}^s_\epsilon(\gamma_\epsilon)$, and $\mathcal{W}^u_\epsilon(\gamma_\epsilon)$.

Choose a point, say $x_0(0)$, on the homoclinic orbit for the unperturbed system. Choose a codimension 1 hyperplane H transverse to the homoclinic orbit at $x_0(0)$. Since $\mathcal{W}^{ss}_\epsilon(p_\epsilon(t_0))$ is C^r close to $x_0(0)$, it intersects H in a unique point, say $x^s_\epsilon(t_0, t_0)$. Define $(x^s_\epsilon(t, t_0), t)$ to be the unique integral curve of the suspended system (4) with initial condition $x^s_\epsilon(t_0, t_0)$. Define $x^u_\epsilon(t, t_0)$ in a similar way. We have

$$x^s_\epsilon(t_0, t_0) = x_0(0) + \epsilon v^s + O(\epsilon^2),$$

$$x^u_\epsilon(t_0, t_0) = x_0(0) + \epsilon v^u + O(\epsilon^2),$$

by construction, where $\|O(\epsilon^2)\| \le \text{const.}\cdot\epsilon^2$ and v^s and v^u are fixed vectors. Notice that

$$(P^\epsilon_{t_0})^n x^s_\epsilon(t_0, t_0) = x^s_\epsilon(t_0 + nT, t_0) \to p_\epsilon(t_0) \quad \text{as} \quad n \to \infty.$$

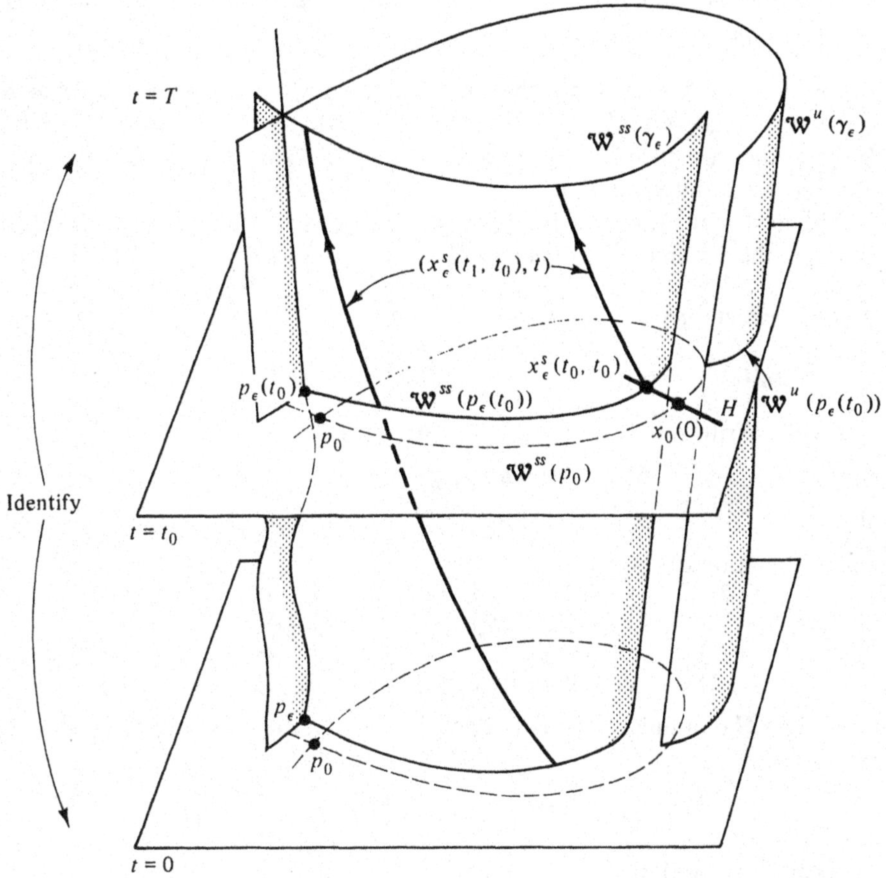

Figure 7.6.5

Since $x_\epsilon^s(t, t_0)$ is an integral curve of a perturbation, we can write

$$x_\epsilon^s(t, t_0) = x_0(t - t_0) + \epsilon x_1^s(t, t_0) + O(\epsilon^2),$$

where $x_1^s(t, y_0)$ is the solution of the first variation equation

$$\frac{d}{dt} x_1^s(t, t_0) = Df_0(x_0(t - t_0)) \cdot x_1^s(t, t_0) + f_1(x_0(t - t_0), t), \tag{9}$$

with $x_1^s(t_0, {}_{00}) = v^s$.

Define the *Melnikov function* by

$$\Delta_\epsilon(t, t_0) = \Omega(f_0(x_0(t - t_0)), x_\epsilon^s(t, t_0) - x_\epsilon^u(t, t_0))$$

and set

$$\Delta_\epsilon(t_0) = \Delta_\epsilon(t_0, t_0).$$

6.5 Lemma *If ϵ is sufficiently small and $\Delta_\epsilon(t_0)$ has a simple zero at some t_0 and maxima and minima that are at least $O(\epsilon)$, then $\mathcal{W}^u_\epsilon(p_\epsilon(t_0))$ and $\mathcal{W}^s_\epsilon(p_\epsilon(t_0))$ intersect transversally near $x_0(0)$.*

The idea is that if $\Delta_\epsilon(t_0)$ changes sign, then $x^s_\epsilon(t_0, t_0) - x^u_\epsilon(t_0, t_0)$ changes orientation relative to $f_0(x_0(0))$. Indeed, this is what symplectic forms measure. If this is the case, then as t_0 increases, $x^s_\epsilon(t_0, t_0)$ and $x^u_\epsilon(t_0, t_0)$ "cross," producing the transversal intersection.

The next lemma gives a remarkable formula that enables one to explicitly compute the leading order terms in $\Delta_\epsilon(t_0)$ in examples.

6.6 Lemma *The following formula holds:*

$$\Delta_\epsilon(t_0) = -\epsilon \int_{-\infty}^{\infty} \Omega(f_0(x_0(t - t_0)), f_1(x_0(t - t_0), t))\, dt + O(\epsilon^2).$$

Proof Write $\Delta_\epsilon(t, t_0) = \Delta^+_\epsilon(t, t_0) - \Delta^-_\epsilon(t, t_0) + O(\epsilon^2)$,

where $$\Delta^+_\epsilon(t, t_0) = \Omega(f_0(x_0(t - t_0)), \epsilon x^s_1(t, t_0))$$

and $$\Delta^-_\epsilon(t, t_0) = \Omega(f_0(x_0(t - t_0)), \epsilon x^u_1(t, t_0)).$$

Using Equation (9), we get

$$\frac{d}{dt}\Delta^+_\epsilon(t, t_0) = \Omega(Df_0(x_0(t, t_0))\cdot f_0(x_0(t - t_0)), \epsilon x^s_1(t, t_0))$$
$$+ \Omega(f_0(x_0(t - t_0)), \epsilon\{Df_0(x_0(t - t_0))\cdot x^s_1(t, t_0) + f_1(x_0(t - t_0), t)\}).$$

Since f_0 is Hamiltonian, Df_0 is Ω-skew. Therefore, the terms involving x^s_1 drop out, leaving

$$\frac{d}{dt}\Delta^+_\epsilon(t, t_0) = \Omega(f_0(x_0(t - t_0)), \epsilon f_1(x_0(t - t_0), t)).$$

Integrating, we have

$$-\Delta^+_\epsilon(t_0, t_0) = \epsilon \int_{t_0}^{\infty} \Omega(f_0(x_0(t - t_0)), f_1(x_0(t - t_0), t))\, dt,$$

since

$$\Delta^+_\epsilon(\infty, t_0) = \Omega(f_0(p_0), \epsilon f_1(p_0, \infty)) = 0, \quad \text{because} \quad f_0(p_0) = 0.$$

Similarly, we obtain

$$\Delta^-_\epsilon(t_0, t_0) = \epsilon \int_{-\infty}^{t_0} \Omega(f_0(x_0(t - t_0)), f_1(x_0(t - t_0), t))\, dt$$

and adding gives the stated formula. ∎

We summarize the situation as follows:

6.7 Theorem *Let hypotheses (H1)–(H5) hold. Let*

$$M(t_0) = \int_{-\infty}^{\infty} \Omega(f_0(x_0(t - t_0)), f_1(x_0(t - t_0), t)\, dt.$$

Suppose that $M(t_0)$ has a simple zero as a function of t_0. Then for $\epsilon > 0$ sufficiently small, the stable manifold $\mathcal{W}^s_\epsilon(p_\epsilon(t_0))$ of p_ϵ for $P^\epsilon_{t_0}$ and the unstable manifold $\mathcal{W}^u_\epsilon(p_\epsilon(t_0))$ intersect transversally.

Having established the transversal intersection of the stable and unstable manifolds, one can now substitute into known results in dynamical systems (going back to Poincaré) to deduce that the dynamics must indeed be complex. In particular, Theorem 6.1 concerning Equation (1) may be deduced. The calculations needed for the examples are outlined in the following problems.

Problem 6.2 (Holmes [1980b]) Consider Equation (2). Show that the Melnikov function is given by

$$M(t_0) = \int_{-\infty}^{\infty} (\dot{x}\gamma \cos \omega t - \delta w)\, dt,$$

where x stands for $x_0(t - t_0) = (\sqrt{2\beta/\alpha}) \operatorname{sech} \sqrt{\beta}\,(t - t_0)$. Evaluate the integral using residues:

$$M(t_0) = -2\gamma\pi\omega\sqrt{\frac{2}{\alpha}} \frac{\sin \omega t_0}{\cosh(\pi\omega/2\sqrt{\beta}\,)} - \frac{4\delta\beta^{3/2}}{3\alpha}.$$

Thus, show that the critical value γ_c above which transversal intersection occurs is

$$\gamma_c = \frac{4\delta\beta^{3/2}}{3\omega\sqrt{2\alpha}} \cosh\left(\frac{\pi\omega}{2\sqrt{\beta}}\right).$$

Problem 6.3 Show that a homoclinic orbit for (1) is given by

$$w_0(z, t) = \frac{2}{\pi}\sqrt{\frac{\Gamma - \pi^2}{\kappa}} \sin(\pi z) \operatorname{sech}(t\pi\sqrt{\Gamma - \pi^2}).$$

Use Problem 6.2 to compute the Melnikov function and hence arrive at the bifurcation value given on p. 506.

REFERENCES

The following list contains only references actually cited in the text and is not intended to be a complete bibliography.

ABRAHAM, R. and J. E. MARSDEN, 1978. *Foundations of Mechanics*, 2d ed. Reading, MA: Addison-Wesley Publishing Co., Inc.

ABRAHAM, R., J. MARSDEN, and T. RATIU, 1982. *Manifolds, Tensor Analysis and Applications*. Reading, MA: Addison-Wesley Publishing Co., Inc. (Second edition: Springer-Verlag, 1988.)

ABRAHAM R. and J. ROBBIN, 1967. *Transversal Mappings and Flows*, Reading, MA: Addison-Wesley Publishing Co., Inc.

ABRAHAM, R. and S. SMALE, 1960. *Lectures on Global Analysis, Notes*. New York: Columbia University Press.

ADAMS, R. A., 1975. *Sobolev Spaces*. New York: Academic Press, Inc.

ADELEKE, S. A., 1980. Stability of some states of plane deformation. *Arch. Rat. Mech. An.* 72:243-263.

AGMON, S., 1965. *Lectures on Elliptic Boundary Value Problems*. Princeton, NJ: D. Van Nostrand Company.

ALEXANDER, J. C. and S. S. ANTMAN, 1981. Global and local behavior of bifurcating multidimensional continua of solutions for multiparameter nonlinear eigenvalue problems. *Arch. Rat. Mech. An.* 76:339-354.

ALEXANDER, J. C. and J. A. YORKE, 1978. Global bifurcation of periodic orbits. *Amer. J. Math.* 100:263-292.

ALTMAN, S., 1982. Quasilinear evolution equations in nonreflexive Banach spaces. *J. Int. Eqns.* (to appear).

AMICK, J. and J. TOLAND, 1981. On solitary water waves of finite amplitude. *Arch. Rat. Mech. An.* 76:9-95.

ANDREWS, G., 1980. On the existence of solutions to the equation $u_{tt} = u_{xxt} + \sigma(u_x)_x$, *J. Diff. Eqns.* 35:200-231.

ANDREWS, G. and J. M. BALL, 1982. Asymptotic behaviour and changes of phase in one-dimensional viscoelasticity. *J. Diff. Eqns.* 44:206-341.

ANDRONOV, A. A. and L. S. PONTRYAGIN, 1937. Coarse systems. *Dokl. Akad. Nauk.* SSSR. 14:247-251.

ANTMAN, S. S., 1972a. The Theory of Rods. *In Handbuch der Physik*, Vol. VI a/2, ed. C. A. Truesdell. Berlin-Gottingen-Heidelberg: Springer-Verlag. 641-703.

—— 1972b. Qualitative theory of the ordinary differential equations on nonlinear elasticity. in Nemat-Nasser, 1972.

—— 1973a. Nonuniqueness of Equilibrium States for Bars in Tension. *J. Math. Anal. Appl.* 44:333-349.

—— 1973b. Monotonicity and invertibility conditions in one dimensional nonlinear elasticity, in Dickey, 1976.

—— 1974a. Stability of Nonlinearly Elastic Structures, *Proc. 11th Annual Meeting, Society of Engineering Science*, ed. G. J. Dvorak, Durham, NC: Duke University Press. 218-219.

—— 1974b. Kirchhoff's problem for nonlinearly elastic rods. *Quart. J. of Appl. Math.* 32:221-240.

—— 1976a. Ordinary Differential Equations of One-Dimensional Nonlinear Elasticity II: Foundations of the Theories of Nonlinearly Elastic Rods and Shells, *Arch. Rat. Mech. An.* 61:307-351.

—— 1976b. Ordinary Differential Equations of One-Dimensional Nonlinear Elasticity II: Existence and Regularity Theory for Conservative Problems. *Arch. Rat. Mech. An.* 61:353-393.

—— 1977. Bifurcation problems for nonlinearly elastic structures, in Rabinowitz 1977.

—— 1978a. Buckled states of nonlinearly elastic plates. *Arch. Rat. Mech. An.* 67: 111-149.

—— 1978b. A family of semi-inverse problems of nonlinear elasticity. *In Contemporary Developments in Continuum Mechanics and Partial Differential Equations.* G. M. de la Penha and L. A. Medeiros, eds. Amsterdam, North-Holland Publishing Co., a Div. of Elsevier North Holland.

—— 1979a. Multiple equilibrium states of nonlinearly elastic strings, *SIAM J. Appl. Math.* 37:588-604.

—— 1979b. The eversion of thick spherical shells. *Arch. Rat. Mech. An.* 70: 113-123.

—— 1980a. Nonlinear eigenvalue problems for the whirling of heavy elastic strings. In *Proc. Roy. Soc. Edinburgh* 85A:59-85.

—— 1980b. The equations for large vibrations of strings. *Am. Math. Monthly.* 87:359-370.

—— 1980c. Geometric aspects of global bifurcation in nonlinear elasticity. In *Springer Lecture Notes*, 775:1-29.

—— 1983. *Nonlinear Problems of Elasticity* (book in preparation).

ANTMAN, S. S. and E. R. CARBONE, 1977. Shear and necking instabilities in nonlinear elasticity. *J. Elasticity* 7:125-151.

ANTMAN, S. S. and J. E. DUNN, 1980. Qualitative behavior of buckled nonlinearly elastic arches. *J. Elasticity* 10:225-239.

ANTMAN, S. S. and C. S. KENNEY, 1981. Large buckled states of nonlinearly elastic rods under torsion, thrust and gravity. *Arch. Rat. Mech. An.* 76:289-388.

ANTMAN, S. S. and A. NACHMAN, 1979. Large buckled states of rotating rods. *Nonlinear analysis, theory, methods and appl.* 4:303-327.

ANTMAN, S. S. and J. E. OSBORNE, 1979. The principle of virtual work and integral laws of motion. *Arch. Rat. Mech. An.* 69:231-262.

ANTMAN, S. S. and G. ROSENFELD, 1978. Global behavior of buckled states of nonlinearly elastic rods, *SIAM Review* 20:513-566, (1980) and 186-187.

APOSTOL, T., 1974. *Mathematical Analysis.* Reading, MA: Addison-Wesley Publishing Co., Inc.

ARMS, J. and J. MARSDEN, 1979. The absence of Killing fields is necessary for linearization stability of Einstein's equations, *Ind. Univ. Math. J.*, 28:119-125.

ARMS, J., J. MARSDEN and V. MONCRIEF, 1981. Symmetry and bifurcation of momentum maps, *Comm. Math. Phys.* 78:455-478.

——— 1982. The structure of the space of solutions of Einstein's equations; II Several Killings Fields and the Einstein-Yang-Mills equations. *Ann. of Phys.* 144:81-106.

ARNOLD, V., 1964. Instability of dynamical systems with several degrees of freedom, *Dokl. Akad. Nauk. SSSR* 156:9-12.

———, 1966. Sur la géometrie différentielle des groupes de Lie de dimension infinie et ses applications à l'hydrodynamique des fluides parfaits, *Ann. Inst. Fourier,* Grenoble 16:319-361.

——— 1972. Bifurcations in versal families, *Russian Math. Surveys,* 27:54-123.

——— 1978. *Mathematical Methods of Classical Mechanics,* Graduate Texts in Mathematics 60. New York: Springer-Verlag.

ARNOLD, V. and A. AVEZ, 1971. *Ergodic Theory of Dynamical Systems.* Reading, MA: Addison-Wesley Publishing Co., Inc.

AUBERT, G., and R. TAHRAOUI, 1979. Theorèmes d'existence en calcul des variations, *J. Diff. Eqns.* 33:1-15.

AUBIN, T., 1976. Espaces de Sobolev sur les variétés riemanniennes, *Bull. Sci. Math. France.* 100:149-173.

BAJAJ, A. K., P. R. SETHNA and T. S. LUNDGREN, 1980. Hopf bifurcation phenomena in tubes carrying a fluid, *SIAM J. Appl. Math.* 39:213-230.

BAKER, M., and J. L. ERICKSEN, 1954. Inequalities restricting the form of the stress deformation relations for isotropic elastic solids and Reiner-Rivlin fluids, *J. Wash. Acad. Sci.* 44:33-35 (reprinted in *Foundations of Elasticity Theory.* New York: Gordon and Breach, Scientific Publishers, Inc., 1965).

BALAKRISHNAN, A. V., 1976. *Applied Functional Analysis,* Applications of Math. 3. New York: Springer-Verlag.

BALL, J. M., 1973. Saddle point analysis for an ordinary differential equation in a Banach space and an application to dynamic buckling of a beam, in *Nonlinear Elasticity.* ed. J. Dickey. New York: Academic Press, Inc.

——— 1977a. Constitutive equalities and existence theorems in elasticity, in Knops (1977).

——— 1977b. Convexity conditions and existence theorems in nonlinear elasticity, *Arch. Rat. Mech. An.* 63:337-403.

—— 1977c. Strongly continuous semigroups, weak solutions and the variations of constants formula, *Proc. AMS,* 63:370-373.

—— 1980. Strict convexity, strong ellipticity and regularity in the calculus of variations, *Math. Proc. Camb. Phil. Soc.* 87:501-513.

—— 1981. Global invertibility of Sobolev functions and the interpenetration of matter, *Proc. Roy. Soc. Edinburgh,* 88A:315-328.

—— 1982. Discontinuous equilibrium solutions and cavitation in nonlinear elasticity. *Phil. Trans. Roy. Soc. London* 306:557-611.

BALL, J. M., and J. CARR, 1976. Decay to zero in critical cases of second order ordinary differential equations of Duffing type. *Arch. Rat. Mech. An.* 63:47-57.

BALL, J. M., R. J. KNOPS and J. E. MARSDEN, 1978. Two examples in nonlinear elasticity, *Springer Lecture Notes in Math.* 665:41-49.

BALL, J. M. and J. E. MARSDEN, 1984. On potential wells in elasticity, *Arch. Rat. Mech. An.* 86:251-277.

BALL, J. M., J. E. MARSDEN and M. SLEMROD, 1982. Controllability of distributed bilinear systems, *SIAM J. on Control and Optimization.* 20:575-597.

BALL, J. M. and D. SCHAEFFER, 1983. Bifurcation and stability of homogeneous equilibrium configurations of an elastic body under dead load tractions, *Math. Proc. Camb. Phil. Soc.* 94:315-339.

BALL, J. M. and M. SLEMROD, 1979. Nonharmonic Fourier Series and the Stabilization of Distributed Semi-Linear Control Systems, *Comm. Pure Appl. Math.* 32:555-587.

BAUER, L., and E. L. REISS, 1965. Nonlinear buckling of rectangular plates, *SIAM Rev.* 13:603-626.

—— 1975. Multiple eigenvalues lead to secondary bifurcations, *SIAM Rev.* 17: 101-122.

BAUER, L., E. L. REISS, 1965. Nonlinear buckling of rectangular plates, *SIAM Rev.* 13:603-626.

BEATTY, M. F., 1965. Some static and dynamic implications of the general theory of elastic stability, *Arch. Rat. Mech. An.* 19:167-188.

—— 1967a. On the foundation principles of general classical mechanics, *Arch. Rat. Mech.* 27:264-273.

—— 1967b. A theory of elastic stability of incompressible, hyperelastic bodies, *Int. J. Solids Struct.* 3:23-37.

—— 1967c. A reciprocal theorem in the linearized theory of couple-stresses, *Acta Mech.,* III:154-166.

—— 1968. Stability of the undistorted states of an isotropic elastic body, *Int. J. Nonlinear Mech.,* 3:337-349.

BEATTY, M. F. and D. E. HOOK, 1968. Some experiments on the stability of circular rubber bars under end thrust, *Int. J. Solids Struct.,* 4:623-635.

BEJU, I., 1971. Theorems on existence, uniqueness and stability of the place boundary value problem, in statics, for hyperelastic materials, *Arch. Rat. Mech. An.* 42: 1-23.

BENJAMIN, T. B., 1961. Dynamics of a system of articulated pipes conveying fluid, *Proc. Roy. Soc. London* A:457-486 and 487-499.

—— 1978. Bifurcation phenomena in steady flows of a viscous fluid, *Proc. Roy. Soc. London,* A, 359:1-26 and 27-43.

BENNOUN, J. F., 1965. Etude des milieux continus élastiques et thermodynamiques en relativité générale, *Ann. Inst. H. Poincaré* 3:41.

BERGER, M. and D. EBIN, 1969. Some decompositions of the space of symmetric tensors on a Riemannian manifold, *J. Diff. Geom.* 3:379-392.

BERGER, M. S., 1974. New applications of the calculus of variations in the large to nonlinear elasticity, *Comm. Math. Phys.* 35:141-150.

—— 1977. *Nonlinearity and Functional Analysis.* New York: Academic Press, Inc.

BERGER, M. S. and P. C. FIFE, 1967. On von Karman's equations and the buckling of a thin elastic plate, *Comm. Pure Appl. Math.* 20:687-719 and 21:227-241.

BHARATHA S. and M. LEVINSON, 1978. Signorini's perturbation scheme for a general reference configuration in finite elastostatics, *Arch. Rat. Mech. An.* 67:365-394.

BIALYNICKI-BIRULA, I. and J. C. HUBBARD, 1982. Gauge-independent canonical formulation of relativistic plasma theory (preprint).

BISHOP, R. and S. GOLDBERG, 1968. *Tensor Analysis on Manifolds,* Englewood Cliffs, NJ: Prentice-Hall, Inc. Reprinted by Dover (1980).

BISPLINGHOFF, R. L. and H. ASHLEY, 1962. *Principles of Aeroelasticity,* New York: John Wiley & Sons, Inc.

BLATZ, P. J. and W. L. KO, 1962. Application of finite elasticity theory to the deformation of rubbery materials, *Trans. Soc. Rheology* 6:223-251.

BLEVINS, R. D., 1977. *Flow-induced Vibration,* New York: D. Van Nostrand Company.

BLOOM, F., 1979. Modern Differential Geometric Techniques in the Theory of Continuous Distributions of Dislocations, *Springer Lecture Notes in Math.* 733.

BOLZA, O., 1904. *Lectures on the Calculus of Variations,* Chicago Univ. Press. Reprinted 1973, New York: Chelsea House Publishers.

BRESSAN, A., 1978. *Relativistic Theories of Materials.* New York: Springer-Verlag.

BROCKETT, R., 1982. *A geometrical framework for nonlinear control and estimation,* (SIAM) CBMS Conf. Series.

BROCKWAY, G. S., 1972. On the uniqueness of singular solutions to boundary-initial value problems in linear elastodynamics, *Arch. Rat. Mech. An.* 48:213-244.

BROWN, G. H., ed., 1976. *Advances in Liquid Crystals,* Vol. 1-3. New York: Academic Press, Inc.

BUCHNER, M., J. MARSDEN and S. SCHECTER, 1982. Applications of the blowing up construction and algebraic geometry to bifurcation problems, *J. Diff. Eqns.* (to appear).

BUCHNER, H., J. MARSDEN and S. SCHECTER, 1983. Examples for the infinite dimensional Morse Lemma, *SIAM, J. Math. An.* 14:1045-1055.

BUZANO E., and M. GOLUBITSKY, 1982. Bifurcation involving the hexagonal lattice, *Phil. Trans. Roy. Soc.* (to appear).

CANTOR, M., 1975a. Sobolev inequalities for Riemannian bundles, *Proc. Symp. Pure Math.* AMS 27:171-184.

—— 1975b. Perfect fluid flows over $\mathbb{R}^{l}$ with asymptotic conditions, *J. Funct. An.* 18:73-84.

—— 1979. Some Problems of Global Analysis on Asymptotically Simple Manifolds, *Comp. Math.* 38:3-35.

CAPRIZ, G., and P. PODIO GUIDUGLI, 1974. On Signorini's perturbation method in finite elasticity, *Arch. Rat. Mech. An.* 57:1-30.

—— 1979. The role of Fredholm conditions in Signorini's perturbation method, *Arch. Rat. Mech. An.* 70:261-288.

CARR, J., 1981. *Applications of Center Manifolds.* New York: Springer-Verlag Applied Math. Sciences series, 35.

CARROLL, R. W., 1969. *Abstract methods in partial differential equations,* New York: Harper and Row, Publishers, Inc.

CARTAN, E, 1922. *Leçons sur les invariants intégraux.* Paris: Hermann.

—— 1923. Sur les variétés à connexion affine et la théorie de la relativité généralisée, *Ann. Ecole Normale, Sup.* 40:325-412; and 41:1-25.

CARTER, B., 1973. Elastic perturbation theory in general relativity and a variational principle for a rotating solid star, *Comm. Math. Phys.* 30:261-286.

CARTER, B., and H. QUINTANA, 1972. Foundations of general relativistic high-pressure elasticity theory, *Proc. Roy. Soc. London* A331:57-83.

CHEN, P. J., 1972. Growth and Decay of Waves in Solids, *Handbuch der Physik,* VIa/3. ed., C. A. Truesdell. Berlin: Springer-Verlag.

CHERNOFF, P., and J. MARSDEN, 1974. *Properties of Infinite Dimensional Hamiltonian Systems,* Springer Lecture Notes in Math 425.

CHILLINGWORTH, D.R.J., 1975. The catastrophe of a buckling beam, *Springer Lecture Notes in Math.* 488:88-91.

—— 1976. *Differential topology with a view to applications,* Research Notes in Math. 9. London: Pitman, Ltd.

CHILLINGWORTH, D.R.J., J. E. MARSDEN, and Y.H. WAN, 1982a. Symmetry and bifurcation in three dimensional elasticity. Part I. *Arch. Rat. Mech. An.* 80:295-331.

—— 1982b. Symmetry and bifurcation in three dimensional elasticity. Part II *Ibid.* 83:363-395; and 84:203-233.

CHOQUET, G., 1976. *Lectures on Analysis,* Reading, MA: Addison-Wesley Publishing Co., Inc. (3 volumes).

CHOQUET-BRUHAT, Y. (Y. Foures-Bruhat), 1952. Théorème d'existence pour certain système d'équations aux dérivées partielles non linéaires, *Acta. Math.* 88:141-225.

CHOQUET-BRUHAT, Y., C. DEWITTE, MORETTE and M. DILLARD-BLEICK, 1977. *Analysis, Manifolds and Physics.* Amsterdam: North Holland Publishing Co., Div. of Elsevier North Holland.

CHORIN, A., T. J. R. HUGHES, M. McCRACKEN and J. E. MARSDEN, 1978. Product formulas and numerical algorithms, *Comm. Pure Appl. Math.* 31:205-256.

CHORIN, A., and J. MARSDEN, 1979. *A Mathematical Introduction to Fluid Mechanics:* Berlin: Springer-Verlag, Universitext. (Third edition, 1993.)

CHOSSAT, P., 1979. Bifurcation and stability of convective flows in a rotating and non-rotating spherical shell, *SIAM J. Appl. Math.* 37:624-647.

CHOW, S. and J. HALE, 1982. *Methods of Bifurcation Theory.* New York: Springer-Verlag.

CHOW, S., J. HALE and J. MALLET-PARET, 1975. Applications of generic bifurcation I, *Arch. Rat. Mech. An.* 59:159-188.

—— 1976. Applications of generic bifurcation II, *Arch. Rat. Mech. An.* 62.

—— 1978. The Fuller index and global Hopf bifurcation, *J. Diff. Eqns.* 29:66-85.

—— 1980. An example of bifurcation to homoclinic orbits, *J. Diff. Eqns.* 37: 351-373.

CIARLET, P. G., 1983. *Topics in Mathematical Elasticity,* North Holland Publishing Co., Div. of Elsevier North Holland.

CIARLET, P. G., and P. RABIER, 1980. *Les équations de von Karman,* Springer Lecture Notes in Math. 826.

COHEN, D. S., 1977. Bifurcation from multiple complex eigenvalues, *J. Math. Anal. Appl.* 57:505-521.

COLEMAN, B. D., 1973. On the energy criterion for stability, in Dickey (1973).

COLEMAN, B. D., and M. E. GURTIN, 1965. Waves in materials with memory III. Thermodynamic influences on the growth and decay of accleration waves, *Arch. Rat. Mech. An.* 19:266-298.

COLEMAN, B. D., and V. J. MIZEL, 1966. Norms and semi-groups in the theory of fading memory, *Arch. Rat. Mech. An.* 23:87-123.

COLEMAN, B. D., and W. NOLL, 1959. On the thermostatics of continuous media, *Arch. Rat. Mech. An.* 4:97-128.

COLLET, P., and J. P. ECKMANN, 1980. *Iterated maps on the interval as dynamical systems.* Boston: Birkhauser.

COSSERAT, E. and F. COSSERAT, 1898. Sur les équations de la théorie de l'élasticité, *Comptes Rendus de l'Acad. Sci. Paris* 126:1129-1132.

—— 1909. *Théories des Corps Déformables.* Paris: Hermann.

COURANT, R. and K. FRIEDRICHS, 1976. *Supersonic Flow and Shock Waves.* New York: Springer-Verlag.

COURANT, R. and D. HILBERT, 1962. *Methods of Mathematical Physics,* Vol. II, John Wiley & Sons, Inc., Interscience.

CRANDALL, M. G., 1972. The semigroup approach to first order quasi-linear equations in several space variables, *Israel J. Math.* 12:108-132.

CRANDALL, M. G. and J. A. NOHEL, 1978. An abstract functional differential equation and a related nonlinear Volterra equation, *Israel J. Math.* 29:313-328.

CRANDALL, M. G. and P. H. RABINOWITZ, 1971. Bifurcation from simple eigenvalues, *J. Funct. An.* 8:321-340.

—— 1973. Bifurcation, perturbation of simple eigenvalues and linearized stability, *Arch. Rat. Mech. An.* 52:161-180.

—— 1978. The Hopf bifurcation in infinite dimensions, *Arch. Rat. Mech. An.* 67:53-72.

DAFERMOS, C. M., 1968. On the existence and asymptotic stability of solutions to the equations of linear thermo-elasticity, *Arch. Rat. Mech. An.* 29:241-271.

—— 1969. The mixed initial-boundary value problem for the equations of nonlinear one-dimensional viscoelasticy, *J. Diff. Eqns.* 6:71-86.

—— 1973. Solutions of the Riemann problem for a class of hyperbolic systems of conservation laws by the viscosity method, *Arch. Rat. Mech. An.* 52:1-9.

—— 1976. Contraction semigroups and trend to equilibrium in continuum mechanics, *Springer Lecture Notes in Math.* 503:295-306.

—— 1982. Global smooth solutions to the initial-boundary value problem for the equations of one-dimensional nonlinear thermoviscoelasticity, *SIAM J. Math. An.* (to appear).

DAFERMOS, C. M. and M. SLEMROD, 1973. Asymptotic behavior of nonlinear contraction semigroups, *J. Funct. An.* 13:97-106.

DANCER, E., 1973. Global solution branches for positive mappings, *Arch. Rat. Mech. An.* 52:181-192.

DAY., W. A. and M. SILHAVY, 1977. Efficiency and the existence of entropy in classical thermodynamics, *Arch. Rat. Mech. An.* 64:205-219.

DICKEY, R. W., 1976. *Bifurcation problems in nonlinear elasticity,* Research Notes in Mathematics, No. 3, London: Pitman, Ltd.

DIEUDONNE, J., 1960. *Foundations of Modern Analysis.* New York: Academic Press, Inc.

DiPERNA, R. J., 1982. Convergence of approximate solutions to conservation laws (preprint).

DOMBROWSKI, H. D., 1966. Eine Charakterisierung der Differentialoperatoren, *Nachr. Akad. Wiss. Gottingen,* II:19-43.

DORROH, J. R. and R. A. GRAFF, 1979. Integral Equations in Banach spaces, a general approach to the linear Cauchy problem, and applications to the non-linear problem, *J. Integral Eqns.* 1:309-359.

DOWELL, E. H., 1975. *Aeroelasticity of plates and shells.* Groningen, NL: Noordhoff Int. Publishing Co.

——— 1980. *Nonlinear aeroelasticity,* in Holmes (1980a), 147-172.

DOYLE, T. C. and J. L. ERICKSEN, 1956. Nonlinear Elasticity, in *Advances in Appl. Mech.* IV. New York: Academic Press, Inc.

DUHEM, P., 1906. *Recherches sur l'élasticité.* Paris: Gauthier-Villars.

DUNFORD, N. and J. SCHWARTZ, 1963. *Linear Operators,* Vol. II, John Wiley & Sons, Inc.

DUVAL, C. and H. P. KUNZLE, 1978. Dynamics of continua and particles from general covariance of Newtonian gravitation theory, *Rep. Math. Phys.* 13:351-368.

DUVAUT, G. and J. L. LIONS, 1972. *Les Inéquations en Mécanique et en Physique.* Paris: Dunod (Translated by Springer, 1976).

DYSON, F. J., 1972. Missed opportunities, *Bull. Am. Math. Soc.* 78:635-652.

EARDLEY, D. and V. MONCRIEF, 1981. Global existence of Yang-Mills-Higgs fields in four dimensional Minkowski space, *Comm. Math. Phys.* 83:171-211.

EBIN, D., 1970. The Manifold of Riemannian Metrics, *Proc. Symp. Pure Math.,* AMS XV, 11-40.

——— 1977. The motion of slightly compressible fluids viewed as a motion with strong constraining force, *Ann. Math.* 105:141-200, and *Comm. Pure Appl. Math.* Vol 35:451-485.

EBIN, D. G. and J. MARSDEN, 1970. Groups of diffeomorphisms and the motion of an incompressible fluid, *Ann Math.* 92:102-163.

EELLS, J. and L. LEMAIRE, 1978. A report on harmonic maps, *Bull. London Math. Soc.* 10:1-68.

EISENHART, L. P., 1926. *Riemannian Geometry.* Princeton, NJ: Princeton University Press.

EKELAND, I. and R. TEMAM, 1974. *Analyse convexe et problèmes variationnels.* Paris: Dunod.

ELIASSON, I. E., 1967. Geometry of manifolds of maps, *J. Diff. Geom.* 1:169-194.

EPSTEIN, D. and W. THURSTON, 1979. Transformation groups and natural bundles, *Proc. London Math. Soc.* III, 38:219-238.

EPSTEIN, M. and R. SEGEV, 1980. Differentiable manifolds and the principle of virtual work in continuum mechanics, *J. Math. Phys.* 21:1243-1245.

ERICKSEN, J. L., 1954. Deformations possible in every isotropic, incompressible perfectly elastic body, *Z.A.M.P.* 5:466-486; and *J. Math. Phys.* (1955) 34: 126-128.

—— 1960. Tensor fields, appendix in Truesdell and Toupin (1960).

—— 1966a. Thermoelastic stability, *Proc. Fifth U.S. Cong. on Appl. Mech.* 187-193.

—— 1966b. A thermokinetic view of elastic stability theory, *Int. Journal Solid. Struct.* 2:573-580.

—— 1975. Equilibrium of Bars, *J. Elasticity* 5:191-201.

ERINGEN, A. C., 1975. *Continuum Physics, Volume II – Continuum Mechanics of Single-Substance Bodies.* New York: Academic Press, Inc.

ERINGEN, A. C. and E. S. SUHUBI, 1974. *Elasto-Dynamics, Volume I – Finite Motions.* New York: Academic Press, Inc.

—— 1975. *Elasto-Dynamics, Volume II – Linear Theory.* New York: Academic Press, Inc.

FICHERA, G., 1972a. Existence theorems in elasticity, *Handbuch der Physik,* Bd. V1a/2, 347-389. Berlin: Springer-Verlag.

—— 1972b. Boundary value problems of elasticity with unilateral constraints, *Handbuch der Physik,* Bd. V1a/2, 391-424. Berlin: Springer-Verlag.

——, ed., 1976. *Trends in applications of pure mathematics to mechanics,* London: Pitman.

FISCHER, A. and J. E. MARSDEN, 1972b. The Einstein evolution equations as a first-order quasi-linear symmetric hyperbolic system, I, *Comm. Math. Phys.* 28:1-38.

—— 1973. Linearization stability of the Einstein equations, *Bull. Am. Math. Soc.* 79:995-1001.

—— 1975a. Deformations of the scalar curvature. Durham, NC: *Duke Math. J.* 42:519-547.

—— 1975b. Linearization stability of non-linear partial differential equations, *Proc. Symp. Pure Math.* AMS 27:219-263.

—— 1979a. Topics in the dynamics of general relativity in *Isolated gravitating systems in general relativity.* ed., J. Ehlers. Italian Physical Society, 322-395.

—— 1979b. Initial Value Problem and Canonical Formalism, in *Einstein centenary volume,* eds., S. Hawking and W. Israel. Cambridge, Cambridge Univ. Press.

FISCHER, A., J. MARSDEN, and V. MONCRIEF, 1980. The structure of the space of solutions of Einstein equations, I. One Killing field, *Ann. Inst. H. Poincare* 33:147-194.

FLANDERS, H., 1963. *Differential Forms.* New York: Academic Press, Inc.

FLETCHER, D. C., 1976. Conservation laws in linear elastodynamics, *Arch. Rat. Mech. An.* 60:329-353.

FLUGGE, W., 1972. *Tensor Analysis and Continuum Mechanics.* Berlin: Springer-Verlag.

FOSDICK, R. L. and J. SERRIN, 1979. On the impossibility of linear Cauchy and Piola-Kirchhoff constitutive theories for stress in solids, *J. Elasticity,* 9: 83-89.

FRANKEL, T., 1965. Critical submanifolds of the classical groups and Stiefel manifolds, in *Differential and Combinatorial Topology,* S. S. Cairns. Princeton, NJ: Princeton University Press.

FRIEDMAN, A., 1969. *Partial Differential Equations.* New York: Holt, Rinehart and Winston.

FRIEDRICHS, K. O., 1941. On the minimum buckling load for spherical shells, von Karman anniversary volume, 258-272.

——— 1947. On the boundary value problems of the theory of elasticity and Korn's inequality, *Ann. Math.* 48:441-471.

——— 1954. Symmetric hyperbolic linear differential equations. *Comm. Pure Appl. Math.* 7:345-392.

——— 1958. Symmetric positive linear differential equations. *Comm. Pure Appl. Math.* 11:333-418.

FUNG, Y. C., 1955. *An Introduction to the Theory of Aeroelasticity.* New York: John Wiley & Sons, Inc.

GALLISSOT, F., 1958. Formes extérieures et la mécanique des milieux continus, *Ann. Inst. Fourier* 8:291-335.

GLIMM, J., 1965. Solutions in the large for nonlinear hyperbolic systems of equations, *Comm. Pure Appl. Math.* 18:697-715.

GOLDSTEIN, H., 1980. *Classical mechanics,* (2nd ed.) Reading, MA: Addison-Wesley Publishing Co., Inc.

GOLDSTEIN, J., 1969. Semigroups and second order differential equations, *J. Funct. Anal.* 4:50-70.

GOLUBITSKY, M. and V. GUILLEMIN, 1973. *Stable mappings and their singularities.* New York: Springer-Verlag.

GOLUBITSKY, M. and W. F. LANGFORD, 1981. Classification and unfoldings of degenerate Hopf bifurcations, *J. Diff. Eqns.* 41:375-415.

GOLUBITSKY, M. and J. MARSDEN, 1983. The Morse Lemma in infinite dimensions via singularity theory. *SIAM J. Math. An.* 14:1037-1044.

GOLUBITSKY, M., J. MARSDEN, and D. SCHAEFFER, 1983. *Bifurcation problems with hidden symmetries,* New York: Plenum (W. Fitzgibbon, ed.).

GOLUBITSKY, M. and D. SCHAEFFER, 1979a. A theory for imperfect bifurcation via singularity theory, *Comm. Pure Appl. Math.,* 32:21-98.

——— 1979b. Imperfect bifurcation in the presence of symmetry, *Comm. Math. Phys.* 67:205-232.

——— 1981. Bifurcation with O(3) symmetry including applications to the Benard problem, *Comm. Pure Appl. Math.* 35:81-109.

——— 1982. A discussion of symmetry and symmetry breaking. *Proc. A.M.S. Summer Institute on Singularity Theory,* New York: Springer-Verlag (to appear).

——— 1984. *Singularities and Bifurcation Theory,* Springer-Verlag Appl. Math. Sciences.

GRAFF, R., 1978. *Elements of non-linear functional analysis.* Memoirs of AMS. No. 206.

——— 1981. Existence, uniqueness and smoothness for nonlinear evolution equations (preprint).

GREEN, A. E. and J. E. ADKINS, 1970. *Large Elastic Deformations,* (2d. ed.). London: Oxford University Press.

GREEN, A. E. and P. M. NAGHDI, 1977. On thermodynamics and the nature of the second law, *Proc. Roy. Soc. London,* 357:253-270.

——— 1978a. The second law of thermodynamics and cyclic processes, *J. Appl. Mech.,* 45:487-492.

——— 1978b. On non-local continuum mechanics, *Math. Proc. Camb. Phil. Soc.* 83:307-319.

GREEN, A. E., P. M. NAGHDI, and W. L. WAINWRIGHT, 1965. A general theory of a Cosserat Surface, *Arch. Rat. Mech. An.* 20:287-308.

GREEN, A. M. and R. S. RIVLIN, 1964a. On Cauchy's equations of motion, *J. Appl. Math. and Physics (ZAMP)* 15:290-292.

—— 1964b. Multipolar continuum mechanics, *Arch. Rat. Mech. An.* 17:113-147.

GREEN, A. E. and W. ZERNA, 1968. *Theoretical Elasticity,* (2d. ed.). London: Oxford University Press.

GREENBERG, J. M., 1967. On the equilibrium configurations of compressible slender bars, *Arch. Rat. Mech. An.* 27:181-194.

GREENSPAN, B. and P. J. HOLMES, 1982. Homoclinic Orbits, Subharmonics and Global Bifurcations in Forced Oscillations, in *Non-linear Dynamics and Turbulence* ed. G. Barenblatt, G. Iooss and D. D. Joseph. London: Pitman, Ltd.

GRIOLI, G., 1962. *Mathematical Theory of Elastic Equilibrium,* Ergebnisse der Ang. Mat. 7. Berlin: Springer-Verlag.

GUCKENHEIMER, J., 1980. Patterns of bifurcation, in Holmes (1980a).

GUCKENHEIMER, J. and P. HOLMES, 1983. *Bifurcations and Nonlinear Oscillations,* Springer Applied Math. Series.

GUILLEMIN, V. and S. STERNBERG, 1977. *Geometric Asymptotics,* Providence, RI: American Mathematics Society, Survey 14.

GUO, ZHONG-HENG, 1963. Time derivatives of tensor fields in non-linear continuum mechanics, *Arch. Mech. Stosowanej. 1 15,* 131-163.

GUREL, O. and O. ROSSLER, eds., 1979. Bifurcation theory and applications in scientific disciplines, *Ann. of N.Y. Acad. Sciences,* Vol. 316.

GURTIN, M. E., 1972a. The linear theory of elasticity, *Handbuch der Physik,* Vol. IVa/2, ed., C. A. Truesdell. Berlin: Springer-Verlag.

—— 1972b. Modern continuum thermodynamics, in *Mechanics Today,* Vol. I, ed., S. Nemat-Nasser.

—— 1973. Thermodynamics and the energy criterion for stability, *Arch. Rat. Mech. An.* 52:93-103.

—— 1975. Thermodynamics and stability, *Arch. Rat. Mech. An.* 59:63-96.

—— 1981a. *Topics in finite elasticity,* CBMS Reg. Conf. Series 35 (SIAM).

—— 1981b. *An introduction to continuum mechanics.* NY: Academic Press, Inc.

GURTIN, M. and L. MARTINS, 1976. Cauchy's theorem in classical physics, *Arch. Rat. Mech. An.* 60:305-328.

GURTIN, M. E. and S. J. SPECTOR, 1979. On stability and uniqueness in finite elasticity, *Arch. Rat. Mech. An.* 70:153-166.

HADAMARD, J., 1902. Sur une question de calcul des variations, *Bull. Soc. Math. France* 30:253-256.

—— 1905. Sur quelques questions de calcul des variations, *Bull. Soc. Math. France* 33:73-80.

HAKEN, H., ed., 1979. *Pattern formation by dynamic systems and pattern recognition,* New York: Springer-Verlag.

HALE, J. K., 1963. *Oscillations in nonlinear systems.* New York: McGraw-Hill Book Co.

—— 1969. *Ordinary differential equations.* New York: John Wiley & Sons, Inc.

—— 1977. Lectures on generic bifurcation theory, in Knops (1977b).

—— 1981. *Topics in dynamic bifurcation theory,* CBMS, Reg. Conf. Series No. 47 AMS.

HARTMAN, P., 1973. *Ordinary Differential Equations,* New York: John Wiley & Sons, Inc., and P. Hartman, reprinted by Birkhauser, Boston, 1982.

HASSARD, B. D., 1979. Numerical evaluations of Hopf bifurcation formulae, *Proc. of the first annual workshop on the information linkage between applied mathematics and industry,* Monterey (1978), P. Wang (ed.), New York: Academic Press, Inc.

HASSARD, B., N. KAZARINOFF, and Y. H. WAN, 1981. *Theory and Applications of Hopf bifurcation.* London Math. Soc. Lecture Note Series 41. Cambridge, England: Cambridge Univ. Press.

HAWKING, S. and G. ELLIS, 1973. *The Large Scale Structure of Spacetime,* Cambridge, England: Cambridge Univ. Press.

HAUGHTON, D. M. and R. W. OGDEN, 1978. Instabilities and symmetry in the pressure problem, *J. Mech. Phys. Solids* 26:111-138, 1982 *J. of Elasticity* 12:239-250.

HAYASHI, C., 1964. *Nonlinear Oscillations in Physical Systems.* New York: McGraw-Hill Book Co.

HELLEMAN, H. G., ed., 1980. *Nonlinear Dynamics,* Ann. of N.Y. Acad. Sci., 357.

HENRY, D., 1981. *Geometric Theory of Semilinear Parabolic Equations,* Springer Lecture Notes in Math,, 840.

HERRMANN, G., 1966. *Dynamic Stability of Structures.* Elmsford, NY: Pergamon Press.

HERMANN, R., 1968. *Differential Geometry and the Calculus of Variations.* New York: Academic Press, Inc., 2d. ed., Mathematics and Science Press.

—— 1973. *Geometry, Physics and Systems.* New York: Marcel Dekker, Inc.

—— 1975. *Interdisciplineary Mathematics,* Volumes 4, 5, and 11. Brookline, MA: Math-Sci Press.

—— 1980. Differential geometry of Engineering-Mechanics systems: a modernized version of the work of Kron and Kondo. (preprint).

HERMES, H., 1979. Local controllability of observables in finite and infinite dimensional nonlineara control systems, *Appl. Math. and Optim.,* 5:117-125.

HILL, R., 1957. On uniqueness and stability in the theory of finite elastic strain, *J. Mech. Phys. Solids* 5:229-241.

HILL, R., 1968. On constitutive inequalities for simple materials, I, *J. Mech. Phys. Solids.* 16:229-242.

HILL, R., 1970. Constitutive inequalities for isotropic elastic solids under finite strain, *Proc. Roy. Soc. London* A314:457-472.

HILLE, E. and R. PHILLIPS, 1957. *Functional Analysis and Semi-Groups,* AMS Colloquia Publishing.

HILTON, P. J. and G. S. YOUNG, eds., 1982. *New Directions in Applied Mathematics,* New York: Springer Publishing Co., Inc.

HOLM, D. and B. KUPERSCHIMIDT, 1982. Poisson brackets and Clebsch representations for magnetohydrodynamics, multifluid plasmas and elasticity (preprint).

HOLMES, P., 1977a. Bifurcations to divergence and flutter in flow-induced oscillations—a finite dimensional analysis, *J. Sound and Vib.* 53, 471-503.

—— 1977b. Strange phenomena in dynamical systems and their physical implications, *Appl. Math. Modelling* 1 (7), 362.

—— 1979a. Global bifurcations and chaos in the forced oscillations of buckled structures, *Proc. IEEE Conference on Decision and Control,* Paper WA7. Jan. 10–12, 1979. San Diego, CA.

—— 1979b. A nonlinear oscillator with a strange attractor, *Phil. Trans. Roy Soc.* A292:419-448.

—— 1979c. Domains of stability in a wind induced oscillation problem, *J. Appl. Mech,* 46:672-676.

——— ed., 1980a. *New approaches to nonlinear problems in Dynamics,* Phildelphia: SIAM.

——— 1980b. Averaging and chaotic motions in forced oscillations, *SIAM J. Appl. Math.* 38:68-80; and 40:167-168.

——— 1980c. Unfolding a degenerate nonlinear oscillator: a codimension two bifurcation, *Ann. N.Y. Acad. Sci.* 357:473-488.

——— 1980d. Pipes supported at both ends cannot flutter, *J. Appl. Mech.* 45, 619-622.

——— 1981. Space and time-periodic perturbations of the sine-Gordon equation, *Springer Lecture Notes in Math* 898:174-191.

——— 1982. Center manifolds, normal forms and bifurcations of vector fields with applications to coupling between periodic and steady motions, *Physica* D 2: 449-481.

HOLMES, P. and Y. K. LIN, 1978. Deterministic stability analysis of a wind loaded structure, *Trans. A.S.M.E. J. Appl. Mech.* 45: 165-169.

HOLMES, P. and J. MARSDEN, 1978a. Bifurcation to divergence and flutter in flow induced oscillations; an infinite dimensional analysis, *Automatica* 14:367-384.

——— 1978b. Bifurcations of Dynamical Systems and Nonlinear Oscillations in Engineering Systems, *Springer Lecture Notes in Math.,* 648:163-206.

——— 1979. Qualitative techniques for bifurcation analysis of complex systems, in Gurel and Rossler (1979), 608-622.

——— 1981. A partial differential equation with infinitely many periodic orbits: chaotic oscillations of a forced beam, *Arch. Rat. Mech. An.* 76:135-166.

——— 1982a. Horseshoes in perturbations of Hamiltonian systems with two degrees of freedom, *Comm. Math. Phys.* 82:523-544.

——— 1982b. Melnikov's method and Arnold diffusion for perturbations of integrable Hamiltonian systems, *J. Math. Phys.* 23:669-675.

——— 1982c. Horseshoes and Arnold diffusion for Hamiltonian systems on Lie groups, *Ind. Univ. Math. J.* 32:273-310.

HOLMES, P. and D. RAND, 1976. The Bifurcations of Duffing's Equation: An Application of Catastrophe Theory, *J. Sound and Vib.* 44:237-253.

——— 1978. Bifurcations of the forced van der Pol oscillator, *Quart. Appl. Math.* 35:495-509.

HOPF, E., 1942. Abzweigung einer periodischen Lösung von einer stationären Lösung eines Differentialsystems. *Ber. Math-Phys. Kl. Sachs. Acad. Wiss. Leipzig* 94:1-22; *Ber. Verh. Sachs. Acad. Wiss. Leipzig Math-Nat. Kl.* 95(1):3-22 (English Translation in Marsden and McCracken (1976).

HOPKINS, H. G. and M. J. SEWELL, 1981. *Mechanics of Solids; the Rodney Hill 60th Anniversary Volume.* Elmwood, NY: Pergamon Press, Inc.

——— 1980. Bifurcation of finitely-deformed rotating elastic cylinders, *Quart J. Mech. Appl. Maths.* 33:251-265.

HSU, C. S., 1977. On nonlinear parametric excitation problems, *Advances in Appl. Mech.* 17:245-301.

HUGHES, T.J.R.., T. KATO, and J. E. MARSDEN, 1977. Well-posed quasi-linear hyperbolic systems with applications to nonlinear elastodynamics and general relativity, *Arch. Rat. Mech. An* 63:273-294.

HUGHES, T.J.R. and J. LUBLINER, 1973. On the one-dimensional theory of blood flow in the larger vessels, *Math. Biosciences* 18:161-170.

HUGHES, T.J.R. and J. E. MARSDEN, 1976. *A Short Course on Fluid Mechanics,* Berkeley, CA: Publish or Perish, Inc.

—— 1977. Some applications of geometry in continuum mechanics, *Rep. on Math. Phys.* 12:35-44.

—— 1978. Classical elastodynamics as a symmetric hyperbolic system, *J. of Elasticity* 8:97-110.

HUNT, G., 1982. Symmetries of elastic buckling, *Eng. Struct.* 4:21-28.

HUSEYIN, K., 1974. *Nonlinear Theory of Elastic Stability.* Groningen, NL: Noordhoff, Int. Publ. Co.

—— 1978. *Vibrations and Stability of Multiple Parameter Systems.* Groningen, ND: Nordhoff Int. Publ. Co.

—— 1978. *Stability of Elastic Structures,* CISM Courses 238, New York: Springer-Verlag.

IOOSS, G. and D. JOSEPH, 1980. *Elementary Stability and Bifurcation Theory.* New York: Springer-Verlag.

IWAN, W. D. and R. D. BLEVINS, 1974. A model for vortex-induced oscillation of structures, *J. Appl. Mech* 41:581-586.

IZE, J., 1976. Bifurcation theory for Fredholm operators, *Memoirs of Amer. Math. Soc.* 174.

JAMES, R. D., 1980. The propagation of phase boundaries in elastic bars, *Arch. Rat. Mech. An.* 73:125-158.

JAUMANN, G., 1911. Geschlossenes System physikalischer und chemischer Differenzialgesetze, *Sitzber. Akad. Wiss. Wien.* (IIa) 120:385-530.

JAUNZEMIS, W., 1967. *Continuum Mechanics,* New York: MacMillan, Inc.

JOHN, F., 1972. Uniqueness of non-linear elastic equilibrium for prescribed boundary displacements and sufficiently small strains, *Comm. Pure Appl. Math.* 25: 617-634.

—— 1977. Finite amplitude waves in a homogeneous isotropic elastic solid, *Comm. Pure Appl. Math.* 30:421-446.

—— 1979. Blow-up of solutions of nonlinear wave equations in three space dimensions, *Manuscripta Math.* 28:235-268.

JORGENS, K., 1961. Das Anfangswertproblem in Grössen fur eine Klasse nichtlinearer Wellengleichungen, *Math. Zeit.* 77:295-308.

JOST, R. and E. ZEHNDER, 1972. A generalization of the Hopf bifurcation theorem. *Helv. Phys. Acta.* 45:258-276.

JURDJEVIC, V. and J. QUINN, 1978. Controllability and Stability, *Journal of Diff. Eqns.* 28:281-289.

KATO, T., 1966. *Perturbation Theory for Linear Operators,* New York: Springer-Verlag, 2d ed. (1977).

—— 1970. Linear evolution equations of hyperbolic type I, *J. Fac. Sci. Univ. Tokyo* 17:241-258.

—— 1973. Linear evolution equations of hyperbolic type, II, *J. Math. Soc. Japan* 25:648-666.

—— 1975a. The Cauchy problem for quasi-linear symmetric hyperbolic systems, *Arch. Rat. Mech. An.* 58:181-205.

—— 1975b. Quasi-linear equations of evolution with applications to partial differential equations. *Springer Lecture Notes in Math.,* 448:25-70.

—— 1977. *Linear quasilinear equations of evolution of hyperbolic type.* CIME lectures, Italian Math Soc.

—— 1979. On the Korteweg-deVries equation. *Manuscripta Math.* 28:89-99.

KAUFMAN, A. and R. SPENCER, 1982. Hamiltonian formulation of a charged two fluid model, *Phys. Rev.* A25:2437-2439.

KEADY, G. and J. NORBURY, 1978. On the existence theory for irrotational water waves, *Math. Proc. Comb. Phil Soc.* 83:137-157.

KEENER, J. P., 1974. Perturbed bifurcation theory at multiple eigenvalues, *Arch. Rat. Mech. An.* 56:348-366.

KEENER, J. P. and H. B. KELLER, 1972. Perturbed bifurcation and buckling of circular plates, *Springer Lect. Notes in Math.* 280:286-293.

—— 1973. Perturbed bifurcation theory, *Arch. Rat. Mech. An.* 50:159-175.

KELLER, H. B., 1973. Buckling of complete spherical shells under slightly nonuniform loads, in Dickey (1973).

KELLER, J. B. and S. ANTMAN, 1969. *Bifurcation theory and nonlinear eigenvalue problems.* Menlo Park, CA: W.A. Benjamin, Inc.

KELLER, H. B., J. B. KELLER, and E. REISS, 1962. Buckled states of circular plates, *Quart. Appl. Math.* 20:55-65.

KELLOGG, R. B. and J. E. OSBORNE, 1976. A regularity result for the Stokes problem in a convex polygon, *J. Funct. An.* 21:397-431.

KEYFITZ, B. L. and H. C. KRANZER, 1980. A system of non-strictly hyperbolic conservation laws arising in elasticity theory, *Arch. Rat. Mech. An.* 72:219-241.

KIELHOFER, H., 1976. On the Lyapunov-Stability of stationary solutions of semilinear parabolic differential equations, *J. Diff. Eqns.* 22:193-208.

KIJOWSKY, J. and W. TULCZYJEW, 1979. *A symplectic framework for field theories,* Springer Lecture Notes in Physics 121.

KILLING, W., 1982. Über die Grundlagen der Geometrie, *J. Reine Angew. Math.* 109:121-186.

KIRCHHOFF, G., 1852. Über die Gleichungen des Gleichgewichts eines elastischen Körpers bei nicht unendlich kleinen Verschrebungen seiner Theile. *Sitzsber. Akad. Wiss. Wien* 9:762-773.

KLAINERMAN, S., 1978. Global existence for nonlinear wave equations, *Comm. Pure Appl. Math.* 33:43-101.

KLAINERMAN, S. and R. KOHN, 1982. The incompressible limit of nonlinear elastodynamics (preprint).

KLAINERMAN, S. and A. MAJDA, 1980. Formation of singularities for wave equations including the nonlinear vibrating string, *Comm. Pure Appl. Math.* 33:241-263.

—— 1981. Singular limits of quasi-linear hyperbolic systems with large parameters and the incompressible limit of compressible fluids, *Comm. Pure Appl. Math.* 34:449-524, and 35:629-651.

KNIGHTLY, G. H., 1967. An existence theorem for the Von Karman equations, *Arch. Rat. Mech. An.* 27:233-242.

KNIGHTLY, G. H. and D. SATHER, 1970. On nonuniqueness of solutions of the von Karmen equations, *Arch. Rat. Mech. An.* 36:65-78.

—— 1974. Nonlinear buckled states of rectangular plates, *Arch. Rat. Mech. An.* 54:356-372.

—— 1975. Nonlinear axisymmetric buckled states of shallow spherical caps, *SIAM J. Math. An.* 6:913-924.

—— 1980. Buckled states of a spherical shell under uniform external pressure, *Arch. Rat. Mech. An.* 72:315-380.

KNOPS, R. J., ed., 1973. *Symposium on Non-well Posed Problems and Logarithmic Convexity,* Springer Lecture Notes in Math. 316.

——, ed., 1977. *Nonlinear Analysis and Mechanics;* Vol. I, Pitman, Ltd.

——, ed., 1978. *Nonlinear Analysis and Mechanics,* Vol. II, London: Pitman, Ltd.

——, ed., 1979a. *Nonlinear Analysis and Mechanics,* Vol. III, London: Pitman, Ltd.

—— 1979b. *Nonlinear Analysis and Mechanics,* IV, London: Pitman. Ltd.

—— 1981. *Trends in Applications of Pure Mathematics to Mechanics,* Vol. III, London: Pitman, Ltd.

KNOPS, R. J., H. A. LEVINE, and L. E. PAYNE, 1974. Non-existence, instability and growth theorems for solutions of a class of abstract nonlinear equations with applications to nonlinear elastodynamics, *Arch. Rat. Mech. An.* 55:52-72.

KNOPS, R. J. and L. E. PAYNE, 1971. *Uniqueness Theorems in Linear Elasticity,* Springer Tracts in Natural Philosophy, 19. New York: Springer-Verlag.

—— 1978. On potential wells and stability in nonlinear elasticity, *Math. Proc. Camb. Phil. Soc.* 84:177-190.

KNOPS, R. and E. WILKES, 1973. Theory of elastic stability, in *Handbuch der Physik,* VIa/3, C. Truesdell, ed., Berlin: Springer-Verlag.

KNOWLES, J. K. and E. STERNBERG, 1972. On a class of conservation laws in linearized and finite elasticity, *Arch. Rat. Mech. An.* 44:187-211.

—— 1975. On the ellipticity of the equations of nonlinear elastostatics for a special material, *J. Elasticity* 5:334-341.

—— 1977. On the failure of ellipticity of the equations for finite elastic plane strain, *Arch. Rat. Mech. An.* 63:321-336.

—— 1978. On the failure of ellipticity and the emergence of discontinuous deformation gradients in plane finite elastostatics, *J. Elasticity* 8:329-379.

KOBAYASHI, S. and K. NOMIZU, 1963. *Foundations of Differential Geometry.* New York: John Wiley & Sons, Inc.

KOITER, W. T., 1945. *On the stability of elastic equilibrium,* Dissertation, Delft, Holland (English translation NASA Tech. Trans. F10:833 (1967)).

—— 1965a. On the instability of equilibrium in the absence of a minimum of the potential energy, *Proc. Kon. Ned. Akad. Wet.* B68:107-113.

—— 1965b. The energy criterion of stability for continuous elastic bodies, *Proc. Kon. Ned. Akad. Wet.* B68:178-202.

—— 1976a. A basic open problem in the theory of elastic stability, *Springer Lecture Notes in Math.* 503:366-373.

—— 1976. On the complementary energy theorem in non-linear elasticity theory, in Fichera (1976).

——, ed., 1977. *Theoretical and Applied Mechanics.* Amsterdam: North Holland Publishing Co., Div. of Elsevier North Holland.

KOLODNER, I., 1955. Heavy rotating string—a nonlinear eigenvalue problem, *Comm. Pure. Appl. Math.* 8:395-408.

KONDO, K., 1955. *RAAG Memoirs of the Unifying study of basic problems in Engineering and Physical Sciences by means of geometry* (3 vols.), Gakujutsu Bunken Fukyu-Kai, Tokyo.

KOPELL, N. and R. B. WASHBURN, 1981. Chaotic motions in the two degree of freedom swing equations (preprint).

KORN, A., 1914. Über die Lösung des Grundproblems der Elastizitätstheorie, *Math. Ann.* 75:497-549.

KRASNOSELSKII, M. A., 1964. *Topological Methods in the Theory of Nonlinear Integral Equations,* Elmsford, NY: Pergamon Press, Inc.

LADYZHENSKAYA, O. A., 1969. *The mathematical theory of viscous incompressible flow* (2d. ed.), New York: Gordon and Breach Science Publishers, Inc.

LANDAU, L. D. and E. M. LIFSHITZ, NY: Pergamon Press, Inc., 1970. *Theory of Elasticity* (2d. ed.) Elmsford.

LANDL, R., 1975. A Mathematical Model for Vortex-Excited Vibrations of Bluff Bodies, *J. Sound and Vib.* 42:219-234.

LANG, S., 1971. *Analysis I.* Reading, MA: Addison-Wesley Publishing Co., Inc.

—— 1972. *Differential Manifolds,* Reading, MA: Addison-Wesley Publishing Co. Inc.

LANGENBACH, A., 1961. On some nonlinear operators of elasticity theory in Hilbert space. *Vest. Leningrad Univ.* 1:38-50.

LANGFORD, W. F., 1979. Periodic and steady state mode interactions lead to tori, *SIAM J. Appl. Math.* 37:22-48.

LANGFORD, W. F. and G. IOOSS, 1980. Interactions of Hopf and pitchfork bifurcations, in *Bifurcation Problems and their Numerical Solution,* H. D. Mittlemann and H. Weber (eds.), Boston, MA: Birkhauser.

LAWRUK, B. and W. M. TULCZYJEW, 1977. Criteria for partial differential equations to be Euler-Lagrange equations. *J. Diff. Eqns.* 24:211-225.

LAX, P. D., 1964. Development of singularities of solutions of nonlinear hyperbolic differential equations, *J. Math. Phys.* 5:611-613.

LeTALLEC, P. and J. T. ODEN, 1981. Existence and characterization of hydrostatic pressure in finite deformations of incompressible elastic bodies *J. Elasticity* 11:341-358.

LEVI, M., F. C. HOPPENSTADT, and W. L. MIRANKER, 1978. Dynamics of the Josephson Junction, *Quart. Appl. Math.* 36:167-198.

LIANIS, G. and R. S. RIVLIN, 1972. Relativistic equations of balance in continuum mechanics, *Arch. Rat. Mech. An.* 48:64-82.

LICHNEROWICZ, A., 1967. *Relativistic Hydrodynamics and Magnetohydrodynamics,* Menlo Park, CA: W. A. Benjamin, Inc.

LIN, Y. K. and P. J. HOLMES, 1968. Stochastic analysis of wind loaded structure, *Proc. A.S.C.E. J. Eng. Mech.* 104:421-440.

LIONS, J. L. and E. MAGENES, 1972. *Nonhomogeneous boundary value problems and applications.* New York: Springer-Verlag.

LITTMAN, W., 1973. L^p-L^q estimates for singular integral operators arising from hyperbolic equations. *Proc. Symp. Pure Math. AMS* 23:477-482.

LORENZ, E. N., 1979. On the prevalence of aperiodicity in simple systems, *Springer Lecture Notes* 755:53-76.

LOVE, A.E.H., 1927. *A Treatise on the Mathematical Theory of Elasticity.* Cambridge, England: Cambridge University Press (Reprinted by Dover (1944)).

LUENBERGER, D. G., 1969. *Optimization by Vector Space Methods.* New York: John Wiley & Sons, Inc.

LUMER, G. and R. S. PHILLIPS, 1961. Dissipative Operators in a Banach space, *Pac. J. Math.* 11:679-698.

LUSTERNIK, L. A. and V. J. SOBOLEV, 1974. *Elements of Functional Analysis.* New York: Halsted Press (John Wiley).

MacCAMY, R. C., 1977. A model for one-dimensional, nonlinear viscoelasticity, *Quart. J. of Appl. Math.* 35:21-33.

MADDOCKS, J., 1982. On non-planar bifurcations of the elastica (preprint).

MAGNUS, R. J., 1976. On the local structure of the zero set of a Banach space valued mapping, *J. Funct. An.* 22:58-72.

—— 1976b. A generalization of multiplicity and a problem of bifurcation, *Proc. London Math. Soc.* 32:251-278.

MAGNUS, R. and T. POSTON, 1977. On the full unfolding of the von Karman equations at a double eigenvalue, *Battelle Report* 109.

MALLET-PARET, J., 1977. Buckling of cylindrical shells with small curvature, *Quart. Appl. Math.* 35:383-400.

MALVERN, L., 1969. *Introduction to the mechanics of a continuous medium.* Englewood Cliffs, NJ: Prentice-Hall, Inc.

MARIETTA, M. G., 1976. An isoperimetric problem for continuous systems: the Aeolian vibration of a conductor span. *J. Franklin Inst.*, 301:317-333.

MARSDEN, J. E., 1968a. Generalized Hamiltonian Mechanics, *Arch. Rat. Mech. An.* 28:326-362.

—— 1968b. Hamiltonian one-parameter groups, *Arch. Rat. Mech. An.* 28:363-396.

—— 1972. Darboux's theorem fails for weak symplectic forms, *Proc. Am. Math. Soc.* 32:590-592.

—— 1973a. A proof of the Calderon extension theorem, *Can. Math. Bull.* 16: 133-136.

—— 1973b. On global solutions for nonlinear Hamiltonian evolution equations, *Comm. Math. Phys.* 30:79-81.

—— 1973c. On product formulas for nonlinear semigroups, *J. Funct. An.* 13:51-72.

—— 1974a. *Elementary Classical Analysis.* San Francisco, CA: W. H. Freeman & Company, Publishers Inc. (Second edition, 1992.)

—— 1974b. *Applications of Global Analysis in Mathematical Physics.* Berkeley, CA: Mathematics Department Lecture Series.

—— 1976. Well-posedness of the equations of a non-homogeneous perfect fluid, *Comm. Part. Diff. Eqns.* 1:215-230.

—— 1978. Qualitative methods in bifurcation theory, *Bull. Am. Math. Soc.* 84: 1125-1148.

—— 1981. *Lectures on Geometric Methods in Mathematical Physics,* CBMS-NSF Regional Conference Series 37, Phildelphia, PA: SIAM.

—— 1982. Four applications of nonlinear analysis to physics and engineering, in Hilton and Young (1982).

MARSDEN, J. E., D. EBIN, and A. FISCHER, 1972. Diffeomorphism Groups, Hydrodynamics and General Relativity, *Proc. 13th Biennial Seminar of the Canadian Math. Congress,* (ed. J. Vanstone). Montreal: 135-279.

MARSDEN, J. E. and T.J.R. HUGHES, 1978. Topics in the Mathematical Foundations of Elasticity, in Knops (1978).

MARSDEN, J. E. and M. McCRACKEN, 1976. *The Hopf bifurcation and its applications,* Applied Math. Sciences No. 19, Springer.

MARSDEN, J. E., T. RATIU, and A. WEINSTEIN, 1984. Semidirect products and reduction in mechanics, *Trans. Am. Math. Soc.* 281:147-177.

MARSDEN, J. E. and A. WEINSTEIN, 1974. Reduction of symplectic manifolds with symmetry, *Rep. Math. Phys.* 5:121-130.

—— 1982. The Hamiltonian structure of the Maxwell-Vlasov equations, *Physica D,* 4:394-406.

MARSDEN, J. E., A. WEINSTEIN, R. SCHMIDT, and R. SPENCER, 1983. Hamiltonian Systems and Symmetry Groups with Applications to Plasma Physics, *Proc. IUTAM Symp. Torino* 289-340.

MARSDEN, J. E. and Y. H. WAN, 1983. Linearization stability and Signorini series for the traction problem in elastostatics, *Proc. Roy. Soc. Ed.* 95:171-180.

MARTINI, R., 1979. On the Frechet differentiability of certain energy functionals, *Proc. Kon. Ned. Akad. Wet.* B82:42-45.

MASLOV, V., 1965. *Théorie des Perturbations et Méthodes Asymptotiques,* Paris: Dunod.

MATKOWSKY, B. J., L. J. PUTNICK, and E. L. REISS, 1980. Secondary buckled states of rectangular plates, *SIAM J. Appl. Math.* 38:38-51.

MATKOWSKY, B. J. and L. J. PUTNIK, 1974. Multiple buckled states of rectangular plates, *Inter. J. Non-Linear Mech.* 9:89-103.

MATKOWSKY, B. J. and E. L. REISS, 1977. Singular perturbations of bifurcations, *SIAM J. Appl. Math.* 33:230-255.

MATSUMURA, A. and T. NISHIDA, 1980. The initial value problem for the equations of motion of viscous and heat-conductive gases. *J. Math. Kyoto U.* 20:67-104.

MAUGIN, G., 1975. *Sur la formulation des lois de comportement en mécanique relativiste des milieux continus,* Thèse, Univ. Paris VI, also *Ann. Inst. H. Poincaré* 15 (1971), 275-302, *J. Gen. Rel. Grav.* (1973) 4:241-272, and *Comm. Math. Phys.* (1977), 53:233-356.

McLEOD, J. B. and D. H. SATTINGER, 1973. Loss of Stability and Bifurcation at a Double Eigenvalue, *J. Funct. An.* 14:62-84.

MEDEIROS, L. A., and G. M. DE LA PENNA, eds., 1978. *Contemporary developments in Continuum Mechanics and Partial Differential Equations,* Amsterdam: North Holland Publishing Co., Inc. Div. of Elsevier North Holland.

MEISTERS, G. H. and C. OLECH, 1963. Locally one-to-one mappings and a classical theorem on Schlicht functions, *Duke Math. J.* 30:63-80.

MEJLBRO, L., 1976. On the energy criterion of stability for continuous elastic bodies, *Report of Math. Institute,* Danmarks Tekniske Hojskole.

MELNIKOV, V. K., 1963. On the stability of the center for time periodic perturbations, *Trans. Moscow Math. Soc.* 12:1-57.

MISNER, C., K. THORNE, and J. WHEELER, 1973. *Gravitation.* San Francisco, CA: W. H. Freeman & Company, Publishers, Inc.

MOON, F. C., 1980a. Experiments on chaotic motions of a forced nonlinear oscillator: strange attractors, *J. Appl. Mech.* 47:638-644.

—— 1980b. Experimental models for strange attractor vibrations in elastic systems, in Holmes (1980a).

MOON, F. and P. HOLMES, 1979. A magneto-elastic strange attractor, *J. Sound Vib.* 65:275-296, and 69:339.

MOONEY, M., 1940. A theory of large elastic deformation, *J. Appl. Phys.* 11: 582-592.

MORREY, C. B., Jr., 1952. Quasi-convexity and the lower semicontinuity of multiple integrals, *Pac. J. of Math.* 2:25-53.

MORREY, C. B., Jr., 1966. *Multiple Integrals in the Calculus of Variations.* New York: Springer-Verlag.

MORRISON, P. J., 1980. The Maxwell-Vlasov equations as a continuous Hamiltonian system, *Phys. Lett.* 80A:383-386.

MORRISON, P. J. and J. M. GREENE, 1980. Noncanonical Hamiltonian density formulation of hydrodynamics and ideal magnetohydrodynamics, *Phys. Rev. Lett.* 45:790-794.

MOSER, J., 1965. On the volume elements on a manifold, *Trans. Am. Math. Soc.* 120:286-294.

—— 1973. *Stable and Random Motions in Dynamical Systems,* Ann. of Math. Studies No. 77. Princeton, NJ: Princeton Univ. Press.

NAGHDI, P. M., 1972. The Theory of Shells, *Handbuch der Physik* VIa/2 (ed. C. Truesdell) New York: Springer-Verlag.

—— 1981. On the role of the second law of thermodynamics in mechanics of materials. *Energy,* 5:771-781.

NAGHDI, P. M. and J. TRAPP, 1973. On the general theory of stability for elastic bodies, *Arch. Rat. Mech. An.* 51:165-191.

NAUDASCHER, E., ed, 1974. *Flow Induced Structural Vibrations* (Proc. I.U.T. A.M. - I.A.H.R. Symposium, Karlsruhe, 1972) Berlin: Springer-Verlag.

NAVARRO, C., 1978. Asymptotic stability in linear thermovisco-elasticity, *J. Math. Anal. Appl.* 65:399-431.

NELSON, E., 1959. Analytic vectors, *Ann. of Math.* 70:572-615.

NEMAT-NASSER, S., ed., 1972. *Mechanics Today I.* Elmsford, NY: Pergamon Press, Inc.

—— ed., 1978. *Mechanics Today IV.* Elmsford: NY: Pergamon Press, Inc.

—— ed., 1980. *Variational Methods in the Mechanics of Solids,* Elmsford, NY: Pergamon Press, Inc.

NIRENBERG, L., 1959. On elliptic partial differential equations, *Ann. de Scuola. Norm. Sup. Pisa.* 13:115-162.

—— 1974. *Topics in Nonlinear Analysis,* Courant Institute Lecture Notes.

NOLL, W., 1958. A mathematical theory of the mechanical behavior of continuous media, *Arch. Rat. Mech. An.* 2:197-226.

—— 1963. La mécanique classique basée sur un axiome d'objectivité in *La Méthode axiomatique dans les mécaniques classiques et nouvelles,* Paris, 47-56.

—— 1972. A new mathematical theory of simple materials. *Arch. Rat. Mech. An.* 48:1-50.

—— 1973. Lectures on the foundations of continuum mechanics and thermodynamics, *Arch. Rat. Mech. An.* 52:62-92.

NOVAK, M., 1969. Aeroelastic Galloping of Prismatic Bodies, *Proc. A.S.C.E. J. Eng. Mech.* 115-142.

—— 1971. Galloping and vortex induced oscillations of structures, *Proc. 3rd Inter. Conf. on Wind Effects on Bldgs. and Structures,* Tokyo.

ODEN, J. T., 1972. *Finite elements of Nonlinear Continua,* New York: McGraw-Hill Book Company, Inc.

—— 1979. *Applied Functional Analysis,* Englewood Cliffs, NJ: Prentice-Hall, Inc.

ODEN, J. T. and J. N. REDDY, 1976a. *Mathematical Theory of Finite Elements,* New York: Wiley-Interscience.

—— 1976b. *Variational Methods in Theoretical Mechanics,* New York: Springer-Verlag.

OGDEN, R. W., 1970. Compressible isotropic elastic solids under finite strain-constitutive inequalities, *Quart. J. Mech. Appl. Math.* 23:457-468.

—— 1972. Large deformation isotropic elasticity, *Proc. Roy. Soc. London.* A326: 565-584 and A328:567-583.

—— 1977. Inequalities associated with the inversion of elastic stress-deformation relations and their implications. *Math. Proc. Camb. Phil. Soc.* 81:313-324.

OLDROYD, J. G., 1950. On the formulation of rheological equations of state, *Proc. Roy. Soc. London* A200:523-541.

OLVER, P. J., 1982. Conservation laws in elasticity: I general principles, II linear homogeneous isotropic elastostatics (preprint).

ONSAGER, L., 1931. Reciprocal relations in irreversible processes. *Phys. Rev.* 37:405-426.

OSTER, G. and A. S. PERELSON, 1973. Systems, circuits and thermodynamics, *Israel J. Chem.* 11:445-478 (1974), Chemical reaction dynamics, *Arch. Rat. Mech. An.* 55:230-274 and 57:31-98.

PAIDOUSSIS, M. P., 1966. Dynamics of flexible slender cylinders in axial flow, *J. Fluid Mech.* 26:717-736, 737-751.

PAIDOUSSIS, M. P. and N. Y. ISSID, 1974. Dynamic stability of pipes conveying fluid, *J. Sound and Vib.* 33:267-294.

PALAIS R., 1954. Definition of the exterior derivative in terms of the Lie derivative, *Proc. Am. Math. Soc.* 5:902-908.

—— 1965. *Seminar on the Atiyah-Singer Index Theorem.* Ann. of Math. Studies, No. 57. Princeton, NJ: Princeton University Press.

—— 1968. *Foundations of Global Nonlinear Analysis.* Reading, MA: Addison-Wesley Publishing Co., Inc.

PALAIS, R. and C. L. TERNG, 1977. Natural bundles have finite order, *Topology.* 16:271-277.

PARKINSON, G. V. and N.P.H. BROOKS, 1961. On the aeroelastic instability of bluff cylinders, *J. Appl. Mech.* 28:252-260.

PARKINSON, G. V. and J. D. SMITH, 1964. The square prism as an aeroelastic nonlinear oscillator, *Quart. J. Mech. and Appl. Math.* 17:225-230.

PARKS, P. C., 1966. A stability criterion for a panel flutter problem via the second Method of Liapunov, in *Differential Equations and Dynamical Systems,* J. K. Hale and J. P. LaSalle (eds.) New York: Academic Press.

PAYNE, L. E. and D. H. SATTINGER, 1975. Saddle points and instability of nonlinear hyperbolic equations, *Israel J. Math.* 22:273-303.

PAYNE, L. E. and H. F. WEINBERGER, 1961. On Korn's inequality, *Arch. Rat. Mech. An.* 8:89-98.

PAZY, A., 1974. *Semi-groups of linear operators and applications to partial differential equations.* Univ. of Maryland, Lect. Notes, No. 10.

PEETRE, J., 1959. Une caractérisation abstraite des opérateurs différentiels, *Math. Scand.* 7:211-218.

PETROVSKII, I., 1937. Uber das Cauchysche Problem fur lineaere und nichtlineaere hyperbolische partielle Differential-gleichungen, *Mat. Sb.* (44) 2:814-868.

PIERCE, J. R. and A. P. WHITMAN, 1980. Topological properties of the manifolds of configurations for several simple deformable bodies, *Arch. Rat. Mech. An.* 74:101-113.

PIOLA, G., 1845. Nuova analysis per tutte le questioni della meccanica *Mem. Mat. Fis. Soc. Ital. Moderna* 24:1-186.

PLAUT, R. H. and K. HUSEYIN, 1975. Instability of fluid-conveying pipes under axial load, *J. of Appl. Mech.* 42:889-890.

POINCARE, H., 1885. Sur l'Equilibre d'Une Masse Fluide Animée d'un Mouvement de Rotation, *Acta. Math.,* 7:259.

POTIER-FERRY, M., 1978a. *Fondements mathématiques de la théorie de la stabilité élastique.* Thesis, Univ. of Paris VI.

—— 1978b. Bifurcation et stabilité pour des systèmes dérivant d'un potentiel, *J. de Mécanique* 17:579-608.

—— 1979. Perturbed Bifurcation Theory, *J. Diff. Eqns.* 33:112-146.

—— 1980. An existence and stability theorem in nonlinear viscoelasticity, in Nemat-Nasser (1980).

—— 1981. The linearization principle for the stability of solutions of quasilinear parabolic equations, I. *Arch. Rat. Mech. An.* 77:301-320.

—— 1982. On the mathematical foundations of elastic stability theory, I. *Arch. Rat. Mech. An.* 78:55-72.

PRODI, G., 1962. Teoremi di tipo locale per il sistema di Navier-Stokes e stabilità della soluzioni stazionarie. *Rend. Sem. Mat. Univ. Padova.* 32.

QUINN, B., 1971. Solutions with shocks, an example of an L_1-contractive semigroups, *Comm. Pure Appl. Math.* 24:125-132.

RABINOWITZ, P. H., 1971. Some global results for nonlinear eigenvalue problems, *J. Funct. An.* 7:487-513.

—— 1977a. A bifurcation theorem for potential operators, *J. Funct. An.* 25: 412-424.

—— ed., 1977b. *Applications of Bifurcation Theory.* NY: Academic Press, Inc.

—— 1978. Free vibrations for a semilinear wave equation, *Comm. Pure. Appl. Math.* 31:31-68.

RAMANUJAN, S., 1969. Morse Theory of certain symmetric spaces. *J. Diff. Geom.* 3:213-229.

RAND, D., 1982. Dynamics and symmetry. Predictions for modulated waves in rotating fluids, *Arch. Rat. Mech. An.* 79:1-37.

RATIU, T., 1979. On the smoothness of the time t-maps of the KdV equations and the bifurcation of eigenvalues of Hill's operator, *Springer Lecture Notes in Math.* 775. 248-294.

—— 1981. Euler-Poisson equations on Lie algebras and the N-dimensional heavy rigid body, *Proc. National Acad. Sci. U.S.A.* 78:1327-1328.

RATIU, T. and P. BERNARD, eds., 1977. *Seminar on Turbulence Theory*, Springer Lecture Notes in Math. 615.

RAUCH, J. B. and F. J. MASSEY III, 1974. Differentiability of solutions to hyperbolic initial-boundary value problems. *Trans. Amer. Math. Soc.* 189:303-318.

REED, M., 1976. *Abstract Nonlinear Wave Equations,* Springer Lecture Notes in Math. 507.

REED, M. and B. SIMON, 1975. *Methods of Modern Mathematical Physics II Fourier Analysis and Self adjointness.* New York: Academic Press, Inc.

REEKEN, M., 1973. Stability of critical points under small perturbations, *Manuscripta Math.* 8:69-72.

REISS, E. L. and B. J. MATKOWSKY, 1971. Nonlinear dynamic buckling of a compressed elastic column, *Appl. Math.* 29:245-260.

RESHETNYAK, Y. G., 1968. Stability theorems for mappings with bounded excursion, *Sibirskii Math.* 9:667-684.

RIESZ, F. and B. Sz. NAGY, 1955. *Functional Analysis.* New York: Frederick Ungar Publishing Co., Inc.

RIVLIN, R. S., 1947. Torsion of a rubber cylinder, *J. Appl. Phys.* 18:444-449.

—— 1948a. Some applications of elasticity theory to rubber engineering, *Proc. 2nd Tech. Conf.*, Cambridge.

—— 1948b. Large elastic deformations of isotropic materials, *Phil. Trans. Roy. Soc. London* 240:491-508; 241:379-397, *Proc. Royal Soc. London* (1949) A242:173-195.

—— 1960a. Some topics in finite elasticity, in *Structural Mechanics.* Elmsford, NY: Pergamon Press, Inc.

—— 1960b. The formulation of constitutive equations in continuum physics II, *Arch. Rat. Mech. An.* 3:262-272.

—— 1966a. The fundamental equations of nonlinear continuum mechanics, in *Dynamics of Fluids and Plasmas.* New York: Academic Press, Inc.

—— 1966b. Stability of pure homogeneous deformations of an elastic cube under dead loading, *Quart. of Appl. Math.*, Oct., 265-271.

—— 1973. Some restrictions on constitutive equations, *Proc. Int. Symp. on the Foundations of Continuum Thermodynamics.* Bussaco.

—— 1974a. The elements of non-linear continuum mechanics, in *Continuum Mechanics Aspects of Geodynamics and Rock Fracture Mechanics* (ed. Thoft-Christensen). Hingham, MA: D. Reidel Publishing Co., Inc.

—— 1974b. Stability of Pure Homogeneous Deformations of an Elastic Cube under Dead Loading, *Quart. of Appl. Math.* Oct., 265-271.

—— 1977. Some research directions in finite elasticity, *Rheo. Acta.* 16:101-112.

RODRIGUES, H. M. and A. L. VANDERBAUWHEDE, 1978. Symmetric perturbations of nonlinear equations: symmetry of small solutions, *Nonlinear Analysis, Theory, Methods and Appl.* 2:27-46.

ROH, H., 1982. Dissipative operators with finite dimensional damping, *Proc. Roy. Soc. Edinburg.* 91:243-263.

ROORDA, J., 1965. Stability of structures with small imperfections, *A.S.C.E., Proc. J. Eng. Mech.* 1:87-106.

ROSENBERG, R. M., 1966. On Nonlinear Vibrations of Systems with Many Degrees of Freedom, *Adv. in Appl. Mech.* 9:159-242.

ROSTAMIAN, R., 1978. Transition from unconstrained to constrained materials in continuum mechanics, *Indiana Univ. Math. J.* 27:637-656.

RUBIN, H. and P. UNGAR, 1957. Motion under a strong constraining force, *Comm. Pure Appl. Math.* 10:65-87.

RUELLE, D., 1969. *Statistical Mechanics. Rigorous Results.* Menlo Park, CA: W. A. Benjamin Publishing, Inc.

—— 1973. Bifurcations in the presence of a symmetry group, *Arch. Rat. Mech. An.* 51:136-152.

—— 1978. *Thermodynamic Formalism,* Encyclopedia of Math. Vol. 5. Reading, MA: Addison-Wesley Publishing Co., Inc.

RUELLE, D. and F. TAKENS, 1971. On the nature of turbulence. *Comm. Math. Phys.* 20:167-192.

RUSSELL, D. L., 1979. *Mathematics of finite dimensional control systems. Theory and Design.* New York: Marcel Dekker, Inc.

SAPIR, M. H. and E. L. REISS, 1979. Dynamic buckling of a nonlinear Timoshenko beam, *SIAM J. Appl. Math.* 37:290-301.

SATHER, D., 1973. Branching of solutions of nonlinear equations, *Rocky Mountain J. Math.* 3:203-250.

—— 1977. Bifurcation and stability for a class of shells, *Arch. Rat. Mech. An.* 63:295-304.

SATTINGER, D. H., 1973. *Topics in stability and bifurcation theory,* Springer Lecture Notes in Math. 309.

—— 1979. *Group theoretic methods in bifurcation theory,* Springer Lecture Notes in Math, 762. New York: Springer-Verlag.

—— 1980. Bifurcation and symmetry breaking in applied mathematics, *Bull. Am. Math. Soc.* 3:779-819.

SAWYERS, K. N., 1976. Stability of an elastic cube under dead loading: two equal forces, *Inter. J. Non-Linear Mech.* 11:11-23.

SAWYERS, K. N. and R. S. RIVLIN, 1982. Stability of a thick elastic plate under thrust, *J. Elasticity.* 12:101-126.

SCHAEFFER, D., 1980. Qualitative analysis of a model for boundary effects in the Taylor Problem, *Math. Proc. Camb. Phil. Soc.* 87:307-337.

SCHAEFFER, D. and M. GOLUBITSKY, 1980. Boundary Conditions and mode jumping in the buckling of a rectangular plate, *Comm. Math. Phys.* 69:209-236.

—— 1981. Bifurcation analysis near a double eigenvalue of a model chemical reaction. *Arch. Rat. Mech. An.* 75:315-347.

SCHAUDER, J., 1935. Das Anfangswertproblem einer quasi-linearen hyperbolischen Differentialgleichung zweiter Ordnung in beliebiger Anzahl von unabhängigen Veränderlichen, *Fund. Math.* 24:213-246.

SCHOUTEN, J. A., 1954. *Ricci Calculus.* Berlin: Springer-Verlag.

SCHUTZ, B., 1980. *Geometrical Methods of mathematical Physics.* Cambridge, England: Cambridge Univ. Press.

SCHWARTZ, J. T., 1967. *Nonlinear Functional Analysis.* New York: Gordon and Breach, Science Publishers, Inc.

SCHWARZ, G., 1975. Smooth functions invariant under the action of a compact Lie group. *Topology* 14:63-68.

SEGAL, I., 1962. Nonlinear Semigroups, *Ann. of Math.* 78:334-362.

SEGEV, R., 1981. *Differentiable manifolds and some basic notions of continuum mechanics,* Ph.D. thesis, Department of Mechanical Engin., University of Calgary.

SELIGER, R. L. and G. B. WHITHAM, 1968. Variational principles in continuum mechanics, *Proc. Roy. Soc. London* 305:1-25.

SERRIN, J., 1975. *Foundations of Classical Thermodynamics,* Lecture Notes, Univ. of Chicago.

—— 1979. Conceptual analysis of the classical second laws of thermodynamics, *Arch. Rat. Mech. An.* 70:355-371.

SETHNA, P. R., 1980. Bifurcation theory and averaging in mechanical systems, in Holmes (1980a).

SETHNA, P. R. and S. M. SHAPIRO, 1977. Nonlinear behavior of flutter unstable systems with gyroscopic and circulatory forces, *Trans. ASME* 44:755-762.

SETHNA, P. R. and A. K. BAJAJ, 1978. Bifurcations in Dynamical Systems with Internal Resonance, *J. Appl. Mech.* 45: No. 4, 895-902.

SEWELL, M. J., 1966a. On configuration-dependent loading, *Arch. Rat. Mech. An.* 23:327-351.

—— 1966b. On the Connexion Between Stability and The Shape of the Equilibrium Surface, *J. Mech. Phys. Solids* 14:203-230.

—— 1969. A method of post buckling analysis, *J. Mech. Phys. Solids* 17:219-233.

SHATAH, J., 1982. Global existence of small solutions to nonlinear evolution equations. *J. Diff. Eqns.* (to appear).

SHEARER, M., 1976. Bifurcation from a multiple eigenvalue, *Springer Lecture Notes in Math.* 564:417-424.

—— 1977. Small solutions of a nonlinear equation on Banach space for a degenerate case, *Proc. Roy. Soc. Edinburgh* 79:35-49.

SHIELD, R., 1971. Deformations possible in every compressible, isotropic, perfectly elastic material, *J. Elasticity,* 1:91-92.

SIDOROFF, R., 1974. Sur les restrictions à imposer à l'énergie de déformation d'un matérial hyperélastique, *C. R. Acad. Sc. Paris.* 279:379-382.

SIGNORINI, A., 1930. Sulle deformazioni termoelastiche finite, *Proc. 3rd Int. Cong. Appl. Mech.* 2:80-89.

—— 1943. Transformazioni termoelastiche finite, *Mem 1, Ann de Mat.* 22:33-143.

SIMPSON, H. and S. SPECTOR, 1985. On failure of the complementing condition and nonuniqueness in linear elastostatics, *J. of Elasticity* 15:229-231.

SLEBODZINSKI, W., 1931. Sur les équations de Hamilton, *Bull Acad. Roy. de Belg.* 17:864-870.

SLEMROD, M., 1976. Asymptotic behavior of C_0 semigroups as determined by the spectrum of the generator, *Indiana Univ. Math. J.* 25:783-792.

—— 1981. Global existence uniqueness and asymptotic stability of classical smooth solutions in one dimensional non-linear thermoelasticity *Arch. Rat. Mech. An.* 76:97-134.

SMALE, S., 1963. Diffeomorphisms with many periodic points, in *Differential and Combinatorial Topology.* S. S. Cairns, ed. Princeton, NJ: Princeton Univ. Press.

—— 1967. Differentiable dynamical systems, *Bull. Am. Math. Soc.* 73:747-817.

SNEDDON, I. N., 1980. Review of "Three-Dimensional problem of the mathematical theory of elasticity and thermoelasticy," *Bull. Am. Math. Soc.* 3:870-878.

SNIATYCKI, J. and W. TULCZYJEW, 1972. Generating forms of Lagrangian submanifolds, *Ind. Univ. Math. J.* 22:267-275.

SOBOLEV, S. S., 1939. On the theory of hyperbolic partial differential equations, *Mat. Sb.* 5 (47):71-99.

—— 1963. *Applications of Functional Analysis in Mathematical Physics,* Translations of Math. Monographs 7. Philadelphia, PA: American Mathematics Society.

SOKOLNIKOFF, I. S., 1956. *The Mathematical Theory of Elasticity,* (2d. ed.) New York: McGraw-Hill Book Company.

SOMMERFIELD, A., 1964a. *Mechanics of Deformable Bodies,* New York: Academic Press, Inc.

—— 1964b. *Thermodynamics and Statistical Mechanics,* New York: Academic Press, Inc.

SOURIAU, J. M., 1970. *Structure des Systèmes Dynamiques,* Paris: Dunod.

—— 1975. Mécanique statistique, groupes de Lie et cosmologie, in *Géométrie Symplectique et Physique Mathématique,* Publ. No. 235 du CNRS.

SPECTOR, S., 1980. On uniqueness in finite elasticity with general loading, *J. Elasticity* 10:145-161.

SPENCER, R., 1982. Hamiltonian formulation of a charged two fluid model (in preparation).

SPIVAK, M., 1975. *Differential Geometry,* I-V. Berkeley, CA: Publish or Perish.

ST. HILAIRE, A. O., 1976. Analytical Prediction of the Non-Linear Response of a Self-Excited Structure. *J. Sound Vib.* 47:185-205.

STAKGOLD, I., 1971. Branching of solutions of non-linear equations, *SIAM Review.* 13:289-332.

STEIN, E., 1970. *Singular Integrals and Differentiability Properties of Functions.* Princeton, NJ: Princeton Univ. Press.

STEIN, M., 1959. The phenomenon of change in buckle patterns in elastic structures, *NASA Technical Reports* R-39 and R-40.

STERNBERG, S., 1963. *Lectures on Differential Geometry.* Englewood Cliffs, NJ: Prentice-Hall, Inc.

STEWART, I., 1981. Applications of catastrophe theory to the physical sciences, *Physica D* 2:245-305.

STOKER, J. J., 1968. *Nonlinear Elasticity.* New York: Gordon and Breach, Science Publishers, Inc.

STOPPELLI, F., 1954. Un teorema di esistenza e di unicità relativo alle equazioni dell'elastostatica isoterma per deformazioni finite, *Ricerche Mat.,* 3:247-267.

—— 1955. Sulla svilluppibilità in serie di potenze di un parametro delle soluzioni delle equazioni dell'elastostatica isoterma, *Ricerche Mat.* 4:58-73.

—— 1958. Sull' esistenza di soluzioni delle equazioni dell' elastostatica isoterma nel caso di sollecitazioni dotate di assi di equilibrio, *Ricerche Mat.* (1957) 6: 244-282 and (1958) 7:71-101, 138-152.

STUART, C. A., 1979. An introduction to bifurcation theory based on differential calculus, in Knops (1979b), 76-135.

SUSSMAN, H. J., 1977. Existence and uniqueness of minimal realizations of nonlinear systems, *Math. Systems Theory,* 10:263-284.

SYNGE, J. L. and A. SCHILD, 1956. *Tensor Calculus,* University of Toronto Press.

TAKENS, F., 1973. Unfoldings of Certain Singularities of Vector Fields: Generalized Hopf Bifurcations. *J. Diff. Eqns.* 14:476-493.

—— 1974a. Introduction to Global Analysis, *Comm. 2, Math. Inst. Utrecht.*

—— 1974b. Forced Oscillations and Bifurcations, *Comm. 3, Math. Inst. Utrecht.*

—— 1977. Symmetries, conservation laws and variational principles, *Springer Lecture Notes in Math.* 597:101-135.

TANYI, G. E., 1978. On the critical points of the classical elastic energy functional, *Afrika Matematika* 1:35-43.

TARTAR, L., 1979. Compensated compactness and applications to partial differential equations, in Knops (1979b), 136-212.

TAUB, A. H., 1949. On Hamilton's Principle for perfect compressible fluids, in *Proc. Symp. Appl. Math. I.* 148-157. Providence, RI: American Mathematical Society.

TEMAM, R., ed., 1976. *Turbulence and Navier-Stokes Equation,* Springer Lecture Notes Math 565.

THOMPSON, J.M.T., 1982. *Instabilities and Catastrophes in Science and Engineering.* New York: John Wiley & Sons, Inc.

THOMPSON, J.M.T. and G. W. HUNT, 1973. *A General Theory of Elastic Stability.* London: John Wiley & Sons, Inc.

—— 1975. Towards a unified bifurcation theory, *J. Appl. Math. Phys. (ZAMP)* 26:581.

—— 1982. *Static and Dynamic Instability Phenomena.* New York: John Wiley & Sons, Inc.

THOMPSON, J.M.T. and T. S. LUNN, 1981a. Resonance-sensitivity in dynamic Hopf bifurcations under fluid loading, *Appl. Math. Modelling* 5:143-150.

—— 1981b. Static elastica formulations of a pipe conveying fluid, *J. Sound Vib.* 77:127-132.

TIMOSHENKO, S., et al, 1974. *Vibration Problems in Engineering,* (4th ed.) New York: John Wiley & Sons, Inc.

TOLOTTI, C., 1943. Deformazioni elastiche finite: onde ordinarie di discontinuità e caso tipico di solidi elastici isotropi, *Rend. Mat. e Applic.* 4:33-59.

TOUPIN, R. A., 1962. Elastic materials with couple stresses, *Arch. Rat. Mech. An.* 11:385-414.

—— 1964. Theories of Elasticity with Couple Stress, *Arch. Rat. Mech. An.* 17: 85-112.

TRAUTMAN, A., 1965. Foundations and current problems of general relativity theory, in *Brandeis 1964 summer institute on theoretical physics.* Englewood Cliffs, NJ: Prentice-Hall, Inc.

TROMBA, A. J., 1976. Almost Riemannian structures on Banach manifolds, the Morse lemma and the Darboux theorem. *Can J. Math.* 28:640-652.

TROTTER, H. F., 1959. On the product of semigroups of operators, *Proc. AMS* 10:545-551.

TRUESDELL, C., 1955a. The simplest rate theory of pure elasticity, *Comm. Pure. Appl. Math.* 8:123-132.

—— 1955b. Hypo-elasticity, *J. Rat. Mech. An.* 4:83-133, 1019-1020.

—— 1960. Invariant and Complete Stress Functions for General Continua, *Arch. Rat. Mech. An.* 4:1-29.

—— 1966a. *The Elements of Continuum Mechanics.* New York: Springer-Verlag.

—— 1966b. *Six Lectures on Modern Natural Philosophy.* New York: Springer-Verlag.

—— 1968. *Essays in the History of Mechanics.* New York: Springer-Verlag.

—— 1969. *Rational Thermodynamics,* New York: Springer-Verlag.

—— 1977. *A First Course in Rational Continuum Mechanics,* Vol. 1, New York: Academic Press.

TRUESDELL, C. and W. NOLL, 1965. The Non-Linear Field Theories of Mechanics, *Handbuch der Physik* III/3, Berlin: ed. S. Flugge. Springer-Verlag.

TRUESDELL, C. and R. TOUPIN, 1960. The Classical Field Theories. *Handbuch der Physik* III/1. ed. S. Flugge, Berlin: Springer-Verlag.

TSENG, W. Y. and J. DUGUNDJI, 1971. Nonlinear vibrations of a cuckled beam under harmonic excitation, *J. Appl. Mech.* 38:467-476.

TULCZYJEW, W. M., 1974. Hamiltonian systems, Lagrangian systems and the Legendre transformation, *Symp. Math.,* 14:247-258.

—— 1976. Simple models of Lagrangian submanifolds of symplectic manifolds, *Rend. Sem. Mat. Torino.* 35:98-112.

TURNER, R.E.L., 1971. Transversality in nonlinear eigenvalue problems, in *Contributions to Nonlinear Functional Analysis* (A. Zerantello, ed.) Academic Press.

VACAITIS, R., M. SHINOZUKA, and M. TAKENO, 1973. Parametric Study of Wind Loading on Structures, *Proc. A.S.C.E. J. Struct. Dev.* 3:453-468.

VAINBERG, M. M., 1964. *Variational Methods in the Theory of Nonlinear Operators.* San Francisco, CA: Holden-Day, Inc.

VAINBERG, M. M. and V. A. TRENOGIN, 1974. *Theory of branching of solutions of nonlinear equations,* Groningen, NL: Noordhoff.

VAN BUREN, W., 1968. On the existence and uniqueness of solutions to boundary value problems in finite elasticity, *Westinghouse Research Report* 68-1D7-Mekmari.

VAN DER SCHAFT, A., 1981. Symmetries and conservation laws for Hamiltonian systems with inputs and outputs: A generalization of Noether's theorem, *Systems and Control Letters,* 1:108-115.

VANDERBAUWHEDE, A., 1978. Generic bifurcation and symmetry with an application to the Von Karmen equations, *Proc. Roy. Soc. Edinburgh* 81A: 211-235.

—— 1980. Symmetry and bifurcation near families of solutions, *J. Diff. Eqns* 36:173-187.

VERONA, A., 1970. Bundles over manifolds of maps and connections, *Rev. Roum. Math. Pures. Appl.* 15: 1097-1112.

VILLAGGIO, P., 1977. *Qualitative methods in elasticity,* Leyden, NL: Noordhoff Int. Publ. Co.

WALKER, J. A., 1980. *Dynamical systems and evolution equations,* New York. Plenum Press.

WAN, Y. H. and J. E. MARSDEN, 1983. Symmetry and bifurcation in three dimensional elasticity, III, *Arch. Rat. Mech. An.* 84:203-233.

WANG, C. C. and C. TRUESDELL, 1973. *Introduction to rational elasticity.* Groningen, NL: Noordhoff.

WARNER, F., 1971. *Foundations of Differentiable Manifolds and Lie Groups.* Glenview, IL: Scott-Forseman & Co.

WASHIZU, K., 1982. *Variational Methods in Elasticity and Plasticity,* 3rd ed.) Elmsford, NY: Pergamon Press, Inc.

WEINSTEIN, A., 1977. *Lectures on symplectic manifolds,* CBMS Conf. Series No. 27. Philadelphia, PA: American Mathematical Society.

—— 1978. Bifurcations and Hamilton's principle, *Math. Zeit.* 159:235-248.

WEINSTEIN, A. and P. MORRISON, 1981. Poisson brackets for the Maxwell-Vlasov equations, *Phys. Lett.* 86A:235-236.

WEISS, B., 1967. Abstract vibrating systems, *J. Math. and Mech.* 17:241-255.

WELLS, R. O., 1980. *Differential Analysis on Complex Manifolds,* New York: Springer-Verlag.

WHEELER, L., 1977. A uniqueness theorem for the displacement problem in finite elastodynamics, *Arch. Rat. Mech. An.* 63:183-190.

WHEELER, L. and R. R. NACHLINGER, 1974. Uniqueness theorems for finite elastodynamics, *J. Elasticity* 4:27-36.

WILKES, N. S., 1980. On the nonexistence of semigroups for some equations of continuum mechanics, *Proc. Roy. Soc. Edinburgh* 86:303-306.

WOLKOWSKI, J. H., 1967. Existence of buckled states of circular plates, *Comm. Pure Appl. Math.* 20:549-560.

WU, F. and C. DESOER, 1972. Global inverse function theorem; *IEEE Trans. on Circuit Theory,* 19:199-201.

YANO, K., 1970. *Integral formulas in Riemannian geometry.* New York: Marcel Dekker, Inc.

YANOWITCH, M., 1956. Nonlinear buckling of thin elastic plates, *Comm. Pure Appl. Math.* 9:661-672.

YOSIDA, K., 1971. *Functional Analysis.* New York: Springer-Verlag.

ZEEMAN, E. C., 1976. Euler Buckling, *Springer Lecture Notes in Math.* 525: 373-395.

ZIEGLER, H., 1968. *Principles of Structural Stability.* Lexington, MA: Ginn-Blaisdell.

ZORSKI, H., ed., 1979. *Trends in applications of pure mathematics to mechanics,* Vol. II. London: Pitman, Ltd.

INDEX

N

T

U